Advanced Lectures in Mathematics
Volume XII

# Cohomology of Groups and Algebraic *K*-theory

Editors: Lizhen Ji, Kefeng Liu, and Shing-Tung Yau

International Press
www.intlpress.com

高等教育出版社
HIGHER EDUCATION PRESS

Lizhen Ji
University of Michigan

Kefeng Liu
University of California at Los Angeles

Shing-Tung Yau
Harvard University



ISBN 978-1-57146-144-5

Typeset using the LaTeX system.
Printed in the USA on acid-free paper.

ADVANCED LECTURES IN MATHEMATICS

# Preface

Cohomology of groups is a fundamental tool in many subjects in modern mathematics. One important generalized cohomology theory is the algebraic $K$-theory, and algebraic $K$-groups of rings such as rings of integers and group rings are important invariants of the rings. They have played important roles in algebra, geometric and algebraic topology, number theory, representation theory etc. Cohomology of groups and algebraic $K$-groups are also closely related. For example, algebraic $K$-groups of rings of integers in number fields can be effectively studied by using cohomology of arithmetic groups.

To introduce these fundamental concepts to graduate students and non-experts, an international summer school titled "Cohomology of groups and algebraic $K$-theory" was organized and held at the Center of Mathematical Sciences (CMS), Zhejiang University, from July 1 to July 13, 2007. Besides lecture series by several distinguished speakers on algebraic $K$-theory, algebraic $K$-groups of rings of integers in number fields and group rings which are fundamental in geometric topology, there were also some research talks.

This volume consists of expanded lecture notes of some talks at the summer school, two fundamental papers of D.Quillen[1], a set of lercture notes of M.Karoubi[2] and several other papers on related topics. Together, they present a coherent picture of the topics under discussion. We hope and expect that this book will be a very valuable reference on and introduction to these topics for both graduate students and mathematicians who want to learn some basics in cohomology of groups and algebraic $K$-groups, and related recent developments.

All the papers in this proceedings are carefully refereed. We would like to thank the referees for their kind help. We would also like to thank the staff members of CMS, Zhejiang University, for their help during the summer school

---

[1]D.Quillen, (1) *Higher algebraic $K$-theory. I*, in *Algebraic $K$-theory, I: Higher $K$-theories* (Proc. Conf., Battelle Memorial Inst., Seattle, Wash., 1972), pp. 85–147. Lecture Notes in Math., Vol. 341, Springer, 1973, and (2) *Finite generation of the groups $K_i$ of rings of algebraic integers*, in *Algebraic $K$-theory, I: Higher $K$-theories* (Proc. Conf., Battelle Memorial Inst., Seattle, Wash., 1972), pp. 179–198. Lecture Notes in Math., Vol. 341, Springer, 1973.

[2]M.Karoubi, *Lectures on $K$-theory*, in *Contemporary developments in algebraic $K$-theory* , pp. 1–95 (electronic), ICTP Lect. Notes, XV, Abdus Salam Int. Cent. Theoret. Phys., Trieste, 2004.

and the preparation of this proceedings. Finally, we would like to thank Springer for their permission for us to reprint the two papers of D.Quillen and ICTP, Trieste, for their permission to reprint the paper of M.Karoubi mentioned above.

Lizhen Ji (the Managing Editor)
Kefeng Liu
Shing-Tung Yau
November 2008

# Contents

*Cohomology of Groups and*
*Algebraic K-theory*
ALM 12, pp. 1–54



# On Crossed Product Rings With Twisted Involutions, Their Module Categories and $L$-Theory

Arthur Bartels* and Wolfgang Lück**

**Abstract**

We study the Farrell-Jones Conjecture with coefficients in an additive $G$-category with involution. This is a variant of the $L$-theoretic Farrell-Jones Conjecture which originally deals with group rings with the standard involution. We show that this formulation of the conjecture can be applied to crossed product rings $R{*}G$ equipped with twisted involutions and automatically implies the a priori more general fibered version.

**2000 Mathematics Subject Classification:** 18F25, 57R67

**Key words and phrases:** $L$-theoretic Farrell-Jones Conjecture of group rings with arbitrary coefficients, additive categories with involution

## Introduction

The *Farrell-Jones Conjecture for algebraic L-theory* predicts for a group $G$ and a ring $R$ with involution $r \mapsto \overline{r}$ that the so called *assembly map*

$$\operatorname{asmb}_n^{G,R}\colon H_n^G\big(E_{\mathcal{VC}\text{yc}}(G);\mathbf{L}_R^{\langle -\infty\rangle}\big) \to L_n^{\langle -\infty\rangle}(RG) \tag{0.1}$$

is bijective for all $n \in \mathbb{Z}$. Here the target is the *L-theory* of the group ring $RG$ with the standard involution sending $\sum_{g\in G} r_g \cdot g$ to $\sum_{g \in G} \overline{r_g} \cdot g^{-1}$. This is the group one wants to understand. It is a crucial ingredient in the surgery program for the classification of closed manifolds. The source is a much easier to handle term, namely, a $G$-homology theory applied to the *the classifying space $E_{\mathcal{VC}yc}(G)$ of the*

*Imperial College London, Huxley Building, London SW7 2AZ, UK. URL: http://ma.ic.ac.uk/abartels/ E-mail: a.bartels@imperial.ac.uk

**Westfälische Wilhelms-Universität Münster, Mathematisches Institut, Einsteinstr. 62, D-48149 Münster, Germany. URL: http://www.math.uni-muenster.de/u/lueck E-mail: lueck@math.uni-muenster.de

*family $\mathcal{VC}yc$ of virtually cyclic subgroups of* $G$. There is also a $K$-theory version of the Farrell-Jones Conjecture. The original source for the (Fibered) Farrell-Jones Conjecture is the paper by Farrell-Jones [6, 1.6 on page 257 and 1.7 on page 262]. More information can be found for instance in the survey article [10].

In this paper we study the Farrell-Jones Conjecture with coefficients in an additive $G$-category with involution. We show that this more general formulation of the conjecture allows to consider instead of the group ring $RG$ the crossed product ring with involution $R *_{c,\tau,w} G$ (see Section 4), which is a generalization of the twisted group ring, and to use more general involutions, for instance the one given by twisting the standard involution with a group homomorphism $w_1 \colon G \to \{\pm 1\}$. The data describing $R *_{c,\tau} G$ and more general involutions are pretty complicated. It turns out that it is convenient to put these into a more general but easier to handle context, where the coefficients are given by an additive $G$-categories $\mathcal{A}$ with involution (see Definition 4.22).

**Definition 0.2** ($L$-theoretic Farrell-Jones Conjecture). A group $G$ together with an additive $G$-category with involution $\mathcal{A}$ satisfies the *$L$-theoretic Farrell-Jones Conjecture with coefficients in $\mathcal{A}$* if the assembly map

$$(0.3)\quad \operatorname{asmb}_n^{G,\mathcal{A}} \colon H_n^G\big(E_{\mathcal{VC}\mathrm{yc}}(G); \mathbf{L}_{\mathcal{A}}^{\langle -\infty\rangle}\big) \to H_n^G\big(\mathrm{pt}; \mathbf{L}_{\mathcal{A}}^{\langle -\infty\rangle}\big) = L_n^{\langle -\infty\rangle}\left(\int_G \mathcal{A}\right).$$

induced by the projection $E_{\mathcal{VC}\mathrm{yc}}(G) \to \mathrm{pt}$ is bijective for all $n \in \mathbb{Z}$..

A group $G$ satisfies the *$L$-theoretic Farrell-Jones Conjecture* if for any additive $G$-category with involution $\mathcal{A}$ the *$L$-theoretic Farrell-Jones Conjecture with coefficients in $\mathcal{A}$* is true.

Here $\int_G \mathcal{A}$ is a certain homotopy colimit which yields an additive category with involution and we use the $L$-theory associated to an additive category with involution due to Ranicki (see [12], [13] and [14]). The $G$-homology theory $H_n^G(-; \mathbf{L}_{\mathcal{A}}^{\langle -\infty\rangle})$ is briefly recalled in Section 9. If $R$ is a ring with involution, $\mathcal{A}$ is the additive category with involution given by finitely generated free $R$-modules and we equip $\mathcal{A}$ with the trivial $G$-action, then the assembly map (0.3) agrees with the one for $RG$ in (0.1) (see Theorem 0.4 below). This general setup is also a very useful framework when one is dealing with categories appearing in controlled topology, which is an important tool for proving the Farrell-Jones Conjecture for certain groups.

Next we state the main results of this paper.

**Theorem 0.4.** *Suppose that $G$ satisfies the $L$-theoretic Farrell-Jones Conjecture in the sense of Definition 0.2. Let $R$ be ring with the data $(c, \tau, w)$ and $R *_{c,\tau,w} G$ be the associated crossed product ring with involution as explained in Section 4. Then the assembly map*

$$(0.5)\qquad \operatorname{asmb}_n^{G,R_{c,\tau,w}} \colon H_n^G\big(E_{\mathcal{VC}\mathrm{yc}}(G); \mathbf{L}_{R,c,\tau,w}^{\langle -\infty\rangle}\big) \to L_n^{\langle -\infty\rangle}(R *_{c,\tau,w} G)$$

*is bijective.*

*Here $\mathbf{L}_{R,c,\tau,w}^{\langle -\infty\rangle}$ is a functor from the orbit category $\mathsf{Or}(G)$ to the category of spectra such that $\pi_n\big(\mathbf{L}_{R,c,\tau,w}^{\langle -\infty\rangle}(G/H)\big)$ for $H \subseteq G$ agrees with $L_n^{\langle -\infty\rangle}(R *_{c|_H,\tau|_N,w|_H} H)$.*

Another important feature is that in this setting the (unfibered) Farrell-Jones Conjecture does already imply the fibered version.

**Definition 0.6** (Fibered $L$-theoretic Farrell-Jones Conjecture). A group $G$ satisfies the *fibered $L$-theoretic Farrell-Jones Conjecture* if for any group homomorphism $\phi\colon K \to G$ and additive $G$-category with involution $\mathcal{A}$ the assembly map

$$\mathrm{asmb}_n^{\phi,\mathcal{A}}\colon H_*^K\left(E_{\phi^*\mathcal{VC}\mathrm{yc}}(G);\mathbf{L}_{\phi^*\mathcal{A}}^{\langle-\infty\rangle}\right) \to L_n^{\langle-\infty\rangle}\left(\int_K \phi^*\mathcal{A}\right).$$

is bijective for all $n \in \mathbb{Z}$, where the family $\phi^*\mathcal{VC}\mathrm{yc}$ of subgroups of $K$ consists of subgroups $L \subseteq K$ with $\phi(L)$ virtually cyclic and $\phi^*\mathcal{A}$ is the additive $K$-category with involution obtained from $\mathcal{A}$ by restriction with $\phi$.

Obviously the fibered version for the group $G$ of Definition 0.6 implies the version for the group $G$ of Definition 0.2, take $\phi = \mathrm{id}$ in Definition 0.6. The converse is also true.

**Theorem 0.7.** *Let $G$ be a group. Then $G$ satisfies the fibered $L$-theoretic Farrell-Jones Conjecture if and only if $G$ satisfies the $L$-theoretic Farrell-Jones Conjecture of Definition 0.2.*

A general statement of a *Fibered Isomorphism Conjecture* and the discussion of its inheritance properties under subgroups and colimits of groups can be found in [1, Section 4] (see also [6, Appendix], [7, Theorem 7.1]). These very useful inheritance properties do not hold for the unfibered version of Definition 0.2. The next three corollaries are immediate consequences of Theorem 0.7 and [1, Theorem 3.3, Lemma 4.4, Lemma 4.5 and Lemma 4.6].

**Corollary 0.8.** *Let $\{G_i \mid i \in I\}$ be a directed system (with not necessarily injective) structure maps and let $G$ be its colimit $\mathrm{colim}_{i\in I} G_i$. Suppose that $G_i$ satisfy the Farrell-Jones Conjecture of Definition 0.2 for every $i \in I$.*

*Then $G$ satisfies the Farrell-Jones Conjecture of Definition 0.2.*

**Corollary 0.9.** *Let $1 \to K \to G \xrightarrow{p} Q \to 1$ be an extension of groups. Suppose that the group $Q$ and for any virtually cyclic subgroup $V \subseteq Q$ the group $p^{-1}(V)$ satisfy the Farrell-Jones Conjecture of Definition 0.2.*

*Then the group $G$ satisfies the Farrell-Jones Conjecture of Definition 0.2.*

**Corollary 0.10.** *If $G$ satisfies the Farrell-Jones Conjecture of Definition 0.2, then any subgroup $H \subseteq G$ satisfies the Farrell-Jones Conjecture of Definition 0.2.*

Corollary 0.9 and Corollary 0.10 have also been proved in [8].

**Remark 0.11.** Suppose that the Farrell-Jones Conjecture of Definition 0.2 has been proved for the product of two virtually cyclic subgroups.

Then Corollary 0.9 and Corollary 0.10 imply that $G \times H$ satisfies the Farrell-Jones Conjecture of Definition 0.2 if and only if both $G$ and $H$ satisfy the Farrell-Jones Conjecture of Definition 0.2.

It is sometimes useful to have strict structures on $\mathcal{A}$, e.g., the involution is desired to be strict and there should be a (strictly associative) functorial direct sum. The functorial direct sum is actually needed in some proofs in order to guarantee good functoriality properties of certain categories arising from controlled topology. We will show

**Theorem 0.12.** *The group $G$ satisfies the $L$-theoretic Farrell-Jones Conjecture of Definition 0.2 if it satisfies the obvious version of it, where one only considers additive $G$-category with (strictly associative) functorial direct sum and strict involution (see Definition 10.6).*

The Farrell-Jones Conjecture with coefficients (in $K$- and $L$-theory) has been introduced in [4]. Our treatment here is more general in that we allow involutions that are not necessarily strict and also deal with twisted involutions on the crossed product ring.

All results mentioned here have obvious analogues for $K$-theory whose proof is actually easier since one does not have to deal with the involutions.

The work was financially supported by the Sonderforschungsbereich 478 – Geometrische Strukturen in der Mathematik – and the Max-Planck-Forschungspreis of the second author.

The paper is organized as follows:



# 1. Additive categories with involution

In this section we will review the notion of an additive category with involution as it appears and is used in the literature. This will be one of our main examples.

Let $\mathcal{A}$ be an *additive category*, i.e., a small category $\mathcal{A}$ such that for two objects $A$ and $B$ the morphism set $\operatorname{mor}_{\mathcal{A}}(A, B)$ has the structure of an abelian group and the direct sum $A \oplus B$ of two objects $A$ and $B$ exists and the obvious compatibility conditions hold. A covariant *functor of additive categories* $F\colon \mathcal{A}_0 \to \mathcal{A}_1$ is a covariant functor such that for two objects $A$ and $B$ in $\mathcal{A}_0$ the map $\operatorname{mor}_{\mathcal{A}_0}(A, B) \to \operatorname{mor}_{\mathcal{A}_1}(F(A), F(B))$ sending $f$ to $F(f)$ respects the abelian group structures and $F(A \oplus B)$ is a model for $F(A) \oplus F(B)$. The notion of a contravariant functor of additive categories is defined analogously.

An *involution* $(I, E)$ *on an additive category* $\mathcal{A}$ is contravariant functor

$$I \colon \mathcal{A} \to \mathcal{A} \tag{1.1}$$

of additive categories together with a natural equivalence of such functors

$$E \colon \operatorname{id}_{\mathcal{A}} \to I^2 := I \circ I \tag{1.2}$$

such that for every object $A$ we have the equality of morphisms

$$E(I(A)) = I(E(A)^{-1}) \colon I(A) \to I^3(A). \tag{1.3}$$

In the sequel we often write $I(A) = A^*$ and $I(f) = f^*$ for a morphism $f \colon A \to B$ in $\mathcal{A}$. If $I^2 = \operatorname{id}_{\mathcal{A}}$ and $E(A) = \operatorname{id}_A$ for all objects $A$, then we call $I = (I, \operatorname{id})$ a *strict involution*.

**Definition 1.4** (Additive category with involution). An *additive category with involution* is an additive category together with an involution $(I, E)$.

An *additive category with strict involution* is an additive category together with a strict involution $I$.

The following example is a key example and illustrates why one cannot expect in concrete situation that the involution is strict.

**Example 1.5.** Let $R$ be a ring. Let $R$-$\mathsf{FGP}$ be the category of finitely generated projective $R$-modules. This becomes an additive category by the direct sum of $R$-modules and the elementwise addition of $R$-homomorphisms.

A *ring with involution* is a ring $R$ together with a map $R \to R$, $r \mapsto \overline{r}$ satisfying $\overline{1} = 1$, $\overline{r+s} = \overline{r} + \overline{s}$ and $\overline{r \cdot s} = \overline{s} \cdot \overline{r}$ for $r, s \in R$. Given a ring with involution $R$, define an involution $I$ on the additive category $R$-$\mathsf{FGP}$ as follows. Given a finitely generated projective $R$-module $P$, let $I(P) = P^*$ be the finitely generated projective $\hom_R(P, R)$,where for $r \in R$ and $f \in \hom_R(P, R)$ the element $rf \in \hom_R(P, R)$ is defined by $rf(x) = f(x) \cdot \overline{r}$ for $x \in P$. The desired natural transformation

$$E \colon \operatorname{id}_{R\text{-}\mathsf{FGP}} \to I^2$$

assigns to a finitely generated projective $R$-module $P$ the $R$-isomorphism $P \xrightarrow{\cong} (P^*)^*$ sending $x \in P$ to $\hom_R(P, R) \to R$, $f \mapsto \overline{f(x)}$.

A *functor of additive categories with involution* $(F, T) \colon \mathcal{A} \to \mathcal{B}$ consists of a covariant functor $F$ of the underlying additive categories together with a natural equivalence $T \colon F \circ I_{\mathcal{A}} \to I_{\mathcal{B}} \circ F$ such that for every object $A$ in $\mathcal{A}$ the following diagram commutes

$$\begin{array}{ccc} F(A) & \xrightarrow{F(E_{\mathcal{A}}(A))} & F(A^{**}) \\ \Big\downarrow {\scriptstyle E_{\mathcal{B}}(F(A))} & & \Big\downarrow {\scriptstyle T(A^*)} \\ F(A)^{**} & \xrightarrow{T(A)^*} & F(A^*)^* \end{array} \tag{1.6}$$

If $T(A) = \mathrm{id}_A$ for all objects $A$, then we call $F$ a *strict* functor of additive categories with involution.

The *composite* of functors of additive categories with involution $(F_1, T_1)\colon \mathcal{A}_1 \to \mathcal{A}_2$ and $(F_2, T_2)\colon \mathcal{A}_2 \to \mathcal{A}_3$ is defined to be $(F_2 \circ F_1, T_2 \circ T_1)$, where $F_2 \circ F_1$ is the composite of functors of additive categories and the natural equivalence $T_2 \circ T_1$ assigns to an object $A \in \mathcal{A}_1$ the isomorphism in $\mathcal{A}_3$

$$F_2 \circ F_1 \circ I_{\mathcal{A}_1}(A) \xrightarrow{F_2(T_1(A))} F_2 \circ I_{\mathcal{A}_2} \circ F_1(A) \xrightarrow{T_2(F_1(A))} I_{\mathcal{A}_3} \circ F_2 \circ F_1(A).$$

A *natural transformation* $S\colon (F_1, T_1) \to (F_2, T_2)$ *of functors* $\mathcal{A}_1 \to \mathcal{A}_2$ *of additive categories with involutions* is a natural transformation $S\colon F_1 \to F_2$ of functors of additive categories such that for every object $A$ in $\mathcal{A}$ the following diagram commutes

(1.7)
$$\begin{array}{ccc} F_1(I_{\mathcal{A}_1}(A)) & \xrightarrow{T_1(A)} & I_{\mathcal{A}_2}(F_1(A)) \\ \downarrow{\scriptstyle S(I_{\mathcal{A}_1}(A))} & & \uparrow{\scriptstyle I_{\mathcal{A}_2}(S(A))} \\ F_2(I_{\mathcal{A}_1}(A)) & \xrightarrow{T_2(A)} & I_{\mathcal{A}_2}F_2(A)) \end{array}$$

## 2. Additive categories with weak $(G, v)$-action

In the sequel $G$ is a group and $v\colon G \to \{\pm 1\}$ is a group homomorphism to the multiplicative group $\{\pm 1\}$. In this section we want to introduce the notion of an additive category with weak $(G, v)$-action such that the notion of an additive category with involution is the special case of an additive category with weak $(\mathbb{Z}/2, v)$-action for $v\colon \mathbb{Z}/2 \to \{\pm 1\}$ the unique group isomorphism and we can also treat $G$-actions up to natural equivalence. Notice that this will force us to deal with covariant and contravariant functors simultaneously. The homomorphism $v$ will take care of that.

We call a functor $+1$-variant if it is covariant and $-1$-variant if it is contravariant. If $F_1\colon \mathcal{C}_0 \to \mathcal{C}_1$ is an $\epsilon_1$-variant functor and $F_2\colon \mathcal{C}_1 \to \mathcal{C}_2$ is an $\epsilon_2$-variant functor, then the composite $F_2 \circ F_1\colon \mathcal{C}_0 \to \mathcal{C}_2$ is $\epsilon_1\epsilon_2$-variant functor. If $f\colon x_0 \to x_1$ is an isomorphism and $\epsilon \in \{\pm 1\}$, then define $f^\epsilon\colon x_0 \to x_1$ to be $f$ if $\epsilon = 1$ and $f^\epsilon\colon x_1 \to x_0$ to be the inverse of $f$ if $\epsilon = -1$. If $F\colon \mathcal{C}_0 \to \mathcal{C}_1$ is $\epsilon$-variant and $f\colon x_0 \xrightarrow{\cong} y_0$ is an isomorphism in $\mathcal{C}_0$, then $F(f)^\epsilon\colon F(x_0) \to F(x_1)$ is an isomorphism in $\mathcal{C}_1$.

**Definition 2.1** (Additive category with weak $(G, v)$-action)**.** Let $G$ be a group together with a group homomorphism $v\colon G \to \{\pm 1\}$. An *additive category with weak $(G, v)$-action* $\mathcal{A}$ is an additive category together with the following data:

- For every $g \in G$ we have a $v(g)$-variant functor $R_g\colon \mathcal{A} \to \mathcal{A}$ of additive categories;
- For every two elements $g, h \in G$ there is a natural equivalence of $v(gh)$-variant functors of additive categories

$$L_{g,h}\colon R_{gh} \to R_h \circ R_g.$$

We require:

(i) $R_e = \mathrm{id}$ for $e \in G$ the unit element;

(ii) $L_{g,e} = L_{e,g} = \mathrm{id}$ for all $g \in G$;

(iii) The following diagram commutes for all $g, h, k \in G$ and objects $A$ in $\mathcal{A}$

$$\begin{CD}
R_{ghk}(A) @>{L_{gh,k}(A)}>> R_k(R_{gh}(A)) \\
@V{L_{g,hk}(A)}VV @VV{R_k(L_{g,h}(A))^{v(k)}}V \\
R_{hk}(R_g(A)) @>{L_{h,k}(R_g(A))}>> R_k(R_h(R_g(A)))
\end{CD}$$

If for every two elements $g, h \in G$ we have $L_{g,h} = \mathrm{id}$ and in particular $R_{gh} = R_h R_g$, we call $\mathcal{A}$ with these data an *additive category with strict $(G,v)$-action* or briefly a *additive $(G,v)$-category.* If $v$ is trivial, we will omit it from the notation.

Let $\mathcal{A}$ and $\mathcal{B}$ be two additive categories with weak $(G,v)$-action and let $\epsilon \in \{\pm 1\}$. An *$\epsilon$-variant functor $(F,T)\colon \mathcal{A} \to \mathcal{B}$ of additive categories with weak $(G,v)$-action* is an $\epsilon$-variant functor $F\colon \mathcal{A} \to \mathcal{B}$ of additive categories together with a collection $\{T_g \mid g \in G\}$ of natural transformations of $\epsilon v(g)$-variant functors of additive categories $T_g\colon F \circ R_g^{\mathcal{A}} \to R_g^{\mathcal{B}} \circ F$. We require that for all $g, h \in G$ and all objects $A$ in $\mathcal{A}$ the following diagram commutes

$$\begin{CD}
F(R_{hg}^{\mathcal{A}}(A)) @>{F(L_{h,g}^{\mathcal{A}}(A))^{\epsilon}}>> F(R_g^{\mathcal{A}}(R_h^{\mathcal{A}}(A))) @>{T_g(R_h(A))}>> R_g^{\mathcal{B}}(F(R_h^{\mathcal{A}}(A))) \\
@V{T_{hg}(A)}VV @. @VV{R_g^{\mathcal{B}}(T_h(A))^{v(g)}}V \\
R_{hg}^{\mathcal{B}}(F(A)) @>{L_{hg}^{\mathcal{B}}(F(A))}>> \cdots @>>> R_g^{\mathcal{B}}(R_h^{\mathcal{B}}(F(A)))
\end{CD} \tag{2.2}$$

The composite $(F_2, T_2) \circ (F_1, T_1)\colon \mathcal{A}_1 \to \mathcal{A}_3$ of an $\epsilon_1$-variant functor of additive categories with weak $(G,v)$-action $(F_1, T_1)\colon \mathcal{A}_1 \to \mathcal{A}_2$ and an $\epsilon_2$-variant functor of additive categories with weak $(G,v)$-action $(F_2, T_2)\colon \mathcal{A}_2 \to \mathcal{A}_3$ is the $\epsilon_1\epsilon_2$-variant functor of additive categories with weak $(G,v)$-action whose underlying $\epsilon_1\epsilon_2$-variant functor of additive categories is $F_2 \circ F_1\colon \mathcal{A}_1 \to \mathcal{A}_3$ and whose required natural transformations for $g \in G$ are given for an object $A$ in $\mathcal{A}_1$ by

$$F_2 \circ F_1 \circ R_g^{\mathcal{A}_1}(A) \xrightarrow{F_2((T_2)_g(A))^{\epsilon_2}} F_2 \circ R_g^{\mathcal{A}_2} \circ F_1(A) \xrightarrow{(T_2)_g(F_1(A))} R_g^{\mathcal{A}_3} \circ F_2 \circ F_1(A).$$

A *natural transformation $S\colon (F_1, T_1) \to (F_2, T_2)$ of functors $\mathcal{A}_1 \to \mathcal{A}_2$ of additive categories with weak $(G,v)$-action* is a natural transformation $S\colon F_1 \to F_2$ of functors of additive categories such that for all $g \in G$ and objects $A$ in $\mathcal{A}_1$ the following diagram commutes

$$\begin{CD}
F_1(R_g^{\mathcal{A}_1}(A)) @>{(T_1)_g(A)}>> R_g^{\mathcal{A}_2}(F_1(A)) \\
@V{S(R_g^{\mathcal{A}_1}(A))}VV @VV{\left(R_g^{\mathcal{A}_2}(S(A))\right)^{v(g)}}V \\
F_2(R_g^{\mathcal{A}_1}(A)) @>{(T_2)_g(A)}>> R_g^{\mathcal{A}_2}(F_2(A))
\end{CD} \tag{2.3}$$

An *$\epsilon$-variant functor $F\colon \mathcal{A} \to \mathcal{B}$ of additive categories with strict $(G,v)$-action* is an $\epsilon$-variant functor $F\colon \mathcal{A} \to \mathcal{B}$ of additive categories satisfying $F \circ R_g^{\mathcal{A}_1} = R_g^{\mathcal{A}_2} \circ F$ for all $g \in G$. A *natural transformation $S\colon F_1 \to F_2$ of $\epsilon$-variant functors $\mathcal{A}_1 \to \mathcal{A}_2$ of additive categories with strict $(G,v)$-action* is a natural transformation $S\colon F_1 \to F_2$ of additive categories satisfying $S(R_g^{\mathcal{A}_1}(A)) = R_g^{\mathcal{A}_2}(S(A))^{v(g)}$ for all $g \in G$ and objects $A$ in $\mathcal{S}_1$.

**Example 2.4** (Additive categories with involution)**.** Given an additive category $\mathcal{A}$, the structure of an additive category with weak $(\mathbb{Z}/2, v)$-action for $v\colon \mathbb{Z}/2 \to \{\pm 1\}$ the unique group isomorphism is the same as an additive category with involution. Namely, let $t \in \mathbb{Z}/2$ be the generator. Given an involution $(I, E)$ in the sense of Definition 1.4, define the structure of an additive category with weak $(\mathbb{Z}/2, v)$-action in the sense of Definition 2.1 by putting $R_e = \mathrm{id}$, $R_t = I$, $L_{e,e} = L_{t,e} = L_{t,e} = \mathrm{id}$ and $L_{t,t} = E$. Condition (iii) in Definition 2.1 follows from condition (1.3). Given the structure of an additive category with weak $(\mathbb{Z}/2, v)$-action, define the involution $(E, I)$ by $E = R_t$ and $I = L_{t,t}$. The corresponding statement is true for functors of additive categories with weak $(\mathbb{Z}/2, v)$-action and natural transformations between them, where diagram (1.6) corresponds to diagram (2.2).

Analogously we get that the structure of an additive category with strict $(\mathbb{Z}/2, v)$-action is the same as an additive category with strict involution.

# 3. Making an additive categories with weak $(G, v)$-action strict

Many interesting examples occur as additive categories with weak $(G, v)$-action which are not necessarily strict. On the other hand additive categories with strict $(G, v)$-action are easier to handle. We explain how we can turn an additive category with weak $(G, v)$-action $\mathcal{A}$ to an additive category with strict $(G, v)$-action which we will denote by $\mathcal{S}(\mathcal{A})$.

**Definition 3.1** ($\mathcal{S}(\mathcal{A})$)**.** An object in $\mathcal{S}(\mathcal{A})$ is a pair $(A, g)$ consisting of an object $A \in \mathcal{A}$ and an element $g \in G$. A morphism $(A, g)$ to $(B, h)$ is a morphism $\phi\colon R_g(A) \to R_h(B)$ in $\mathcal{A}$. The composition of morphisms is given by the one in $\mathcal{A}$. The category $\mathcal{S}(\mathcal{A})$ inherits the structure of an additive category from $\mathcal{A}$ in the obvious way.

Next we define the structure of an additive category with strict $(G, v)$-action on $\mathcal{S}(\mathcal{A})$. Define for $g \in G$ a functor $R_g^{\mathcal{S}}\colon \mathcal{S}(\mathcal{A}) \to \mathcal{S}(\mathcal{A})$ of additive categories as follows. Given an object $(A, h)$, define

$$R_g^{\mathcal{S}}(A, h) = (A, hg).$$

Given a morphism $\phi\colon (A, h) \to (B, k)$ define

$$R_g^{\mathcal{S}}(\phi)\colon R_g^{\mathcal{S}}(A, h) = (A, hg) \to R_g^{\mathcal{S}}(B, k) = (B, kg)$$

by the composite of morphisms in $\mathcal{A}$

$$R_{hg}(A) \xrightarrow{L_{h,g}(A)} R_g\left(R_h(A)\right) \xrightarrow{R_g(\phi)} R_g\left(R_k(B)\right) \xrightarrow{L_{k,g}(B)^{-1}} R_{kg}(B).$$

if $v(g) = 1$ and

$$R_g^{\mathcal{S}}(\phi)\colon R_g^{\mathcal{S}}(B,k) = (B,kg) \to R_g^{\mathcal{S}}(A,h) = (A,hg)$$

by the composite of morphisms in $\mathcal{A}$

$$R_{kg}(B) \xrightarrow{L_{k,g}(B)} R_g\left(R_k(B)\right) \xrightarrow{R_g(\phi)} R_g\left(R_h(A)\right) \xrightarrow{L_{h,g}(A)^{-1}} R_{hg}(A)$$

if $v(g) = -1$.

A direct computation shows that $R_g^{\mathcal{S}}$ is indeed a functor of additive categories. We conclude $R_e^{\mathcal{S}} = \mathrm{id}_{\mathcal{S}(\mathcal{A})}$ from the conditions $R_e = \mathrm{id}$ and $L_{g,e} = L_{e,g} = \mathrm{id}$. We have to check $R_{g_2}^{\mathcal{S}} \circ R_{g_1}^{\mathcal{S}} = R_{g_1g_2}^{\mathcal{S}}$. We will do this for simplicity only in the case $v(g_1) = v(g_2) = 1$, the other cases are analogous. Given a morphism $\phi\colon (A,h) \to (B,k)$, the morphism $R_{g_1g_2}^{\mathcal{S}}(\phi)$ is given by the composite in $\mathcal{A}$

$$R_{hg_1g_2}(A) \xrightarrow{L_{h,g_1g_2}(A)} R_{g_1g_2}\left(R_hA)\right) \xrightarrow{R_{g_1g_2}(\phi)} R_{g_1g_2}\left(R_k(B)\right) \xrightarrow{L_{k,g_1g_2}(B)^{-1}} R_{kg_1g_2}(B).$$

The morphism $R_{g_2}^{\mathcal{S}} \circ R_{g_1}^{\mathcal{S}}(\phi)$ is given by the composite in $\mathcal{A}$

$$R_{hg_1g_2}(A) \xrightarrow{L_{hg_1,g_2}(A)} R_{g_2}\left(R_{hg_1}(A)\right) \xrightarrow{R_{g_2}(L_{h,g_1}(A))} R_{g_2}\left(R_{g_1}(R_h(A))\right) \xrightarrow{R_{g_2}\left(R_{g_1}(\phi)\right)} R_{g_2}\left(R_{g_1}(R_k(B))\right) \xrightarrow{R_{g_2}\left(L_{k,g_1}(B)^{-1}\right)} R_{g_2}\left(R_{kg_1}(B)\right) \xrightarrow{L_{kg_1,g_2}(B)^{-1}} R_{kg_1g_2}(B).$$

Next we compute that these two morphisms agree. Because of condition (iii) in Definition 2.1 have

$$R_{g_2}(L_{h,g_1}(A)) \circ L_{hg_1,g_2}(A) = L_{g_1,g_2}(R_h(A)) \circ L_{h,g_1g_2}(A);$$
$$R_{g_2}(L_{k,g_1}(B)) \circ L_{kg_1,g_2}(B) = L_{g_1,g_2}(R_k(B)) \circ L_{k,g_1g_2}(B).$$

Hence it suffices to show that the composite

$$R_{g_1g_2}\left(R_h(A)\right) \xrightarrow{L_{g_1,g_2}(R_h(A))} R_{g_2}\left(R_{g_1}(R_h(A))\right) \xrightarrow{R_{g_2}\left(R_{g_1}(\phi)\right)} R_{g_2}\left(R_{g_1}(R_k(B))\right) \xrightarrow{L_{g_1,g_2}(R_k(B))^{-1}} R_{g_1g_2}\left(R_k(B)\right)$$

agrees with

$$R_{g_1g_2}\left(R_k(B)\right) \xrightarrow{R_{g_1g_2}(\phi)} R_{g_1g_2}\left(R_k(B)\right).$$

This follows from the fact that $L_{g_1,g_2}\colon R_{g_1g_2} \to R_{g_2} \circ R_{g_2}$ is a natural equivalence.

Let $(F,T)\colon \mathcal{A} \to \mathcal{B}$ be an $\epsilon$-variant functor of additive categories with weak $(G,v)$-action. It induces an $\epsilon$-variant functor $\mathcal{S}(F,T)\colon \mathcal{S}(\mathcal{A}) \to \mathcal{S}(\mathcal{B})$ of additive categories with strict $(G,v)$-action as follows. For simplicity we will only treat the case $\epsilon = 1$, the other case $\epsilon = -1$ is analogous. The functor $\mathcal{S}(F,T)$ sends an object $(A,h)$ in $\mathcal{S}(\mathcal{A})$ to the object $(F(A),h)$ in $\mathcal{S}(\mathcal{B})$. It sends a morphism $\phi\colon (A,h) \to (B,k)$ in $\mathcal{S}(\mathcal{A})$ which is given by a morphism $\phi\colon R_h^{\mathcal{A}}(A) \to R_k^{\mathcal{A}}(B)$ in $\mathcal{A}$ to the morphism $\mathcal{S}(F,T)(\phi)\colon (F(A),h) \to (F(B),k)$ in $\mathcal{S}(\mathcal{B})$ which is given by the following composite of morphisms in $\mathcal{B}$

$$R_h^{\mathcal{B}}(F(A)) \xrightarrow{T_h(A)^{-1}} F(R_h^{\mathcal{A}}(A)) \xrightarrow{F(\phi)} F(R_k^{\mathcal{A}}(B)) \xrightarrow{T_k(B)} R_k^{\mathcal{A}}(F(B)).$$

We have to show $R_g^{\mathcal{S}(\mathcal{B})} \circ \mathcal{S}(F) = \mathcal{S}(F) \circ R_g^{\mathcal{S}(\mathcal{A})}$ for every $g \in G$. We only treat the case $v(g) = 1$. This is obvious on objects since both composites send an object $(A,h)$ to $(F(A),hg)$. Let $\phi\colon (A,h) \to (B,k)$ be a morphism in $\mathcal{S}(\mathcal{A})$ which is given by a morphism $\phi\colon R_h^{\mathcal{A}}(A) \to R_k^{\mathcal{A}}(B)$ in $\mathcal{A}$. Then $R_g^{\mathcal{S}(\mathcal{B})} \circ \mathcal{S}(F)(\phi)$ is the morphism $(F(A),hg) \to (F(B),kg)$ in $\mathcal{S}(\mathcal{B})$ which is given by the composite in $\mathcal{B}$

$$\begin{aligned} R_{hg}^{\mathcal{B}}(F(A)) &\xrightarrow{L_{h,g}^{\mathcal{B}}(F(A))} R_g^{\mathcal{B}}(R_h^{\mathcal{B}}(F(A))) \xrightarrow{R_g^{\mathcal{B}}(T_h(A)^{-1})} R_g^{\mathcal{B}}(F(R_h^{\mathcal{A}}(A))) \\ &\xrightarrow{R_g^{\mathcal{B}}(F(\phi))} R_g^{\mathcal{B}}(F(R_k^{\mathcal{A}}(B))) \xrightarrow{R_g^{\mathcal{B}}(T_k(B))} R_g^{\mathcal{B}}(R_k^{\mathcal{B}}(F(B))) \\ &\xrightarrow{L_{k,g}^{\mathcal{B}}(F(B))^{-1}} R_{kg}^{\mathcal{B}}(F(B)) \end{aligned}$$

and $\mathcal{S}(F) \circ R_g^{\mathcal{S}(\mathcal{A})}(\phi)$ is the morphism $(F(A),hg) \to (F(B),kg)$ in $\mathcal{S}(\mathcal{B})$ which is given by the composite in $\mathcal{B}$

$$\begin{aligned} R_{hg}^{\mathcal{B}}(F(A)) &\xrightarrow{T_{hg}(A)^{-1}} F(R_{hg}^{\mathcal{A}}(A)) \xrightarrow{F(L_{h,g}^{\mathcal{A}}(A))} F(R_g^{\mathcal{A}}(R_h^{\mathcal{A}}(A))) \\ &\xrightarrow{F(R_g^{\mathcal{A}}(\phi))} F(R_g^{\mathcal{A}}(R_k^{\mathcal{A}}(B))) \xrightarrow{F(L_{k,g}^{\mathcal{A}}(B)^{-1})} F(R_{kg}^{\mathcal{A}}(B)) \\ &\xrightarrow{T_{kg}(A)} R_{kg}^{\mathcal{B}}(F(B)). \end{aligned}$$

Since $T_g\colon F \circ R_g^{\mathcal{A}} \to R_g^{\mathcal{B}} \circ F$ is a natural transformation, the following diagram commutes

$$\begin{array}{ccc} F(R_g^{\mathcal{A}}(R_h^{\mathcal{A}}(A))) & \xrightarrow{F(R_g^{\mathcal{A}}(\phi))} & F(R_g^{\mathcal{A}}(R_k^{\mathcal{A}}(B))) \\ \downarrow{\scriptstyle T_g(R_h^{\mathcal{A}}(A))} & & \downarrow{\scriptstyle T_g(R_k^{\mathcal{A}}(B))} \\ R_g^{\mathcal{B}}(F(R_h^{\mathcal{A}}(A))) & \xrightarrow{R_g^{\mathcal{B}}(F(\phi))} & R_g^{\mathcal{B}}(F(R_k^{\mathcal{A}}(B))) \end{array}$$

Hence it suffices to show that the composite

$$\begin{aligned} R_{hg}^{\mathcal{B}}(F(A)) &\xrightarrow{L_{h,g}^{\mathcal{B}}(F(A))} R_g^{\mathcal{B}}(R_h^{\mathcal{B}}(F(A))) \xrightarrow{R_g^{\mathcal{B}}(T_h(A)^{-1})} R_g^{\mathcal{B}}(F(R_h^{\mathcal{A}}(A))) \\ &\xrightarrow{T_g(R_h^{\mathcal{A}}(A))^{-1}} F(R_g^{\mathcal{A}}(R_h^{\mathcal{A}}(A))) \end{aligned}$$

agrees with the composite

$$R^{\mathcal{B}}_{hg}(F(A)) \xrightarrow{T_{hg}(A)^{-1}} F(R^{\mathcal{A}}_{hg}(A)) \xrightarrow{F(L^{\mathcal{A}}_{h,g}(A))} F(R^{\mathcal{A}}_g(R^{\mathcal{A}}_h(A)))$$

and that the composite

$$F(R^{\mathcal{A}}_g(R^{\mathcal{A}}_k(B))) \xrightarrow{T_g(R^{\mathcal{A}}_k(B))} R^{\mathcal{B}}_g(F(R^{\mathcal{A}}_k(B))) \xrightarrow{R^{\mathcal{B}}_g(T_k(B))} R^{\mathcal{B}}_g(R^{\mathcal{B}}_k(F(B)))$$
$$\xrightarrow{L^{\mathcal{B}}_{k,g}(F(B))^{-1}} R^{\mathcal{B}}_{kg}(F(B))$$

agrees with the composite

$$F(R^{\mathcal{A}}_g(R^{\mathcal{A}}_k(B))) \xrightarrow{F(L^{\mathcal{A}}_{k,g}(B)^{-1})} F(R^{\mathcal{A}}_{kg}(B)) \xrightarrow{T_{kg}(A)} R^{\mathcal{B}}_{kg}(F(B)).$$

This follows in both cases from the commutativity of the diagram (2.2). This finishes the proof that $\mathcal{S}(F,T)$ is a functor of additive categories with strict $(G,v)$-action.

Let $S\colon (F_1,T_1) \to (F_2,T_2)$ be a natural transformation of $\epsilon$-variant functors of additive categories with weak $(G,v)$-action $(F_1,T_1)\colon \mathcal{A}_1 \to \mathcal{A}_2$ and $(F_2,T_2)\colon \mathcal{A}_1 \to \mathcal{A}_2$. It induces a natural transformation $\mathcal{S}(S)\colon \mathcal{S}(F_1,T_1) \to \mathcal{S}(F_2,T_2)$ of functors of additive categories with strict $(G,v)$-action as follows. Given an object $(A,g)$ in $\mathcal{S}(\mathcal{A})$, we have to specify a morphism in $\mathcal{S}(\mathcal{A})$

$$\mathcal{S}(S)(A)\colon \mathcal{S}(F_1,T_1)(A,g) = (F_1(A),g) \to \mathcal{S}(F_2,T_2)(A,g) = (F_2(A),g),$$

i.e., a morphism $R^{\mathcal{A}}_g(F_1(A)) \to R^{\mathcal{A}}_g(F_2(A))$ in $\mathcal{A}$. We take $R^{\mathcal{A}}_g(S(A))^{v(g)}$. We leave it to the reader to check that this is indeed a natural transformation of $\epsilon$-variant functors of additive categories with strict $(G,v)$-action using the commutativity of the diagram (2.3).

Let $(G,v)\text{-}\mathsf{Add}\text{-}\mathsf{Cat}^{\epsilon}$ be the category of additive categories with weak $(G,v)$-action with $\epsilon$-variant functors as morphisms and let $\mathsf{strict}\text{-}(G,v)\text{-}\mathsf{Add}\text{-}\mathsf{Cat}^{\epsilon}$ be the category of additive categories with strict $(G,v)$-action with $\epsilon$-variant functors as morphisms. There is the forgetful functor

$$\text{forget}\colon \mathsf{strict}\text{-}(G,v)\text{-}\mathsf{Add}\text{-}\mathsf{Cat}^{\epsilon} \to (G,v)\text{-}\mathsf{Add}\text{-}\mathsf{Cat}^{\epsilon}$$

and the functor constructed above

$$\mathcal{S}\colon (G,v)\text{-}\mathsf{Add}\text{-}\mathsf{Cat}^{\epsilon} \to \mathsf{strict}\text{-}(G,v)\text{-}\mathsf{Add}\text{-}\mathsf{Cat}^{\epsilon}.$$

**Lemma 3.2.** (i) *We obtain an adjoint pair of functors* $(\mathcal{S}, \text{forget})$.

(ii) *We get for every additive category* $\mathcal{A}$ *with weak* $(G,v)$*-action a functor of additive categories with weak* $(G,v)$*-action*

$$P_{\mathcal{A}}\colon \mathcal{A} \to \text{forget}(\mathcal{S}(\mathcal{A}))$$

*which is natural in* $\mathcal{A}$ *and whose underlying functor of additive categories is an equivalence of additive categories.*

*Proof.* We will only treat the case, where $v$ is trivial and $\epsilon = 1$, the other cases are analogous.

(i) We have to construct for any additive category $\mathcal{A}$ with weak $G$-action and any additive category $\mathcal{B}$ with strict $G$-action to one another inverse maps

$$\alpha\colon \operatorname{func}_{\text{strict-}(G,v)\text{-Add-Cat}}(\mathcal{S}(\mathcal{A}), \mathcal{B}) \to \operatorname{func}_{(G,v)\text{-Add-Cat}}(\mathcal{A}, \operatorname{forget}(\mathcal{B}))$$

and

$$\beta\colon \operatorname{func}_{(G,v)\text{-Add-Cat}}(\mathcal{A}, \operatorname{forget}(\mathcal{B})) \to \operatorname{func}_{\text{strict-}(G,v)\text{-Add-Cat}}(\mathcal{S}(\mathcal{A}), \mathcal{B}).$$

For a functor of additive categories with strict $G$-action $F\colon \mathcal{S}(\mathcal{A}) \to \mathcal{B}$, the functor of additive categories with weak $G$-action, $\alpha(F)\colon \mathcal{A} \to \operatorname{forget}(\mathcal{B})$ is given by a functor $\alpha(F)\colon \mathcal{A} \to \operatorname{forget}(\mathcal{B})$ of additive categories and a collection of natural transformations $T(F)_g\colon \alpha(F) \circ R_g^{\mathcal{A}} \to R_g^{\mathcal{B}} \circ \alpha(F)$ satisfying certain compatibility conditions. We first explain the functor $\alpha(F)\colon \mathcal{A} \to \operatorname{forget}(\mathcal{B})$. It sends a morphism $f\colon A \to B$ in $\mathcal{A}$ to the morphism in $\mathcal{B}$ which is given by the value of $F$ on the morphism $(A,e) \to (B,e)$ in $\mathcal{S}(\mathcal{A})$ defined by $f$. For $g \in G$ the transformation $T(F)_g$ evaluated at an object $A$ in $\mathcal{A}$ is the morphism

$$\alpha(F)(R_g^{\mathcal{A}}(A)) = F(R_g^{\mathcal{A}}(A), e) \to R_g^{\mathcal{B}}(\alpha(F)(A)) = R_g^{\mathcal{B}}(F(A,e))$$

defined as follows. It is given by the composite of the image under $F$ of the morphism $(R_g^{\mathcal{A}}(A), e) \to R_g^{\mathcal{S}(\mathcal{A})}(A,e) = (A,g)$ in $\mathcal{S}(\mathcal{A})$ which is defined by the identity morphism $\operatorname{id}\colon R_g^{\mathcal{A}}(A) \to R_g^{\mathcal{A}}(A)$ in $\mathcal{A}$ and the identity $F(R_g^{\mathcal{S}(\mathcal{A})}(A,e)) = R_g^{\mathcal{B}}(F(A,e))$ which comes from the assumption that $F$ is a functor of strict additive $G$-categories. One easily checks that $\alpha(F)$ satisfies condition (2.2) since it is satisfied for $F$.

Given a functor of additive categories with weak $G$-action $(F,T)\colon \mathcal{A} \to \operatorname{forget}(\mathcal{B})$, the functor of additive categories with strict $G$-action $\beta(F,T)\colon \mathcal{S}(\mathcal{A}) \to \mathcal{B}$ is defined as follows. It sends an object $(A,h)$ to $R_h^{\mathcal{B}}(F(A))$. A morphism $\phi\colon (A,h) \to (B,k)$ in $\mathcal{S}(A)$ which is given by a morphism $\phi\colon R_h^{\mathcal{A}}A \to R_k^{\mathcal{A}}B$ in $\mathcal{A}$ is sent to morphism in $\mathcal{B}$ given by the composite

$$R_h^{\mathcal{B}}(F(A)) \xrightarrow{T_h(A)^{-1}} F(R_h^{\mathcal{A}}(A)) \xrightarrow{F(\phi)} F(R_k^{\mathcal{A}}(B)) \xrightarrow{T_k(B)} R_k^{\mathcal{B}}(F(B)).$$

The following calculation shows that $\beta(F,T)$ is indeed a functor of additive categories with strict $G$-action. Given an element $g \in G$ the morphism $R_g^{\mathcal{S}(\mathcal{A})}(\phi)\colon (A,hg) \to (B,kg)$ in $\mathcal{S}(\mathcal{A})$ is given by the morphism in $\mathcal{A}$

$$R_{hg}^{\mathcal{A}}(A) \xrightarrow{L_{h,g}^{\mathcal{A}}(A)} R_g^{\mathcal{A}}(R_h^{\mathcal{A}}(A)) \xrightarrow{R_g^{\mathcal{A}}(\phi)} R_g^{\mathcal{A}}(R_k^{\mathcal{A}}(B)) \xrightarrow{L_{k,g}^{\mathcal{A}}(B)^{-1}} R_{kg}^{\mathcal{A}}(B).$$

Hence $\beta(F,T) \circ R_g^{\mathcal{S}(\mathcal{A})}(\phi)$ is the morphism in $\mathcal{B}$ given by the composite

$$\begin{aligned} R_{hg}^{\mathcal{B}}(F(A)) &\xrightarrow{T_{hg}(A)^{-1}} F(R_{hg}^{\mathcal{A}}(A)) \xrightarrow{F(L_{h,g}^{\mathcal{A}}(A))} F(R_g^{\mathcal{A}}(R_h^{\mathcal{A}}(A))) \\ &\xrightarrow{F(R_g^{\mathcal{A}}(\phi))} F(R_g^{\mathcal{A}}(R_k^{\mathcal{A}}(B))) \xrightarrow{F(L_{k,g}^{\mathcal{A}}(B)^{-1})} F(R_{kg}^{\mathcal{A}}(B))) \\ &\xrightarrow{T_{kg}(B)} R_{kg}^{\mathcal{A}}(F(B))). \end{aligned}$$

The morphism $R_g^{\mathcal{B}}(B) \circ \beta(F,T)(\phi)$ in $\mathcal{B}$ is given by the composite

$$R_g^{\mathcal{B}}(R_h^{\mathcal{B}}(F(A))) \xrightarrow{R_g^{\mathcal{B}}(T_h(A)^{-1})} R_g^{\mathcal{B}}(F(R_h^{\mathcal{A}}(A))) \xrightarrow{R_g^{\mathcal{B}}(F(\phi))} R_g^{\mathcal{B}}(F(R_k^{\mathcal{A}}(B))) \xrightarrow{R_g^{\mathcal{B}}(T_k(B))} R_g^{\mathcal{B}}(R_k^{\mathcal{B}}(F(B))).$$

Since $\mathcal{B}$ is a additive category with strict $G$-action by assumption, we have the equalities $R_g^{\mathcal{B}}(R_h^{\mathcal{B}}(F(A))) = R_{hg}^{\mathcal{B}}(F(A))$ and $R_g^{\mathcal{B}}(R_k^{\mathcal{B}}(B)) = R_{kg}^{\mathcal{A}}(F(B)))$. We must show that under these identifications the two morphisms in $\mathcal{B}$ above agree. Since $T_g$ is a natural transformation $F \circ R_g^{\mathcal{A}} \to R_g^{\mathcal{B}} \circ F$, the following diagram commutes

$$\begin{array}{ccc} F(R_g^{\mathcal{A}}(R_h^{\mathcal{A}}(A))) & \xrightarrow{F(R_g^{\mathcal{A}}(\phi))} & F(R_g^{\mathcal{A}}(R_k^{\mathcal{A}}(B))) \\ \downarrow{\scriptstyle T_g(R_h^{\mathcal{A}}(A))} & & \downarrow{\scriptstyle T_g(R_k^{\mathcal{A}}(B))} \\ R_g^{\mathcal{B}}(F(R_h^{\mathcal{A}}(A))) & \xrightarrow{R_g^{\mathcal{B}}(F(\phi))} & R_g^{\mathcal{B}}(F(R_k^{\mathcal{A}}(B))) \end{array}$$

Hence it suffices to show that the composites

$$R_g^{\mathcal{B}}(R_h^{\mathcal{B}}(F(A))) = R_{hg}^{\mathcal{B}}(F(A)) \xrightarrow{T_{hg}(A)^{-1}} F(R_{hg}^{\mathcal{A}}(A)) \xrightarrow{F(L_{h,g}^{\mathcal{A}}(A))} F(R_g^{\mathcal{A}}(R_h^{\mathcal{A}}(A)))$$

and

$$R_g^{\mathcal{B}}(R_h^{\mathcal{B}}(F(A))) \xrightarrow{R_g^{\mathcal{B}}(T_h(A)^{-1})} R_g^{\mathcal{B}}(F(R_h^{\mathcal{A}}(A))) \xrightarrow{T_g(R_h^{\mathcal{A}}(A))^{-1}} F(R_g^{\mathcal{A}}(R_h^{\mathcal{A}}(A)))$$

agree and that the composites

$$F(R_g^{\mathcal{A}}(R_k^{\mathcal{A}}(B))) \xrightarrow{F(L_{k,g}^{\mathcal{A}}(B)^{-1})} F(R_{kg}^{\mathcal{A}}(B))) \xrightarrow{T_{kg}(B)} R_{kg}^{\mathcal{A}}(F(B))) = R_g^{\mathcal{B}}(R_k^{\mathcal{B}}(F(B)))$$

and

$$F(R_g^{\mathcal{A}}(R_k^{\mathcal{A}}(B))) \xrightarrow{T_g(R_k^{\mathcal{A}}(B))} R_g^{\mathcal{B}}(F(R_k^{\mathcal{A}}(B))) \xrightarrow{R_g^{\mathcal{B}}(T_k(B))} R_g^{\mathcal{B}}(R_k^{\mathcal{B}}(F(B))$$

agree. This follows in both cases from the commutativity of the diagram (2.2). This finishes the proof that $\beta(F)$ is a functor of additive categories with strict $G$-action. We leave it to the reader to check that both composites $\beta \circ \alpha$ and $\alpha \circ \beta$ are the identity.

(ii) The in $\mathcal{A}$ natural functor of additive categories with weak $(G,v)$-action

$$P_{\mathcal{A}} \colon \mathcal{A} \to \operatorname{forget}(\mathcal{S}(\mathcal{A}))$$

is defined to be the adjoint of the identity functor $\mathrm{id} \colon \mathcal{S}(\mathcal{A}) \to \mathcal{S}(\mathcal{A})$. Explicitly it sends an object $A$ to the object $(A,e)$ and a morphism $\phi \colon A \to B$ to the morphism $(A,e) \to (B,e)$ given by $\phi$. Obviously $P_{\mathcal{A}}$ induces a bijection $\operatorname{mor}_{\mathcal{A}}(A,B) \to \operatorname{mor}_{\mathcal{S}(\mathcal{A})}(P_{\mathcal{A}}(A), P_{\mathcal{A}}(B))$ and for every object $(A,g)$ in $\mathcal{S}(\mathcal{A})$ there is an object in the image of $P_{\mathcal{A}}$ which is isomorphic to $(A,g)$, namely, $P_{\mathcal{A}}(R_g^{\mathcal{A}}(A)) = (R_g^{\mathcal{A}}(A), e)$. Hence the underlying functor $R_{\mathcal{A}}$ is an equivalence of additive categories. □

# 4. Crossed product rings and involutions

In this subsection we will introduce the concept of a crossed product ring. Let $R$ be a ring and let $G$ be a group. Let $e \in G$ be the unit in $G$ and denote by 1 the multiplicative unit in $R$. Suppose that we are given maps of sets

$$c\colon G \to \operatorname{aut}(R), \quad g \mapsto c_g; \tag{4.1}$$

$$\tau\colon G \times G \to R^{\times}. \tag{4.2}$$

We require

$$c_{\tau(g,g')} \circ c_{gg'} = c_g \circ c_{g'}; \tag{4.3}$$

$$\tau(g,g') \cdot \tau(gg',g'') = c_g(\tau(g',g'')) \cdot \tau(g,g'g''); \tag{4.4}$$

$$c_e = \mathrm{id}_R; \tag{4.5}$$

$$\tau(e,g) = 1; \tag{4.6}$$

$$\tau(g,e) = 1, \tag{4.7}$$

for $g, g', g'' \in G$, where $c_{\tau(g,g')}\colon R \to R$ is conjugation with $\tau(g,g')$, i.e., it sends $r$ to $\tau(g,g')r\tau(g,g')^{-1}$. Let $R * G = R *_{c,\tau} G$ be the free $R$-module with the set $G$ as basis. It becomes a ring with the following multiplication

$$\left(\sum_{g\in G} \lambda_g g\right) \cdot \left(\sum_{h \in G} \mu_h h\right) = \sum_{g \in G} \left( \sum_{\substack{g',g'' \in G,\\ g'g''=g}} \lambda_{g'} c_{g'}(\mu_{g''}) \tau(g',g'') \right) g.$$

This multiplication is uniquely determined by the properties $g \cdot r = c_g(r) \cdot g$ and $g \cdot g' = \tau(g,g') \cdot (gg')$. The conditions (4.3) and (4.4) relating $c$ and $\tau$ are equivalent to the condition that this multiplication is associative. The other conditions (4.5), (4.6) and (4.7) are equivalent to the condition that the element $1 \cdot e$ is a multiplicative unit in $R * G$. We call

$$R * G = R *_{c,\tau} G \tag{4.8}$$

the *crossed product* of $R$ and $G$ with respect to $c$ and $\tau$.

**Example 4.9.** Let $1 \to H \xrightarrow{i} G \xrightarrow{p} Q \to 1$ be an extension of groups. Let $s\colon Q \to G$ be a map satisfying $p \circ s = \mathrm{id}$ and $s(e) = e$. We do not require $s$ to be a group homomorphism. Define $c\colon Q \to \operatorname{aut}(RH)$ by $c_q(\sum_{h \in H} \lambda_h h) = \sum_{h\in H} \lambda_h s(q) h s(q)^{-1}$. Define $\tau\colon Q \times Q \to (RH)^{\times}$ by $\tau(q,q') = s(q)s(q')s(qq')^{-1}$. Then we obtain a ring isomorphism $RH * Q \to RG$ by sending $\sum_{q \in Q} \lambda_q q$ to $\sum_{q \in Q} i(\lambda_q) s(q)$, where $i\colon RH \to RG$ is the ring homomorphism induced by $i\colon H \to G$. Notice that $s$ is a group homomorphism if and only if $\tau$ is constant with value $1 \in R$.

Next we consider the additive category with involution $R$-$\mathsf{FGP}$ of finitely generated projective $R$-modules. For $g \in G$ we obtain a functor $\operatorname{res}_{c_g}\colon R\text{-}\mathsf{FGP} \to$

$R$-FGP by restriction with the ring automorphism $c_g\colon R \to R$. Define natural transformation of functors $R$-FGP $\to$ $R$-FGP

$$L_{\tau(g,h)}\colon \operatorname{res}_{c_{gh}} \to \operatorname{res}_{c_h} \circ \operatorname{res}_{c_g}$$

by assigning to a finitely generated projective $R$-module the $R$-homomorphism

$$\operatorname{res}_{c_{gh}} P \to \operatorname{res}_{c_h} \operatorname{res}_{c_g} P, \quad p \mapsto \tau(g,h)p.$$

This is indeed an $R$-linear map because of the following computation for $r \in R$ and $p \in P$

$$\begin{aligned}\tau(g,h)c_{gh}(r) &= \tau(g,h)c_{gh}(r)\tau(g,h)^{-1}\tau(g,h) = c_{\tau(g,h)} \circ c_{gh}(r)\tau(g,h)\\ &= c_g \circ c_h(r)\tau(g,h).\end{aligned}$$

**Lemma 4.10.** *We get from the collections* $\{\operatorname{res}_{c_g} \mid g \in G\}$ *and* $\{L_{\tau(g,h)} \mid g,h \in G\}$ *the structure of an additive category with weak $G$-action on $R$-FGP.*

*Proof.* Condition (4.4) implies that for every finitely generated projective $R$-module the composites

$$\operatorname{res}_{c_{gg'g''}} P \xrightarrow{L_{\tau(g,g'g'')}} \operatorname{res}_{c_{g'g''}} \operatorname{res}_{c_g} P \xrightarrow{L_{c_g(\tau(g',g''))}} \operatorname{res}_{c_{g''}} \operatorname{res}_{C_{g'}} \operatorname{res}_{c_g} P$$

and

$$\operatorname{res}_{c_{gg'g''}} P \xrightarrow{L_{\tau(gg',g'')}} \operatorname{res}_{c_{g''}} \operatorname{res}_{c_{gg'}} P \xrightarrow{L_{\tau(g,g')}} \operatorname{res}_{c_{g''}} \operatorname{res}_{C_{g'}} \operatorname{res}_{c_g} P$$

agree. This takes care of condition (iii) in Definition 2.1. We conclude $(\operatorname{res}_{c(e)} = \mathrm{id}$, $L_{\tau(g,e)} = \mathrm{id}$ and $L_{\tau(e,g)} = \mathrm{id}$ for all $g \in G$ from (4.5), (4.6) and (4.7). □

Because of Lemma 4.10 we obtain two additive categories with strict $G$-action from the constructions of Section 3

(4.11) $$R\text{-FGP}_{c,\tau} := \mathcal{S}(R\text{-FGP});$$

From now on assume that $R$ comes with an involution of rings $r \mapsto \overline{r}$. We want to consider extensions of it to an involution on $R * G$. Suppose that additionally we are given a map

(4.12) $$w\colon G \to R.$$

We require the following conditions for $g,h \in G$ and $r \in R$

(4.13) $$w(e) = 1;$$

(4.14) $$w(gh) = w(h)c_{h^{-1}}(w(g))\tau(h^{-1},g^{-1})c_{(gh)^{-1}}\left(\overline{\tau(g,h)}\right)^{-1};$$

(4.15) $$\overline{w(g)} = w(g)c_g^{-1}\left(\tau(g,g^{-1})\overline{\tau(g,g^{-1})}^{-1}\right);$$

(4.16) $$\overline{c_g(r)} = c_g\left(\left(w(g)\tau(g^{-1},g)\right)^{-1}\overline{r}\left(w(g)\tau(g^{-1},g)\right)\right).$$

We claim that there is precisely one involution on $R * G$ with the properties that it extends the involution on $R$ and sends $g$ to $w(g) \cdot g^{-1}$. The candidate for the involution is

$$\overline{\sum_{g \in G} r_g \cdot g} := \sum_{g \in G} w(g) c_{g^{-1}}(\overline{r_g}) \cdot g^{-1}. \tag{4.17}$$

One easily concludes from the requirements and the axioms of an involution that this is the only possible formula for such an involution. Namely,

$$\begin{aligned}\overline{\sum_{g \in G} r_g \cdot g} &= \sum_{g \in G} \overline{r_g \cdot g} = \sum_{g \in G} \overline{g} \cdot \overline{r_g} = \sum_{g \in G} w(g) \cdot g^{-1} \cdot \overline{r_g} \\ &= \sum_{g \in G} w(g) \cdot (g^{-1} \cdot \overline{r_g} \cdot g) \cdot g^{-1} = \sum_{g \in G} w(g) c_{g^{-1}}(\overline{r_g}) \cdot g^{-1}.\end{aligned}$$

Before we explain that this definition indeed satisfies the axioms for an involution, we show that the conditions about $w$ above are necessary for this map to be an involution on $R * G$. So assume that we have an involution on $R * G$ that extends the involution on $R$ and sends $g$ to $w(g) \cdot g^{-1}$ for a given map $w \colon G \to R$. Denote by 1 the multiplicative unit in both $R$ and $R * G$. From

$$1 \cdot e = 1 = \overline{1} = \overline{1 \cdot e} = w(e) \cdot e$$

we conclude (4.13). The equality

$$\begin{aligned} w(gh) c_{(gh)^{-1}} \left( \overline{\tau(g,h)} \right) \cdot (gh)^{-1} &= \overline{\tau(g,h) \cdot gh} = \overline{g \cdot h} \\ &= \overline{h} \cdot \overline{g} = w(h) \cdot h^{-1} \cdot w(g) \cdot g^{-1} \\ &= w(h) \left( h^{-1} \cdot w(g) \cdot h \right) \cdot h^{-1} \cdot g^{-1} \\ &= w(h) c_{h^{-1}}(w(g)) \tau(h^{-1}, g^{-1}) \cdot (gh)^{-1} \end{aligned}$$

implies (4.14). If we take $h = g^{-1}$ in (4.14) and use (4.13), we get

$$1 = w(e) = w(gg^{-1}) = w(g^{-1}) c_g(w(g)) \tau(g, g^{-1}) \overline{\tau(g, g^{-1})}^{-1}. \tag{4.18}$$

This implies that for all $g \in G$ the element $w(g)$ is a unit in $R$ with inverse

$$w(g)^{-1} = c_{g^{-1}}(w(g^{-1})) \tau(g^{-1}, g) \overline{\tau(g^{-1}, g)}^{-1}.$$

The equality

$$\begin{aligned} g = \overline{\overline{g}} = \overline{w(g) \cdot g^{-1}} &= \overline{g^{-1}} \cdot \overline{w(g)} = w(g^{-1}) \cdot g \cdot \overline{w(g)} \\ &= w(g^{-1}) \cdot \left( g \cdot \overline{w(g)} \cdot g^{-1} \right) \cdot g = w(g^{-1}) c_g \left( \overline{w(g)} \right) \cdot g \end{aligned}$$

together with (4.18) implies

$$w(g^{-1}) c_g \left( \overline{w(g)} \right) = 1 = w(g^{-1}) c_g(w(g)) \tau(g, g^{-1}) \overline{\tau(g, g^{-1})}^{-1}.$$

If we multiply this equation with $w(g^{-1})^{-1}$ and apply the inverse $c_g^{-1}$ of $c_g$, we derive condition (4.15). The equality

$$\begin{aligned}\overline{r}\cdot w(g)\cdot g^{-1} &= \overline{r}\cdot\overline{g} = \overline{g\cdot r} = \overline{(g\cdot r\cdot g^{-1})\cdot g} = \overline{c_g(r)\cdot g} = \overline{g}\cdot\overline{c_g(r)}\\ &= w(g)\cdot g^{-1}\cdot\overline{c_g(r)} = w(g)\cdot\left(g^{-1}\cdot\overline{c_g(r)}\cdot g\right)\cdot g^{-1}\\ &= w(g)\cdot c_{g^{-1}}\left(\overline{c_g(r)}\right)\cdot g^{-1}\end{aligned}$$

implies that for all $g\in G$ and $r\in R$ we have $\overline{r}\cdot w(g) = w(g)\cdot c_{g^{-1}}\left(\overline{c_g(r)}\right)$ and hence

$$\overline{c_g(r)} = c_{g^{-1}}^{-1}\left(w(g)^{-1}\overline{r}w(g)\right).$$

From the relation (4.3) we conclude $c_{\tau(g^{-1},g)} = c_{g^{-1}}\circ c_g$ and hence $c_{g^{-1}}^{-1} = c_g\circ c_{\tau(g^{-1},g)}^{-1}$. Now condition (4.16) follows.

Finally we show that the conditions (4.13), (4.14), (4.15) and (4.16) on $w$ do imply that we get an involution of rings on $R*G$ by the formula (4.17). Obviously this formula is compatible with the additive structure on $R*G$ and sends 1 to 1. In order to show that it is an involution and compatible with the multiplicative structure we have to show $\overline{g\cdot h} = \overline{h}\cdot\overline{g}$, $\overline{rs} = \overline{s}\cdot\overline{r}$, $\overline{r\cdot g} = \overline{g}\cdot\overline{r}$, $\overline{g\cdot r} = \overline{r}\cdot\overline{g}$, $\overline{\overline{r}} = r$ and $\overline{\overline{g}} = g$ for $r,s\in R$ and $g,h\in G$. We get $\overline{rs} = \overline{s}\cdot\overline{r}$ and $\overline{\overline{r}} = r$ from the fact that we start with an involution on $R$. The other equations follow from the proofs above that (4.17) is the only possible candidate and that the conditions about $w$ are necessary for the existence of the desired involution on $R*G$, just read the various equations and implications backwards. We will denote the resulting ring with involution by

(4.19) $$R *_{c,\tau,w} G.$$

**Example 4.20.** Suppose that we are in the situation of Example 4.9. Suppose that we are additionally given a group homomorphism $w_1\colon G\to \operatorname{cent}(R)^\times$ to the abelian group of invertible central elements in $R$ satisfying $\overline{w_1(g)} = w_1(g)$ for all $g\in G$. The $w_1$-twisted involution on $RG$ is defined by $\overline{\sum_{g\in G} r_g\cdot g} = \sum_{g\in G}\overline{r_g}w_1(g)\cdot g^{-1}$. It extends the $w_1|_H$-involution on $RH$. We obtain an involution on $RH*Q$ if we conjugate the $w_1$-twisted involution with the isomorphism $RH*Q\xrightarrow{\cong} RG$ which we have introduced in Example 4.9. This involution on $RH*Q$ sends $q\in Q$ to the element $w_1(s(q))\tau(q^{-1},q)^{-1}\cdot q^{-1}$ because of the following calculation in $RG$ for $q\in Q$

$$\begin{aligned}\overline{s(q)} &= w_1(s(q))\cdot s(q)^{-1} = w_1(s(q))\cdot s(q)^{-1}\cdot s(q^{-1})^{-1}\cdot s(q^{-1})\\ &= w_1(s(q))\cdot\left(s(q^{-1})\cdot s(q)\right)^{-1}\cdot s(q^{-1})\\ &= w_1(s(q))\cdot\left(\tau(q^{-1},q)s(q^{-1}q)\right)^{-1}\cdot s(q^{-1})\\ &= w_1(s(q))\tau(q^{-1},q)^{-1}\cdot s(q^{-1}).\end{aligned}$$

Define

$$w\colon Q\to RH,\quad q\mapsto w_1(s(q))\tau(q^{-1},q)^{-1}.$$

Then $w$ satisfies the conditions (4.13), (4.14), (4.15) and (4.16) and the involution on $RH * Q$ determined by $w$ corresponds under the isomorphism $RH * Q \xrightarrow{\cong} RG$ to the $w_1$-twisted involution on $RG$.

Let

$$t_g\colon \operatorname{res}_{c_g} \circ I_{R\text{-FGP}} \to I_{R\text{-FGP}} \circ \operatorname{res}_{c_g} \tag{4.21}$$

be the natural transformation which assigns to a finitely generated projective $R$-module $P$ the $R$-isomorphism $t_g(P)\colon \operatorname{res}_{c_g} P^* \to (\operatorname{res}_{c_g} P)^*$ which sends the $R$-linear map $f\colon P \to R$ to the $R$-linear map

$$t_g(P)(f)\colon \operatorname{res}_{c_g} P \to R, \quad p \mapsto c_g^{-1}(f(p))\left(w(g)\tau(g^{-1},g)\right)^{-1}.$$

We firstly check that $t_g(P)(f)\colon \operatorname{res}_{c_g} P \to R$ is $R$-linear by the following computation

$$\begin{aligned}
t_g(P)(f)(c_g(r)p) &= c_g^{-1}\left(f(c_g(r)p)\right)\left(w(g)\tau(g^{-1},g)\right)^{-1}\\
&= c_g^{-1}\left(c_g(r)f(p)\right)\left(w(g)\tau(g^{-1},g)\right)^{-1}\\
&= c_g^{-1}\left(c_g(r)\right)c_g^{-1}(f(p))\left(w(g)\tau(g^{-1},g)\right)^{-1}\\
&= rc_g^{-1}(f(p))\left(w(g)\tau(g^{-1},g)\right)^{-1}\\
&= rt_g(P)(f)(p).
\end{aligned}$$

Finally we check that $t_g(P)\colon \operatorname{res}_{c_g} P^* \to (\operatorname{res}_{c_g} P)^*$ is $R$-linear by the following calculation for $f \in P^*$ and $p \in P$

$$\begin{aligned}
&t_g(P)\left((c_g(r)f)\right)(p)\\
&= c_g^{-1}\left((c_g(r)f)(p)\right)\left(w(g)\tau(g^{-1},g)\right)^{-1}\\
&= c_g^{-1}\left(f(p)\overline{c_g(r)}\right)\left(w(g)\tau(g^{-1},g)\right)^{-1}\\
&= c_g^{-1}(f(p))c_g^{-1}\left(\overline{c_g(r)}\right)\left(w(g)\tau(g^{-1},g)\right)^{-1}\\
&= c_g^{-1}(f(p))c_g^{-1}\left(c_g\left(\left(w(g)\tau(g^{-1},g)\right)^{-1}\overline{r}\left(w(g)\tau(g^{-1},g)\right)\right)\right)\left(w(g)\tau(g^{-1},g)\right)^{-1}\\
&= c_g^{-1}(f(p))\left(w(g)\tau(g^{-1},g)\right)^{-1}\overline{r}\left(w(g)\tau(g^{-1},g)\right)\left(w(g)\tau(g^{-1},g)\right)^{-1}\\
&= c_g^{-1}(f(p))\left(w(g)\tau(g^{-1},g)\right)^{-1}\overline{r}\\
&= t_g(P)(f)(p)\overline{r}\\
&= \left(rt_g(P)\right)(f)(p).
\end{aligned}$$

**Definition 4.22.** An *additive $G$-category with involution* $\mathcal{A}$ is an additive $G$-category, which is the same as an additive category with strict $G$-action (see Definition 2.1), together with an involution $(I, E)$ of additive categories (see (1.1) and (1.2)) with the following properties: $I\colon \mathcal{A} \to \mathcal{A}$ is a contravariant functor of additive $G$-categories, i.e., $R_g \circ I = I \circ R_g$ for all $g \in G$, and $E\colon \mathrm{id}_{\mathcal{A}} \to I \circ I$ is a natural transformation of functors of additive $G$-categories, i.e., for every $g \in G$ and every object $A$ in $\mathcal{A}$ the morphisms $E(R_g(A))$ and $R_g(E(A))$ from $R_g(A)$ to $I^2 \circ R_g(A) = R_g \circ I^2(A)$ agree.

**Lemma 4.23.** *The additive category with strict $G$-action $R\text{-}\mathsf{FGP}_{c,\tau}$ of (4.11) inherits the structure of an additive $G$-category with involution in the sense of Definition 4.22.*

*Proof.* We firstly show that

$$I_{R\text{-}\mathsf{FGP}} \colon R\text{-}\mathsf{FGP} \to R\text{-}\mathsf{FGP}$$

together with the collection of the $\{t_g^{-1} \colon I_{R\text{-}\mathsf{FGP}} \circ \operatorname{res}_{c_g} \to \operatorname{res}_{c_g} \circ I_{R\text{-}\mathsf{FGP}} \mid g \in G\}$ (see (4.21)) is a contravariant functor of additive categories with weak $G$-action. We have to verify that the diagram (2.2) commutes. This is equivalent to show for every finitely generated projective $R$-module $P$ and $g, h \in G$ that the following diagram commutes

$$\begin{array}{ccccc}
\operatorname{res}_{c_{gh}} P^* & & \xrightarrow{\quad t_{gh}(P) \quad} & & (\operatorname{res}_{c_{gh}} P)^* \\
\Big\downarrow{\scriptstyle L_{\tau(g,h)}(P^*)} & & & & \Big\uparrow{\scriptstyle L_{\tau(g,h)}(P)^*} \\
\operatorname{res}_{c_h} \operatorname{res}_{c_g} P^* & \xrightarrow{\operatorname{res}_{c_h} t_g(P)} & \operatorname{res}_{c_h} (\operatorname{res}_{c_g} P)^* & \xrightarrow{t_h(\operatorname{res}_g P)} & (\operatorname{res}_{c_h} \operatorname{res}_{c_g} P)^*
\end{array}$$

We start with an element $f \colon P \to R$ in the left upper corner. Its image under the upper horizontal arrow is $p \mapsto c_{gh}^{-1}(f(p)) \left(w(gh)\tau((gh)^{-1}, gh)\right)^{-1}$. Next we list successively how its image looks like if we go in the anticlockwise direction from the left upper corner to the right upper corner. We first get $p \mapsto f(p)\overline{\tau(g,h)}$. After the second map we get $p \mapsto c_g^{-1}\left(f(p)\overline{\tau(g,h)}\right)\left(w(g)\tau(g^{-1},g)\right)^{-1}$. After applying the third map we obtain $p \mapsto c_h^{-1}\left(c_g^{-1}\left(f(p)\overline{\tau(g,h)}\right)\left(w(g)\tau(g^{-1},g)\right)^{-1}\right)\left(w(h)\tau(h^{-1},h)\right)^{-1}$. Finally we get $p \mapsto c_h^{-1}\left(c_g^{-1}\left(f(\tau(g,h)p)\overline{\tau(g,h)}\right)\left(w(g)\tau(g^{-1},g)\right)^{-1}\right)\left(w(h)\tau(h^{-1}, h)\right)^{-1}$. Since $f$ lies in $P^*$, we have $f(\tau(g,h)p) = \tau(g,h)f(p)$. Hence it suffices to show for all $r \in R$

$$\begin{aligned}
&c_h^{-1}\left(c_g^{-1}\left(\tau(g,h)r\overline{\tau(g,h)}\right)\left(w(g)\tau(g^{-1},g)\right)^{-1}\right)\left(w(h)\tau(h^{-1},h)\right)^{-1} \\
&\quad = c_{gh}^{-1}(r)\left(w(gh)\tau((gh)^{-1},gh)\right)^{-1}.
\end{aligned}$$

(Notice that now $f$ has been eliminated.) By applying $c_{gh}$ we see that this is equivalent to showing

$$\begin{aligned}
&c_{gh}\left(c_h^{-1}\left(c_g^{-1}\left(\tau(g,h)r\overline{\tau(g,h)}\right)\right)\right) \\
&\quad = rc_{gh}\left(\left(w(gh)\tau((gh)^{-1},gh)\right)^{-1}\left(w(h)\tau(h^{-1},h)\right)c_h^{-1}\left(w(g)\tau(g^{-1},g)\right)\right).
\end{aligned}$$

From the relation (4.3) we conclude that $c_{gh} \circ c_{h^{-1}} \circ c_{g^{-1}}(s) = \tau(g,h)^{-1}s\tau(g,h)$ holds for all $s \in R$. Hence it remains to show

$$\begin{aligned}
&\tau(g,h)^{-1}\left(\tau(g,h)r\overline{\tau(g,h)}\right)\tau(g,h) \\
&\quad = rc_{gh}\left(\left(w(gh)\tau((gh)^{-1},gh)\right)^{-1}w(h)\tau(h^{-1},h)c_h^{-1}\left(w(g)\tau(g^{-1},g)\right)\right).
\end{aligned}$$

This reduces to proving for $g, h \in G$

$$\begin{aligned}
&\overline{\tau(g,h)}\tau(g,h) \\
&\quad = c_{gh}\left(\tau((gh)^{-1},gh)^{-1}w(gh)^{-1}w(h)\tau(h^{-1},h)c_h^{-1}\left(w(g)\tau(g^{-1},g)\right)\right).
\end{aligned}$$

(Notice that now $r$ has been eliminated.) By inserting condition (4.14) and the conclusions $c_{\tau(h^{-1},h)} \circ c_h^{-1} = c_{h^{-1}}$ and $c_{\tau((gh)^{-1},gh)} \circ c_{gh}^{-1} = c_{(gh)^{-1}}$ from conditions (4.3) and (4.5) we get

$$\begin{aligned}
&w(gh)^{-1}w(h)\tau(h^{-1},h)c_h^{-1}\left(w(g)\tau(g^{-1},g)\right) \\
&= \left(w(h)c_{h^{-1}}(w(g))\tau(h^{-1},g^{-1})c_{(gh)^{-1}}\left(\overline{\tau(g,h)}\right)^{-1}\right)^{-1} w(h) \\
&\qquad \tau(h^{-1},h)c_h^{-1}\left(w(g)\tau(g^{-1},g)\right)\tau(h^{-1},h)^{-1}\tau(h^{-1},h) \\
&= c_{(gh)^{-1}}\left(\overline{\tau(g,h)}\right)\tau(h^{-1},g^{-1})^{-1}c_{h^{-1}}(w(g))^{-1}w(h)^{-1}w(h) \\
&\qquad c_{h^{-1}}\left(w(g)\tau(g^{-1},g)\right)\tau(h^{-1},h) \\
&= c_{(gh)^{-1}}\left(\overline{\tau(g,h)}\right)\tau(h^{-1},g^{-1})^{-1}c_{h^{-1}}(w(g))^{-1}c_{h^{-1}}(w(g)) \\
&\qquad c_{h^{-1}}\left(\tau(g^{-1},g)\right)\tau(h^{-1},h) \\
&= c_{(gh)^{-1}}\left(\overline{\tau(g,h)}\right)\tau(h^{-1},g^{-1})^{-1}c_{h^{-1}}\left(\tau(g^{-1},g)\right)\tau(h^{-1},h) \\
&= \tau((gh)^{-1},gh)c_{gh}^{-1}\left(\overline{\tau(g,h)}\right)\tau((gh)^{-1},gh)^{-1}\tau(h^{-1},g^{-1})^{-1} \\
&\qquad c_{h^{-1}}\left(\tau(g^{-1},g)\right)\tau(h^{-1},h).
\end{aligned}$$

This implies

$$\begin{aligned}
&c_{gh}\left(\tau((gh)^{-1},gh)^{-1}w(gh)^{-1}w(h)\tau(h^{-1},h)c_h^{-1}\left(w(g)\tau(g^{-1},g)\right)\right) \\
&\quad = c_{gh}\left(\tau((gh)^{-1},gh)^{-1}\tau((gh)^{-1},gh)c_{gh}^{-1}\left(\overline{\tau(g,h)}\right)\tau((gh)^{-1},gh)^{-1}\right. \\
&\qquad\qquad \left.\tau(h^{-1},g^{-1})^{-1}c_{h^{-1}}\left(\tau(g^{-1},g)\right)\tau(h^{-1},h)\right) \\
&\quad = \overline{\tau(g,h)}c_{gh}\left(\tau((gh)^{-1},gh)^{-1}\tau(h^{-1},g^{-1})^{-1}c_{h^{-1}}\left(\tau(g^{-1},g)\right)\tau(h^{-1},h)\right).
\end{aligned}$$

Hence it remains to show

$$\tau(g,h) = c_{gh}\left(\tau((gh)^{-1},gh)^{-1}\tau(h^{-1},g^{-1})^{-1}c_{h^{-1}}\left(\tau(g^{-1},g)\right)\tau(h^{-1},h)\right).$$

(Notice that we have eliminated any expression involving the involution.) From condition (4.3), (4.4) and (4.5) we conclude

$$\begin{aligned}
\tau(h^{-1},g^{-1})\tau((gh)^{-1},g) &= c_{h^{-1}}(\tau(g^{-1},g)); \\
\tau((gh)^{-1},g)\tau(h^{-1},h) &= c_{(gh)^{-1}}(\tau(g,h))\tau((gh)^{-1},gh); \\
c_{gh}^{-1} &= c_{\tau((gh)^{-1},gh)^{-1}} \circ c_{(gh)^{-1}}.
\end{aligned}$$

Hence

$$\begin{aligned}
&\tau((gh)^{-1},gh)^{-1}\tau(h^{-1},g^{-1})^{-1}c_{h^{-1}}\left(\tau(g^{-1},g)\right)\tau(h^{-1},h)\\
&= \tau((gh)^{-1},gh)^{-1}\tau(h^{-1},g^{-1})^{-1}\tau(h^{-1},g^{-1})\tau((gh)^{-1},g)\tau(h^{-1},h)\\
&= \tau((gh)^{-1},gh)^{-1}\tau((gh)^{-1},g)\tau(h^{-1},h)\\
&= \tau((gh)^{-1},gh)^{-1}c_{(gh)^{-1}}(\tau(g,h))\tau((gh)^{-1},gh)\\
&= c_{gh}^{-1}(\tau(g,h)).
\end{aligned}$$

This finishes the proof of the commutativity of the diagram (2.2).

Next we show that $E_{R\text{-FGP}}\colon \mathrm{id}_{R\text{-FGP}} \to I_{R\text{-FGP}} \circ I_{R\text{-FGP}}$ is a natural transformation of contravariant functors of additive categories with weak $G$-action. We have to show that the diagram (2.3) commutes. This is equivalent to show for every finitely generated projective $R$-module $P$ the following diagram commutes

$$\begin{array}{ccc}
\mathrm{res}_{c_g} P & \xrightarrow{E_{R\text{-FGP}}(\mathrm{res}_{c_g} P)} & (\mathrm{res}_{c_g} P)^{**}\\
\Big\downarrow{\scriptstyle \mathrm{res}_{c_g} E_{R\text{-FGP}}(P)} & & \Big\downarrow{\scriptstyle t_g(P)^*}\\
\mathrm{res}_{c_g}(P^{**}) & \xrightarrow{t_g(P^*)} & (\mathrm{res}_{c_g} P^*)^*
\end{array}$$

We start with an element $p \in P$ in the left upper corner. It is sent under the left vertical arrow to the element given by $f \mapsto f(p)$. The image of this element under the lower horizontal is given by $f \mapsto c_g^{-1}(f(p))\left(w(g)\tau(g^{-1},g)\right)^{-1}$. The image of $p \in P$ under the upper horizontal arrow is $f \mapsto f(p)$. The image of this element under the right vertical arrow sends $f$ to $f \circ t_g(P)(p) = c_g^{-1}(f(p))\left(w(g)\tau(g^{-1},g)\right)^{-1}$.

From the naturality of the construction of the additive category with strict $G$-action $R\text{-FGP}_{c,\tau} := \mathcal{S}(R\text{-FGP})$ (see Section 3) we conclude that $(I_{R\text{-FGP}}, \{t_g \mid g \in G\})$ induces a functor of additive categories with strict $G$-action

$$I_{R\text{-FGP}_{c,\tau}}\colon R\text{-FGP}_{c,\tau} \to R\text{-FGP}_{c,\tau}$$

and $E_{R\text{-FGP}}$ induces a natural transformation of functors of additive categories with strict $G$-action

$$E_{R\text{-FGP}_{c,\tau}}\colon \mathrm{id}_{R\text{-FGP}} \to I_{R\text{-FGP}_{c,\tau}} \circ I_{R\text{-FGP}_{c,\tau}}.$$

It remains to prove that condition (1.3) holds for $(I_{R\text{-FGP}_{c,\tau}}, E_{R\text{-FGP}_{c,\tau}})$. But this follows easily from the fact that condition (1.3) holds for $(I_{R\text{-FGP}}, E_{R\text{-FGP}})$. □

The additive $G$-category with involution constructed in Lemma 4.23 will be denoted in the sequel by

(4.24) $$R\text{-FGP}_{c,\tau,w}.$$

# 5. Connected groupoids and additive categories

Groupoids are always to be understood to be small. A groupoid is called *connected* if for two objects $x$ and $y$ there exists a morphism $f\colon x \to y$. Let $\mathcal{G}$ be a connected groupoid. Let $\mathsf{Add\text{-}Cat}$ be the category of small additive categories. Given a contravariant functor $F\colon \mathcal{G} \to \mathsf{Add\text{-}Cat}$, we define a new small additive category, which we call its *homotopy colimit* (see for instance [15])

$$\int_{\mathcal{G}} F \tag{5.1}$$

as follows. An object is a pair $(x, A)$ consisting of an object $x$ in $\mathcal{G}$ and an object $A$ in $F(x)$. A morphism in $\int_{\mathcal{G}} F$ from $(x, A)$ to $(y, B)$ is a formal sum

$$\sum_{f \in \operatorname{mor}_{\mathcal{G}}(x,y)} f \cdot \phi_f$$

where $\phi_f\colon A \to F(f)(B)$ is a morphism in $F(x)$ and only finitely many coefficients $\phi_f$ are different from zero. The composition of a morphism $\sum_{f \in \operatorname{mor}_{\mathcal{G}}(x,y)} f \cdot \phi_f\colon (x, A) \to (y, B)$ and a morphism $\sum_{g \in \operatorname{mor}_{\mathcal{G}}(y,z)} g \cdot \phi_g\colon (y, B) \to (z, C)$ is given by the formula

$$\sum_{h \in \operatorname{mor}_{\mathcal{G}}(x,z)} h \cdot \left( \sum_{\substack{f \in \operatorname{mor}_{\mathcal{G}}(x,y) \\ g \in \operatorname{mor}_{\mathcal{G}}(y,z) \\ h = g \circ f}} F(f)(\psi_g) \circ \phi_f \right).$$

The decisive special case is

$$(g \cdot \psi) \circ (f \cdot \phi) = (g \circ f) \cdot (F(f)(\psi) \circ \phi).$$

The $\mathbb{Z}$-module structure on $\operatorname{mor}_{\int_{\mathcal{G}} F}(x, y)$ is given by

$$\left( \sum_{f \in \operatorname{mor}_{\mathcal{G}}(x,y)} f \cdot \phi_f \right) + \left( \sum_{f \in \operatorname{mor}(\mathcal{G})} f \cdot \psi_f \right) = \sum_{f \in \operatorname{mor}_{\mathcal{G}}(x,y)} f \cdot (\phi_f + \psi_f).$$

A model for the sum of two objects $(x, A)$ and $(x, B)$ is $(x, A \oplus B)$ if $A \oplus B$ is a model for the sum of $A$ and $B$ in $F(x)$. Since $\mathcal{G}$ is by assumption connected, we can choose for any object $(y, B)$ in $\int_{\mathcal{G}} F$ and any object $x$ in $\mathcal{G}$ an isomorphism $f\colon x \to y$ and the objects $(x, F(f)(B))$ and $(y, B)$ in $\int_{\mathcal{G}} F$ are isomorphic. Namely $f \cdot \mathrm{id}_{F(f)(B)}$ is an isomorphism $(x, F(f)(B)) \xrightarrow{\cong} (y, B)$ whose inverse is $f^{-1} \cdot \mathrm{id}_B$. Hence the direct sum of two arbitrary objects $(x, A)$ and $(y, B)$ exists in $\int_{\mathcal{G}} F$.

Notice that we need the connectedness of $\mathcal{G}$ only to show the existence of a direct sum. This will become important later when we deal with non-connected groupoids.

This construction is functorial in $F$. Namely, if $S\colon F_0 \to F_1$ is a natural transformation of contravariant functors $\mathcal{G} \to \mathsf{Add\text{-}Cat}$, then it induces a functor

$$\int_{\mathcal{G}} S\colon \int_{\mathcal{G}} F_0 \to \int_{\mathcal{G}} F_1 \tag{5.2}$$

of additive categories as follows. It sends an object $(x, A)$ in $\int_{\mathcal{G}} F_0$ to the object $(x, S(x)(A))$ in $\int_{\mathcal{G}} F_1$. A morphism $\sum_{f\in \mathrm{mor}_{\mathcal{G}}(x,y)} f \cdot \phi_f\colon (x, A) \to (y, B)$ is sent to the morphism

$$\sum_{f\in \mathrm{mor}_{\mathcal{G}}(x,y)} f \cdot S(x)(\phi_f)\colon (x, s(x)(A)) \to (y, s(y)(B)).$$

This makes sense since $S(x)(\phi_f)$ is a morphism in $F_1(x)$ from $S(x)(A)$ to $S(x)(F_0(f)(B)) = F_1(f)(S(y)(B))$. The decisive special case is that $\int_{\mathcal{G}} S$ sends $(f\colon x \to y) \cdot \phi$ to $(f\colon x \to y) \cdot S(x)(\phi)$. One easily checks that $\int_{\mathcal{G}} S$ is compatible with the structures of additive categories and we have

$$\left(\int_{\mathcal{G}} S_2\right) \circ \left(\int_{\mathcal{G}} S_1\right) = \int_{\mathcal{G}} (S_2 \circ S_1); \tag{5.3}$$

$$\int_{\mathcal{G}} \mathrm{id}_F = \mathrm{id}_{\int_{\mathcal{G}} F}. \tag{5.4}$$

The construction is also functorial in $\mathcal{G}$. Namely, let $W\colon \mathcal{G}_1 \to \mathcal{G}_2$ be a covariant functor of groupoids. Then we obtain a covariant functor

$$W_*\colon \int_{\mathcal{G}_1} F \circ W \to \int_{\mathcal{G}_2} F \tag{5.5}$$

of additive categories as follows. An object $(x_1, A)$ in $\int_{\mathcal{G}_1} F \circ W$ is sent to the object $(W(x_1), A)$ in $\int_{\mathcal{G}_2} F$. A morphism $\sum_{f\in \mathrm{mor}_{\mathcal{G}_1}(x_1,y_1)} f \cdot \phi_f\colon (x_1, A) \to (y_1, B)$ in $\int_{\mathcal{G}_1} F \circ W$ is sent to the morphism

$$\sum_{f\in \mathrm{mor}_{\mathcal{G}_2}(W(x_1),W(y_1))} f \cdot \left(\sum_{\substack{f_1\in \mathrm{mor}_{\mathcal{G}_1}(x_1,y_1)\\ W(f_1)=f}} \phi_{f_1}\right)\colon (W(x_1), A) \to (W(y_1), B)$$

in $\int_{\mathcal{G}_2} F$. Here the decisive special case is that $W_*$ sends the morphism $f \cdot \phi$ to $W(f) \cdot \phi$. One easily checks that $W_*$ is compatible with the structures of additive categories and we have for covariant functors $W_1\colon \mathcal{G}_1 \to \mathcal{G}_2$, $W_2\colon \mathcal{G}_2 \to \mathcal{G}_3$ and a contravariant functor $F\colon \mathcal{G} \to \mathsf{Add\text{-}Cat}$

$$(W_2)_* \circ (W_1)_* = (W_2 \circ W_1)_*; \tag{5.6}$$

$$(\mathrm{id}_{\mathcal{G}})_* = \mathrm{id}_{\int_{\mathcal{G}} F}. \tag{5.7}$$

These two constructions are compatible. Namely, given a natural transformation $S_1\colon F_1 \to F_2$ of contravariant functors $\mathcal{G} \to \mathsf{Add\text{-}Cat}$ and a covariant functor $W\colon \mathcal{G}_1 \to \mathcal{G}$, we get

$$\left(\int_{\mathcal{G}} S\right) \circ W_* = W_* \circ \left(\int_{\mathcal{G}_1} (S \circ W)\right). \tag{5.8}$$

A functor $F\colon \mathcal{C}_0 \to \mathcal{C}_1$ of categories is called an *equivalence* if there exists a functor $F'\colon \mathcal{C}_1 \to \mathcal{C}_0$ with the property that $F' \circ F$ is naturally equivalent to the identity functor $\mathrm{id}_{\mathcal{C}_0}$ and $F \circ F'$ is naturally equivalent to the identity functor $\mathrm{id}_{\mathcal{C}_1}$. A functor $F$ is a natural equivalence if and only if it is *full* and *faithful*, i.e., it induces a bijection on the isomorphism classes of objects and for any two objects $c, d$ in $\mathcal{C}_0$ the induced map $\mathrm{mor}_{\mathcal{C}_0}(c,d) \to \mathrm{mor}_{\mathcal{C}_1}(F(c), F(d))$ is bijective. If $\mathcal{C}_0$ and $\mathcal{C}_1$ come with an additional structure such as of an additive category (with involution) and $F$ is compatible with this structure, we require that $F'$ and the two natural equivalences $F' \circ F \simeq \mathrm{id}_{\mathcal{C}_0}$ and $F \circ F' \simeq \mathrm{id}_{\mathcal{C}_1}$ are compatible with these. In this case it still true that $F$ is an equivalence of categories with this additional structure if and only if $F$ is full and faithful.

One easily checks

**Lemma 5.9.** (i) *Let $W\colon \mathcal{G}_1 \to \mathcal{G}$ be an equivalence of connected groupoids. Let $F\colon \mathcal{G} \to \mathsf{Add\text{-}Cat}$ be a contravariant functor. Then*

$$W_*\colon \int_{\mathcal{G}_1} F \circ W \to \int_{\mathcal{G}} F$$

*is an equivalence of additive categories.*

(ii) *Let $\mathcal{G}$ be a connected groupoid. Let $S\colon F_1 \to F_2$ be a transformation of contravariant functors $\mathcal{G} \to \mathsf{Add\text{-}Cat}$ such that for every object $x$ in $\mathcal{G}$ the functor $S(x)\colon F_0(x) \to F_1(x)$ is an equivalence of additive categories. Then*

$$\int_{\mathcal{G}} S\colon \int_{\mathcal{G}} F_1 \to \int_{\mathcal{G}} F_2$$

*is an equivalence of additive categories.*

# 6. From crossed product rings to additive categories

**Example 6.1.** Here is our main example of a contravariant functor $\mathcal{G} \to \mathsf{Add\text{-}Cat}$. Notice that a group $G$ is the same as a groupoid with one object and hence a contravariant functor from a group $G$ to $\mathsf{Add\text{-}Cat}$ is the same as an additive $G$-category what is the same as an additive category with strict $G$-action (see Definition 2.1). Let $R$ be a ring together with maps of sets

$$c\colon G \to \mathrm{aut}(R), \quad g \mapsto c_g;$$
$$\tau\colon G \times G \to R^\times.$$

satisfying (4.3), (4.4), (4.5), (4.6) and (4.7). We have introduced the additive $G$-category $R\text{-}\mathsf{FGP}_{c,\tau}$ in (4.11). All the construction restrict to the subcategory $R\text{-}\mathsf{FGF} \subseteq R\text{-}\mathsf{FGP}$ of finitely generated free $R$-modules and lead to the additive $G$-category

(6.2) $$R\text{-}\mathsf{FGF}_{c,\tau} := \mathcal{S}(R\text{-}\mathsf{FGP});$$

**Lemma 6.3.** *Consider the data $(R, c, \tau)$ and the additive category $R$-$\mathsf{FGF}_{c,\tau}$ appearing in Example 6.1. Let $\int_G R$-$\mathsf{FGF}_{c,\tau}$ be the additive category defined in* (5.1). *Since $G$ regarded as a groupoid has precisely one object, we can (and will) identify the set of objects in $\int_G R$-$\mathsf{FGF}_{c,\tau}$ with the set of objects in $R$-$\mathsf{FGF}_{c,\tau}$ which consists of pairs $(M, g)$ for $M$ a finitely generated free $R$-module and $g \in G$. Denote by $\left(\int_G R\text{-}\mathsf{FGF}_{c,\tau}\right)_e$ the full subcategory of $\int_G R$-$\mathsf{FGF}_{c,\tau}$ consisting of objects of the shape $(M, e)$ for $e \in G$ the unit element. Denote by $R * G = R *_{c,\tau} G$ the crossed product ring (see* (4.8)*). Then*

(i) *There is an equivalence of additive categories*

$$\alpha\colon \left(\int_G R\text{-}\mathsf{FGF}_{c,\tau}\right)_e \to R *_{c,\tau} G\text{-}\mathsf{FGF};$$

(ii) *The inclusion*

$$\left(\int_G R\text{-}\mathsf{FGF}_{c,\tau}\right)_e \to \int_G R\text{-}\mathsf{FGF}_{c,\tau}$$

*is an equivalence of additive categories.*

*Proof.* (i) An object $(M, e)$ in $\left(\int_G R\text{-}\mathsf{FGF}_{c,\tau}\right)_e$ is sent under $\alpha$ to the finitely generated free $R *_{c,\tau} G$-module $R *_{c,\tau} G \otimes_R M$. A morphism $\phi = \sum_{g \in G} g \cdot (\phi_g\colon M \to \operatorname{res}_{c_g}(N))$ from $(M, e)$ to $(N, e)$ is sent to the $R *_{c,\tau} G$-homomorphism

$$\alpha(\phi)\colon R *_{c,\tau} G \otimes_R M \to R *_{c,\tau} G \otimes_R N, \quad u \otimes x \mapsto \sum_{g \in G} u \cdot \tau(g^{-1}, g)^{-1} \cdot g^{-1} \otimes \phi_g(x)$$

for $u \in R *_{c,\tau} G$ and $x \in M$. This is well-defined, i.e., compatible with the tensor relation, by the following calculation for $r \in R$ using (4.3) and (4.5).

$$\begin{aligned}
&u \cdot \tau(g^{-1}, g)^{-1} \cdot g^{-1} \otimes \phi_g(rx) \\
&= u \cdot \tau(g^{-1}, g)^{-1} \cdot g^{-1} \otimes c_g(r)\phi_g(x) \\
&= u \cdot \tau(g^{-1}, g)^{-1} \cdot g^{-1} \cdot c_g(r) \otimes \phi_g(x) \\
&= u \cdot \tau(g^{-1}, g)^{-1} c_{g^{-1}}(c_g(r)) \cdot g^{-1} \otimes \phi_g(x) \\
&= u \cdot \tau(g^{-1}, g)^{-1} c_{g^{-1}}(c_g(r)) \tau(g^{-1}, g) \tau(g^{-1}, g)^{-1} \cdot g^{-1} \otimes \phi_g(x) \\
&= u \cdot c_{g^{-1}g}(r) \tau(g^{-1}, g)^{-1} \cdot g^{-1} \otimes \phi_g(x) \\
&= u \cdot c_e(r) \tau(g^{-1}, g)^{-1} \cdot g^{-1} \otimes \phi_g(x) \\
&= (u \cdot r) \tau(g^{-1}, g)^{-1} \cdot g^{-1} \otimes \phi_g(x).
\end{aligned}$$

Next we show that $\alpha$ is a covariant functor. Obviously $\alpha(\mathrm{id}_{(M,e)}) = \mathrm{id}_{\alpha(M,e)}$. Consider morphisms $\phi = \sum_{g \in G} g \cdot \phi_g\colon (M, e) \to (N, e)$ and $\psi = \sum_{g \in G} g \cdot \psi_g\colon (N, e) \to$

$(P, e)$ in $\left(\int_G R\text{-FGF}_{c,\tau}\right)_e$. A direct computation shows for $u \in R *_{c,\tau} G$ and $x \in M$

$$\begin{aligned}
&\alpha(\psi)\left(\alpha(\phi)(u \otimes x)\right) \\
&= \alpha(\psi)\left(\sum_{k \in G} u \cdot \tau(k^{-1}, k)^{-1} \cdot k^{-1} \otimes \phi_k(x)\right) \\
&= \sum_{h \in G} \sum_{k \in G} u \cdot \tau(k^{-1}, k)^{-1} \cdot k^{-1} \cdot \tau(h^{-1}, h)^{-1} \cdot h^{-1} \otimes \psi_h \circ \phi_k(x) \\
&= \sum_{h,k \in G} u \cdot \tau(k^{-1}, k)^{-1} c_{k^{-1}}(\tau(h^{-1}, h)^{-1}) \cdot k^{-1} \cdot h^{-1} \otimes \psi_h \circ \phi_k(x) \\
&= \sum_{h,k \in G} u \cdot \tau(k^{-1}, k)^{-1} c_{k^{-1}}(\tau(h^{-1}, h)^{-1}) \tau(k^{-1}, h^{-1}) \cdot (hk)^{-1} \otimes \psi_h \circ \phi_k(x)
\end{aligned}$$

and

$$\begin{aligned}
&\alpha(\psi \circ \phi)(u \otimes x) \\
&= \sum_{g \in G} u \cdot \tau(g^{-1}, g)^{-1} \cdot g^{-1} \otimes (\psi \circ \phi)_g(x) \\
&= \sum_{g \in G} u \cdot \tau(g^{-1}, g)^{-1} \cdot g^{-1} \otimes \left(\sum_{\substack{hk \in G, \\ hk = g}} r_k(\psi_h) \circ \phi_k(x)\right) \\
&= \sum_{g \in G} u \cdot \tau(g^{-1}, g)^{-1} \cdot g^{-1} \otimes \left(\sum_{\substack{hk \in G, \\ hk = g}} \tau(h, k)^{-1} \psi_h \circ \phi_k(x)\right) \\
&= \sum_{g \in G} \sum_{\substack{hk \in G, \\ hk = g}} u \cdot \tau(g^{-1}, g)^{-1} \cdot g^{-1} \cdot \tau(h, k)^{-1} \otimes \psi_h \circ \phi_k(x) \\
&= \sum_{g \in G} \sum_{\substack{hk \in G, \\ hk = g}} u \cdot \tau(g^{-1}, g)^{-1} c_{g^{-1}}(\tau(h, k)^{-1}) \cdot g^{-1} \otimes \psi_h \circ \phi_k(x) \\
&= \sum_{hk \in G} u \cdot \tau((hk)^{-1}, hk)^{-1} c_{(hk)^{-1}}(\tau(h, k)^{-1}) \cdot (hk)^{-1} \otimes \psi_h \circ \phi_k(x).
\end{aligned}$$

Hence it remains to show for $h, k \in G$

$$\tau(k^{-1}, k)^{-1} c_{k^{-1}}(\tau(h^{-1}, h)^{-1}) \tau(k^{-1}, h^{-1}) = \tau((hk)^{-1}, hk)^{-1} c_{(hk)^{-1}}(\tau(h, k)^{-1}),$$

or, equivalently,

$$\tau(k^{-1}, h^{-1}) c_{(hk)^{-1}}(\tau(h, k)) \tau((hk)^{-1}, hk) = c_{k^{-1}}(\tau(h^{-1}, h)) \tau(k^{-1}, k).$$

Since (4.4) yields

$$\tau((hk)^{-1}, h) \tau(k^{-1}, k) = c_{(hk)^{-1}} \tau(h, k) \tau((hk)^{-1}, hk),$$

it suffices to show

$$\tau(k^{-1}, h^{-1})\tau((hk)^{-1}, h) = c_{k^{-1}}(\tau(h^{-1}, h)).$$

But this follows from (4.4) and (4.7). This finishes the proof that $\alpha$ is a covariant functor. Obviously it is compatible with the structures of an additive category. One easily checks that $\alpha$ induces a bijection between the isomorphism classes of objects. In order to show that $\alpha$ is a weak equivalence, we have to show for two objects $(M, e)$ and $(N, e)$ that $\alpha$ induces a bijection

$$\mathrm{mor}_{\left(\int_G R\text{-}\mathsf{FGF}_{c,\tau}\right)_e}((M, e), (N, e)) \xrightarrow{\cong} \hom_{R*_{c,\tau}G}(R *_{c,\tau} \otimes_R M, R *_{c,\tau} \otimes_R N).$$

Since $\alpha$ is compatible with the structures of an additive category, it suffices to check this in the special case $M = N = R$, where it is obvious.

(ii) An object of the shape $(M, g)$ in $\int_G R\text{-}\mathsf{FGF}_{c,\tau}$ is isomorphic to the object $(M, e)$, namely an isomorphism $(M, g) \xrightarrow{\cong} (M, e)$ in $\int_G R\text{-}\mathsf{FGF}_{c,\tau}$ is given by $g \cdot \mathrm{id}_{\mathrm{res}_{c_g}(M)}$. □

# 7. Connected groupoids and additive categories with involutions

Next we want to enrich the constructions of Section 5 to additive categories with involutions. Let $\mathsf{Add\text{-}Cat}_{\mathrm{inv}}$ be the category of additive categories with involution. Given a contravariant functor $(F, T) \colon \mathcal{G} \to \mathsf{Add\text{-}Cat}_{\mathrm{inv}}$, we want to define on the additive category $\int_{\mathcal{G}} F$ the structure of an additive category with involution. Here the pair $(F, T)$ means that we assign to every object $x$ in $\mathcal{G}$ an additive category with involution $F(x)$ and for every morphism $f \colon x \to y$ in $\mathcal{G}$ we have a functor of additive categories with involution $(F(f), T(f)) \colon F(y) \to F(x)$.

Next we construct for a functor $\mathcal{G} \to \mathsf{Add\text{-}Cat}_{\mathrm{inv}}$ an involution of additive categories

(7.1) $$(I_{\int_{\mathcal{G}} F}, E_{\int_{\mathcal{G}} F})$$

on the additive category $\int_{\mathcal{G}} F$ which we have introduced in (5.1). On objects we put

$$I_{\int_{\mathcal{G}} F}(x, A) := (x, I_{\mathcal{G}}(A)) = (x, A^*).$$

Let $\phi = \sum_{f \in \mathrm{mor}_{\mathcal{G}}(x,y)} f \cdot \phi_f \colon (x, A) \to (y, B)$ be a morphism in $\int_{\mathcal{G}} F$. Define $I_{\int_{\mathcal{G}} F}(\phi) \colon B^* \to A^*$ to be the morphism $\phi^* = \sum_{f \in \mathrm{mor}_{\mathcal{G}}(y,x)} f \cdot (\phi^*)_f \colon (y, B^*) \to (x, A^*)$ in $\int_{\mathcal{G}} F$ whose component for $f \in \mathrm{mor}_{\mathcal{G}}(y, x)$ is given by the composite

$$\begin{aligned}(\phi^*)_f \colon B^* = F(f)\left(F(f^{-1})(B^*)\right) &\xrightarrow{F(f)(T(f^{-1})(B))} F(f)\left(F(f^{-1})(B)^*\right) \\ &\xrightarrow{F(f)\left((\phi_{f^{-1}})^*\right)} F(f)(A^*).\end{aligned}$$

Next we show that $I_{\int_{\mathcal{G}} F}$ is a contravariant functor. Obviously $I_{\int_{\mathcal{G}} F}$ sends the identity $\mathrm{id}_A$ to $\mathrm{id}_{I_{\int_{\mathcal{G}} F}(A)}$. We have to show

$$I_{\int_{\mathcal{G}} F}(\psi \circ \phi) = I_{\int_{\mathcal{G}} F}(\phi) \circ I_{\int_{\mathcal{G}} F}(\psi)$$

for morphisms $\phi = \sum_{h \in \mathrm{mor}_{\mathcal{G}}(x,y)} h \cdot \phi_h \colon (x, A) \to (y, B)$ and $\psi \colon \sum_{k \in \mathrm{mor}_{\mathcal{G}}(y,z)} k \cdot \psi_k \colon (y, B \to (z, C)$, or in short notation $(\psi \circ \phi)^* = \phi^* \circ \psi^*$.

By definition $(\phi^* \circ \psi^*) = \sum_{g \in \mathrm{mor}_{\mathcal{G}}(z,x)} g \cdot (\phi^* \circ \psi^*)_g$ for

$$(\phi^* \circ \psi^*)_g := \sum_{\substack{k \in \mathrm{mor}_{\mathcal{G}}(z,y), \\ h \in \mathrm{mor}_{\mathcal{G}}(y,x), \\ hk = g}} F(k)((\phi^*)_h) \circ (\psi^*)_k.$$

By definition

$$(\psi^*)_k \colon C^* = F(k)(F(k^{-1})(C^*)) \xrightarrow{F(k)(T(k^{-1})(C))} F(k)(F(k^{-1})(C)^*) \xrightarrow{F(k)((\psi_{k^{-1}})^*)} F(k)(B^*)$$

and

$$(\phi^*)_h \colon B^* = F(h)(F(h^{-1})(B^*)) \xrightarrow{F(h)(T(h^{-1})(B))} F(h)(F(h^{-1})(B)^*) \xrightarrow{F(h)((\phi_{h^{-1}})^*)} F(h)(A^*).$$

Hence the component $(\phi^* \circ \psi^*)_g$ of $(\phi^* \circ \psi^*)$ at $g \colon z \to x$ is given by the sum of morphisms from $C^*$ to $F(g)(A^*)$

$$\sum_{\substack{k \in \mathrm{mor}_{\mathcal{G}}(z,y), \\ h \in \mathrm{mor}_{\mathcal{G}}(y,x), \\ hk = g}} F(k)\left(F(h)((\phi_{h^{-1}})^*)\right) \circ F(k)\left(F(h)(T(h^{-1})(B))\right) \circ F(k)((\psi_{k^{-1}})^*) \circ F(k)(T(k^{-1})(C)).$$

The component of $(\psi \circ \phi)^*_g$ of $(\psi \circ \phi)^*$ at $g \colon z \to x$ is given by

$$C^* = F(g)\left(F(g^{-1})(C^*)\right) \xrightarrow{F(g)(T(g^{-1})(C))} F(g)\left(F(g^{-1})(C)^*\right) \xrightarrow{F(g)\left(((\psi \circ \phi)_{g^{-1}})^*\right)} F(g)(A^*).$$

Since for $g \colon z \to x$ we have

$$(\psi \circ \phi)_{g^{-1}} = \sum_{\substack{h \in \mathrm{mor}_{\mathcal{G}}(y,z), \\ k \in \mathrm{mor}_{\mathcal{G}}(x,y), \\ hk = g^{-1}}} F(k)(\psi_h) \circ \phi_k,$$

the component of $(\psi\circ\phi)^*_g$ of $(\psi\circ\phi)^*$ at $g\colon z\to x$ is given by the sum of morphisms $C^*$ to $F(g)(A^*)$

$$\sum_{\substack{h\in\mathrm{mor}_{\mathcal{G}}(y,z),\\ k\in\mathrm{mor}_{\mathcal{G}}(x,y),\\ hk=g^{-1}}} F(g)\left((\phi_k)^*\right)\circ F(g)\left(F(k)(\psi_h)^*\right)\circ F(g)(T(g^{-1})(C)).$$

By changing the indexing by replacing $h$ with $k^{-1}$ and $k$ by $h^{-1}$, this transforms to

$$\sum_{\substack{k\in\mathrm{mor}_{\mathcal{G}}(z,y),\\ h\in\mathrm{mor}_{\mathcal{G}}(y,x),\\ hk=g}} F(g)\left((\phi_{h^{-1}})^*\right)\circ F(g)\left(F(h^{-1})(\psi_{k^{-1}})^*\right)\circ F(g)(T(g^{-1})(C)).$$

Hence we have to show for every $k\colon z\to y$ and $h\colon y\to x$ with $hk=g$ that the two composites

$$F(k)\left(F(h)((\phi_{h^{-1}})^*)\right)\circ F(k)\left(F(h)(T(h^{-1})(B))\right)\circ F(k)((\psi_{k^{-1}})^*)\circ F(k)(T(k^{-1})(C))$$

and

$$(F(g)((\phi_{h^{-1}})^*))\circ F(g)\left(F(h^{-1})(\psi_{k^{-1}})^*\right)\circ F(g)(T(g^{-1})(C))$$

agree. We compute for the first one

$$\begin{aligned}
&F(k)\left(F(h)((\phi_{h^{-1}})^*)\right)\circ F(k)\left(F(h)(T(h^{-1})(B))\right)\circ F(k)((\psi_{k^{-1}})^*)\\
&\qquad\circ F(k)(T(k^{-1})(C))\\
&=(F(g)((\phi_{h^{-1}})^*))\circ F(g)(T(h^{-1})(B))\circ F(g)\left(F(h^{-1})((\psi_{k^{-1}})^*)\right)\\
&\qquad\qquad\circ F(g)\left(F(h^{-1})(T(k^{-1})(C))\right).
\end{aligned}$$

Hence it remains to show that the composites

$$F(g^{-1})(C^*)\xrightarrow{T(g^{-1})(C)}F(g^{-1})(C)^*\xrightarrow{F(h^{-1})(\psi_{k^{-1}})^*}F(h^{-1})(B)^*$$

and

$$\begin{aligned}
F(g^{-1})(C^*)=F(h^{-1})\left(F(k^{-1})(C^*)\right)&\xrightarrow{F(h^{-1})(T(k^{-1})(C))}F(h^{-1})(F(k^{-1})(C)^*)\\
&\xrightarrow{F(h^{-1})((\psi_{k^{-1}})^*)}F(h^{-1})(B^*)\xrightarrow{T(h^{-1})(B)}F(h^{-1})(B)^*
\end{aligned}$$

agree. The second one agrees with the composite

$$\begin{aligned}
F(g^{-1})(C^*)=F(h^{-1})\left(F(k^{-1})(C^*)\right)&\xrightarrow{F(h^{-1})(T(k^{-1})(C))}F(h^{-1})(F(k^{-1})(C)^*)\\
&\xrightarrow{T(h^{-1})(F(k^{-1})(C))}F(h^{-1})(F(k^{-1})(C))^*\\
&\xrightarrow{F(h^{-1})(\psi_{k^{-1}})^*}F(h^{-1})(B)^*
\end{aligned}$$

since $T(h^{-1})$ is a natural transformation $F(h^{-1}) \circ I_{F(y)} \to I_{F(x)} \circ F(h^{-1})$. Since

$$\begin{aligned}(F(h^{-1}), T(h^{-1})) \circ (F(k^{-1}), T(k^{-1})) &= (F(k^{-1}h^{-1}), T(k^{-1}h^{-1}))\\ &= (F(g^{-1}), T(g^{-1}))\end{aligned}$$

the map $T(g^{-1})(C)$ can be written as the composite

$$\begin{aligned}T(g^{-1})(C)\colon F(g^{-1})(C^*) &= F(h^{-1})\left(F(k^{-1})(C^*)\right)\\ &\xrightarrow{F(h^{-1})\left(T(k^{-1})(C)\right)} F(h^{-1})\left(F(k^{-1})(C)^*\right)\\ &\xrightarrow{T(h^{-1})\left(F(k^{-1})(C)\right)} F(h^{-1})\left(F(k^{-1})(C)\right)^* = F(g)(C)^*.\end{aligned}$$

This finishes the proof that $I_{\int_{\mathcal{G}} F}$ is a contravariant functor.

The natural equivalence

$$E_{\int_{\mathcal{G}} F}\colon \mathrm{id}_{\int_{\mathcal{G}} F} \to I_{\int_{\mathcal{G}} F} \circ I_{\int_{\mathcal{G}} F}$$

assigns to an object $(x, A)$ in $\int_{\mathcal{G}} F$ the isomorphism

$$\mathrm{id}_x \cdot \left(E_{\mathcal{G}}(A)\colon A \xrightarrow{\cong} A^{**}\right) : (x, A) \to (x, A^{**}).$$

We have to check that $E_{\int_{\mathcal{G}} F}$ is a natural equivalence. Consider a morphism $\phi = \sum_{f \in \mathrm{mor}_{\mathcal{G}}(x,y)} f \cdot (\phi_f\colon (x, A) \to (y, B))$ in $\int_{\mathcal{G}} F$. Then $\left(I_{\int_{\mathcal{G}} F} \circ I_{\int_{\mathcal{G}} F}\right)(\phi)$ has as the component for $f\colon x \to y$ the composite

$$\begin{aligned}A^{**} = F(f)\left(F(f^{-1})(A^{**})\right) &\xrightarrow{F(f)\left(T(f^{-1})(A^*)\right)} F(f)\left(F(f^{-1})(A^*)^*\right)\\ &\xrightarrow{F(f)\left(F(f^{-1})((\phi_f)^*)^*\right)} F(f)\left(F(f^{-1})(F(f)(B)^*)^*\right)\\ &\xrightarrow{F(f)\left(F(f^{-1})(T(f)(B))^*\right)} F(f)\left(F(f^{-1})(F(f)(B^*))^*\right) = F(f)(B^{**}).\end{aligned}$$

Hence $\left(I_{\int_{\mathcal{G}} F} \circ I_{\int_{\mathcal{G}} F}\right)(\phi) \circ E_{\int_{\mathcal{G}} F}(x, A)$ has as component for $f\colon x \to y$ the composite

$$\begin{aligned}A \xrightarrow{E_{\mathcal{A}}(A)} A^{**} = F(f)\left(F(f^{-1})(A^{**})\right) &\xrightarrow{F(f)\left(T(f^{-1})(A^*)\right)} F(f)\left(F(f^{-1})(A^*)^*\right)\\ &\xrightarrow{F(f)\left(F(f^{-1})(\phi_f^*)^*\right)} F(f)\left(F(f^{-1})(F(f)(B)^*)^*\right)\\ &\xrightarrow{F(f)\left(F(f^{-1})(T(f)(B))^*\right)} F(f)\left(F(f^{-1})(F(f)(B^*))^*\right) = F(f)(B^{**}).\end{aligned}$$

The component of $E_{\int_{\mathcal{G}} F}(y, B) \circ \phi$ at $f\colon x \to y$ is the composite

$$A \xrightarrow{\phi_f} F(f)(B) \xrightarrow{F(f)(E_{\mathcal{A}}(B))} F(f)(B^{**}).$$

It remains to show that these two morphisms $A \to F(f)(B^{**})$ agree. The following two diagrams commute since $E_{\mathcal{A}}$ and $T(f^{-1})$ are natural transformations

$$\begin{array}{ccc} A & \xrightarrow{\quad E_{\mathcal{A}}(A) \quad} & A^{**} = F(f)\big(F(f^{-1})(A^{**})\big) \\ \Big\downarrow{\scriptstyle \phi_f} & & \Big\downarrow{\scriptstyle \phi_f^{**} = F(f)\left(F(f^{-1})(\phi_f^{**})\right)} \\ F(f)(B) & \xrightarrow{E_{\mathcal{A}}(F(f)(B))} & F(f)(B)^{**} = F(f)\big(F(f^{-1})(F(f)(B)^{**})\big) \end{array}$$

and

$$\begin{array}{ccc} F(f)\big(F(f^{-1})(A^{**})\big) & \xrightarrow{\quad F(f)(T(f^{-1})(A^{*})) \quad} & F(f)\big(F(f^{-1})(A^{*})^{*}\big) \\ \Big\downarrow{\scriptstyle F(f)\left(F(f^{-1})(\phi_f^{**})\right)} & & \Big\downarrow{\scriptstyle F(f)\left(F(f^{-1})(\phi_f^{*})^{*}\right)} \\ F(f)\big(F(f^{-1})(F(f)(B)^{**})\big) & \xrightarrow{F(f)\left(T(f^{-1})(F(f)(B)^{*})\right)} & F(f)\big(F(f^{-1})(F(f)(B)^{*})^{*}\big). \end{array}$$

Hence we have to show that

$$F(f)(B) \xrightarrow{F(f)(E_{\mathcal{A}}(B))} F(f)(B^{**})$$

agrees with the composite

$$\begin{aligned} F(f)(B) &\xrightarrow{E_{\mathcal{A}}(F(f)(B))} F(f)(B)^{**} = F(f)\big(F(f^{-1})(F(f)(B)^{**})\big) \\ &\xrightarrow{F(f)\left(T(f^{-1})(F(f)(B)^{*})\right)} F(f)\big(F(f^{-1})(F(f)(B)^{*})^{*}\big) \\ &\xrightarrow{F(f)\left(F(f^{-1})(T(f)(B))^{*}\right)} F(f)\big(F(f^{-1})(F(f)(B^{*}))^{*}\big) = F(f)(B^{**}). \end{aligned}$$

(Notice that $\phi$ is not involved anymore.) The following diagram commutes by the axioms (see (1.6))

$$\begin{array}{ccc} F(f)(B) & \xrightarrow{\quad E_{\mathcal{A}}(F(f)(B)) \quad} & F(f)(B)^{**} \\ \Big\downarrow{\scriptstyle F(f)(E_{\mathcal{A}}(B))} & & \Big\downarrow{\scriptstyle T(f)(B)^{*}} \\ F(f)(B^{**}) & \xrightarrow{\quad T(f)(B^{*}) \quad} & F(f)(B^{*})^{*} \end{array}$$

Hence it remains to show the commutativity of the following diagram (which does not involve $\phi$ and $E_{\mathcal{A}}$ anymore).

$$\begin{array}{ccc} \begin{array}{l} F(f)(B)^{**} \\ = F(f)\big(F(f^{-1})(F(f)(B)^{*})\big)^{*} \end{array} & \xrightarrow{F(f)\left(T(f^{-1})(F(f)(B)^{*})\right)} & F(f)\big(F(f^{-1})(F(f)(B)^{*})^{*}\big) \\ \Big\downarrow{\scriptstyle T(f)(B)^{*}} & & \Big\downarrow{\scriptstyle F(f)\left(F(f^{-1})(T(f)(B))^{*}\right)} \\ F(f)(B^{*})^{*} & \xleftarrow[\quad T(f)(B^{*}) \quad]{} & \begin{array}{l} F(f)(B^{**}) \\ = F(f)\big(F(f^{-1})(F(f)(B^{*}))^{*}\big) \end{array} \end{array}$$

Since $(F(f),T(f)) \circ (F(f^{-1}),T(f^{-1})) = \mathrm{id}$, we have

$$T(f)\left(F(f^{-1})(F(f)(B)^*)\right) \;\circ\; F(f)\left(T(f^{-1})(F(f)(B)^*)\right) = \mathrm{id}\,.$$

Hence it suffices to prove the commutativity of the following diagram

$$\begin{array}{ccc}
\begin{array}{l} F(f)(B)^{**} \\ = F(f)\left(F(f^{-1})(F(f)(B)^*)\right)^* \end{array} & \xleftarrow{T(f)\left(F(f^{-1})(F(f)(B)^*)\right)} & F(f)\left(F(f^{-1})(F(f)(B)^*)^*\right) \\
\Big\downarrow {\scriptstyle T(f)(B)^*} & & \Big\downarrow {\scriptstyle F(f)\left(F(f^{-1})(T(f)(B))^*\right)} \\
F(f)(B^*)^* & \xleftarrow[T(f)(B^*)]{} & \begin{array}{l} F(f)(B^{**}) \\ = F(f)\left(F(f^{-1})(F(f)(B^*))^*\right). \end{array}
\end{array}$$

This follows because this diagram is obtained by applying the natural transformation $T(f)$ to the morphism

$$F(f^{-1})(F(f)(B^*)) \xrightarrow{F(f^{-1})(T(f)(B))} F(f^{-1})(F(f)(B)^*).$$

The condition (1.3) is satisfied for $(I_{\int_{\mathcal{G}} F}, E_{\int_{\mathcal{G}} F})$ since it holds for $(I_{\mathcal{A}}, E_{\mathcal{A}})$.

We will denote the resulting additive category $\int_{\mathcal{A}} F$ with involution $(I_{\int_{\mathcal{G}} F}, E_{\int_{\mathcal{G}} F})$ by

$$\int_{\mathcal{G}} (F,T). \tag{7.2}$$

Let $(F_0,T_0)$ and $(F_1,T_1)$ be two contravariant functors $\mathcal{G} \to \mathsf{Add\text{-}Cat}_{\mathrm{inv}}$. Let $(S,U)\colon (F_0,T_0) \to (F_1,T_1)$ be a natural transformation of such functors. This means that we for each object $x$ in $\mathcal{G}$ we have an equivalence $(S(x),U(x))\colon F_0(x) \to F_1(y)$ of additive categories with involution such that for all $f\colon x \to y$ in $\mathcal{G}$ the following diagram of functors of additive categories with involution commutes

$$\begin{array}{ccc}
F_0(y) & \xrightarrow{(S(y),U(y))} & F_1(y) \\
\Big\downarrow {\scriptstyle (F_0(f),T_0(f))} & & \Big\downarrow {\scriptstyle (F_1(f),T_1(f))} \\
F_0(x) & \xrightarrow{(S(x),U(x))} & F_1(x)
\end{array} \tag{7.3}$$

Then both $\int_{\mathcal{G}}(F_0,T_0)$ and $\int_{\mathcal{G}}(F_1,T_1)$ are additive categories with involutions. The functor of additive categories $\int_{\mathcal{G}} S\colon \int_{\mathcal{G}} F_0 \to \int_{\mathcal{G}} F_1$ defined in (5.2) extends to a functor of additive categories with involution

$$\int_{\mathcal{G}} (S,U)\colon \int_{\mathcal{G}} (F_0,T_0) \to \int_{\mathcal{G}} (F_1,T_1) \tag{7.4}$$

as follows. We have to specify a natural equivalence

$$\widehat{U}\colon \left(\int_{\mathcal{G}} S\right) \circ I_{\int_{\mathcal{G}}(F_0,T_0)} \to I_{\int_{\mathcal{G}}(F_1,T_1)} \circ \int_{\mathcal{G}} S.$$

For an object $(x, A)$ in $\int_{\mathcal{G}} F_0$ the isomorphism

$$\widehat{U}(x,A)\colon \left(\int_{\mathcal{G}} S\right) \circ I_{\int_{\mathcal{G}}(F_0,T_0)}(x,A) \to I_{\int_{\mathcal{G}}(F_1,T_1)} \circ \int_{\mathcal{G}} S(x,A)$$

is given by the isomorphism

$$\mathrm{id}_x \cdot U(x)(A)\colon (x, S(x)(A^*)) \to (x, S(x)(A)^*)$$

in $\int_{\mathcal{G}} F_1$. Next we check that $\widehat{U}$ is a natural equivalence.

Let $\sum_{f \in \mathrm{mor}_{\mathcal{G}}(x,y)} f \cdot \phi_f\colon (x,A) \to (y,B)$ be a morphism in $\int_{\mathcal{G}} F_0$, where by definition $\phi_f\colon A \to F(f)(B)$ is a morphism in the additive category $F_0(x)$. We have to show the commutativity of the following diagram in the additive category $\int_{\mathcal{G}} F_1$

$$\begin{array}{ccc}
\left(\int_{\mathcal{G}} S\right) \circ I_{\int_{\mathcal{G}}(F_0,T_0)}(y,B) & \xrightarrow{\left(\int_{\mathcal{G}} S\right)\circ I_{\int_{\mathcal{G}}(F_0,T_0)}(\phi)} & \left(\int_{\mathcal{G}} S\right) \circ I_{\int_{\mathcal{G}}(F_0,T_0)}(x,A) \\
\Big\downarrow \widehat{U}(y,B) & & \Big\downarrow \widehat{U}(x,A) \\
I_{\int_{\mathcal{G}}(F_1,T_1)} \circ \int_{\mathcal{G}} S(y,B) & \xrightarrow{I_{\int_{\mathcal{G}}(F_1,T_1)}\circ \int_{\mathcal{G}} S(\phi)} & I_{\int_{\mathcal{G}}(F_1,T_1)} \circ \int_{\mathcal{G}} S(x,A)
\end{array}$$

The morphism $I_{\int_{\mathcal{G}}(F_0,T_0)}(\phi)$ in $\int_{\mathcal{G}} F_0$ is given by

$$\phi^* = \sum_{f \in \mathrm{mor}_{\mathcal{G}}(x,y)} f \cdot (\phi^*)_f\colon (y,B^*) \to (x,A^*),$$

where the component $(\phi^*)_f$ is the composite

$$(\phi^*)_f\colon B^* = F_0(f)\left(F_0(f^{-1})(B^*)\right) \xrightarrow{F_0(f)\left(T_0(f^{-1})(B)\right)} F_0(f)\left(F_0(f^{-1})(B)^*\right) \xrightarrow{F_0(f)((\phi_{f^{-1}})^*)} F_0(f)\left(A^*\right).$$

The morphism $\left(\int_{\mathcal{G}} S\right) \circ I_{\int_{\mathcal{G}}(F_0,T_0)}(\phi)\colon (y,B^*) \to (x,A^*)$ in $\int_{\mathcal{G}} F_1$ is given by $\sum_{f \in \mathrm{mor}_{\mathcal{G}}(x,y)} f \cdot \psi_f\colon (y,B^*) \to (x,A^*)$, where $\psi_f$ is the composite

$$\begin{aligned}
\psi_f\colon S(y)(B^*) &= S(y)\left(F_0(f)\left(F_0(f^{-1})(B^*)\right)\right) \\
&\xrightarrow{S(y)\left(F_0(f)\left(T_0(f^{-1})(B)\right)\right)} S(y)\left(F_0(f)\left(F_0(f^{-1})(B)^*\right)\right) \\
&\xrightarrow{S(y)(F_0(f)((\phi_{f^{-1}})^*))} S(y)\left(F_0(f)\left(A^*\right)\right) = F_1(f)\left(S(x)\left(A^*\right)\right).
\end{aligned}$$

Hence the morphism $\widehat{U}(x,A) \circ \left(\int_{\mathcal{G}} S\right) \circ I_{\int_{\mathcal{G}}(F_0,T_0)}(\phi)\colon (y,B^*) \to (x,A^*)$ in $\int_{\mathcal{G}} F_1$ is given by $\sum_{f \in \mathrm{mor}_{\mathcal{G}}(x,y)} f \cdot \mu_f\colon (y,B^*) \to (x,A^*)$, where $\mu_f$ is the composite in

$F_1(y)$

$$\begin{aligned}\mu_f\colon S(y)(B^*) &= S(y)\left(F_0(f)\left(F_0(f^{-1})(B^*)\right)\right)\\ &\xrightarrow{S(y)\left(F_0(f)\left(T_0(f^{-1})(B)\right)\right)} S(y)\left(F_0(f)\left(F_0(f^{-1})(B)^*\right)\right)\\ &\xrightarrow{S(y)(F_0(f)((\phi_{f^{-1}})^*))} S(y)\left(F_0(f)\left(A^*\right)\right)\\ &= F_1(f)\left(S(x)\left(A^*\right)\right) \xrightarrow{F_1(f)(U(x)(A))} F_1(f)\left(S(x)(A)^*\right).\end{aligned}$$

The morphism $\int_{\mathcal{G}} S(\phi)\colon (x, S(x)(A)) \to (y, S(y)(B))$ in $\int_{\mathcal{G}} F_1$ is given by

$$\sum_{f\in\mathrm{mor}_{\mathcal{G}}(x,y)} f\cdot (S(x)(\phi_f))\colon S(x)(A) \to S(x)(F_0(f)(B)) = F_1(f)(S(y)(B)).$$

The morphism $I_{\int_{\mathcal{G}}(F_1,T_1)}\circ\int_{\mathcal{G}} S(\phi)\colon (y, S(y)(B)^*) \to (x, S(x)(A)^*)$ in $\int_{\mathcal{G}} F_1$ is given by $\sum_{f\in\mathrm{mor}_{\mathcal{G}}(y,x)} f\cdot\nu_f$, where $\nu_f$ is the composite in $F_1(y)$.

$$\begin{aligned}\nu_f\colon S(y)(B)^* &= F_1(f)\left(F_1(f^{-1})\left(S(y)(B)^*\right)\right)\\ &\xrightarrow{F_1(f)\left(T_1(f^{-1})(S(y)(B))\right)} F_1(f)\left(F_1(f^{-1})\left(S(y)(B)\right)^*\right)\\ &= F_1(f)\Big(S(x)\left(F_0(f^{-1})(B)\right)^*\Big)\\ &\xrightarrow{F_1(f)\left((S(x)(\phi_{f^{-1}}))^*\right)} F_1(f)\left(S(x)(A)^*\right).\end{aligned}$$

The morphism $I_{\int_{\mathcal{G}}(F_1,T_1)}\circ\int_{\mathcal{G}} S(\phi)\circ\widehat{U}(y,B)\colon (y, S(y)(B^*)) \to (x, S(x)(A)^*)$ in $\int_{\mathcal{G}} F_1$ is given by $\sum_{f\in\mathrm{mor}_{\mathcal{G}}(y,x)} f\cdot\omega_f$, where $\omega_f$ is the composite in $F_1(y)$.

$$\begin{aligned}\omega_f\colon S(y)(B^*) &\xrightarrow{U(y)(B)} S(y)(B)^*\\ &= F_1(f)\left(F_1(f^{-1})\left(S(y)(B)^*\right)\right)\\ &\xrightarrow{F_1(f)\left(T_1(f^{-1})(S(y)(B))\right)} F_1(f)\left(F_1(f^{-1})\left(S(y)(B)\right)^*\right)\\ &= F_1(f)\Big(S(x)\left(F_0(f^{-1})(B)\right)^*\Big) \xrightarrow{F_1(f)\left((S(x)(\phi_{f^{-1}}))^*\right)} F_1(f)\left(S(x)(A)^*\right).\end{aligned}$$

Hence we have to show for all $f\colon y\to x$ in $\mathrm{mor}_{\mathcal{G}}(y,x)$ that the two composites in $F_1(y)$

$$\begin{aligned}S(y)(B^*) &= S(y)\left(F_0(f)\left(F_0(f^{-1})(B^*)\right)\right)\\ &\xrightarrow{S(y)\left(F_0(f)\left(T_0(f^{-1})(B)\right)\right)} S(y)\left(F_0(f)\left(F_0(f^{-1})(B)^*\right)\right)\\ &\xrightarrow{S(y)(F_0(f)((\phi_{f^{-1}})^*))} S(y)\left(F_0(f)\left(A^*\right)\right) = F_1(f)\left(S(x)\left(A^*\right)\right)\\ &\xrightarrow{F_1(f)(U(x)(A))} F_1(f)\left(S(x)(A)^*\right)\end{aligned}$$

and

$$
\begin{aligned}
&S(y)(B^*) \xrightarrow{U(y)(B)} S(y)(B)^* \\
&\quad = F_1(f)\left(F_1(f^{-1})\left(S(y)(B)^*\right)\right) \\
&\quad \xrightarrow{F_1(f)\left(T_1(f^{-1})(S(y)(B))\right)} F_1(f)\left(F_1(f^{-1})\left(S(y)(B)\right)^*\right) \\
&\quad = F_1(f)\Big(S(x)\big(F_0(f^{-1})(B)\big)^*\Big) \xrightarrow{F_1(f)\left((S(x)(\phi_{f^{-1}}))^*\right)} F_1(f)\left(S(x)(A)^*\right).
\end{aligned}
$$

agree. Since $S$ is a natural transformation from $F_0 \to F_1$, the first composite can be rewritten as the composite

$$
\begin{aligned}
S(y)(B^*) = {} & F_1(f)\left(S(x)\left(F_0(f^{-1})(B^*)\right)\right) \\
& \xrightarrow{F_1(f)\left(S(x)\left(T_0(f^{-1})(B)\right)\right)} F_1(f)\left(S(x)\left(F_0(f^{-1})(B)^*\right)\right) \\
& \xrightarrow{F_1(f)\left(S(x)((\phi_{f^{-1}})^*)\right)} F_1(f)\left(S(x)\left(A^*\right)\right) \\
& \xrightarrow{F_1(f)(U(x)(A))} F_1(f)\left(S(x)(A)^*\right).
\end{aligned}
$$

Since $U(x)$ is a natural transformation from $S(x) \circ I_{F_0(x)}$ to $I_{F_1(x)} \circ S(x)$, this agrees with the composite

$$
\begin{aligned}
S(y)(B^*) = {} & F_1(f)\left(S(x)\left(F_0(f^{-1})(B^*)\right)\right) \\
& \xrightarrow{F_1(f)\left(S(x)\left(T_0(f^{-1})(B)\right)\right)} F_1(f)\left(S(x)\left(F_0(f^{-1})(B)^*\right)\right) \\
& \xrightarrow{F_1(f)\left(U(x)\left(F_0(f^{-1})(B)\right)\right)} F_1(f)\Big(S(x)\left(F_0(f^{-1})(B)\right)^*\Big) \\
& \xrightarrow{F_1(f)\left((S(x)(\phi_{f^{-1}}))^*\right)} F_1(f)\left(S(x)(A)^*\right).
\end{aligned}
$$

Hence it suffices to show that the following two composites agree

$$
\begin{aligned}
S(y)(B^*) = {} & F_1(f)\left(S(x)\left(F_0(f^{-1})(B^*)\right)\right) \\
& \xrightarrow{F_1(f)\left(S(x)\left(T_0(f^{-1})(B)\right)\right)} F_1(f)\left(S(x)\left(F_0(f^{-1})(B)^*\right)\right) \\
& \xrightarrow{F_1(f)\left(U(x)\left(F_0(f^{-1})(B)\right)\right)} F_1(f)\Big(S(x)\left(F_0(f^{-1})(B)\right)^*\Big)
\end{aligned}
$$

and

$$
\begin{aligned}
&S(y)(B^*) \xrightarrow{U(y)(B)} S(y)(B)^* \\
&\quad = F_1(f)\left(F_1(f^{-1})\left(S(y)(B)^*\right)\right) \\
&\quad \xrightarrow{F_1(f)\left(T_1(f^{-1})(S(y)(B))\right)} F_1(f)\left(F_1(f^{-1})\left(S(y)(B)\right)^*\right) \\
&\quad = F_1(f)\Big(S(x)\big(F_0(f^{-1})(B)\big)^*\Big)
\end{aligned}
$$

(Notice that $\phi_{f^{-1}}$ has been eliminated.) This will follow by applying $F_1(f)$ to the following diagram, provided we can show that it does commute.

$$\begin{array}{ccc} S(x)\left(F_0(f^{-1})(B^*)\right) = F_1(f^{-1})\left(S(y)(B^*)\right) & \xrightarrow{S(x)\left(T_0(f^{-1})(B)\right)} & S(x)\left(F_0(f^{-1})(B)^*\right) \\ \Big\downarrow {\scriptstyle F_1(f^{-1})(U(y)(B))} & & \Big\downarrow {\scriptstyle U(x)\left(F_0(f^{-1})(B)\right)} \\ F_1(f^{-1})\left(S(y)(B)^*\right) & \xrightarrow{T_1(f^{-1})(S(y)(B))} & S(x)\left(F_0(f^{-1})(B)\right)^* \end{array}$$

But the latter diagram commutes because we require the following equality of functors of additive categories with involution for $f^{-1}\colon x \to y$ (see (7.3))

$$(F_0(f^{-1}), T_0(f^{-1})) \circ (S(x), U(x)) = (S(y), U(y)) \circ (F_1(f^{-1}), T_1(f^{-1})).$$

This finishes the proof that $\widehat{U}$ is a natural equivalence. One easily checks that condition (1.6) is satisfied by $\widehat{U}$ since it holds for $U(x)$ for all objects $x$ in $\mathcal{G}$. This finishes the construction of the functor of additive categories with involution $(S, U)$ (see (7.4)).

One easily checks

$$\left(\int_{\mathcal{G}} (S_2, U_2)\right) \circ \left(\int_{\mathcal{G}} (S_1, U_1)\right) = \int_{\mathcal{G}} (S_2, U_2) \circ (S_1, U_1) \tag{7.5}$$

$$\int_{\mathcal{G}} \mathrm{id}_F = \mathrm{id}_{\int_{\mathcal{G}} F}. \tag{7.6}$$

Given a functor of groupoids $W\colon \mathcal{G}_1 \to \mathcal{G}$ and a functor $(F, T)\colon \mathcal{G} \to$ Add-Cat$_{\mathrm{inv}}$, the composition with $W$ a yields a functor $(F \circ W, T \circ W)$. Hence both $\int_{\mathcal{G}_1} (F, T) \circ W$ and $\int_{\mathcal{G}} (F, T)$ are additive categories with involutions. One easily checks that $I_{\int_{\mathcal{G}} F} \circ W_* = W_* \circ I_{\int_{\mathcal{G}_1} F \circ W}$ holds for the functor $W_*$ defined in (5.5). Hence

$$(W_*, \mathrm{id})\colon \int_{\mathcal{G}_1} (F, T) \circ W \to \int_{\mathcal{G}} (F, T). \tag{7.7}$$

is a functor of additive categories with involution. One easily checks

$$((W_2)_*, \mathrm{id}) \circ ((W_1)_*, \mathrm{id}) = ((W_2 \circ W_1)_*, \mathrm{id}); \tag{7.8}$$

$$(\mathrm{id}_{\mathcal{G}})_* = \mathrm{id}_{\int_{\mathcal{G}} F}. \tag{7.9}$$

These two constructions are compatible. Namely, we get

$$\left(\int_{\mathcal{G}} (S, U)\right) \circ (W_*, \mathrm{id}) = (W_*, \mathrm{id}) \circ \left(\int_{\mathcal{G}_1} (S \circ W, U \circ W)\right). \tag{7.10}$$

One easily checks

**Lemma 7.11.** (i) *Let* $W\colon \mathcal{G}_1 \to \mathcal{G}$ *be an equivalence of connected groupoids. Let* $(F,T)\colon \mathcal{G} \to \mathsf{Add\text{-}Cat}_{\mathrm{inv}}$ *be a contravariant functor. Then*

$$W_*\colon \int_{\mathcal{G}_1} (F,T)\circ W \to \int_{\mathcal{G}} (F,T)$$

*is an equivalence of additive categories with involution.*

(ii) *Let* $\mathcal{G}$ *be a connected groupoid. Let* $S\colon (F_1,T_1) \to (F_2,T_2)$ *be a transformation of contravariant functors* $\mathcal{G} \to \mathsf{Add\text{-}Cat}_{\mathrm{inv}}$ *such that for every object* $x$ *in* $\mathcal{G}$ *the functor* $S(x)\colon F_0(x) \to F_1(x)$ *is an equivalence of additive categories. Then*

$$\int_{\mathcal{G}} S\colon \int_{\mathcal{G}} (F_1,T_1) \to \int_{\mathcal{G}} (F_2,T_2)$$

*is an equivalence of additive categories with involution.*

# 8. From crossed product rings with involution to additive categories with involution

Next we want to extend Example 6.1 and Lemma 6.3 to rings and additive categories with involutions. Let $R$ be a ring and let $G$ be a group. Suppose that we are given maps of sets

$$\begin{aligned} c\colon G &\to \mathrm{aut}(R), \quad g \mapsto c_g;\\ \tau\colon G\times G &\to R^\times;\\ w\colon G &\to R, \end{aligned}$$

satisfying conditions (4.3), (4.4), (4.5), (4.6), (4.7), (4.13), (4.14), (4.15), and (4.16). We have constructed in Section 4 an involution on the crossed product $R * G = R *_{c,\tau} G$. We have denoted this ring with involution by $R * G = R *_{c,\tau,w} G$ (see (4.19)). The additive category $R*G\text{-}\mathsf{FGF}$ inherits the structure of an additive category with involution (see Example 1.5).

We have introduced notion of an additive $G$-category with involution in Definition 4.22 and constructed an explicit example $R\text{-}\mathsf{FGP}_{c,\tau,w}$ in (4.24). All these constructions restrict to the subcategory $R\text{-}\mathsf{FGF} \subseteq R\text{-}\mathsf{FGP}$ of finitely generated free $R$-modules. Thus we obtain the additive $G$-category with involution

$$R\text{-}\mathsf{FGP}_{c,\tau,w} \tag{8.1}$$

**Lemma 8.2.** *Consider the data* $(R,c,\tau,w)$ *and the additive* $G$*-category with involution* $R\text{-}\mathsf{FGF}_{c,\tau,w}$ *of* (8.1)*. Let* $\int_G R\text{-}\mathsf{FGF}_{c,\tau,w}$ *be the additive category with involution defined in* (7.1)*. Since* $G$ *regarded as a groupoid has precisely one object, we can (and will) identify the set of objects in* $\int_G R\text{-}\mathsf{FGF}_{c,\tau,w}$ *with the set of objects in* $R\text{-}\mathsf{FGF}_{c,\tau,w}$ *which consists of pairs* $(M,g)$ *for* $M$ *a finitely generated free* $R$*-module and* $g \in G$*. Denote by* $\left(\int_G R\text{-}\mathsf{FGF}_{c,\tau,w}\right)_e$ *the full subcategory of* $\int_G FGFR_{c,\tau,w}$ *consisting of objects of the shape* $(M,e)$ *for* $e \in G$ *the unit element. Denote by* $R * G = R *_{c,\tau,w} G$ *the ring with involution given by the crossed product ring (see 4.19). Then*

(i) *There is an equivalence of additive categories with involution*

$$(\alpha, \beta)\colon \left(\int_G R\text{-}\mathsf{FGF}_{c,\tau,w}\right)_e \to R *_{c,\tau,w} G\text{-}\mathsf{FGF};$$

(ii) *The inclusion*

$$\left(\int_G R\text{-}\mathsf{FGF}_{c,\tau,w}\right)_e \to \int_G R\text{-}\mathsf{FGF}_{c,\tau,w}$$

*is an equivalence of additive categories with involution.*

*Proof.* (i) We have already constructed an equivalence of categories

$$\alpha\colon \left(\int_G R\text{-}\mathsf{FGF}_{c,\tau}\right)_e \to R *_{c,\tau} G\text{-}\mathsf{FGF};$$

in Lemma 6.3 (i). We want to show that $\alpha$ is compatible with the involution, i.e., there is a functor of categories with involutions

$$(\alpha, \beta)\colon \left(\int_G R\text{-}\mathsf{FGF}_{c,\tau,w}\right)_e \to R *_{c,\tau,w} G\text{-}\mathsf{FGF}.$$

The natural equivalence $\beta\colon \alpha \circ I_{\left(\int_G R\text{-}\mathsf{FGF}_{c,\tau,w}\right)_e} \to I_{R *_{c,\tau,w} G\text{-}\mathsf{FGF}} \circ \alpha$ assigns to an object $(M, e)$ in $\left(\int_G R\text{-}\mathsf{FGF}_{c,\tau,w}\right)_e$ the $R *_{c,\tau} G$-isomorphism

$$\beta(M, e)\colon R *_{c,\tau,w} G \otimes_R M^* \xrightarrow{\cong} (R *_{c,\tau,w} G \otimes_R M)^*$$

given by $\beta(M, e)(u \otimes f)(v \otimes m) = vf(m)\overline{u}$ for $f \in M^*$, $u \in R *_{c,\tau} G$ and $m \in M$. Obviously $\beta$ is compatible with the structures of additive categories.

Next we check that $\beta$ is a natural transformation. We have to show for a morphism $\phi\colon (M, e) \to (N, e)$ in $\left(\int_G R\text{-}\mathsf{FGF}_{c,\tau,w}\right)_e$ that the following diagram commutes

$$\begin{array}{ccc} R * G \otimes_R N^* & \xrightarrow{\alpha(\phi^*)} & R * G \otimes_R M^* \\ \downarrow{\scriptstyle \beta(N,e)} & & \downarrow{\scriptstyle \beta(M,e)} \\ (R * G \otimes_R N)^* & \xrightarrow{\alpha(\phi)^*} & (R * G \otimes_R M)^* \end{array}$$

Recall that a morphism

$$\phi = \sum_{g \in G} g \cdot \phi_g\colon (M, e) \to (N, e)$$

in $\left(\int_G R\text{-}\mathsf{FGF}_{c,\tau,w}\right)_e$ is given by a collection of morphisms $\phi_g\colon (M, e) \to R_g(N, e) = (N, g)$ in $R\text{-}\mathsf{FGF}_{c,\tau,w}$ for $g \in G$, where $\phi_g$ is a $R$-homomorphism $M \to \operatorname{res}_{c_g} N$. We want to unravel what the dual morphism

$$\phi^* = \sum_{g \in G} g \cdot (\phi^*)_g\colon (N, e)^* = (N^*, e) \to (M, e)^* = (M^*, e)$$

in $\left(\int_G R\text{-}\mathsf{FGF}_{c,\tau,w}\right)_e$ is. It is given by a collection of morphisms $\{(\phi^*)_g\colon (N^*,e) \to R_g(M^*,e) = (M^*,g) \mid g \in G\}$ in $R\text{-}\mathsf{FGF}_{c,\tau,w}$, where $(\phi^*)_g$ is a $R$-homomorphism $N^* \to \operatorname{res}_{c_g} M^*$. In $R\text{-}\mathsf{FGF}_{c,\tau,w}$ the morphism $(\phi^*)_g$ is given by the composite

$$(N^*,e) = (N^*,g^{-1}) \cdot g = R_g(N^*,g^{-1}) \xrightarrow{R_g((\phi_{g^{-1}})^*)} R_g(M^*,e) = (M^*,g).$$

The morphism $(\phi_{g^{-1}})^*$ is given by the composite

$$\operatorname{res}_{c_{g^{-1}}} N^* \xrightarrow{t_{g^{-1}}(N)} \left(\operatorname{res}_{c_{g^{-1}}} N\right)^* \xrightarrow{(\phi_{g^{-1}})^*} M^*.$$

Explicitly this is the map

$$N^* \to M^*, \quad f(x) \mapsto c_{g^{-1}}^{-1}\left(f \circ \phi_{g^{-1}}(x)\right)\left(w(g^{-1})\tau(g,g^{-1})\right)^{-1}.$$

The morphism $R_g((\phi_{g^{-1}})^*)$ is the composite

$$N^* = \operatorname{res}_{c_{g^{-1}g}} N^* \xrightarrow{L_{\tau(g^{-1},g)}} \operatorname{res}_g \operatorname{res}_{c_{g^{-1}}} N^* \xrightarrow{\operatorname{res}_{c_g}(\phi_{g^{-1}})^*} \operatorname{res}_{c_g} M^*.$$

Hence the $R$-linear map $(\phi^*)_g\colon N^* \to \operatorname{res}_{c_g} M^*$ sends $f \in N^*$ to the element in $M^*$ given by

$$x \;\mapsto\; c_{g^{-1}}^{-1}\left(f \circ \phi_{g^{-1}}(x)\overline{\tau(g^{-1},g)}\right)\left(w(g^{-1})\tau(g,g^{-1})\right)^{-1}.$$

This implies that the $R*G$-homomorphism

$$\alpha(\phi^*)\colon R*G \otimes_R N^* \to R*G \otimes_R M^*$$

sends $u \otimes f$ for $u \in R*G$ and $f \in N^*$ to the $R$-linear map $M \to R$ given by

$$\sum_{g\in G} u\cdot\tau(g^{-1},g)^{-1}\cdot g^{-1} \otimes \left(c_{g^{-1}}^{-1} \circ f \circ \phi_{g^{-1}}\right) c_{g^{-1}}^{-1}\left(\overline{\tau(g^{-1},g)}\right)\left(w(g^{-1})\tau(g,g^{-1})\right)^{-1}.$$

We conclude that the composite $\beta(M,e)\circ\alpha(\phi^*)$ sends $u \otimes f$ for $u \in R*G$ and $f \in N^*$ to the $R$-linear map $R*G \otimes_R M \to R*G$ which maps $v \otimes x$ for $v \in R*G$ and $x \in M$ to the element in $R*G$

$$\sum_{g\in G} v\cdot\left(c_{g^{-1}}^{-1} \circ f \circ \phi_{g^{-1}}\right)(x) c_{g^{-1}}^{-1}\left(\overline{\tau(g^{-1},g)}\right)\left(w(g^{-1})\tau(g,g^{-1})\right)^{-1} \cdot \overline{u\cdot\tau(g^{-1},g)^{-1}\cdot g^{-1}}.$$

We compute that the composite $\alpha(\phi)^* \circ \beta(N,e)$ sends $u \otimes f$ for $u \in R*G$ and $f \in N^*$ to the $R$-linear map $R*G \otimes_R M \to R*G$ which maps $v \otimes x$ for $v \in R*G$ and $x \in M$ to the element in $R*G$

$$\begin{aligned}
&\beta(N,e)(u\otimes f)\left(\alpha(\phi)(v\otimes x)\right)\\
&= \beta(N,e)(u\otimes f)\left(\sum_{g\in G} v\cdot\tau(g^{-1},g)^{-1}\cdot g^{-1}\otimes\phi_g(x)\right)\\
&= \sum_{g\in G}\beta(N,e)(u\otimes f)\left(v\cdot\tau(g^{-1},g)^{-1}\cdot g^{-1}\otimes\phi_g(x)\right)\\
&= \sum_{g\in G} v\cdot\tau(g^{-1},g)^{-1}\cdot g^{-1}\cdot f\left(\phi_g(x)\right)\cdot\overline{u}.
\end{aligned}$$

Hence it suffices to show for each $g \in G$, $u, v \in R * G$ and $x \in M$

$$\begin{aligned} v\left(c_{g^{-1}}^{-1} \circ f \circ \phi_{g^{-1}}\right)(x) c_{g^{-1}}^{-1}\left(\overline{\tau(g^{-1},g)}\right)\left(w(g^{-1})\tau(g,g^{-1})\right)^{-1} & \\ \cdot \overline{u \cdot \tau(g^{-1},g)^{-1} \cdot g^{-1}} & \\ = v \cdot \tau(g,g^{-1})^{-1} \cdot g \cdot f\left(\phi_{g^{-1}}(x)\right) \cdot \overline{u}. & \end{aligned}$$

Since

$$\begin{aligned} \overline{u \cdot \tau(g^{-1},g)^{-1} \cdot g^{-1}} &= w(g^{-1}) c_g(\overline{\tau(g^{-1},g)^{-1}}) \cdot g \cdot \overline{u}; \\ g \cdot f\left(\phi_{g^{-1}}(x)\right) \cdot \overline{u} &= c_g\left(f\left(\phi_{g^{-1}}(x)\right)\right) \cdot g \cdot \overline{u}, \end{aligned}$$

it remains to show for all $g \in G$ and $x \in M$

$$\begin{aligned} \left(c_{g^{-1}}^{-1} \circ f \circ \phi_{g^{-1}}\right)(x) c_{g^{-1}}^{-1}\left(\overline{\tau(g^{-1},g)}\right)\left(w(g^{-1})\tau(g,g^{-1})\right)^{-1} & \\ \cdot w(g^{-1}) c_g(\overline{\tau(g^{-1},g)^{-1}}) & \\ = \tau(g,g^{-1})^{-1} \cdot c_g\left(f\left(\phi_{g^{-1}}(x)\right)\right). & \end{aligned}$$

If we put $r = f \circ \phi_{g^{-1}}(x)$, this becomes equivalent to showing for all $g \in G$ and $r \in R$

$$\begin{aligned} & c_{g^{-1}}^{-1}(r) c_{g^{-1}}^{-1}\left(\overline{\tau(g^{-1},g)}\right)\left(w(g^{-1})\tau(g,g^{-1})\right)^{-1} \cdot w(g^{-1}) c_g(\overline{\tau(g^{-1},g)^{-1}}) \\ & \quad = \tau(g,g^{-1})^{-1} \cdot c_g(r). \end{aligned}$$

This is equivalent to showing

$$\tau(g,g^{-1}) c_{g^{-1}}^{-1}\left(r\overline{\tau(g^{-1},g)}\right)\tau(g,g^{-1})^{-1} = c_g(r\overline{\tau(g^{-1},g)}).$$

From (4.3) and (4.5) we conclude for $x \in R$

$$\tau(g,g^{-1}) c_{g^{-1}}^{-1}(x) \tau(g,g^{-1})^{-1} = c_g(x),$$

and the claim follows, i.e., $\beta$ is a natural equivalence.

It remains to check that the following diagram (see (1.6)) commutes for every object $(M,e)$ in $R\text{-}\mathsf{FGF}_{c,\tau}[G]^e$.

$$\begin{array}{ccc} R * G \otimes_R M & \xrightarrow{E_{R*_{c,\tau,w}G\text{-}\mathsf{FGF}}(R*G\otimes_R M)} & (R * G \otimes_R M)^{**} \\ \Big\downarrow {\scriptstyle \alpha(E_{R\text{-}\mathsf{FGF}_{c,\tau,w}}(M,e))} & & \Big\downarrow {\scriptstyle (\beta(M,e))^*} \\ R * G \otimes_R M^{**} & \xrightarrow{\beta((M,e)^*)} & (R * G \otimes_R M^*)^* \end{array}$$

We consider an element $u \otimes x$ in the left upper corner for $u \in R * G$ and $x \in M$. It is sent by the upper horizontal arrow to the element in $(R * G \otimes_R M)^{**}$ which maps $h \in (R * G \otimes_R M)^*$ to $\overline{h(u \otimes x)}$. This element is mapped by the right vertical arrow to the element in $(R * G \otimes_R M^*)^*$ which sends $v \otimes f$ for $v \in R * G$ and $f \in M^*$ to

$$\overline{\beta(M,e)(v \otimes f)(u \otimes x)} = \overline{u f(x) \overline{v}} = v \overline{f(x)}\, \overline{u}.$$

The left vertical arrow sends $u \otimes x$ to $u \otimes I_{R\text{-FGF}}(x)$, where $I_{R\text{-FGF}}(x)$ sends $f \in M^*$ to $f(x)$. This element is mapped by the lower horizontal arrow to the element in $(R * G \otimes_R M^*)^*$ which sends $v \otimes f$ for $v \in R * G$ and $f \in M^*$ to

$$vI_{R\text{-FGF}}(x)(f)\overline{u} = v\overline{f(x)}\overline{u}.$$

This finishes the proof that

$$(\alpha, \beta)\colon \left(\int_G R\text{-FGF}_{c,\tau,w}\right)_e \to R *_{c,\tau,w} G\text{-FGF}$$

is an equivalence of additive category with involutions.

(ii) This has already been proved in Lemma 6.3 (ii). □

## 9. $G$-homology theories

In this section we construct $G$-homology theories and discuss induction.

**Definition 9.1** (Transport groupoid). Let $G$ be a group and let $\xi$ be a $G$-set. Define the *transport groupoid* $\mathcal{G}^G(\xi)$ to be the following groupoid. The set of objects is $\xi$ itself. For $x_1, x_2 \in \xi$ the set of morphisms from $x_1$ to $x_2$ consists of those elements $g$ in $G$ for which $gx_1 = x_2$ holds. Composition of morphisms comes from the group multiplication in $G$.

A $G$-map $\alpha\colon \xi \to \eta$ of $G$-sets induces a covariant functor $\mathcal{G}^G(\alpha)\colon \mathcal{G}^G(\xi) \to \mathcal{G}^G(\eta)$ by sending an object $x \in \xi$ to the object $\alpha(x) \in \eta$. A morphism $g\colon x_1 \to x_2$ is sent to the morphism $g\colon \alpha(x_1) \to \alpha(x_2)$.

Fix a functor

$$\mathbf{E}\colon \mathsf{Add\text{-}Cat_{inv}} \to \mathsf{Spectra}$$

which sends weak equivalences of additive categories with involutions to weak homotopy equivalences of spectra.

Let $G$ be a group. Let $\mathsf{Groupoids} \downarrow G$ be the category of connected groupoids over $G$ considered as a groupoid with one object, i.e., objects a covariant functors $F\colon \mathcal{G} \to G$ with a connected groupoid as source and $G$ as target and a morphism from $F_0\colon \mathcal{G}_0 \to G$ to $F_1\colon \mathcal{G}_1 \to G$ is a covariant functor $W\colon \mathcal{G}_0 \to \mathcal{G}_1$ satisfying $F_1 \circ W = F_0$. For a $G$-set $S$ let

$$\mathrm{pr}_G\colon \mathcal{G}^G(S) \to \mathcal{G}(G/G) = G$$

be the functor induced by the projection $S \to G/G$. The transport category yields a functor

$$\mathcal{G}^G\colon \mathsf{Or}G \to \mathsf{Groupoids} \downarrow G$$

by sending $G/H$ to $\mathrm{pr}_G\colon \mathcal{G}^G(G/H) \to \mathcal{G}^G(G/G) = G$.

Let $\mathcal{A}$ be an additive $G$-category with involution in the sense of Definition 4.22. We obtain a functor

$$(9.2) \qquad \mathbf{E}_{\mathcal{A}} \colon \mathsf{Or}G \to \mathsf{Spectra}, \quad G/H \mapsto \mathbf{E}\left(\textstyle\int_{\mathcal{G}(G/H)} \mathcal{A} \circ \mathrm{pr}_G\right).$$

Associated to it there is a $G$-homology theory in the sense of [9, Section 1]

$$(9.3) \qquad H_*^G(-;\mathbf{E}_{\mathcal{A}})$$

such that $H_*^G(G/H;\mathbf{E}_{\mathcal{A}}) \cong \pi_n\left(\mathbf{E}_{\mathcal{A}}(G/H)\right)$ holds for every $n \in \mathbb{Z}$ and every subgroup $H \subseteq G$. Namely, define for a $G$-$CW$-complex $X$

$$H_n^G(X;\mathbf{E}_{\mathcal{A}}) = \pi_n\left(\mathrm{map}_G(G/?,X)_+ \wedge_{\mathsf{Or}(G)} \mathbf{E}_{\mathcal{A}}(G/?)\right).$$

For more details about spectra and spaces over a category and associated homology theories we refer to [5]. (Notice that there $\wedge_{\mathsf{Or}(G)}$ is denoted by $\otimes_{\mathsf{Or}(G)}$.)

**Lemma 9.4.** *Let $f\colon \mathcal{A} \to \mathcal{B}$ be a weak equivalence of additive $G$-categories with involution. Then the induced map*

$$H_n^G(X;\mathbf{E}_f)\colon H_n^G(X;\mathbf{E}_{\mathcal{A}}) \xrightarrow{\cong} H_n^G(X;\mathbf{E}_{\mathcal{A}})$$

*is a bijection for all $n \in \mathbb{Z}$.*

*Proof.* This follows from Lemma 7.11 and [5, Lemma 4.6]. □

Let $\phi\colon K \to G$ be a group homomorphism. Given a $K$-$CW$-complex $X$, let $G \times_\phi X$ be the $G$-$CW$-complex obtained from $X$ by induction with $\phi$. If $\mathcal{H}_*^G(-)$ is a $G$-homology theory, then $\mathcal{H}^G(\phi_*(-))$ is a $K$-homology theory. The next result is essentially the same as the proof of the existence of an induction structure in [1, Lemma 6.1].

**Lemma 9.5.** *Let $\phi\colon K \to G$ be a group homomorphism. Let $\mathcal{A}$ be an additive $G$-category with involution in the sense of Definition 4.22. Let $\mathrm{res}_\phi \mathcal{A}$ be the additive $K$-category with involution obtained from $\mathcal{A}$ by restriction with $\phi$.*

*Then there is a transformation of $K$-homology theories*

$$\sigma_*\colon H_*^K(-;\mathbf{E}_{\mathrm{res}_\phi \mathcal{A}}) \to H_*^G(\phi_*(-);\mathbf{E}_{\mathcal{A}})$$

*If $X$ is a $K$-$CW$-complex on which $\ker(\phi)$ acts trivially, then*

$$\sigma_n\colon H_n^K(X;\mathbf{E}_{\mathrm{res}_\phi \mathcal{A}}) \xrightarrow{\cong} H_n^G(\phi_* X;\mathbf{E}_{\mathcal{A}})$$

*is bijective for all $n \in \mathbb{Z}$.*

*Proof.* We have to construct for every $K$-$CW$-complex $X$ a natural transformation

$$(9.6) \quad \mathrm{map}_K(K/?,X)_+ \wedge_{\mathsf{Or}(K)} \mathbf{E}\left(\int_{\mathcal{G}^K(K/?)} \mathrm{res}_\phi \mathcal{A} \circ \mathrm{pr}_K\right) \to \mathrm{map}_G(G/?,\phi_* X)_+ \wedge_{\mathsf{Or}(G)} \mathbf{E}\left(\int_{\mathcal{G}^G(G/?)} \mathcal{A} \circ \mathrm{pr}_G\right).$$

The group homomorphism $\phi$ induces for every transitive $K$-set $\xi$ a functor, natural in $\xi$,

$$\mathcal{G}^\phi(\xi)\colon \mathcal{G}^K(\xi) \to \mathcal{G}^G(\phi_*\xi)$$

which sends an object $x \in \xi$ to the object $(e, x)$ in $G \times_\phi \xi$ and sends a morphism given by $k \in K$ to the morphism given by $\phi(k)$. We obtain for every transitive $K$-set $\xi$ a functor of additive categories with involutions, natural in $\xi$ (see (7.7))

$$\mathcal{G}^\phi(\xi)_*\colon \int_{\mathcal{G}^K(\xi)} \operatorname{res}_\phi \mathcal{A} \circ \operatorname{pr}_K = \int_{\mathcal{G}^K(\xi)} \mathcal{A} \circ \operatorname{pr}_G \circ \mathcal{G}^\phi(\xi) \to \int_{\mathcal{G}^G(\phi_*\xi)} \mathcal{A} \circ \operatorname{pr}_G .$$

Thus we obtain a map of spectra

$$\begin{aligned} &\operatorname{map}_K(K/?, X)_+ \wedge_{\mathsf{Or}(K)} \mathbf{E}\left(\int_{\mathcal{G}^K(K/?)} \operatorname{res}_\phi \mathcal{A} \circ \operatorname{pr}_K\right) \\ &\qquad \to \operatorname{map}_K(K/?, X)_+ \wedge_{\mathsf{Or}(K)} \mathbf{E}\left(\int_{\mathcal{G}^G(\phi_*(K/?))} \mathcal{A} \circ \operatorname{pr}_G\right). \end{aligned}$$

From the adjunction of induction and restriction with the functor

$$\mathsf{Or}(\phi)\colon \mathsf{Or}(K) \to \mathsf{Or}(G), \quad K/H \mapsto \phi_* K/H,$$

and the canonical map of contravariant $\mathsf{Or}(G)$-spaces

$$\mathsf{Or}(\phi)_* \left(\operatorname{map}_K(K/?, X)\right) \to \operatorname{map}_G(G/?, \phi_* X),$$

which is an isomorphism for a $K$-$CW$-complexes $X$, we obtain maps of spectra

$$\begin{aligned} &\operatorname{map}_K(K/?, X)_+ \wedge_{\mathsf{Or}(K)} \mathbf{E}\left(\int_{\mathcal{G}^G(\phi_*(K/?))} \mathcal{A} \circ \operatorname{pr}_G\right) \\ &\cong \operatorname{map}_K(K/?, X)_+ \wedge_{\mathsf{Or}(K)} \mathsf{Or}(\phi)^*\left(\mathbf{E}\left(\int_{\mathcal{G}^G(G/?)} \mathcal{A} \circ \operatorname{pr}_G\right)\right) \\ &\cong \mathsf{Or}(\phi)_* \left(\operatorname{map}_K(K/?, X)\right)_+ \wedge_{\mathsf{Or}(G)} \mathbf{E}\left(\int_{\mathcal{G}^G(G/?)} \mathcal{A} \circ \operatorname{pr}_G\right) \\ &\cong \operatorname{map}_G(G/?, \phi_* X)_+ \wedge_{\mathsf{Or}(G)} \mathbf{E}\left(\int_{\mathcal{G}^G(G/?)} \mathcal{A} \circ \operatorname{pr}_G\right). \end{aligned}$$

Now the desired map of spectra (9.6) is the composite of the two maps above.

The proof that $\tau_n(X)$ is bijective if $\ker(\phi)$ acts freely on $X$ is the same as the one of [1, Lemma 1.5]. □

# 10. $\mathbb{Z}$-categories and additive categories with involutions

For technical reason it will be useful that $\mathcal{A}$ comes with a (strictly associative) functorial direct sum. It will be used in the definition of the category $\operatorname{ind}_\phi \mathcal{A}$ in (11.5) and in functorial constructions about categories arising in controlled topology. (See for instance [2, Section 2.2], [3, Section 3].)

**Definition 10.1** ($\mathbb{Z}$-category (with involution)). A *$\mathbb{Z}$-category* $\mathcal{A}$ is an additive category except that we drop the condition that finite direct sums do exists. More precisely, a $\mathbb{Z}$-category $\mathcal{A}$ is a small category such that for two objects $A$ and $B$ the morphism set $\operatorname{mor}_{\mathcal{A}}(A, B)$ has the structure of an abelian group and composition yields bilinear maps $\operatorname{mor}_{\mathcal{A}}(A, B) \times \operatorname{mor}_{\mathcal{A}}(B, C) \to \operatorname{mor}_{\mathcal{A}}(B, C)$.

The notion of a *$\mathbb{Z}$-category with involution* $\mathcal{A}$ is defined analogously. Namely, we require the existence of the pair $(I_{\mathcal{A}}, E_{\mathcal{A}})$ with the same axioms as in Section 1 except that we forget everything about finite direct sums.

Of course an additive category (with involution) is a $\mathbb{Z}$-category (with involution), just forget the existence of the direct sum of two objects.

Given a $\mathbb{Z}$-category $\mathcal{A}$, we can enlarge it to an additive category $\mathcal{A}_\oplus$ with a functorial direct sums as follows. The objects in $\mathcal{A}_\oplus$ are $n$-tuples $\underline{A} = (A_1, A_2, \ldots, A_n)$ consisting of objects $A_i$ in $\mathcal{A}$ for $i = 1, 2, \ldots, n$ and $n = 0, 1, 2, \ldots$, where we think of the empty set as 0-tuple which we denote by 0. The $\mathbb{Z}$-module of morphisms from $\underline{A} = (A_1, \ldots, A_m)$ to $\underline{B} = (B_1, \ldots, B_n)$ is given by

$$\operatorname{mor}_{\mathcal{A}_\oplus}(\underline{A}, \underline{B}) := \bigoplus_{1 \le i \le m, 1 \le j \le n} \operatorname{mor}_{\mathcal{A}}(A_i, B_j).$$

Given a morphism $f \colon \underline{A} \to \underline{B}$, we denote by $f_{i,j} \colon A_i \to B_j$ the component which belongs to $i \in \{1, \ldots, m\}$ and $j \in \{1, \ldots, n\}$. If $A$ or $B$ is the empty tuple, then $\operatorname{mor}_{\mathcal{A}_\oplus}(A, B)$ is defined to be the trivial $\mathbb{Z}$-module. The composition of $f \colon \underline{A} \to \underline{B}$ and $g \colon \underline{B} \to \underline{C}$ for objects $\underline{A} = (A_1, \ldots, A_m)$, $\underline{B} = (B_1, \ldots, B_n)$ and $\underline{C} = (C_1, \ldots, C_p)$ is defined by

$$(g \circ f)_{i,k} := \sum_{j=1}^{n} g_{j,k} \circ f_{i,j}.$$

The sum on $\mathcal{A}_\oplus$ is defined on objects by sticking the tuples together, i.e., for $\underline{A} = (A_1, \ldots, A_m)$ and $\underline{B} = (B_1, \ldots, B_n)$ define

$$\underline{A} \oplus \underline{B} := (A_1, \ldots, A_m, B_1, \ldots, B_n).$$

The definition of the sum of two morphisms is now obvious. The zero object is given by the empty tuple 0. The construction is strictly associative. These data define the structure of an additive category with functorial direct sum on $\mathcal{A}_\oplus$. Notice that this is more than an additive category since for an additive category the existence of the direct sum of two objects is required but not a functorial model.

In the sequel functorial direct sum is always to be understood to be strictly associative, i.e., we have for three objects $A_1$, $A_2$ and $A_3$ the equality $(A_1 \oplus A_2) \oplus A_3 = A_1 \oplus (A_2 \oplus A_3)$ and we will and can omit the brackets from now on in the notion. We have constructed a functor from the category of $\mathbb{Z}$-categories to the category of additive categories with functorial direct sum

$$\oplus\colon \mathbb{Z}\text{-}\mathsf{Cat} \to \mathsf{Add\text{-}Cat}_{\oplus}, \quad \mathcal{A} \mapsto \mathcal{A}_{\oplus}.$$

Let

$$\text{forget}\colon \mathsf{Add\text{-}Cat}_{\oplus} \to \mathbb{Z}\text{-}\mathsf{Cat}$$

be the forgetful functor.

**Lemma 10.2.** (i) *We obtain an adjoint pair of functors* $(\oplus, \text{forget})$.

(ii) *We get for every* $\mathbb{Z}$*-category* $\mathcal{A}$ *a functor of* $\mathbb{Z}$*-categories*

$$Q_{\mathcal{A}}\colon \mathcal{A} \to \text{forget}(\mathcal{A}_{\oplus})$$

*which is natural in* $\mathcal{A}$.

*If* $\mathcal{A}$ *is already an additive category,* $Q_{\mathcal{A}}$ *is an equivalence of additive categories.*

*Proof.* (i) We have to construct for every $\mathbb{Z}$-category $\mathcal{A}$ and every additive category $\mathcal{B}$ with functorial direct sum to one another inverse maps

$$\alpha\colon \text{func}_{\mathsf{Add\text{-}Cat}_{\oplus}}(\mathcal{A}_{\oplus}, \mathcal{B}) \to \text{func}_{\mathbb{Z}\text{-}\mathsf{Cat}}(\mathcal{A}, \text{forget}(\mathcal{B}))$$

and

$$\beta\colon \text{func}_{\mathbb{Z}\text{-}\mathsf{Cat}}(\mathcal{A}, \text{forget}(\mathcal{B})) \to \text{func}_{\mathsf{Add\text{-}Cat}_{\oplus}}(\mathcal{A}_{\oplus}, \mathcal{B}).$$

Given $F\colon \mathcal{A}_{\oplus} \to \mathcal{B}$, define $\alpha(F)\colon \mathcal{A} \to \mathcal{B}$ to be the composite of $F$ with the obvious inclusion $Q_{\mathcal{A}}\colon \mathcal{A} \to \mathcal{A}_{\oplus}$ which sends $A$ to $(A)$. Given $F\colon \mathcal{A} \to \text{forget}(\mathcal{B})$, define $\beta(F)\colon \mathcal{A}_{\oplus} \to \mathcal{B}$ by sending $(A_1, A_2, \dots, A_n)$ to $F(A_1) \oplus F(A_2) \oplus \dots \oplus F(A_n)$.

(ii) We have defined $Q_{\mathcal{A}}$ already above. It is the adjoint of the identity on $\mathcal{A}_{\oplus}$. Obviously $Q_{\mathcal{A}}$ induces a bijection $\text{mor}_{\mathcal{A}}(A, B) \to \text{mor}_{\mathcal{A}_{\oplus}}(Q_{\mathcal{A}}(A), Q_{\mathcal{A}}(B))$ for every objects $A, B \in \mathcal{A}$. Suppose that $\mathcal{A}$ is an additive category. Then every object $(A_1, A_2, \dots, A_n)$ in $\mathcal{A}_{\oplus}$ is isomorphic to an object in the image of $P_{\mathcal{A}}$, namely to $P_{\mathcal{A}}(A_1 \oplus A_2 \oplus \dots \oplus A_n) = (A_1 \oplus A_2 \oplus \dots \oplus A_n)$. Hence $Q_{\mathcal{A}}$ is an equivalence of additive categories. □

**Definition 10.3** (Additive category with functorial direct sum and involution). An *additive category with functorial sum and involution* is an additive category with (strictly associative) functorial sum $\oplus$ and involution $(I, E)$ which are strictly compatible with one another, i.e., if $A_1$ and $A_2$ are two objects in $\mathcal{A}$, then $I(A_1 \oplus A_2) = I(A_1) \oplus I(A_2)$ and $E(A_1 \oplus A_2) = E(A_1) \oplus E(A_2)$ hold.

One easily checks that if the $\mathbb{Z}$-category $\mathcal{A}$ comes with an involution $(I_{\mathcal{A}}, E_{\mathcal{A}})$, the additive category $\mathcal{A}_{\oplus}$ constructed above inherits the structure of an additive category with functorial direct sum and involution in the sense of Definition 10.3. Namely, define

$$I_{\mathcal{A}_\oplus}((A_1, A_2, \dots, A_n)) = (I_{\mathcal{A}}(A_1), I_{\mathcal{A}}(A_1), \dots, I_{\mathcal{A}}(A_1));$$
$$E_{\mathcal{A}_\oplus}((A_1, A_2, \dots, A_n)) = E_{\mathcal{A}}(A_1) \oplus E_{\mathcal{A}}(A_2) \oplus \cdots \oplus E_{\mathcal{A}}(A_1).$$

We obtain a functor from the category of $\mathbb{Z}$-categories with involution to the category of additive categories with functorial direct sum and involution

$$\oplus\colon \mathbb{Z}\text{-}\mathsf{Cat}_{\mathrm{inv}} \to \mathsf{Add}\text{-}\mathsf{Cat}_{\mathrm{inv}\oplus}, \quad \mathcal{A} \mapsto \mathcal{A}_\oplus.$$

Let

$$\mathrm{forget}\colon \mathsf{Add}\text{-}\mathsf{Cat}_{\mathrm{inv}\oplus} \to \mathbb{Z}\text{-}\mathsf{Cat}_{\mathrm{inv}}$$

be the forgetful functor. One easily extends the proof of Lemma 10.2 to the case with involution.

**Lemma 10.4.** (i) *We obtain an adjoint pair of functors* $(\oplus, \mathrm{forget})$.

(ii) *We get for every $\mathbb{Z}$-category with involution $\mathcal{A}$ a functor of $\mathbb{Z}$-categories with involution*

$$Q_{\mathcal{A}}\colon \mathcal{A} \to \mathrm{forget}(\mathcal{A}_\oplus)$$

*which is natural in $\mathcal{A}$.*

*If $\mathcal{A}$ is already an additive category with involution, then $Q_{\mathcal{A}}$ is an equivalence of additive categories with involution.*

**Definition 10.5.** A *$\mathbb{Z}$-$G$-category with involution $\mathcal{A}$* is the same as an additive $G$-category in the sense of Definition 4.22 except that one forgets about the direct sum.

**Definition 10.6** (Additive $G$-category with functorial sum and (strict) involution)**.** An *additive $G$-category with functorial sum and involution* is an additive $G$-category with (strictly associative) functorial sum $\oplus$ and involution $(I, E)$ which are strictly compatible with one another, i.e., we have:

(i) If $A_1$ and $A_2$ are two objects in $\mathcal{A}$, then $I(A_1 \oplus A_2) = I(A_1) \oplus I(A_2)$ and $E(A_1 \oplus A_2) = E(A_1) \oplus E(A_2)$ hold;
(ii) If $A_1$ and $A_2$ are two objects in $\mathcal{A}$ and $g \in G$, then $R_g(A_1) \oplus R_g(A_2) = R_g(A_1 \oplus A_2)$ holds;
(iii) If $A$ is an object in $\mathcal{A}$, then $I(R_g(A)) = R_g(I(A))$ and $E(R_g(A)) = R_g(E(A))$ hold.

If the involution is strict in the sense of Section 1, i.e., $E = \mathrm{id}$ and $I \circ I = \mathrm{id}$, we call $\mathcal{A}$ an *additive $G$-category with functorial sum and strict involution.*

Define a $\mathbb{Z}$-$G$-category with (strict) involution analogously, just forget the direct sum.

We obtain a functor from the category of $\mathbb{Z}$-$G$-categories with involution to the category of additive categories with functorial direct sum and involution

$$\oplus\colon \mathbb{Z}\text{-}\mathsf{G}\text{-}\mathsf{Cat}_{\mathrm{inv}} \to \mathsf{Add}\text{-}\mathsf{G}\text{-}\mathsf{Cat}_{\mathrm{inv}\oplus}, \quad \mathcal{A} \mapsto \mathcal{A}_\oplus.$$

Let

$$\mathrm{forget}\colon \mathsf{Add\text{-}G\text{-}Cat}_{\mathrm{inv}\oplus} \to \mathbb{Z}\text{-}\mathsf{Cat}_{\mathrm{inv}}$$

be the forgetful functor. One easily extends the proof of Lemma 10.2 to the case with $G$-action and involution.

**Lemma 10.7.** (i) *We obtain an adjoint pair of functors* $(\oplus, \mathrm{forget})$.

(ii) *We get for every $\mathbb{Z}$-$G$-category with involution $\mathcal{A}$ a functor of $\mathbb{Z}$-categories with involution*

$$Q_{\mathcal{A}}\colon \mathcal{A} \to \mathrm{forget}(\mathcal{A}_{\oplus})$$

*which is natural in $\mathcal{A}$.*

*If $\mathcal{A}$ is already an additive $G$-category with involution, then $Q_{\mathcal{A}}$ is an equivalence of additive $G$-categories with involution;*

(iii) *The corresponding definitions and results carry over to the case of strict involutions.*

**Remark 10.8.** Given an additive $G$-category $\mathcal{A}$ and a $G$-set $T$, we have constructed the additive $G$-category $\left(\int_{\mathcal{G}(T)} \mathcal{A} \circ \mathrm{pr}_G\right)_{\oplus}$. Let $\mathcal{A} *_G T$ be the additive $G$-category defined in [4, Definition 2.1]. We obtain a functor of $\mathbb{Z}$-categories

$$\rho(T)\colon \int_{\mathcal{G}(T)} \mathcal{A} \circ \mathrm{pr}_G \to \mathcal{A} *_G T$$

by sending an object $(x, A)$ to the object $\{B_t \mid t \in T\}$ for which $B_x = A$ if $x = t$ and $B_x = 0$ if $x \neq t$. It induces a functor of additive categories with functorial direct sum

$$\rho(T)_{\oplus}\colon \left(\int_{\mathcal{G}(T)} \mathcal{A} \circ \mathrm{pr}_G\right)_{\oplus} \to (\mathcal{A} *_G T)_{\oplus}.$$

Recall that we have the functor of $\mathbb{Z}$-categories

$$Q_{\mathcal{A} *_G T}\colon \mathcal{A} *_G T \to (\mathcal{A} *_G T)_{\oplus}.$$

One easily checks that both $\rho(T)_{\oplus}$ and $Q_{\mathcal{A} *_G T}$ are equivalences of additive categories and natural in $T$.

If $\mathcal{A}$ is an additive $G$-category with strict involution, then we obtain on the source and the target of $\rho(T)_{\oplus}$ and of $Q_{\mathcal{A} *_G T}$ strict involutions such that both $\rho(T)_{\oplus}$ and $Q_{\mathcal{A} *_G T}$ are equivalences of additive categories with strict involution.

This implies that the $G$-homology theories constructed for $K$- and $L$-theory here and in [4, Definition 2.1] are naturally isomorphic and lead to isomorphic assembly maps.

## 11. $G$-homology theories and restriction

Fix a functor

(11.1) $$\mathbf{E}\colon \mathsf{Add\text{-}Cat}_{\mathrm{inv}} \to \mathsf{Spectra}$$

which sends weak equivalences of additive categories with involutions to weak homotopy equivalences of spectra. We call it *compatible with direct sums* if for any family of additive categories with involutions $\{\mathcal{A}_i \mid i \in I\}$ the map induced by the canonical inclusions $\mathcal{A}_i \to \bigoplus_{i\in I} \mathcal{A}_i$ for $i \in I$

$$\bigvee_{i\in I} \mathbf{E}(\mathcal{A}_i) \to \mathbf{E}\left(\bigoplus_{i\in I} \mathcal{A}_i\right)$$

is a weak homotopy equivalence of spectra.

**Example 11.2.** The most important examples for $\mathbf{E}$ will be for us the functor which sends an additive category $\mathcal{A}$ to its non-connective algebraic $K$-theory spectrum $\mathbf{K}_{\mathcal{A}}$ in the sense of Pedersen-Weibel [11], and the functor which sends an additive category with involution $\mathcal{A}$ to its algebraic $L^{-\infty}$-spectrum $\mathbf{L}_{\mathcal{A}}^{-\infty}$ in the sense of Ranicki (see [12], [13] and [14]). Both functors send weak equivalences to weak homotopy equivalences and are compatible with direct sums. The latter follows from the fact that they are compatible with finite direct sums and compatible with directed colimits. This is proven for rings in [1, Lemma 5.2], the proof carries over to additive categories with involution.

Given a $G$-$CW$-complex $X$ and a group homomorphism $\phi\colon K \to G$, let $\phi^*X$ be the $K$-$CW$-complex obtained from $X$ by restriction with $\phi$. Given a $K$-homology theory $\mathcal{H}_*^K$, we obtain a $G$-homology theory by sending a $G$-$CW$-complex $X$ to $\mathcal{H}_*^K(\phi^*X)$. Recall that we have assigned to an additive $G$-category $\mathcal{A}$ with involution a $G$-homology theory $H_*^G(-;\mathbf{E}_{\mathcal{A}})$ in (9.3). The main result of this section is

**Theorem 11.3.** *Suppose that the functor* $\mathbf{E}$ *of* (11.1) *is compatible with direct sums. Let* $\phi\colon K \to G$ *be a group homomorphism. Let* $\mathcal{A}$ *be a* $\mathbb{Z}$-$K$*-category with involution in the sense of Definition 10.5. Let* $\operatorname{ind}_\phi \mathcal{A}$ *be the* $G$-$\mathbb{Z}$*-category with involution defined in* (11.5).

*Then there is a natural equivalence of* $G$*-homology theories*

$$\tau_*\colon \mathcal{H}^K\big(\phi^*(-);\mathbf{E}_{\mathcal{A}_\oplus}\big) \xrightarrow{\cong} \mathcal{H}^G\big(-;\mathbf{E}_{(\operatorname{ind}_\phi \mathcal{A})_\oplus}\big).$$

Its proof needs some preparation.

Given a contravariant functor $F\colon \mathcal{G} \to \mathsf{Add\text{-}Cat}_{\mathrm{inv}}$ from a groupoid into the category $\mathsf{Add\text{-}Cat}_{\mathrm{inv}}$ of additive categories with involution, we have defined an additive category with involution $\int_{\mathcal{G}} F$ in (7.2), provided that $\mathcal{G}$ is connected. We want to drop the assumption that $\mathcal{G}$ is connected. The connectedness of $\mathcal{G}$ was only used in the construction of the direct sum of two objects in $\int_{\mathcal{G}} F$. Hence everything goes through if we refine us to the construction of $\mathbb{Z}$-categories with involution. Namely, if we drop the connectivity assumption on $\mathcal{G}$, all constructions and all the functoriality properties explained in Section 7 remain true if we work within the category $\mathbb{Z}\text{-}\mathsf{Cat}_{\mathrm{inv}}$ instead of $\mathsf{Add\text{-}Cat}_{\mathrm{inv}}$.

Let $G$ and $K$ be groups. Consider a (left) $K$-set $\xi$ and a $K$-$G$-biset $\eta$. Then $G$ acts from the right on the transport groupoid $\mathcal{G}^K(\eta)$. Namely, for an element

$g \in G$ the map $R_g \colon \eta \to \eta$, $x \mapsto xg$ is $K$-equivariant and induces a functor $\mathcal{G}^K(R_g) \colon \mathcal{G}^K(\eta) \to \mathcal{G}^K(\eta)$.

Consider a $K$-$\mathbb{Z}$-category with involution $\mathcal{A}$. Let $\mathrm{pr}_K \colon \mathcal{G}^K(\eta) \to \mathcal{G}^K(K/K) = K$ be the functor induced by the projection $\eta \to K/K$. Then $\mathcal{A} \circ \mathrm{pr}_K$ is a contravariant functor $\mathcal{G}^K(\eta) \to \mathbb{Z}\text{-}\mathsf{Cat}_{\mathrm{inv}}$. We obtain a $\mathbb{Z}$-category with involution $\int_{\mathcal{G}^K(\eta)} \mathcal{A} \circ \mathrm{pr}_K$ (compare (7.2)). Given $g \in G$, the functor $\mathcal{G}^G(R_g) \colon \mathcal{G}^K(\eta) \to \mathcal{G}^K(\eta)$ induces a functor of $\mathbb{Z}$-categories with involution (compare (7.7))

$$\mathcal{G}^G(R_g) \colon \int_{\mathcal{G}^K(\eta)} \mathcal{A} \circ \mathrm{pr}_K = \int_{\mathcal{G}^K(\eta)} \mathcal{A} \circ \mathrm{pr}_K \circ \mathcal{G}^K(R_g) \to \int_{\mathcal{G}^K(\eta)} \mathcal{A} \circ \mathrm{pr}_K,$$

which strictly commutes with the involution. Thus $\int_{\mathcal{G}^K(\eta)} \mathcal{A} \circ \mathrm{pr}_K$ becomes a $\mathbb{Z}$-$G$-category with involution in the sense of Definition 10.5. We conclude that $\left(\int_{\mathcal{G}^K(\eta)} \mathcal{A} \circ \mathrm{pr}_K\right) \circ \mathrm{pr}_G$ is a contravariant functor $\mathcal{G}^G(\xi) \to \mathbb{Z}\text{-}\mathsf{Cat}_{\mathrm{inv}}$. We obtain a $\mathbb{Z}$-category with involution (compare (7.2))

$$\int_{\mathcal{G}^G(\xi)} \left(\int_{\mathcal{G}^K(\eta)} \mathcal{A} \circ \mathrm{pr}_K\right) \circ \mathrm{pr}_G .$$

Consider $\eta \times \xi$ as a left $G \times K$ set by $(g,k) \cdot (y,x) = (kyg^{-1}, gx)$. Then $\mathcal{A} \circ \mathrm{pr}_{G \times K}$ is a contravariant functor $\mathcal{G}^{G \times K}(\eta \times \xi) \to \mathbb{Z}\text{-}\mathsf{Cat}_{\mathrm{inv}}$. We obtain a $\mathbb{Z}$-category with involution (compare (7.2))

$$\int_{\mathcal{G}^{G \times K}(\eta \times \xi)} \mathcal{A} \circ \mathrm{pr}_{G \times K} .$$

**Lemma 11.4.** *There is an isomorphism of $\mathbb{Z}$-categories with involution*

$$\omega \colon \int_{\mathcal{G}^G(\xi)} \left(\int_{\mathcal{G}^K(\eta)} \mathcal{A} \circ \mathrm{pr}_K\right) \circ \mathrm{pr}_G \xrightarrow{\cong} \int_{\mathcal{G}^{G \times K}(\eta \times \xi)} \mathcal{A} \circ \mathrm{pr}_{G \times K}$$

*which is natural in both $\xi$ and $\eta$.*

*Proof.* An object in $\int_{\mathcal{G}^G(\xi)} \left(\int_{\mathcal{G}^K(\eta)} \mathcal{A} \circ \mathrm{pr}_K\right) \circ \mathrm{pr}_G$ is given by $(x,(y,A))$, where $x \in \xi$ is an object in $\mathcal{G}^G(\xi)$ and $(y,A)$ is an object in $\int_{\mathcal{G}^K(\eta)} \mathcal{A} \circ \mathrm{pr}_K$ which is given by an object $y \in \eta$ in $\mathcal{G}^K(\eta)$ and an object $A$ in $\mathcal{A}$. The object $(x,(y,A))$ is sent under $\omega$ to the object $((y,x),A)$ given by the object $(y,x)$ in $\mathcal{G}^{G \times K}(\eta \times \xi)$ and the object $A \in \mathcal{A}$.

A morphism $\phi$ in $\int_{\mathcal{G}^G(\xi)} \left(\int_{\mathcal{G}^K(\eta)} \mathcal{A} \circ \mathrm{pr}_K\right) \circ \mathrm{pr}_G$ from $(x_1,(y_1,A_1))$ to $(x_1,(y_2,A_2))$ is given by $g \cdot \psi$ for a morphism $g \colon x_1 \to x_2$ in $\mathcal{G}^G(\xi)$ and a morphism $\psi \colon (y_1,A_1) \to \mathcal{G}^K(R_g)_*(y_2,A_2)$. The morphism $\psi$ itself is given by $k \cdot \nu$ for a morphism $k \colon y_1 \to y_2 g$ in $\mathcal{G}^K(\eta)$ and a morphism $\nu \colon A \to r_k(A)$ in $\mathcal{A}$. Define the image of $\phi$ under $\omega$ to be the morphism in $\int_{\mathcal{G}^{G \times K}(\eta \times \xi)} \mathcal{A} \circ \mathrm{pr}$ given by the morphism $(g^{-1},k) \colon (y_1,x_1) \to (y_2,x_2)$ in $\mathcal{G}^{G \times K}(\eta \times \xi)$ and the morphism $\phi \colon A \to r_k(A)$. This makes sense since $r_k(A)$ is the image of $A$ under the functor $\mathcal{A} \circ \mathrm{pr}(g^{-1},k)$.

One easily checks that $\omega$ is an isomorphism of $\mathbb{Z}$-categories with involution and natural with respect to $\xi$ and $\eta$. $\square$

Let $\phi\colon K \to G$ be a group homomorphism and $\xi$ be a $G$-set. Let $\phi^*\xi$ be the $K$-set obtained from the $G$-set $\xi$ by restriction with $\phi$. Consider a $K$-$\mathbb{Z}$-category with involution $\mathcal{A}$ in the sense of Definition 10.5. Let $\phi^*G$ be the $K$-$G$-biset for which multiplication with $(k,g) \in K \times G$ sends $x \in G$ to $\phi(k)xg^{-1}$. We have explained above how $\int_{\mathcal{G}^K(\phi^*G)} \mathcal{A}$ can be considered as a $G$-$\mathbb{Z}$-category with involution. We will denote it by

$$\operatorname{ind}_\phi \mathcal{A} := \int_{\mathcal{G}^K(\phi^*G)} \mathcal{A}. \tag{11.5}$$

**Lemma 11.6.** *For every $G$-set $\xi$ there is a natural equivalence of $\mathbb{Z}$-categories with involutions*

$$\tau\colon \int_{\mathcal{G}^G(\xi)} \operatorname{ind}_\phi \mathcal{A} \xrightarrow{\simeq} \int_{\mathcal{G}^K(\phi^*\xi)} \mathcal{A} \circ \operatorname{pr}_K .$$

*It is natural in $\xi$.*

*Proof.* Because of Lemma 11.4 it suffices to construct a natural equivalence

$$\tau\colon \int_{\mathcal{G}^{G\times K}(\phi^*G\times\xi)} \mathcal{A} \circ \operatorname{pr} \xrightarrow{\simeq} \int_{\mathcal{G}^K(\phi^*\xi)} \mathcal{A} \circ \operatorname{pr}_K .$$

Consider the functor

$$W\colon \mathcal{G}^{G\times K}(\phi^*G \times \xi) \to \mathcal{G}^K(\phi^*\xi)$$

sending an object $(x,y) \in G\times\xi$ in $\mathcal{G}^{G\times K}(\phi^*G\times\xi)$ to the object $xy \in \xi$ in $\mathcal{G}^K(\phi^*\xi)$ and a morphism $(g,k)\colon (x_1,y_1) \to (x_2,y_2)$ to the morphism $k\colon xy_1 \to xy_2$. Now define $\tau$ to be

$$W_*\colon \int_{\mathcal{G}^{G\times K}(\phi^*G\times\xi)} \mathcal{A} \circ \operatorname{pr} = \int_{\mathcal{G}^{G\times K}(\phi^*G\times\xi)} \mathcal{A} \circ \operatorname{pr}_K \circ W \xrightarrow{\simeq} \int_{\mathcal{G}^K(\phi^*\xi)} \mathcal{A}$$

(see (7.7)). Since $W$ is a weak equivalence of groupoids, $\tau$ is a weak equivalence of additive categories with involution by Lemma 5.9 (i). One easily checks that this construction is natural in $\xi$. $\square$

Now we can give the proof of Theorem 11.3.

*Proof.* In the sequel we write $E_\oplus\colon \mathbb{Z}\text{-}\mathsf{Cat}_{\mathrm{inv}} \to \mathsf{Spectra}$ for the composite of the functor $\mathbf{E}$ of (11.1) and the functor $\mathbb{Z}\text{-}\mathsf{Cat}_{\mathrm{inv}} \to \mathsf{Add\text{-}Cat}_{\mathrm{inv}}$ sending $\mathcal{A}$ to $\mathcal{A}_\oplus$. Given a $G$-$CW$-complex $X$, we have to define a weak equivalence of spectra

$$\begin{aligned} &\operatorname{map}_K(K/?,\phi^*X)_+ \wedge_{\mathsf{Or}(K)} \mathbf{E}_\oplus\left(\int_{\mathcal{G}^K(K/?)} \mathcal{A} \circ \operatorname{pr}_K\right) \\ &\qquad \to \operatorname{map}_G(G/?,X)_+ \wedge_{\mathsf{Or}(G)} \mathbf{E}_\oplus\left(\int_{\mathcal{G}^G(G/?)} \operatorname{ind}_\phi \mathcal{A} \circ \operatorname{pr}_G\right). \end{aligned}$$

The left hand side can be rewritten as

$$\begin{aligned}
&\operatorname{map}_K(K/?, \phi^* X)_+ \wedge_{\mathrm{Or}(K)} \mathbf{E}_\oplus\left(\int_{\mathcal{G}^K(K/?)} \mathcal{A} \circ \mathrm{pr}_K\right) \\
&= \operatorname{map}_G(\phi_*(K/?), X)_+ \wedge_{\mathrm{Or}(K)} \mathbf{E}_\oplus\left(\int_{\mathcal{G}^K(K/?)} \mathcal{A} \circ \mathrm{pr}_K\right) \\
&= \operatorname{map}_G(G/?, X)_+ \wedge_{\mathrm{Or}(G)} \operatorname{map}_K(\phi_*(K/??), G/?)_+ \\
&\quad \wedge_{\mathrm{Or}(K)} \mathbf{E}_\oplus\left(\int_{\mathcal{G}^K(K/??)} \mathcal{A} \circ \mathrm{pr}_K\right) \\
&= \operatorname{map}_G(G/?, X)_+ \wedge_{\mathrm{Or}(G)} \operatorname{map}_K(K/??, \phi^* G/?)_+ \\
&\quad \wedge_{\mathrm{Or}(K)} \mathbf{E}_\oplus\left(\int_{\mathcal{G}^K(K/??)} \mathcal{A} \circ \mathrm{pr}_K\right).
\end{aligned}$$

Because of Lemma 11.6 the right hand side can be identified with

$$\begin{aligned}
&\operatorname{map}_G(G/?, X)_+ \wedge_{\mathrm{Or}(G)} \mathbf{E}_\oplus\left(\int_{\mathcal{G}^G(G/?)} \operatorname{ind}_\phi \mathcal{A} \circ \mathrm{pr}_G\right) \\
&\quad = \operatorname{map}_G(G/?, X)_+ \wedge_{\mathrm{Or}(G)} \mathbf{E}_\oplus\left(\int_{\mathcal{G}^K(\phi^* G/?)} \mathcal{A} \circ \mathrm{pr}_K\right).
\end{aligned}$$

Hence we need to construct for every $K$-set $\xi$ a weak homotopy equivalence, natural in $\xi$

$$\rho(\xi)\colon \operatorname{map}_K(K/??, \xi)_+ \wedge_{\mathrm{Or}(K)} \mathbf{E}_\oplus\left(\int_{\mathcal{G}^K(K/??)} \mathcal{A} \circ \mathrm{pr}_K\right) \to \mathbf{E}_\oplus\left(\int_{\mathcal{G}^K(\xi)} \mathcal{A} \circ \mathrm{pr}_K\right).$$

The map $\rho(\xi)$ sends an element in the source given by $(\phi, z)$ for a $K$-map $\phi\colon K/?? \to \xi$ and $z \in \mathbf{E}_\oplus\left(\int_{\mathcal{G}^K(K/??)} \mathcal{A} \circ \mathrm{pr}_K\right)$ to $\mathbf{E}_\oplus\left(\mathcal{G}^K(\phi)_*\right)(z)$, where

$$\mathcal{G}^K(\phi)_*\colon \int_{\mathcal{G}^K(K/??)} \mathcal{A} \circ \mathrm{pr}_K = \int_{\mathcal{G}^K(K/??)} \mathcal{A} \circ \mathrm{pr}_K \circ \mathcal{G}^K(\phi) \to \int_{\mathcal{G}^K(\xi)} \mathcal{A} \circ \mathrm{pr}_K$$

has been defined in (7.7). Obviously it is natural in $\xi$ and is an isomorphism if $\xi$ is a $K$-orbit. For a family of $K$-sets $\{\xi_i \mid i \in I\}$ there is a natural isomorphism of spectra

$$\begin{aligned}
&\bigvee_{i \in I}\left(\operatorname{map}_K(K/??, \xi_i)_+ \wedge_{\mathrm{Or}(K)} \mathbf{E}_\oplus\left(\int_{\mathcal{G}^K(K/??)} \mathcal{A} \circ \mathrm{pr}_K\right)\right) \\
&\qquad \xrightarrow{\cong} \operatorname{map}_K\left(K/??, \coprod_{i \in I} \xi_i\right)_+ \wedge_{\mathrm{Or}(K)} \mathbf{E}_\oplus\left(\int_{\mathcal{G}^K(K/??)} \mathcal{A} \circ \mathrm{pr}_K\right).
\end{aligned}$$

We have

$$\coprod_{i\in I} \mathcal{G}^K(\xi_i) \cong \mathcal{G}^K\left(\coprod_{i\in I}\xi_i\right);$$

$$\coprod_{i\in I}\int_{\mathcal{G}^K(\xi_i)} \mathcal{A}\circ \mathrm{pr}_k \cong \int_{\coprod_{i\in I}\mathcal{G}^K(\xi_i)} \mathcal{A}\circ \mathrm{pr}_k;$$

$$\bigoplus_{i\in I}\left(\int_{\mathcal{G}^K(\xi_i)} \mathcal{A}\circ \mathrm{pr}_k\right)_{\oplus} \cong \left(\coprod_{i\in I}\int_{\mathcal{G}^K(\xi_i)} \mathcal{A}\circ \mathrm{pr}_k\right)_{\oplus}.$$

By assumption $\mathbf{E}$ is compatible with direct sums. Hence we obtain a weak equivalence

$$\bigvee_{i\in I}\mathbf{E}_{\oplus}\left(\int_{\mathcal{G}^K(\xi_i)} \mathcal{A}\circ \mathrm{pr}_K\right) \xrightarrow{\simeq} \mathbf{E}_{\oplus}\left(\int_{\mathcal{G}^K(\coprod_{i\in I}\xi_i)} \mathcal{A}\circ \mathrm{pr}_K\right).$$

We conclude that $\rho\left(\coprod_{i\in I}\xi_i\right)$ is a weak homotopy equivalence if and only if $\bigvee_{i\in I}\rho(\xi_i)$ is a weak homotopy equivalence. Since a $K$-set is the disjoint union of its $K$-orbits and a wedge of weak homotopy equivalences of spectra is again a weak homotopy equivalence, $\rho(\xi)$ is a weak homotopy equivalence for every $K$-set $\xi$. This finishes the proof of Theorem 11.3. □

# 12. Proof of the main theorems

In this section we can finally give the proofs of Theorem 0.4, Theorem 0.7 and Theorem 0.12.

*Proof of Theorem 0.4.* This follows from Lemma 7.11 and Lemma 8.2. □

*Proof of Theorem 0.7.* Let $\phi\colon K\to G$ be a group homomorphism and let $\mathcal{B}$ be a additive $K$-category with involution. We have to show that the following assembly map is bijective

$$\mathrm{asmb}_n^{K,\mathcal{B}}\colon H_*^K(E_{\phi^*\mathcal{VC}\mathrm{yc}}(K);\mathbf{L}_{\mathcal{B}}) \to H_n^K(\mathrm{pt};\mathbf{L}_{\mathcal{B}}) = L_n\left(\textstyle\int_K \mathcal{B}\right).$$

Since $\phi^*E_{\mathcal{VC}\mathrm{yc}}(G)$ is a model for $E_{\phi^*\mathcal{VC}\mathrm{yc}}(K)$, this follows from the commutative diagram

$$\begin{array}{ccc}
H_n^K(E_{\phi^*\mathcal{VC}\mathrm{yc}}(K);\mathbf{L}_{\mathcal{B}}) & \xrightarrow{H_n^K(\mathrm{pr};\mathbf{L}_{\mathcal{B}})} & H_n^K(\mathrm{pt};\mathbf{L}_{\mathcal{B}}) = L_n\left(\int_K \mathcal{B}\right) \\
\cong\downarrow H_n^K(\mathrm{id};\mathbf{L}_{Q_{\mathcal{B}}}) & & \cong\downarrow H_n^K(\mathrm{id};\mathbf{L}_{Q_{\mathcal{B}}}) \\
H_n^K(E_{\phi^*\mathcal{VC}\mathrm{yc}}(K);\mathbf{L}_{\mathcal{B}_\oplus}) & \xrightarrow{H_n^K(\mathrm{pr};\mathbf{L}_{\mathcal{B}_\oplus})} & H_n^K(\mathrm{pt};\mathbf{L}_{\mathcal{B}_\oplus}) = L_n\left(\int_K \mathcal{B}_\oplus\right) \\
\cong\downarrow \tau_n^{\phi}(E_{\mathcal{VC}\mathrm{yc}}(G)) & & \cong\downarrow \tau_n^{\phi}(\mathrm{pt}) \\
H_n^G(E_{\mathcal{VC}\mathrm{yc}}(G);\mathbf{L}_{(\mathrm{ind}_\phi \mathcal{B})_\oplus}) & \xrightarrow{H_n^G(\mathrm{pr};\mathbf{L}_{(\mathrm{ind}_\phi \mathcal{B})_\oplus})} & H_n^G(\mathrm{pt};\mathbf{L}_{(\mathrm{ind}_\phi \mathcal{B})_\oplus}) = L_n\left(\int_G (\mathrm{ind}_\phi \mathcal{B})_\oplus\right)
\end{array}$$

where pr denotes the projection onto the one-point-space pt and $Q_{\mathcal{B}}\colon \mathcal{B} \to \mathcal{B}_{\oplus}$ is the natural equivalence coming from Lemma 10.7 and the vertical arrows are isomorphisms because of Lemma 9.4 and Theorem 11.3. □

*Proof of Theorem 0.12.* Given an additive category $\mathcal{A}$ with involution $\mathcal{A}$, we can consider it as an additive category with $(\mathbb{Z}/2, v)$-operation as explained in Example 2.4. If we apply Lemma 3.2, we obtain an additive category with strict involution $\mathcal{S}^{\mathbb{Z}/2}(\mathcal{A})$ together with a weak equivalence of additive categories with involutions

$$P_{\mathcal{A}}\colon \mathcal{A} \to \mathcal{S}^{\mathbb{Z}/2}(\mathcal{A})$$

If $\mathcal{A}$ is an additive $G$-category with involution in the sense of Definition 4.22, then $P_{\mathcal{A}}$ is an equivalence of additive $G$-categories with involution.

If we apply Lemma 10.7, we obtain an additive $G$-category with functorial direct sum and strict involution $\mathcal{S}^{\mathbb{Z}/2}(\mathcal{A})_{\oplus}$ in the sense of Definition 10.6 and an equivalence of additive $G$-categories with strict involution

$$Q_{\mathcal{S}^{\mathbb{Z}/2}(\mathcal{A})}\colon \mathcal{S}^{\mathbb{Z}/2}(\mathcal{A}) \to \mathcal{S}^{\mathbb{Z}/2}(\mathcal{A})_{\oplus}$$

The composite

$$f := Q_{\mathcal{S}^{\mathbb{Z}/2}(\mathcal{A})} \circ P_{\mathcal{A}}\colon \mathcal{A} \to \mathcal{S}^{\mathbb{Z}/2}(\mathcal{A})_{\oplus}$$

is a weak equivalence of additive $G$-category with involution. Now the claim follows from the following commutative diagram

$$\begin{array}{ccc}
H_n^G(E_{\mathcal{VC}\mathrm{yc}}(G);\mathbf{L}_{\mathcal{A}}) & \xrightarrow{H_n^K(\mathrm{pr};\mathbf{L}_{\mathcal{A}})} & H_n^G(\mathrm{pt};\mathbf{L}_{\mathcal{A}}) = L_n\left(\int_G \mathcal{A}\right) \\
{\scriptstyle \cong}\downarrow{\scriptstyle H_n^K(\mathrm{id};\mathbf{L}_f)} & & {\scriptstyle \cong}\downarrow{\scriptstyle H_n^K(\mathrm{id};\mathbf{L}_f)} \\
H_n^G\left(E_{\mathcal{VC}\mathrm{yc}}(G);\mathbf{L}_{\mathcal{S}^{\mathbb{Z}/2}(\mathcal{A})_{\oplus}}\right) & \xrightarrow{H_n^K\left(\mathrm{pr};\mathbf{L}_{\mathcal{S}^{\mathbb{Z}/2}(\mathcal{A})_{\oplus}}\right)} & H_n^G(\mathrm{pt};\mathbf{L}_{\mathcal{A}}) = L_n\left(\int_G \mathcal{S}^{\mathbb{Z}/2}(\mathcal{A})_{\oplus}\right)
\end{array}$$

whose vertical arrows are isomorphisms by Lemma 9.4. □

# References


[1] A. Bartels, S. Echterhoff, and W. Lück. Inheritance of isomorphism conjectures under colimits. Preprintreihe SFB 478 — Geometrische Strukturen in der Mathematik, Heft 452, Münster, arXiv:math.KT/0702460, 2007.

[2] A. Bartels, T. Farrell, L. Jones, and H. Reich. On the isomorphism conjecture in algebraic $K$-theory. *Topology*, 43(1):157–213, 2004.

[3] A. Bartels, W. Lück, and H. Reich. The $K$-theoretic Farrell-Jones Conjecture for hyperbolic groups. Preprintreihe SFB 478 — Geometrische Strukturen in der Mathematik, Heft 434, Münster, arXiv:math.GT/0609685, 2007.

[4] A. Bartels and H. Reich. Coefficients for the Farrell-Jones conjecture. Preprintreihe SFB 478 — Geometrische Strukturen in der Mathematik, Heft 402, Münster, arXiv:math.KT/0510602, to appear in Advances, 2005.

[5] J. F. Davis and W. Lück. Spaces over a category and assembly maps in isomorphism conjectures in $K$- and $L$-theory. *$K$-Theory*, 15(3):201–252, 1998.

[6] F. T. Farrell and L. E. Jones. Isomorphism conjectures in algebraic $K$-theory. *J. Amer. Math. Soc.*, 6(2):249–297, 1993.

[7] F. T. Farrell and P. A. Linnell. $K$-theory of solvable groups. *Proc. London Math. Soc. (3)*, 87(2):309–336, 2003.

[8] I. Hambleton, E. K. Pedersen, and D. Rosenthal. Assembly maps for group extensions in $K$- and $L$-theory. Preprint, arXiv: math.KT/0709.0437v1, 2007.

[9] W. Lück. Chern characters for proper equivariant homology theories and applications to $K$- and $L$-theory. *J. Reine Angew. Math.*, 543:193–234, 2002.

[10] W. Lück and H. Reich. The Baum-Connes and the Farrell-Jones conjectures in $K$- and $L$-theory. In *Handbook of $K$-theory. Vol. 1, 2*, pages 703–842. Springer, Berlin, 2005.

[11] E. K. Pedersen and C. A. Weibel. A non-connective delooping of algebraic $K$-theory. In *Algebraic and Geometric Topology; proc. conf. Rutgers Uni., New Brunswick 1983*, volume 1126 of *Lecture notes in mathematics*, pages 166–181. Springer, 1985.

[12] A. A. Ranicki. Additive $L$-theory. *$K$-Theory*, 3(2):163–195, 1989.

[13] A. A. Ranicki. *Algebraic $L$-theory and topological manifolds.* Cambridge University Press, Cambridge, 1992.

[14] A. A. Ranicki. *Lower $K$- and $L$-theory.* Cambridge University Press, Cambridge, 1992.

[15] R. W. Thomason. Homotopy colimits in the category of small categories. *Math. Proc. Cambridge Philos. Soc.*, 85(1):91–109, 1979.

*Cohomology of Groups and*
*Algebraic K-theory*
ALM 12, pp. 55–129



# Deformation Spaces for Affine Crystallographic Groups

Oliver Baues*

**Abstract**

We develop the foundations of the deformation theory of compact complete affine space forms and affine crystallographic groups. Using methods from the theory of linear algebraic groups we show that these deformation spaces inherit an algebraic structure from the space of crystallographic homomorphisms. We also study the properties of the action of the homotopy mapping class groups on deformation spaces. In our context these groups are arithmetic groups, and we construct examples of flat affine manifolds where every finite group of mapping classes admits a fixed point on the deformation space. We also show that the existence of fixed points on the deformation space is equivalent to the realisation of finite groups of homotopy equivalences by finite groups of affine diffeomorphisms. Extending ideas of Auslander we relate the deformation spaces of affine space forms with solvable fundamental group to deformation spaces of manifolds with nilpotent fundamental group. We give applications concerning the classification problem for affine space forms.

**2000 Mathematics Subject Classification:** Primary 57S30 22E40; Secondary 22E40 22E45 20H15 58D29

**Keyword and Phrases:** Affine space form, affine crystallographic group, proper actions, discrete group, locally homogeneous manifold, deformation space, moduli space, mapping classes, homotopy equivalences, realisation problem, linear algebraic group, arithmetic group, solvable group, polycyclic group, nilpotent group

# Introduction

This work is devoted to some aspects of the study of the fundamental groups of affine space forms. An affine space form is a *compact, complete, affinely flat manifold*. In modern terminology, an affinely flat manifold is a *locally homogeneous manifold* which is modelled on the action of the affine group $\mathrm{Aff}(V)$ on some real vector space $V$. One defines more restricted kinds of affinely flat manifolds by

*Institut für Algebra und Geometrie, Universität Karlsruhe, D-76128 Karlsruhe, Deutschland. E-mail: baues@math.uni-karlsruhe.de

considering any transitive subgroup $A$ of $\mathrm{Aff}(V)$. For example, if $A = E(V)$ is the group of Euclidean isometries of $V$, then one obtains the concept of a *Riemannian flat manifold*, and similarly one obtains geometrically interesting cases like *Lorentz-flat* manifolds or *flat symplectic* manifolds.

It is well known that a compact complete affine manifold arises as a quotient $\Gamma\backslash V$, where $\Gamma \leq \mathrm{Aff}(V)$ is a properly discontinuously and cocompactly acting subgroup of $\mathrm{Aff}(V)$. These groups are called *affine crystallographic groups*. An affine crystallographic subgroup $\Gamma \leq \mathrm{E}(V)$ is called an *Euclidean* crystallographic group. The study of affine crystallographic groups has a long history which goes back to Bieberbach's famous three theorems and Hilbert's 18th problem on Euclidean crystallographic groups.

Much of the recent work on affine crystallographic groups is an effort to generalize the satisfactory theory of Euclidean crystallographic groups to the general case. This approach has encountered serious obstacles. A long standing conjecture ("Auslander's conjecture") states that any affine crystallographic group $\Gamma$ is a *virtually polycyclic group*. Under the assumption that Auslander's conjecture holds certain generalizations of Bieberbach's theory are possible. Still, for example, the question which virtually polycyclic groups are isomorphic to an affine crystallographic group remains mysterious.[1]

In this article, we study the deformation theory of complete affinely flat structures on a fixed compact manifold $M$. Let $\Gamma$ be a group. An injective homomorphism $\rho : \Gamma \to \mathrm{Aff}(V)$, such that $\rho$ gives rise to a proper action of $\Gamma$ on $V$ with compact quotient $\Gamma\backslash V$ is called a *crystallographic homomorphism*. By the deformation theory of locally homogeneous spaces, developed by Thurston, the description of the topology of the *deformation space* of complete affinely flat structures on $M$ roughly amounts to an analysis of the space of crystallographic homomorphisms of $\pi_1(M)$ up to conjugacy by $\mathrm{Aff}(V)$. This approach, of course, has a long and fruitful history in geometry. One particular important example is the Teichmüller theory of surfaces.

The phenomena one could expect for the topology (and possibly geometry) of the deformation spaces of complete affine manifolds are, however, quite different from the situation of constant curvature geometric surfaces. The main theme here is that the deformation spaces of complete affine manifolds with virtually polycyclic fundamental group are of an *algebraic* and *arithmetic* nature.

We provide a conceptual framework for the study of the set of affine crystallographic actions of a given virtually polycylic group $\Gamma$. This builds on the strong interaction of affine crystallographic with the theory of representations of Lie- and algebraic groups. A basic result we obtain is that for any virtually polycyclic group $\Gamma$ the space $\mathrm{Hom}_c(\Gamma, \mathrm{Aff}(V))$ of crystallographic embeddings into $\mathrm{Aff}(V)$ has a natural structure of a real algebraic variety defined over the rational numbers. We prove that $\mathrm{Hom}_c(\Gamma, \mathrm{Aff}(V))$ is Zariski-open in a certain closed subspace of the space of all homomorphisms of $\Gamma$ into the affine group. This result may be seen as an analogue of a classic theorem of Weil on discrete subgroups of Lie

[1] Benoist [17] gave a striking example of a finitely generated nilpotent group which is not isomorphic to an affine crystallographic group.

groups.

The *deformation space* $\mathcal{D}_c(\Gamma, \mathrm{Aff}(V))$ arises as a quotient of $\mathrm{Hom}_c(\Gamma, \mathrm{Aff}(V))$ by the action of $\mathrm{Aff}(V)$. The algebraic structure of $\mathrm{Hom}_c(\Gamma, \mathrm{Aff}(V))$ carries over to the deformation space which in some cases turns out to be homeomorphic to a semi-algebraic set, and in particular it is then a Hausdorff-space. This turns out to be true, for example, for any manifold which is finitely covered by a torus and also for certain solvmanifolds. In general, the structure of the quotient spaces $\mathcal{D}_c(\Gamma, \mathrm{Aff}(V))$ seems particular hard to understand because the group actions which define the moduli problem are not reductive, and the usual techniques of geometric invariant theory do not apply.

Of particular interest in this context is furthermore the action of the mapping class group $\mathrm{Out}(\pi_1(M))$ on the deformation space of complete flat affine structures, and its quotient, the moduli space of affine structures on $M$. Compared with the Teichmüller theory of conformal structures for surfaces, the general situation in our subject resembles more the case of a genus 1 surface than the situation for higher genus. In fact, if $\Gamma$ is a polycyclic group, then $\mathrm{Out}(\Gamma)$ is an arithmetic group, similar to the model situation $\mathrm{Out}(\mathbb{Z}^2) = \mathrm{GL}_2(\mathbb{Z})$.[2] Moreover, we show that the action of $\mathrm{Out}(\pi_1(M))$ extends to an action of an algebraic group on the deformation space. In contrast to the situation for surfaces, however, the action of $\mathrm{Out}(\pi_1(M))$ on the deformation space is not proper, and the moduli space may be highly singular.

We also consider the following type of question: *Let $\Gamma$ be an affine crystallographic group and let $\Gamma \leq \Delta$ be a finite extension. Is $\Delta$ also isomorphic to an affine crystallographic group?* Questions like this are known as *realization problems*. From classic work of Dehn, Nielsen and Poincare in surface theory it is known that there is a close interplay between such a question and Teichmüller theory.

We show in detail that, for virtually nilpotent groups, the solution of the above realization problem depends only on the action of the group of mapping classes for $\Gamma$ on the deformation space. This leads us in turn to a description of the deformation space $\mathcal{D}_c(\Delta, \mathrm{Aff}(V))$ of $\Delta$ as a fixed point set in $\mathcal{D}_c(\Gamma, \mathrm{Aff}(V))$ for the finite group of mapping classes which belongs to the extension $\Delta$. We mention here that our contribution to the realization problem may be understood as a suitable generalization of the classic Burckhardt-Zassenhaus theorem on Euclidean crystallographic groups. This is because there exists a fixed point for every finite subgroup of $\mathrm{GL}_n(\mathbb{Z})$ on the deformation space of Euclidean tori. In this realm, we exhibit large classes of infranilmanifolds where the finite group of mapping classes have fixed points in $\mathcal{D}_c(\Gamma, \mathrm{Aff}(V))$. Therefore the answer to the affine crystallographic realization problem for these manifolds is positive.

Another type of realization problem enters the stage, coming from Auslander's *nilshadow construction* for solvmanifolds. *Let $\Gamma$ be the fundamental group of a solvmanifold, and let $\Theta$ be its nilshadow. If $\Theta$ is an affine crystallographic group, is it true that $\Gamma$ is isomorphic to an affine crystallographic group?* This question admits a treatment which is similar to the solution of the problem for

[2]This result is however highly non-trivial, see [15].

finite extensions. The solution for this realization problem paves the way for a description of the deformation spaces for affine solvmanifolds, and enables us to construct new examples of such manifolds.

**Broader context of the article** Let us mention briefly related work in the context of our article. The deformation theory of locally homogeneous spaces was formulated and developed by Thurston, see [76, 30, 26] for exposition. In similar spirit, but with different methods, is the local and global deformation theory of complex structures on compact manifolds, as pioneered by Kodaira and Spencer [47]. For compact surfaces both approaches lead to the same mathematical object, namely *Teichmüller space*, which is the analogue of our deformation space.

Traditionally locally homogeneous spaces are modeled on homogeneous spaces $G/H$ with $H$ compact. In this setup, discrete subgroups $\Gamma \leq G$ give rise to proper and, if $\Gamma$ is a uniform lattice, cocompact actions on $G/H$ in abundance. The case $H$ non-compact has been more or less ignored for quite some time, although its obvious relevance for relativity and Pseudo-Riemannian geometry. Motivated by phenomena discussed in the seminal paper [48], there has been lots of recent activity to understand locally homogeneous spaces modeled on $G/H$, where $G$ is reductive and $H$ is non-compact. (See, [40, 18, 64, 69] for important contributions.)

Kobayashi [44] more generally formulated the program to classify the deformation and moduli spaces of proper actions of a discrete group $\Gamma$ on a homogeneous space $G/H$ with non-compact $H$. The subject of this article with $G = \mathrm{Aff}(V)$ and $H = \mathrm{GL}(V)$, and $\Gamma$ crystallographic is an important special case.[3]

Other recent activity in this program concerns the situation, where $G$ is nilpotent and $\Gamma$ is acting properly but not necessarily cocompact on $G/H$. (See, for example [6, 7, 46, 81].)

In complex geometry, there is related recent work of Catanese et al. [21, 22, 36] to understand and classify the deformation of complex structures on compact nilmanifolds. The latter subject bears certain similarities and interactions with the deformation of affine structures on these manifolds.

**Structure of the article** There are three main parts. In Part I we introduce and develop the basic techniques in the study of crystallographic actions of solvable groups on affine space. The main result is the description of the space of crystallographic homomorphism as an algebraic variety defined over the rational numbers.

In Part II we analyse the basic properties of the associated deformation spaces. We provide examples of such spaces which have interesting properties with respect to the action of the group of mapping classes. In particular, we study convexity and fixed point properties of deformation spaces.

Part III is technically most demanding and concerns the realisation questions for affine crystallographic groups, and the extension of the action of the group of mapping classes to an algebraic group action on the deformation space. These constructions are carried out in the context of nilmanifolds.

[3]Margulis constructed proper non-uniform affine actions of finitely generated free groups on a three-dimensional vector space, see [54, 23].

**Acknowledgement** The author enjoyed many fruitful conversations with colleagues concerning the contents of this article. He profited a lot from the insight and interest in the subject shared by Bill Goldman, Fritz Grunewald, Dave Morris, Yves Benoist. He wishes to thank further Herbert Abels, Richard Pink, Karel Dekimpe, Yoshinobu Kamishima, Bernd Siebert, Mark Gross, Vicente Cortes, Anders Karlsson for helpful comments and discussions.

## Notational conventions

### Groups

$Z(G)$, the center of $G$
$C_G(H)$, the centralizer of $H$ in $G$
$\mathrm{Fitt}(\Gamma)$, Fitting subgroup, the maximal nilpotent normal subgroup of $\Gamma$
$H \leq_f G$, $H$ is a subgroup of finite index in $G$
$[A, B] \leq G$, the commutator subgroup of $A, B \subseteq G$
$G$ is said to be *virtually* $\mathcal{P}$, if $G$ has a finite index subgroup which satisfies $\mathcal{P}$
$G$ is called a wfn-*group* if $G$ is without a non-trivial finite normal subgroup
f.t.n.-group, a finitely generated torsionfree nilpotent group

### Algebraic sets

We will frequently consider affine algebraic varieties defined over the rationals or reals. Our special interest lies on the set of real points of these varieties. We call these sets real algebraic varieties.
$\bar{M}$, the real Zariski closure of $M \subset X$, $X$ a real algebraic variety

### Linear algebraic groups

A $\mathbb{Q}$-defined linear algebraic group $\mathbf{G}$ is a subgroup $\mathbf{G} \leq \mathrm{GL}_n(\mathbb{C})$ defined by polynomials with rational coefficients. For a subring $R$ of $\mathbb{C}$ we put $\mathbf{G}_R = \mathbf{G} \cap \mathrm{GL}_n(R)$.
$g = g_s g_u$, the Jordan decomposition for $g \in \mathbf{G}$.
$\mathrm{u}(\Gamma)$, the maximal normal unipotent subgroup of $\Gamma \leq \mathbf{G}$
$M_s = \{m_s \mid m \in M\}$
$M_u = \{m_u \mid m \in M\}$, where $M \subset \mathbf{G}$ is a subset
$\overline{M}$, the Zariski-closure of $M$ in $\mathbf{G}$ $\rho_u$, the map $\gamma \mapsto \rho(\gamma)_u$, where $\rho : \Gamma \to \mathbf{G}$ is a homomorphism

### Lie groups, real algebraic groups

A Lie group $G$ is called real algebraic, if $G$ is (a connected component of) the set of real points $\mathbf{G}_{\mathbb{R}}$ of a linear algebraic group $\mathbf{G}$
$N(G)$, nilpotent radical of $G$, the maximal nilpotent normal connected subgroup
$G^0$, identity component in the (real) Zariski-topology, if $G$ is real algebraic
$G_0$, identity component in the Hausdorff-topology
$\overline{G}$, the real Zariski-closure of a linear group $G \leq \mathrm{GL}_n(\mathbb{R})$
$\mathrm{u}(G)$, the unipotent radical, $G \subset \mathrm{GL}_n$ a linear group

**Affine crystallographic groups**
$V$, a finite dimensional real vector space
$\mathrm{Aff}(V) = V \cdot \mathrm{GL}(V)$, the group of affine transformations of $V$
$A \leq \mathrm{Aff}(V)$, a Zariski closed subgroup of $\mathrm{Aff}(V)$, transitive on $V$
$\mathrm{GL}_A = A \cap \mathrm{GL}(V)$
$A_x = \{a \in A \mid ax = x\}$, for $x \in V$
$\rho : \Gamma \to A$ a homomorphism, $\rho^a(\gamma) = a\rho(\gamma)a^{-1}$

**Definition** Let $\Gamma \leq \mathrm{Aff}(V)$ be a subgroup. $\Gamma$ is called *properly discontinuous* if, for all compact subsets $K \subset V$, the set $\{\gamma \in \Gamma \mid \gamma K \cap K \neq \emptyset\}$ is finite
$\Gamma \leq A$ is called an *affine crystallographic group of type* $A$, if $\Gamma$ is properly discontinuous, and if the quotient space ${}_\Gamma \backslash V$ is compact.

ACG, affine crystallographic group

# Part I
# Crystallographic Homomorphisms

Let $V$ denote a finite dimensional real vector space, and $A \leq \mathrm{Aff}(V)$ a Zariski closed subgroup, which acts transitively on $V$. A subgroup $\Gamma \leq \mathrm{Aff}(V)$ is called *affine crystallographic* if $\Gamma$ acts properly discontinuously and with a compact quotient on $V$. If $\Gamma \leq A$ then $\Gamma$ is called an affine crystallographic group of type $A$. We define the *space of crystallographic homomorphisms* as

$$\mathrm{Hom}_c(\Gamma, A) = \{\rho : \Gamma \to A \mid \rho \text{ is crystallographic}\} .$$

A homomorphism $\rho : \Gamma \to \mathrm{Aff}(V)$ is called a *crystallographic homomorphism* if $\rho$ is injective and the image $\rho(\Gamma) \leq \mathrm{Aff}(V)$ is a crystallographic subgroup. In this part we study the space of crystallographic homomorphisms

$$\mathrm{Hom}_c(\Gamma, A) \subset \mathrm{Hom}(\Gamma, A)$$

as a subset of the real algebraic variety $\mathrm{Hom}(\Gamma, A)$ of all homomorphisms of $\Gamma$ into $A$. We establish that the space $\mathrm{Hom}_c(\Gamma, A)$ may be described by algebraic equalities and inequalities, and carries itself a natural structure as a real algebraic variety.

## 1. Polycyclic groups, crystallographic groups

In this section we develop the foundations of affine crystallographic virtually polycyclic groups. We will need these results to build our later arguments on. Our basic references are [27] and [35] which are, however, written from different points of view. We put the notion of algebraic hull for a torsionfree polycyclic group (as developed by Mostow [58], see also, for example, [66, Chapter IV]) at the center of our considerations. We hope that our presentation clarifies some aspects of the theory of polycyclic affine crystallographic groups.[4]

**Solvable crystallographic groups** First some further definitions and introductory remarks. A long standing conjecture of Louis Auslander states that affine crystallographic groups are virtually solvable groups. See Milnor's paper [23] for an introduction to this problem, and related results. The conjecture is verified in certain special cases, in particular in low dimensions. No counter example is known. See [1] for details. By a result of Mostow [57], a discrete solvable subgroup of a Lie group is a polycyclic group. Hence, a solvable affine crystallographic group is, in particular, a polycyclic group.[5] Virtually polycyclic affine crystallographic groups are therefore the objects of our study.

We add that there are a few obvious mild restrictions for an abstract finitely generated group $\Delta$ to be realised as a crystallographic subgroup of $\mathrm{Aff}(V)$: Let $\Delta \leq \mathrm{Aff}(V)$ be crystallographic. Then the quotient space $\Delta\backslash V$ is a manifold if and only if $\Delta$ is torsionfree. Moreover, by Selberg's lemma every affine crystallographic group contains a torsionfree normal subgroup of finite index. Since $V$ is contractible, this implies that, the *virtual cohomological dimension* of $\Delta$ (see [20] for definition) satisfies $\operatorname{vcd}\Delta = \dim V$.

**Lemma 1.1.** *Let $\Delta \leq \mathrm{Aff}(V)$ be an ACG. Then every finite normal subgroup of $\Delta$ is trivial.*

*Proof.* Let $N$ be a finite normal subgroup of $\Delta$. Since $N$ is finite, there exist a fixed point $x \in V$ for $N$. The set of fixed points for $N$ is an affine subspace $H$ of $V$, and it contains the orbit $\Delta x$. Since $\Delta$ acts as an ACG on $H$, we have $\dim H = \operatorname{vcd}\Delta = \dim V$. Therefore $H = V$, and $N = \{1\}$. □

A group which satisfies the conclusion of the lemma will be called a wfn-*group*.

## Hulls, splittings and shadows

We start by providing some foundational material on algebraic and analytic hulls for polycyclic groups. Let us first recall the definition of polycyclic groups. Namely, a group $\Gamma$ is called *polycyclic* if $\Gamma$ admits a finite normal series

$$\Gamma = \Gamma_0 \geq \Gamma_1 \geq \cdots \geq \Gamma_k = \{1\}$$

such that each quotient $\Gamma_i/\Gamma_{i+1}$ is cyclic.We let $\operatorname{rank}\Gamma$ denote its rank which is the number of infinite cyclic quotients $\Gamma_i/\Gamma_{i+1}$. (By some authors the rank of $\Gamma$ is called the Hirsch-length.) Then it is true that $\operatorname{rank}\Gamma = \operatorname{vcd}\Gamma$. The rank of a virtually polycyclic group is defined as the rank of any finite index polycyclic subgroup.

[4] A generalisation in the context of affine actions on solvable Lie groups is obtained in [13]. The importance of the *algebraic hull functor* for arbitrary (that is not necessarily affine crystallographic virtually) polycyclic groups is now particularly evident from [15].

[5] There is a rich and well developed theory of polycyclic groups, as is documented in the book [72]. For recent developments see also [15].

**The algebraic hull for a virtually polycyclic group** Let $\mathbf{H}$ be a linear algebraic group. We put $\mathrm{u}(\mathbf{H})$ for its maximal unipotent normal subgroup. The group $\mathrm{u}(\mathbf{H})$ is Zariski-closed in $\mathbf{H}$ and is called the *unipotent radical* of $\mathbf{H}$. It is customary to say that $\mathbf{H}$ has a *strong unipotent radical* if $\mathrm{C}_{\mathbf{H}}(\mathrm{u}(\mathbf{H})) = \mathrm{Z}(\mathrm{u}(\mathbf{H}))$.

**Definition 1.2.** A $\mathbb{Q}$-defined linear algebraic group $\mathbf{H}$ with a strong unipotent radical is called an *algebraic hull for* $\Gamma$ if $\Gamma$ is a Zariski-dense subgroup of $\mathbf{H}$, $\Gamma \leq \mathbf{H}_{\mathbb{Q}}$, and $\dim \mathrm{u}(\mathbf{H}) = \operatorname{rank} \Gamma$.

If $\Gamma$ is a virtually polycyclic wfn-group then algebraic hulls for $\Gamma$ exist and are unique up to $\mathbb{Q}$-isomorphism. Also $\Gamma \cap \mathbf{H}_{\mathbb{Z}}$ has finite index in $\Gamma$. This fact was proved by Mostow in the torsionfree polycyclic case, see also [66, Proposition 4.40]. In Section 7 of this article the case of finite extensions of f.t.n.-groups will be developed in detail. The more general construction for virtually polycyclic wfn-groups is given in [13]. Further applications of the theory of algebraic hulls are developed in [15].

Let us put $\mathbf{H}_\Gamma$ for the algebraic hull of $\Gamma$. We will also need algebraic hulls over the real numbers.

**Definition 1.3.** The group $H_\Gamma = \mathbf{H}_{\Gamma,\mathbb{R}}$ is called a *real algebraic hull for* $\Gamma$.

Note that $\Gamma$ is a discrete subgroup in its real hull $H_\Gamma$ since $\Gamma \cap \mathbf{H}_{\mathbb{Z}}$ has finite index in $\Gamma$. Let us henceforth write f.t.n.-group for a finitely generated torsionfree nilpotent group. We understand the algebraic hull for a torsionfree polycyclic group $\Gamma$ as a substitute for the Malcev-completion[6] of an f.t.n.-group. In fact, if $\Gamma$ is an f.t.n.-group then $\mathbf{H}_\Gamma$ is the Malcev completion of $\Gamma$. Also the algebraic hull $\mathbf{H}_\Gamma$ of a virtually polycyclic wfn-group $\Gamma$ satisfies the following rigidity property:

**Proposition 1.4.** *Let* $\mathbf{H}_\Gamma$ *be an algebraic hull for* $\Gamma$, $\mathbf{G}$ *a* $\mathbb{Q}$*-defined linear algebraic group with a strong unipotent radical. Let* $\rho : \Gamma \longrightarrow \mathbf{G}$ *be a homomorphism so that* $\rho(\Gamma)$ *is Zariski-dense in* $\mathbf{G}$. *Then* $\rho$ *extends uniquely to a homomorphism* $\rho_{\mathbf{H}_\Gamma} : \mathbf{H}_\Gamma \longrightarrow \mathbf{G}$. *If* $\rho(\Gamma) \leq \mathbf{G}_k$, *where* $k$ *is a subfield of* $\mathbb{C}$, *then* $\rho_{\mathbf{H}_\Gamma}$ *is defined over* $k$.

*Proof.* We will use the diagonal argument. Therefore we consider the subgroup

$$D = \{(\gamma, \rho(\gamma)) \mid \gamma \in \Gamma\} \leq \mathbf{H} \times \mathbf{G} \ .$$

Let $\pi_1, \pi_2$ denote the projection morphisms on the factors of the product $\mathbf{H} \times \mathbf{G}$. Let $\mathbf{D}$ be the Zariski-closure of $D$, and $\mathbf{U} = \mathrm{u}(\mathbf{D})$ the unipotent radical of $\mathbf{D}$. Now $\mathbf{D}$ is a solvable algebraic group, hence $\mathbf{U} = \mathbf{D}_u$. Let $\alpha = \pi_1|_{\mathbf{D}}$. Since $\alpha$ is onto it follows that $\alpha$ maps $\mathbf{U}$ onto $\mathrm{u}(\mathbf{H})$. By [66, Lemma 4.36] we have $\dim \mathbf{U} \leq \operatorname{rank} \Gamma = \dim \mathrm{u}(\mathbf{H})$, and hence $\dim \mathbf{U} = \dim \mathrm{u}(\mathbf{H})$. In particular it follows that the restriction $\alpha : \mathbf{U} \to \mathrm{u}(\mathbf{H})$ is an isomorphism. Thus the kernel of $\alpha$ consists only of semi-simple elements. Therefore every $x \in \ker \alpha$ centralizes $\mathbf{U}$, and since $\pi_2(\mathbf{U}) = \mathrm{u}(\mathbf{G})$, $\pi_2(x)$ centralizes $\mathrm{u}(\mathbf{G})$. Since $\mathbf{G}$ has a strong unipotent radical, $x$

[6] See [53, 33].

is in the kernel of $\pi_2$. Hence $x = 1$, and it follows that $\alpha$ is an isomorphism. We put $\rho_{\mathbf{H}_\Gamma} = \pi_2 \circ \alpha^{-1}$ to get the required unique extension. If $\rho(\Gamma) \leq \mathbf{G}_k$ then $\mathbf{D}$ is $k$-defined, and hence also the morphism $\rho_{\mathbf{H}_\Gamma}$. □

*Remark* The proposition shows that the condition that the Zariski-closure $\overline{\rho(\Gamma)}$ has a strong unipotent radical forces the homomorphism $\rho$ to be well behaved. For example, $\rho$ must be unipotent on the Fitting subgroup of $\Gamma$. See Proposition 1.6 below.

**Corollary 1.5.** *The algebraic hull $\mathbf{H}_\Gamma$ of $\Gamma$ is unique up to $\mathbb{Q}$-isomorphism. In particular, every automorphism $\phi$ of $\Gamma$ extends uniquely to a $\mathbb{Q}$-defined automorphism $\Phi$ of $\mathbf{H}_\Gamma$.*

Let us put $\mathrm{Fitt}(\Gamma)$ for the Fitting subgroup of $\Gamma$, that is, the unique maximal nilpotent normal subgroup of $\Gamma$. We note another property of the algebraic hull:

**Proposition 1.6.** *Let $\Gamma \leq \mathbf{H}_\Gamma$. Then* $\mathrm{Fitt}(\Gamma) \leq \mathrm{u}(\mathbf{H}_\Gamma)$. *In particular,* $\mathrm{u}(\Gamma) = \mathrm{Fitt}(\Gamma)$.

*Proof.* Let $F$ be the maximal nilpotent normal subgroup of $\mathbf{H}_\Gamma$. Clearly, $F = \mathbf{F}$ is a Zariski-closed subgroup. Therefore $\mathrm{u}(\mathbf{F}) = \mathrm{u}(\mathbf{H}_\Gamma)$. Now since $\mathbf{F}$ is nilpotent, $\mathbf{F}_s$ is a subgroup, $\mathbf{F}_u = \mathrm{u}(\mathbf{F})$ and $\mathbf{F} = \mathbf{F}_s \cdot \mathrm{u}(\mathbf{F})$ is a direct product of groups. Since $\mathbf{H}_\Gamma$ has a strong unipotent radical, $\mathbf{F}_s$ must be trivial, and it follows that $\mathbf{F} = \mathrm{u}(\mathbf{H}_\Gamma)$. The Zariski-closure of $\mathrm{Fitt}(\Gamma)$ is a nilpotent normal subgroup of $\mathbf{H}_\Gamma$, and therefore $\mathrm{Fitt}(\Gamma)$ is contained in $\mathbf{F}$, hence $\mathrm{Fitt}(\Gamma) \leq \mathrm{u}(\mathbf{H}_\Gamma)$. □

Recall that a *Cartan subgroup* of $\mathbf{H}_\Gamma$ is the centralizer of a maximal torus in $\mathbf{H}_\Gamma^0$. Let $\mathbf{C}$ be a Cartan-subgroup and $\mathbf{N} = \overline{\mathrm{Fitt}(\Gamma)}$. We just proved that $\mathbf{N} \leq \mathrm{u}(\mathbf{H}_\Gamma)$.

**Lemma 1.7.** *Any Cartan subgroup $\mathbf{C}$ is nilpotent, $\mathbf{C} = \mathbf{C}_s \cdot \mathrm{u}(\mathbf{C})$. Moreover,*

$$\mathrm{u}(\mathbf{H}_\Gamma) = \mathrm{u}(\mathbf{C})\mathbf{N} \ .$$

*Proof.* For the first statement, see [19, 11.7]. Let $\mathbf{U} = \mathrm{u}(\mathbf{H}_\Gamma)$. Since $\mathbf{H}_\Gamma$ is solvable, $(\mathbf{H}_\Gamma)_u = \mathbf{U}$. Therefore $\mathrm{u}(\mathbf{C}) = \mathbf{C} \cap \mathbf{U}$. Since $[\Gamma,\Gamma] \leq \mathrm{Fitt}(\Gamma)$, $[\mathbf{H}_\Gamma, \mathbf{H}_\Gamma] \leq \mathbf{N}$. It is known ([19, 10.6]) that all maximal tori are conjugate by elements of $[\mathbf{H}_\Gamma, \mathbf{H}_\Gamma]$. Hence they are conjugate by elements of $\mathbf{N}$. Therefore, if $\mathbf{T}$ is a maximal torus, $\mathrm{u}(\mathbf{H}_\Gamma) = \mathbf{N}C_{\mathbf{U}}(\mathbf{T}) = \mathbf{N}\mathrm{u}(\mathbf{C})$. □

**The algebraic hull and the semisimple splitting of a solvable Lie group**
The notion of algebraic hull applies also to connected, simply connected solvable Lie groups $G$. We summarize, cf. [66, Proposition 4.40]:

**Proposition 1.8.** *There exists a linear algebraic group $\mathbf{H}_G$ with a strong unipotent radical, so that $G \leq \mathbf{H}_\mathbb{R}$ is a Zariski-dense subgroup, and* $\dim \mathrm{u}(\mathbf{H}) = \dim G$.

As for discrete groups, the algebraic hull $\mathbf{H}_G$ is unique up to isomorphism defined over $\mathbb{R}$, and it has analogous rigidity properties. Let $N$ be the nilpotent radical of $G$, i.e., the maximal, connected nilpotent normal subgroup of $G$. Then it follows (with the same proof as for Proposition 1.6) that $N \leq \mathrm{u}(\mathbf{H}_G)$. Note that $N$ is the connected component of $\mathrm{u}(G) = G \cap \mathrm{u}(\mathbf{H}_G)$.

**Definition 1.9.** The group $H_G = \mathbf{H}_{G,\mathbb{R}}$ is called the *real algebraic hull* of $G$, and $U_G = \mathrm{u}(H_G)$ is called the *unipotent shadow* of $G$. If $N = \mathrm{u}(G)$ then $G$ is called *u-connected.*

We remark further that $G$ is a normal subgroup of $H_G$. In fact, $N \leq U_G$ is Zariski-closed in $H_G$, and $[G, G] \leq N$ implies therefore that $[H_G, H_G] \leq N$. The existence of the algebraic hull for $G$ allows for the following:

**Definition 1.10.** A subset $X \leq G$ is called Zariski-dense in $G$, if $X$ is Zariski-dense in the algebraic hull of $G$.

In particular, a subgroup $\Gamma \leq G$ is Zariski-dense in $G$ if it is so in $H_G$. We consider next the (real) *semisimple splitting* construction. (Compare [24, Chapter 2,§3.6].) This idea goes back to Malcev's and Auslander's work on solvmanifolds.

**Definition 1.11.** Let $M = S \cdot U$ be a splittable, simply connected solvable Lie group, so that $U$ is the nilradical of $M$, and $S$ acts by semisimple automorphisms. $M$ is called a *semisimple splitting for* $G$, if $G$ is a normal subgroup of $M$, so that $M = S \cdot G$ (semidirect product), and $M = GN$ (product of subgroups).

Note that it is said that a group of automorphisms of $U$ is semisimple if the corresponding induced automorphism group of the Lie algebra acts by semisimple linear maps. Auslander (see [3]) proved that semisimple splittings for $G$ exist and are unique. We briefly show the existence of $M$ by realizing it inside the algebraic hull $H_G$.

**Proposition 1.12.** *There exists a compatible embedding of the semisimple splitting $M$ into the real algebraic hull $H_G$, so that $U$ coincides with the unipotent shadow $U_G$.*

*Proof.* Since $H_G$ is (real) algebraic and Zariski-connected, there exists a semidirect product decomposition $H_G = T \cdot U_G$, where $T$ is a maximal torus. We denote $\psi : H \to U_G$ the projection map on $U_G$ which is defined by the splitting. Let $C$ be a Cartan subgroup of $G$, so that $G = CN$, where $N \leq U_G$ is the nilradical of $G$. We put $S = C_s = \{g_s \mid g \in C\}$, so that $C \leq SC_u$. $S$ is a Zariski-connected abelian subgroup of $H_G$, centralized by $C$. By the conjugacy of maximal tori, we may assume that $S \leq T$. Since $H_G = \overline{G} \leq TC_uN$, we conclude that $U_G = C_uN$. Therefore $U_G \leq SG$, and also $G \leq SU_G$. It follows that the crossed homomorphism $\psi : G \to U_G$ is surjective, in fact since $\dim U = \dim G$ it is a covering. Since $U$ is simply connected $\psi$ is a diffeomorphism. Therefore $S \cap G = \{1\}$ and we put $M = S \cdot U_G = S \cdot G = SC_uF = GU_G$. Note that $S$ acts faithfully on the strong unipotent radical $U_G$. Therefore $M \leq H_G$ is a semisimple splitting for $G$. □

**Syndetic hulls and standard groups** The notion of *syndetic hull* of a solvable subgroup of a linear group is due to Fried and Goldman, cf. [27, §1.6 ], and is an important tool in the study of discrete solvable groups, compare also [79, §5]. Fried and Goldman introduced this notion in the context of affine crystallographic groups. We use the slightly modified definition for the syndetic hull which is given in [35]. We then carry out some known constructions for ACGs in the algebraic hull of a general virtually polycyclic wfn-group.

**Definition 1.13.** Let $\Gamma$ be a polycyclic subgroup of $\mathrm{GL}(V)$, and $G$ a closed, connected subgroup of $\mathrm{GL}(V)$. $G$ is called a *syndetic hull* of $\Gamma$ if $\Gamma$ is a Zariski-dense uniform lattice in $G$, and $\dim G = \operatorname{rank}\Gamma$.

We remark that a syndetic hull for $\Gamma$ is necessarily a connected, simply connected solvable Lie group. (One has to show that $G$ has no compact subgroups.) The construction of a syndetic hull for $\Gamma$ may take place in any (real) linear algebraic group $H$ which contains $\Gamma$, provided $H$ satisfies certain conditions.

**Definition 1.14.** A polycyclic group $\Gamma \leq H$ is called *standard in H* if $\Gamma$ is Zariski-dense, $\Gamma \leq H_0$, $\Gamma' \leq \mathrm{u}(H)$, and $\Gamma/\mathrm{u}(\Gamma)$ is torsionfree. $\Gamma$ is called a *standard polycyclic group* if $\Gamma$ is standard in its real algebraic hull $H_\Gamma$.

Note that for $\Gamma \leq H_\Gamma$ to be standard it is enough to assume that $\Gamma \leq (H_\Gamma)_0$, and that $\Gamma/\mathrm{u}(\Gamma)$ is torsionfree. We recall from Proposition 1.6 that for the embedding $\Gamma \leq H_\Gamma$ we have $\mathrm{Fitt}(\Gamma) = \mathrm{u}(\Gamma)$. So for a standard polycyclic group $\Gamma/\mathrm{Fitt}(\Gamma)$ is torsionfree.

**Proposition 1.15.** *If $\Gamma$ is standard then $\Gamma$ has a u-connected syndetic hull $G$ in its real algebraic hull $H_\Gamma$.*

*Proof.* The proposition is in fact true if $\Gamma$ is a discrete standard subgroup in a real algebraic group $H$. Since $\Gamma \leq \mathbf{H}_{\Gamma,\mathbb{Z}}$ has finite index in $\Gamma$, $\Gamma$ is a discrete subgroup of $H_\Gamma = \mathbf{H}_{\Gamma,\mathbb{R}}$. Now $\Gamma \leq H_\Gamma$ satisfies all the assumptions which are needed to carry out the construction given in [35, Proposition 4.1, Lemma 4.2]. □

We remark that $\Gamma$ *is a standard polycyclic group if and only if* $\Gamma$ *may be realized as a Zariski-dense lattice in a u-connected simply connected solvable Lie group G.* One half of this statement is implied by the previous proposition. The other is the following proposition.

**Proposition 1.16.** *Let $G$ be a connected simply connected solvable Lie group, and $\Gamma \leq G$ a Zariski-dense lattice. Then the real algebraic hull $H_G$ is a real algebraic hull for $\Gamma$. If $G$ is u-connected (in $H_G$) then $\Gamma$ is standard.*

*Proof.* The Zariski-denseness of $\Gamma$ in $G$ implies by definition that $\Gamma$ is Zariski-dense in $\mathbf{H}_G$. Moreover, $\operatorname{rank}\Gamma = \dim G$ since $\Gamma$ is cocompact. Therefore $\mathbf{H}_G$ is a $\mathbb{R}$-defined algebraic hull for $\Gamma$. By an application of Propositition 1.4, $\mathbf{H}_G$ is isomorphic over $\mathbb{R}$ to $\mathbf{H}_\Gamma$. In particular, $H_G$ is isomorphic to $H_\Gamma$. Since $\Gamma$ is contained in the connected group $G \leq H_\Gamma$, $\Gamma \leq (H_\Gamma)_0$.

Let $N$ be the nilpotent radical of $G$. It is known that $\Gamma/(\Gamma \cap N) \leq A = G/N$ is torsionfree. Now if $G$ is $u$-connected, $N = G \cap \mathrm{u}(H_\Gamma)$, and by Proposition 1.6 $\mathrm{u}(\Gamma) = \Gamma \cap \mathrm{u}(H_\Gamma) = \Gamma \cap N$. We conclude that $\Gamma/\,\mathrm{u}(\Gamma)$ is torsionfree. In particular, $\Gamma$ is standard. □

The following fact is easy to see.

**Proposition 1.17.** *Let $\Delta$ be a virtually polycyclic group. Then $\Delta$ has a normal polycyclic subgroup $\Gamma$ of finite index which is standard.*

**Discrete shadows** The semisimple splitting construction for a discrete subgroup $\Gamma$ of a connected, simply connected solvable Lie group $G$ was introduced in the study of solvmanifolds. In particular, as Auslander showed (see [3]), the construction allows to associate to $\Gamma$ certain lattices in the nilshadow of $G$. A semisimple splitting for certain polycyclic groups $\Gamma$ may be obtained with the help of *nilpotent supplements.* (See for example [72].) Maximal nilpotent supplements provide an analogue for the Cartan-subgroups in a solvable Lie-group $G$. The use of nilpotent supplements allows to use the Jordan-decomposition in an algebraic group to construct the splitting. At this point, we only need the most elementary features of this construction.

Let $\Gamma \leq \mathbf{G}$ be a torsionfree polycyclic subgroup of some linear algebraic group $\mathbf{G}$. We further assume that $\Gamma = CF$, where $F$ is a normal subgroup of $\Gamma$, and $C$ is a nilpotent subgroup. $C$ will be called a *nilpotent supplement for $F$*. Let $h : C \to \mathbf{G}$ be a homomorphism, where $C$ is nilpotent. For $\gamma \in C$, we put $h_u(\gamma) = h(\gamma)_u$. This defines a homomorphism $h_u : C \to \mathbf{G}$. Now we associate with $\Gamma \leq \mathbf{G}$ the group

$$\Gamma_C^u = < C_u, F_u > \ .$$

We call $\Gamma_C^u$ the *unipotent shadow of $\Gamma$ in $\mathbf{G}$ (with respect to $C$).*

**Lemma 1.18.** *The group $\Gamma_C^u \leq \mathbf{G}_\mathbb{Q}$ is a finitely generated, Zariski-dense subgroup of* $\mathrm{u}(\overline{\Gamma}) \leq \mathbf{G}$.

*Proof.* By nilpotency, we get from the Jordan-decomposition $C \leq C_s C_u$ and $F \leq F_s F_u$. $F_s$ is a central subgroup of $F_s F_u$ and consists of semisimple elements. Therefore $\overline{F} = \overline{F}_s$. Since $C$ normalizes $F$ it normalizes $\overline{F}$. Now $\overline{\Gamma} = \overline{C}\,\overline{F} = (\overline{C})_s(\overline{F})_s(\overline{C})_u(\overline{F})_u$. Since $\overline{\Gamma}$ is solvable, $\mathrm{u}(\overline{\Gamma}) = \overline{\Gamma}_u$. Therefore $\mathrm{u}(\overline{\Gamma}) = (\overline{C})_u(\overline{F})_u = \overline{C_u}\,\overline{F_u}$, and clearly $\Gamma_C^u = < C_u, F_u >$ is finitely generated and Zariski-dense. □

Usually, we consider $\Gamma \leq \mathbf{H}_\Gamma$ as a subgroup of its algebraic hull, and $\Gamma_C^u \leq \mathbf{H}_\Gamma$ will be called *the* unipotent shadow of $\Gamma$.

**Proposition 1.19.** *Let $\Gamma = CF$ be a torsionfree polycyclic group, where $F =$* $\mathrm{Fitt}(\Gamma)$ *and $C$ is a nilpotent supplement for $F$. Then*

$$\mathrm{u}(\mathbf{H}_\Gamma) = (\overline{C})_u \overline{F} \ .$$

*The unipotent shadow $\Gamma_C^u \leq$* $\mathrm{u}(\mathbf{H}_\Gamma)_\mathbb{Q}$ *is a Zariski-dense subgroup of* $\mathrm{u}(\mathbf{H}_\Gamma)$. *In particular,* $\mathrm{rank}\,\Gamma = \mathrm{rank}\,\Gamma_C^U$.

*Proof.* Note first that by the properties of the Jordan decomposition $\Gamma \leq \mathbf{H}_{\Gamma,\mathbb{Q}}$ implies that $\Gamma^u_C \leq \mathrm{u}(\mathbf{H}_\Gamma)_\mathbb{Q}$. Since $\overline{C}$ is nilpotent, $\overline{C} = S\,\overline{C}_u$, where $S = \overline{C}_s$ centralizes the unipotent group $\mathrm{u}(\overline{C}) = (\overline{C})_u$. By Proposition 1.6, $\mathrm{Fitt}(\Gamma) \leq \mathrm{u}(\mathbf{H}_\Gamma)$. Therefore $\overline{F}$ is a unipotent normal subgroup of $\overline{\Gamma} = \mathbf{H}_\Gamma$. By the proof of the preceding lemma, $\overline{\Gamma} = \overline{C}_u\overline{F}_u = \overline{C}_u\overline{F}$, and hence the first part of the proposition follows. Since $\Gamma^u_C$ is Zariski-dense in $\mathrm{u}(\mathbf{H}_\Gamma)$ it is known (c.f. [66, Theorem 2.10]) that $\operatorname{rank}\Gamma = \dim \mathrm{u}(\mathbf{H}_\Gamma) \leq \operatorname{rank}\Gamma^u_C$. On the other hand for any f.t.n.-group $\Theta \leq \mathrm{u}(\mathbf{H}_\Gamma)_\mathbb{Q}$, $\operatorname{rank}\Theta \leq \dim \mathrm{u}(\mathbf{H}_\Gamma)$. Hence, $\operatorname{rank}\Gamma = \operatorname{rank}\Gamma^u_C$. □

Put $F = \mathrm{Fitt}(\Gamma)$. Let $\mathbf{G}$ be a linear algebraic group.

**Definition 1.20.** We say that a homomorphism $\rho : \Gamma \to \mathbf{G}$ *descends to the unipotent shadow* $\Gamma^u_C$ if there is a homomorphism $\rho^u : \Gamma^u_C \to \mathbf{G}$ so that $\rho^u|_F = \rho_u$, and $\rho^u|_{C_u}(\gamma_u) = \rho_u(\gamma)$, for all $\gamma \in C$.

In this sense, the unipotent shadow $\Gamma^u_C \leq \mathbf{H}_\Gamma$ has a rather strong functorial property:

**Proposition 1.21.** *Let $\rho : \Gamma \to \mathbf{G}$ be a homomorphism. Then $\rho$ descends to a unique homomorphism $\rho^u : \Gamma^u_C \to \mathbf{G}$. If $\rho(\Gamma) \leq \mathbf{G}_k$, where $k$ is a subfield of $\mathbb{C}$, then $\rho(\Gamma^u_C) \leq \mathbf{G}_k$.*

*Proof.* We may assume that $\rho(\Gamma)$ is Zariski-dense in $\mathbf{G}$. As in the proof of Proposition 1.4 we use the diagonal construction for the homomorphism $\rho$. We proved there that, since $\dim \mathrm{u}(\mathbf{H}_\Gamma) = \operatorname{rank}\Gamma$, the projection map $\alpha : \mathbf{D} \to \mathbf{H}_\Gamma$ induces a $k$-defined isomorphism $\alpha : \mathrm{u}(\mathbf{D}) \to \mathrm{u}(\mathbf{H}_\Gamma)$ of the unipotent radicals. Since $\Gamma^u_C \leq \mathrm{u}(\mathbf{H}_\Gamma)$ we can define $\rho^u = \pi_2\alpha^{-1}$. We have to show that $\rho$ descends to $\rho^u$ in the above sense. Since $F \leq \mathrm{u}(\mathbf{H}_\Gamma)$, it is enough to verify that $\pi_2\alpha^{-1}(\gamma_u) = \rho(\gamma)_u$, for all $\gamma \in \Gamma$. But, since $d = (\gamma, \rho(\gamma))_u = (\gamma_u, \rho(\gamma)_u) \in \mathrm{u}(\mathbf{D})$ it is clear that $\alpha^{-1}(\gamma_u) = d$, and $\pi_2(\alpha^{-1}(\gamma_u)) = \rho(\gamma)_u$. The rationality statement follows from well known properties of the Jordan-decomposition. □

Proposition 1.21 implies that the correspondence $\rho \longmapsto \rho^u$ defines a map

$$s_u : \mathrm{Hom}(\Gamma, \mathbf{G}) \longrightarrow \mathrm{Hom}(\Gamma^u_C, \mathbf{G}) .$$

**Definition 1.22.** We call $s_u : \mathrm{Hom}(\Gamma, \mathbf{G}) \to \mathrm{Hom}(\Gamma^u_C, \mathbf{G})$ the *shadow map for* $\Gamma$ *(and* $\mathbf{G}$*).*

We will need the following result (compare [72, §7, Theorem 2]):

**Proposition 1.23.** *If $\Delta$ is a polycyclic group then $\Delta$ has a characteristic subgroup $\Gamma$ of finite index which admits a nilpotent supplement $C$ for* $\mathrm{Fitt}(\Gamma)$.

Let $\Delta \leq \mathbf{H}_\Delta$ be a virtually polycyclic wfn-group, and $\Gamma \leq \Delta$ a characteristic subgroup of finite index which admits a nilpotent supplement $C$ for $\mathrm{Fitt}(\Gamma)$. Since $\mathrm{Fitt}(\Gamma) \leq F = \mathrm{Fitt}(\Delta)$, the group $C\,\mathrm{Fitt}(\Delta)$ is of finite index in $\Delta$. $C$ is then an almost nilpotent supplement for $\Delta$. In any case, we can associate to $\Delta$ the group

$$\Delta^u_C = < C_u, F_u > \leq \mathbf{H}_\Delta ,$$

and we call it a *unipotent shadow for* $\Delta$.

## Crystallographic groups and simply transitive groups

In this subsection, we collect, and sometimes reformulate, basic results about virtually polycyclic ACGs. Our basic references are [27] and [35]. First, we explain the role of simply transitive groups.

### Simply transitive hulls

The importance of syndetic hulls in our setting comes from the following central result, (compare [27],[35, §4]):

**Theorem 1.24.** *A virtually polycyclic subgroup $\Delta \subset \mathrm{Aff}(V)$ is an ACG if and only if there exists a simply transitive subgroup $G \subset \mathrm{Aff}(V)$ and $\Gamma \leq_f \Delta$ such that $G$ is a syndetic hull for $\Gamma$.*

The theorem describes the link between virtually polycyclic ACGs and certain simply transitive subgroups $G \leq \mathrm{Aff}(V)$. We will call $G$ a *simply transitive hull* for the ACG $\Gamma$. We remark that the drawback in considering the syndetic hull $G$ is that $G$ is, in general, not uniquely determined by $\Gamma$.

### Extension property of simply transitive groups

The next proposition was noticed in [27]. Our aim in this section is to provide a converse.

**Proposition 1.25.** *Let $G \leq \mathrm{Aff}(V)$ be a simply transitive group of affine motions. Then the embedding of $G$ into its algebraic closure $\overline{G} \leq \mathrm{Aff}(V)$ is a real algebraic hull for $G$.*

*Proof.* Let $U$ denote the unipotent radical of $\overline{G}$. Auslander [4, §3] proved that $U$ is simply transitive on $V$. Hence $\dim U = \dim V = \dim G$.

$\mathrm{C}_{\overline{G}}(U)$ is an algebraic subgroup of $\overline{G}$. Since $U$ is transitive, the elements of $\mathrm{C}_{\overline{G}}(U)$ act without fixed points on $V$. Therefore $\mathrm{C}_{\overline{G}}(U)$ contains no semisimple elements. Hence $\mathrm{C}_{\overline{G}}(U)$ is a Zariski-closed normal unipotent subgroup of $\overline{G}$. Hence $\mathrm{C}_{\overline{G}}(U) \leq \mathrm{u}(\overline{G}) = U$. So $\overline{G}$ has a strong unipotent radical. □

Let us call a homomorphism $i : H \to \mathrm{Aff}(V)$ of real linear algebraic groups *u-simply transitive* if the unipotent radical $\mathrm{u}(i(H))$ acts simply transitively on $V$. We can characterise simply transitive groups now as follows[7]:

**Theorem 1.26.** *A connected, simply connected, solvable Lie subgroup $G \leq \mathrm{Aff}(V)$ is simply transitive on $V$ if and only if the embedding of $G$ into $\mathrm{Aff}(V)$ extends to a u-simply transitive embedding of algebraic groups $H_G \to \mathrm{Aff}(V)$.*

*Proof.* The uniqueness of the real algebraic hull implies that the "only if" part of the theorem is just the previous proposition. So let us prove that $G$ is simply transitive if $\overline{G} \leq \mathrm{Aff}(V)$ is a $u$-simply transitive algebraic hull for $G$. By Proposition 1.12, there exists a torus $S \leq \overline{G}$ such that $U \leq GS$. The torus $S$ has a fixed point $x \in V$, so that $Gx \supset Ux = V$. Therefore $G$ acts transitively, and, since $\dim G = \dim U = \dim V$, $G$ acts a fortiori simply transitively on $V$. □

[7] Auslander [4] observed that the nilpotent shadow of $G$ acts simply transitively.

**Extension property of crystallographic groups**

We come now to the main result of this subsection.[8]

**Theorem 1.27.** *Let $\Delta \leq \mathrm{Aff}(V)$ be a virtually polycyclic* wfn-*group. Then $\Delta$ is an ACG if and only if the embedding of $\Delta$ into* $\mathrm{Aff}(V)$ *extends to a u-simply transitive embedding of algebraic groups $H_\Delta \to \mathrm{Aff}(V)$.*

*Proof.* Let us assume that $\Delta$ is an ACG. By Theorem 1.24, $\overline{\Delta}$ contains a simply transitive hull $G$ for a finite index subgroup $\Gamma$ of $\Delta$. Since $G$ is a syndetic hull for $\Gamma$, the algebraic closure of $\Gamma$ coincides with $\overline{G}$, and $\operatorname{rank}\Gamma = \dim G$. Therefore, by Theorem 1.26, $\dim \mathrm{u}(\overline{\Delta}) = \operatorname{rank}\Gamma$. Since $\overline{G}$ is $u$-simply transitive, $\overline{\Delta}$ is $u$-simply transitive. Since $\overline{\Delta}^0 = \overline{G}$ has a strong unipotent radical, we can conclude that $\overline{\Delta}$ is a real algebraic hull for $\Delta$.

For the converse, let us assume that $\Gamma \leq \mathrm{Aff}(V)$ is a subgroup such that $\overline{\Gamma}$ is a $u$-simply transitive real algebraic hull for $\Gamma$. Since $\Gamma$ has a standard subgroup of finite index, we may as well assume that $\Gamma$ is standard. Now if $\Gamma$ is standard, there exists, by Proposition 1.15, a syndetic hull $G \leq \overline{\Gamma}$ for $\Gamma$. By Proposition 1.16, $\overline{G} = \overline{\Gamma}$ is an algebraic hull for $G$, and $u$-simply transitive by assumption. By Theorem 1.26, $G$ acts simply transitively on $V$. In particular, $\Gamma$ is an ACG. □

*Remark* The theorem, together with Proposition 1.6, implies that any ACG $\Gamma \leq \mathrm{Aff}(V)$ satisfies $\mathrm{Fitt}(\Gamma) = \mathrm{u}(\Gamma)$. This is the content of Lemma C in [35].

# 2. The variety of crystallographic homomorphisms

Let $\Gamma$ be a virtually polycyclic wfn-group, and $A \leq \mathrm{Aff}(V)$ a Zariski closed subgroup. The purpose of this section is to study the space of crystallographic homomorphisms

$$\mathrm{Hom}_c(\Gamma, A) \subset \mathrm{Hom}(\Gamma, A)$$

as a subset of the real algebraic variety $\mathrm{Hom}(\Gamma, A)$. The main result of this section, Theorem 2.15, establishes that the space $\mathrm{Hom}_c(\Gamma, A)$ is described by algebraic equalities and inequalities, and carries itself a natural structure as a real algebraic variety which is defined over the rational numbers. We first prove the result for finitely generated torsionfree nilpotent groups. Our strategy is to use the unipotent shadow construction to extend from finitely generated torsionfree nilpotent groups to general virtually polycyclic groups.

## Spaces of homomorphisms

Let us introduce here the spaces which we want to study. We start with some preliminary remarks.

---

[8]The result was announced in [11, Theorem 3.2].

**Topology on the space of homomorphisms** Let $H, G$ be locally compact groups, and $R(H, G)$ the space of all continuous homomorphisms from $H$ to $G$. We equip the space $R(H, G)$ with the *compact open topology*, meaning that a fundamental system of neighbourhoods in $R(H, G)$ is specified by all sets of the form

$$\mathcal{U}_{K,U} = \{\, \rho \in R(H, G) \mid \rho(K) \subset U \,\},$$

where $K \subset H$ is compact, and $U \subset G$ is open.

Let us assume next that $\Gamma$ is a discrete group. We equip $\operatorname{Hom}(\Gamma, G)$ with the subspace topology which is inherited from the *product topology* on

$$\operatorname{Map}(\Gamma, G) = G^{\Gamma} .$$

It is easy to see that this topology coincides with the compact open topology on $\operatorname{Hom}(\Gamma, G) = R(\Gamma, G)$. If $\Gamma$ is finitely generated, and $S = \{\gamma_1, \ldots, \gamma_n\}$ is a system of generators, then $\operatorname{Hom}(\Gamma, G)$ is a subset of $G^n$ in a natural way. In fact, the inclusion

$$\jmath : \operatorname{Hom}(\Gamma, G) \longrightarrow G^n$$

which is given by $\rho \mapsto (\rho(\gamma_1), \ldots, \rho(\gamma_n))$ is a homeomorphism onto a closed subspace of $G^n$, see for example [61, 2.23]. Therefore, if $\Gamma$ is finitely generated, $\operatorname{Hom}(\Gamma, G)$ carries also the subspace topology from $G^n$.

If $G = \mathbf{G}_{\mathbb{R}}$ is a real (linear) algebraic group, $G$ carries the locally compact Hausdorff-topology and the Zariski-topology. So does $G^n$, and the embedding $\jmath$ identifies $\operatorname{Hom}(\Gamma, G)$ with a Zariski-closed subset. Aside from the Hausdorff-topology, the space $\operatorname{Hom}(\Gamma, G)$ carries a natural structure of a real algebraic variety which is independent of the embedding $\jmath$. This follows from:

**Proposition 2.1.** *Let $\Gamma$ be a finitely generated group, and $\mathbf{G}$ a $\mathbb{Q}$-defined linear algebraic group. Then* $\operatorname{Hom}(\Gamma, \mathbf{G})$ *has a natural structure of affine algebraic variety defined over $\mathbb{Q}$ and* $\operatorname{Hom}(\Gamma, \mathbf{G}_k) = \operatorname{Hom}(\Gamma, \mathbf{G})_k$ *for any subfield $k \leq \mathbb{C}$.*

See [52] for a proof. The space $\operatorname{Hom}(\Gamma, \mathbf{G})$ is called a representation variety. Note that $\mathbf{G}$ acts on homomorphisms by conjugation. This turns $\operatorname{Hom}(\Gamma, \mathbf{G})$ into a $\mathbf{G}$-variety.

**Crystallographic homomorphisms** Recall that a homomorphism $\rho : \Gamma \to \operatorname{Aff}(V)$ is an *affine crystallographic homomorphism*, if $\rho$ is an isomorphism onto its image $\rho(\Gamma)$ and $\rho(\Gamma)$ is a crystallographic subgroup of $\operatorname{Aff}(V)$. We put

$$\operatorname{Hom}_c(\Gamma, \operatorname{Aff}(V)) = \{\, \rho : \Gamma \to \operatorname{Aff}(V) \mid \rho \text{ is crystallographic}\}$$

for the *space of crystallographic homomorphisms.* We consider then $\operatorname{Hom}_c(\Gamma, \operatorname{Aff}(V))$ as a Hausdorff-topological space with the subspace topology inherited from the space $\operatorname{Hom}(\Gamma, \operatorname{Aff}(V))$. The space $\operatorname{Hom}_c(\Gamma, \operatorname{Aff}(V))$ has a natural action of $\operatorname{Aff}(V)$, which is induced by conjugation on homomorphisms.

## Nilpotent crystallographic groups

We prove here that for finitely generated torsionfree nilpotent groups the space of crystallographic homomorphisms is a Zariski-open subset of the space of unipotent homomorphisms. This result was developed in [10].

**Nilpotent crystallographic groups** Let $\Gamma$ be an f.t.n.-group. Our aim is to describe the inclusion

$$\mathrm{Hom}_c(\Gamma, \mathrm{Aff}(V)) \subset \mathrm{Hom}(\Gamma, \mathrm{Aff}(V)) .$$

*Remark* Using the fact that rank and cohomological dimension of $\Gamma$ coincide, it is evident that $\mathrm{Hom}_c(\Gamma, \mathrm{Aff}(V)) = \emptyset$, for $\mathrm{rank}\,\Gamma \neq \dim V$.

Let us briefly specialise some facts of Section 1. Since $\Gamma$ is an f.t.n.-group, the real algebraic hull $U_\Gamma$ is just the real Malcev hull. The group $\Gamma$ is then a discrete, cocompact lattice in $U_\Gamma$. The following extension property is a special case of Proposition 1.4:

**Proposition 2.2.** *Let $\Gamma$ be an* f.t.n.-*group. Then every homomorphism $\rho : \Gamma \to U$ into a unipotent Zariski-closed subgroup $U \leq \mathrm{Aff}(V)$ uniquely extends to a homomorphism of algebraic groups $\rho_{U_\Gamma} : U_\Gamma \to U$.*

From Theorem 1.27 we deduce the following characterization of crystallographic f.t.n.-groups:

**Theorem 2.3.** *Let $\Gamma \subset \mathrm{Aff}(V)$ be an* f.t.n.-*group. Then $\Gamma$ acts crystallographically on $V$ if and only if the algebraic closure $\overline{\Gamma} \subset \mathrm{Aff}(V)$ is a unipotent simply transitive (real) Malcev hull for $\Gamma$.*

The theorem will be useful together with

**Lemma 2.4.** *Let $\Gamma \subset \mathrm{Aff}(V)$ be an* f.t.n.-*group, and* $\mathrm{rank}\,\Gamma \leq \dim V$. *If the algebraic closure $U = \overline{\Gamma} \subset \mathrm{Aff}(V)$ is a unipotent simply transitive group then $U$ is a real Malcev hull for $\Gamma$.*

*Proof.* Let $U_\Gamma$ be the Malcev hull of $\Gamma$. Let $j : \Gamma \hookrightarrow \overline{\Gamma} = U$ be the inclusion homomorphism. Now $\dim \mathrm{u}(\overline{\Gamma}) \leq \mathrm{rank}\,\Gamma = \dim U_\Gamma$. By Proposition 2.2, $j$ extends to a surjective homomorphism $j_{U_\Gamma} : U_\Gamma \to U$ of unipotent algebraic groups. Since $\dim U_\Gamma = \mathrm{rank}\,\Gamma = \dim V = \dim U$, $j_{U_\Gamma}$ is an isomorphism. □

Before stating the main result, we introduce some further notation. A function $\delta$ on a $G$-variety $V$, where $G$ is a group, is called a *relative $G$-invariant* if there exists a character $\chi$ of $G$, so that, for all $g \in G$ and $v \in V$,

$$\delta(g \cdot v) = \chi(g)\, \delta(v) .$$

For any continuous function $\delta$, the set $V_\delta$ defined as

$$V_\delta = \{v \in V \mid \delta(v) \neq 0\}$$

is called a *special open subset* of $V$.

Let us put

$$\mathrm{Hom}_u(\Gamma, \mathrm{Aff}(V)) = \{\rho \mid \rho(\Gamma) \text{ is unipotent } \} \subset \mathrm{Hom}(\Gamma, \mathrm{Aff}(V))$$

for the variety of unipotent representations.

**Lemma 2.5.** *The space* $\mathrm{Hom}_u(\Gamma, \mathrm{Aff}(V))$ *is a Zariski-closed in* $\mathrm{Hom}(\Gamma, \mathrm{Aff}(V))$, *and it is defined over* $\mathbb{Q}$.

*Proof.* If $\Gamma \leq \mathrm{GL}(V)$ is nilpotent then $\Gamma_s$ and $\Gamma_u$ are commuting subgroups, $\Gamma \leq \Gamma_s \times \Gamma_u$ and the projection maps are homomorphisms. (See for example [72, Chapter 7, Proposition 3].) Therefore $\rho(\Gamma)$ is unipotent if and only if $\rho$ is unipotent on a set of generators, and we conclude that $\mathrm{Hom}_u(\Gamma, \mathrm{Aff}(V))$ is a Zariski-closed subset of $\mathrm{Hom}(\Gamma, \mathrm{Aff}(V))$. □

Theorem 2.3 implies that

$$\mathrm{Hom}_c(\Gamma, \mathrm{Aff}(V)) \subset \mathrm{Hom}_u(\Gamma, \mathrm{Aff}(V)) \,. \tag{1}$$

Surprisingly, $\mathrm{Hom}_c(\Gamma, \mathrm{Aff}(V))$ is a special open subset of $\mathrm{Hom}_u(\Gamma, \mathrm{Aff}(V))$:

**Theorem 2.6.** *Let* $\Gamma$ *be an* f.t.n.*-group which satisfies* $\mathrm{rank}\,\Gamma = \dim V$. *Then there exists a* $\mathbb{Q}$*-defined polynomial function* $\delta$ *on the variety of unipotent representations* $\mathrm{Hom}_u(\Gamma, \mathrm{Aff}(V))$, *such that*

$$\mathrm{Hom}_c(\Gamma, \mathrm{Aff}(V)) = \mathrm{Hom}_u(\Gamma, \mathrm{Aff}(V))_\delta \,.$$

*Moreover,* $\delta$ *is a relative invariant with respect to the natural* $\mathrm{Aff}(V)$*-action on* $\mathrm{Hom}_u(\Gamma, \mathrm{Aff}(V))$.

*Remark* The character which belongs to the relative invariant function $\delta$ is the determinant on $\mathrm{Aff}(V)$, cf. [10] .

We need some more preparations for the proof of the theorem.

**The Malcev hull of an f.t.n.-group** Let $\mathbf{U}_\Gamma$ be the Malcev hull of $\Gamma$. This means that $\mathbf{U}_\Gamma$ is a $\mathbb{Q}$-defined unipotent linear algebraic group, and $\Gamma \leq (\mathbf{U}_\Gamma)_\mathbb{Q}$ is a Zariski-dense subgroup. Let $\mathfrak{u}$ be the Lie algebra of $\mathbf{U}_\Gamma$. The exponential map

$$\exp : \mathfrak{u} \longrightarrow \mathbf{U}_\Gamma$$

is a polynomial map which is an isomorphism of algebraic varieties. The inverse map is $\log : \mathbf{U}_\Gamma \to \mathfrak{u}$. It is known that there exists a $\mathbb{Q}$-structure on the complex Lie algebra $\mathfrak{u}$, so that $\log \Gamma \leq \mathfrak{u}_\mathbb{Q}$, and, in fact, $\mathfrak{u}_\mathbb{Q} = \mathbb{Q} \log \Gamma$ is the $\mathbb{Q}$-span of $\log \Gamma$. Moreover, exp and log are $\mathbb{Q}$-defined maps. (See [33] for more details.)

**Definition 2.7.** Let $S = \{\gamma_1, \ldots, \gamma_n\}$ be a system of generators for $\Gamma$. $\Gamma$ is called a *Malcev-basis of* $\Gamma$, if the series of subgroups

$$1 \leq <\gamma_1> \leq <\gamma_1, \gamma_2> \leq \cdots \leq <\gamma_1, \ldots, \gamma_n> = \Gamma$$

is a central series for $\Gamma$ with infinite cyclic factors.

By results of Malcev [53] , every f.t.n.-group $\Gamma$ admits a Malcev-basis.

The following facts can be derived from the Baker-Campbell-Hausdorff formula: Put $(\mathfrak{u}_i)_{\mathbb{Q}} = \mathbb{Q} \log \Gamma_i$, where $\Gamma_i = <\gamma_1, \ldots, \gamma_i>$. The complex span $\mathfrak{u}_i$ of $\log \Gamma_i$ is an ideal in the Lie algebra $\mathfrak{u}$ and the series

$$0 \subset \mathfrak{u}_1 \subset \mathfrak{u}_2 \cdots \subset \mathfrak{u}$$

is a central series for $\mathfrak{u}$ with one-dimensional factors. In particular,

$$\mathfrak{u}_{\mathbb{Q}} = < \log \gamma_1, \ldots, \log \gamma_n >_{\mathbb{Q}} .$$

**Simply transitive and étale unipotent actions** We have to consider simply transitive unipotent actions. Let $U \leq \mathrm{Aff}(V)$ be a unipotent algebraic subgroup which satisfies $\dim U = \dim V$. $U$ is called *étale on* $V$ if it has an open orbit on $V$. By a result of Rosenlicht [68] the orbits of the unipotent group $U$ on $V$ are all closed. Therefore $U$ is simply transitive if and only if $U$ is étale.

*Proof of Theorem 2.6* Let $\mathfrak{a}(V)$ denote the Lie algebra of $\mathrm{Aff}(V)$, and let $\mathfrak{a}_n(V)$ denote the (algebraic) subset of nilpotent elements in $\mathfrak{a}(V)$. Recall that the exponential map

$$\exp : \mathfrak{a}(V) \longrightarrow \mathrm{Aff}(V)$$

defines a polynomial map which induces a *polynomial* equivalence from $\mathfrak{a}_n(V)$ onto the (algebraic) subset $\mathrm{Aff}_u(V)$ of unipotent elements in $\mathrm{Aff}(V)$. On $\mathrm{Aff}_u(V)$, the map exp has a well defined polynomial inverse

$$\log : \mathrm{Aff}_u(V) \longrightarrow \mathfrak{a}_n(V) .$$

Let $S = \{\gamma_1, \ldots, \gamma_n\}$ be a Malcev-basis for $\Gamma$, where $n = \dim V$. If $\rho \in \mathrm{Hom}_u(\Gamma, \mathrm{Aff}(V))$, then $\rho(\Gamma) \leq \mathrm{Aff}(V)$ is a unipotent f.t.n.-group. The Lie algebra $\mathfrak{u}_\rho$ of $U_\rho = \overline{\rho(\Gamma)}$ is therefore contained in $\mathfrak{a}_n(V)$. Since $\mathfrak{u}_\rho$ is a homomorphic image of $\mathfrak{u}_\Gamma$ it is spanned by the set $\{\log \rho(\gamma_i) \mid i = 1 \ldots n\}$. It follows from Proposition 2.3, together with Lemma 2.4, that, $\rho \in \mathrm{Hom}_c(\Gamma, \mathrm{Aff}(V))$ if and only if the algebraic closure $U_\rho$ is a simply transitive unipotent subgroup of $\mathrm{Aff}(V)$.

Choose an arbitrary base point $x \in V$. The differential of the orbit map $A \mapsto A \cdot x$ of the action of $\mathrm{Aff}(V)$ on $V$ defines a linear map $o_x : \mathfrak{a}(V) \to V$. Therefore $U_\rho \leq \mathrm{Aff}(V)$ is étale in $x$ if and only if the restriction of $o_x$ to the Lie-algebra $\mathfrak{u}_\rho \subset \mathfrak{a}(V)$ is an isomorphism of vector spaces. It follows a fortiori that $U$ is simply transitive on $V$ if and only if the restriction $o_x : \mathfrak{u}_\rho \to V$ is an isomorphism.

Let us next consider the linear map $\tau_x(\rho) : \mathbb{R}^n \to \mathbb{R}^n$ which is defined by

$$\tau_x(\rho) :\ (\alpha_1, \dots, \alpha_n) \longmapsto o_x\big( \sum_{i=1}^{n} \alpha_i \log \rho(\gamma_i) \big) \ .$$

(After choosing an arbitrary basis in $V$, we may as well view $o_x$ as a map with values in $\mathbb{R}^n$.) We easily see that $\tau_x(\rho)$ is an isomorphism if and only if $o_x : \mathfrak{u}_\rho \to V$ is. We therefore define

$$\delta(\rho) = \det \tau_x(\rho) \ ,$$

and, clearly, $\delta$ is a polynomial function in the matrix entries of the $\rho(\gamma_i)$. Therefore $\delta$ is a polynomial on $\mathrm{Hom}_u(\Gamma, \mathrm{Aff}(V))$ with the property that $\delta(\rho) \neq 0$ if and only if $U_\rho = \overline{\rho(\Gamma)} \leq \mathrm{Aff}(V)$ is a simply transitive subgroup. In this case, the extension $\rho_{U_\Gamma} : U_\Gamma \to U_\rho$ is an isomorphism of algebraic groups, and $\rho(\Gamma)$ is a crystallographic group. Up to now we proved that

$$\mathrm{Hom}_c(\Gamma, \mathrm{Aff}(V)) = \mathrm{Hom}_u(\Gamma, \mathrm{Aff}(V))_\delta \ .$$

Next we show that the function $\delta$ is independent of $x \in V$. To see this, we remark that our previous reasoning is also valid over the field of complex numbers. In particular if $\delta(\rho) \neq 0$ the corresponding unipotent algebraic group $\mathbf{U}_\Gamma$ acts simply transitively on $\mathbb{C}^n$. If $\rho$ is fixed and $\delta(\rho) \neq 0$ then the function

$$x \mapsto \det \tau_x(\rho)$$

is a polynomial on $\mathbb{C}^n$ which does not vanish. Hence $\det \tau_x(\rho)$ is constant. To show that $\delta(\rho) = \det \tau_x(\rho)$ is a relative invariant on $\mathrm{Hom}_c(\Gamma, \mathrm{Aff}(V))$ we remark that by direct calculation, for all $g \in \mathrm{Aff}(V)$, the formula

$$\det \tau_x(\rho^g) = (\det g) \det \tau_{g^{-1}x}(\rho)$$

holds. It follows therefore that $\det \tau_x(\rho^g) = (\det g) \det \tau_x(\rho)$. □

*Remark* We may also define $\tilde{\delta}(\rho) = \delta(\rho_u)$. If $\rho \in \mathrm{Hom}(\Gamma, \mathrm{Aff}(V))$, such that $\tilde{\delta}(\rho) \neq 0$, then, by the Theorem, $\rho_u$ is in $\mathrm{Hom}_c(\Gamma, \mathrm{Aff}(V))$. In particular $U = \overline{\rho_u(\Gamma)}$ is a simply transitive subgroup. But since $\rho(\Gamma)_s$ consists of semisimple elements, and is contained in the centralizer of $U$, it must be trivial. Therefore $\rho = \rho_u$, and a fortiori $\rho \in \mathrm{Hom}_c(\Gamma, \mathrm{Aff}(V))$. Hence,

$$\mathrm{Hom}_c(\Gamma, \mathrm{Aff}(V)) = \mathrm{Hom}(\Gamma, \mathrm{Aff}(V))_{\tilde{\delta}} \ .$$

Note however that the function $\tilde{\delta}$ is *not* continuous on $\mathrm{Hom}(\Gamma, \mathrm{Aff}(V))$.

The next example illustrates the theorem in the simplest possible situation:

**Example 2.8.** Let $V = \mathbb{R}$. Then

$$\mathrm{Hom}(\mathbb{Z}, \mathrm{Aff}(\mathbb{R})) = \mathrm{Aff}(V) = \left\{ \begin{pmatrix} \epsilon & v \\ 0 & 1 \end{pmatrix} \middle| \, \epsilon \neq 0 \right\} \leq \mathrm{GL}(2, \mathbb{R})$$

$$\mathrm{Hom}_u(\mathbb{Z}, \mathrm{Aff}(\mathbb{R})) = V = \left\{ \begin{pmatrix} 1 & v \\ 0 & 1 \end{pmatrix} \right\}$$

and

$$\mathrm{Hom}_c(\mathbb{Z}, \mathrm{Aff}(\mathbb{R})) = \left\{ \begin{pmatrix} 1 & v \\ 0 & 1 \end{pmatrix} \middle| v \neq 0 \right\} .$$

## Virtually polycyclic crystallographic groups

After some more preparations we come to the statement and proof of our main result on the space of crystallographic homomorphisms, Theorem 2.15.

**The crystallographic shadow map** Let $\Gamma$ be a virtually polycyclic wfn-group. The main step in our description of the space $\mathrm{Hom}_c(\Gamma, A)$ is to characterise the elemens $\rho \in \mathrm{Hom}_c(\Gamma, A)$ by properties which may be verified on the unipotent shadow of $\Gamma$. Let us assume here that $\Gamma$ is polycyclic, and $\Gamma = CF \leq \mathrm{Aff}(V)$ admits a nilpotent supplement for $F = \mathrm{Fitt}(\Gamma)$. Recall then from Proposition 1.19 that there is a certain unipotent group $\Gamma^u_C \leq \mathrm{Aff}(V)$, the unipotent shadow of $\Gamma$ in $\mathrm{Aff}(V)$, which is associated with $\Gamma$.

**Proposition 2.9.** *Let $\Gamma \leq \mathrm{Aff}(V)$ be a torsionfree polycyclic group which satisfies* $\mathrm{rank}\,\Gamma = \dim V$*. Moreover, we assume that $\Gamma = CF$ admits a nilpotent supplement $C$ for $F =$* $\mathrm{Fitt}(\Gamma)$*. Then $\Gamma$ is an ACG if and only if the unipotent shadow $\Gamma^u_C$ of $\Gamma$ in* $\mathrm{Aff}(V)$ *is an ACG.*

*Proof.* Let us first assume that $\Gamma$ is an ACG. Then, by Theorem 1.27, the algebraic closure $\overline{\Gamma}$ is a $u$-simply transitive real algebraic hull for $\Gamma$. By Proposition 1.19, $\Gamma^u_C = C_u F$ is a lattice in $U = \mathrm{u}(\overline{\Gamma})$. Since $U$ is simply transitive, $\Gamma^u_C$ is an ACG.

Now let us assume that $\Gamma^u_C = C_u F_u$ is an unipotent ACG. It follows that $U = \overline{\Gamma^u_C} = \overline{C_u F_u}$ is a unipotent simply transitive subgroup of $\overline{\Gamma}$. Since $\overline{\Gamma}$ is solvable, $U \leq \mathrm{u}(\overline{\Gamma})$. By a lemma of Mostow (cf. [66, 4.36]) it is known that $\dim \mathrm{u}(\overline{\Gamma}) \leq \mathrm{rank}\,\Gamma = \dim V$, and hence it follows that $\mathrm{u}(\overline{\Gamma}) = U$. So $\overline{\Gamma}$ is in fact $u$-simply transitive, and $\dim \mathrm{u}(\overline{\Gamma}) = \mathrm{rank}\,\Gamma$. The centralizer $C_{\overline{\Gamma}}(U)$ of the simply transitive normal subgroup $U$ is a unipotent subgroup of $\overline{\Gamma}$. Therefore $C_{\overline{\Gamma}}(U) \leq \mathrm{u}(\overline{\Gamma}) = U$. Hence $\overline{\Gamma}$ has a strong unipotent radical, and is a $u$-simply transitive algebraic hull for $\Gamma$. By Theorem 1.27, $\Gamma$ is an ACG. □

Recall from Definition 1.20 that, for $\Gamma = CF$ as above, a homomorphism $\rho : \Gamma \to \mathbf{G}$ is said to descend to the unipotent shadow $\Gamma^u_C \leq H_\Gamma$ if there is a homomorphism $\rho^u : \Gamma^u_C \to \mathbf{G}$ so that $\rho^u|_F = \rho_u$, and $\rho^u|_{C_u}(\gamma_u) = \rho_u(\gamma)$, for all $\gamma \in C$.

**Proposition 2.10.** *Let $\Gamma$ be a torsionfree polycyclic group which satisfies* $\mathrm{rank}\,\Gamma = \dim V$*. We assume also that $\Gamma = CF$ admits a nilpotent supplement $C$ for $F =$* $\mathrm{Fitt}(\Gamma)$*. Let $\rho : \Gamma \to$* $\mathrm{Aff}(V)$ *be a homomorphism. Then $\rho$ is a crystallographic homomorphism if and only if $\rho$ descends to a crystallographic homomorphism $\rho^u$ on the unipotent shadow $\Gamma^u_C$ of $\Gamma$.*

*Proof.* Let us first assume that $\rho$ is crystallographic. By Theorem 1.27, $\rho$ extends to a $u$-simply transitive morphism of (real) algebraic groups $\rho_{H_\Gamma} : H_\Gamma \hookrightarrow \mathrm{Aff}(V)$.

As in the proof of Proposition 1.21, $\rho$ descends to the restriction of $\rho_{H_\Gamma}$ to $\Gamma^u_C \leq \mathrm{u}(H_\Gamma)$. Since $\mathrm{u}(\rho(H_\Gamma))$ is simply transitive, this homomorphism is crystallographic.

Conversely, let us assume that $\rho$ descends to a crystallographic homomorphism $\rho^u$ on $\Gamma^u_C \leq \mathrm{u}(H_\Gamma)$. Then, since $\rho^u(\Gamma^u_C)$ is a unipotent shadow for $\rho(\Gamma)$ in $\mathrm{Aff}(V)$, Proposition 2.9 implies that $\rho(\Gamma)$ is crystallographic on $V$. In particular, $\operatorname{rank}\rho(\Gamma) = \dim V = \operatorname{rank}\Gamma$. It follows that the kernel of $\rho$ is finite. Since $\Gamma$ is torsionfree, $\rho$ is injective, hence $\rho$ is a crystallographic homomorphism. □

Proposition 2.10 implies that the correspondence $\rho \longmapsto \rho^u$ defines a map

$$s_u : \mathrm{Hom}_c(\Gamma, A) \longrightarrow \mathrm{Hom}_c(\Gamma^u_C, A) \ .$$

**Definition 2.11.** We call $s_u : \mathrm{Hom}_c(\Gamma, A) \to \mathrm{Hom}_c(\Gamma^u_C, A)$ the *crystallographic shadow map for* $\Gamma$.

In fact, the crystallographic shadow map for $\Gamma$ is just the restriction of the shadow map $s_u : \mathrm{Hom}(\Gamma, A) \to \mathrm{Hom}(\Gamma^u_C, A)$, see Proposition 1.21.

**Adapted systems of generators** Let $\Gamma = CF \leq \mathbf{H}_\Gamma$ be a torsionfree polycyclic group which admits a nilpotent supplement $C$ for $F = \mathrm{Fitt}(\Gamma)$.

**Definition 2.12.** Let $S = \{\gamma_i \in \Gamma,\ i = 1, \dots, l\}$, be a system of generators for $\Gamma$. $S$ is called *adapted to the supplement* $C$, if $C = <\gamma_i,\ i = 1, \dots, s>$, and $F = <\gamma_i,\ i = s+1, \dots, l>$.

Let $S$ be an adapted system of generators for $\Gamma = C\,\mathrm{Fitt}(\Gamma)$. Then it follows that

$$S_u = \{\, \theta_i = (\gamma_i)_u \in \mathrm{u}(H_\Gamma) \,\}$$

is a set of generators for $\Gamma^u_C$. Assume now also that $\operatorname{rank}\Gamma = \dim V$.

**Proposition 2.13.** *Let $\rho : \Gamma \to \mathrm{Aff}(V)$ be a homomorphism, $S$ an adapted system of generators. Then $\rho$ is a crystallographic homomorphism if and only if the assignment*

$$\theta_i \longmapsto \rho(\gamma_i)_u \quad , \ \theta_i \in S_u$$

*defines a crystallographic homomorphism $\rho^u \in \mathrm{Hom}_c(\Gamma^u_C, \mathrm{Aff}(V))$.*

*Proof.* By Proposition 1.21, there exists a unique homomorphism

$$\rho^u \in \mathrm{Hom}(\Gamma^u_C, \mathrm{Aff}(V)) \ ,$$

so that $\rho$ descends to $\rho^u$, and moreover $\rho^u(\theta_i) = \rho(\gamma_i)_u$. The proposition then follows from Proposition 2.10. □

*Remark* By the remark following Theorem 1.27, every crystallographic group has a unipotent Fitting subgroup. Astonishingly enough, we do not have to *assume* this necessary condition on $\rho$ to ensure that $\rho$ is crystallographic. In fact, the proof of Proposition 2.9 shows that if $\rho^u$ is crystallographic then also $\rho(\gamma) = \rho(\gamma)_u$, for all $\gamma \in \mathrm{Fitt}(\Gamma)$.

**The crystallographic restriction** A nilpotent crystallographic group $\Gamma$ is necessarily unipotent. This lead us to consider the space of unipotent homomorphisms $\mathrm{Hom}_u(\Gamma, \mathrm{Aff}(V))$. For a general virtually polycyclic wfn-group $\Gamma$, a corresponding restriction arises from the internal structure of the algebraic hull.

For $g \in H_\Gamma$, we let $c_U(g) \in \mathrm{Aut}(U_\Gamma)$ denote conjugation with $g$ on $U_\Gamma = \mathrm{u}(H_\Gamma)$. Taking the differential of $c_U(g)$ in the identity defines a representation

$$\alpha_U : H_\Gamma \longrightarrow \mathrm{Aut}(\mathfrak{u}_\Gamma)$$

on the Lie algebra $\mathfrak{u}_\Gamma$ of $U_\Gamma$. Since $H_\Gamma$ has a strong unipotent radical, the kernel of $\alpha_U$ is contained in $U$. For $\gamma \in H_\Gamma$, we define the characteristic polynomial

$$\chi_U(\gamma, T) = \det(T\mathrm{id} - \alpha_U(\gamma)) ,$$

and, correspondingly for $g \in \mathrm{Aff}(V)$, we let $\chi(g, T)$ denote the characteristic polynomial of $g$ as an element of $\mathrm{GL}(V \oplus \mathbb{R})$. The crystallographic restrictions for $\Gamma$ are described by the following:

**Lemma 2.14.** *Let $\Gamma$ be virtually polycyclic* wfn*-group. If $\rho : \Gamma \to \mathrm{Aff}(V)$ is a crystallographic homomorphism then, for all $\gamma \in \Gamma$,*

$$\chi(\rho(\gamma), T) = \chi_U(\gamma, T)\,(T - 1) . \tag{2}$$

*Proof.* Since $\rho$ is crystallographic there exists a $u$-simply transitive homomorphism $\rho_\Gamma : H_\Gamma \to \mathrm{Aff}(V)$ which extends $\rho$, so that $U = \rho_\Gamma(\mathrm{U}_\Gamma)$ is a simply transitive subgroup of $\mathrm{Aff}(V)$. $\rho(H_\Gamma)$ acts by conjugation on $U$, as well as on the Lie algebra $\mathfrak{u} \subset \mathrm{End}(V \oplus \mathbb{R})$ of $U$. Let $\phi_\Gamma : \mathfrak{u}_\Gamma \to \mathfrak{u}$ be the differential of the homomorphism $\rho_\Gamma$. The isomorphism $\phi_\Gamma$ is equivariant with respect to conjugation, i.e., $\phi_\Gamma$ satisfies, for all $X \in \mathfrak{u}_\Gamma$, $g \in H_\Gamma$,

$$\phi_\Gamma(\alpha(g)X) = \rho(g)\phi_\Gamma(X)\rho(g)^{-1} .$$

Since $\chi_U(\gamma, T) = \chi_U(\gamma_s, T)$ we are allowed to assume that $\gamma = \gamma_s$ is a semisimple element. We may also assume that $\rho(\gamma_s) = \rho(\gamma)_s \in \mathrm{GL}(V)$. Let $o_0 : \mathfrak{u} \to V$ be the differential of the orbit map of $U$ in 0 (the evaluation map in 0). An immediate calculation shows that the isomorphism $o_0\, \phi_\Gamma : \mathfrak{u}_\Gamma \to V$ is equivariant with respect to the action of $\gamma_s$, i.e., for all $X \in \mathfrak{u}_\Gamma$,

$$o_0\, \phi_\Gamma\, (\alpha(\gamma_s)X) = \rho(\gamma_s)\, o_0\, \phi_\Gamma(X).$$

Hence, $\chi_U(\gamma_s, T) = \chi_{\mathrm{GL}(V)}(\rho(\gamma)_s, T)$, and the lemma follows. □

We define now a certain Zariski-closed subspace of $\mathrm{Hom}(\Gamma, A)$. Namely, we put

$$\mathrm{Hom}_\chi(\Gamma, A) = \{\rho \mid \chi(\rho(\gamma), T) = \chi_U(\gamma, T)(T - 1) \text{ , for all } \gamma \in \Gamma\}$$

for the subvariety of homomorphisms of type $A$ which satisfy the crystallographic restrictions. If $\Gamma$ is nilpotent, then $\mathrm{Hom}_\chi(\Gamma, A) = \mathrm{Hom}_u(\Gamma, A)$. From the lemma we infer that

$$\mathrm{Hom}_c(\Gamma, A) \subset \mathrm{Hom}_\chi(\Gamma, A) .$$

We will show next that $\mathrm{Hom}_c(\Gamma, A)$ is Zariski-open in $\mathrm{Hom}_\chi(\Gamma, A)$.

**The variety of crystallographic homomorphisms** We are ready now to generalize Theorem 2.6 from nilpotent groups to virtually polycyclic groups. Also we consider the situation now for any Zariski-closed subgroup $A \leq \mathrm{Aff}(V)$. If $\Gamma$ is a polycyclic finite index subgroup of $\Delta$, we let $\mathrm{Hom}_\chi(\Delta, A) \subset \mathrm{Hom}(\Delta, A)$ be the Zariski-closed subset of homomorphisms which satisfy the crystallographic restrictions (2) on $\Gamma$.

**Theorem 2.15.** *Let $\Delta$ be a virtually polycyclic* wfn-*group which satisfies* $\operatorname{rank}\Delta = \dim V$. *Then there exists a unique $\mathbb{Q}$-defined polynomial function $\delta$ on the real algebraic variety* $\mathrm{Hom}_\chi(\Delta, A)$, *such that*

$$\mathrm{Hom}_c(\Delta, A) = \mathrm{Hom}_\chi(\Delta, A)_\delta .$$

*Moreover, $\delta$ is a relative invariant for the conjugation action of $A$ on* $\mathrm{Hom}_\chi(\Delta, A)$.

*Proof.* Since $\mathrm{Hom}(\Delta, A)$ is Zariski-closed in $\mathrm{Hom}(\Delta, \mathrm{Aff}(V))$, it is clearly enough to prove the result in the case $A = \mathrm{Aff}(V)$. Observe that $\rho(\Delta)$ is a crystallographic group if and only if a finite index subgroup of $\rho(\Delta)$ is a crystallographic group. Since $\Delta$ is a wfn-group, $\rho$ is crystallographic (in particular injective) if and only if it is so on a finite index subgroup $\Gamma$. Since the restriction map $\mathrm{Hom}(\Delta, \mathrm{Aff}(V)) \to \mathrm{Hom}(\Gamma, \mathrm{Aff}(V))$ is algebraic, it is therefore enough to show the theorem for $\Gamma$. By Proposition 1.23, there exists a finite index subgroup $\Gamma$ of $\Delta$ which is torsionfree and admits a nilpotent supplement $C$ for $\mathrm{Fitt}(\Gamma)$. We now show the result for $\Gamma$.

By Theorem 2.6, there exists a polynomial function $\delta_u$ on $\mathrm{Hom}_u(\Gamma^u_C, \mathrm{Aff}(V))$ which is a relative invariant and tests crystallography for $\mathrm{Hom}(\Gamma^u_C, \mathrm{Aff}(V))$. The shadow map for $\Gamma$ is a map

$$s_u : \mathrm{Hom}(\Gamma, \mathrm{Aff}(V)) \longrightarrow \mathrm{Hom}_u(\Gamma^u_C, \mathrm{Aff}(V)) .$$

Therefore (compare Proposition 2.13) the function $\delta = \delta_u \circ s_u$ tests crystallography on $\mathrm{Hom}(\Gamma, \mathrm{Aff}(V))$, and in particular on $\mathrm{Hom}_\chi(\Gamma, \mathrm{Aff}(V))$.

We now show that $s_u$ is continuous on $\mathrm{Hom}_\chi(\Gamma, \mathrm{Aff}(V))$. Let $S = \{\gamma_i\}$ be an adapted system of generators for $\Gamma$. By Proposition 2.13, the shadow map $s_u$ is obtained by computing $\rho(\gamma_i)_u$, for each generator $\theta_i = (\gamma_i)_u$ of the unipotent shadow $\Gamma^u_C$. Now, for $g \in \mathrm{GL}(W)$, we may compute the unipotent part of the multiplicative Jordan decomposition of $g$ by the formula

$$g_u = I + g_s^{-1} g_n ,$$

where $g = g_s + g_n$ is the additive Jordan-decomposition in $\mathrm{End}(W)$. Moreover $g_s = P(g)$, $g_n = Q(g)$, where $P(T), Q(T) \in \mathbb{Q}[T]$ are certain polynomials which depend only on the characteristic polynomial of $g \in \mathrm{GL}(W)$. If $\rho \in \mathrm{Hom}_\chi(\Gamma, \mathrm{Aff}(V))$ satisfies the crystallographic restriction then $\rho(\gamma_i)$ has characteristic polynomial $\chi_U(\gamma, T)\,(T-1)$ which does not depend on $\rho$. Therefore, there exist polynomials $P_i(T)$, $Q_i(T)$ such that, for every $\rho \in \mathrm{Hom}_\chi(\Gamma, \mathrm{Aff}(V))$,

$$\rho(\gamma_i)_u = I + P_i(\rho(\gamma_i))^{-1} Q_i(\rho(\gamma_i)) .$$

This means that $s_u$ is a $\mathbb{Q}$-defined polynomial map on $\mathrm{Hom}_\chi(\Gamma, \mathrm{Aff}(V))$, and so is $\delta = \delta_u \circ s_u$. It is easy to see that $s_u$ is $\mathrm{Aff}(V)$-equivariant. Hence, $\delta$ is a $\mathbb{Q}$-defined polynomial function which is a relative invariant, and the theorem is proved. □

*Remark* Let $\Delta$ be as in Theorem 2.15. The theorem implies that the existence problem for crystallographic subgroups of $\mathrm{Aff}(V)$ which are isomorphic to $\Delta$ may be reduced to the existence problem for real solutions of a certain system of algebraic equations with rational coefficients. Therefore, by the Tarski-Seidenberg Theorem this decision problem admits an effective solution.

In particular, Theorem 2.15 shows that $\mathrm{Hom}_c(\Delta, A)$ is a Zariski-open subset of $\mathrm{Hom}_\chi(\Delta, A)$. The theorem therefore also establishes a certain *rigidity property* for affine crystallographic homomorphisms: *If* $\rho \in \mathrm{Hom}_c(\Gamma, \mathrm{Aff}(V))$ *then any nearby* $\rho' \in \mathrm{Hom}_\chi(\Gamma, \mathrm{Aff}(V))$ *is in* $\mathrm{Hom}_c(\Gamma, \mathrm{Aff}(V))$. In his celebrated paper [78], A. Weil proved the following[9]:

Let $\Gamma \leq G$ be a discrete cocompact subgroup of the Lie group $G$. Then the space

$$R(\Gamma, G) = \{\rho : \Gamma \to G \mid \rho \text{ is injective and } \rho(\Gamma) \text{ is discrete and cocompact}\}$$

is open in the space of all homomorphisms from $\Gamma$ to $G$. (The space $R(\Gamma, G)$ is now called the Weil space.)

Our Theorem 2.15 thus may be interpreted as an analogue of the result of Weil in the specific situation of crystallographic affine actions for virtually polycyclic group.

We state some first consequences now. The main applications, however, concern the structure and topology of the deformation spaces of affine crystallographic actions. This will be described in the following part of this article.

**Corollary 2.16.** *Let* $\Gamma$ *be a virtually polycyclic group. Then* $\mathrm{Hom}_c(\Gamma, A)$ *is a locally closed subset with respect to the Zariski-topology on* $\mathrm{Hom}(\Gamma, A)$.

As a special open subset of an affine real algebraic variety the set of crystallographic homomorphisms has a natural algebraic structure itself:

**Corollary 2.17.** $\mathrm{Hom}_c(\Gamma, \mathrm{Aff}(V))$ *has the structure of a* $\mathbb{Q}$*-defined real affine algebraic variety. In particular,* $\mathrm{Hom}_c(\Gamma, \mathrm{Aff}(V))$ *has only finitely many connected components also in the Hausdorff-topology.*

Another important application of the reasoning in the proof of Theorem 2.15 is:

**Corollary 2.18.** *Let* $\Theta$ *be a unipotent shadow for* $\Gamma$*. Then the shadowmap*

$$s_u : \mathrm{Hom}_c(\Gamma, A) \longrightarrow \mathrm{Hom}_c(\Theta, A)$$

[9] Abels (see [1]) gave generalizations of Weil's theorem in the general context of proper actions on homogeneous spaces of Lie groups..

*is a morphism of real algebraic varieties defined over* $\mathbb{Q}$. *In particular,* $s_u$ *is a continuous map.*

*Proof.* In fact, in the course of the proof of Theorem 2.15 we proved that the shadow map

$$s_u : \mathrm{Hom}_\chi(\Gamma, A) \longrightarrow \mathrm{Hom}_u(\Theta, A)$$

is given by certain rational polynomials with respect to an adapted system of generators of $\Gamma$. Therefore $s_u$ is algebraic also on $\mathrm{Hom}_c(\Gamma, A) \subset \mathrm{Hom}_\chi(\Gamma, A)$. □

# Part II
# Deformation Spaces

In this part we study properties of the deformation spaces of affine crystallographic actions of virtually polycyclic groups. These spaces are defined in a purely algebraic way as quotients of spaces of homomorphisms with an appropiate topology. If $M$ is a fixed compact smooth manifold then the deformation space $\mathcal{D}_c(M, A)$ of complete affine $A$-structures on $M$ is the space of such structures up to diffeomorphism, equipped with the $C^\infty$-topology from the space of developing maps. If $\Gamma$ is the fundamental group of $M$ the *deformation theorem* of Thurston links the spaces $\mathcal{D}_c(M, A)$ and $\mathcal{D}_c(\Gamma, A)$ via a continuous and open map, the *holonomy map*

$$hol : \mathcal{D}_c(M, A) \longrightarrow \mathcal{D}_c(\Gamma, A) .$$

Note that the theory of Thurston covers the general situation of locally homogeneous manifolds $M$, modelled on a homogeneous space $G/H$, see [75], and, in particular, [26, 30] for further explanation and proofs. In [13] implications are discussed for the deformations of compact complete afffine manifolds.

In this realm, Kobayashi [44], more generally formulates the program to determine the deformation spaces for *proper*, not necessarily cocompact, actions of a group $\Gamma$ on a homogeneous space $G/H$. The situation $G = \mathrm{Aff}(V)$, and $G/H = V$ is an important special case in this program.[10]

If $\Gamma$ is a torsionfree virtually polycyclic ACG of type $A$ then the quotient space $M = \Gamma\backslash V$ is a compact complete affine manifold which admits an affine atlas with coordinate changes in the group $A$. It is also a smooth aspherical compact manifold with fundamental group $\Gamma$. Results of the previous part (cf. Proposition1.15) imply that the smooth manifold $M$ falls into the class of infrasolvmanifolds. In this class of smooth manifolds the diffeomorphism type of $M$ is determined by the fundamental group $\Gamma$ alone. This is proved in [13].[11] This *smooth rigidity* for compact complete affine manifolds implies that the holononomy map is a homeomorphism, compare [10]. Therefore, the algebraic viewpoint on the deformation space $\mathcal{D}_c(M, A)$ captures the whole picture. In this sense, we allow ourselves at

[10] See [6, 7, 46, 81] for recent contributions on this problem.

[11] A result in [27] shows that any two compact complete affine manifolds with virtually solvable fundamental group are (polynomially) diffeomorphic.

some points to speak about the deformation spaces of affine structures on certain manifolds, although we only give results on the affine crystallographic actions of their fundamental groups.

# 3. The geometry of the deformation space

Let $\Gamma$ be a group, and let $\mathrm{Hom}_c(\Gamma, A)$ be the space of crystallographic homomorphisms of type $A$. The group $A$ acts by conjugation on the space of homomorphisms. Our principal object of interest is the deformation space of crystallographic homomorphisms of type $A$ which is defined as

$$\mathcal{D}_c(\Gamma, A) = \mathrm{Hom}_c(\Gamma, A) \,/_{A} \,.$$

In this situation, the group $\mathrm{Aut}(\Gamma)$ of automorphisms of $\Gamma$ acts freely on $\mathrm{Hom}_c(\Gamma, A)$ commuting with the action of $A$. The quotient topological space

$$\mathcal{C}_c(\Gamma, A) = {}_{\mathrm{Aut}(\Gamma)}\backslash \mathrm{Hom}_c(\Gamma, A)$$

is the space of crystallographic subgroups of $A$ which are isomorphic to $\Gamma$. This space is sometimes called the *Chabauty space*. The group $A$ acts by conjugation on $\mathcal{C}_c(\Gamma, A)$, and the action of the group $\mathrm{Aut}(\Gamma)$ on $\mathrm{Hom}_c(\Gamma, A)$ factorizes to an action of the outer automorphism group $\mathrm{Out}(\Gamma) = \mathrm{Aut}(\Gamma)/\mathrm{Inn}(\Gamma)$ on $\mathcal{D}_c(\Gamma, A)$. The quotient space

$$\mathcal{M}_c(\Gamma, A) = {}_{\mathrm{Out}(\Gamma)}\backslash \mathcal{D}_c(\Gamma, A)$$

is called the *moduli space*. The situation is described in the following commutative diagram of maps and spaces:

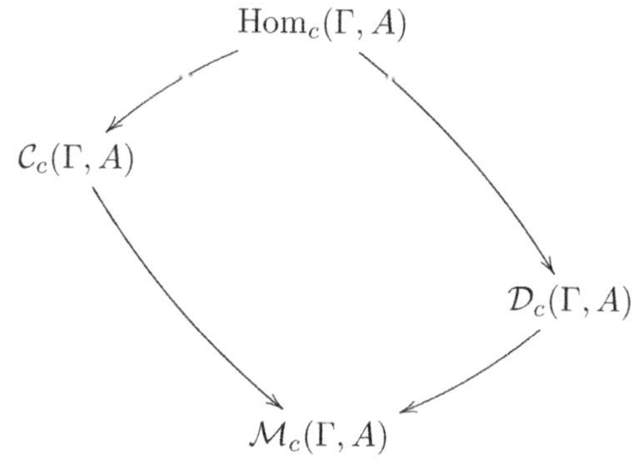


. (3)

Basic problems in the deformation theory for $\Gamma$ are to understand the topology and geometry of the spaces involved in this diagram.

We restrict ourselves here to the case that $\Gamma$ is a virtually polycyclic group. Recall from Part I that then the affine crystallographic Weil space $\mathrm{Hom}_c(\Gamma, A)$ carries a Hausdorff-topology, and also a Zariski-topology which is induced from $\mathrm{Hom}(\Gamma, A)$. This is the essence of our Theorem 2.15. The next step is to learn more about the remaining spaces. Then the following interesting picture emerges:

**Algebraic and arithmetic nature of deformation spaces** The deformation space arises as a quotient of the real points of the $\mathbb{Q}$-defined algebraic variety $\mathrm{Hom}_c(\Gamma, A)$ by a compatible $\mathbb{Q}$-defined algebraic action of $A$. The moduli space $\mathcal{M}_c(\Gamma, A)$ arises as a quotient of the deformation space by a compatible $\mathbb{Q}$-defined algebraic action of the *arithmetic group*[12] $\mathrm{Out}(\Gamma)$. In fact, as is proved recently in [15], the outer automorphism group of a virtually polycyclic group is an arithmetic group.

Moreover, the action of $\mathrm{Aut}(\Gamma)$ on $\mathrm{Hom}_c(\Gamma, A)$ extends to an algebraic $\mathbb{Q}$-defined action of the *algebraic automorphism group* $\mathrm{Aut}_a(H_\Gamma)$, and this action factors over the algebraic inner group

$$\mathrm{Out}_a(H_\Gamma) = \mathrm{Aut}_a(H_\Gamma)/\mathrm{Inn}(H_\Gamma)$$

to an action on $\mathcal{D}_c(\Gamma, A)$, which extends the $\mathrm{Out}(\Gamma)$-action on $\mathcal{D}_c(\Gamma, A)$. These facts are announced in [11]. In Part III of this article we will give a proof for (virtually) f.t.n.-groups $\Gamma$. (The proof for the general fact needs more of the machinery which relates the groups $\mathrm{Aut}(\Gamma)$ and $\mathrm{Aut}_a(H_\Gamma)$, as is developed in [13, 15].)

**Topology of deformation spaces** The deformation space, as well as the moduli space have nice topological properties in some well known geometric situations. For example they are Hausdorff-manifolds in some cases. This is in particular true for the deformation spaces of constant curvature Riemannian metrics, and the related Teichmüller theory of complex analytic structures on surfaces, see for example [67]. Such properties were suspected to be no longer true for the deformation spaces of affine crystallographic groups, see [30].

To the contrary it was shown in [9] that the deformation space $\mathcal{D}_c(\mathbb{Z}^2, \mathrm{Aff}(\mathbb{R}^2))$ is a Hausdorff-space, in fact, homeomorphic to $\mathbb{R}^2$. Whereas the moduli space

$$\mathcal{M}_c(\mathbb{Z}^2, \mathbb{R}^2) = {}_{\mathrm{GL}_2(\mathbb{Z})}\backslash \mathbb{R}^2$$

is highly non Hausdorff.

Note that via the holonomy map the space $\mathcal{D}_c(\mathbb{Z}^2, \mathrm{Aff}(\mathbb{R}^2))$ is homeomorphic to the deformation space of complete affine structures on the two torus $T^2$. A geometric interpretation of the coordinates of $\mathcal{D}_c(\mathbb{Z}^2, \mathrm{Aff}(\mathbb{R}^2))$ in terms of periods of developing maps for affine structures on $T^2$ is given in [14].

This example also nicely illustrates the *arithmetic and algebraic nature* of the deformation space. For an interpretation of rational points in $\mathbb{R}^2$ in terms of geometric properties of the corresponding affine structures on $T^2$, see [9]. It is shown there that rationality of an affine structure is equivalent to the existence of closed geodesics.

In this part we concentrate our effort to a further study of the deformation spaces $\mathcal{D}_c(\Gamma, A)$. We will show that in some cases the deformation spaces for

[12] See [73] for a discussion of the notion of arithmetic group.

affine crystallographic groups are in fact Hausdorff and homeomorphic to a semi-algebraic set. Let us mention first the following basic result on the separation properties of deformation spaces:

**Theorem 3.1.** *Let $\Gamma$ be a virtually polycyclic group, and $A \leq \mathrm{Aff}(V)$ a Zariski-closed subgroup. Then the deformation space $\mathcal{D}_c(\Gamma, A)$ is a $T1$-topological space*[13] *with finitely many connected components.*

*Proof.* By Corollary 2.17, the space $\mathrm{Hom}_c(\Gamma, A)$ has only finitely many components. Therefore, the same holds for the quotient space $\mathcal{D}_c(\Gamma, A)$.

To see that points in $\mathcal{D}_c(\Gamma, A)$ are closed, we shall need some of the machinery, which will be developed in the sequel. We can argue as follows: By the realisation theorem, Theorem 9.5, it is enough to prove separation properties for deformation spaces of f.t.n.-groups. If $\Gamma$ is an f.t.n.-group then, by Proposition 11.4, the deformation space $\mathcal{D}_c(\Gamma, A)$ may be represented as the quotient space of a real algebraic variety by a unipotent action of the real Malcev hull $U_\Gamma$. By the Rosenlicht theorem [68] every orbit of $U_\Gamma$ is closed. Therefore, the points in $\mathcal{D}_c(\Gamma, A)$ are closed. □

**Geometric notions for deformation spaces** Geometric notions are introduced by the natural group actions on our spaces. Let $\rho \in \mathrm{Hom}_c(\Gamma, A)$ be a crystallographic homomorphism. An important concept concerns the dimension of the deformation space.

**Definition 3.2.** The action $\rho$ is called *(locally) A-rigid* if the orbit $A\rho$ is an open set in $\mathrm{Hom}_c(\Gamma, A)$.

Local rigidity of $\rho$ is equivalent to the fact that the point $[\rho]$ constitutes a connected component of the deformation space $\mathcal{D}_c(\Gamma, A)$. The action $\rho$ is *rigid* if the deformation space is a point. It is well known that rigidity occurs for certain geometric structures, for example hyperbolic structures in dimension $n > 2$. We expect that local rigidity fails for most affine crystallographic actions.

Another important notion is

**Definition 3.3.** The deformation space $\mathcal{D}_c(\Gamma, A)$ is called *convex* if every finite subgroup of $\mathrm{Out}(\Gamma)$ has a fixed point. A special case occurs if the group $\mathrm{Out}(\Gamma)$ has fixed points. Then we will call $\mathcal{D}_c(\Gamma, A)$ *fixed pointed.*

The convexity of the deformation space of $\Gamma$ is in particular important because of the role it plays in the realization problem for finite extensions, see Part III. In Theorem 5.3 below we show that natural classes of f.t.n.-groups, and as a consequence many f.t.n.-by finite and torsionfree polycyclic groups have fixed-pointed or convex deformation spaces.

[13] meaning that every point is closed

# 4. Models for deformation spaces

The purpose of this section is to provide some examples of deformation spaces for affine crystallographic groups which are Hausdorff topological spaces. We first give some details on the structure of deformation spaces for tori. The results of Part III allow then to construct more examples of deformation spaces which are Hausdorff. In particular we show that the deformation spaces of complete affine structures of certain three-manifolds are Hausdorff spaces.

## Deformation spaces for affine tori

In [10] it was proved that the deformation spaces $\mathcal{D}_c(\mathbb{Z}^n, \mathrm{Aff}(V))$ of affine tori are homeomorphic to a semi-algebraic set in some Euclidean space. Here we want to study more closely these deformation spaces and the structure of the algebraic varieties associated to them.

Theorem 2.15 asserts that the space $\mathrm{Hom}_c(\mathbb{Z}^n, A)$ has a natural structure as a real algebraic variety. Here we want to describe the algebraic variety $\mathrm{Hom}_c(\mathbb{Z}^n, A)$, that is, the space of crystallographic homomorphisms of type $A$, and the associated deformation space $\mathcal{D}_c(\mathbb{Z}^n, A)$ in more detail. A basic result on these deformation spaces is:

**Theorem 4.1.** *The deformation space $\mathcal{D}_c(\mathbb{Z}^n, A)$ is homeomorphic to a semi algebraic set, and in particular $\mathcal{D}_c(\mathbb{Z}^n, A)$ is a Hausdorff space.*

*Proof.* The proof of [10, Corollary 2.9] for $A = \mathrm{Aff}(V)$ carries over almost verbatim to a more general $A$ in the case that $\mathrm{GL}_A = A \cap \mathrm{GL}(V)$ is a real reductive group. Let us briefly recall the argument. Since $\mathbb{Z}^n$ is abelian the centralizer of every cystallographic action is a simply transitive subgroup of $A$, and $\mathcal{D}_c(\mathbb{Z}^n, A)$ is homeomorphic to the orbit space $\mathrm{Hom}_c(\mathbb{Z}^n, A)/\mathrm{GL}_A$. It follows from Lemma 11.7 that $\mathrm{GL}_A$ acts freely on $\mathrm{Hom}_c(\mathbb{Z}^n, A)$, and in particular all orbits are closed. It is known that the space of closed orbits of the reductive group $\mathrm{GL}_A$ is homeomorphic to a semi-algebraic set. For a more general $A$ the theorem will follow from Proposition 4.3 below. $\square$

Using a fixed choice of basis we embed $\mathbb{Z}^n$ as a lattice in $V$. As a real algebraic group, $V$ identifies then with the real Malcev hull of $\mathbb{Z}^n$. As shown in Section 11 the automorphism group of $V$, that is the group $\mathrm{GL}(V)$, acts freely on the space $\mathrm{Hom}_c(\mathbb{Z}^n, A)$. For $g \in \mathrm{GL}(V)$ the action is given by

$$\rho \longmapsto \rho g = \bar{\rho} g^{-1} j \,,$$

where $\bar{\rho} : V \to A$ is the unique simply transitive representation of $V$ which extends $\rho$, and $j : \mathbb{Z}^n \to V$ is the fixed embedding of $\mathbb{Z}^n$ in $V$. This action of $\mathrm{GL}(V)$ commutes with the conjugation action of $A$. The quotient space

$$G_{st}(V, A) \;=\; \mathrm{GL}(V) \backslash \mathrm{Hom}_c(\mathbb{Z}^n, A)$$

is the space of simply transitive abelian subgroups of $A$. We describe the structure of the algebraic variety $\mathrm{Hom}_c(\mathbb{Z}^n, A)$ as follows:

**Proposition 4.2.** *The space $G_{st}(V, A)$ is homeomorphic to a real algebraic variety and there is an isomorphism of algebraic varieties*

$$\mathrm{Hom}_c(\Gamma, A) = \mathrm{GL}(V) \times G_{st}(V, A) ,$$

*such that the action of* $\mathrm{GL}(V)$ *on* $\mathrm{Hom}_c(\Gamma, A)$ *is given by right multiplication on the first factor. The action of* $\mathrm{GL}_A$ *by conjugation on* $\mathrm{Hom}_c(\Gamma, A)$ *is given by its action on* $G_{st}(V, A)$ *and by left multiplication on* $\mathrm{GL}(V)$.

*Proof.* Each $\rho \in \mathrm{Hom}_c(\mathbb{Z}^n, A)$ extends to a homomorphism $U_{\mathbb{Z}^n} = V \to A$. The translational components of its derivative define a linear isomorphism $\bar{t}(\rho) : V \to V$, since $V$ acts simply transitively. Also it is easy to see that the $\mathrm{GL}(V)$- (change of basis) action on $\mathrm{Hom}_c(\mathbb{Z}^n, A)$ corresponds to (transposed) right multiplication on $\mathrm{GL}(V)$, i.e., $\bar{t}(\rho g) = \bar{t}(\rho)(g^{-1})^t$. In particular, each element of $G_{st}(V, A)$ has a unique representative $\rho_{id} \in \mathrm{Hom}_c(\mathbb{Z}^n, A)$ which satisfies $\bar{t}(\rho_{id}) = id_V$, and this representative is computed by the formula $\rho_{id} = \rho\bar{t}(\rho)^{-1}$. Therefore $G_{st}(V, A)$ identifies with a Zariski-closed subspace $\mathrm{Hom}_{id}(\mathbb{Z}^n, A)$ of $\mathrm{Hom}_c(\mathbb{Z}^n, A)$, and the natural map

$$\mathrm{GL}(V) \times \mathrm{Hom}_{id}(\mathbb{Z}^n, A) \longrightarrow \mathrm{Hom}_c(\mathbb{Z}^n, A)$$

given by $(g, \rho) \mapsto \rho g$ is an algebraic isomorphism. The equivariance statements are easy to verify. Hence the proposition follows. □

We let $\mathrm{GL}_A$ act on $\mathrm{GL}(V)$ by left multiplication and put

$$X_A = {}_{\mathrm{GL}_A}\backslash \mathrm{GL}(V) .$$

The following result reveals the structure of the deformation spaces $\mathcal{D}_c(\mathbb{Z}^n, A)$ and the nature of the $\mathrm{GL}(V)$-action on $\mathcal{D}_c(\mathbb{Z}^n, A)$ more precisely:

**Proposition 4.3.** *Let $F = G_{st}(V, A)$ be the algebraic variety of simply transitive abelian subgroups of $A$ with the natural action of* $\mathrm{GL}(A)$. *Then*

$$\mathcal{D}_c(\mathbb{Z}^n, A) = \mathrm{GL}(V) \underset{\mathrm{GL}_A}{\times} F$$

*is homeomorphic to a fiber product, and in particular a bundle over the homogeneous space $X_A$ with fiber $F$. The* $\mathrm{GL}(V)$*-action on* $\mathcal{D}_c(\mathbb{Z}^n, A)$ *corresponds to the natural* $\mathrm{GL}(V)$*-action on the fiber product which is induced by right multiplication of* $\mathrm{GL}(V)$ *on itself.*

*Proof.* We already remarked in the proof of Theorem 4.1 that $\mathcal{D}_c(\mathbb{Z}^n, A)$ is homeomorphic to the quotient by $\mathrm{GL}_A$. Therefore the proposition follows from Proposition 4.2. □

As a consequence we see that $\mathcal{D}_c(\mathbb{Z}^n, \mathrm{Aff}(V))$ is actually homeomorphic to a real algebraic variety.

**Corollary 4.4.** *The deformation space* $\mathcal{D}_c(\mathbb{Z}^n, \mathrm{Aff}(V))$ *is homeomorphic to the real algebraic variety* $G_{st}(V, \mathrm{Aff}(V))$.

*Proof.* The group $\mathrm{GL}_{\mathrm{Aff}(V)}(V) = \mathrm{GL}(V)$ acts by conjugation on $\mathrm{Hom}_c(\mathbb{Z}^n, \mathrm{Aff}(V))$, so that $t(\rho^g) = gt(\rho)$. Therefore each orbit intersects $\mathrm{Hom}_{id}(\mathbb{Z}^n, \mathrm{Aff}(V))$ precisely once. The corollary now follows from Proposition 4.3. □

**Example 4.5.** The deformation space $\mathcal{D}_c(\mathbb{Z}^2, \mathrm{Aff}(\mathbb{R}^2))$ is homeomorphic to $\mathbb{R}^2$ and the actions of $\mathrm{GL}(\mathbb{R}^2)$ and $\mathrm{GL}_2(\mathbb{Z})$ on $\mathcal{D}_c(\mathbb{Z}^2, \mathrm{Aff}(\mathbb{R}^2))$ correspond to the canonical linear actions. The fixed point $o \in \mathcal{D}_c(\mathbb{Z}^2, \mathrm{Aff}(\mathbb{R}^2))$ corresponds to the natural action of $\mathbb{Z}^2$ by translations. (See [9])

We next cover a few more special cases:

**Example 4.6.** Let $q$ be an inner product on $V$, $\mathrm{O}(q)$ the group of linear $q$-isometries, and $\mathrm{A}(q)$ the group of affine isometries for $q$. It is easy to see and well known that $G_{st}(V, A(q))$ consists of the groups of translations only. (A proof may be found in [12]) It follows that $\mathcal{D}_c(\mathbb{Z}^n, \mathrm{A}(q))$ is homeomorphic to $X_{A(q)}$.

The symplectic case is already more interesting. Let $V$ be a vector space with a nondegenerate alternating product $\omega$. We let $\mathrm{Sp}(\omega)$ denote the group of linear $\omega$-isometries of $V$, and $\mathrm{A}(\omega)$ the group of affine isometries for $\omega$. Let $\mathcal{U}_k$ denote the tautological vector bundle over the Grassmanian of $k$-dimensional isotropic subspaces of the symplectic vector space $(V, \omega)$. It is a natural $\mathrm{Sp}(\omega)$-variety. Let $2n = \dim V$. The following is proved in [12]:

**Proposition 4.7.** *The variety $G_{st}(V, A(\omega))$ admits a $\mathrm{Sp}(\omega)$-invariant stratification*

$$G_{st}(V, A(\omega)) = \bigcup_{k=0}^{n} \mathcal{G}_{st}(V, A(\omega))_k \,.$$

*Each stratum is an open subbundle of the third symmetric power $S^3\mathcal{U}_k$ of $\mathcal{U}_k$ .*

Together with Proposition 4.3 we obtain a model for the space $\mathcal{D}_c(\mathbb{Z}^n, A(\omega))$.

*Remark* As is observed in [12] the study of simply transitive abelian symplectic affine groups plays a role in the construction of flat models for *special Kähler geometry*, a particular geometry which arises in supersymmetric quantum field theory.

## Semi-algebraic deformation spaces

We proved up to now that the spaces $\mathcal{D}_c(\mathbb{Z}^n, A)$ are homeomorphic to semi algebraic sets. The results of Part III allow us to further generalize this result. A finite effective extension group of $\mathbb{Z}^n$ is traditionally called a *Bieberbach group.*

**Corollary 4.8.** *Let $\Delta$ be a Bieberbach group. Then the deformation space $\mathcal{D}_c(\Delta, A)$ is homeomorphic to a semi algebraic set and is a Hausdorff space.*

This implies corresponding results on the deformation spaces of affine space forms which are finitely covered by a torus. More details and the proofs will be given in Section 8. A further result is:

**Corollary 4.9.** *Let $\Gamma$ be a virtually torsionfree polycyclic group such that the unipotent shadow of $\Gamma$ is abelian. Then the deformation space $\mathcal{D}_c(\Gamma, A)$ is homeomorphic to a semi algebraic set and is a Hausdorff space.*

The proof will be given in Section 9.

**Deformation spaces of complete affine three manifolds** The following question was raised by Bill Goldman[14]:

*Let $M$ be a closed 3-manifold which is a torus bundle over $S^1$. Is the deformation space of complete affine structures on $M$ Hausdorff?*

As is proved in [27] if $M$ is a closed complete affine 3-manifold then $M$ is finitely covered by a torus bundle over $S^1$. The previous results imply a partial answer to the above question:

**Theorem 4.10.** *Let $M$ be a closed complete affine three manifold, and $A \leq \mathrm{Aff}(3)$ a Zariski-closed subgroup.*

i) *If $M$ is finitely covered by a torus, then the deformation space $\mathcal{D}_c(M, A)$ is Hausdorff.*
ii) *If $M$ is a torus bundle over $S^1$ where the attaching map is hyperbolic with a positive trace then $\mathcal{D}_c(M, A)$ is Hausdorff.*

*Proof.* The first claim follows from Corollary 4.8 above, keeping in mind the nontrivial fact that the deformation space of complete affine structures on $M$ is homeomorphic to the deformation space of crystallographic actions of the fundamental group.

For the second claim, we remark that if the condition on the bundle is satisfied, the fundamental group $\pi_1(M)$ of $M$ is a Zariski-dense lattice in the 3-dimensional Lie group Sol. Therefore, the unipotent shadow of $\Gamma = \pi_1(M)$ is abelian, and Corollary 4.9 applies. □

*Remark* It would be further interesting to understand the separation properties of f.t.n.-groups which are lattices in the 3-dimensional Heisenberg group.

# 5. Convexity properties of deformation spaces

The purpose of this section is to provide some examples of deformation spaces for affine crystallographic groups which are convex in the sense of Definition 3.3.

**Strong convexity** Let $\Gamma$ be an f.t.n.-group and let $U_\Gamma$ be a real Malcev hull for $\Gamma$. The automorphism group of $U_\Gamma$ naturally acts on the space $\mathrm{Hom}_c(\Gamma, A)$ and on the deformation space $\mathcal{D}_c(\Gamma, A)$. Moreover, there is an induced action

[14] private communication

of $\mathrm{Out}(U_\Gamma) = \mathrm{Aut}(U_\Gamma)/\mathrm{Inn}(U_\Gamma)$ on $\mathcal{D}_c(\Gamma, A)$ which extends the natural $\mathrm{Out}(\Gamma)$-action. This is explained in Section 11.

We may as well interpret the fixed point properties of the $\mathrm{Aut}(U_\Gamma)$-action as a *convexity property* of the deformation space $\mathcal{D}_c(\Gamma, A)$. For an f.t.n.-group $\Gamma$, we will henceforth understand Definition 3.3 in the following stronger sense:

**Definition 5.1.** Let $\Gamma$ be an f.t.n.-group. The space $\mathcal{D}_c(\Gamma, A)$ is called *strongly convex* if every reductive subgroup of $\mathrm{Aut}(U_\Gamma)$ has a fixed point in $\mathcal{D}_c(\Gamma, A)$. If the group $\mathrm{Aut}(U_\Gamma)$ has a fixed point, $\mathcal{D}_c(\Gamma, A)$ will be called *fixed pointed.*

Note, if $\mathcal{D}_c(\Gamma, A)$ is fixed pointed in the sense of Definition 5.1 then every finite subgroup of $\mathrm{Out}(\Gamma)$ has a fixed point on $\mathcal{D}_c(\Gamma, A)$. In fact, every finite subgroup $\mu \leq \mathrm{Out}(\Gamma)$ may be lifted to a finite subgroup of $\mathrm{Aut}(U_\Gamma)$.[15]

**Example 5.2.** Let $\Gamma = \mathbb{Z}^n \leq V$ be a lattice, so that $U_\Gamma = V$. The space $\mathcal{D}_c(\mathbb{Z}^n, \mathrm{Aff}(V))$ has a natural fixed point for all automorphisms. The fixed point for $\mathrm{Out}(U_{\mathbb{Z}^n}) = \mathrm{GL}(V)$ is given by the natural action of $\mathbb{Z}^n$ on $V$ via translations. (Compare also Proposition 4.3)

For the moment being, we restrict our interest to the principal case $A = \mathrm{Aff}(V)$. Our aim here is to construct examples of groups with fixed pointed deformation spaces $\mathcal{D}_c(\Gamma, \mathrm{Aff}(V))$, and more generally with convex deformation spaces. The first step is to consider f.t.n.-groups.

**Conditions for strong convexity** Let $\Gamma$ be an f.t.n.-group. Recall that the *nilpotency class* of $\Gamma$ is the length of a shortest central series for $\Gamma$. (For example, if $\Gamma$ is abelian then $\Gamma$ is of class 1.) By abuse of language we say that $\Gamma$ *admits an invariant grading* if the Lie algebra of the Malcev hull $U_\Gamma$ has an invariant grading. (See Definition 5.12.) Let us consider now the following conditions for $\Gamma$:

i) $\Gamma$ is of nilpotency class $\leq 2$,
ii) $\Gamma$ is of nilpotency class $\leq 3$,
iii) $\Gamma$ admits a positive invariant grading.

It is known from constructions given by Scheuneman that if one of the conditions i)-iii) is satisfied $\Gamma$ admits affine crystallographic actions. This will be explained further below.

We will prove below:

**Theorem 5.3.** *Let $\Gamma$ be an* f.t.n.*-group. If $\Gamma$ satisfies condition* i) *then the deformation space $\mathcal{D}_c(\Gamma, \mathrm{Aff}(V))$ is fixed pointed. If one of the conditions* ii) *or* iii) *is satisfied by $\Gamma$ then $\mathcal{D}_c(\Gamma, \mathrm{Aff}(V))$ is strongly convex.*

The proof of the theorem will show that is possible to construct affine crystallographic actions which are fixed by reductive groups if the algebraic structure

[15] using Theorem 7.23

of the f.t.n.-group $\Gamma$ is not too complicated. In particular if the rank of $\Gamma$ is small, for example, $\operatorname{rank}\Gamma \leq 5$, conditions ii) or iii) are satisfied. This follows from known classification results for nilpotent Lie algebras. On the other hand, it is known that for a "sufficiently generic" nilpotent group connected reductive groups of automorphisms do not exist at all, compare [31]. This leaves us with the open problem:

*Does there exist an affine crystallographic* f.t.n.*-group* $\Gamma$ *with a non-convex deformation space?*

The answer to this question is in particular important in the light of the solution to the realization problems in Part III. For example, by Theorem 6.4 the convexity of $\mathcal{D}_c(\Gamma, A)$ implies that $\Gamma$ satisfies the realization property for finite extensions. In particular, every finite effective extension $\Delta$ of $\Gamma$ admits an affine crystallographic action. Therefore the question above is more or less equivalent to the question:

*Does there exist an affine crystallographic* f.t.n.*-group* $\Gamma$ *with a finite effective extension* $\Delta$ *which is not isomorphic to an affine crystallographic group?*

By Theorem 8.9 the convexity of the deformation space is inherited to finite extensions of $\Gamma$. Therefore Theorem 5.3 also yields:

**Corollary 5.4.** *Let* $\Delta$ *be a finite extension of an* f.t.n.*-group,* $\Gamma = \operatorname{Fitt}(\Delta)$. *Then the deformation space* $\mathcal{D}_c(\Delta, \operatorname{Aff}(V))$ *is fixed pointed if condition* i) *is satisfied by* $\Gamma$. *The deformation space* $\mathcal{D}_c(\Delta, \operatorname{Aff}(V))$ *is convex if the group* $\Gamma$ *satisfies one of* ii) *or* iii).

We remark that the corollary in particular establishes the existence of affine crystallographic actions for $\Delta$. (See also Section 10) By Theorem 9.10 the convexity of the deformation space is also inherited from unipotent shadows. This yields:

**Corollary 5.5.** *Let* $\Gamma$ *be a torsionfree polycyclic group,* $\Theta$ *a unipotent shadow for* $\Gamma$. *Then the deformation space* $\mathcal{D}_c(\Gamma, \operatorname{Aff}(V))$ *is fixed pointed if condition* i) *is satisfied by* $\Theta$. $\mathcal{D}_c(\Gamma, \operatorname{Aff}(V))$ *is convex if the shadow* $\Theta$ *satisfies one of* ii) *or* iii).

The previous corollaries are an immediate consequence of Theorem 5.3 and the above mentioned inheritance results of Part III. Theorem 5.3 follows from the existence of certain well understood simply transitive unipotent actions constructed by Scheuneman, see below.

*Proof of Theorem 5.3* For a simply transitive subgroup $U \leq \operatorname{Aff}(V)$ we let $\operatorname{Aut}_{\operatorname{Aff}(V)}(U) \leq \operatorname{Aut}(U)$ be the image of the normalizer of $U$. (cf. Section 11).

Let $\Gamma$ be of nilpotency class $\leq 2$, and $U_\Gamma$ its real Malcev hull. By Proposition 5.10, $U_\Gamma$ embeds into $\operatorname{Aff}(V)$ so that the image $U \leq \operatorname{Aff}(V)$ is a simply transitive subgroup with the property that $\operatorname{Aut}_{\operatorname{Aff}(V)}(U) = \operatorname{Aut}(U)$. This representation corresponds to an element $\rho \in \operatorname{Hom}_c(\Gamma, \operatorname{Aff}(V))$.

Recall now that $\mathrm{Aut}(U_\Gamma)$ acts on the deformation space $\mathcal{D}_c(\Gamma, \mathrm{Aff}(V))$, extending the action of $\mathrm{Aut}(\Gamma)$. By Proposition 11.8, the subgroup $\mathrm{Aut}_{\mathrm{Aff}(V)}(U) \leq \mathrm{Aut}(U)$ corresponds to a subgroup of the stabiliser $\mathrm{Aut}(U_\Gamma)_{[\rho]}$ of $[\rho] \in \mathcal{D}_c(\Gamma, \mathrm{Aff}(V))$. Therefore, $\mathrm{Aut}_{\mathrm{Aff}(V)}(U) = \mathrm{Aut}(U)$ implies that $\mathrm{Aut}(U_\Gamma)_{[\rho]} = \mathrm{Aut}(U_\Gamma)$. This means that $[\rho]$ is a fixed point for $\mathrm{Aut}(U_\Gamma)$.

The second statement is proved similarly. If $\Gamma$ satisfies condition ii) or condition iii), Scheuneman's examples (cf. Theorem 5.15, Theorem 5.14) provide us with crystallographic homomorphisms $\rho$ for $\Gamma$ so that $\mathrm{Aut}_{\mathrm{Aff}(V)}(U)$ contains a Levi subgroup of $\mathrm{Aut}(U)$. This shows that $\mathrm{Aut}(U_\Gamma)_{[\rho]}$ contains a Levi subgroup $L$ of $\mathrm{Aut}(U_\Gamma)$. Therefore $[\rho]$ is a fixed point for $L$. Hence any reductive subgroup of $\mathrm{Aut}(U_\Gamma)$ has a fixed point in the $\mathrm{Aut}(U_\Gamma)$-orbit of $[\rho]$. In particular, this holds for finite subgroups of $\mathrm{Aut}(U_\Gamma)$. □

## Scheuneman's examples

Scheuneman (cf. [70, 71]) constructed unipotent simply transitive actions on affine space to give new examples of compact complete affine manifolds. We analyze here the automorphism groups of these actions. Some of the examples we study were known already to Elie Cartan (cf. [37]). Scheuneman's examples provide the only known general method to establish the existence of affine crystallographic actions on large classes of (reasonably well behaved) torsionfree nilpotent groups.

To present the constructions, we need to establish first the infinitesimal picture of simply transitive actions.

### Simply transitive affine actions of Lie algebras

Let $\mathfrak{g}$ be a Lie algebra, $\phi : \mathfrak{g} \to \mathfrak{gl}(V)$ a representation of $\mathfrak{g}$ on a vector space $V$. A map $D : \mathfrak{g} \to V$ is called a derivation for $\phi$ if

$$D([X,Y]) = \phi_X DY - \phi_Y DX \quad , \quad \text{for all } X, Y \in \mathfrak{g}.$$

$D$ will be called *nonsingular* if $D$ is an isomorphism of vector spaces. We call the pair $(\phi, D)$ an *affine representation on* $V$. Let $\mathfrak{a}(V) = \mathfrak{gl}(V) \oplus V$ be the Lie algebra of $\mathrm{Aff}(V)$. Every Lie algebra homomorphism $\varphi : \mathfrak{g} \to \mathfrak{a}(V)$ is, with respect to the splitting of $\mathfrak{a}(V)$, a sum $\varphi = \phi + D$, such that $(\phi, D)$ is an affine representation.

Let $U$ denote a simply connected nilpotent Lie group with Lie algebra $\mathfrak{u}$. The next proposition is well known (compare [39]):

**Proposition 5.6.** *An affine representation $(\phi, D) : \mathfrak{u} \to \mathfrak{a}(V)$ is the differential of a unipotent simply transitive representation of $U$ if and only if $\phi$ is a representation of nilpotent linear operators and $D$ is a nonsingular derivation for $\phi$.*

If $U$ is a unipotent simply transitive subgroup of $\mathrm{Aff}(V)$, we call the Lie algebra $\mathfrak{u} \subset \mathfrak{a}(V)$ a *simply transitive subalgebra* of $\mathfrak{a}(V)$.

**Normalisers of simply transitive actions** Let $U$ be a unipotent (real) linear algebraic group with Lie algebra $\mathfrak{u}$. Recall that the automorphism group $\mathrm{Aut}(U)$ of $U$ identifies with the group $\mathrm{Aut}(\mathfrak{u})$ of Lie algebra automorphisms of $\mathfrak{u}$ by taking the differential in the identity. (This identification also defines the natural algebraic group structure on $\mathrm{Aut}(U)$.)

Let $U \leq A$ be a subgroup of the affine group. We define (see also Section 11) the group

$$\mathrm{Aut}_A(U) \leq \mathrm{Aut}(U)$$

of $A$-automorphisms of $U$ as the natural image of $N_{\mathrm{GL}_A}(U)$ in $\mathrm{Aut}(U)$, and

$$\mathrm{Aut}_A(\mathfrak{u}) \leq \mathrm{Aut}(\mathfrak{u})$$

as the corresponding group of Lie algebra automorphisms of $\mathfrak{u}$.

Note that $\mathrm{Aut}_A(\mathfrak{u})$ may also be computed as the image of $N_{\mathrm{GL}_A}(\mathfrak{u})$ in $\mathrm{Aut}(\mathfrak{u})$, since $\mathfrak{u}$ is a subalgebra of $\mathfrak{a}$, where $\mathrm{GL}_A$ acts by conjugation on $\mathfrak{a}$. Now let $\varphi : \mathfrak{u} \to \mathfrak{a}$ be a simply transitive representation of $\mathfrak{u}$, $\varphi = (\phi, D)$. We put

$$\mathrm{Aut}_A(\mathfrak{u}, \varphi) = \mathrm{Aut}_A(\varphi(\mathfrak{u}))^{\varphi^{-1}} .$$

To compute $\mathrm{Aut}_A(\mathfrak{u}, \varphi)$ it is convenient to identify $V$ with the underlying vector space of $\mathfrak{u}$, i.e., we consider simply transitive representations $\varphi$ of $\mathfrak{u}$ on itself. For elements $D, g \in \mathrm{GL}(\mathfrak{u})$ let us define $(D, g) := g^{-1}DgD^{-1}$. The group $\mathrm{Aut}_A(\mathfrak{u})$ may now be characterized in terms of $\phi$ as follows:

**Proposition 5.7.** *Let $A \leq \mathrm{Aff}(\mathfrak{u})$ be a transitive subgroup. Let $\varphi : \mathfrak{u} \to \mathfrak{a}(\mathfrak{u})$ be a a unipotent simply transitive representation of $\mathfrak{u}$ on itself. Let $g \in \mathrm{Aut}(\mathfrak{u})$. Then $g \in \mathrm{Aut}_A(\mathfrak{u}, \varphi)$ if and only if*

$$DgD^{-1} \in N_{GL_A}(\phi(\mathfrak{u}))$$

*and*

$$(D, g)\, \phi_X\, (D, g)^{-1} = g^{-1}\phi_{gX}\, g \ , \quad \textit{for all } X \in \mathfrak{g} \ .$$

*Proof.* Let $h \in N_{GL_A}(\phi(\mathfrak{u}))$. If $\varphi = (\phi, D)$ then the conjugate representation is $\varphi^h = (\phi^h, hD)$. The image $\Phi_h$ of $h$ in $\mathrm{Aut}_A(\mathfrak{u}, \varphi)$ is defined by the condition that

$$\varphi\, \Phi_h \ = \ \varphi^h \ .$$

This is equivalent to $\phi^h = \phi\, \Phi_h$, and $hD = D\Phi_h$. By the latter condition it follows that

$$D\Phi_h D^{-1} = h \in N_{GL_A}(\phi(\mathfrak{u})) \ .$$

The proposition is immediate with the first condition. □

This gives rise to the following

**Definition 5.8.** Let $\phi : \mathfrak{u} \to \mathfrak{gl}(\mathfrak{u})$ be a representation of Lie algebras, and $g \in \mathrm{Aut}(\mathfrak{u})$. Then $g$ is called *compatible* with $\phi$ if

$$g^{-1}\phi_{gX}\, g = \phi_X \ , \quad \text{for all } X \in \mathfrak{g} \ .$$

We remark that a particular special case is the adjoint representation of $\mathfrak{u}$ on itself, $\phi = \mathrm{ad}$ . By its very definition, $\mathrm{Aut}(\mathfrak{u})$ is compatible with $\mathrm{ad}$ .

**Basic examples of unipotent simply transitive actions**

We come now to a first family of examples which was known to Cartan already. These examples are fully invariant and may be seen as a generalization of the natural simply transitive representation of an abelian Lie algebra.

**Definition 5.9.** We call a simply transitive representation $\varphi$ of $\mathfrak{u}$ *invariant* if $\mathrm{Aut}(\mathfrak{u},\varphi) = \mathrm{Aut}(\mathfrak{u})$. We call $\varphi$ invariant by the subgroup $H \leq \mathrm{Aut}(\mathfrak{u})$ if $H \leq \mathrm{Aut}(\mathfrak{u},\varphi)$.

**Proposition 5.10.** *Let $\mathfrak{u}$ be a 2-step nilpotent Lie algebra. Then $\mathfrak{u}$ has a natural invariant simply transitive representation.*

*Proof.* We take $\phi = \frac{1}{2}\,\mathrm{ad}$, and $D = \mathrm{id}_{\mathfrak{g}}$, where ad denotes the adjoint representation. □

This construction was generalized by Scheuneman ([70]) as follows:

**Proposition 5.11.** *Let $\mathfrak{u}$ be a nilpotent Lie algebra, which admits a nonsingular derivation $D$. Then $\mathfrak{u}$ has a simply transitive affine representation $\varphi = (\mathrm{ad}\,, D)$. The representation $\varphi$ is invariant by the centralizer of $D$ in $\mathrm{Aut}(U)$.*

*Proof.* Since ad is compatible with $\mathrm{Aut}(\mathfrak{u})$, the centralizer

$$\mathrm{C}_{Aut(\mathfrak{u})}(D) = \{g \in \mathrm{Aut}(\mathfrak{u}) \mid (g, D) = id_{\mathfrak{u}}\}$$

is contained in $\mathrm{Aut}(\mathfrak{u},\varphi)$, by Proposition 5.7. □

**Construction of derivations** The proposition raises the particular problem to study the centralizers of certain elements in $\mathrm{Aut}(\mathfrak{u})$. A semisimple element $g \in \mathrm{Aut}(\mathfrak{u})$ is called expanding if all eigenvalues have absolute value $> 1$. We construct now examples of Lie algebras which contain expanding elements in the center of the Levi subgroups of $\mathrm{Aut}(\mathfrak{u})$. (We call a maximal reductive subgroup of the linear algebraic group a Levi subgroup, cf. Theorem 7.23.)

For preparation we have to introduce a new concept.

**Definition 5.12.** A (positive) *filtration* on a Lie algebra $\mathfrak{g}$ is a nested sequence of subspaces

$$\mathfrak{g} = \mathfrak{g}_1 \supseteq \mathfrak{g}_2 \supseteq \mathfrak{g}_3 \supseteq \cdots$$

such that $[\mathfrak{g}_i, \mathfrak{g}_j] \subseteq \mathfrak{g}_{i+j}$. The filtration is called *invariant* if it is preserved by $\mathrm{Aut}(\mathfrak{g})$.

For each positive filtration on $\mathfrak{g}$ there is an associated graded Lie algebra

$$\bar{\mathfrak{g}} = \bigoplus_{i=1,\dots} \mathfrak{g}_i/\,\mathfrak{g}_{i+1}$$

where the Lie product on $\overline{\mathfrak{g}}$ is defined by setting, for $x \in \mathfrak{g}_i$, $y \in \mathfrak{g}_j$,

$$[\overline{x}, \overline{y}] = \overline{[x, y]} \in \mathfrak{g}_{i+j} / \mathfrak{g}_{i+j+1} .$$

A grading

$$\mathfrak{g} = \bigoplus_{l=1,\dots,k} V_l$$

is called invariant by a subgroup $G \leq \mathrm{Aut}(\mathfrak{g})$ if every $g \in G$ preserves the decomposition. If $\mathfrak{g}$ has a positive grading $\mathfrak{g} = \oplus_{l=1,\dots,k} V_l$ then the nested sequence of ideals $\mathfrak{g}_i = \oplus_{l=i,\dots,k} V_l$ defines an associated positive filtration.

Note that every positive grading is preserved by a one parameter group $l_\lambda$ of expanding automorphisms in $\mathrm{Aut}(\mathfrak{g})$. In fact, for $\lambda > 1$, $l_\lambda$ is given by

$$l_\lambda(v) = \lambda^i v , \ v \in V_i .$$

**Proposition 5.13.** *Let* $\mathfrak{u}$ *be a nilpotent Lie algebra which has a positive grading with an invariant associated filtration. Let* $L$ *be a Levi-subgroup of the linear algebraic group* $\mathrm{Aut}(\mathfrak{u})$. *Then* $L$ *contains a one parameter group of expanding automorphisms in its center.*

*Proof.* Since the filtration $\mathfrak{u}_i$ of $\mathfrak{u}$ associated with the grading is invariant, $L$ acts on the factor spaces $\mathfrak{u}_i / \mathfrak{u}_{i+1}$. Therefore $L$ acts by automorphisms on the associated graded Lie algebra $\overline{u}$ such that the grading is preserved. Since $L$ is reductive the action of $L$ on $\overline{u}$ is also faithful.

Now the fact that $\mathfrak{u}$ is graded with respect to the filtration $\mathfrak{u}_i$ implies that there is an isomorphism of filtered Lie algebras $\pi : \mathfrak{u} \to \overline{\mathfrak{u}}$. Hence it follows that $\mathrm{Aut}(\mathfrak{u})$ has a Levi subgroup which preserves a grading compatible which is compatible with the original filtration. Since all Levi subgroups are conjugate in $\mathrm{Aut}(\mathfrak{u})$, every Levi subgroup $L$ preserves a corresponding positive grading of $\mathfrak{u}$. The one parameter family $l_\lambda$ of expanding automorphisms which belongs to this positive grading commutes with $L$. □

As a consequence we obtain

**Theorem 5.14.** *Let* $u$ *be a nilpotent Lie algebra which has an invariant positive grading, and let* $L$ *be a Levi-subgroup of* $\mathrm{Aut}(U)$. *Then there exists an affine simply transitive representation which is invariant by* $L$.

*Remark* It is easily checked from certain lists of nilpotent Lie algebras that every nilpotent Lie algebra of dimension less or equal to five has an invariant positive grading.

**Scheuneman representations for 3-step nilpotent Lie algebras** A little more refined construction which builds on the previous cases was given by Scheuneman in [71]. The construction is compelling because it works for every 3-step nilpotent Lie algebra.

**Theorem 5.15.** *Let* $u$ *be a 3-step nilpotent Lie algebra, and let* $L$ *be a Levi-subgroup of* $\mathrm{Aut}(U)$. *Then there exists an affine simply transitive representation which is invariant by* $L$.

*Proof.* Following [71], we consider the filtration

$$\mathfrak{u} = \mathfrak{u}_1 \supseteq \mathfrak{u}_2 \supseteq \mathfrak{u}_3 \supseteq \{0\}$$

which is given by the descending central series, i.e., $\mathfrak{u}_2 = [\mathfrak{u}, \mathfrak{u}]$, $\mathfrak{u}_3 = [\mathfrak{u}, \mathfrak{u}_2]$. We choose a compatible decomposition of $\mathfrak{u} = \oplus_{l=1,\dots,3} V_l$ as a direct sum of vector spaces, $V_3 = \mathfrak{u}_3$. We define a diagonal element $g = g_{\alpha,\beta,\gamma} \in \mathrm{GL}(\mathfrak{u})$ which preserves the grading by $gv = \alpha v$, for $v \in V_1$, $gv = \beta v$, for $v \in V_2$, and $gv = \gamma v$, for $v \in V_3$. Correspondingly, we define $D = D_{r,s,t}$. It is easy to see (compare [71]) that there exist $\alpha, \beta, \gamma > 0$, and $r, s, t > 0$ so that $D$ is a derivation for the representation $\phi = \mathrm{ad}^{\,g}$ of $\mathfrak{u}$ on itself. The pair $(\phi, D)$ defines then a simply transitive representation $\varphi$ of $\mathfrak{u}$. Since the filtration we chose is invariant, we may assume that the spaces $V_i$ are invariant by automorphisms of a Levi subgroup $L$ of $\mathrm{Aut}(\mathfrak{u})$. In this case $g$ commutes with $L$, and the representation $\phi$ is compatible with $L$ in the sense of Definition 5.8. Since also $D$ commutes with $L$, it follows that $L \leq \mathrm{Aut}(\mathfrak{u}, \varphi)$, by Proposition 5.7. □

*Remark* When applied to the 2-step nilpotent case the method recovers just the natural invariant simply transitive action.

# Part III
# Geometric Realization Problems

## 6. Nielsen's realization problem

Nielsen considered in [60] the following question which is now called the *Nielsen realization problem.* Namely, *can any finite group $G$ of isotopy classes of self-homeomorphisms of a surface be realized as a finite group of self-homeomorphisms?* As it turned out, Nielsen's question is equivalent to the following: *Let $\Delta$ be a finite effective extension of the fundamental group $\pi_1(S)$ of a surface $S$. Is it true, that $\Delta$ acts discontinuously on the plane?*

Geometry gives a positive answer to this question. For example, every effective finite extension of $\mathbb{Z}^2$ acts as an Euclidean crystallographic group. These are the well known so called wall-paper groups. Therefore the realization problem for Euclidean surfaces has a positive solution. For hyperbolic surfaces the problem was settled not long ago. Kerckhoff's [38] celebrated fixed point theorem states: Any finite group of mapping classes of a closed orientable surface $M_g$ of genus $g \geq 2$ has a fixed point in the Teichmüller space of hyperbolic structures. A particular consequence of the fixed point theorem is: Let $\Delta$ be a finite effective extension of $\pi_1(M_g)$, then $\Delta$ is isomorphic to a discontinuous group of hyperbolic motions. These results give a beautiful example of the significance of geometric methods in topology.

Recall, that a homomorphism $\alpha : F \to \mathrm{Out}(\Gamma)$ from a group $F$ to the outer automorphism group of a group $\Gamma$ is called an *abstract kernel for* $\Gamma$. Every group

$F$ of homeomorphisms of a manifold naturally defines an abstract kernel $\alpha : F \to \pi_1(M)$. If this is the case, we say that $\alpha$ is realized by a group of homeomorphisms. The *topological realization problem* may be posed as follows: *Let $M$ be a manifold and let $\alpha$ be a finite abstract kernel for $\pi_1(M)$. Is it possible to realize $\alpha$ by a group of homeomorphisms of $M$?*

Under suitable topological conditions on the manifold $M$, for example if $M$ is aspherical, a necessary condition for $\alpha$ to be realizable as a group of homeomorphisms is the existence of a finite *effective* extension group $\Delta$ for $\Gamma = \pi_1(M)$ which induces $\alpha$. The group $\Delta$ is then an *algebraic realization* of the kernel $\alpha$. As illustrated above in the case of surfaces, geometry may help to realize $\Delta$ as properly discontinuous group of homeomorphisms of the universal cover $\tilde{M}$ of $M$. Still, *geometric realization questions* are interesting in their own right. An answer to such a question may show for example how well a particular geometry is adapted to the topology of a manifold.

A classic example which extends beyond the surface case is the following: Bieberbach proved in 1912 his famous theorems on the structure of the Euclidean crystallographic groups. Later Burckhardt and Zassenhaus showed that Euclidean crystallographic groups are characterized by their algebraic properties. In particular, their remark implies (see [83][§25]) a positive answer to the geometric realization problem for *Euclidean crystallographic groups*:

**Theorem 6.1.** *Let $\Gamma$ be isomorphic to an Euclidean crystallographic group. If $\Delta$ is a finite effective extension of $\Gamma$ then $\Delta$ is also isomorphic to an Euclidean crystallographic group.*

Since, by Bieberbach's second theorem, flat Euclidean manifolds are determined by their fundamental group up to affine diffeomorphism this implies:

**Theorem 6.2.** *Let $M$ be compact Euclidean space form, and $\alpha$ an injective abstract kernel for $\pi_1(M)$. Then the following are equivalent:*

i) *$\alpha$ has an algebraic realization $\Delta$,*
ii) *$\alpha$ can be realized by a group of affine diffeomorphisms of $M$.*

**The realization problem for affine crystallographic groups** In our work we are concerned with the realization problem for affine crystallographic groups. We restrict ourselves here to the case that $\Gamma$ is a virtually nilpotent ACG.[16]

Every affine space form $M$ with fundamental group isomorphic to $\Gamma$ is finitely covered by a nilmanifold. (A nilmanifold is diffeomorphic to the quotient of a nilpotent Lie group by a closed subgroup. $M$ itself has an infranilmanifold structure.) For certain aspherical manifolds the topological realization question was solved by Lee and Raymond (cf. [50]) with methods from the theory of Seifert fiber spaces. One particular result is:

[16] Most of our results here generalise to virtually polycyclic ACGs in the context of infrasolvmanifolds. See [11, 13] for some of the general constructions.

**Theorem 6.3.** *Let $M$ be an infranilmanifold, $\alpha$ an abstract kernel for $\pi_1(M)$. If there exists an algebraic realization of $\alpha$ then $\alpha$ may be realized as a group of homeomorphisms of $M$.*

However, the following *geometric question* remains: *Which of the finite quotient manifolds of an affine space form $M$ are again an affine space form?* Our basic result shows that if $M$ has a virtually nilpotent fundamental group then the solution to the geometric realization problem is a question about the fixed points of abstract kernels on the deformation space.

**Theorem 6.4.** *Let $\Lambda$ be virtually nilpotent and isomorphic to an affine crystallographic group of type $A$. Then a finite effective extension $\Delta$ of $\Lambda$ is isomorphic to an affine crystallographic group of type $A$ if and only if the kernel associated to the extension has a fixed point in the deformation space $\mathcal{D}_c(\Lambda, A)$.*

We will also consider the question for which crystallographic groups the analogue of Theorem 6.1 holds. These are precisely those affine crystallographic groups $\Gamma$ which satisfy the following *realization property*.

**Definition 6.5.** We say that $\Gamma$ has the affine realization property if every finite effective extension $\Delta$ of $\Gamma$ is isomorphic to an affine crystallographic group.

We remark that if the fundamental group $\pi_1(M)$ of an affine space form $M$ satisfies the realization property then, by some standard reasoning[17], the following generalization of Theorem 6.2 holds:

**Theorem 6.6.** *Let $M$ be a compact affine space form with a virtually nilpotent fundamental group $\pi_1(M)$, and $\alpha$ an injective abstract kernel for $\pi_1(M)$. If $\pi_1(M)$ has the realization property then the following are equivalent:*

i) *$\alpha$ has an algebraic realization $\Delta$.*
ii) *$\alpha$ can be realized by a group of polynomial diffeomorphisms of $M$.*

Theorem 6.4 implies that $\Gamma$ satisfies the realization property if the deformation space $\mathcal{D}_c(\Gamma, \mathrm{Aff}(V))$ is convex in the sense of Definition 3.3. In particular, compare Corollary 10.1, the examples of Part II show that some natural classes of virtually nilpotent crystallographic groups satisfy the realization property.

**Corollary 6.7.** *Let $\Gamma$ be a virtually nilpotent affine crystallographic group. If the deformation space $\mathcal{D}_c(\Gamma, \mathrm{Aff}(V))$ is convex then $\Gamma$ has the realization property.*

This fact should provide us with enough motivation to study the deformation spaces for affine space forms.

---

[17]using that compact complete affine manifolds with isomorphic fundamental group are polynomially diffeomorphic, see [27].

# 7. The Realization of finite extensions for f.t.n.-groups

Let $\Gamma$ be a group which is isomorphic to an ACG of type $A$, and let $\beta : F \longrightarrow \mathrm{Out}(\Gamma)$ be an abstract kernel for $\Gamma$. As a subgroup of $\mathrm{Out}(\Gamma)$, $\beta(F)$ acts on the deformation space $\mathcal{D}_c(\Gamma, A)$. Let us assume that there exists a finite normal extension $\Gamma \leq_f \Delta$ which realizes the abstract kernel $\beta$. This means that $\beta$ coincides with the natural homomorphism $F = \Delta/\Gamma \longrightarrow \mathrm{Out}(\Gamma)$ which is associated to the extension. The following is a fundamental observation:

**Proposition 7.1.** *Let $\beta : F \longrightarrow \mathrm{Out}(\Gamma)$ be the kernel associated to the normal extension $\Gamma \leq_f \Delta$. Assume that there exists a crystallographic homomorphism $\rho_\Delta \in \mathrm{Hom}_c(\Delta, A)$. Let $\rho$ denote the restriction of $\rho_\Delta$ to $\Gamma$. Then $\rho$ is in $\mathrm{Hom}_c(\Gamma, A)$, and $[\rho] \in \mathcal{D}_c(\Gamma, A)$ is a fixed point for the action of $F$ on $\mathcal{D}_c(\Gamma, A)$.*

*Proof.* Clearly, the restriction of $\rho_\Delta$ to $\Gamma$ remains crystallographic. For $g \in F$, we have $\beta(g)[\rho] = [\rho^{\phi_g}]$, where $\phi_g \in \mathrm{Aut}(\Gamma)$ represents $\beta(g) \in \mathrm{Out}(\Gamma)$. But, since $\Delta$ realizes the kernel $\beta$, there exists $\delta \in \Delta$ such that $\beta(g)$ is represented by conjugation with $\delta$. Therefore, we may assume that $\phi_g(\gamma) = \delta\gamma\delta^{-1}$, for all $\gamma \in \Gamma$. It follows that $\rho^{\phi_g}(\gamma) = \rho_\Delta(\delta)\rho(\gamma)\rho_\Delta(\delta)^{-1}$. Since $\rho_\Delta(\delta) \in A$, we conclude $\beta(g)[\rho] = [\rho^{\phi_g}] = [\rho]$. □

From now on let us assume that $\Gamma$ is an f.t.n.-group.

**Definition 7.2.** A normal extension group $\Delta$ of $\Gamma$ is called *effective* if the associated kernel $\beta : \Delta/\Gamma \longrightarrow \mathrm{Out}(\Gamma)$ is an injective homomorphism.

We are going to show that the converse of Proposition 7.1 holds for all finite effective extensions $\Delta$ of $\Gamma$. Namely, the existence of fixed points for $\beta$ on the deformation space of $\Gamma$ is the only obstruction to realize $\Delta$ as an affine crystallographic group.

**Theorem 7.3.** *Let $\Gamma$ be an* f.t.n.*-group which is isomorphic to an affine crystallographic group of type $A$, and let $\Gamma \leq_f \Delta$ be an effective finite extension with associated abstract kernel $\beta : F \longrightarrow \mathrm{Out}(\Gamma)$. Then $\Delta$ may be realized as an affine crystallographic group of type $A$ if and only if the kernel $\beta$ has a fixed point in $\mathcal{D}_c(\Gamma, A)$.*

*Remark* In the course of the proof of Theorem 7.3, we will show a stronger statement: Every $\rho \in \mathrm{Hom}_c(\Gamma, A)$ which is a fixed point for $\beta$ may be extended to a homomorphism $\rho_\Delta \in \mathrm{Hom}_c(\Delta, A)$.

Theorem 6.4 follows from Theorem 7.3. The proof of Theorem 6.4 will be given below in Section 8. The purpose of the remainder of this section is to prove Theorem 7.3. We will proceed as follows:

The first step is, to consider the related problem of realizing a finite extension $\Delta$ of $\Gamma$ as a Zariski dense subgroup of a linear algebraic group. The solution to this problem (Proposition 7.8) leads us to the construction of an algebraic hull for $\Delta$ which satisfies functorial properties with respect to the representations of $\Delta$. As one application of the construction we deduce a splitting result for the extension $\Gamma \leq \Delta$ (Proposition 7.17). Another application is a certain characterization of the finite effective nilpotent extensions of $\Gamma$. (Corollary 7.11)

The second step is the proof of the affine crystallographic realization for the finite effective extensions $\Delta$ of $\Gamma$. This breaks into two distinct parts: The case that $\Delta$ itself is an f.t.n.-group, and the case that the extension $\Gamma \leq \Delta$ splits.

We will use frequently certain results from the theory of f.t.n.-groups. For a general reference on the relevant aspects of the theory, see [24, Chapter 2], [66, Chapter II], and in particular the book [72]. We start with some preparatory material.

## The algebraic hull of an extension

**Automorphisms and the Malcev hull** Let $\Gamma$ be an f.t.n.-group. It is known that there exists a $\mathbb{Q}$-defined unipotent algebraic group $\mathbf{U}$ (called the Malcev completion of $\Gamma$), and an embedding $j : \Gamma \to \mathbf{U}_{\mathbb{Q}}$ such that $\Gamma$ is a Zariski-dense subgroup in $\mathbf{U}$. Every automorphism of $\Gamma$ extends to a unique $\mathbb{Q}$-defined automorphism of $\mathbf{U}$. In fact, the automorphism group $\mathrm{Aut}(\mathbf{U})$ of $\mathbf{U}$ is a $\mathbb{Q}$-defined linear algebraic group, in such a way that $\mathrm{Aut}(\mathbf{U})_{\mathbb{Q}} = \mathrm{Aut}(\mathbf{U}_{\mathbb{Q}})$, and the extension defines an injective homomorphism

$$\epsilon_j : \mathrm{Aut}(\Gamma) \longrightarrow \mathrm{Aut}(\mathbf{U})_{\mathbb{Q}}$$

which satisfies $\epsilon_j(\phi)\, j = j\, \phi$, for all $\phi \in \mathrm{Aut}(\Gamma)$. By factorization there exists a homomorphism $o_j : \mathrm{Aut}(\Gamma) \to \mathrm{Out}(\mathbf{U})$ such that the diagram

$$\begin{array}{ccc} \mathrm{Aut}(\Gamma) & \xrightarrow{\epsilon_j} & \mathrm{Aut}(\mathbf{U}) \\ \downarrow & & \downarrow \\ \mathrm{Out}(\Gamma) & \xrightarrow{o_j} & \mathrm{Out}(\mathbf{U}) \end{array}$$

is commutative. Note that $o_j$ is, in general, not injective. The preimage of its kernel is $\epsilon_j^{-1}(\mathrm{Inn}(\mathbf{U}))$. Let us therefore denote

$$\mathrm{Inn}_{\mathbb{Q}}(\Gamma) = \epsilon_j^{-1}(\mathrm{Inn}(\mathbf{U})) \text{ and } \mathrm{Out}_{\mathbb{Q}}(\Gamma) = \mathrm{Aut}(\Gamma)/\mathrm{Inn}_{\mathbb{Q}}(\Gamma)\ .$$

**Definition 7.4.** Let $\beta : F \to \mathrm{Out}(\Gamma)$ be an abstract kernel. $\beta$ is called *radicably effective* if $\beta$ induces an embedding $F \hookrightarrow \mathrm{Out}_{\mathbb{Q}}(\Gamma)$. $\beta$ is called *radicably trivial* if $\beta$ induces the trivial homomorphism to $\mathrm{Out}_{\mathbb{Q}}(\Gamma)$.

Let $\Gamma \leq \Delta$ be a finite extension with $\Gamma$ normal in $\Delta$, and let $c : \Delta \to \mathrm{Aut}(\Gamma)$ be the homomorphism which is induced by conjugation. We put $\Gamma^r = c^{-1}(\mathrm{Inn}_{\mathbb{Q}}(\Gamma))$. So $\Gamma \leq \Delta$ is radicably effective if and only if $\Gamma^r = \Gamma$. We remark that if the extension is finite and effective $\Gamma^r$ is a nilpotent normal subgroup in $\Delta$.

In fact, it follows (Corollary 7.10) that $\Gamma^r$ is the Fitting subgroup of $\Delta$.

Later we will need the following lemma:

**Lemma 7.5.** *Let $\mu \leq \mathrm{Aut}(\mathbf{U}_{\mathbb{Q}})$ be a finite subgroup, and $\Gamma \leq \mathbf{U}_{\mathbb{Q}}$ a finitely generated subgroup. Then there exists $\Gamma_f \leq \mathbf{U}_{\mathbb{Q}}$ such that $\Gamma \leq_f \Gamma_f$, and $\Gamma_f$ is normalized by $\mu$.*

*Proof.* Let $\mathfrak{u}$ be the Lie algebra of $\mathbf{U}$. $\mathfrak{u}$ is defined over $\mathbb{Q}$, and the exponential map $\exp : \mathfrak{u} \to \mathbf{U}$ identifies $\mathfrak{u}_{\mathbb{Q}}$ with $\mathbf{U}_{\mathbb{Q}}$. The group $\mu$ acts as a subgroup of automorphisms in $\mathrm{GL}(\mathfrak{u}_{\mathbb{Q}})$ on $\mathfrak{u}_{\mathbb{Q}}$, and to show that $\mu$ normalizes $\Gamma \leq \mathbf{U}_{\mathbb{Q}}$ is equivalent to show that $\mu$ normalizes the set $\log \Gamma \subseteq \mathfrak{u}_{\mathbb{Q}}$. Replacing, if necessary, $\Gamma$ with a finite extension group, we may assume ([72][§6 B, Theorem 3]) that $\Gamma$ is a lattice subgroup, such that $\Lambda = \log \Gamma = \mathbb{Z} \log \Gamma$ is a full lattice in $\mathfrak{u}_{\mathbb{Q}}$. Since $\mu \leq \mathrm{GL}(\mathfrak{u}_{\mathbb{Q}})$ is finite there exists an integer $m$ sucht that, for all $g \in \mu$, $g\,\frac{1}{m}\Lambda \subseteq \frac{1}{m}\Lambda$. Let $\Gamma^{\frac{1}{m}}$ be the group which is generated by the set $\{\exp u \mid u \in \frac{1}{m}\Lambda\}$. Now $\Gamma_f = \Gamma^{\frac{1}{m}}$ is a finite extension of $\Gamma$, and is normalized by $\mu$. □

**The Malcev extension functor** As before let the $\mathbb{Q}$-defined algebraic group $\mathbf{U}$ denote the Malcev completion of $\Gamma$. We need to consider the extension functor (*Malcev-rigidity*) from the unipotent representations of $\Gamma$ to representations of $\mathbf{U}$. Remark, that this is just a special case of Propostion 1.4.

**Proposition 7.6.** *Let $\rho : \Gamma \to \mathbf{H}$ be a homomorphism of $\Gamma$ to a unipotent $\mathbb{Q}$-defined linear algebraic group $\mathbf{H}$. Then there exists a unique morphism of algebraic groups*

$$\rho_{\mathbf{U}} : \mathbf{U} \longrightarrow \mathbf{H}$$

*which extends $\rho$. If $\rho(\Gamma) \leq \mathbf{H}_k$, where $k$ is a subfield of $\mathbb{C}$, then $\rho_{\mathbf{U}}$ is defined over $k$.*

*Proof.* Consider the subgroup $D = \{(\gamma, \rho(\gamma)) \mid \gamma \in \Gamma\}$ of the product $\mathbf{U} \times \mathbf{H}$. Let $\pi_1, \pi_2$ denote the projection morphisms on the factors of the product. Let $\mathbf{D}$ be the Zariski-closure of $D$. By [66, Theorem 2.10] we have $\dim \mathbf{D} = \operatorname{rank} \Gamma = \dim \mathbf{U}$, and it follows that $\pi_1 : \mathbf{D} \to \mathbf{U}$ is an isomorphism. Hence $\rho_{\mathbf{U}} = \pi_2 \circ \pi_1^{-1}$ is the unique extension. If $\rho(\Gamma) \leq \mathbf{H}_k$ then $\mathbf{D}$ is $k$-defined, and hence also the morphism $\rho_{\mathbf{U}}$. □

At some point later we consider the spaces $\mathrm{Hom}(\Gamma, \mathbf{H}_{\mathbb{R}})$ as topological spaces equipped with the compact-open topology. (Compare Section 2) Now let $\Gamma^*$ be an f.t.n.-group which is a finite extension of $\Gamma$. There exists a unique embedding of $\Gamma^*$ into $\mathbf{U}_{\mathbb{Q}}$ which is the identity on $\Gamma$, see [72][§6A]. We note:

**Corollary 7.7.** *Let $\mathbf{H}$ be a unipotent $\mathbb{Q}$-defined linear algebraic group. Then every $\rho \in \mathrm{Hom}(\Gamma, \mathbf{H}_{\mathbb{R}})$ has a unique extension $\rho_{\Gamma^*} \in \mathrm{Hom}(\Gamma^*, \mathbf{H}_{\mathbb{R}})$. The extension functor induces a homeomorphism*

$$e : \mathrm{Hom}(\Gamma, \mathbf{H}_{\mathbb{R}}) \longrightarrow \mathrm{Hom}(\Gamma^*, \mathbf{H}_{\mathbb{R}}) .$$

*Proof.* We use the embedding $\Gamma^* \hookrightarrow \mathbf{U}_{\mathbb{Q}}$. Malcev rigidity implies that every $\rho \in \mathrm{Hom}(\Gamma, \mathbf{H}_{\mathbb{R}})$ extends to a homomorphism $\rho_{\mathbf{U}} : \mathbf{U}_{\mathbb{R}} \to \mathbf{H}_{\mathbb{R}}$. The restriction of $\rho_{\mathbf{U}}$ to $\Gamma^*$ is the unique extension $e(\rho) = \rho_{\Gamma^*} \in \mathrm{Hom}(\Gamma^*, \mathbf{H}_{\mathbb{R}})$. The map $e$ is then clearly a bijection.

Let $G$ be a discrete group. For any $V \subset \mathbf{H}_{\mathbb{R}}$ which is open in the Hausdorff-topology on $\mathbf{H}_{\mathbb{R}}$, and any $\gamma \in G$ we put

$$\mathcal{U}_G(\gamma, V) = \{\rho \in \mathrm{Hom}(G, \mathbf{H}_{\mathbb{R}}) \mid \rho(\gamma) \in V\} .$$

The compact open topology on $\mathrm{Hom}(G, \mathbf{H}_{\mathbb{R}})$ is then generated by all sets $\mathcal{U}_G(\gamma, V)$. It is immediate that the restriction map $e^{-1}$ is continuous. To prove that $e$ is continuous we take the following approach: We can assume that $\Gamma^* = \Gamma^{\frac{1}{m}}$, for some $m \in \mathbb{N}$. By [72][§6B, Proposition 2] there exists $s \in \mathbb{N}$ such that for every $\tilde{\gamma} \in \Gamma^*$, $\tilde{\gamma}^s \in \Gamma$. We put $V^s = \{v^s \mid v \in V\}$, $\gamma = \tilde{\gamma}^s$. Now

$$e^{-1}(\mathcal{U}_{\Gamma^*}(\tilde{\gamma}, V)) = \{\rho \mid \rho(\gamma) \in V^s\} = \mathcal{U}_{\Gamma}(\gamma, V^s) .$$

Hence, $e$ is continuous. □

**The algebraic hull of an effective extension** Let $\Gamma$ be an f.t.n.-group. Then $\Gamma \leq \mathbf{U}_{\mathbb{Q}}$, where $\mathbf{U}$ is the Malcev completion of $\Gamma$. In fact, we may assume that $\Gamma \leq \mathbf{U}_{\mathbb{Z}}$, i.e., $\Gamma$ is contained in the subgroup $\mathbf{U}_{\mathbb{Z}}$ of integer points of $\mathbf{U}$. Now let $\Delta$ be a finite normal extension group of $\Gamma$, i.e., $\Gamma$ is a normal subgroup $|\Delta/\Gamma| < \infty$. We can also realize the group $\Delta$ as a subgroup of a linear algebraic group (in fact, the construction realizes $\Delta$ as an arithmetic group, as follows from [34]):

**Proposition 7.8.** *There exists a $\mathbb{Q}$-defined linear algebraic group $I(\mathbf{U}, \Delta)$ with $\mathbf{U}$ its component of identity, and an embedding $\psi : \Delta \to I(\mathbf{U}, \Delta)_{\mathbb{Q}}$ which is the identity on $\Gamma$, such that $I(\mathbf{U}, \Delta) = \psi(\Delta)\mathbf{U}$ and $\psi(\Delta) \cap \mathbf{U} = \Gamma$.*

*Proof.* We may assume that $\Gamma \leq \mathbf{U}_{\mathbb{Z}}$, i.e., $\Gamma$ is contained in the subgroup $\mathbf{U}_{\mathbb{Z}}$ of integer points of $\mathbf{U}$. Let $\Delta = \Gamma r_1 \cup \cdots \cup \Gamma r_m$ be a decomposition of $\Delta$ into left cosets. By Malcev-rigidity, conjugation with $r_i$ on $\Gamma$ extends to $\mathbb{Q}$-defined rational homomorphisms $f_i$ of $\mathbf{U}$. A straightforward application of the construction used to prove Proposition 2.2 in [34] implies the result. □

In the case that the extension $\Gamma \leq \Delta$ is effective we may further refine Proposition 7.8. Let $\mathrm{Fitt}(\Delta)$ be the Fitting subgroup of $\Delta$, i.e., its maximal nilpotent normal subgroup. We show that $\mathrm{Fitt}(\Delta)$ is an f.t.n.-group and embeds into $\mathbf{U}_{\mathbb{Q}}$.

**Proposition 7.9.** *If the finite extension $\Gamma \leq \Delta$ is effective then there exists a $\mathbb{Q}$-defined linear algebraic group $I^*(\mathbf{U}, \Delta)$ with $\mathbf{U}$ its component of identity, and an embedding $\psi : \Delta \to I^*(\mathbf{U}, \Delta)_{\mathbb{Q}}$ which is the identity on $\Gamma$, such that $I^*(\mathbf{U}, \Delta) = \psi(\Delta)\mathbf{U}$ and $\psi(\Delta) \cap \mathbf{U} = \psi(\mathrm{Fitt}(\Delta))$. Moreover the centralizer of $\mathbf{U}$ in $I^*(\mathbf{U}, \Delta)$ is contained in $\mathbf{U}$.*

*Proof.* By Proposition 7.8 we may assume that $\Delta$ is a subgroup of $\mathbf{G}_{\mathbb{Q}}$, where $\mathbf{G}$ is a linear algebraic group with $\mathrm{u}(\mathbf{G}) = \mathbf{U}$, and $\Gamma \leq \mathbf{U}$. Let $\Gamma^* = \mathrm{Fitt}(\Delta)$.

Since $\Gamma$ is a subgroup of finite index in $\Gamma^*$, the group $\Gamma^*_s = \{\gamma_s \mid \gamma \in \Gamma^*\}$ is finite. Since $\Gamma^*$ is normal in $\Delta$, $\Gamma^*_s$ is normalized by $\Gamma$. The centralizer of $\Gamma^*_s$ is a Zariski-closed subgroup of $\mathbf{G}$ which contains a finite index subgroup of $\Gamma$. Therefore the centralizer $\mathrm{C}_{\mathbf{G}}(\Gamma^*_s)$ contains $\mathbf{U}$, in particular $\Gamma^*_s$ centralizes $\Gamma$. Consider the homomorphism $\psi_u : \Gamma^* \to \mathbf{U}_{\mathbb{Q}}$ given by $\gamma \mapsto \gamma_u$. Since the extension $\Gamma \leq \Gamma^*$ is effective, the homomorphism $\psi_u$ is injective. Therefore $\Gamma^*$ is an f.t.n.-group, and embeds as a subgroup of $\mathbf{U}_{\mathbb{Q}}$ containing $\Gamma$. We obtain $I^*(\mathbf{U}, \Delta)$ by applying Proposition 7.8 to the extension $\Gamma^* \leq \Delta$.

The centralizer of $\mathbf{U}$ in $I^*(\mathbf{U}, \Delta)$ splits as a direct product $\mathrm{C}_{I^*(\mathbf{U},\Delta)}(\mathbf{U}) = H \times \mathrm{Z}(\mathbf{U})$ where $H$ is a finite group of semisimple elements. It follows that the set $X_\Delta = \{\gamma \in \Delta \mid \gamma_s \in H\}$ is a normal subgroup of $\Delta$, and the map $\psi_u : \gamma \mapsto \gamma_u$ is a homomorphism on $X_\Delta$. Since the extension $\Gamma \leq \Delta$ is effective and $\Gamma \leq \mathbf{U}$, $\psi_u$ is injective on $X_\Delta$. Therefore $X_\Delta$ is nilpotent and hence contained in $\mathrm{Fitt}(\Delta)$. By construction $\mathrm{Fitt}(\Delta)$ is unipotent in $I^*(\mathbf{U}, \Delta)$ and hence no $\gamma_s \neq 1$ centralizes $\mathbf{U}$. Since $I^*(\mathbf{U}, \Delta) = \Delta\mathbf{U}$, $H$ is trivial. □

Let $c : \Delta \to \mathrm{Aut}(\Gamma)$ be the homomorphism which is induced by conjugation, and $\Gamma^r = c^{-1}(\mathrm{Inn}_{\mathbb{Q}}(\Gamma))$. Then there are finite extensions

$$\Gamma \leq \Gamma^r \leq \Delta\,.$$

Recall that the extension $\Gamma \leq \Delta$ is called radicably effective (see Definition 7.4) if and only if $\Gamma^r = \Gamma$.

**Corollary 7.10.** *Let $\Gamma$ be an* f.t.n.*-group and $\Gamma \leq \Delta$ an effective finite extension. Then $\Gamma^r = \mathrm{Fitt}(\Delta)$, and in particular, the extension $\mathrm{Fitt}(\Delta) \leq \Delta$ is radicably effective.*

*Proof.* We consider $\Gamma^r \leq I^*(\mathbf{U}, \Delta)$. Since $c(\Gamma^r) < \mathrm{Inn}_{\mathbb{Q}}(\Gamma)$ consists of unipotent automorphisms the semisimple parts of the elements of $\Gamma_r$ centralize $\mathbf{U}$. Hence, by Proposition 7.9, $\Gamma^r \leq U$ and therefore $\Gamma^r = \Delta \cap \mathbf{U} = \mathrm{Fitt}(\Delta)$.

Since $\Gamma \leq \mathrm{Fitt}(\Delta)$, the extension $\mathrm{Fitt}(\Delta) \leq \Delta$ is effective. Since $\mathrm{Fitt}(\Delta)$ is an f.t.n.-group, the first part of the corollary implies that $\mathrm{Fitt}(\Delta)^r = \mathrm{Fitt}(\Delta)$. □

**Corollary 7.11.** *Let $\Gamma$ be an* f.t.n.*-group and $\Gamma \leq \Delta$ an effective finite extension. Then $\Delta$ is a nilpotent group if and only if the extension is radicably trivial. If $\Delta$ is nilpotent then $\Delta$ is an* f.t.n.*-group.*

The following lemma shows that the class of finite effective extensions of f.t.n.-groups is closed under the operation of taking effective extensions.

**Lemma 7.12.** *Let $\Lambda$ be a finite effective extension group of the* f.t.n.*-group $\Theta$, and $\Delta$ a finite effective extension of $\Lambda$. Let $\Gamma = \mathrm{Fitt}(\Delta)$. Then $\Gamma$ is an* f.t.n.*-group and the extension $\Gamma \leq \Delta$ is effective.*

*Proof.* We may assume that $\Theta = \mathrm{Fitt}(\Lambda)$ and embed $\Delta$ as a subgroup of a linear algebraic group $I(\mathbf{U}, \Delta)$ such that $\mathbf{U} \cap \Delta = \Theta$. It follows that $\mathrm{Z}_\Delta(\Theta)_s = \mathrm{Z}_\Delta(\mathbf{U})_s$ is a finite subgroup of $I(\mathbf{U}, \Delta)$ which is normalized by the Zariski-closure $\overline{\Lambda}$ of $\Lambda$.

Since $\Lambda$ is a normal subgroup of $\Delta$, $\overline{\Lambda}$ is also normalized by $Z_\Delta(\Theta)_s$. Therefore $C = [Z_\Delta(\Theta)_s, \Lambda] \leq \overline{\Lambda} \cap Z_\Delta(\Theta)_s$. Since $C \leq \overline{\Lambda}$ centralizes $\mathbf{U}$, Proposition 7.9 implies that $C$ is unipotent. Hence $C$ is trivial, and $Z_\Delta(\Theta)_s$ centralizes $\Lambda$. Since $\Delta$ is an effective extension of $\Lambda$ the map $\gamma \mapsto \gamma_u$ is an injective homomorphism on $Z_\Delta(\Theta)$. Therefore $Z_\Delta(\Theta)$ is torsionfree, and also $\tilde{\Gamma} = Z_\Delta(\Theta)\Theta$ is a f.t.n.-group. Since the extension $\tilde{\Gamma} \leq \Delta$ is effective, $\Delta$ is a finite effective extension of an f.t.n.-group. The lemma follows. □

**Definition 7.13.** Let $\Delta$ be an effective finite extension of some f.t.n.-group. A $\mathbb{Q}$-defined linear algebraic group $\mathbf{U}_\Delta$ is called an *algebraic hull for* $\Delta$, if the following hold: $\Delta$ is a Zariski-dense subgroup of $\mathbf{U}_\Delta$, so that $\Delta \leq \mathbf{U}_{\Delta,\mathbb{Q}}$, $\mathrm{Fitt}(\Delta)$ is contained in the unipotent radical $\mathrm{u}(\mathbf{U}_\Delta)$, and $\dim \mathrm{u}(\mathbf{U}_\Delta) = \mathrm{rank}\,\mathrm{Fitt}(\Delta)$.[18]

**Corollary 7.14.** *Every $\Delta$ as above admits an algebraic hull, and the algebraic hull is unique up to $\mathbb{Q}$-isomorphism.*[19]

The existence of the algebraic hull for $\Delta$ is proved by Proposition 7.9. In fact $\mathbf{U}_\Delta = I^*(\mathbf{U}, \Delta)$, where $\mathbf{U}$ is the Malcev completion of $\mathrm{Fitt}(\Gamma)$. The uniqueness of the algebraic hull follows from the following functorial property.

**Proposition 7.15.** *Let $\mathbf{U}_\Delta$ be an algebraic hull for $\Delta$, $\mathbf{G}$ a $\mathbb{Q}$-defined linear algebraic group. Then every homomorphism $\rho : \Delta \to \mathbf{G}_\mathbb{Q}$, such that $\rho(\mathrm{Fitt}(\Delta))$ is contained in a $\mathbb{Q}$-defined unipotent subgroup $\mathbf{H}$ of $\mathbf{G}$, extends uniquely to a $\mathbb{Q}$-defined homomorphism $\rho_{\mathbf{U}_\Delta} : \mathbf{U}_\Delta \to \mathbf{G}$.*

*Proof.* Let $\Gamma = \mathrm{Fitt}(\Delta)$, and let $\mathbf{U}$ be the unipotent radical of $\mathbf{U}_\Delta$. By the Malcev extension property the induced homomorphism $\rho : \Gamma \to \mathbf{H}_\mathbb{Q}$ extends to a $\mathbb{Q}$-defined homomorphism $\rho_\mathbf{U} : \mathbf{U} \to \mathbf{H}$. Now every $g \in \mathbf{U}_\Delta$ may be written as a product $g = u\,\delta$, where $u \in \mathbf{U}$, $\delta \in \Delta$. Therefore any extension $\rho_{\mathbf{U}_\Delta}$ of $\rho$ must satisfy

$$\rho_{\mathbf{U}_\Delta}(g) = \rho_\mathbf{U}(u)\rho(\delta)\ .$$

We claim that this expression also defines a homomorphism from $\mathbf{U}_\Delta$ to $\mathbf{G}$. We have to verify that the expression is well defined on $\mathbf{U}_\Delta$: Assume therefore that $u\delta = u'\delta'$. Then $\delta(\delta')^{-1} = u'u^{-1} \in \mathbf{U} \cap \Delta$. Since $\mathbf{U} \cap \Delta = \Gamma$ it follows that $\rho_\mathbf{U}(u')\rho_\mathbf{U}(u^{-1}) = \rho(u'u^{-1}) = \rho(\delta)\rho(\delta')^{-1}$. Hence, $\rho_\mathbf{U}(u)\rho(\delta) = \rho_\mathbf{U}(u')\rho(\delta')$.

Similarly, we note that, by Zariski-denseness of $\Gamma$ in $\mathbf{U}$, for all $\delta \in \Delta$, $u \in \mathbf{U}$, the identitity

$$\rho_\mathbf{U}(\,\delta u \delta^{-1}) = \rho(\delta)\rho_\mathbf{U}(u)\rho(\delta^{-1})$$

holds. This shows that the expression for $\rho_{\mathbf{U}_\Delta}$ defines a homomorphism. We see that $\rho_{\mathbf{U}_\Delta}$ is a $\mathbb{Q}$-defined morphism on $\mathbf{U}_\Delta$, by computing $\rho_{\mathbf{U}_\Delta}$ on the product of varieties $\mathbf{U}_\Delta = \mu \times \mathbf{U}$, where $\mu \leq \mathbf{U}_\Delta$ is a finite subgroup of $\mathbf{U}_\Delta$. □

In summary, the algebraic hull for $\Delta$ satisfies the same rigidity properties as the Malcev-completion for $\Gamma$ does. The proposition implies in particular:

[18] See Definition 1.2, for generalisation in the more general context of finite extensions of polycyclic groups.

[19] In fact, every virtually polycyclic wfn-group has an algebraic hull. See [13], and Part I of this article.

**Corollary 7.16.** *Every automorphism of $\Delta$ extends to a unique $\mathbb{Q}$-defined automorphism of $\mathbf{U}_\Delta$.*

**Splitting extensions** We use the algebraic hull of $\Delta$ to show that the extension $\Gamma \leq \Delta$ splits in a finite extension. Let us first introduce some terminology.

Let $\tilde{\Delta}$ be a group, $\tilde{\Gamma} \leq \tilde{\Delta}$ a normal subgroup, and assume there is a subgroup $\mu \leq \tilde{\Delta}$ such that $\tilde{\Delta} = \tilde{\Gamma} \cdot \mu$ is a semidirect product. Let us put $F = \Delta/\Gamma$. The normal extension $\Gamma \leq \Delta$ is said to *split* (in the extension $\tilde{\Gamma} \leq \tilde{\Delta}$) if there exists an embedding $\psi : \Delta \longrightarrow \tilde{\Delta}$ and a commutative diagram of homomorphisms

$$\begin{array}{ccccccccc} 1 & \longrightarrow & \Gamma & \longrightarrow & \Delta & \longrightarrow & F & \longrightarrow & 1 \;. \\ & & \downarrow & & \downarrow{\scriptstyle \psi} & & \downarrow{\scriptstyle \mathrm{id}_F} & & \\ 1 & \longrightarrow & \tilde{\Gamma} & \longrightarrow & \tilde{\Delta} & \longrightarrow & F & \longrightarrow & 1 \end{array} \qquad (4)$$

Now consider $\Delta$ as a subgroup of its algebraic hull $\mathbf{U}_\Delta$. We recall that $\mathbf{U}_\Delta = \Delta\mathbf{U}$, where $\mathbf{U}$ is the unipotent radical of $\mathbf{U}_\Delta$. It is known (c.f. Theorem 7.23) that that there exists a finite subgroup $\mu \leq \mathbf{U}_{\Delta,\mathbb{Q}}$ such that $\mathbf{U}_{\Delta,\mathbb{Q}} = \mu \cdot \mathbf{U}_\mathbb{Q}$ is a semi-direct product. This implies, in fact, that the extension $\Gamma \leq \Delta$ splits in the extension $\mathbf{U}_\mathbb{Q} \leq \mathbf{U}_{\Delta,\mathbb{Q}}$.

**Proposition 7.17.** *There exists a finite normal extension $\Gamma_f \leq \Delta_f$, where $\Gamma_f$ is an* f.t.n.*-group, and an embedding $\Delta \leq_f \Delta_f$, such that the extension $\Gamma \leq \Delta$ splits in the extension $\Gamma_f \leq \Delta_f$.*

*Proof.* Let $\Delta = \Gamma r_1 \cup \cdots \cup \Gamma r_m$ be a decomposition of $\Delta$ in left cosets. It follows from the splitting of $\mathbf{U}_{\Delta,\mathbb{Q}}$ that $r_i = u_i g_i$ with $u_i \in \mathbf{U}_\mathbb{Q}$, $g_i \in \mu$, $i = 1, \ldots, m$. Let $\tilde{\Gamma}$ be the subgroup of $\mathbf{U}_\mathbb{Q}$ generated by the $u_i$ and $\Gamma$. Since $u_i \in \mathbf{U}_\mathbb{Q}$, $\tilde{\Gamma}$ is a finite extension of $\Gamma$. Moreover, since the finite group $\mu$ normalizes $\mathbf{U}_\mathbb{Q}$, Lemma 7.5 implies that there exists a subgroup $\Gamma_f \leq \mathbf{U}_\mathbb{Q}$ which is a finite extension of $\Gamma$ and which is normalized by $\mu$. A finitely generated subgroup of $\mathbf{U}_\mathbb{Q}$ is an f.t.n.-group, hence $\Gamma_f$ is an f.t.n.-group. Let $\Delta_f = \Gamma_f\,\mu$ then it follows from the construction that $\Delta$ is contained in $\Delta_f$. □

The splitting group $\Delta_f$ inherits the functorial properties of $\mathbf{U}_\Delta$.

**Proposition 7.18.** *Let $\rho : \Delta \to \mathbf{G}_\mathbb{Q}$ be a homomorphism of $\Delta$ to a $\mathbb{Q}$-defined linear algebraic group $\mathbf{G}$ such that $\rho(\Gamma)$ is contained in $\mathbf{H}_\mathbb{Q}$, where $\mathbf{H}$ is a $\mathbb{Q}$-defined unipotent subgroup of $\mathbf{G}$. Then $\rho$ extends to a homomorphism $\rho_f : \Delta_f \to \mathbf{G}_\mathbb{Q}$.*

*Proof.* By Proposition 7.15, $\rho$ extends to a homomorphism $\rho_{\mathbf{U}_\Delta} : \mathbf{U}_\Delta \to \mathbf{G}$. Now by construction of $\Delta_f$ we have inclusions $\Delta \leq \Delta_f \leq \mathbf{U}_\Delta$, and the restriction of $\rho_{\mathbf{U}_\Delta}$ to $\Delta_f$ induces the required extension $\rho_f$. □

## Realization as affine crystallographic group

We are now dealing with the proof of Theorem 7.3. Let $\Delta$, $\Gamma$ satisfy the assumptions of Theorem 7.3, i.e., $\Gamma$ is an f.t.n.-group, $\Gamma \leq_f \Delta$ is an effective normal extension with associated kernel $\beta$. We assume that $\Gamma$ is isomorphic to an ACG of type $A$.

Let us first consider the special case, where the extension $\Gamma \leq \Delta$ is radicably trivial. Proposition 11.1 implies that, in this case, the kernel of the extension acts trivially on $\mathcal{D}_c(\Gamma, A)$. Also, by Proposition 7.9, $\Delta$ is an f.t.n.-group. Therefore let us consider now a (not necessarily normal) finite extension $\Gamma \leq_f \Gamma^*$, where $\Gamma^*$ is an f.t.n.-group. The following result is an immediate consequence of Proposition 7.7.

**Proposition 7.19.** *Every $\rho \in \mathrm{Hom}_c(\Gamma, A)$ has an extension $\rho_{\Gamma^*} \in \mathrm{Hom}_c(\Gamma^*, A)$, and this extension is unique.*

Now let us consider a tower of finite extensions $\Gamma \leq \Gamma^* \leq \Delta$, where the extensions $\Gamma \leq \Delta$, $\Gamma^* \leq \Delta$ are normal. For later purposes we note:

**Lemma 7.20.** *If $[\rho] \in \mathcal{D}_c(\Gamma, A)$ is a fixed point for $\Delta/\Gamma$ then $[\rho_{\Gamma^*}] \in \mathcal{D}_c(\Gamma^*, A)$ is a fixed point for $\Delta/\Gamma^*$.*

*Proof.* Let $c_g : \Gamma^* \to \Gamma^*$ denote conjugation with $g \in \Delta$. Since $[\rho] \in \mathcal{D}_c(\Gamma, A)$ is a fixed point for $\Delta/\Gamma$ there exists $a \in A$, such that, for all $\gamma \in \Gamma$,

$$\rho_{\Gamma^*}(c_g(\gamma)) = \rho\,(c_g(\gamma)) = \rho^a(\gamma) = \rho^a_{\Gamma^*}(\gamma)\,.$$

By the uniqueness of extensions we must have $\rho_{\Gamma^*} \circ c_g = \rho^a_{\Gamma^*}$. Therefore $[\rho_{\Gamma^*}]$ is a fixed point for $\Delta/\Gamma^*$. □

We turn now to the realization of radicably effective extensions. We start by considering split extensions $\Delta = F \cdot \Gamma$, where $F \leq \Delta$ is a finite subgroup with $F \cap \Gamma = \{1\}$. Recall from Section 11, Proposition 11.6, that there exists a natural homomorphism $c_\rho : \mathrm{N}_A(\rho(\Gamma)) \longrightarrow \mathrm{Aut}(\Gamma)_{[\rho]}$ which is onto.

**Proposition 7.21.** *Let $F \leq \mathrm{Aut}(\Gamma)$ be a finite subgroup. If $[\rho] \in \mathcal{D}_c(\Gamma, A)$ is a fixed point for $F$, then there exists an embedding*

$$\iota : F \longrightarrow \mathrm{N}_A(\,\rho(\Gamma))$$

*which satisfies $c_\rho(\,\iota(g)) = g$, for all $g \in F$. Any two embeddings $\iota, \iota' : F \to \mathrm{N}_A(\,\rho(\Gamma))$ with the latter property are conjugate by an element of $\mathrm{C}_A(\overline{\rho(\Gamma)})$.*

*Proof.* We will identify $\mathrm{Aut}(\Gamma)$ with a subgroup of $\mathrm{Aut}(U_\Gamma)$, where $U_\Gamma$ is a real hull of $\Gamma$, as in Section 11. From the assumption that $[\rho]$ is a fixed point we deduce that $F \leq \mathrm{Aut}(\Gamma)_{[\rho]}$. We are going to apply some of the facts which are collected in Section 11.

The group $U = \overline{\rho(\Gamma)}$ is a simply transitive subgroup of $A$. By Proposition 11.5 and Proposition 11.6, the map $c : N_A(U) \to \mathrm{Aut}(U_\Gamma)_{[\rho]}$ is onto with kernel $\mathrm{C}_A(U)$, and restricts to a surjective map $c_\rho : N_A(\rho(\Gamma)) \to \mathrm{Aut}(\Gamma)_{[\rho]}$. Now let $H = c^{-1}(F)$. $H$ is a Zariski-closed subgroup of $N_A(U)$. Since $F \leq \mathrm{Aut}(\Gamma)_{[\rho]}$ it follows that $H \leq \mathrm{N}_A(\rho(\Gamma))$. $H$ contains $\mathrm{C}_A(U)$ as a normal subgroup of finite index. In fact, $H/\mathrm{C}_A(U)$ is isomorphic to $F$. Since $\mathrm{C}_A(U)$ is unipotent, $\mathrm{C}_A(U)$ is the unipotent radical of $H$. Now by splitting of algebraic groups there exists a subgroup $\mu \leq H$ such that $H = \mu \cdot \mathrm{C}_A(U)$. Since the restriction of $c$ to $\mu$ is an isomorphism onto $F$, we can set $\iota = c^{-1}$. We thus proved the existence of $\iota$.

We prove now the conjugacy statement. Let us remark first that every homomorphism $\iota : F \to \mathrm{N}_A(\rho(\Gamma))$ which satisfies the assumption maps $F$ into the group $H$. Moreover $\iota$ is uniquely determined by its image $\iota(F) \leq H$. Since both $\iota(F)$ and $\iota'(F)$ are Levi-subgroups in $H$, $\iota'(F)$ is conjugated to $\iota(F)$ by an element of $\mathrm{C}_A(U)$. Hence also the homomorphisms $\iota$ and $\iota'$ are conjugate. □

**Corollary 7.22.** *Let $F \leq \mathrm{Aut}(\Gamma)$ be a finite subgroup, and $\Delta = F \cdot \Gamma$ be the corresponding split extension of $\Gamma$. Then $\rho \in \mathrm{Hom}_c(\Gamma, A)$ extends to a homomorphism $\rho_\Delta \in \mathrm{Hom}_c(\Delta, A)$, if and only if $[\rho] \in \mathcal{D}_c(\Gamma, A)$ is a fixed point for $F$. Any two extensions $\rho_\Delta, \rho'_\Delta \in \mathrm{Hom}_c(\Delta, A)$ of $\rho$ are conjugate by an element of $\mathrm{C}_A(\overline{\rho(\Gamma)})$.*

*Proof.* By Proposition 7.21 there exists an embedding $\iota : F \to \mathrm{N}_A(\rho(\Gamma))$ such that $c_\rho(\iota(g)) = g$, for all $g \in F$. For $g \in F$, $\gamma \in \Gamma$ we can define

$$\rho_\Delta(g\gamma) = \iota(g)\rho(\gamma) ,$$

to get the required extension $\rho_\Delta$. Now, if $\rho_\Delta$ is any extension of $\rho$ to $\Delta$, then $\iota(g) = \rho_\Delta(g)$ defines a homomorphism $\iota : F \to \mathrm{N}_A(\rho(\Gamma))$ which satisfies $c_\rho(\iota(g)) = g$, for all $g \in F$. Therefore the conjugacy statement of Proposition 7.21 implies that any two extensions of $\rho$ to $\Delta$ are conjugate by an element $u \in \mathrm{C}_A(\overline{\rho(\Gamma)})$. □

We are now ready to finish the proof of Theorem 7.3.

*Proof of Theorem 7.3* Let $\rho \in \mathrm{Hom}_c(\Gamma, A)$ be such that $[\rho] \in \mathcal{D}_c(\Gamma, A)$ is a fixed point for $\beta$. Let $\Gamma^* = \mathrm{Fitt}(\Delta)$. By Proposition 7.19, $\rho$ extends to a unique homomorphism $\tilde{\rho} \in \mathrm{Hom}_c(\Gamma^*, A)$. By Lemma 7.20, $[\tilde{\rho}] \in \mathcal{D}_c(\Gamma^*, A)$ is a fixed point for the kernel of the extension $\Gamma^* \leq \Delta$.

So we may assume now that $\Gamma = \mathrm{Fitt}(\Delta)$, or equivalently that the extension $\Gamma \leq \Delta$ is radicably effective. Let $F = \Delta/\Gamma$. We apply Proposition 7.17 to embed $\Delta$ in an extension group $\Delta_f = \mu \cdot \Gamma_f$ which splits. $\Gamma$ embeds as a subgroup of finite index in the f.t.n.-group $\Gamma_f$, and $\mu$ is isomorphic to $F$. By Proposition 7.19, $\rho$ uniquely extends to $\rho_f \in \mathrm{Hom}_c(\Gamma_f, A)$. We remark that, by a similar argument as given in the proof of Lemma 7.20, $[\rho_f] \in \mathcal{D}_c(\Gamma_f, A)$ is a fixed point for $\mu$. Moreover, since $\Gamma \leq \Delta$ is radicably effective, the extension $\Gamma_f \leq \Delta_f$ is a (radicably) effective extension. So we may view $\mu$ as a subgroup of $\mathrm{Aut}(\Gamma_f)$. By Corollary 7.22, the affine crystallographic representation of $\Gamma_f$ extends to $\Delta_f$. Since $\Delta$ is a subgroup of $\Delta_f$ we have, in particular, extended the original representation $\rho$ of $\Gamma$ to a representation $\rho_\Delta$ of $\Delta$. □

**Appendix: Splitting of algebraic groups** In the preceding proofs we made crucial use of the existence of Levi subgroups in linear algebraic groups and the fact that the Levi subgroups are all conjugated, a result which is originally due to Mostow. Let us recall the precise statement of the result, cf. [16][5.1]:

**Theorem 7.23.** *Let* $\mathbf{G}$ *be a* $\mathbb{Q}$*-defined linear algebraic group. Then there exists a reductive* $\mathbb{Q}$*-defined subgroup* $\mathbf{L}$ *of* $\mathbf{G}$ *such that* $\mathbf{G} = \mathbf{L}\mathrm{u}(\mathbf{G})$ *and* $\mathbf{L} \cap \mathrm{u}(\mathbf{G}) = \{1\}$. *Any two such subgroups* $\mathbf{L}$ *and* $\mathbf{L}'$ *are conjugate by an element of* $\mathrm{u}(\mathbf{G})_{\mathbb{Q}}$. *Moreover, each* $\mathbb{Q}$*-defined reductive subgroup* $\mathbf{H}$ *of* $\mathbf{G}$ *is conjugate to a subgroup of* $L$ *by an element of* $\mathrm{u}(\mathbf{G})_{\mathbb{Q}}$.

From the theorem corresponding rational splitting and conjugacy statements are easily deduced for any subfield $k \leq \mathbb{C}$. In fact, by the theorem, the variety $\mathbf{G}$ is a product of varieties $\mathbf{L}$ and $\mathrm{u}(\mathbf{G})$, where $\mathbf{L}$ is isomorphic to $\mathbf{G}/\mathrm{u}(\mathbf{G})$. The same statement holds then for the rational points of these varieties. In particular we deduce that $(\mathbf{G}/\mathrm{u}(\mathbf{G}))_{\mathbb{Q}} = \mathbf{G}_{\mathbb{Q}}/\mathrm{u}(\mathbf{G})_{\mathbb{Q}}$, and

$$\mathbf{G}_{\mathbb{Q}} = \mathbf{L}_{\mathbb{Q}}\mathrm{u}(\mathbf{G})_{\mathbb{Q}} \ , \text{ with } \mathbf{L}_{\mathbb{Q}} \cap \mathrm{u}(\mathbf{G})_{\mathbb{Q}} = \{1\} \ .$$

The application of this result replaces in our setting the explicit use of *cohomological reasoning* which is pervasive in most treatments of the classical Bieberbach theory.

# 8. Deformation spaces of finite extensions

Let $\Gamma$ be an f.t.n.-group which is isomorphic to an ACG of type $A$. The extension results of the previous section are particular well suited to understand the deformation spaces of the finite extensions of $\Gamma$. We continue to use the notational conventions of the previous section.

Let $\beta : F \longrightarrow \mathrm{Out}(\Gamma)$ be an abstract kernel for $\Gamma$ which is realized by a finite extension $\Gamma \leq_f \Delta$. The restriction map $r_\Gamma : \mathrm{Hom}_c(\Delta, A) \longrightarrow \mathrm{Hom}_c(\Gamma, A)$ factorizes to a continuous map

$$\bar{r}_\Gamma : \mathcal{D}_c(\Delta, A) \longrightarrow \mathcal{D}_c(\Gamma, A)$$

on the deformation spaces. We denote the set of fixed points for the kernel $\beta$ on $\mathcal{D}_c(\Gamma, A)$ with $\mathcal{D}_c(\Gamma, A)^F$, and we put $\mathrm{Hom}_c(\Gamma, A)^F$ for its preimage in $\mathrm{Hom}_c(\Gamma, A)$. We topologize $\mathcal{D}_c(\Gamma, A)^F$ with the induced subspace topology from $\mathcal{D}_c(\Gamma, A)$. It follows from Proposition 7.1 that the image of $\bar{r}_\Gamma$ is contained in $\mathcal{D}_c(\Gamma, A)^F$. The following is then a commutative diagram of (continuous) maps:

$$\begin{array}{ccc} \mathrm{Hom}_c(\Delta, A) & \xrightarrow{r_\Gamma} & \mathrm{Hom}_c(\Gamma, A)^F \\ \downarrow & & \downarrow \\ \mathcal{D}_c(\Delta, A) & \xrightarrow{\bar{r}_\Gamma} & \mathcal{D}_c(\Gamma, A)^F \end{array} .$$

We apply the results of the previous section to study the properties of the map $\bar{r}_\Gamma$. Let $\Gamma$ be an f.t.n.-group and $\mathbf{U}$ its Malcev completion. Recall from Section 11 that $\mathrm{Out}(\mathbf{U}_{\mathbb{R}})$ acts naturally on $\mathcal{D}_c(\Gamma, A)$. We consider first the case of commensurable f.t.n.-groups. Their deformation spaces are essentially identical.

**Proposition 8.1.** *Let* $\Gamma$ *and* $\Gamma'$ *be commensurable* f.t.n.*-groups. Then there exists an* $\mathrm{Out}(\mathbf{U}_{\mathbb{R}})$*-equivariant homeomorphism* $h : \mathcal{D}_c(\Gamma, A) \xrightarrow{\approx} \mathcal{D}_c(\Gamma', A)$.

*Proof.* $\Gamma$ and $\Gamma'$ contain a common subgroup of finite index which is an f.t.n.-group. It is therefore enough to prove the statement in the case of a finite extension $\Gamma \leq \Gamma'$. Now by Proposition 7.19 and Proposition 7.7 the restriction map from $\Gamma'$ to $\Gamma$ is a homeomorphism from $\mathrm{Hom}_c(\Gamma', A)$ to $\mathrm{Hom}_c(\Gamma, A)$. This homeomorphism is easily seen to be compatible with the actions of $\mathrm{Aut}(\mathbf{U}_{\mathbb{R}})$ on these spaces. □

However, this phenomenon is not restricted to commensurability. More generally the following holds:

**Proposition 8.2.** *Let* $\Gamma$ *and* $\Gamma'$ *be lattices in the real Lie group* $\mathbf{U}_{\mathbb{R}}$*. Then there exists an* $\mathrm{Out}(\mathbf{U}_{\mathbb{R}})$*-equivariant homeomorphism* $h : \mathcal{D}_c(\Gamma, A) \xrightarrow{\approx} \mathcal{D}_c(\Gamma', A)$.

We leave the proof to the reader. In fact, Theorem 11.3 implies that the properties of the deformation space $\mathcal{D}_c(\Delta, A)$ depend only on the algebraic hull $\mathbf{U}_\Delta$ of $\Delta$.

**Corollary 8.3.** *Let* $\Gamma$ *be an* f.t.n.*-group and* $\Gamma \leq_f \Delta$*, where* $\Gamma$ *is normal in* $\Delta$*, be a finite, effective extension,* $F = \Delta / \Gamma$*. Then there is a continuous bijection*

$$\bar{r}_\Gamma : \mathcal{D}_c(\Delta, A) \longrightarrow \mathcal{D}_c(\Gamma, A)^F .$$

*Remark* We do not expect in general that the map $\bar{r}_\Gamma$ will be a homeomorphism. Instead, we expect that there is a stratification of $\mathcal{D}_c(\Delta, A)$ such that $\bar{r}_\Gamma$ will induce a homeomorphism on the strata.

Before we give the proof of the corollary, let us exemplify the previous remark by considering a particular important stratum in the deformation space. Recall from Section 11, that $\rho \in \mathrm{Hom}_c(\Gamma, A)$ is called *symmetric* if the centralizer of $\rho(\Gamma)$ in $A$ acts transitively on $V$. Let $\mathrm{Hom}_c(\Gamma, A)_s$ denote the set of all symmetric crystallographic homomorphisms of $\Gamma$. $\mathrm{Hom}_c(\Gamma, A)_s$ is a closed $A$-invariant subspace of $\mathrm{Hom}_c(\Gamma, A)$, and the deformation space $\mathcal{D}_c(\Gamma, A)_s$ embeds as a closed subspace into $\mathcal{D}_c(\Gamma, A)$. Let us denote $\mathcal{D}_c(\Delta, A)_s = \bar{r}_\Gamma^{-1}(\mathcal{D}_c(\Gamma, A)_s)$.

**Theorem 8.4.** *The map* $\bar{r}_\Gamma$ *induces a homeomorphism*

$$\mathcal{D}_c(\Delta, A)_s \xrightarrow{\approx} \mathcal{D}_c(\Gamma, A)_s^F .$$

*Proof of Corollary 8.3* We want to show that the map $\bar{r}_\Gamma : \mathcal{D}_c(\Delta, A) \longrightarrow \mathcal{D}_c(\Gamma, A)^F$ is a bijection. Now $\Delta$ is a subgroup of $\Delta_f$, where $\Delta_f = \Gamma_f \cdot \tilde{F}$ is as in Proposition 7.17, and it is shown in the proof of Theorem 7.3 that there is a commutative diagram of restriction maps

$$\begin{array}{ccc} \mathcal{D}_c(\Gamma, A)^F & \xleftarrow[\bar{r}_\Gamma]{} & \mathcal{D}_c(\Delta, A) \\ \cong \uparrow & & \uparrow \\ \mathcal{D}_c(\Gamma_f, A)^{\tilde{F}} & \xleftarrow[\bar{r}_{\Gamma_f}]{\cong} & \mathcal{D}_c(\Delta_f, A) \end{array} .$$

By Proposition 7.19 the left upgoing arrow is a bijection. The map $\bar{r}_{\Gamma_f}$ is a bijection by the conjugacy statement of Corollary 7.22, and the right upgoing arrow is onto by Proposition 7.18. It follows that $\bar{r}_\Gamma$ is a bijection. □

For the proof of the theorem we need a lemma.

**Lemma 8.5.** *Let $\tilde{F} \leq \mathrm{Aut}(\Gamma_f)$ be a finite subgroup, and $\Delta_f = \tilde{F} \cdot \Gamma_f$ be the corresponding split extension. Then the restriction map $r_{\Gamma_f} : \mathrm{Hom}_c(\Delta_f, A) \to \mathrm{Hom}_c(\Gamma_f, A)$ admits a (semi-algebraic) continuous cross section $s : \mathrm{Hom}_c(\Gamma_f, A)^F \to \mathrm{Hom}_c(\Delta_f, A)$.*

*Proof.* By Proposition 11.12, there exists a continuous map $s_x : \mathrm{Hom}_c(\Gamma, A)_s^F \to \mathrm{Hom}(F, A_x)$ such that $c_\rho(s_x(\rho, g)) = g$, for all $g \in F$. Let $\rho \in \mathrm{Hom}_c(\Gamma, A)_s^F$. By Corollary 7.22, there exists a corresponding $\rho_{\Delta_f} = s(\rho) \in \mathrm{Hom}_c(\Delta_f, A)$ which restricts to $\rho$. □

*Proof of Theorem 8.4* We consider again the diagram in the proof of Corollary 8.3. By Proposition 7.7 the restriction map from $\mathrm{Hom}_c(\Gamma_f, A)$ to $\mathrm{Hom}_c(\Gamma, A)$ is a homeomorphism, and consequently the induced map from $\mathcal{D}_c(\Gamma_f, A)_s$ to $\mathcal{D}_c(\Gamma, A)_s$ is a homeomorphism. The same is true for the bijection from $\mathcal{D}_c(\Delta_f, A)$ to $\mathcal{D}_c(\Delta, A)$. Now it follows from the lemma that $\bar{r}_{\Gamma_f} : \mathcal{D}_c(\Delta_f, A)_s \to \mathcal{D}_c(\Gamma_f, A)$ is a homeomorphism. □

## Some applications

**Moduli spaces of finite extensions** Let $\Gamma$ be an f.t.n.-group. Recall that the moduli space $\mathcal{M}_c(\Gamma, A)$ is just the space $\mathcal{D}_c(\Gamma, A)/\mathrm{Out}(\Gamma)$ with the quotient topology. Let $\Delta$ be a finite effective extension group of $\Gamma$. There is a natural induced restriction map

$$\tilde{r} : \mathcal{M}_c(\Delta, A) \longrightarrow \mathcal{M}_c(\Gamma, A)$$

of moduli spaces. In general, this map is not injective and does not give a precise picture for the space $\mathcal{M}_c(\Delta, A)$.

We give now a description of $\mathcal{M}_c(\Delta, A)$ in the case that the extension is radicably effective, i.e., we assume that $\Gamma = \mathrm{Fitt}(\Delta)$. We may view $F = \Delta/\Gamma$ as a subgroup of $\mathrm{Out}(\Gamma)$. Since $\Gamma$ is an invariant normal subgroup of $\Delta$, there is a natural homomorphism

$$\bar{\nu} : \mathrm{Out}(\Delta) \longrightarrow \mathrm{N}_{\mathrm{Out}(\Gamma)}(F)/F$$

which comes associated with the extension. Note that the group $\mathrm{N}_{\mathrm{Out}(\Gamma)}(F)/F$ acts as a group of homeomorphisms on $\mathcal{D}_c(\Gamma, A)^F$. In fact, this action desribes the action of $\mathrm{Out}(\Delta)$ on $\mathcal{D}_c(\Delta, A)$, and it is possible to recover $\mathcal{M}_c(\Delta, A)$ from $\mathcal{D}_c(\Gamma, A)$ and the action of $F \leq \mathrm{Out}(\Gamma)$.

**Theorem 8.6.** *Let $\Delta$ be an effective finite extension of the* f.t.n.*-group $\Gamma$, and assume that $\Gamma = \mathrm{Fitt}(\Delta)$. Let $\mathrm{Out}(\Delta)$ act on $\mathcal{D}_c(\Delta, A)$ with respect to the homomorphism $\nu$. Then the embedding $\bar{r}_\Gamma : \mathcal{D}_c(\Delta, A) \longrightarrow \mathcal{D}_c(\Gamma, A)^F$ is $\mathrm{Out}(\Delta)$-equivariant.*

*In particular, there is a continuous bijection from the moduli space $\mathcal{M}_c(\Delta, A)$ to the quotient space $\mathcal{D}_c(\Gamma, A)^F / \bar{\nu}(\mathrm{Out}(\Delta))$. The bijection induces a homeomorphism on the moduli space $\mathcal{M}_c(\Delta, A)_s$ of symmetric crystallographic homomorphisms to $\mathcal{D}_c(\Gamma, A)^F_s / \bar{\nu}(\mathrm{Out}(\Delta))$.*

*Proof.* That the map $\bar{r}_\Gamma$ is $\mathrm{Out}(\Delta)$-equivariant is a routine calculation from the definitions. We omit it therefore. The remaining statements follow then from the previous results on deformation spaces. □

**Inheritance of the Hausdorff-property** Though the topology on $\mathcal{D}_c(\Delta, A)$ is potentially finer than the topology induced from $\mathcal{D}_c(\mathrm{Fitt}(\Delta), A)^F$ we still can deduce an immediate useful consequence from Corollary 8.3.

**Corollary 8.7.** *Let $\Delta$ be a virtually nilpotent ACG, and let $\Gamma = \mathrm{Fitt}(\Delta)$ be the Fitting subgroup. If $\mathcal{D}_c(\Gamma, A)$ is a Hausdorff topological space, then $\mathcal{D}_c(\Delta, A)$ is a Hausdorff space too.*

**Inheritance of convexity to finite extensions** If $\Delta$ is a virtually nilpotent ACG then the Fitting subgroup $\Gamma = \mathrm{Fitt}(\Delta)$ is a crystallographic f.t.n.-group, and is also a characteristic subgroup of $\Delta$. Therefore there is a natural restriction homomorphism

$$\nu : \mathrm{Aut}(\Delta) \to \mathrm{Aut}(\Gamma) ,$$

and, evidently, the map $r_\Gamma : \mathrm{Hom}_c(\Delta, A) \to \mathrm{Hom}_c(\Gamma, A)$ is equivariant with respect to $\nu$. It follows

**Proposition 8.8.** *Let $\Delta$ be virtually nilpotent ACG, and let $\Gamma = \mathrm{Fitt}(\Delta)$. Then the induced inclusion on deformation spaces*

$$\mathcal{D}_c(\Delta, A) \longrightarrow \mathcal{D}_c(\Gamma, A)$$

*is equivariant with respect to the actions of $\mathrm{Aut}(\Delta)$ and $\mathrm{Aut}(\Gamma)$.*

As a particular consequence we see that convexity properties of $\mathcal{D}_c(\mathrm{Fitt}(\Delta), A)$ will be inherited to $\Delta$.

**Theorem 8.9.** *Let $\Delta$ be a virtually nilpotent ACG, and let $\Gamma = \mathrm{Fitt}(\Delta)$. Then the following hold:*

i) *If $\mathcal{D}_c(\Gamma, A)$ is fixed pointed then $\mathcal{D}_c(\Delta, A)$ is fixed pointed too.*
ii) *If $\mathcal{D}_c(\Gamma, A)$ is convex then $\mathcal{D}_c(\Delta, A)$ is convex too.*

*Proof.* Let $F = \Delta/\Gamma$. If $\mathcal{D}_c(\Gamma, A)$ has a fixed point $[\rho_o]$ for $\mathrm{Aut}(\Gamma)$ then, by Corollary 8.3, $\bar{r}_\Gamma(\mathcal{D}_c(\Delta, A))$ meets $[\rho_o]$. By equivariance and the injectivity of the map $\bar{r}_\Gamma$, $[\rho_o]$ is a fixed point for $\mathrm{Aut}(\Delta)$ on $\mathcal{D}_c(\Delta, A)$.

We assume now that $\mathcal{D}_c(\Gamma, A)$ is convex. The restriction homomorphism $\nu$ induces a map

$$\bar{\nu} : \mathrm{Out}(\Delta) \longrightarrow \mathrm{N}_{\mathrm{Out}(\Gamma)}(F)/F .$$

Let $\mu$ be a finite subgroup of $\mathrm{Out}(\Delta)$, and let $\tilde{\mu} \leq N_{\mathrm{Out}(\Gamma)}(F)$ be the preimage of $\bar{\nu}(\mu)$. Since $F$ is finite so is $\tilde{\mu}$ which is a finite normal extension of $F$. If $[\rho_o] \in \mathcal{D}_c(\Gamma, A)$ is a fixed point for $\tilde{\mu}$ then, since $F \leq \tilde{\mu}$, $\bar{r}_\Gamma(\mathcal{D}_c(\Delta, A))$ meets $[\rho_o]$. By the equivariance properties of the embedding of $\mathcal{D}_c(\Delta, A)$ into $\mathcal{D}_c(\Gamma, A)$, $[\rho_o]$ is a fixed point for $\mu$ too. □

**The realization theorem** We now come to the proof of the realization theorem for finite extensions of a virtually nilpotent affine crystallographic group:

*Proof of Theorem 6.4* Since the virtually nilpotent group $\Lambda$ is isomorphic to an affine crystallographic group, $\Theta = \mathrm{Fitt}(\Lambda)$ is an f.t.n.-group, and the centralizer of $\Theta$ is contained in $\Theta$. Hence the extension $\Theta \leq \Lambda$ is effective. It follows from Lemma 7.12 that $\Delta$ is an effective extension of $\Gamma = \mathrm{Fitt}(\Delta)$, and also that $\Gamma = \mathrm{Fitt}(\Delta)$ is an f.t.n.-group. Since we already proved the realization theorem for effective extensions of f.t.n.-groups, to show that $\Delta$ is isomorphic to a crystallographic group, it is enough to show that the kernel $F \leq \mathrm{Out}(\Gamma)$ which is associated to the extension $\Gamma \leq \Delta$ has a fixed point in $\mathcal{D}_c(\Gamma, A)$.

We remark that $\Theta = \mathrm{Fitt}(\Lambda)$ has finite index in $\Gamma = \mathrm{Fitt}(\Delta)$. Therefore (compare Proposition 8.1) $\mathrm{Out}(\Theta)$ and the kernel $F_\Theta \leq \mathrm{Out}(\Theta)$ which is associated to $\Theta \leq \Lambda$ act also on $\mathcal{D}_c(\Gamma, A)$, and the set of fixed points of $F_\Theta$ corresponds to $\mathcal{D}_c(\Lambda, A)$. The action of $\mathrm{Out}(\Theta)$ on $\mathcal{D}_c(\Gamma, A)$ factorizes over $\mathrm{Out}(\mathbf{U}_\mathbb{R})$. The action of $\mathrm{Out}(\Lambda)$ on $\mathcal{D}_c(\Lambda, A) \subseteq \mathcal{D}_c(\Gamma, A)$ is then encoded in the homomorphism

$$\bar{\nu} : \mathrm{Out}(\Lambda) \longrightarrow \mathrm{N}_{\mathrm{Out}(\mathbf{U}_\mathbb{R})}(F_\Theta)/ F_\Theta \ .$$

Let $\mu \leq \mathrm{Out}(\Lambda)$ be the kernel which is associated to the extension $\Lambda \leq \Delta$, and let $\tilde{\mu} \leq \mathrm{Out}(\mathbf{U}_\mathbb{R})$ be the preimage of $\bar{\nu}(\mu)$ in $\mathrm{N}_{\mathrm{Out}(\mathbf{U}_\mathbb{R})}(F_\Theta)$. By our assumption, $\mu$ has a fixed point in $\mathcal{D}_c(\Lambda, A)$ and this implies that $\tilde{\mu}$ has a fixed point in $\mathcal{D}_c(\Gamma, A)$. It is easy to see that the kernel $F \leq \mathrm{Out}(\mathbf{U}_\mathbb{R})$ is contained in $\tilde{\mu}$. Therefore $F$ has a fixed point in $\mathcal{D}_c(\Gamma, A)$. □

## Manifolds of Euclidean type

Let $M$ be a closed manifold which admits a Riemannian metric of constant curvature zero, i.e., a flat Riemannian metric. In the spirit of the theory of geometrization, $M$ is called a *manifold of Euclidean type*. By Bieberbach theory $M$ is finitely covered by a torus $T^n$. The fundamental group $\pi_1(M)$ is isomorphic to a finite extension of $\mathbb{Z}^n$. It was remarked previously in [10] that the deformation space $\mathcal{D}_c(T^n, \mathrm{Aff}(\mathbb{R}^n))$ of affine strucutures on $T^n$ is a Hausdorff-space and homeomorphic to a semi-algebraic set. We obtain the following generalization of this result:

**Corollary 8.10.** *Let $M$ be a closed manifold of Euclidean type. Then the deformation space of complete affine structures of type $A$ is a Hausdorff space and homeomorphic to a semi-algebraic set.*

*Proof.* First let us remark that, by Theorem 4.1, $\mathcal{D}_c(\mathbb{Z}^n, A)$ is a semi-algebraic set. Now $\pi_1(M)$ is a normal extension of $\mathbb{Z}^n$ with some finite group $F$. We remark next that the crystallographic actions of $\mathbb{Z}^n$ are all symmetric. By Corollary 8.4 it follows that $\mathcal{D}_c(M, A)$ is homeomorphic to $\mathcal{D}_c(\mathbb{Z}^n, A)^F$. Therefore $\mathcal{D}_c(M, A)$ is homeomorphic to a closed subspace of $\mathcal{D}_c(\mathbb{Z}^n, A)$, and also it is Hausdorff. Since the action of $\mathrm{GL}(n, \mathbb{Z})$ on $\mathcal{D}_c(\mathbb{Z}^n, A)$ is algebraic $\mathcal{D}_c(\mathbb{Z}^n, A)^F$ is a semi-algebraic set. □

**The Burckhardt-Zassenhaus theorem** We let $\mathrm{GL}_A$ act on $\mathrm{GL}(V)$ by left multiplication and put

$$X_A = {}_{\mathrm{GL}_A}\backslash \mathrm{GL}(V)\,.$$

Let $F = G_{st}(V, A)$ be the algebraic variety of simply transitive abelian subgroups of $A$. $\mathrm{GL}(A)$ acts by conjugation on $F$. It was proved in Proposition 4.3 that

$$\mathcal{D}_c(\mathbb{Z}^n, A) = \mathrm{GL}(V) \underset{\mathrm{GL}_A}{\times} F$$

is a fiber product, and in particular a bundle over the homogeneous space $X_A$ with fiber $F$. The $\mathrm{GL}(V)$-action on $\mathcal{D}_c(\mathbb{Z}^n, A)$ corresponds to the natural $\mathrm{GL}(V)$-action on the fiber product which is induced by right multiplication of $\mathrm{GL}(V)$ on itself.

If $A$ contains the subgroup $V$ of translations we note that $\mathrm{GL}_A$ fixes $V$ as an element of $G_{st}(V, A)$. Therefore, provided $A$ contains the subgroup $V$ of translations, $\mu \leq \mathrm{GL}(V)$ fixes a point in $\mathcal{D}_c(\mathbb{Z}^n, A)$ if and only if it fixes a point on $X_A$. It follows that, in this case, the existence problem for fixed points in the deformation spaces of tori is only a matter of the action of $\mathrm{GL}(V)$ on $X_A$. In particular we recover from Theorem 7.3 the Burckhardt-Zassenhaus theorem (see [82]) on Euclidean crystallographic groups:

**Corollary 8.11.** *Every finite effective extension of $\mathbb{Z}^n$ acts as an Euclidean crystallographic group on affine space $\mathbb{A}^n$.*

*Proof.* Every finite subgroup $\mu$ of $\mathrm{GL}_n$ has a fixed point in $X = O_n \backslash \mathrm{GL}_n$, as is well known. Since $G_{st}(V, \mathrm{E}(n))$ contains the subgroup of translations, $\mu$ has a fixed point on $\mathcal{D}_c(\mathbb{Z}^n, E(n))$. □

In general, it seems to be a nontrivial problem to give a precise description of the variety $G_{st}(V, A)$, and therefore also of the deformation spaces for complete affine structures on manifolds of Euclidean type. Next we cover some accessible special cases.

**Complete affine surfaces** It is well known that a closed affine surface is diffeomorphic to the two-torus or the Klein bottle. In [9] the following result is proved:

**Proposition 8.12.** *The deformation space $\mathcal{D}_c(T^2, \mathrm{Aff}(\mathbb{R}^2))$ of the two-torus is homeomorphic to $\mathbb{R}^2$. In these coordinates the action of $\mathrm{GL}_2(\mathbb{Z})$ corresponds to the natural linear action. The fixed point $0 \in \mathbb{R}^2$ represents the flat Euclidean structure on $T^2$.*

It is easy to see from [9] that also the natural $GL_2(\mathbb{R})$-action is linear in the above model for $\mathcal{D}_c(T^2, \mathrm{Aff}(\mathbb{R}^2))$. In a sense, our model encodes all information on the affine crystallographic groups in dimension two. We apply our method to study the two dimensional affine crystallographic groups up to affine conjugacy: Recall that there exist up to isomorphism seventeen finite effective extensions of $\mathbb{Z}^2$. All these groups act as Euclidean crystallographic groups on the plane. These are the famous *wall paper groups*. As crystallographic subgroups of the Euclidean group, the wall paper groups are unique up to conjugacy with affine transformations. Let us call an affine crystallographic group *non-Euclidean* if it is not conjugated to an Euclidean crystallographic group by an affine transformation. So which of the finite effective extensions of $\mathbb{Z}^2$ admit non-Euclidean affine crystallographic actions?

**Proposition 8.13.** *Let $\Gamma$ be a finite effective extension of $\mathbb{Z}^2$ which acts as a non-Euclidean affine crystallographic group. If $\Gamma$ preserves orientation then $\Gamma$ is isomorphic to $\mathbb{Z}^2$. If $\Gamma$ does not preserve orientation then $\Gamma$ contains $\mathbb{Z}^2$ as a translation subgroup of index two. There are three isomorphism classes of the latter type and for these groups the deformation space $\mathcal{D}_c(\Gamma, \mathrm{Aff}(\mathbb{R}^2))$ is homeomorphic to the real line.*

*Proof.* Let $\mu$ be a finite subgroup of $GL_2(\mathbb{Z})$ which fixes a line in $\mathbb{R}^2$. Every nonidentity element of $\mu$ has eigenvalues 1 and $-1$ and is of order two. It follows that $\mu$ is conjugate in $GL_2(\mathbb{Z})$ to

$$\mu_1 = < \begin{pmatrix} 1 & 0 \\ 0 & -1 \end{pmatrix} > \quad \text{or} \quad \mu_2 = < \begin{pmatrix} 0 & 1 \\ 1 & 0 \end{pmatrix} > \quad .$$

From the classification (see for example [83][§22, §23]) it follows that there are two (effective) extensions of $\mathbb{Z}^2$ by $\mu_1$, and only one by $\mu_2$. The torsionfree extension of $\mathbb{Z}^2$ by $\mu_1$ is the fundamental group of the Klein bottle. □

**Corollary 8.14.** *The deformation space $\mathcal{D}_c(\mathcal{K}, \mathrm{Aff}(\mathbb{R}^2))$ of complete affine structures on the Klein bottle $\mathcal{K}$ is homeomorphic to the real line. The moduli space $\mathcal{M}_c(\mathcal{K}, \mathrm{Aff}(\mathbb{R}^2))$ coincides with the deformation space.*

*Proof.* Let $\Delta$ be the Klein-bottle group, $q \in \mu_1$ the non-identity element of $\mu_1$, $a, b$ the standard generators for the lattice $\mathbb{Z}^2$. Then $\Delta$ is generated by $q, a, b$ with the relations $qaq^{-1} = a$, $qbq^{-1} = -b$ and $q^2 = a$. The group $\mu_1$ fixes a line in $\mathbb{R}^2$ which is the deformation space for $\Delta$. It is easy to see that the image of the map from $\mathrm{Out}(\Delta)$ to the normalizer of $\mu_1$ in $GL_2(\mathbb{Z})$ is $\mu_1$ itself. By Theorem 8.6 the fixed line in $\mathbb{R}^2$ is therefore also the moduli space. □

**Metric and symplectic affine structures on tori** We turn our attention to some particular interesting subgeometries, namely complete affine structures with a parallel inner product (flat pseudo-Riemannian manifolds) or a parallel nondegenerate alternating product (symplectic affine manifolds). In these cases we can give a precise description of the spaces $G_{st}(V, A)$, see Section 4. This amounts to a good understanding of the deformation spaces for these geometries on manifolds of Euclidean type.

# 9. The realization problem for unipotent shadows

Every torsionfree polycyclic group $\Gamma$ has an unipotent shadow $\Theta$ in its real algebraic hull $H_\Gamma$. Recall from Proposition 2.10, that if $\Gamma$ is an affine crystallographic group then the unipotent shadow $\Theta$ is an affine crystallographic group as well. Therefore the following *realization problem* makes sense:

*Let $\Gamma$ be a torsionfree polycyclic group with unipotent shadow $\Theta$. Assume that $\Theta$ is isomorphic to an affine crystallographic group. Is it true that $\Gamma$ is isomorphic to an affine crystallographic group?*

We need the following

**Definition 9.1.** Let $H_\Gamma$ be a real algebraic hull for $\Gamma$. We choose a splitting $H_\Gamma = T{\cdot}U_\Gamma$, where $T$ is a Levi subgroup of $H_\Gamma$ and put $U_\Gamma = \mathrm{u}(H_\Gamma)$. Then $\Theta \leq U_\Gamma$, and $U_\Gamma$ is a real Malcev hull for $\Theta$. Since $H_\Gamma$ has a strong unipotent radical, conjugation embeds $T$ as a subgroup of $\mathrm{Aut}(U_\Gamma)$. We call the corresponding image $T_\Gamma \leq \mathrm{Out}(U_\Gamma)$ of $T$ the *semisimple kernel associated to* $\Gamma$.

Recall (see Section 11) that, since $U_\Gamma$ is a Malcev-hull for $\Theta$, $\mathrm{Out}(U_\Gamma)$, and hence also $T_\Gamma$, act on the deformation space $\mathcal{D}_c(\Theta, A)$ of $\Theta$. The main result of ths section is the following theorem. It shows that the answer to the question above lies in the action of the semisimple kernel $T_\Gamma$ of $\Gamma$ on the deformation space of $\Theta$.

**Theorem 9.2.** *Let $\Gamma$ be a torsionfree polycyclic group with unipotent shadow $\Theta$. Assume that $\Theta$ is an ACG and let $\tau \in \mathrm{Hom}_c(\Theta, A)$. If $[\tau] \in \mathcal{D}_c(\Theta, A)$ is a fixed point for the semisimple kernel $T_\Gamma$ of $\Gamma$ then there exist $\rho \in \mathrm{Hom}_c(\Gamma, A)$ which descends to $\tau$.*

An immediate consequence is the following realization result. It gives an answer to the question which polycyclic groups can act as affine crystallographic groups in terms of the set of crystallographic actions of the unipotent shadow.

**Corollary 9.3.** *Let $\Gamma$ be a torsionfree polycyclic group, $\Theta \leq H_\Gamma$ a unipotent shadow for $\Gamma$. Then $\Gamma$ may be realized as an affine crystallographic group of type $A$ if and only if there exists a fixed point for the action of the semisimple kernel $T_\Gamma$ in the deformation space $\mathcal{D}_c(\Theta, A)$.*

*Remark* Auslander (compare [4]) remarked that the unipotent shadow of a crystallographic group is crystallographic. The realization criterion of the Corollary may be seen as providing a converse to his result.

Applications to the existence problem for crystallographic actions on large classes of polycyclic groups follow. (See Section 10)

## The shadow map on deformation spaces

Let $\Gamma$ be a torsionfree polycyclic group, $\Gamma \leq \mathbf{H}_\Gamma$ and $\Theta = \Gamma_C^u \leq \mathrm{u}(\mathbf{H}_\Gamma)$ a unipotent shadow for $\Gamma$ with respect to some (almost) nilpotent supplement $C$ for $\mathrm{Fitt}(\Gamma)$. Recall from Proposition 2.10, that there exists a canonical crystallographic shadow map

$$s_u : \mathrm{Hom}_c(\Gamma, A) \longrightarrow \mathrm{Hom}_c(\Theta, A) .$$

We remark first (see Proposition 9.4) that the crystallographic shadow map induces a shadow map

$$\sigma_u : \mathcal{D}_c(\Gamma, A) \longrightarrow \mathcal{D}_c(\Theta, A)$$

on the deformation spaces. We study here the properties of the shadow map $\sigma_u$ in order to obtain information about the deformation spaces $\mathcal{D}_c(\Gamma, A)$. The results and methods are analoguous to those in Section 7 on the realization of finite extension for f.t.n.-groups. We can interpret them also as a solution to a suitable realization problem for polycyclic groups with a prescribed shadow.

**The induced shadow map** Let $H_\Gamma$ be a real algebraic hull for $\Gamma$, and let $\Theta$ be a unipotent shadow. We choose a splitting $H_\Gamma = T \cdot U_\Gamma$ of $H_\Gamma$. We may view $T$ as a subgroup of $\mathrm{Aut}(U_\Gamma)$, and $T$ acts on $\mathcal{D}_c(\Theta, A)$ via the semisimple kernel $T_\Gamma \leq \mathrm{Out}(U_\Gamma)$.

**Proposition 9.4.** *The crystallographic shadow map $s_u$ induces a continuous map*

$$\sigma_u : \mathcal{D}_c(\Gamma, A) \longrightarrow \mathcal{D}_c(\Theta, A)^T .$$

*Proof.* To see that there is an induced map $\sigma_u$ on deformation spaces which is given by $\sigma_u([\rho]) = [s_u(\rho)]$, it is enough to verify that, for all $g \in A$, $\rho \in \mathrm{Hom}_c(\Gamma, A)$,

$$s_u(\rho^g) = s_u(\rho)^g .$$

To verify this, we have to recall the construction of the homomorphism $s_u(\rho) = \rho^u$. If $\Gamma = C\,\mathrm{Fitt}(\Gamma)$ then, by Proposition 1.21, $\rho^u : \Gamma_C^u \to A$ is determined by the conditions $\rho^u(\gamma_u) = \rho(\gamma)_u$, for all $\gamma \in C$, $\gamma \in \mathrm{Fitt}(\Gamma)$. Therefore the above formula is immediate since the Jordan-decomposition in $A$ is preserved by conjugation. The map $\sigma_u$ is continuous since (Corollary 2.18) $s_u$ is continuous.

Next we have to prove that $\sigma_u(\rho) \in \mathcal{D}_c(\Theta, A)^T$. But this is an immediate consequence of Theorem 1.27: The homomorphism $\rho : \Gamma \to A$ extends to a $u$-simply transitive embedding $\rho_{H_\Gamma} : H_\Gamma \to A$. By Proposition 1.21, $\rho_{H_\Gamma}$ restricts to $s_u(\rho) = \rho^u$ on $\Theta = \Gamma_C^u \leq U_\Gamma$. Then the action of $T \leq \mathrm{Aut}(U_\Gamma)$ on $\mathrm{Hom}_c(\Theta, A)$ is induced by conjugation via $\rho_{H_\Gamma}(T)$. Hence, $T \leq \mathrm{Aut}(U_\Gamma)_{\sigma_u([\rho])}$. □

We are going to prove here

**Theorem 9.5.** *Let $\Gamma$ be a torsionfree polycyclic group, $\Theta \leq U_\Gamma$ a unipotent shadow for $\Gamma$, and $T$ the semisimple kernel associated to $\Gamma$. Then the induced shadow map*

$$\sigma_u : \mathcal{D}_c(\Gamma, A) \longrightarrow \mathcal{D}_c(\Theta, A)^T$$

*is a continuous bijection onto $\mathcal{D}_c(\Theta, A)^T$.*

It does not seem to be clear whether the map $\sigma_u$ is also a homeomorphism onto its image. However, this is the case on a certain stratum of $\mathcal{D}_c(\Gamma, A)$. Let $\mathcal{D}_c(\Theta, A)_s$ be the closed stratum of symmetric structures in the deformation space $\mathcal{D}_c(\Theta, A)$. (Compare Theorem 8.4). We define a stratum $\mathcal{D}_c(\Gamma, A)_s = \sigma_u^{-1}(\mathcal{D}_c(\Theta, A)_s)$ in $\mathcal{D}_c(\Gamma, A)$.

**Theorem 9.6.** *The map $\sigma_u$ induces a homeomorphism*

$$\mathcal{D}_c(\Gamma, A)_s \xrightarrow{\approx} \mathcal{D}_c(\Theta, A)_s^T .$$

We postpone the proof of Theorem 9.2, Theorem 9.5, and Theorem 9.6 for a moment in order to derive some consequences.

**Inheritance of the Hausdorff-property** Though the topology on $\mathcal{D}_c(\Gamma, A)$ is potentially finer than the topology induced from $\mathcal{D}_c(\Theta, A)_s^T$ we still can deduce an immediate useful consequence.

**Corollary 9.7.** *Let $\Gamma$ be a torsionfree polycyclic ACG, and let $\Theta$ be a unipotent shadow for $\Gamma$. If $\mathcal{D}_c(\Theta, A)$ is a Hausdorff topological space, then $\mathcal{D}_c(\Gamma, A)$ is Hausdorff too.*

Some applications of this fact were already described in Part II. If $\Theta$ is abelian then $\mathcal{D}_c(\Theta, A) = \mathcal{D}_c(\Theta, A)_s$. Therefore it follows from Theorem 9.6:

**Corollary 9.8.** *Let $\Gamma$ be a torsionfree polycyclic ACG, so that the unipotent shadow $\Theta$ is abelian. Then $\mathcal{D}_c(\Gamma, A)$ is homeomorphic to a semi-algebraic set.*

**Inheritance of convexity** Let us recall that every automorphism $\phi \in \mathrm{Aut}(\Gamma)$ extends to a unique algebraic automorphism $\Phi$ of $H_\Gamma$. Since $\Phi$ is algebraic it restricts to an automorphism $\Phi^u$ of the unipotent radical $U_\Gamma$ of $H_\Gamma$. Therefore there is a natural induced homomorphism

$$\nu^u : \mathrm{Aut}(\Gamma) \longrightarrow \mathrm{Aut}(U_\Gamma) .$$

Since $U_\Gamma = U_\Theta$ is a Malcev-hull for $\Theta$, the group $\mathrm{Aut}(U_\Gamma)$ acts on the Deformation space $\mathcal{D}_c(\Theta, A)$ and it is easy to see that the shadow map $s_u : \mathrm{Hom}_c(\Gamma, A) \to \mathrm{Hom}_c(\Theta, A)$ is equivariant with respect to $\nu^u$. We summarize:

**Proposition 9.9.** *Let $\Gamma$ be a polycyclic ACG, $\Theta$ a unipotent shadow for $\Gamma$. Then the shadow map on deformation spaces*

$$\sigma_u : \mathcal{D}_c(\Gamma, A) \hookrightarrow \mathcal{D}_c(\Theta, A)$$

*is equivariant with respect to the actions of* $\mathrm{Aut}(\Gamma)$ *and* $\mathrm{Aut}(U_\Gamma)$ *and the homomorphism $\nu^u$.*

As a particular consequence we see that the convexity properties of the unipotent shadow $\Theta$ will be inherited.

**Theorem 9.10.** *Let $\Gamma$ be a polycyclic ACG and $\Theta$ a unipotent shadow for $\Gamma$. Then the following hold:*

i) *If $\mathcal{D}_c(\Theta, A)$ is fixed pointed then $\mathcal{D}_c(\Gamma, A)$ is fixed pointed too.*
ii) *If $\mathcal{D}_c(\Theta, A)$ is convex then $\mathcal{D}_c(\Gamma, A)$ is convex too.*

*Proof.* If $\mathcal{D}_c(\Theta, A)$ has a fixed point $[\rho_o]$ for $\mathrm{Aut}(U_\Gamma)$, Theorem 9.5 implies that $\sigma_u(\mathcal{D}_c(\Gamma, A))$ meets $[\rho_o]$. By equivariance and injectivity of the map $\sigma_u$, $[\rho_o]$ is a fixed point for $\mathrm{Aut}(\Gamma)$ in $\mathcal{D}_c(\Theta, A)$.

We assume now that $\mathcal{D}_c(\Theta, A)$ is convex. Let $T_\Gamma \leq \mathrm{Out}(U_\Gamma)$ be the semisimple kernel for the shadow $\Theta$. It is easy to see that there is a well defined homomorphism

$$\bar{\nu}^u : \mathrm{Out}(\Delta) \longrightarrow \mathrm{N}_{\mathrm{Out}(U_\Gamma)}(T_\Gamma)/\,T_\Gamma\ .$$

Let $\mu \leq \mathrm{Aut}(\Gamma)$ be a finite subgroup, and let $\tilde{\mu} \leq N_{\mathrm{Out}(\Gamma)}(T_\Gamma)$ be the preimage of $\bar{\nu}^u(\mu)$. Since $T_\Gamma$ is reductive so is $\tilde{\mu}$ which is a finite normal extension of $T_\Gamma$. If $[\rho_o]$ is a fixed point for $\tilde{\mu}$ then, since $T_\Gamma \leq \tilde{\mu}$, $\sigma_u(\mathcal{D}_c(\Gamma, A))$ meets $[\rho_o]$. Moreover $[\rho_o]$ is a fixed point for $\mu$ too. Therefore $\mathcal{D}_c(\Gamma, A)$ is convex. □

**Surjectivity of the shadow map** We start now with the proofs of the main theorems. Let $\Theta$ be an f.t.n.-group, $U_\Theta$ a real Malcev-hull for $\Theta$. For $\tau \in \mathrm{Hom}_c(\Theta, A)$ we put $U = \overline{\tau(\Theta)}$ to denote the unipotent simply transitive hull of $\tau(\Theta) \leq A$. Recall from Proposition 11.5 that in this situation conjugation in $A$ defines a natural map

$$c_\tau : N_A(U) \longrightarrow \mathrm{Aut}(U_\Theta)_{[\tau]}\ .$$

The map $c_\tau$ is then onto with kernel $\mathrm{C}_A(U)$.

**Proposition 9.11.** *Let $T \leq \mathrm{Aut}(U_\Theta)$ be a reductive algebraic subgroup. If $[\tau] \in \mathcal{D}_c(\Theta, A)$ is a fixed point for $T$, then there exists an embedding of (real-) linear algebraic groups*

$$\jmath : T \longrightarrow \mathrm{N}_A(U)$$

*which satisfies $c(\jmath(t)) = t$, for all $t \in T$. Any two embeddings $\jmath, \jmath' : T \to \mathrm{N}_A(U)$ with the latter property are conjugate by an element of $\mathrm{C}_A(U)$.*

*Proof.* By assumption $T \leq \mathrm{Aut}(U_\Theta)_{[\tau]}$. Let $H = c^{-1}(T)$. $H$ is a Zariski-closed subgroup of $N_A(U)$, and $H$ contains the unipotent group $\mathrm{C}_A(U)$ as a normal subgroup. In fact, since $H/\,\mathrm{C}_A(U)$ is isomorphic to $T$, $\mathrm{C}_A(U)$ is the unipotent radical of $H$. Now by splitting of algebraic groups there exists a subgroup $T_H \leq H$ such that $H = T_H \cdot \mathrm{C}_A(U)$. Since the restriction of $c$ to $T_H$ is an isomorphism of algebraic groups onto $T$, we can set $\jmath = c^{-1} : T \to T_H \leq H$. We thus proved the existence of $\jmath$.

We prove now the conjugacy statement. Let us remark first that every homomorphism $\jmath : T \to \mathrm{N}_A(U)$ which satisfies the assumptions maps $T$ into the group $H$. Moreover $\jmath$ is uniquely determined by its image $\jmath(T) \leq H$. Since $\jmath(T)$ and $\jmath'(T)$ are Levi-subgroups in $H$, $\jmath'(T)$ is conjugated to $\jmath(T)$ by an element of $\mathrm{C}_A(U)$. Hence also the homomorphisms $\jmath$ and $\jmath'$ are conjugate. □

Let $\Gamma$ be a torsionfree polycyclic group with real algebraic hull $H_\Gamma$, $\Theta \leq U_\Gamma$ a unipotent shadow for $\Gamma$. We say that $\rho_{H_\Gamma} : H_\Gamma \to A$ extends a homomorphism $\tau \in \mathrm{Hom}_c(\Theta, A)$ if $\tau$ is the restriction of $\rho_{H_\Gamma}$ to $\Theta$.

**Corollary 9.12.** *A homomorphism $\tau \in \mathrm{Hom}_c(\Theta, A)$ extends to a $u$-simply transitive embedding of linear algebraic groups $\rho_{H_\Gamma} : H_\Gamma \to A$ if and only if $[\tau] \in \mathcal{D}_c(\Theta, A)$ is a fixed point for the semisimple kernel $\overline{T_\Gamma}$. Any two such extensions $\rho_{H_\Gamma}, \rho'_{H_\Gamma}$ of $\tau$ are conjugate by an element of $\mathrm{C}_A(\tau(\Theta))$.*

*Proof.* Since $H_\Gamma = T{\cdot}U_\Gamma$ has a strong unipotent radical, we may view the reductive group $T$ as a subgroup of $\mathrm{Aut}(U_\Gamma)$. Let us assume that $[\tau] \in \mathcal{D}_c(\Theta, A)$ is a fixed point for $\overline{T_\Gamma}$. By Proposition 9.11, there exists an embedding $\jmath : T \to \mathrm{N}_A(U)$, $U = \overline{\tau(\Theta)} \leq A$, such that $c(\jmath(t)) = t$, for all $t \in T$. We already used implicitly that, since $U_\Gamma$ is a real Malcev hull for $\Theta$, the homomorphism $\tau \in \mathrm{Hom}_c(\Theta, A)$ extends to a $u$-simply transitive homomorphism $\tau_{U_\Gamma} : U_\Gamma \to A$. For $t \in T$, $u \in U_\Gamma$ we can then define

$$\rho_{H_\Gamma}(t\,u) \;=\; \jmath(t)\,\tau_{U_\Gamma}(u)\;.$$

By the properties of $\jmath$, this defines a homomorphism $\rho_{H_\Gamma} : H_\Gamma \to A$ of algebraic groups which is clearly injective and $u$-simply transitive. Thus we proved the "if" part of the first statement. The "only if" we saw in Proposition 9.4.

Now, if $\rho_{H_\Gamma}$ is any $u$-simply transitive algebraic group embedding which extends $\tau$ to $H_\Gamma$, then $\jmath(t) = \rho_{H_\Gamma}(t)$ defines a homomorphism $\jmath : T \to \mathrm{N}_A(U)$, $U = \overline{\tau(\Theta)}$, which satisfies $c(\jmath(t)) = t$, for all $t \in T$. Therefore the conjugacy statement of Proposition 9.11 implies that any two $u$-simply transitive extensions of $\tau$ to $H_\Gamma$ are conjugate by an element $u \in \mathrm{C}_A(U)$. □

*Proof of Theorem 9.2* By the previous corollary, $\tau \in \mathrm{Hom}_c(\Theta, A)$ extends to a $u$-simply transitive algebraic group homomorphism $\rho_{H_\Gamma} : H_\Gamma \to A$. Let $\rho$ be the restriction of $\rho_{H_\Gamma}$ to $\Gamma$. Let $\gamma \in \Gamma$ be so that $\gamma_u \in \Theta$. Since $\rho_{H_\Gamma}$ preserves the Jordan-decomposition, $\tau(\gamma_u) = \rho_{H_\Gamma}(\gamma_u) = \rho(\gamma)_u$. Therefore $\tau$ descends to $\rho$ in the sense of Definition 2.2. By Theorem 1.27, $\rho$ is a crystallographic homomorphism for $\Gamma$. □

*Proof of Theorem 9.5* It follows from Theorem 9.2 that the map $\sigma_u$ is onto $\mathcal{D}_c(\Theta, A)^T$. Let $\tau \in \mathrm{Hom}_c(\Theta, A)$. By Theorem 1.27, every extension $\rho \in \mathrm{Hom}_c(\Gamma, A)$ of $\tau$ extends *uniquely* to a $u$-simply transitive homomorphism $\rho_{H_\Gamma} : H_\Gamma \to A$ which then is also an extension of $\tau$. Therefore the conjugacy statement of Corollary 9.12 implies that $\sigma_u$ is injective. By Corollary 2.18, the shadow map $s_u : \mathrm{Hom}_c(\Gamma, A) \to \mathrm{Hom}_c(\Theta, A)$ is algebraic, in particular $s_u$ is continuous, hence also $\sigma_u$. □

*Proof of Theorem 9.6* To prove that $\sigma_u$ is an open mapping on $\mathcal{D}_c(\Gamma, A)_s$ we show that $s_u : \mathrm{Hom}_c(\Gamma, A)_s \to \mathrm{Hom}_c(\Theta, A)_s$ admits a continuous section. This is done in a manner completely analoguous to the proof of Lemma 8.5. We leave the details to the reader. □

# 10. Applications to the existence of affine crystallographic actions

One of the main questions of the subject is to decide whether a given virtually polycyclic group $\Delta$ may act as an affine crystallographic group. Not many general results on this question seem to be known. (But see [32] for the classification of torsionfree polycyclic groups which act crystallographically by affine Lorentz-transformations.) Our methods here apply when some information on the unipotent shadow of $\Delta$ is available. In general, this suggests that the main difficulty of the problem lies in the existence and structure of crystallographic actions of torsionfree nilpotent groups. In Section 5 we exhibited some classes of well understood f.t.n.-groups. Since these examples have strongly convex deformation spaces our realization results provide a positive answer for the existence of crystallographic actions for their finite extensions and also certain associated polycyclic groups. (See Corollary 5.4, Corollary 5.5)

**Crystallographic actions of finite extensions** Let $\Gamma$ be an f.t.n.-group. We say that $\Gamma$ admits an invariant grading if the Lie algebra of the real Malcev hull $U_\Gamma$ has a positive grading which is preserved by a Levi subgroup of $\mathrm{Aut}(U_\Gamma)$. (Compare Definition 5.12.)

**Corollary 10.1.** *Let $\Gamma$ be an* f.t.n.*-group which satisfies one of the following conditions*

i) $\Gamma$ *is of nilpotency class* $\leq 3$,
ii) $\Gamma$ *has rank* $\leq 5$,
iii) $\Gamma$ *admits an invariant grading.*

*Then $\Gamma$ is isomorphic to an affine crystallographic group which satisfies the realization property. Moreover, if $\Delta$ is a finite effective extension of $\Gamma$ then $\Delta$ is isomorphic to an affine crystallographic group, and $\Delta$ has the realization property.*

*Proof.* By Theorem 5.3, the deformation space of $\Gamma$ is convex, in particular, it is not empty. If $\Delta$ is a finite effective extension of $\Gamma$, then the associated kernel $\beta : \Delta/\Gamma \to \mathrm{Out}(\Gamma)$ has a fixed point in $\mathcal{D}_c(\Gamma, \mathrm{Aff}(V))$. By Theorem 7.3, $\Delta$ admits an affine crystallographic action. Since the convexity of the deformation space is inherited to $\Delta$, also $\Delta$ has the realization property. □

*Remark* Our corollary includes the result of Lee ([49]) who proved that a finitely generated torsionfree virtually nilpotent group $\Gamma$ with $\mathrm{rank}\,\Gamma \leq 3$ acts as an affine crystallographic group.

**Examples of affine crystallographic polycyclic groups** Let $\Gamma$ be a torsionfree polycyclic group. We associated to $\Gamma$ its unipotent shadow $\Theta \leq U_\Theta$, and the semisimple kernel $T_\Gamma \leq \mathrm{Out}(U_\Theta)$. Let $\mathfrak{u}_\Theta$ be the Lie algebra of $U_\Theta$. We say that $T_\Gamma$ centralizes a nonsingular derivation $D$ of $\mathfrak{u}_\Theta$ if there exists a subgroup $T \leq \mathrm{Aut}(U_\Theta)$ which projects onto $T_\Gamma$ and centralizes $D$.

**Corollary 10.2.** *Let $\Gamma$ be a torsionfree polycyclic group with unipotent shadow $\Theta$. We assume that $\Theta$ satisfies one of the following conditions*

i) *$\Theta$ is of nilpotency class $\leq 3$,*
ii) *$\Theta$ admits an invariant grading,*
iii) *The semisimple kernel $T_\Gamma$ of $\Gamma$ centralizes a nonsingular derivation $D$ of $\mathfrak{u}_\Theta$.*

*Then $\Gamma$ is isomorphic to an affine crystallographic group.*

*Proof.* By Theorem 5.3 conditions i) and ii) imply that $\mathcal{D}_c(\Theta, \mathrm{Aff}(V))$ is strongly convex. In particular, the semisimple kernel $T_\Gamma \leq \mathrm{Out}(U_\Theta)$ has a fixed point in $\mathcal{D}_c(\Theta, \mathrm{Aff}(V))$. If condition iii) is satisfied it follows from Proposition 5.11 that there exists an affine crystallographic action $\rho$ of $\Theta$ so that $T_\Gamma \leq \mathrm{Out}(U_\Theta)_{[\rho]}$, that is, $T_\Gamma$ fixes the point $[\rho] \in \mathcal{D}_c(\Theta, \mathrm{Aff}(V))$. Therefore, in all three cases, it follows from Theorem 9.2 that $\Gamma$ admits an affine crystallographic action. □

To illustrate the corollary, we consider the particular case that $\Gamma$ is a Zariski-dense lattice in a semi-direct product. Then the conditions on the unipotent shadow $\Theta$ of $\Gamma$ become in particular transparent.

**Example 10.3 (semi-direct products).** Let $G$ be a connected simply connected solvable Lie group, and $\Gamma \leq G$ a Zariski-dense lattice. We assume that $G$ is a semisimple semi-direct product, that is, the Lie algebra $\mathfrak{g}$ of $G$ splits as a direct sum

$$\mathfrak{g} = \mathfrak{a} \oplus \mathfrak{n}\,,$$

where $\mathfrak{n}$ is the nilpotent radical of $\mathfrak{g}$, and $\mathfrak{a}$ is an abelian subalgebra which acts on $\mathfrak{n}$ by semisimple transformations. The unipotent shadow $U$ of $G$ is obtained by killing the torus action. That is, the Lie algebra $\mathfrak{u}$ of $U$ is just the direct product

$$\mathfrak{u} = \mathfrak{b} \oplus \mathfrak{n}\,,$$

of an abelian ideal $\mathfrak{b}$ and the ideal $\mathfrak{n}$. The Lie algebra $\mathfrak{a}$ acts on $\mathfrak{u}$ by centralizing $\mathfrak{b}$, and by the adjoint action on $\mathfrak{n}$. Moreover, the Zariski closure of the semisimple abelian subgroup $T_A \leq \mathrm{Aut}(U)$, which belongs to $\mathfrak{a}$, is a torus $T \leq \mathrm{Aut}(U)$ which projects onto the semisimple kernel $T_\Gamma \leq \mathrm{Out}(U)$.

Therefore we get:

**Corollary 10.4.** *Let $G$ be a simply connected solvable Lie group which is a semisimple semi-direct product $G = T_A N$, where $N$ is the nilradical of $G$. Assume that $G$ has a Zariski-dense lattice $\Gamma$, and assume further that one of the following conditions is satisfied:*

i) *$N$ has a positive grading which is invariant by $T_A$.*
ii) *$T_A$ centralizes a nonsingular derivation of $N$.*
iii) *$N$ is of nilpotency class $\leq 3$.*

*Then $\Gamma$ is isomorphic to an affine crystallographic group.*

## 11. Group actions on deformation spaces

Let $\Gamma$ be an f.t.n.-group, and $\Delta$ a finite effective extension group of $\Gamma$. We let $\mathbf{U}$ denote the Malcev completion of $\Gamma$, and $\mathbf{U}_\Delta$ the algebraic hull for $\Delta$. In this section we view $\Gamma$ as a lattice in the connected simply connected (real) Lie group $U_\Gamma = \mathbf{U}_\mathbb{R}$, and $\Delta$ as a subgroup of $U_\Delta = \mathbf{U}_{\Delta,\mathbb{R}}$. The purpose of this final section is to exploit the functorial properties of the real algebraic hull $U_\Delta$ in the study of the deformation space $\mathcal{D}_c(\Delta, A)$, and to provide some auxiliary results which we need at various places in this work.

**The hull functor on deformation spaces** Recall that the group $\mathrm{Aut}(\Delta)$ acts in a natural way on $\mathrm{Hom}_c(\Delta, A)$. In fact, the action of $\phi \in \mathrm{Aut}(\Delta)$ is described for all $\rho \in \mathrm{Hom}_c(\Delta, A)$ by

$$\rho \longmapsto \rho^\phi = \rho \circ \phi^{-1} .$$

The $\mathrm{Aut}(\Delta)$-action on $\mathrm{Hom}_c(\Delta, A)$ induces then an $\mathrm{Out}(\Delta)$-action on the deformation space $\mathcal{D}_c(\Delta, A)$. We fix an embedding $j : \Delta \hookrightarrow U_\Delta$ of $\Delta$ as a lattice in its real algebraic hull $U_\Delta$. By Corollary 7.16 the embedding $j$ induces an embedding

$$\epsilon_j : \mathrm{Aut}(\Delta) \longrightarrow \mathrm{Aut}(U_\Delta) .$$

We show now that the action of $\mathrm{Aut}(\Delta)$ extends to an action of $\mathrm{Aut}(U_\Delta)$. By Proposition 7.15, there exists, for every $\rho \in \mathrm{Hom}_c(\Delta, A)$, a unique homomorphism $\bar\rho = \rho_{U_\Delta} : U_\Delta \to A$ which satisifies $\bar\rho \circ j = \rho$. For $\rho \in \mathrm{Hom}_c(\Delta, A)$, $\Phi \in \mathrm{Aut}(U_\Delta)$, we put then

$$\rho^\Phi = \bar\rho \circ \Phi^{-1} \circ j .$$

Since $\rho^\Phi(\Gamma)$ is a lattice in the unipotent simply transitive group $\bar\rho(U_\Gamma)$ it follows that $\rho^\Phi(\Delta)$ is an ACG, hence $\rho^\Phi \in \mathrm{Hom}_c(\Delta, A)$.

**Proposition 11.1.** *The correspondence $\rho \longmapsto \rho^{\Phi^{-1}}$ extends the $\mathrm{Aut}(\Delta)$-action on $\mathrm{Hom}_c(\Delta, A)$ to an action of $\mathrm{Aut}(U_\Delta)$ on $\mathrm{Hom}_c(\Delta, A)$. The $\mathrm{Aut}(U_\Delta)$-action on $\mathrm{Hom}_c(\Delta, A)$ is free, and induces an action of $\mathrm{Out}(U_\Delta)$ on $\mathcal{D}_c(\Delta, A)$. In particular the action of $\mathrm{Out}(\Delta)$ factorizes over $\mathrm{Out}(U_\Delta)$.*

*Proof.* Let $\Phi, \Psi \in \mathrm{Aut}(U_\Delta)$. We calculate $(\rho^\Phi)^\Psi = \overline{\rho^\Phi} \circ \Psi^{-1} \circ j$, where $\overline{\rho^\Phi} : U_\Delta \to A$ is the extension of $\rho^\Phi : \Delta \to A$. Therefore $\overline{\rho^\Phi} = \bar\rho \circ \Phi^{-1}$, and we get $(\rho^\Phi)^\Psi = \bar\rho \circ (\Phi^{-1} \circ \Psi^{-1}) \circ j = \rho^{\Psi\circ\Phi}$. So, in fact, $\mathrm{Aut}(U_\Delta)$ acts on $\mathrm{Hom}_c(\Delta, A)$. Obviously, the $\mathrm{Aut}(U_\Delta)$-action extends the $\mathrm{Aut}(\Delta)$-action, in the sense that, for $\phi \in \mathrm{Aut}(\Delta)$, $\rho^\phi = \rho^{\epsilon_j(\phi)}$. Since $\Delta$ is Zariski-dense in $U_\Delta$, $\mathrm{Aut}(U_\Delta)$ acts freely.

For $h \in U_\Delta$, $\gamma \in \Delta$, let $c_h : U_\Delta \to U_\Delta$ denote conjugation with $h$. We get $\rho^{c_h}(\delta) = \bar\rho(h)\rho(\delta)\bar\rho(h)^{-1}$. Therefore $\mathrm{Out}(U_\Delta)$ acts on $\mathcal{D}_c(\Delta, A)$. □

*Remark* The $\mathrm{Aut}(U_\Delta)$-action depends on the embedding $j : \Delta \to U_\Delta$. Nevertheless, the image of $\mathrm{Aut}(U_\Delta)$ as a group of morphisms of $\mathrm{Hom}_c(\Delta, A)$ is independent from the choice of embedding. In particular, $\mathrm{Out}(U_\Delta)$ maps to a well defined subgroup of homeomorphisms on the deformation space $\mathcal{D}_c(\Delta, A)$.

The presence of the $\mathrm{Aut}(U_\Delta)$-action on the deformation space $\mathcal{D}_c(\Delta, A)$ comes from a correspondence of the affine crystallographic representations of $\Delta$ with certain representations of its real hull $U_\Delta$. We want to briefly explain this now:

Let $H$ be a real algebraic group which contains a normal unipotent subgroup $U$ of finite index, and assume that the centralizer of $U$ is contained in $U$. Note that the assumptions on $H$ are satisfied by the algebraic hull $U_\Delta$.

**Definition 11.2.** A homomorphism $\bar{\rho} : H \to A$ is called *crystallographic* if $\bar{\rho}$ is injective and if $U$ acts simply transitively on $V$.

We then define the space $\mathrm{Hom}_c(H, A)$ of crystallographic homomorphisms of $H$ and the corresponding deformation space $\mathcal{D}_c(H, A) = \mathrm{Hom}_c(H, A)/A$. The space $\mathrm{Hom}(H, A)$ is a topological space with the compact open topology. Also by the results of Part I there is a natural structure of real algebraic variety on $\mathrm{Hom}_c(H, A)$.

We fix an embedding $j : \Delta \hookrightarrow U_\Delta$ of $\Delta$ in its real algebraic hull $U_\Delta$. Recall that every $\rho \in \mathrm{Hom}_c(\Delta, A)$ is unipotent on the Fitting subgroup $\Gamma$ of $\Delta$. By Proposition 7.15, $\rho \in \mathrm{Hom}_c(\Delta, A)$ extends to a representation $\rho_{U_\Delta}$ of $U_\Delta$. Without a proof we state:

**Theorem 11.3.** *The correspondence $\rho \longmapsto \rho_{U_\Delta}$ defines a homeomorphism*

$$h_j : \mathrm{Hom}_c(\Delta, A) \xrightarrow{\approx} \mathrm{Hom}_c(U_\Delta, A)$$

*which commutes with the natural actions of automorphism groups on both spaces. In particular, $h_j$ induces a homeomorphism $\bar{h}_j : \mathcal{D}_c(\Delta, A) \xrightarrow{\approx} \mathcal{D}_c(U_\Delta, A)$ which commutes with the natural outer automorphism actions.*

The theorem implies that the topology and structure of the deformation space $\mathcal{D}_c(\Delta, A)$ for $\Delta$, and also of the moduli space $\mathcal{M}_c(\Delta, A)$, depend only on the properties of the algebraic hull for $\Delta$.

Here is another application. Look at the variety $\mathcal{C} = \mathrm{Hom}_c(U_\Delta, A) \times V$. Then $A$ acts on $\mathcal{C}$, where $g(\rho, v) = (\rho^g, gv)$. Moreover, $\mathrm{Aut}(U)$ acts on the first factor of $\mathcal{C}$, commuting with the action of $A$. Define $\mathcal{LC}(U) = \mathcal{C}/A$.

**Proposition 11.4.** *The deformation space $\mathcal{D}_c(U, A)$ is a quotient of the variety $\mathcal{LC}(U)$ by the induced unipotent action of* $\mathrm{Inn}(U)$ *on $\mathcal{LC}(U)$.*

*Proof.* Since $A$ acts transitively on $V$, we can deduce that

$$\mathcal{D}_c(U, A) = \mathcal{LC}(U)/\mathrm{Inn}(U) .$$

Furthermore, $\mathcal{LC}(U)$ is the set of $A$-conjugacy classes of étale representation with basepoint, and this corresponds to the algebraic variety of left-symmetric algebra products on the Lie algebra $\mathfrak{u}$. (See, for example, [12], for discussion.) □

**Stabilizers in deformation space** For a moment we restrict our considerations to f.t.n.-groups. Let $\Gamma \leq U_\Gamma$ be an f.t.n.-group and assume that $\Gamma$ is an ACG. Let $\rho \in \mathrm{Hom}_c(\Gamma, A)$ be an affine crystallographic representation, and $[\rho] \in \mathcal{D}_c(\Gamma, A)$ the corresponding point in the deformation space. We let $U = \bar{\rho}(U_\Gamma) \leq A$ be the unipotent simply transitive hull for $\rho(\Gamma)$. The group $\mathrm{Aut}(\mathrm{U}_\Gamma)$ acts on $\mathcal{D}_c(\Gamma, A)$. The stabilizer $\mathrm{Aut}(U_\Gamma)_{[\rho]} \leq \mathrm{Aut}(U_\Gamma)$ of $[\rho]$ may be desribed in terms of the normalizer $\mathrm{N}_A(U)$ of the unipotent simply transitive hull $U$.

**Proposition 11.5.** *There are canonical isomorphisms*

$$\mathrm{N}_A(U)/\,\mathrm{C}_A(U) \xrightarrow{\cong} \mathrm{Aut}(U_\Gamma)_{[\rho]}\ ,$$
$$\mathrm{N}_A(U)/\,(\mathrm{C}_A(U)U) \xrightarrow{\cong} \mathrm{Out}(U_\Gamma)_{[\rho]}\ .$$

*Proof.* $\mathrm{N}_A(U)/\,\mathrm{C}_A(U)$ acts by conjugation as a group of automorphisms of $U$. Since $\bar{\rho}$ is an isomorphism, there exists for each $g \in \mathrm{N}_A(U)$ an element $\Phi_g \in \mathrm{Aut}(U_\Gamma)$ such that $\rho^g = \rho^{\Phi_g}$. This is, precisely, the condition that $[\rho^{\Phi_g}] = [\rho]$. Therefore the map $g \mapsto \Phi_g$ defines a homomorphism of $\mathrm{N}_A(U)$ onto $\mathrm{Aut}(U_\Gamma)_{[\rho]}$ with kernel $\mathrm{C}_A(U)$. This factorizes to an isomorphism $\mathrm{N}_A(U)/\,(\mathrm{C}_A(U)\,U) \longrightarrow \mathrm{Out}(U_\Gamma)_{[\rho]}$. □

We consider next the $\mathrm{Aut}(\Gamma)$-action on $\mathcal{D}_c(\Gamma, A)$. Let us put

$$\mathrm{Out}_{A,\rho}(\Gamma) = \mathrm{N}_A(\rho(\Gamma))/\,(\mathrm{C}_A(U)\rho(\Gamma))\ .$$

We describe the stabilizer of $[\rho]$ in terms of the normalizer $\mathrm{N}_A(\rho(\Gamma)) \leq A$.

**Proposition 11.6.** *There are canonical isomorphisms*

$$\mathrm{N}_A(\rho(\Gamma))/\,\mathrm{C}_A(U) \xrightarrow{\cong} \mathrm{Aut}(\Gamma)_{[\rho]}\ ,$$
$$\mathrm{Out}_{A,\rho}(\Gamma) \xrightarrow{\cong} \mathrm{Out}(\Gamma)_{[\rho]}\ .$$

*Proof.* As in Proposition 11.1 we view $\mathrm{Aut}(\Gamma)$ as a subgroup of $\mathrm{Aut}(U_\Gamma)$. With this identification $\mathrm{Aut}(\Gamma)_{[\rho]} = \mathrm{Aut}(\Gamma) \cap \mathrm{Aut}(U_\Gamma)_{[\rho]}$. By Proposition 11.5 conjugation on $U$ induces a surjective homomorphism $c : \mathrm{N}_A(U) \longrightarrow \mathrm{Aut}(U_\Gamma)_{[\rho]}$. Therefore

$$\mathrm{Aut}(\Gamma)_{[\rho]} = \mathrm{Aut}(\Gamma) \cap c(\mathrm{N}_A(U)) = c(\mathrm{N}_A(\rho(\Gamma)))\ .$$

This shows that $\mathrm{N}_A(\rho(\Gamma))$ is mapped onto $\mathrm{Aut}(\Gamma)_{[\rho]}$. The first isomorphism follows.

Now $\mathrm{Out}(\Gamma)_{[\rho]}$ is just the image of $\mathrm{Aut}(\Gamma)_{[\rho]}$ in $\mathrm{Out}(\Gamma)$. Hence, $\mathrm{Out}(\Gamma)_{[\rho]}$ is a quotient of $\mathrm{N}_A(\rho(\Gamma))$, and clearly the kernel is $\mathrm{C}_A(U)\rho(\Gamma)$. □

**The normalizer of a unipotent simply transitive group** Let $A \leq \mathrm{Aff}(V)$ be a Zariski-closed subgroup, and $U \leq A$ a simply transitive unipotent group. We discuss some of the structure of the normalizer $\mathrm{N}_A(U)$ of $U$ in $A$.

**Lemma 11.7.** *The centralizer* $\mathrm{C}_A(U) \leq A$ *is a unipotent normal subgroup of* $\mathrm{N}_A(U)$ *and acts freely on* $V$. *The group* $\mathrm{C}_A(U)U$ *is a unipotent normal subgroup in* $\mathrm{N}_A(U)$.

*Proof.* $\mathrm{C}_A(U)$ acts freely since $U$ is simply transitive, and clearly $\mathrm{C}_A(U)$ is a normal Zariski-closed subgroup of $A$. Since any reductive subgroup of $\mathrm{C}_A(U)$ has fixed points on $V$, $\mathrm{C}_A(U)$ is unipotent. □

There are some subgroups of $\mathrm{Aut}(U)$ which are associated to the normalizer $N_A(U)$, and which are important in our context. (These groups also carry some (differential-) geometric interpretations in terms of the simply transitive group action of $U$ and the associated flat left invariant connection on the real Lie group $U$. We do not go into these details here, but see a related discussion in [27][§3.11].) We consider the homomorphism

$$c_U : \mathrm{N}_A(U) \longrightarrow \mathrm{Aut}(U) ,$$

where $c_U(g) : U \to U$ is given by conjugation with $g \in \mathrm{N}_A(U)$. Let us define subgroups $\mathrm{Aff}_A(U)$, $\mathrm{Aut}_A(U)$, $\mathrm{Inn}_A(U) \leq \mathrm{Aut}(U)$ as follows

$$\begin{array}{lll}
\mathrm{Aff}_A(U) &= c(\mathrm{N}_A(U)) & (\cong \mathrm{N}_A(U)/\mathrm{C}_A(U)) \\
\mathrm{Aut}_A(U) &= c(\mathrm{N}_{\mathrm{GL}_A}(U)) & (\cong \mathrm{N}_A(U)/U) \\
\mathrm{Inn}_A(U) &= c(\mathrm{N}_{\mathrm{GL}_A}(U) \cap \mathrm{C}_A(U)U) & (\cong (\mathrm{C}_A(U)U)/U)
\end{array}$$

Finally, we put $\mathrm{Out}_A(U)$ for the image of $\mathrm{N}_A(U)$ in $\mathrm{Out}(U)$. Note that the group $\mathrm{Aut}_A(U)$ projects onto $\mathrm{Out}_A(U)$. Proposition 11.5 implies now

**Proposition 11.8.** *Let* $\Gamma$ *be an* f.t.n.*-group and* $\rho \in \mathrm{Hom}_c(\Gamma, A)$. *If* $U \leq A$ *is the simply transitive hull for* $\rho(\Gamma)$ *then there are natural isomorphisms*

$$\mathrm{Aff}_A(U) \xrightarrow{\cong} \mathrm{Aut}(U_\Gamma)_{[\rho]}$$
$$\mathrm{Out}_A(U) \xrightarrow{\cong} \mathrm{Out}(U_\Gamma)_{[\rho]} \quad .$$

*Proof.* In fact, the representation $\bar{\rho}$ defines an isomorphism $\varrho : U_\Gamma \to U$, and the proof of Proposition 11.5 shows that $\mathrm{Aut}(U_\Gamma)_{[\rho]} = \varrho^{-1}\mathrm{Aff}_A(U)\varrho$. The statement for $\mathrm{Out}_A(U)$ follows as well. □

The following are geometrically interesting special cases:

**Definition 11.9.** Let $U \leq A$ be a simply transitive unipotent subgroup. $U$ is called *A-symmetric* if $\mathrm{C}_A(U)$ is transitive on $V$. We call $U$ *fully A-symmetric* if $U$ is $A$-symmetric and $c_U(\mathrm{Out}_A(U)) = \mathrm{Out}(U)$. Finally, we call $U$ *A-invariant* if $c_U(\mathrm{Out}_A(U)) = \mathrm{Out}(U)$, and $U$ is called *A-convex* if $c_U(\mathrm{Out}_A(U))$ contains a Levi subgroup of $\mathrm{Out}_A(U)$.

In our context these conditions are interpreted in terms of group actions on the deformation spaces.

**The normalizer of a unipotent ACG** Let us consider now the normalizer $\mathrm{N}_A(\rho(\Gamma))$ of $\rho(\Gamma)$ in $A$. The Zariski-closure of $U = \overline{\rho(\Gamma)}$ is a simply transitive unipotent subgroup of type $A$. By Zariski-denseness of $\rho(\Gamma)$ in $U$, we have $\mathrm{N}_A(\rho(\Gamma)) \leq \mathrm{N}_A(U)$, as well as $\mathrm{C}_A(\rho(\Gamma)) = \mathrm{C}_A(U)$. For $x \in V$, we put $\mathrm{N}_{A,x}(\rho(\Gamma)) = \mathrm{N}_A(\rho(\Gamma)) \cap A_x$. Our interest will be in the image of $\mathrm{N}_{A,x}$ in $\mathrm{Out}_{A,\rho}(\Gamma)$. In a special case, the projection from $\mathrm{N}_{A,x}$ to $\mathrm{Out}_{A,\rho}(\Gamma)$ is surjective, for all $x \in V$.

**Lemma 11.10.** *If $U$ is an $A$-symmetric simply transitive group then, for all $x \in V$, the natural map*

$$\mathrm{N}_{A,x}(\rho(\Gamma)) \longrightarrow \mathrm{N}_A(\rho(\Gamma))/\mathrm{C}_A(U)$$

*is an isomorphism.*

*Proof.* Put $H = \mathrm{N}_A(\rho(\Gamma))/\mathrm{C}_A(U)$. $H$ acts on the orbit space $\mathrm{C}_A(U)\backslash V$. It is easy to see that $\mathrm{N}_{A,x}(\rho(\Gamma))$ projects onto $H_{[x]}$, where $H_{[x]}$ is the stabilizer of $[x] = \mathrm{C}_A(U)x$. If $U$ is fully symmetric then $\mathrm{Z}_A(U)$ acts transitively on $V$. Hence $H = H_{[x]}$, $\mathrm{N}_{A,x}(\rho) \cong \mathrm{N}_A(\rho(\Gamma))/\mathrm{C}_A(U)$. □

**Definition 11.11.** $\rho \in \mathrm{Hom}_c(\Gamma, A)$ is called symmetric, if $U = \overline{\rho(\Gamma)}$ is $A$-symmetric.

Let $c_\rho : N_A(\rho(\Gamma)) \to \mathrm{Aut}(\Gamma)$ denote the conjugation homomorphism, and $\mathrm{Hom}_c(\Gamma, A)_s$ the set of symmetric crystallographic homomorphisms. The next fact was required in Section 8.

**Proposition 11.12.** *Let $F \leq \mathrm{Aut}(\Gamma)$ be a finite subgroup. There exists a continuous map $s_x : \mathrm{Hom}_c(\Gamma, A)_s^F \longrightarrow \mathrm{Hom}(F, A_x)$ such that $c_\rho(s_x(\rho, g)) = g$, for all $g \in F$.*

*Proof.* For every $\rho \in \mathrm{Hom}_c(\Gamma, A)$, $\bar\rho(U_\Gamma)$ is a simply transitive subgroup of $A$. Therefore there exists, for every $g \in \mathrm{Aut}(\Gamma)$, a unique polynomial diffeomorphism $\Phi_g$ of $V$ which satisfies $\Phi_g \bar\rho(u)x = \bar\rho(gu)x$. In particular, $\Phi_g \bar\rho(u)\Phi_g^{-1} = \bar\rho(gu)$ holds, for all $u \in U_\Gamma$. Now, if $\rho$ is fixed by $F$, and $\rho \in \mathrm{Hom}_c(\Gamma, A)_s$ then by Proposition 11.6, and Lemma 11.10 there exists a $\phi \in N_{A,x}(\rho(\Gamma))$ such that $\phi\bar\rho(u)\phi^{-1} = \bar\rho(gu)$. We remark that, since $\mathrm{C}_A(U)$ is transitive on $V$, it coincides with the centralizer of $U$ in the group of diffeomorphisms of $V$. We conclude that $\Phi_g\phi^{-1} \in \mathrm{C}_A(U)$, and in particular $\Phi_g \in N_{A,x}(U)$. So we set $s_x(\rho, g) = \Phi_g$. The map $s_x$ is easily seen to be continuous, since it is the restriction of the continuous map on $\mathrm{Hom}_c(\Gamma, A)$ which assigns to $\Phi_g(\rho)$ its first derivative in $x$. □

We remark that symmetric simply transitive actions may be constructed for many unipotent Lie groups $U$.

**Example 11.13.** Let $\mathcal{A}$ be a nilpotent associative algebra, finite dimensional over $\mathbb{R}$ and $\mathfrak{u}$ the commutation Lie algebra of $\mathcal{A}$. Then it is known, compare [4, 23], that the corresponding simply connected Lie group $U$ admits a simply transitive unipotent representation. It is possible to show that this representation is also symmetric, and that every symmetric simply transitive representation arises from this construction.

# References


[1] H. Abels, *Über die Erzeugung von eigentlichen Transformationsgruppen*, Math. Z. **103** (1968), 333-357.

[2] H. Abels, *Properly discontinuous groups of affine transformations, a survey*, Geom. Dedicata **87** (2001), 309-333.

[3] L. Auslander, *An exposition of the structure of solvmanifolds. I. Algebraic theory*, Bull. Amer. Math. Soc. **79** (1973), no. 2, 262-285.

[4] L. Auslander, *Simply transitive groups of affine motions*, Amer. J. Math. **99** (1977), no. 4, 809-826.

[5] L. Auslander, R. Tolimieri, *On a conjecture of G. D. Mostow and the structure of solvmanifolds*, Bull. Amer. Math. Soc. **75** (1969), 1330-1333.

[6] A. Baklouti, F. Khlif, *Proper actions on some exponential solvable homogeneous spaces*, Int. J. Math. **16** (2005), 941-955.

[7] A. Baklouti, I. Kédim, T. Yoshino, *On the deformation space of Clifford-Klein forms of Heisenberg groups*, Int. math. Res. Not., to appear.

[8] A. Baklouti, I. Kédim, *On the deformation space of Clifford-Klein forms of some exponential homogeneous spaces*, preprint 2007.

[9] O. Baues, *Gluing affine two-manifolds with polygons*, Geom. Dedicata **75** (1999), no. 1, 33-56.

[10] O. Baues, *Varieties of discontinuous groups*, in *Crystallographic groups and their generalizations*, Contemporary Mathematics **262** (2000), 147-158.

[11] O. Baues, *Finite extensions and unipotent shadows of affine crystallographic groups*, C. R. Math. Acad. Sci. Paris **335** (2002), no. 10, 785-788.

[12] O. Baues, V. Cortés, *Simply transitive abelian groups of symplectic type*, Ann. Inst. Fourier (Grenoble) **52** (2002), no. 6, 1729-1751.

[13] O. Baues, *Infra-solvmanifolds and rigidity of subgroups in solvable linear algebraic groups*, Topology **43** (2004), no. 4, 903-924.

[14] O. Baues, W.M. Goldman, *Is the deformation space of complete affine structures on the 2-torus smooth?*, in *Geometry and dynamics*, Contemp. Math., **389**, 69-89.

[15] O. Baues, F. Grunewald, *Automorphism groups of polycyclic-by-finite groups and arithmetic groups*, Publ. Math. Inst. Hautes Études Sci. No. **104** (2006), 213-268.

[16] A. Borel, J.-P. Serre, *Théorèmes de finitude en cohomologie galoisienne*, Comment. Math. Helv. **39** (1964), 111-164.

[17] Y. Benoist, *Une nilvariété non affine*, J. Diff. Geom. **41** (1995), 21-52.

[18] Y. Benoist, *Actions propres sur les espaces homogènes réductifs*, Ann. of Math. **16** (2005), 941-955.

[19] A. Borel, *Linear algebraic groups*, Second edition, Graduate Texts in Mathematics **126**, Springer-Verlag, (1991).

[20] K.S. Brown, *Cohomology of groups*, Graduate Texts in Mathematics, **87**. Springer-Verlag, New York-Berlin, (1982).

[21] F. Catanese, *Deformation in the large of some complex manifolds. I.* Ann. Mat. Pura Appl. (4) **183** (2004), no. 3, 261-289.

[22] F. Catanese, *Deformation types of Real and Complex Manifolds*, Contempo-

rary Trends in Algebraic Geometry and Algebraic Topology (Tianjin, 2000), pp. 195-238. Nankai Tracts Math., 5, World Sci. Publishing, River Edge, NJ (2002).

[23] T. Drumm, *Linear holonomy of Margulis space-times*, J. Differential Geom. **38** (1993), 679-691.

[24] *Lie groups and Lie algebras. II*, Encyclopaedia of Mathematical Sciences, **21**, Springer-Verlag, Berlin, 2000.

[25] P. E. Conner, F. Raymond, *Deforming homotopy equivalences to homeomorphisms in aspherical manifolds*, Bull. Amer. Math. Soc. **83** (1977), no. 1, 36-85.

[26] R.D. Canary, D.B.A. Epstein, P. Green, *Notes on notes of Thurston*, Analytical and Geometric Aspects of Hyperbolic Space, London Mathematical Society Lecture Note Series **111**, Cambridge University Press, (1984).

[27] D. Fried, W. M. Goldman, *Three-dimensional affine crystallographic groups*, Adv. in Math. **47** (1983), no. 1, 1-49.

[28] P. Furness, D. Arrowsmith, *Locally symmetric spaces*, J. London Math. Soc. (2) **10** (1972), 487-499.

[29] É. Ghys, *Déformations des structures complexes sur les espaces homogènes de* $\mathrm{SL}(2,\mathbb{C})$, J. reine angew. Math., **468** (1995), 113-138.

[30] W.M. Goldman, *Geometric structures on manifolds and varieties of representations*, Geometry of group representations, (Boulder, Colorado, 1987), Contemporary Mathematics **74** (1988), 169-198.

[31] M. Goze, Y. Khakimdjanov, *Nilpotent Lie Algebras*, Mathematics and Its Applications **361**, Kluwer 1996.

[32] F. Grunewald, G. Margulis, *Transitive and quasitransitive actions of affine groups preserving a generalized Lorentz-structure*, J. Geom. Phys. **5** (1988), no. 4, 493-531.

[33] F. Grunewald, J. O'Halloran, *Nilpotent groups and unipotent algebraic groups*, J. Pure Appl. Algebra **37** (1985), no. 3, 299-313.

[34] F. Grunewald, V. Platonov, *Rigidity results for groups with radical, cohomology of finite groups and arithmeticity problems*, Duke Math. J. **100** (1999), no. 2, 321-358.

[35] F. Grunewald, D. Segal, *On affine crystallographic groups*, J. Differential Geom. **40** (1994), no. 3, 563-594.

[36] K. Hasegawa, *Complex moduli and pseudo-Kähler structures on three-dimensional compact complex solvmanifolds*, Electronic Research Announcements in Mathematical Sciences **14**, Pages 3034 (July 13, 2007) S 1935-9179 AIMS (2007).

[37] S. Helgason, *Differential geometry, Lie groups, and symmetric spaces*, Pure and Applied Mathematics **80**, Academic Press, Inc., 1978.

[38] S. P. Kerckhoff, *The Nielsen realization problem*, Ann. of Math. (2) **117** (1983), no. 2, 235-265.

[39] H. Kim, *Complete left-invariant affine structures on nilpotent Lie groups*, J. Diff. Geom. **24** (1986), 373-394.

[40] T. Kobayashi, *Proper action on a homogeneous space of reductive type*, Math. Ann. **285** (1989), no. 2, 249-263.

[41] T. Kobayashi, *Discontinuous groups and Clifford-Klein forms of pseudo-Riemannian homogeneous manifolds*, Algebraic and analytic methods in representation theory (Sonderborg, 1994), 99-165, Perspect. Math., **17**, Academic Press, San Diego, CA, 1997.

[42] T. Kobayashi, *On discontinuous groups acting on homogeneous spaces with noncompact isotropy subgroups*, J. Geom. Phys. **12** (1993), no. 2, 133-144.

[43] T. Kobayashi, *Deformation of compact Clifford-Klein forms of indefinite-Riemannian homogeneous manifolds*, Math. Ann. **310** (1998), no. 3, 395-409.

[44] T. Kobayashi, *Discontinuous groups for non-Riemannian homogeneous spaces*, Mathematics unlimited—2001 and beyond, 723-747, Springer, Berlin, 2001.

[45] T. Kobayashi, *On discontinuous groups acting on non-Riemannian homogeneous spaces*, Sugaku Exposition, arXiv:math/0603319, Amer. Math. soc., to appear.

[46] T. Kobayashi, S. Nasrin, *Deformation of properly discontinuous actions of $\mathbb{Z}^k$ on $\mathbb{R}^{k+1}$*, Internat. J. Math. **17** (2006), no. 10, 1175-1193.

[47] K. Kodaira, D.C. Spencer, *On deformations of complex analytic structures I, II.*, Ann. of Math. (2) **67** (1958), 328-466.

[48] R. Kulkarni, *Proper actions and pseudo-Riemannian space forms*, Adv. Math. **40** (1981), 10-51.

[49] K. B. Lee, *Aspherical manifolds with virtually* 3*-step nilpotent fundamental group*, Amer. J. Math **105** (1983), no. 6, 1435-1453.

[50] K. B. Lee, F. Raymond, *Geometric realization of group extensions by the Seifert construction*, Contributions to group theory, 353-411, Contemp. Math. **33** (1984).

[51] R. Lipsman, *Proper actions and a compactness condition*, J. Lie Theory **5** (1995), no. 1, 25-39.

[52] A. Lubotzky, A. Magid, *Varieties of representations of finitely generated groups*, Mem. Amer. Math. Soc. **58** (1985), no. 336, xi+117 pp.

[53] A. I. Malcev, *On a class of homogeneous spaces*, Amer. Math. Soc. Translation **39** (1951).

[54] G. A. Margulis, *Free completely discontinuous groups of affine transformations*, Dokl. Akad. Nauk SSSR **272** (1983), no. 4, 785-788.

[55] Y. Matsushima, *Affine structures on complex manifolds*, Osaka J. Math. **5** (1968), 215-222.

[56] J. Milnor, *On fundamental groups of complete affinely flat manifolds*, Advances in Math. **25** (1977), no. 2, 178-187.

[57] G.D. Mostow, *On the fundamental group of a homogeneous space*, Ann. of Math. (2) **66** (1957), 249-255.

[58] G.D. Mostow, *Representative functions on discrete groups and solvable arithmetic subgroups*, Amer. J. Math. **92** (1970) 1-32.

[59] S. Nasrin, *Criterion of proper actions for 2-step nilpotent Lie groups*, Tokyo J. Math. **24** (2001) 535-543.

[60] J. Nielsen, *Abbildungsklassen endlicher Ordnung*, Acta Math. **75** (1942), 23-115.

[61] A. M. Macbeath, D. Singerman, *Spaces of subgroups and Teichmüller space*,

Proc. London Math. Soc. (3) **31** (1975), no. 2, 211-256.
[62] D. Johnson, J. Millson, *Deformation spaces associated to compact hyperbolic manifolds*, Discrete groups in geometry and analysis (New Haven, Conn., 1984), Progr. Math. **67** (1987), 48-106.
[63] T. Nagano, K. Yagi, *The affine structures on the real two-torus*, Osaka J. Math. **11** (1974), 181-210.
[64] H. Oh, D. Witte, *New examples of compact Clifford-Klein forms of homogeneous spaces of* $\mathrm{SO}(2,n)$, Internat. Math. Res. Notices **2000**, no. 5, 235-251.
[65] A. Püttmann, *Biquotient actions on unipotent Lie groups*, Internat. J. Math., to appear.
[66] M. S. Raghunathan, *Discrete subgroups of Lie groups*, Ergebnisse der Mathematik und ihrer Grenzgebiete **68**, Springer-Verlag, (1972).
[67] J. G. Ratcliffe, *Foundations of Hyperbolic Manifolds*, Graduate Texts in Mathematics **149**, Springer-Verlag, (1994).
[68] M. Rosenlicht, *On quotient varieties and the affine embedding of certain homogeneous spaces*, Trans. Amer. Math. Soc. **101** (1961), 211-223.
[69] F. Salein, *Variétés anti-de Sitter de dimension 3 exotiques*, Ann. Inst. Fourier (Grenoble) **50** (2000), no. 1, 257-284.
[70] J. Scheuneman, *Examples of compact locally affine spaces, Bull. Amer. Math. Soc.* **77** (1971), 589-592.
[71] J. Scheuneman, *Affine structures on three-step nilpotent Lie algebras*, Proc. Amer. Math. Soc. **46** (1974), 451-454.
[72] D. Segal, *Polycyclic groups*, Cambridge tracts in mathematics **82**, Cambridge University Press, (1983).
[73] J.-P. Serre, *Arithmetic groups*, London Math. Soc. Lecture Notes **36**, (1979), 105-135.
[74] R. W. Richardson, P. Slodowy, *Minimum vectors for real reductive algebraic groups*, J. London Math. Soc. (2) **42** (1990), 409-429.
[75] W. P. Thurston, *The Geometry and Topology of Three-Manifolds*, mimeographed lecture notes, Princeton University (1980), http://www.msri.org/publications/books/gt3m/ (electronic version 1.1)
[76] W. P. Thurston, *Three dimensional geometry and topology*, Vol.1, Princeton University Press, (1997).
[77] V. S. Varadarajan, *Lie groups, Lie algebras, and their representations*, Reprint of the 1974 edition, Graduate Texts in Mathematics **102**, Springer-Verlag, (1984).
[78] A. Weil, *On discrete subgroups of Lie groups*, Ann. of Math. (2) **75** (1962), 578-602.
[79] D. Witte, *Superrigidity of lattices in solvable Lie groups*, Invent. Math. **122** (1995), 147-193.
[80] K. Yagi, *On compact homogeneous affine manifolds*, Osaka J. Math. **7** (1970), 457-475.
[81] T. Yoshino, *Criterion of proper actions for 3-step nilpotent Lie groups*, Internat. J. Math. **18** (2007), no. 7, 783-795.
[82] H. Zassenhaus, *Über einen Algorithmus zur Bestimmung der Raumgruppen*, Comment. Math. Helv. **21** (1948), 117-141.

[83] H. Zieschang, *Finite groups of mapping classes of surfaces*, Lecture Notes in Mathematics **875**, Springer Verlag, (1981).

*Cohomology of Groups and Algebraic K-theory*
ALM 12, pp. 131–166



# Lectures on the Cohomology of Groups

Kenneth S. Brown*

**Abstract**

These lectures give an introduction to homology and cohomology theories of groups and their basic properties, in particular cohomological finiteness properties.

**2000 Mathematics Subject Classification:** 20F32, 20J05

**Keywords and Phrases:** Cohomology of group, equivariant homology, cohomological finiteness

## 0. Historical introduction

The cohomology theory of groups arose from both topological and algebraic sources. The starting point for the topological aspect of the theory was a 1936 paper by Hurewicz [7], in which he introduced aspherical spaces. These are spaces $X$ such that $\pi_n(X) = 0$ for $n \neq 1$. (Hurewicz had introduced higher homotopy groups just one year earlier, and he was now trying to understand the spaces with the simplest possible higher homotopy groups.) Hurewicz proved that such an $X$ is determined up to homotopy equivalence by its fundamental group $\pi := \pi_1(X)$. Thus homotopy invariants of $X$ can be thought of as invariants of the group $\pi$. Examples of such invariants include homology, cohomology, and the Euler characteristic. Thus we can define

$$H_*(\pi) := H_*(X) \tag{0.1}$$

if $X$ is an aspherical space with fundamental group $\pi$, and similarly for cohomology and the Euler characteristic. [We will replace (0.1) with an equivalent algebraic definition in the next section.]

For example, we have $H_0(\pi) = \mathbb{Z}$ for all $\pi$ (so $H_0$ is a boring invariant of $\pi$), and $H_1(\pi) = \pi_{\mathrm{ab}} := \pi/[\pi, \pi]$, the abelianization of $\pi$. This is less boring, but still well known. The next group, $H_2(\pi)$ is more interesting. As a sample calculation, note that $H_2(\pi) = \mathbb{Z}$ if $\pi$ is free abelian of rank 2, since we can take $X$ to be the torus $S^1 \times S^1$.

*Department of Mathematics, Cornell University, Ithaca, NY 14853, USA. Email: kbrown@cornell.edu

The next step was taken by Hopf [6] in 1942. One of the results of his paper is that $H_2(\pi)$ can be used to measure the failure of the Hurewicz map to be surjective in dimension 2. More precisely, consider an arbitrary path-connected space $X$ (not necessarily aspherical), and set $\pi := \pi_1(X)$. The Hurewicz map $h_2 \colon \pi_2(X) \to H_2(X)$ is an isomorphism if $\pi$ is the trivial group (i.e., if $X$ is simply connected). In general, according to Hopf, there is an exact sequence

$$\pi_2(X) \xrightarrow{h_2} H_2(X) \longrightarrow H_2(\pi) \longrightarrow 0.$$

A second result of Hopf's paper is a calculation of $H_2(\pi)$ if $\pi$ is given as a quotient of a free group. Suppose $\pi = F/R$, where $F$ is a free group and $R$ is a normal subgroup. Then Hopf's formula says

$$H_2(\pi) \cong (R \cap [F, F]) / [R, F].$$

Thus $H_2(\pi)$ can be viewed as the group of relations among commutators, modulo those relations that hold trivially.

Following Hopf's paper, the subject developed rapidly in the 1940s, due primarily to the work of Eckmann, Eilenberg–MacLane, Freudenthal, and Hopf. By the end of the decade there was a purely algebraic definition of the homology and cohomology of a group. As a consequence of the algebraic definition, one could make connections with algebra going back to the early 1900s. For example, $H^1$ turned out to be a group of equivalence classes of "derivations" (also called "crossed homomorphisms"). And $H^2$ turned out to be similarly related to "factor sets", which had been studied by Schur [11], Schreier [9], and Brauer [2], while $H_2$ coincided with the "Schur multiplier" (also introduced in [11]). Finally, $H^3$ had appeared in the work of Teichmüller [14]. These are the algebraic sources referred to in the first sentence of this section. Of course, none of this algebra had suggested that derivations, factor sets, and so on were part of a coherent "cohomology theory". This had to wait for the impetus from topology.

We now proceed to the six lectures, which correspond to the six sections that follow. They give a very brief introduction to the homology and cohomology theory of groups, with an emphasis on infinite groups and finiteness properties. The lectures are based on my book [3] and are organized as follows:

1. In the first lecture we will redefine $H_*(G)$ for an arbitrary group $G$, taking the algebraic point of view (homological algebra) that had evolved by the end of the 1940s. Although we are now thinking algebraically, we will always keep the topological interpretation in mind and will use it when convenient.
2. In the second lecture we take a first look at finiteness properties of infinite groups. These properties may be defined either algebraically or topologically.
3. The third lecture is devoted to $H_*(G, M)$ and $H^*(G, M)$, homology and cohomology with coefficients in a $G$-module $M$. These generalize $H_*(G)$, which is $H_*(G, \mathbb{Z})$ (with trivial $G$-action on the coefficient module $\mathbb{Z}$). They are important for both algebraic and topological applications, even if one is primarily interested in the special case $H_*(G)$.

4. With the material of Lecture 3 available, we can treat finiteness properties more seriously in Lecture 4. In particular, we will discuss homological duality. This includes Poincaré duality as a special case, but it is much more general.
5. Lecture 5 is technical. I will attempt to give an introduction to equivariant homology theory and the associated spectral sequences.
6. The final lecture treats the special features of cohomology theory when the group $G$ is finite. Although we are mainly interested in infinite groups in these lectures, anyone learning the cohomology theory of groups for the first time should know the basic facts about the finite case.

# 1. The homology of a group

A reference for this section is [3, Chapter II]. We will give an algebraic approach to homology theory, based on free resolutions. These generalize presentations of modules by generators and relations. If $R$ is a ring and $M$ is an $R$-module with $n$ generators, then we have an exact sequence

$$R^n \longrightarrow M \longrightarrow 0. \tag{1.1}$$

Here $n$ is a cardinal number, possibly infinite. The surjection $R^n \twoheadrightarrow M$ has a kernel $K$, whose elements represent relations among the given generators. If $K$ admits $m$ generators as an $R$-module (so that $M$ is defined by $n$ generators and $m$ relations), then we can map $R^m$ onto $K$, thereby obtaining a continuation of (1.1) to a diagram

$$\begin{array}{ccccccc} R^m & \longrightarrow & R^n & \longrightarrow & M & \longrightarrow & 0 \\ & \searrow & & \nearrow & & & \\ & & K & & & & \end{array} \tag{1.2}$$

with the top row exact. The surjection $R^m \twoheadrightarrow K$ has a kernel $L$ ( consisting of "relations among the relations"). Choosing a free module that maps onto $L$ and continuing *ad infinitum*, we obtain an exact sequence

$$\cdots \to F_2 \to F_1 \to F_0 \to M \to 0,$$

where each $F_i$ is a free $R$-module. Such an exact sequence is called a *free resolution* of $M$.

We will mainly be interested in the case where $R$ is the integral group ring $\mathbb{Z}G$ of a group $G$ and $M = \mathbb{Z}$, with trivial $G$-action. This situation arises naturally in topology, as follows. Suppose $X$ is a $G$-CW-complex, by which we mean a CW-complex with a $G$-action that permutes the cells. Then $G$ acts on the cellular chain complex $C_*(X)$, which therefore becomes a chain complex of $\mathbb{Z}G$-modules. This complex is naturally augmented over $\mathbb{Z}$, so we have a diagram

$$\cdots \to C_2(X) \to C_1(X) \to C_0(X) \to \mathbb{Z} \to 0 \tag{1.3}$$

of $\mathbb{Z}G$-modules. If $G$ acts freely on $X$ (i.e., all cell stabilizers are trivial), then each module $C_n(X)$ is a direct sum of copies of $\mathbb{Z}G$, with one copy for each $G$-orbit of $n$-cells. In particular, $C_n(X)$ is a free $\mathbb{Z}G$-module. If, in addition, $X$ is contractible, then (1.3) is a free resolution of $\mathbb{Z}$ over $\mathbb{Z}G$.

We can reformulate this as follows: Introduce the quotient complex $Y := G\backslash X$. If $G$ acts freely on $X$, then we can view $X$ as a regular covering space of $Y$, with $G$ as the group of deck transformations. If $X$ is simply connected, then $X$ is the universal cover of $Y$, and $G$ can be identified with $\pi_1(Y)$. If, in addition, $X$ is contractible, then elementary homotopy theory implies that the higher homotopy groups $\pi_i(Y)$ are trivial, since these do not change when one passes to a covering space. Thus $Y$ is an aspherical space in the sense of Section 0. Conversely, if we start with an aspherical CW-complex $Y$, then its universal cover $X$ has trivial homotopy groups in all dimensions and hence is contractible by a theorem of Whitehead. So the situation we described in the previous paragraph can be summarized as follows: If $Y$ is an aspherical CW-complex with fundamental group $G$, then its universal cover $X$ is a contractible, free $G$-CW-complex, whose cellular chain complex gives rise to a free resolution of $\mathbb{Z}$ over $\mathbb{Z}G$ as in (1.3). There is some standard notation and terminology associated with this situation:

**Definition 1.1.** Let $Y$ be a CW-complex with fundamental group $G$. We say that $Y$ is an *Eilenberg–MacLane complex* of type $K(G, 1)$ if it is aspherical or, equivalently, if its universal cover is contractible.

In some contexts one also writes $Y = BG$ and calls $Y$ a *classifying space* for $G$, but we will stick to the $K(G, 1)$ notation here.

It is a fact, which is not difficult to prove, that every group $G$ admits a $K(G, 1)$-complex $Y$; moreover, $Y$ is unique up to homotopy equivalence. (The uniqueness part is a theorem of Hurewicz that we quoted in Section 0.) This fact has an algebraic analogue, whose proof is even easier, which is sometimes called the *fundamental lemma of homological algebra*: Given any module $M$ (over an arbitrary ring), free resolutions of $M$ exist and are unique up to chain homotopy equivalence. We already proved existence above. And a proof of uniqueness can be found in any book on homological algebra. It consists of constructing the desired chain maps and homotopies step by step. At each step, one has to solve a mapping problem of the following form:

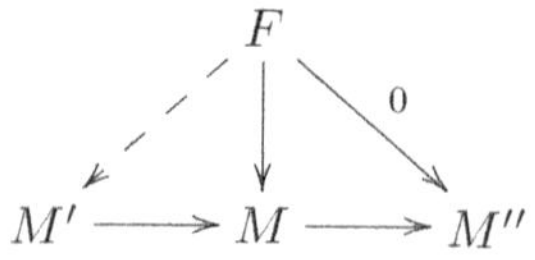


where $F$ is free and the row is exact. The solid arrows represent given maps, with the composite $F \to M \to M''$ equal to the zero map, and the dotted arrow represents a map we want to construct. Note that the image of $F \to M$ lies in the kernel of $M \to M''$ and hence in the image of $M' \to M$; so we can construct the desired map by lifting the images of basis elements of the free module $F$.

Free modules are not the only modules for which such mapping problems can be solved. If $P$ is a module such that all mapping problems as above can be solved (with $F$ replaced by $P$), then $P$ is said to be a *projective* module. There are many equivalent characterizations of projective modules. For example, a module $P$ is projective if and only if it is a direct summand of a free module, i.e., there is a module $Q$ such that $P \oplus Q$ is free.

The upshot of the discussion above is that the fundamental lemma of homological algebra is applicable to *projective resolutions*, not just to free ones. This gives us more flexibility in constructing resolutions of a module, all of which are homotopy equivalent. Even in the setting where we can find a free resolution via topology as above, it might be more convenient to consider projective resolutions that are not necessarily free (and hence, in particular, do not necessarily come from topology).

We proceed now to two examples, where we do in fact use topology to get the desired resolutions.

*Example* 1.2. Let $G = F(S)$, the free group generated by a set $S$. Let $X$ be the Cayley graph of $G$ with respect to $S$. It is a tree with vertex set $G$, with an edge from $g$ to $gs$ for each $g \in G$ and $s \in S$. It is a free, contractible, $G$-CW-complex. There is a single $G$-orbit of vertices, and there is one $G$-orbit of edges for each $s \in S$. Thus the quotient complex $Y$ is a bouquet of circles indexed by $S$. The resulting free resolution is

$$0 \longrightarrow \mathbb{Z}G^{(S)} \xrightarrow{\partial} \mathbb{Z}G \xrightarrow{\varepsilon} \mathbb{Z} \longrightarrow 0, \tag{1.4}$$

where $\mathbb{Z}G^{(S)}$ is a free $\mathbb{Z}G$-module with a basis $(e_s)_{s\in S}$. Here $\varepsilon\colon \mathbb{Z}G \to \mathbb{Z}$ is the *canonical augmentation*, given by $\varepsilon(g) = 1$ for all $g \in G$, and $\partial\colon \mathbb{Z}G^{(S)} \to \mathbb{Z}G$ is given by $\partial(e_s) = s - 1$ for $s \in S$. (This is because the circle corresponding to $s$ lifts to the edge in $X$ from the vertex 1 to the vertex $s$.) Although we have proved the exactness of (1.4) via topology, it is possible to give a purely algebraic proof; see [3, Section IV.4.2, Exercise 3].

In the special case where $S$ is a singleton $\{t\}$ (i.e., $G$ is an infinite cyclic group generated by $t$), we can identify $\mathbb{Z}G$ with the ring $\mathbb{Z}[t, t^{-1}]$ of Laurent polynomials, and (1.4) becomes

$$0 \longrightarrow \mathbb{Z}[t, t^{-1}] \xrightarrow{t-1} \mathbb{Z}[t, t^{-1}] \xrightarrow{\varepsilon} \mathbb{Z} \longrightarrow 0. \tag{1.5}$$

The arrow labeled $t - 1$ is given by multiplication by $t - 1$, and $\varepsilon$ can be described as the evaluation map $f \mapsto f(1)$ for $f = f(t) \in \mathbb{Z}[t, t^{-1}]$. In this special case, it is quite easy to verify the exactness by pure algebra.

*Example* 1.3. Suppose $X$ is a free $G$-CW-complex homeomorphic to an odd-dimensional sphere $S^{2k-1}$. Then $X$ is of course not contractible, but we can still use its cellular chain complex $C_* := C_*(X)$ to get a free resolution of $\mathbb{Z}$ over $\mathbb{Z}G$. Indeed, $X$ has homology only in dimensions 0 and $2k-1$, where it is $\mathbb{Z}$ (with trivial $G$-action by the Lefschetz fixed-point theorem). So we have an exact sequence of $\mathbb{Z}G$-modules

$$0 \to \mathbb{Z} \to C_{2k-1} \to \cdots \to C_1 \to C_0 \to \mathbb{Z} \to 0 \tag{1.6}$$

with each $C_i$ free. We can now obtain the desired free resolution by splicing together infinitely many copies of (1.6):

$$\cdots \longrightarrow C_1 \longrightarrow C_0 \longrightarrow C_{2k-1} \longrightarrow \cdots \longrightarrow C_0 \longrightarrow \mathbb{Z} \longrightarrow 0 \qquad (1.7)$$

with $C_0 \searrow \mathbb{Z} \nearrow C_{2k-1}$.

Note that, in contrast, to our previous examples, this resolution continues forever to the left.

As a simple special case, consider $G = \langle t\,;\, t^n = 1 \rangle$, the finite cyclic group of order $n$, acting on the circle by rotation. We can triangulate the circle by cutting it into $n$ arcs, so that there is a single orbit of 1-cells and a single orbit of 0-cells. It is then easy to check that (1.6) becomes

$$0 \longrightarrow \mathbb{Z} \xrightarrow{\eta} \mathbb{Z}G \xrightarrow{t-1} \mathbb{Z}G \xrightarrow{\varepsilon} \mathbb{Z} \longrightarrow 0, \qquad (1.8)$$

where $\varepsilon$ is again the canonical augmentation and

$$\eta(1) = N := \sum_{i=0}^{n-1} t^i.$$

(Note that $N$ is simply the sum of all the group elements. Such an element $N$ exists for any finite group $G$ and is called the *norm element* of $\mathbb{Z}G$.) The resulting resolution (1.7) is then

$$\cdots \xrightarrow{N} \mathbb{Z}G \xrightarrow{t-1} \mathbb{Z}G \xrightarrow{N} \mathbb{Z}G \xrightarrow{t-1} \mathbb{Z}G \xrightarrow{\varepsilon} \mathbb{Z} \longrightarrow 0, \qquad (1.9)$$

the maps being alternately multiplication by $N$ and by $t-1$. As in (1.5), it is not hard to verify exactness by pure algebra. (Note that $\mathbb{Z}G = \mathbb{Z}[t]/(t^n - 1)$ and that $t^n - 1 = (t-1)N$ in the polynomial ring $\mathbb{Z}[t]$.)

We are now ready to give the (algebraic) definition of the homology of a group $G$. It makes use of the *coinvariants* functor $M \mapsto M_G$, which is the algebraic analogue of forming the quotient by a $G$-action. Here $M$ is a $G$-module (i.e., a $\mathbb{Z}G$-module), and

$$M_G := M/\langle gm - m \mid g \in G,\, m \in M \rangle,$$

where the angle brackets denote the subgroup generated by the given elements. Thus $M_G$ is the largest quotient of $M$ on which $G$ acts trivially. (Compare this to the set of *invariants*

$$M^G := \{m \in M \mid gm = m \text{ for all } g \in G\},$$

which is the largest submodule of $M$ on which $G$ acts trivially.)

**Definition 1.4.** Given a group $G$, choose a projective resolution $P = (P_i)_{i \geq 0}$ of $\mathbb{Z}$ over $\mathbb{Z}G$ and set

$$H_*(G) := H_*(P_G).$$

In words, we apply the coinvariants functor to the acyclic chain complex $P$ and then take homology. The fundamental lemma of homological algebra guarantees that $H_*(G)$ is well-defined (independent of the choice of $P$) up to canonical isomorphism. We also immediately get

$$H_*(G) = H_*(Y)$$

if $Y$ is a $K(G,1)$-complex. Indeed, if $X$ is the universal cover of $Y$, then we can take $P$ to be $C_*(X)$, and then $P_G = C_*(Y)$.

From the algebraic point of view, we can think of $H_*(G)$ as measuring the failure of the coinvariants functor to be exact. (If it were exact, then $P_G$ would again be acyclic and hence would have trivial homology in positive dimensions.) But this algebraic point of view would seem very artificial without the motivation from topology. The latter dictated the choice of $\mathbb{Z}G$ as the ring, the choice of $\mathbb{Z}$ as the module to resolve, and the choice of coinvariants as the functor to apply.

*Exercise* 1.5. Use the resolutions in Examples 1.2 and 1.3 to calculate $H_*(G)$ if $G$ is free or cyclic.

# 2. Finiteness properties (introduction)

There is a canonical free resolution of $\mathbb{Z}$ over $\mathbb{Z}G$, sometimes called the bar resolution, that one can use in principle to compute the homology of any group $G$. It comes from a canonical $K(G,1)$-complex. See [3, Section I.5]. The bar resolution and canonical $K(G,1)$ are useful for theoretical purposes, but they are very big. If one wants to actually compute the homology, or at least discover qualitative properties of it (such as finite generation or vanishing in high dimensions), it is desirable to have a small resolution $P$ or a small $K(G,1)$-complex $Y$. Here "small" might mean that $P_n = 0$ for large $n$ (or that $Y$ is finite dimensional). Alternatively, it might mean that each module $P_n$ is finitely generated (or that $Y$ has only finitely many cells in each dimension). Examples 1.2 and 1.3 illustrate this. We will treat these two notions of smallness in the two subsections that follow. Our treatment is based on [3, Chapter VIII].

## 2.1. Dimension

There are two natural definitions of the dimension of a group, depending on whether we think topologically or algebraically.

**Definition 2.1.** The *geometric dimension* of $G$, denoted $\operatorname{gd} G$, is the smallest non-negative integer $n$ such that there exists an $n$-dimensional $K(G,1)$-complex. (Or, if no such $n$ exists, then we set $\operatorname{gd} G = \infty$.)

**Definition 2.2.** The *cohomological dimension* of $G$, denoted $\operatorname{cd} G$, is the smallest non-negative integer $n$ such that there exists a projective resolution $P = (P_i)_{i\geq 0}$ of $\mathbb{Z}$ over $\mathbb{Z}G$ of length $\leq n$, i.e., satisfying $P_i = 0$ for $i > n$. (Or, if no such $n$ exists, then we set $\operatorname{cd} G = \infty$.)

*Remark* 2.3. It is not clear at this stage why we call cd(−) "cohomological dimension" instead of, for example, "projective dimension". The name comes from a characterization of $\operatorname{cd} G$ that we will explain in Section 4.

Since a $K(G,1)$-complex $Y$ yields a free resolution of length equal to the dimension of $Y$, it is clear that

$$\operatorname{cd} G \leq \operatorname{gd} G. \tag{2.1}$$

Another simple observation is that dimension can only go down if one passes to a subgroup. In other words, if dim(−) denotes either cohomological dimension or geometric dimension, then

$$\dim(H) \leq \dim(G) \quad \text{if } H \leq G. \tag{2.2}$$

For cd this follows from the fact that $\mathbb{Z}G$ is a free $\mathbb{Z}H$-module, which implies that a projective $\mathbb{Z}G$-module is also projective as a $\mathbb{Z}H$-module. Hence any projective resolution of $\mathbb{Z}$ over $\mathbb{Z}G$ is still a projective resolution of the same length when viewed as a complex of $\mathbb{Z}H$-modules. For gd one instead argues using covering spaces. [Any $K(G,1)$-complex has a covering space that is a $K(H,1)$-complex of the same dimension.]

*Examples* 2.4. (a) A 0-dimensional $K(G,1)$ is necessarily a point, so it exists if and only if $G$ is the trivial group. The algebraic analogue of this (which requires a little thought but is still easy) is that $\mathbb{Z}$ is a projective $\mathbb{Z}G$-module if and only if $G$ is the trivial group. Hence

$$\operatorname{cd} G = 0 \iff \operatorname{gd} G = 0 \iff G = \{1\}.$$

(b) A 1-dimensional $K(G,1)$ is necessarily homotopy equivalent to a bouquet of circles, so there exists one if and only if $G$ is free. (For the "if" part, see Example 1.2.) It is also true, but much more difficult, that $\operatorname{cd} G = 1$ only if $G$ is free. This is a deep theorem of Stallings [12] and Swan [13]. Hence

$$\operatorname{cd} G = 1 \iff \operatorname{gd} G = 1 \iff G \text{ is free and nontrivial.}$$

(c) If $G$ is the fundamental group of a closed surface $Y$ other than the sphere or projective plane, then $\operatorname{cd} G = \operatorname{gd} G = 2$. Indeed, $Y$ is a $K(G,1)$ since its universal cover is homeomorphic to $\mathbb{R}^2$, so $\operatorname{cd} G \leq \operatorname{gd} G \leq 2$; equality holds in the orientable case because $H_2(G) = H_2(Y) \neq 0$, implying $\operatorname{cd} G \geq 2$. In the non-orientable case one can give a similar argument based on mod 2 homology, or one can pass to the orientable double cover and apply (2.2).

(d) The free abelian group $G = \mathbb{Z}^n$ has $\operatorname{cd} G = \operatorname{gd} G = n$. This follows as in (c) from the fact that the $n$-torus is a $K(G,1)$ and has nontrivial homology in dimension $n$. More generally, if $G$ is the fundamental group of a closed aspherical $n$-manifold, then $\operatorname{cd} G = \operatorname{gd} G = n$.

(e) If $G$ is a nontrivial finite cyclic group, then $\operatorname{cd} G = \operatorname{gd} G = \infty$ since $G$ has nontrivial homology in arbitrarily high dimensions (see Exercise 1.5). In view of (2.2), it follows that $\operatorname{cd} G = \operatorname{gd} G = \infty$ for any group $G$ with torsion. Equivalently,

$$\operatorname{cd} G < \infty \implies G \text{ is torsion-free.}$$

This can be used to prove, for instance, that knot groups are torsion-free (since knot complements are aspherical).
(f) Torsion-free arithmetic groups have finite cohomological dimension, which can be calculated explicitly by methods that we will explain at the end of Section 4.

It is not known whether equality always holds in (2.1). We have seen in (a) and (b) above that equality holds if $\operatorname{cd} G \leq 1$. It also holds if $\operatorname{cd} G \geq 3$. In fact, for any $n \geq 3$ one can show by fairly straightforward homotopy theory that the existence of a projective resolution of length $n$ implies the existence of an $n$-dimensional $K(G,1)$-complex. The argument breaks down if $n = 2$, however, so it is conceivable that there is a group $G$ with $\operatorname{cd} G = 2$ but $\operatorname{gd} G = 3$. The famous *Eilenberg–Ganea problem* asks whether or not this can happen. In summary, we have:

**Theorem 2.5.** *For any group $G$,*

$$\operatorname{cd} G \leq \operatorname{gd} G,$$

*with equality except possibly if* $\operatorname{cd} G = 2$ *but* $\operatorname{gd} G = 3$.

In view of the close connection between $\operatorname{cd} G$ and $\operatorname{gd} G$, we will often simply write $\dim G$ in cases where they are known to be equal or where it does not matter which one we use. For example, we can speak of *finite-dimensional groups* and write $\dim G < \infty$.

## 2.2. The $F_n$ and $FP_n$ conditions

**Definition 2.6.** We say that a group $G$ is of *type* $F_n$ $(0 \leq n < \infty)$ if there is a $K(G,1)$-complex with a finite $n$-skeleton, i.e., with only finitely many cells in dimensions $\leq n$. We say that $G$ is of *type* $F_\infty$ if there is a $K(G,1)$ with all of its skeleta finite, and that $G$ is of *type* $F$ if there is a finite $K(G,1)$.

One can always build a $K(G,1)$-complex by starting with a single vertex, then attaching 1-cells corresponding to generators of $G$, then attaching 2-cells corresponding to relators, and then attaching 3-cells, 4-cells, and so on, to kill the higher homotopy groups $\pi_2, \pi_3, \ldots$. This leads to the following interpretation of the $\mathrm{F}_n$ property for small $n$:

- Every group is of type $\mathrm{F}_0$.
- $G$ is of type $\mathrm{F}_1$ if and only if it is finitely generated.
- $G$ is of type $\mathrm{F}_2$ if and only if it is finitely presented.

The successively stronger higher finiteness properties $\mathrm{F}_3$, $\mathrm{F}_4$, ..., $\mathrm{F}_\infty$, and F are more subtle and do not have simple group-theoretic interpretations. Groups of type F, of course, are finite dimensional; in particular, only torsion-free groups can have this property.

*Remarks* 2.7. (a) All of the examples mentioned in our discussion of dimension are of type $\mathrm{F}_\infty$ (provided, in the case of free groups, that we require the group to be finitely generated). And all of the torsion-free groups (with the same proviso) are even of type F.

(b) In view of (a), one might wonder whether all torsion-free groups of type $F_\infty$ are in fact of type F. This is not the case. The first counterexample was given by Brown and Geoghegan [4]. That example, however, is infinite dimensional. So one can still ask whether every *finite-dimensional* group of type $F_\infty$ is of type F. In other words, if there is a $K(G,1)$-complex with only finitely many cells in each dimension, and if there is also a $K(G,1)$-complex that is finite-dimensional, is there a $K(G,1)$-complex that satisfies both of these conditions simultaneously? This question is still open. We will understand the difficulty better shortly, after we discuss the $FP_n$ conditions.

**Definition 2.8.** We say that a group $G$ is of *type* $FP_n$ $(0 \leq n < \infty)$ if there is a projective resolution $P$ of $\mathbb{Z}$ over $\mathbb{Z}G$ such that $P_i$ is finitely generated for $i \leq n$. We say that $G$ is of *type* $FP_\infty$ if there is a projective resolution $P$ of $\mathbb{Z}$ over $\mathbb{Z}G$ with $P_i$ finitely generated for all $i$, and that $G$ is of *type FP* if there is a finite projective resolution, i.e., a projective resolution such that $P_i$ is finitely generated for all $i$ and is 0 for sufficiently large $i$.

It is obvious that $F_n \implies FP_n$. Moreover, one can show:

- Every group is of type $FP_0$.
- $G$ is of type $FP_1$ if and only if it is finitely generated (so $FP_1$ is equivalent to $F_1$).
- $G$ is of type $FP_2$ if and only if $G \cong \widetilde{G}/N$, where $\widetilde{G}$ is finitely presented and $N$ is a perfect normal subgroup.

If $G \cong \widetilde{G}/N$ as above, then $G$ is finitely presented if and only if $N$ is finitely generated as a normal subgroup. Examples due to Bestvina and Brady [1] show that this need not be the case. In other words, $FP_2$ is definitely weaker than $F_2$. For higher $n$, however, there is no further difference between $FP_n$ and $F_n$ beyond finite presentability. In other words, $G$ is of type $F_n$ $(2 \leq n \leq \infty)$ if and only if it is finitely presented and of type $FP_n$. The analogous statement for property F is not known:

*Question* 2.9. If $G$ is finitely presented and of type FP, is $G$ of type F?

To understand this better, we introduce one more finiteness condition, which is algebraic in nature but is motivated by topology:

**Definition 2.10.** $G$ is of *type FL* if there is a finite free resolution of $\mathbb{Z}$ over $\mathbb{Z}G$, i.e., a free resolution $F$ such that $F_i$ is finitely generated for all $i$ and is 0 for sufficiently large $i$.

(The "L" in "FL" stands for "libre". Some authors write "FF", for "finite free" instead of "FL".)

Now it is not difficult to show that $G$ is of type F if and only if it is finitely presented and of type FL. So the real question underlying Question 2.9 is:

*Question* 2.11. If $G$ is of type FP, is it of type FL?

*Remarks* 2.12. (a) A group $G$ is of type FP if and only if it is finite dimensional and of type $FP_\infty$. (Thus algebra is easier than topology in this setting, cf. Remark 2.7(b).) In fact, if $G$ is of type $FP_\infty$ and finite dimensional, then one can construct a projective resolution step by step using finitely generated modules, and the kernel will eventually be projective. One can therefore stop at that stage and have a finite projective resolution.

(b) In the step-by-step construction just described, we can use finitely-generated free modules at each stage. The kernels will all be finitely generated, so there is no obstruction to continuing the process. But when we reach the point where the kernel $K$ is projective, we cannot be sure that it will be free. It is therefore not clear that we can stop and have a finite free resolution. In fact, one can show that there exists a finite *free* resolution if and only if $K$ is stably free. So Questions 2.9 and 2.11 would have affirmative answers if we knew that finitely-generated projective $\mathbb{Z}G$-modules are always stably free if $G$ is torsion free. A famous conjecture asserts that this is the indeed the case; see Lück's lectures in this volume.

(c) It can be shown that a group $G$ is finitely presented and of type FP if and only if $K(G,1)$ is finitely dominated (i.e., is a retract up to homotopy of a finite complex). This is a topological analogue of the fact that a module is projective if and only if it is free. So the question we have been discussing is whether a finitely-dominated $K(G,1)$ is homotopy equivalent to a finite complex. Now there are in fact plenty of examples of finitely-dominated spaces that are not homotopy equivalent to finite complexes; but no known examples have torsion-free fundamental groups. In particular, they cannot be $K(G,1)$s.

# 3. Homology with coefficients

This section is based on [3, Chapter III].

## 3.1. Definitions

We defined the homology of a group $G$ by applying the coinvariants functor $(-)_G$ to a projective resolution $P$ of $\mathbb{Z}$ over $\mathbb{Z}G$. There are other functors we could apply.

First, fix a $G$-module $M$ and consider the tensor-product functor $- \otimes_G M$. Applying this to a projective resolution $P$ yields a nonnegative chain complex

$$\cdots \to P_1 \otimes_G M \to P_0 \otimes_G M,$$

and taking homology gives the *homology of $G$ with coefficients in $M$*:

$$H_*(G,M) := H_*(P \otimes_G M).$$

A word of explanation is in order concerning the tensor product. Normally, a tensor product $N \otimes_R M$ is defined when $N$ is a right $R$-module and $M$ is a left $R$-module. It is obtained by forming the tensor product $N \otimes M := N \otimes_{\mathbb{Z}} M$ and introducing the relations $nr \otimes m = n \otimes rm$ for $n \in N$, $r \in R$, and $m \in M$. In case $R = \mathbb{Z}G$, however, we work entirely with left modules and, in order to form the tensor product $N \otimes_G M := N \otimes_{\mathbb{Z}G} M$, we convert the left $G$-module $N$ to a right

$G$-module by setting $ng := g^{-1}n$ for $g \in G$ and $n \in N$. Thus $N \otimes_G M$ is obtained from $N \otimes M$ by introducing the relations $g^{-1}n \otimes m = n \otimes gm$. If we replace $n$ by $gn$, this becomes

$$n \otimes m = gn \otimes gm. \tag{3.1}$$

So we can also describe the tensor product by

$$N \otimes_G M = (N \otimes M)_G\,, \tag{3.2}$$

where the coinvariants on the right are formed with respect to the diagonal $G$-action on $N \otimes M$ (i.e., $g(n \otimes m) := gn \otimes gm$). If we take $M = \mathbb{Z}$, for example, then $N \otimes \mathbb{Z} = N$, so (3.1) gives $N \otimes_G \mathbb{Z} = N_G$. Thus

$$H_*(G, \mathbb{Z}) = H_*(G),$$

and we see that the homology defined in Section 1 is a special case of the homology that we are considering now.

As in that case, there is a topological interpretation, provided one knows about homology with local coefficients. Namely, if $Y$ is a $K(G, 1)$-complex, then the $G$-module $M$ may be viewed as a local coefficient system on $Y$, and

$$H_*(G, M) = H_*(Y, M). \tag{3.3}$$

If $G$ acts trivially on $M$, the right side is just the ordinary homology of $Y$ with coefficients in the abelian group $M$.

Next, we consider the (contravariant) Hom functor $\mathrm{Hom}_G(-, M)$ for a fixed $G$-module $M$. Applying this to a projective resolution $P$ yields a nonnegative cochain complex

$$\mathrm{Hom}_G(P_0, M) \to \mathrm{Hom}_G(P_1, M) \to \cdots\,,$$

and taking cohomology gives the *cohomology of $G$ with coefficients in $M$*:

$$H^*(G, M) := H^*(\mathrm{Hom}_G(P, M)).$$

To better see the analogy with homology, note that, for any $G$-module $N$, there is a natural "diagonal" $G$-action on $\mathrm{Hom}(N, M) := \mathrm{Hom}_{\mathbb{Z}}(N, M)$ by functoriality, simply because $G$ acts on both $N$ and $M$. A moment's thought (taking account of the contravariance of Hom in the first variable), shows that this action should be defined by

$$(gf)(n) := gf(g^{-1}n)$$

for $g \in G$, $f \in \mathrm{Hom}(N, M)$, and $n \in N$. This immediately yields the following analogue of (3.2):

$$\mathrm{Hom}_G(N, M) = \mathrm{Hom}(N, M)^G, \tag{3.4}$$

where the invariants on the right are formed with respect to the $G$-action on $\mathrm{Hom}(N, M)$ that we have just defined. And, as in (3.3), cohomology with coefficients admits a topological interpretation as the cohomology of $Y = K(G, 1)$ with local coefficients in $M$.

In dimension 0, our homology and cohomology functors (viewed as functors of the coefficient module $M$) are familiar ones. Indeed, the tensor-product functor $-\otimes_G M$ is right exact, so we have an exact sequence

$$P_1 \otimes_G M \to P_0 \otimes_G M \to \mathbb{Z} \otimes_G M \to 0;$$

thus we can make the identification

$$H_0(G, M) = \mathbb{Z} \otimes_G M = M_G\,.$$

Similarly, the contravariant functor $\mathrm{Hom}_G(-, M)$ takes right exact sequences to left exact sequences, so we have an exact sequence

$$0 \to \mathrm{Hom}(\mathbb{Z}, M) \to \mathrm{Hom}(P_0, M) \to \mathrm{Hom}(P_1, M);$$

thus we can make the identification

$$H^0(G, M) = \mathrm{Hom}_G(\mathbb{Z}, M) = M^G.$$

As we will see, the existence of nontrivial higher homology and cohomology functors reflects the failure of the invariants and coinvariants functors to be exact. We will also see that homology and cohomology with coefficients are fundamental technical tools, even if we only care about $H_*(G) = H_*(G, \mathbb{Z})$.

## 3.2. Examples

*Example* 3.1. Let $G = \langle t \rangle$ be infinite cyclic, generated by $t$. Recall from Example 1.2 that we have a free resolution

$$0 \longrightarrow \mathbb{Z}G \xrightarrow{t-1} \mathbb{Z}G \longrightarrow \mathbb{Z} \longrightarrow 0.$$

Applying $-\otimes_G M$ as above (for a fixed $G$-module $M$), we obtain the chain complex

$$\cdots \longrightarrow 0 \longrightarrow 0 \longrightarrow M \xrightarrow{t-1} M$$

for computing $H_*(G, M)$. And applying $\mathrm{Hom}_G(-, M)$, we obtain the cochain complex

$$M \xrightarrow{t-1} M \longrightarrow 0 \longrightarrow 0 \longrightarrow \cdots$$

for computing $H^*(G, M)$. (In both cases, the two copies of $M$ are in dimensions 0 and 1.) Thus

$$H_0(G, M) = H^1(G, M) = M_G\,,$$

and

$$H^0(G, M) = H_1(G, M) = M^G,$$

and all higher homology and cohomology functors vanish. The vanishing is of course to be expected, since $G$ is a 1-dimensional group. And the calculations above are perhaps also to be expected. Indeed, they suggest Poincaré duality, which one would predict since the standard $K(G, 1)$ (the circle) is a closed 1-manifold. We will return to this in the next section.

*Example* 3.2. Now let $G = \langle t\,;\, t^n = 1 \rangle$, the finite cyclic group of order $n$. Recall from Example 1.3 that we have a free resolution

$$\cdots \xrightarrow{N} \mathbb{Z}G \xrightarrow{t-1} \mathbb{Z}G \xrightarrow{N} \mathbb{Z}G \xrightarrow{t-1} \mathbb{Z}G \longrightarrow \mathbb{Z} \longrightarrow 0.$$

As in the previous example, this yields the chain complex

$$\cdots \xrightarrow{N} M \xrightarrow{t-1} M \xrightarrow{N} M \xrightarrow{t-1} M$$

for computing homology and the cochain complex

$$M \xrightarrow{t-1} M \xrightarrow{N} M \xrightarrow{t-1} M \xrightarrow{N} \cdots$$

for computing cohomology. We therefore obtain, for $i \geq 1$,

$$H^i(G, M) = \begin{cases} M^G/N \cdot M & \text{if } i \text{ is even} \\ \ker N/(t-1) \cdot M & \text{if } i \text{ is odd.} \end{cases} \tag{3.5}$$

The result for homology is similar, with the roles of even and odd indices reversed. To get a neater statement, note that the norm operator (i.e., multiplication by $N$) induces a map

$$\overline{N} \colon M_G \to M^G,$$

and the quotients in (3.5) are simply the cokernel and kernel of this map.

## 3.3. Properties

Topology motivates thinking of homology and cohomology as functors of a space, and hence of the group $G$ in our setting. Algebra, on the other hand, suggests that we think of homology and cohomology as functors of the coefficient module $M$. Both points of view are useful. Here we explore the second. Thus we fix $G$ and let the module vary.

The first observation is that a short exact sequence of $G$-modules

$$0 \to M' \to M \to M'' \to 0$$

yields a **long exact sequence**

$$\cdots \to H_1(G, M') \to H_1(G, M) \to H_1(G, M'') \to H_0(G, M') \\ \to H_0(G, M) \to H_0(G, M'') \to 0.$$

Thus we can think of $H_1$ as measuring the failure of the coinvariants functor $H_0$ to be left exact. Similarly, $H_2$ measures the failure of $H_1$ to be left exact, and so on. If $\dim G < \infty$, we eventually reach a left-exact functor and the remaining functors are 0.

The derivation of the long exact sequence is straightforward. One simply tensors the given short exact sequence with a projective resolution. This yields a short exact sequence of chain complexes (because projective modules are flat, i.e., tensoring with a projective module is an exact functor). A standard argument therefore yields a long exact sequence in homology.

Similar remarks apply to cohomology. This time we use the fact that the functor $\mathrm{Hom}_G(-,-)$ is exact (and covariant) in the second variable if we fix a projective module in the first variable. Our exact sequence of coefficient modules therefore yields a long exact sequence

$$0 \to H^0(G,M') \to H^0(G,M) \to H^0(G,M'') \to H^1(G,M') \\ \to H^1(G,M) \to H^1(G,M'') \to \cdots .$$

Thus $H^1$ measures the failure of the invariants functor to be right exact, $H^2$ measures the failure of $H^1$ to be right exact, and so on.

The next property is sometimes called **acyclicity**: If $M$ is a projective $\mathbb{Z}G$-module, then

$$H_n(G,M) = 0$$

for $n > 0$. This is again a consequence of the flatness of projectives. [Tensoring a projective resolution with $M$ preserves the exactness of the resolution.] Readers familiar with injective modules will note that, in the same way, injectives are acyclic for cohomology, i.e.,

$$H^n(G,M) = 0$$

for $n > 0$ if $M$ is injective [because then $\mathrm{Hom}_G(-,M)$ is an exact functor].

The significance of acyclicity is that it leads to **effaceability**: Every module is a quotient of a module that is acyclic for homology. Similarly, every module can be embedded in a module that is acyclic for cohomology. This makes it possible to use a technique known as **dimension shifting** for reducing high-dimensional homology and cohomology to low-dimensional homology and cohomology, provided one is willing to change the coefficient module. See [3, Section III.7] for details.

The next property we wish to discuss is often called Shapiro's lemma (or, more accurately, the Eckmann–Shapiro lemma). In order to explain it, we need to digress and talk about induction and coinduction, which are special cases of extension and co-extension of scalars.

Consider an arbitrary inclusion $R \subseteq S$ of rings. Then any $S$-module may be viewed as an $R$-module by restriction of scalars. We wish to go in the other direction. In other words, given an $R$-module, we want to enlarge it so as to obtain an $S$-module. There are two ways to do this, depending on how one interprets "enlarge". Many readers will have seen the first method when $R$ and $S$ are fields. For example, the complexification of a real vector space is an example of the construction that follows.

**Method 1: Extension of scalars.** Given an $R$-module $M$, form the tensor product $S \otimes_R M$. Here we view $S$ as a right $R$-module in order to make sense out of the tensor product, and we then use the left action of $S$ on itself to make the tensor product an $S$-module:

$$s \cdot (s' \otimes m) := ss' \otimes m$$

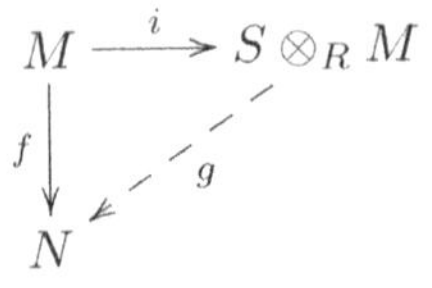


Figure 3.1 The universal property of extension of scalars

for $s, s' \in S$ and $m \in M$. This is legitimate because the left action of $S$ on itself commutes with the right action of $R$ on $S$ that was used in forming the tensor product. We say that $S \otimes_R M$ is the $S$-module obtained from $M$ by *extension of scalars* from $R$ to $S$. It is an "enlargement" of $M$ in the sense that there is a canonical $R$-module map

$$i \colon M \to S \otimes_R M$$

given by $m \mapsto 1 \otimes m$. (This is often injective in concrete examples, so that $M$ is embedded in $S \otimes_R M$.) The map $i$ is universal for $R$-maps of $M$ to an $S$-module, in the following sense: Given an $S$-module $N$ and an $R$-module map $f \colon M \to N$, there is a unique $S$-module map $g \colon S \otimes_R M \to N$ such that $gi = f$, as illustrated in Figure 3.1.

More concisely,

$$\mathrm{Hom}_R(M, N) = \mathrm{Hom}_S(S \otimes_R M, N) \tag{3.6}$$

for any $R$-module $M$ and $S$-module $N$. This says that extension of scalars is left adjoint to restriction of scalars.

**Method 2: Co-extension of scalars.** Given an $R$-module $M$, consider the abelian group $\mathrm{Hom}_R(S, M)$. Here we view $S$ as a left $R$-module in order to make sense out of the Hom, and we then use the right action of $S$ on itself to make $\mathrm{Hom}_R(S, M)$ an $S$-module:

$$(sf)(s') := f(s's)$$

for $s, s' \in S$ and $f \in \mathrm{Hom}_R(S, M)$. We say that $\mathrm{Hom}_R(S, M)$ is the $S$-module obtained from $M$ by *co-extension of scalars* from $R$ to $S$. It is an "enlargement" of $M$ in the sense that there is a canonical $R$-module map

$$p \colon \mathrm{Hom}_R(S, M) \to M$$

given by $f \mapsto f(1)$. (This is often surjective in concrete examples, so that $M$ is a quotient of $\mathrm{Hom}_R(S, M)$.) The map $p$ is universal for $R$-maps of an $S$-module to $M$, in the following sense: Given an $S$-module $N$ and an $R$-module map $f \colon N \to M$, there is a unique $S$-module map $g \colon N \to \mathrm{Hom}_R(S, M)$ such that $pg = f$, as illustrated in Figure 3.2. More concisely,

$$\mathrm{Hom}_R(N, M) = \mathrm{Hom}_S(N, \mathrm{Hom}_R(S, M)) \tag{3.7}$$

for any $R$-module $M$ and $S$-module $N$. This says that co-extension of scalars is right adjoint to restriction of scalars.

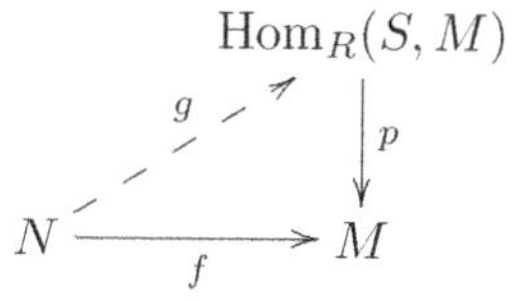


Figure 3.2 The universal property of co-extension of scalars

We will be interested in the case where $R$ and $S$ are group rings, say $R = \mathbb{Z}H$ and $S = \mathbb{Z}G$, where $H \leq G$. In this case extension of scalars is called *induction* and co-extension of scalars is called *coinduction*. Given a $G$-module $M$, we set

$$\mathrm{Ind}_H^G := \mathbb{Z}G \otimes_{\mathbb{Z}H} M$$

and

$$\mathrm{Coind}_H^G := \mathrm{Hom}_{\mathbb{Z}H}(\mathbb{Z}G, M).$$

We will talk mainly about induction, but everything we say has an analogue for coinduction.

The first observation is that the canonical $H$-map $i\colon M \to \mathrm{Ind}_H^G M$ is injective. We may therefore identify $M$ with its image $i(M)$, and $M$ then becomes an $H$-invariant additive subgroup of the $G$-module $\mathrm{Ind}_H^G M$. The transform $gM$ for $g \in G$ therefore depends only on the class of $g$ in $G/H$, and one checks easily that there is an abelian group decomposition

$$\mathrm{Ind}_H^G M = \bigoplus_{g \in G/H} gM. \tag{3.8}$$

This property in fact characterizes induced modules. More precisely, suppose $N$ is a $G$-module which is a direct sum of additive subgroups that are permuted transitively by the $G$-action. If $M$ is one of the summands and $H$ is its stabilizer in $G$, then $N$ is canonically isomorphic to $\mathrm{Ind}_H^G M$. Thus (3.8) really captures the essence of the induction construction.

Similar remarks apply to coinduction, but this time we have a direct product decomposition such that the $G$-action permutes the factors. In case $H$ has finite index in $G$, there is no difference between a direct sum decomposition and a direct product decomposition, and one concludes that

$$\mathrm{Ind}_H^G M \cong \mathrm{Coind}_H^G M \tag{3.9}$$

for any $H$-module $M$ in this case.

We now return to our list of properties of homology and cohomology. The **Eckmann–Shapiro lemma** is the following result:

**Proposition 3.3.** *Given $H \leq G$ and an $H$-module $M$, there are canonical isomorphisms*

$$H_*(H, M) \cong H_*(G, \mathrm{Ind}_H^G M)$$

*and*

$$H^*(H, M) \cong H^*(G, \mathrm{Coind}_H^G M).$$

*Proof.* Take a projective resolution $P$ of $\mathbb{Z}$ over $\mathbb{Z}G$. If we restrict operators from $G$ to $H$, $P$ is still a projective resolution of $\mathbb{Z}$ over $\mathbb{Z}H$, and we have

$$\begin{aligned} H^*(H,M) &= H^*(\operatorname{Hom}_H(P,M)) \\ &= H^*(\operatorname{Hom}_G(P, \operatorname{Coind}_H^G M)) \\ &= H^*(G, \operatorname{Coind}_H^G M), \end{aligned}$$

where the second equality comes from the universal property of coinduction, as formulated in (3.7). For homology, we instead use properties of tensor products:

$$\begin{aligned} H_*(H,M) &= H_*(P \otimes_H M) \\ &= H_*(P \otimes_G (\mathbb{Z}G \otimes_H M)) \\ &= H_*(G, \operatorname{Ind}_H^G M). \end{aligned}$$

□

Taking $M = \mathbb{Z}$, for example, the induced module $\operatorname{Ind}_H^G \mathbb{Z}$ can be identified, in view of (3.8), with the permutation module $\mathbb{Z}[G/H]$. So we have

$$H_*(H) \cong H_*(G, \mathbb{Z}[G/H]).$$

Thus the ordinary homology of a group can be computed as the homology of any supergroup, provided we are willing to introduce coefficients. And if $H$ has finite index in $G$, then we also have

$$H^*(H,\mathbb{Z}) \cong H^*(G, \mathbb{Z}[G/H])$$

in view of (3.9).

The final property we wish to state is the existence of **transfer maps**. We continue to assume that we are given $H \leq G$. One always has the expected covariance of homology with respect to group homomorphisms, giving a map

$$H_*(H,M) \to H_*(G,M)$$

for any $G$-module $M$. Algebraically, this comes from the canonical surjection

$$P \otimes_H M \twoheadrightarrow P \otimes_G M$$

if $P$ is a projective resolution of $\mathbb{Z}$ over $\mathbb{Z}G$. Topologically, it comes from the canonical map $K(H,1) \to K(G,1)$. Similarly, cohomology is contravariant with respect to the group: We have a map

$$H^*(G,M) \to H^*(H,M)$$

for any $G$-module $M$. But if $[G : H] < \infty$, we can also define so-called "transfer" maps that go in the opposite direction.

Consider cohomology for example, where we seek a map

$$H^*(H,M) \to H^*(G,M)$$

for any $G$-module $M$. If we use Proposition 3.3 and the isomorphism (3.9) to identify $H^*(H, M)$ with $H^*(G, \operatorname{Ind}_H^G M)$, the desired map has the form

$$H^*(G, \operatorname{Ind}_H^G M) \to H^*(G, M). \tag{3.10}$$

To define it, simply observe that there is a canonical $G$-module map

$$\operatorname{Ind}_H^G M \to M. \tag{3.11}$$

Indeed, since we started with a $G$-module $M$, restricted it to $H$, and then induced back up, we can use the original $G$-action to define a $G$-map

$$\mathbb{Z}G \otimes_{\mathbb{Z}H} M \to M$$

by $r \otimes m \mapsto rm$ for $r \in \mathbb{Z}G$ and $m \in M$. Formally, this is just an instance of the universal mapping property of induction. [Apply (3.6), and consider the identity map on $M$, viewed as an $H$-map to a $G$-module.] The map (3.10) that we are seeking is now obtained from (3.11) by functoriality of cohomology with respect to the coefficient module. Similar remarks apply to homology.

See [3, Section III.9] for other ways of explaining the existence of the transfer map. The name "transfer" comes from the special case $H_1(G) \to H_1(H)$, which is a map on abelianizations $G_{\text{ab}} \to H_{\text{ab}}$ that goes back to Schur [10], who called it the transfer ("Verlagerung"). The extension of Schur's transfer to homology and cohomology is due to Eckmann [5].

Finally, we remark that the ordinary functorial map

$$H^*(G, M) \to H^*(H, M)$$

can also be explained in terms of the Eckmann–Shapiro lemma and maps of coefficient modules (and similarly for homology). This does not require finite index, but, if we assume finite index for simplicity, the relevant map $M \to \mathbb{Z}G \otimes_{\mathbb{Z}H} M$ turns out to be given by

$$m \mapsto \sum_{g \in G/H} g \otimes g^{-1} m.$$

In particular, this shows that the composite

$$M \to \operatorname{Ind}_H^G M \to M$$

is simply multiplication by the index $[G : H]$, so the same is true of the composite

$$H^*(G, M) \to H^*(H, M) \to H^*(G, M). \tag{3.12}$$

Here the first map is the ordinary functorial map (sometimes called "restriction" in the case of cohomology), and the second map is the transfer. Suppose, for example, that $G$ is finite and $H$ is the trivial subgroup. Then this argument shows that multiplication by $[G : H] = |G|$ in positive-dimensional cohomology is the zero map, and similarly for homology. Consequently:

**Corollary 3.4.** *If $G$ is finite, then $H_n(G, M)$ and $H^n(G, M)$ are annihilated by $|G|$ for every $G$-module $M$ and all $n > 0$.* □

*Remark* 3.5. We noted above that the functorial map $H^*(G) \to H^*(H)$ is sometimes called *restriction*; here we have suppressed the coefficient module from the notation for simplicity. The transfer map $H^*(H) \to H^*(G)$ may then be called *corestriction*. In homology, on the other hand, it is the transfer map $H_*(G) \to H_*(H)$ that is called *restriction*, while the functorial map $H_*(H) \to H_*(G)$ is called *corestriction*. In both cases, then, "restriction" refers to the map $H(G) \to H(H)$, where $H(-)$ denotes either homology or cohomology. This uniform terminology makes it easier to remember certain formulas. For example, restriction followed by corestriction as in (3.12) is multiplication by the index in both homology and cohomology.

# 4. Finiteness properties revisited

This is a continuation of Section 2. The reference is still [3, Chapter VIII]. We begin with dimension.

## 4.1. Dimension

Recall that we defined the cohomological dimension $\operatorname{cd} G$ in terms of projective resolutions. We are now in a position to explain why this is called cohomological dimension. Namely, $\operatorname{cd} G$ is the largest $n$ (if any) for which the cohomology functor $H^n(G, -)$ is not identically 0:

**Proposition 4.1.** *For any group $G$,*

$$\operatorname{cd} G = \sup \{n \mid H^n(G, -) \neq 0\} .$$

*In other words, $\operatorname{cd} G \leq m \iff H^k(G, M) = 0$ for all $k > m$ and all $G$-modules $M$.*

*Proof.* Let's temporarily write $d(G)$ for the supremum on the right side of the equation to be proved. Clearly $d(G) \leq \operatorname{cd} G$. For the opposite inequality, we may assume $d(G) < \infty$. Set $n := d(G)$, and consider an arbitrary projective resolution $P$ of $\mathbb{Z}$ over $\mathbb{Z}G$. Let $L$ and $K$ be the images of the boundary maps $P_{n+1} \to P_n$ and $P_n \to P_{n-1}$:

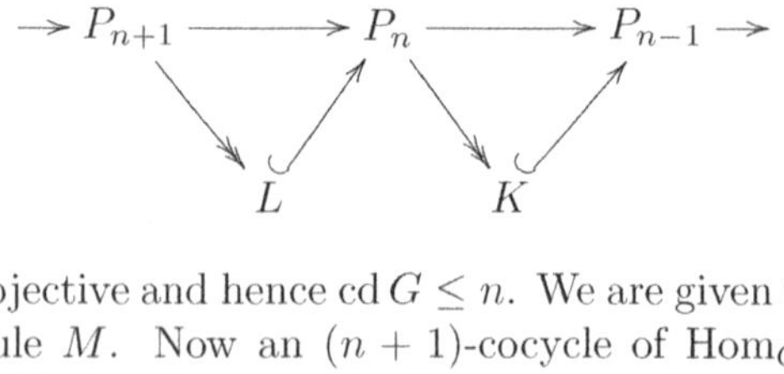

I claim that $K$ is projective and hence $\operatorname{cd} G \leq n$. We are given that $H^{n+1}(G, M) = 0$ for every $G$-module $M$. Now an $(n+1)$-cocycle of $\operatorname{Hom}_G(P, M)$ is a homomorphism $P_{n+1} \to M$ that becomes 0 when composed with the boundary map $P_{n+2} \to P_{n+1}$. Thus the group of $n$-cocycles can be identified with $\operatorname{Hom}_G(L, M)$. Saying that every such cocycle is a coboundary means that every map $L \to M$ extends to a map $P_n \to M$. This being true for every $G$-module $M$, it follows that

$L$ is a direct summand of $P_n$. Hence $P_n \cong L \oplus K$, so $K$ is a direct summand of a projective module and is therefore projective, as claimed. □

Note that every $P_n$ above could be taken to be free. But the proof would still only show that $K$ is projective. That is why we need to work with projective resolutions in developing the theory, in spite of the fact that in all known examples the modules can be taken to be free.

Recall from (2.2) that $\operatorname{cd} H \leq \operatorname{cd} G$ if $H \leq G$. We proved this using projective resolutions, but it also follows immediately from Propositions 3.3 and 4.1. Our next goal is to show that equality holds if $[G : H] < \infty$ and $G$ is finite dimensional. (But trivial examples show that one can have $\operatorname{cd} H < \infty$ and $\operatorname{cd} G = \infty$ if $G$ has torsion.)

**Proposition 4.2.** *If $\operatorname{cd} G < \infty$, then $\operatorname{cd} H = \operatorname{cd} G$ for every subgroup $H \leq G$ of finite index.*

The proof will use the first part of the following lemma:

**Lemma 4.3.** (a) *If $\operatorname{cd} G < \infty$, then*

$$\operatorname{cd} G = \max \{n \mid H^n(G, F) \neq 0 \textit{ for some free } \mathbb{Z}G\textit{-module } F\}.$$

(b) *If $G$ is of type* FP, *then*

$$\operatorname{cd} G = \max \{n \mid H^n(G, \mathbb{Z}G) \neq 0\}.$$

The proposition now follows immediately from the Eckmann–Shapiro lemma, since every free $\mathbb{Z}G$-module is induced from the free $\mathbb{Z}H$-module of the same rank. It remains to prove the lemma.

*Proof of the lemma.* (a) If $\operatorname{cd} G = n < \infty$, then the functor $H^n(G, -)$ is right exact (see Section 3.3). Since this functor is nonzero and every module is a quotient of a free module, it follows that the functor is nonzero on some free module.
(b) It is easy to check that $H^n(G, -)$ preserves direct sums if $G$ is of type FP. So the nonvanishing of this functor on a direct sum of copies of $\mathbb{Z}G$ implies that it is nonvanishing on $\mathbb{Z}G$. □

We noted above that one can have $\operatorname{cd} H \neq \operatorname{cd} G$ if $G$ has torsion. Our next result, which is due to Serre, says that this is the *only* way the equality can fail for subgroups of finite index.

**Theorem 4.4.** *If $G$ is torsion free, then $\operatorname{cd} H = \operatorname{cd} G$ for every subgroup $H \leq G$ of finite index.*

*Sketch of the proof.* In view of Proposition 4.2, it suffices to show that $\operatorname{cd} H < \infty \implies \operatorname{cd} G < \infty$. To this end, one starts with a finite-dimensional, free, contractible $H$-CW-complex $X$. One can then use a topological analogue of coinduction to produce a $G$-CW-complex which, as a CW-complex, is a product of $[G : H]$ copies of $X$, with the $G$-action permuting the factors. This complex is still finite dimensional and contractible, and the $H$-action on it is still free. It follows that the $G$-action has finite stabilizers and hence is free (because $G$ is torsion free). □

We close by using Lemma 4.3(b) to give an extremely useful topological characterization of cohomological dimension for groups of type F. Theorem 2.5, of course, gives a topological characterization for more-or-less arbitrary groups. But we have in mind something quite different. Suppose $G$ is of type F, and let $X$ be a free, contractible $G$-CW-complex with $G\backslash X$ finite. Then $H^*(G, \mathbb{Z}G)$, which occurred in Lemma 4.3, has the following interpretation:

$$H^*(G, \mathbb{Z}G) = H^*_c(X, \mathbb{Z}), \tag{4.1}$$

where $H^*_c$ denotes cohomology with compact supports (based on cellular cochains that vanish on all but finitely many cells of $X$). To prove (4.1), one need only stare at $\mathrm{Hom}_G(C_*(X), \mathbb{Z}G)$ and observe that it can be identified with the group of $\mathbb{Z}$-valued cochains of $X$ with compact supports. Thus Lemma 4.3(b) immediately yields:

**Corollary 4.5.** *Let $G$ be a group of type* F*, and let $X$ be a free, contractible $G$-CW-complex with finite quotient. Then*

$$\mathrm{cd}\, G = \max\{n \mid H^n_c(X) \neq 0\}. \qquad \square$$

## 4.2. Duality

In the topology of manifolds, one proves duality (classically) by using dual cell decompositions; this makes a cochain complex look like a chain complex. From the algebraic point of view, one instead uses duality theory for finitely generated projective modules to achieve the same effect. We begin by reviewing this duality theory. It is probably familiar to many readers for vector spaces, and the general case is no more difficult.

Let $R$ be a ring and $P$ a finitely-generated projective (left) $R$-module. The *dual* of $P$ is defined by

$$P^* := \mathrm{Hom}_R(P, R).$$

Here $R$ is viewed as a left $R$-module in forming the Hom. But there is also a right action of $R$ on itself that commutes with the left action, and this makes $P^*$ a right $R$-module. One easily checks that it is again finitely generated and projective. Dualizing again, one gets back a left $R$-module $P^{**}$ and a canonical isomorphism

$$P \xrightarrow{\simeq} P^{**}$$

defined as in duality theory for finite-dimensional vector spaces. Duality is useful for us because it can convert Hom to tensor product and vice versa: If $P$ is a finitely-generated projective and $M$ is arbitrary, then

$$\mathrm{Hom}_R(P, M) \cong P^* \otimes_R M \quad \text{and} \quad P \otimes_R M \cong \mathrm{Hom}_R(P^*, M). \tag{4.2}$$

(To be precise here, one needs to distinguish between left modules and right modules; I leave this as an exercise for the reader.)

Suppose now that $G$ is a group of type FP with $\mathrm{cd}\, G = n$, and take a finite projective resolution

$$0 \to P_n \to \cdots \to P_1 \to P_0 \to \mathbb{Z} \to 0$$

of $\mathbb{Z}$ over $\mathbb{Z}G$. Form the cochain complex dual to $P$, i.e., apply the functor $\mathrm{Hom}_G(-, \mathbb{Z}G)$, to get

$$P_0^* \to P_1^* \to \cdots \to P_n^* .$$

Note that this is what one uses to compute $H^*(G, \mathbb{Z}G)$. If we re-index by setting $Q_i := P_{n-i}^*$, then we have a chain complex

$$Q_n \to Q_{n-1} \to \cdots \to Q_0 ,$$

and (4.2) yields a canonical identification

$$\mathrm{Hom}_G(P_i, M) = Q_{n-i} \otimes_G M$$

for each $i$. Hence

$$H^i(G, M) = H_{n-i}(Q \otimes_G M). \tag{4.3}$$

It might happen that the dual complex $Q$ is again a resolution of $\mathbb{Z}$. This happens if and only if

$$H^i(G, \mathbb{Z}G) \cong \begin{cases} \mathbb{Z} & \text{for } i = n \\ 0 & \text{for } i \neq n. \end{cases} \tag{4.4}$$

[A technical point: $H^n(G, \mathbb{Z}G)$ is naturally a right $G$-module via the right action of $G$ on $\mathbb{Z}G$, which commutes with the left action that is used in defining $H^n(G, \mathbb{Z}G)$. What we mean in (4.4) is that $H^n(G, \mathbb{Z}G)$ is infinite cyclic with trivial $G$-action.] In this case (4.3) yields a result that looks like Poincaré duality for a closed orientable manifold:

$$H^i(G, M) = H_{n-i}(G, M).$$

More generally, suppose $Q$ is a resolution of $\mathbb{Z}$ with a possibly nontrivial action of $G$. We write $\widetilde{\mathbb{Z}}$ for this $G$-module, and for any $G$-module $M$ we set $\widetilde{M} := M \otimes \widetilde{\mathbb{Z}}$, with the diagonal $G$-action. Thus $\widetilde{M}$ is the same abelian group as $M$, with the $G$ action twisted by a homomorphism $G \to \{\pm 1\}$. Then (4.3) becomes

$$H^i(G, M) = H_{n-i}(G, \widetilde{M}),$$

which looks like Poincaré duality for a (possibly non-orientable) closed manifold. The justification for this is that $Q$ and $\widetilde{P} = P \otimes \widetilde{\mathbb{Z}}$ are both projective resolutions of $\widetilde{\mathbb{Z}}$, so they are homotopy equivalent. The homology $H_*(Q \otimes_G M)$ that occurs in (4.3) is therefore canonically isomorphic to $H_*((P \otimes \widetilde{\mathbb{Z}} \otimes M)_G) = H_*(P \otimes_G \widetilde{M}) = H_*(G, \widetilde{M})$.

Still more generally, suppose $Q$ is a resolution of *some* $G$-module $D$, not necessarily infinite cyclic. In other words, we are simply assuming that

$$H^i(G, \mathbb{Z}G) = 0 \quad \text{for } i \neq n,$$

and we are setting

$$D := H^n(G, \mathbb{Z}G).$$

Assume for simplicity that $D$ is free as an abelian group. Then we can argue as above to obtain

$$H^i(G, M) \cong H_{n-i}(G, D \otimes M), \tag{4.5}$$

where $D \otimes M$ is the tensor product over $\mathbb{Z}$, with the diagonal $G$-action. This is known as *Bieri–Eckmann duality*, and $G$ is said to be a *duality group* if it holds. The $G$-module $D$ is called the *dualizing module*. Borel and Serre proved that all torsion-free arithmetic groups are duality groups, so there is no shortage of examples.

Incidentally, there are also "inverse duality" isomorphisms under the hypotheses above (including the assumption that $D := H^n(G, \mathbb{Z}G)$ is free abelian):

$$H_i(G, M) \cong H^{n-i}(G, \mathrm{Hom}(D, M)),$$

where the Hom is taken over $\mathbb{Z}$ and has the diagonal $G$-action. To see this, note that there is a homotopy equivalence $Q \simeq P \otimes D$ (which is already needed in the proof of (4.5)), whence

$$\begin{aligned} P \otimes_G M &= \mathrm{Hom}_G(P^*, M) \\ &= \mathrm{Hom}_G(Q, M) \\ &\simeq \mathrm{Hom}_G(P \otimes D, M) \\ &= \mathrm{Hom}_G(P, \mathrm{Hom}(D, M)). \end{aligned}$$

## 4.3. Topological interpretation

Suppose $G$ is a group of type F. As usual, let $X$ be a free, contractible $G$-CW-complex with finite quotient. Then the hypothesis (4.4) that led to duality above can be written, in view of (4.1), as

$$H_c^i(X) = \begin{cases} 0 & i \neq n \\ D & i = n \end{cases}$$

with $D$ free abelian. Here the $G$-action on the dualizing module $D$ is induced by the $G$-action on $X$. Suppose, for example, that $X$ is an $m$-manifold, possibly with nonempty boundary $\partial X$. [Necessarily $m \geq n = \mathrm{cd}\, G$.] Then we can use Poincaré–Lefschetz duality to compute $H_c^i(X)$. This yields

$$H_c^i(X) \cong H_{m-i}(X, \partial X) \cong \widetilde{H}_{m-i-1}(\partial X), \tag{4.6}$$

where the second isomorphism comes from the contractibility of $X$, and $\widetilde{H}$ denotes reduced homology. Suppose further that $\partial X$ has the homotopy type of a bouquet of $k$-spheres for some $k$. (This holds with $k = -1$ if $\partial X = \emptyset$.) We then conclude from (4.6) that $H_c^i(X) = 0$ for $i \neq m-k-1$ and is free abelian when $i = m-k-1$. Thus $G$ is a duality group of dimension $n = m - k - 1$.

It is precisely in this way that Borel and Serre proved duality for torsion-free arithmetic groups (and calculated the cohomological dimension).

# 5. Equivariant homology

This section gives a very brief introduction to a technical subject that is treated in more detail in [3, Chapter VII]. Soulé's lectures in this volume give some interesting applications.

## 5.1. Introduction to spectral sequences

The only spectral sequences we will consider here are those that arise from double complexes. A double complex is a commutative diagram of abelian groups as in Figure 5.1 (extending forever in all four directions) in which every row and every column is a chain complex. Thus $\partial'^2 = 0$, $\partial''^2 = 0$, and $\partial''\partial' = \partial'\partial''$. In practice, double complexes are often concentrated in the first quadrant, i.e., $C_{pq} = 0$ unless $p, q \geq 0$.

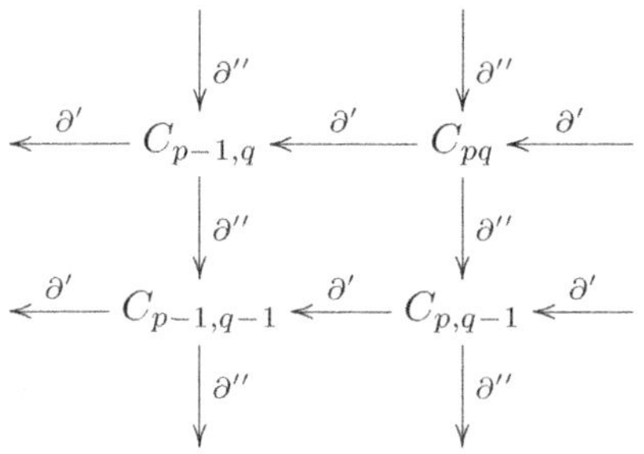


Figure 5.1 A double complex

A simple example of a double complex is provided by the tensor product of two chain complexes. Thus if $D$ and $E$ are chain complexes, then there is a double complex with $C_{pq} = D_p \otimes E_q$. The horizontal boundary operator $\partial'$ is induced by the boundary operator in $D$, and the vertical boundary operator $\partial''$ is induced by the boundary operator in $E$.

There are three ways to think about a double complex:

(a) Each (vertical) column $C_{p,*}$ for fixed $p$ is a chain complex, and the horizontal arrows are chain maps such that the composite of two consecutive ones is 0. Thus we have a "chain complex in the category of chain complexes".

(b) Each (horizontal) row $C_{*,q}$ for fixed $q$ is a chain complex, and the vertical arrows are chain maps such that the composite of two consecutive ones is 0. This is a second way of viewing $C$ as a chain complex in the category of chain complexes.

(c) We can simplify the whole double complex to an ordinary chain complex $TC$ (the *total complex*) by setting

$$(TC)_n := \bigoplus_{p+q=n} C_{pq}\,.$$

It has a boundary operator $\partial$, whose restriction to the summand $C_{pq}$ is given by $\partial = \partial' + (-1)^p \partial''$ as in the familiar case of the tensor product of two chain complexes.

The theory that we wish to sketch here relates all three of these viewpoints.

The basic fact is that there is a machine called a *spectral sequence* for computing $H_*(TC)$ by successive approximation. The spectral sequence is an infinite sequence of bigraded abelian groups

$$E^0_{pq},\ E^1_{pq},\ E^2_{pq},\ \ldots.$$

Under mild hypotheses (which are satisfied in the first-quadrant case), the sequence $E^r_{pq}$ stabilizes for each fixed $p, q$ as $r \to \infty$. Let $E^\infty_{pq}$ denote $E^r_{pq}$ for large $r$. It turns out to be the "$p$th layer" of $H_{p+q}(TC)$, in the following sense.

There is an increasing filtration of $TC$ by subcomplexes $F_pTC$, where $F_pTC$ is given in dimension $n$ by

$$(F_pTC)_n := \bigoplus_{i \le p} C_{i,n-i}\,.$$

Thus we draw a vertical line at the horizontal position $p$, and everything to the left of it is in $F_pTC$. As the line moves to the right, we get bigger and bigger subcomplexes. There is an induced filtration on homology, given by

$$F_pH_*(TC) := \operatorname{Im}\{H_*(F_pTC) \to H_*(TC)\}\,.$$

Thus $F_pH_n(TC)$ consists of homology classes represented by cycles in $\bigoplus_{i \le p} C_{i,n-i}$. We call the subquotient

$$F_pH_n(TC)/F_{p-1}H_n(TC)$$

of $H_n(TC)$ its *pth layer*. Thus our claim at the end of the previous paragraph is that

$$E^\infty_{pq} \cong F_pH_{p+q}(TC)/F_{p-1}H_{p+q}(TC).$$

Informally, then, one has to combine the groups $E^\infty_{pq}$ along a diagonal line $p+q = n$ to compute $H_n(TC)$.

The final ingredient of the spectral sequence is that there is a procedure for passing from each approximation $E^r$ to the next one by taking homology. More precisely, there is a boundary operator $d^r$ on $E^r$, induced by the boundary operator on $TC$, such that $E^{r+1} = H(E^r, d^r)$. This is easy to explain for $r = 0$ and 1. For $r = 0$, we have $E^0_{pq} = C_{pq}$ and $d^0 = \partial''$. Thus $(E^0, d^0)$ is simply the collection of columns of $C$, each viewed as a chain complex as in (a) above, and hence $E^1$ is the vertical homology. Next, $d^1$ is the map induced on the vertical homology by $\partial'$, viewed as a chain map as in (a). So $E^2$ can be described as the horizontal homology of the vertical homology. The differential $d^2$ is more subtle, and all I will say about it is that it maps $E^2_{pq}$ to $E^2_{p-2,q+1}$. So one can visualize it as going 2 steps to the left and 1 step up (decreasing the total degree $p + q$ by 1, as do $d^0$ and $d^1$). In general, $d^r$ goes $r$ steps to the left and $r - 1$ steps up.

Note that one can transpose the double complex $C$ to get a new double complex $C'$ with $C'_{pq} = C_{qp}$. The total complex $TC'$ is canonically isomorphic to $TC$, so we obtain a second spectral sequence for computing $H_*(TC)$ with the roles of $\partial'$ and $\partial''$ reversed.

In summary, there are two spectral sequences converging to $H_*(TC)$, based on the two viewpoints (a) and (b). [Warning: The statement that the two spectral sequences both converge to $H_*(TC)$ is somewhat misleading, since there are two different filtrations on $H_*(TC)$ and hence two different families of "layers" that occur as $E^\infty$.] We have introduced these spectral sequences because they are useful in connection with equivariant homology, to which we turn next.

## 5.2. Equivariant homology

Equivariant homology is the same as what Lück calls "Borel homology" in his lectures in this volume, but I will describe an algebraic approach. For simplicity I will stick to homology, but everything I say has an analogue for cohomology.

**Definition 5.1.** If $X$ is a $G$-CW-complex and $M$ is a $G$-module, then we set

$$H_*^G(X, M) := H_*(P \otimes_G C(X, M)),$$

where $P$ is a projective resolution of $\mathbb{Z}$ over $\mathbb{Z}G$ and $C(X, M)$ is the cellular chain complex of $X$ with coefficients in $M$ (with diagonal $G$-action). We call $H_*^G(X, M)$ the *equivariant homology* of $(G, X)$ with coefficients in $M$.

Heuristically, equivariant homology is a mixture of homology of groups and homology of spaces. For us, the main point is that it is a tool for getting information about $H_*(G)$ from *any* action (not necessarily free) on *any* space (not necessarily contractible).

*Examples* 5.2. (a) If $X$ is a point, then $H_*^G(X, M) = H_*(G, M)$. More generally, this holds if $X$ is contractible. Thus equivariant homology is the same as the homology of the group if the space is trivial. [Sketch of proof: There is a weak equivalence $C(X, M) \to M$, where $M$ is viewed as a chain complex concentrated in dimension 0, and a weak equivalence is a map that induces an isomorphism in homology. Tensoring over $G$ with the complex of projectives $P$, we obtain a weak equivalence $P \otimes_G C(X, M) \to P \otimes_G M$.]
(b) At the other extreme, if $G$ is the trivial group, then $H_*^G(X, M) = H_*(X, M)$.
(c) If $G$ acts freely on $X$, and $Y := G\backslash X$, then

$$H_*^G(X, M) = H_*(Y, M);$$

here $M$ on the right side is viewed as a $\pi_1(Y)$-module (and hence a local coefficient system on $Y$) via the canonical map $\pi_1(Y) \to G$ provided by the theory of covering spaces. [Sketch of proof: If $P$ is a projective resolution of $\mathbb{Z}$ over $\mathbb{Z}G$, then there is a weak equivalence $P \otimes M \to M$. Tensoring over $G$ with the complex of free $\mathbb{Z}G$-modules $C(X)$, we obtain a weak equivalence $P \otimes_G C(X, M) \to C(X) \otimes_G M = C(Y, M)$.]

If $X$ is contractible and the $G$-action is free, then (a) and (c) give back the familiar fact that $H_*(G) = H_*(K(G, 1))$.

Note that $H_*^G(X, M)$ is, by definition, the homology of the total complex associated to the double complex $P_p \otimes_G C_q(X, M)$ or, alternatively, the double complex $P_q \otimes_G C_p(X, M)$. The theory sketched in Section 5.1 therefore gives us two spectral sequences for computing $H_*^G(X)$. The $E^2$ term in each case is the horizontal homology of the vertical homology. Let's start with the second spectral sequence, where the double complex is viewed as $P_q \otimes_G C_p(X, M)$. The vertical homology is obtained by fixing $p$ and taking the homology with respect to $q$; thus

$$E^1_{pq} = H_q(G, C_p(X, M)). \tag{5.1}$$

To analyze this further, recall that $C_p(X)$ is the direct sum of infinite cyclic groups $\mathbb{Z}_\sigma$, one for each $p$-cell $\sigma$, where the two generators of $\mathbb{Z}_\sigma$ correspond to the two possible orientations of $\sigma$. Tensoring with $M$ (over $\mathbb{Z}$) and grouping the summands into $G$-orbits, we recognize $C_p(X, M)$ as a direct sum of induced modules:

$$C_p(X, M) = \bigoplus_\sigma \operatorname{Ind}_{G_\sigma}^G M_\sigma \,. \tag{5.2}$$

Here $\sigma$ ranges over a set of representatives for the $G$-orbits of $p$-cells, $G_\sigma$ is the stabilizer of $\sigma$, and $M_\sigma := \mathbb{Z}_\sigma \otimes M$ with the diagonal $G$-action. Thus $M_\sigma$ is $M$, with the action restricted to $G_\sigma$ and twisted by the "orientation homomorphism" $G_\sigma \to \{\pm 1\}$. [Each element of $G_\sigma$ either preserves the orientation of $\sigma$ or reverses it.] Combining (5.1) and (5.2), and using the Eckmann–Shapiro lemma, we obtain

$$E^1_{pq} = \bigoplus_\sigma H_q(G_\sigma, M_\sigma)$$

where, again, $\sigma$ ranges over representatives for the $p$-cells of $X$ mod $G$. It is customary to summarize what we have done so far by writing

$$E^1_{pq} = \bigoplus_\sigma H_q(G_\sigma, M_\sigma) \implies H^G_{p+q}(X; M). \tag{5.3}$$

This is a short-hand way of saying that we have constructed a spectral sequence with the given $E^1$ term, which converges to $H_*^G(X; M)$ (i.e., $E^\infty_{pq}$ is the $p$th layer of $H^G_{p+q}(X; M)$ with respect to some filtration on the latter).

*Remark* 5.3. Intuitively, $E^1_{pq}$ in (5.3) is the group of $p$-chains of $Y := G\backslash X$ with coefficients in the "system of coefficients" $\sigma \mapsto \{H_q(G_\sigma, M_\sigma)\}$. This suggests that

$$E^2_{pq} = H_p(Y, \{H_q(G_\sigma, M_\sigma)\}). \tag{5.4}$$

We will not attempt to formulate this precisely.

In case $G$ acts freely on $X$, all the stabilizers are trivial. The spectral sequence is then concentrated on the horizontal line $q = 0$, and $E^2 = E^\infty$. One easily recovers the result of Example 5.2(c) in this case. For a more interesting application, suppose that the action is not necessarily free but that $X$ is contractible. Then the equivariant homology can be identified with $H_*(G, M)$, so the spectral sequence converges to the latter. Thus (5.3) becomes

$$E^1_{pq} = \bigoplus_\sigma H_q(G_\sigma, M_\sigma) \implies H_{p+q}(G, M)$$

in this case. See Soulé's lectures in this volume for some illustrations of how this spectral sequence can be used to obtain concrete results.

We turn now to the other spectral sequence, arising from the double complex $P_p \otimes_G C_q(X, M)$. This time if we take the vertical homology (fixing $p$ and taking the homology with respect to $q$), we get $E^1_{pq} = P_p \otimes_G H_q(X, M)$ by the flatness of $P_p$. Taking homology with respect to $p$ now gives $H_p(G, H_q(X, M))$. Thus our spectral sequence has the form

$$E^2_{pq} = H_p(G, H_q(X)) \implies H^G_{p+q}(X)$$

where, to simplify the notation, I have suppressed the coefficient module $M$. If $X$ is contractible, this gives back the result of Example 5.2(a). For a more interesting application, suppose that $X$ is not necessarily contractible but that the $G$-action is free. Then we know that $H^G_*(X) = H_*(G \backslash X)$, so the spectral sequence becomes

$$E^2_{pq} = H_p(G, H_q(X)) \implies H_{p+q}(G \backslash X). \tag{5.5}$$

This is called the *Cartan–Leray* spectral sequence.

As a special case of (5.5) we can derive a spectral sequence associated to a group extension. Suppose we have a normal subgroup $N \lhd G$ and we set $Q := G/N$. Start with a free, contractible $G$-CW-complex $X$. Then we can form the quotient $X_G := G \backslash X$ in two steps, in which we first divide out by the $N$-action to get a $Q$-CW-complex $X_N$, and then go mod $Q$:

$$X_G = (X_N)_Q\,.$$

Now $X_N$ is a $K(N, 1)$-complex and a free $Q$-complex, and the quotient $(X_N)_Q = X_G$ is a $K(G, 1)$-complex. The Cartan–Leray spectral sequence for the $Q$-complex $X_N$ therefore becomes

$$E^2_{pq} = H_p(Q, H_q(N)) \implies H_{p+q}(G). \tag{5.6}$$

This is the *Hochschild–Serre* spectral sequence.

*Remarks* 5.4. (a) The description of the $E^2$ term in (5.6) suggests that there is an action of $Q$ on $H_*(N)$. To see where this comes from, observe that the conjugation action of $G$ on $N$ induces an action of $G$ on $H_*(N)$. One can show that the action of $N$ on its own homology is trivial, so we get an action of $Q = G/N$ on $H_*(N)$.

(b) Even if the suppressed coefficient module $M$ is $\mathbb{Z}$ (with trivial $G$-action), the action of $Q$ on $H_*(N)$ is generally nontrivial. Thus, once again, homology with coefficients arises naturally when one tries to compute ordinary integral homology.

# 6. The cohomology theory of finite groups

In this final lecture we point out some special features of the homology and cohomology theory of finite groups. Even if one is primarily interested in infinite groups, the homology of finite groups will arise whenever one tries to apply the spectral sequence (5.3) to a proper action (where the stabilizers $G_\sigma$ are finite). A reference for this section is [3, Chapter VI]. Assume from now on that $G$ is a finite group.

## 6.1. Functoriality

When there is no need to distinguish between homology and cohomology, we will often write $H(G, M)$ to denote the homology or cohomology of a group $G$ with coefficients in $M$. For brevity, we may even suppress $M$ from the notation and simply write $H(G)$. Recall that, for arbitrary groups, we often have restriction and corestriction maps between $H(G)$ and $H(H)$ if $H$ is a subgroup of $G$; see Remark 3.5. The qualifier "often" refers to the fact that we have to assume $H$ has finite index in $G$ in order to have a restriction map in homology and a corestriction map in cohomology. All of this simplifies if $G$ is finite, since every subgroup has finite index. Thus we always have maps in both directions, and formal differences between homology and cohomology disappear.

A similar (and related) phenomenon is that there is no need to distinguish between induced modules and coinduced modules; see (3.9). Thus we can give a unified statement of the Eckmann–Shapiro lemma (Proposition 3.3):

$$H(H, M) \cong H(G, \operatorname{Ind}_H^G M),$$

with no distinction between homology and cohomology.

## 6.2. Local computation of homology and cohomology

We continue to write $H(G)$ for homology or cohomology with an arbitrary coefficient module. Our starting point is Corollary 3.4, which states that $H(G)$ is annihilated by $|G|$ in positive dimensions. (This is of course false for $H_0$ and $H^0$.) We therefore have a decomposition

$$H(G) = \bigoplus_p H(G)_{(p)} \tag{6.1}$$

in positive dimensions, where $p$ ranges over the primes dividing $|G|$, and $(-)_{(p)}$ denotes the $p$-primary component. This simple observation can be quite useful in practice, since it allows one to localize the computation of $H(G)$ by focusing on one prime at a time. For example, one might be trying to compute $H(G)$ with the aid of a seemingly complicated spectral sequence. If one is lucky, the complications will disappear (or at least become more manageable) after localizing at one prime at a time.

To take (6.1) one step further, we show that the $p$-primary component $H(G)_{(p)}$ can be described in terms of the $p$-subgroups of $G$. There are two versions of the result. For the first, fix a prime $p$, and choose a Sylow $p$-subgroup $S \leq G$. Then $H(S)$ consists entirely of $p$-torsion by Corollary 3.4. The result that we will state below says, roughly speaking, that $H(G)$ can be identified with the subgroup of $H(S)$ invariant under conjugation. This is easy to state precisely if $S$ is normal in $G$. Recall from Remark 5.4(a) that, if $S \lhd G$, there is an action of $G$ [and even of $G/S$] on $H(S)$. So we can certainly talk about the invariants under this action. In general, we have to work a little harder.

Given $g \in G$, conjugation by $g$ induces an isomorphism between $H(S)$ and $H(gSg^{-1})$, which we denote by $z \mapsto gz$ for $z \in H(S)$. Now it does not make sense to ask whether $gz = z$ unless $g$ normalizes $S$. But we can ask whether $gz$ and $z$ restrict to the same class in $H(S \cap gSg^{-1})$. If $z$ has the property that this is true for all $g \in G$, then we will say that $z$ is invariant under conjugation. We can now state:

**Theorem 6.1.** *If $S$ is a Sylow $p$-subgroup of $G$, then the restriction map $H(G) \to H(S)$ is injective in positive dimensions. Its image is the set of elements of $H(S)$ invariant under conjugation.*

The second version of the result is more concise. It simply says that the canonical map

$$H(G)_{(p)} \to \lim_{P} H(P)$$

is an isomorphism in positive dimensions, where the limit (or inverse limit) is taken over the category whose objects are the $p$-subgroups $P$ of $G$ and whose morphisms are the maps $P_1 \to P_2$ induced by conjugation by elements of $G$. It is straightforward to prove that this version of the result is equivalent to the one in Theorem 6.1.

The proof of the theorem is based on formal properties of the restriction and corestriction maps between $H(G)$ and $H(S)$. Recall first that the composite

$$H(G) \to H(S) \to H(G)$$

(restriction followed by corestriction) is simply multiplication by the index $[G : S]$, which is relatively prime to $p$. This composite therefore induces an automorphism of the $p$-primary component $H(G)_{(p)}$; in particular, the restriction map is injective on this component. Next, one checks that the image of the monomorphism $H(G)_{(p)} \hookrightarrow H(S)$ is contained in the conjugation invariants in $H(S)$. Finally, to prove that the image is the set of all invariants, one considers the other composite:

$$H(S) \to H(G)_{(p)} \to H(S)$$

(corestriction followed by restriction). This composite is not as easy to describe as the other one, but there is in fact a formula for it involving conjugation, and the formula allows one to conclude that the composite, when restricted to the invariants, is again multiplication by $[G : S]$ and hence an automorphism. Theorem 6.1 follows at once.

## 6.3. Complete cohomology (Tate)

We have seen in Sections 6.1 and 6.2 that, for finite groups, homology and cohomology have similar properties and can be treated simultaneously. Carrying this idea to the extreme, one finds that there is a unified cohomology theory that combines all of the homology and cohomology functors (after $H_0$ and $H^0$ are modified so that they behave like the others). The resulting theory, called *complete cohomology* or *Tate cohomology*, involves a doubly-infinite family of functors $\widehat{H}^n$

($n \in \mathbb{Z}$), which are related to the usual functors as follows:

$$\begin{array}{cccccccc} \cdots & H_2 & H_1 & H_0 & H^0 & H^1 & H^2 & \cdots \\ & \| & \| & \uparrow & \downarrow & \| & \| & \\ \cdots & \widehat{H}^{-3} & \widehat{H}^{-2} & \widehat{H}^{-1} & \widehat{H}^0 & \widehat{H}^1 & \widehat{H}^2 & \cdots \end{array}$$

The modifications of $H_0$ and $H^0$ that produce $\widehat{H}^{-1}$ and $\widehat{H}^0$ are easy to explain. Recall that, for any $G$-module $M$, there is a *norm operator* $N\colon M \to M$ given by $m \mapsto \sum_{g\in G} gm$ for $m \in M$. It is obvious that the image of this operator is contained in $M^G = H^0(G, M)$. And it is equally easy to check that the norm operator induces a map on the quotient $M_G = H_0(G, M)$ of $M$. Thus we have a map

$$\overline{N}\colon H_0(G, M) \to H^0(G, M),$$

which already arose naturally in Example 3.2. The functors $\widehat{H}^{-1}$ and $\widehat{H}^0$ are simply the kernel and cokernel of $\overline{N}$. Taking $M = \mathbb{Z}$, for example, we have $H_0(G, \mathbb{Z}) = H^0(G, \mathbb{Z}) = \mathbb{Z}$, and $\overline{N}$ is multiplication by $m := |G|$. So $\widehat{H}^{-1}(G, \mathbb{Z}) = 0$, and

$$\widehat{H}^0(G, \mathbb{Z}) = \mathbb{Z}_m\,, \tag{6.2}$$

where the right side is the group of integers mod $m$.

As a first illustration of the usefulness of the complete theory, let's return to Example 3.2, where $G$ is a finite cyclic group. The calculation of homology and cohomology in that example can be restated as follows: $\widehat{H}^n$ is periodic of period 2; it is the kernel of $\overline{N}$ if $n$ is odd and the cokernel of $\overline{N}$ if $n$ is even.

## 6.4. Construction of the complete theory

The definition of the functors $\widehat{H}^*$ above does not explain why they fit together to form a cohomology theory. We will therefore give an alternate definition based on the notion of *complete resolution*. The starting point is the fact that $\mathbb{Z}$ admits a "backwards" projective resolution

$$0 \to \mathbb{Z} \to Q^0 \to Q^1 \to \cdots. \tag{6.3}$$

One can prove this by relative homological algebra [3, Section VI.2] or by duality theory (in the spirit of Section 4.2). The duality proof goes as follows.

Start with an ordinary resolution

$$\cdots \to P_1 \to P_0 \to \mathbb{Z} \to 0$$

by finitely generated projective $\mathbb{Z}G$-modules, and dualize, i.e., set $Q^i := \mathrm{Hom}_G(P_i, \mathbb{Z}G)$. To see that this yields an exact sequence as in (6.3), observe that the cochain complex $Q$ computes $H^*(G, \mathbb{Z}G)$; the latter is the same as the cohomology of the trivial group with $\mathbb{Z}$ coefficients (Eckmann–Shapiro), so it is $\mathbb{Z}$ in dimension 0 and it vanishes elsewhere. [Note that we have used the finiteness of $G$ in this argument: The coefficient module $\mathbb{Z}G$ is an induced module, hence also a coinduced module.]

We now splice together an ordinary projective resolution of $\mathbb{Z}$ with a backwards resolution of $\mathbb{Z}$ to obtain an exact sequence of projectives

$$\cdots \to P_1 \to P_0 \to Q^0 \to Q^1 \to \cdots,$$

where the map $P_0 \to Q^0$ is the composite

$$P_0 \twoheadrightarrow \mathbb{Z} \hookrightarrow Q^0.$$

Setting $P_i := Q^{-1-i}$ for $i < 0$, we obtain a complete resolution in the sense of the following definition:

**Definition 6.2.** A *complete resolution* for the finite group $G$ is an acyclic chain complex of projectives

$$\cdots \to P_1 \to P_0 \to P_{-1} \to P_{-2} \to \cdots$$

such that the map $P_0 \to P_{-1}$ factors as a composite $P_0 \twoheadrightarrow \mathbb{Z} \hookrightarrow P_{-1}$. Here, as usual, $\mathbb{Z}$ is assumed to have trivial $G$-action.

We have seen that complete resolutions exist. Here is a situation in which they arise naturally.

*Example* 6.3. Suppose $G$ acts freely on a CW-complex $X$ homeomorphic to the sphere $S^{2k-1}$ as in Example 1.3. Recall that one can construct an ordinary resolution by splicing together infinitely many copies of the chain complex $C_* := C_*(X)$. Similarly, one can construct a complete resolution, which is periodic of period $2k$, by splicing together a doubly-infinite collection of copies of $C_*$:

$$\cdots \to C_0 \to C_{2k-1} \to \cdots \to C_0 \to C_{2k-1} \to \cdots \to C_0 \to C_{2k-1} \to \cdots$$

One can show that complete resolutions are unique up to homotopy. We use them to give the "right" definition of complete cohomology:

**Definition 6.4.** For any $G$-module $M$,

$$\widehat{H}^*(G,M) := H^*(\mathrm{Hom}_G(P,M)),$$

where $P$ is a complete resolution.

It is not hard to check that this definition is consistent with the *ad hoc* definition given in Section 6.3. The advantage of the present approach is that one can develop all the usual cohomological properties (long exact sequences, restriction and corestriction, cup products, etc.) Here is a nice illustration of cup products: For any integer $i$, there is a cup product

$$\widehat{H}^i(G,\mathbb{Z}) \otimes \widehat{H}^{-i}(G,\mathbb{Z}) \to \widehat{H}^0(G,\mathbb{Z}) = \mathbb{Z}_m, \tag{6.4}$$

where $m := |G|$; see (6.2). Now the groups $\widehat{H}^*(G,M)$ are always annihilated by $m$ as in Corollary 3.4, i.e., they are $\mathbb{Z}_m$-modules. Moreover, there is a good duality theory for finitely generated $\mathbb{Z}_m$-modules. [The dual is given by $\mathrm{Hom}(-,\mathbb{Z}_m)$. Passage to the dual is an exact contravariant functor. A finitely-generated $\mathbb{Z}_m$-module is always non-canonically isomorphic to its dual and canonically isomorphic to its double dual.]

**Proposition 6.5.** *The cup product in* (6.4) *is a duality pairing. In other words, the induced map*

$$\widehat{H}^i(G,\mathbb{Z}) \to \mathrm{Hom}(\widehat{H}^{-i}(G,\mathbb{Z}),\mathbb{Z}_m)$$

*is an isomorphism.*

We omit the proof but simply remark that elementary arguments (involving the universal coefficient theorem) can be used to prove the abstract duality of $\widehat{H}^i(G,\mathbb{Z})$ and $\widehat{H}^{-i}(G,\mathbb{Z})$, based on the definitions of these cohomology groups in Section 6.3. The significance of the proposition, then, is the fact that duality is given by cup product; this takes some work to prove.

## 6.5. Groups with periodic cohomology

Groups that act freely on a $(2k-1)$-sphere have a periodic complete resolution of period $2k$ by Example 6.3, hence

$$\widehat{H}^n(G,M) \cong \widehat{H}^{n+2k}(G,M)$$

for all $G$-modules $M$ and all integers $n$. The complete cohomology theory allows one to formulate this kind of periodicity in several equivalent ways:

**Proposition 6.6.** *The following conditions on the finite group $G$ are equivalent.*

(i) *For some $d \neq 0$ there is an element $u \in \widehat{H}^d(G,\mathbb{Z})$ that is invertible in the ring $\widehat{H}^*(G,\mathbb{Z})$.*
(ii) *For some $d \neq 0$ there is an element $u \in \widehat{H}^d(G,\mathbb{Z})$ such that multiplication by $u$ gives an isomorphism $\widehat{H}^n(G,M) \xrightarrow{\sim} \widehat{H}^{n+d}(G,M)$ for all $n \in \mathbb{Z}$ and all $G$-modules $M$.*
(iii) *There are integers $n$ and $d$, with $d \neq 0$, such that $\widehat{H}^n(G,M) \cong \widehat{H}^{n+d}(G,M)$ for all $G$-modules $M$.*
(iv) *There is an integer $d \neq 0$ such that $\widehat{H}^d(G,\mathbb{Z}) \cong \mathbb{Z}_m$ $(m := |G|)$.*
(v) *There is an integer $d \neq 0$ such that $\widehat{H}^d(G,\mathbb{Z})$ contains an element of order $m := |G|$.*

*Sketch of the proof.* It is obvious that (i) $\Longrightarrow$ (ii) $\Longrightarrow$ (iii) and that (iv) $\Longrightarrow$ (v). To prove (iii) $\Longrightarrow$ (iv), first use dimension-shifting to show that there are isomorphisms as in (iii) for *all* integers $n$; now take $M = \mathbb{Z}$ and $n = 0$. Finally, the implication (v) $\Longrightarrow$ (i), which is perhaps the most surprising one *a priori*, follows from Proposition 6.5. □

Combining Proposition 6.6 with the local calculation of cohomology (Section 6.2) and doing a little group theory, one can prove:

**Theorem 6.7.** *The following conditions are equivalent.*

(i) *$G$ has periodic cohomology.*
(ii) *Every abelian subgroup of $G$ is cyclic.*
(iii) *For every prime $p$, every elementary abelian $p$-subgroup of $G$ has rank at most* 1.

(iv) *The Sylow subgroups of $G$ are cyclic or generalized quaternion groups.*

The condition that $G$ have periodic cohomology is very restrictive, and the groups with this property have been completely classified. A less restrictive (but still quite useful) condition is periodicity of the $p$-primary component $\widehat{H}^*(G)_{(p)}$ for a fixed prime $p$. More briefly, we say that $G$ has *$p$-periodic cohomology.* There are various characterizations of this property analogous to the results stated above. For example, $G$ has $p$-periodic cohomology (for a given prime $p$) if and only if every elementary abelian $p$-subgroup of $G$ has rank at most 1. Roughly speaking, then, the $p$-primary part of the cohomology of $G$ is very simple if and only if the structure of the $p$-subgroups of $G$ is very simple.

We close by stating a beautiful and far-reaching generalization of the last assertion, due to Quillen [8]. In order to motivate it, suppose that $G$ has $p$-periodic cohomology for some prime $p$ dividing $|G|$. Then the ordinary mod $p$ cohomology ring $H^*(G, \mathbb{F}_p)$ is finitely generated over a polynomial subring $\mathbb{F}_p[u]$ and hence has Krull dimension 1. Quillen's generalization asserts, among other things, the following:

**Theorem 6.8.** *For any finite group $G$ and prime $p$, the Krull dimension of $H^*(G, \mathbb{F}_p)$ is the maximal rank of an elementary abelian $p$-subgroup of $G$.*

Thus the complexity of $H^*(G, \mathbb{F}_p)$ correlates precisely with the complexity of the $p$-subgroups of $G$.

# References


[1] M. Bestvina and N. Brady, *Morse theory and finiteness properties of groups*, Invent. Math. **129** (1997), no. 3, 445–470.

[2] R. Brauer, *Über Zusammenhänge zwischen arithmetischen und invariantentheoretischen Eigenschaften von Gruppen linearer Substitutionen*, Sitzungsber. Preuss. Akad. Wiss. (1926), 410–416, Collected papers, vol. 1, 5–11.

[3] K. S. Brown, *Cohomology of groups*, Graduate Texts in Mathematics, vol. 87, Springer-Verlag, New York, 1994, corrected reprint of the 1982 original.

[4] K. S. Brown and R. Geoghegan, *An infinite-dimensional torsion-free $FP_\infty$ group*, Invent. Math. **77** (1984), no. 2, 367–381.

[5] B. Eckmann, *Cohomology of groups and transfer*, Ann. of Math. (2) **58** (1953), 481–493.

[6] H. Hopf, *Fundamentalgruppe und zweite Bettische Gruppe*, Comment. Math. Helv. **14** (1942), 257–309.

[7] W. Hurewicz, *Beiträge zur Topologie der Deformationen. IV. Asphärische Räume*, Nederl. Akad. Wetensch. Proc. **39** (1936), 215–224.

[8] D. Quillen, *The spectrum of an equivariant cohomology ring. I, II*, Ann. of Math. (2) **94** (1971), 549–572; ibid. (2) 94 (1971), 573–602.

[9] O. Schreier, *Über die Erweiterung von Gruppen I*, Monatsh. Math. Phys. **34** (1926), no. 1, 165–180.

[10] I. Schur, *Neuer Beweis eines Satzes über endliche Gruppen*, Sitzungsber. Preuss. Akad. Wiss. (1902), 1013–1019, Gesammelte Abhandlungen, I, 79–85.

[11] ———, *Über die Darstellung der endlichen Gruppen durch gebrochene lineare Substitutionen*, J. Reine Angew. Math. **127** (1904), 20–50, Gesammelte Abhandlungen, I, 86–116.

[12] J. R. Stallings, *On torsion-free groups with infinitely many ends*, Ann. of Math. (2) **88** (1968), 312–334.

[13] R. G. Swan, *Groups of cohomological dimension one*, J. Algebra **12** (1969), 585–610.

[14] O. Teichmüller, *Über die sogenannte nichtkommutative Galoissche Theorie und die Relation* $\xi_{\lambda,\mu,\nu}\xi_{\lambda,\mu\nu,\pi}\xi^{\lambda}_{\mu,\nu,\pi} = \xi_{\lambda,\mu,\nu\pi}\xi_{\lambda,\mu,\nu,\pi}$, Deutsche Math. **5** (1940), 138–149.

*Cohomology of Groups and Algebraic K-theory*
ALM 12, pp. 167–185



# A Brief Introduction to Algebraic $K$-Theory*

Daniel R. Grayson**

**Abstract**

We give a brief survey of higher algebraic $K$-theory and its connection to motivic cohomology. We start with limits and colimits, and then pass to the combinatorial construction of topological spaces by means of "systems of simplices", usefully mediated by simplicial sets. Definitions of $K$-theory are offered, and the main theorems are stated. A definition of motivic cohomology is offered and its major properties are listed. Many references to the literature are provided, especially for the development of motivic cohomology.

**2000 Mathematics Subject Classification:** 19

**Keywords and phrases:** Higher algebraic $K$-theory, motivic cohomology, combinatorial construction of topological spaces

## Introduction

This chapter summarizes a six-lecture course on "Algebraic $K$-Theory" that I gave at the conference and summer school on "Cohomology of Groups and Algebraic $K$-Theory", held at the Center of Mathematical Sciences of Zhejiang University, in Hangzhou, China, July 2 to 12, 2007, with recent references to the literature added. I am grateful to the Center of Mathematical Sciences and to the Higher Education Press for the support that allowed me to visit China for the summer school. I am also grateful to Marco Varisco for taking notes of the lectures, in TEX, upon which this chapter is based.

Other useful surveys of (or textbooks about) algebraic $K$-theory include [3, 33, 24, 5, 43, 17, 1, 41, 31, 18, 39, 22, 11, 20, 26, 62].



*Supported by NSF grant DMS 03-11378.

**URL: http://www.math.uiuc.edu/~dan/ E-mail: dan@math.uiuc.edu

# 1. Limits, colimits, and equations

There are two things you can do with an equation (or a system of equations): you can ask for the solution set, or you can declare the equation to be true. Perhaps surprisingly, from the right point of view, these operations are *dual* to each other. (And what that means is that there are two points of view, symmetrically related to each other, from which the two concepts turn out to be identical.)

**Example 1.1.** *Suppose $F : V \to W$ is a linear map between vector spaces over a field, and you consider the equation $F(v) = 0$. The solution set is the* kernel *of the operator,* $\ker F = \{v \in V \mid F(v) = 0\}$. *On the other hand, declaring the equation to be true leads to the* cokernel *of the operator (where the equation is true), which is defined as the quotient space* $\operatorname{coker} F := W/F(V)$.

**Example 1.2.** *Suppose $F_1, \ldots, F_m$ are polynomials in the variables $x_1, \ldots, x_n$ over an algebraically closed field $K$. The solution set of the system $F(x) = 0$ (where the indices are inferred) is the algebraic set $\{\alpha \in K^n \mid F(\alpha) = 0\}$. On the other hand, declaring the equations to be true leads to the quotient ring $K[x]/(F(x))$, in which the equations are true. (The algebraic variety defined by the system incorporates both the solution set and the quotient ring.)*

The duality between these two operations is especially visible in an abstract category $\mathcal{C}$. The only operation available is the composition $f \circ g$ of arrows $f$ and $g$, so the only equations available have the form $f \circ g = h$. Let $\mathsf{I}$ be a small "index" category, and consider a functor (or *diagram*) $F : \mathsf{I} \to \mathcal{C}$. The limit of $F$ is the universal solution to the problem of finding an object $C \in \mathcal{C}$ and arrows $f_I : C \to F(I)$ for all $I \in \mathsf{I}$ that make the equation $f_J = F(\alpha) \circ f_I$ true, for every arrow $\alpha : I \to J$ in $\mathsf{I}$. On the other hand, the colimit of $F$ is the co-universal solution to the problem of finding an object $C \in \mathcal{C}$ and arrows $g_I : F(I) \to C$ for all $I \in \mathsf{I}$ that make the equation $g_J \circ F(\alpha) = g_I$ true, for every arrow $\alpha : I \to J$ in $\mathsf{I}$.

In the category $\mathsf{Top}$ of topological spaces (and in the category $\mathsf{Sets}$ of sets) limits and colimits always exist. The limit is constructed as the subspace of the product $\prod_I F(I)$ where the equations $f_J = F(\alpha) \circ f_I$ are all true, whereas the colimit is constructed as the quotient space obtained from the disjoint union $\coprod_I F(I)$ by declaring all the equations $g_J \circ F(\alpha) = g_I$ to be true, mirroring the pairs of constructions in each of our first two examples.

Returning to example 1.1, one sees that $\ker F$ arises as

$$\lim \left( \begin{array}{ccc} & & V \\ & & \Big\downarrow {\scriptstyle F} \\ 0 & \longrightarrow & W \end{array} \right),$$

and that $\operatorname{coker} F$ arises as

$$\operatorname{colim} \left( \begin{array}{ccc} & & W \\ & & {\scriptstyle F} \Big\uparrow \\ 0 & \longleftarrow & V \end{array} \right).$$

The two diagrams are dual to each other (reverse the arrows and rename the objects).

The colimit of a diagram of spaces is said to be obtained by *gluing* the spaces of the diagram together.

# 2. Constructing spaces combinatorially

In algebraic $K$-theory we consider spaces that are large, but constructed from simple building blocks glued together in a simple way.

The building blocks are the *topological simplices* $\Delta^n_{\mathsf{Top}} := \{a \in \mathbb{R}^{n+1} \mid \forall i\ a_i \geq 0, \sum a_i = 1\} \in \mathsf{Top}$, for each $n \geq 0$. The maps between them that are usable in our diagrams are those that send vertices[1] to vertices, preserve the numerical ordering of the vertices, and are affine, in the sense that they preserve convex linear combinations[2]. Let $\mathsf{Sim}$, the *category of simplices*, be the subcategory of $\mathsf{Top}$ with those objects and those maps.

The colimit (in $\mathsf{Top}$) of a diagram in $\mathsf{Sim}$ will be called a *combinatorially constructed space*. All the important spaces of topology can be constructed in this way, even very complicated ones.

For example, a model for the topological $n$-sphere $S^n_{\mathsf{Top}}$ can be constructed combinatorially as

$$\operatorname{colim}\left(\begin{array}{ccc} \Delta^{n-1}_{\mathsf{Top}} & \overset{\longrightarrow}{\underset{\longrightarrow}{\vdots}} & \Delta^{n}_{\mathsf{Top}} \\ \Big\downarrow & & \\ \Delta^{0}_{\mathsf{Top}} & & \end{array}\right),$$

where the horizontal maps are the *face maps* (injective maps).

We can replace the category $\mathsf{Sim}$ by something entirely combinatorial, and thus slightly more convenient to use, as follows. Let $\mathsf{Ord}$ denote the category of finite ordered nonempty sets whose objects are of the form $\underline{\underline{n}} := \{0 < 1 < 2 < \cdots < n\}$, for $n \geq 0$; the arrows in the category are the order-preserving functions. The maps in $\mathsf{Sim}$ are determined by their effect on the vertices, so there is an equivalence $\mathsf{Ord} \cong \mathsf{Sim}$, so diagrams in $\mathsf{Sim}$ correspond to diagrams in $\mathsf{Ord}$.

Let $\Delta_{\mathsf{Top}} : \mathsf{Ord} \to \mathsf{Top}$ denote the composite functor $\mathsf{Ord} \cong \mathsf{Sim} \hookrightarrow \mathsf{Top}$. To construct a space combinatorially, we may start with a diagram in $\mathsf{Ord}$, compose it with $\Delta_{\mathsf{Top}}$, and take the colimit. The $n$-sphere arises in this way from the diagram

$$\begin{array}{ccc} \underline{\underline{n-1}} & \overset{\longrightarrow}{\underset{\longrightarrow}{\vdots}} & \underline{\underline{n}}\,. \\ \Big\downarrow & & \\ \underline{\underline{0}} & & \end{array}$$

[1]The vertices of $\Delta^n_{\mathsf{Top}}$ are the points with one nonzero coordinate.

[2]A *convex linear combination* of points $p_i$ in an affine space is a sum $\sum a_i p_i$ where $a_i \geq 0$ for each $i$ and $\sum a_i = 1$.

The horizontal maps used above are the injective order-preserving maps $\underline{n-1} \hookrightarrow \underline{n}$.

We abstract the situation above as follows. We say that a *system of simplices* in a category $\mathcal{C}$ is a functor $\Delta_{\mathcal{C}}\colon \mathsf{Ord} \to \mathcal{C}$. For each $n \geq 0$ we let $\Delta^n_{\mathcal{C}} := \Delta_{\mathcal{C}}(\underline{n})$; we call it a *simplex* of $\mathcal{C}$.

When $\mathcal{C}$ has a system of simplices $\Delta_{\mathcal{C}}$, then for any object $X$ of $\mathcal{C}$ we can construct a diagram of topological simplices from $X$ by using the simplices of $\mathcal{C}$ to probe the structure of $X$, as follows. The index category $\mathrm{I}_X$ of the diagram is the "comma category" where an object is a pair $(\underline{n}, f : \Delta^n_{\mathcal{C}} \to X)$ and where an arrow from $(\underline{n}, f : \Delta^n_{\mathcal{C}} \to X)$ to $(\underline{m}, f : \Delta^m_{\mathcal{C}} \to X)$ is an arrow $\varphi : \underline{n} \to \underline{m}$ in $\mathsf{Ord}$ such that the triangle

$$\begin{array}{ccc} \Delta^n_{\mathcal{C}} & \longrightarrow & X \\ {\scriptstyle \Delta_{\mathcal{C}}(\varphi)}\downarrow & \nearrow & \\ \Delta^m_{\mathcal{C}} & & \end{array}$$

commutes. The diagram is the functor $F_X : \mathrm{I}_X \to \mathsf{Sim}$ that sends $(\underline{n}, f : \Delta^n_{\mathcal{C}} \to X)$ to $\Delta^n_{\mathsf{Top}}$. We let $|X|$ denote the colimit of the diagram $F_X$; it is a topological space (and is a CW-complex); we call it the *geometric realization* of $X$ formed with respect to the system of simplices $\Delta_{\mathcal{C}}$. The construction is natural, and we get a functor $\mathcal{C} \to \mathsf{Top}$ that sends $X$ to $|X|$.

**Exercise 2.1.** *If $\Delta_{\mathcal{C}}$ is fully faithful and $X = \Delta^n_{\mathcal{C}}$, then $\mathrm{I}_X$ has a final object and $|X|$ is homeomorphic to $\Delta^n_{\mathsf{Top}}$.*

**Example 2.2.** *Let $\mathcal{C} = \mathsf{Top}$ equipped with the system of simplices $\Delta_{\mathsf{Top}}$ defined above. For any space $X$ the space $|X|$ is a* CW-approximation *to $X$. For example, the space $|X|$ is connected if and only if $X$ is path-connected. In this case, $\Delta_{\mathsf{Top}}$ is not fully faithful.*

**Example 2.3.** *Let $\mathcal{C}$ be the category $\mathsf{PO}$ of partially ordered sets, and let $\Delta_{\mathsf{PO}}$ be the inclusion $\mathsf{Ord} \hookrightarrow \mathsf{PO}$, so that $\Delta^n_{\mathsf{PO}} = \underline{n}$. In this case, $\Delta_{\mathsf{PO}}$ is fully faithful. Example: the geometric realization of the partially ordered set $\{t \geq x \leq y \leq z\}$ can be visualized (using the exercise) as a 2-simplex $xyz$ and a 1-simplex $xt$ glued together at the vertex $x$.*

**Exercise 2.4.** *Interpreting $\underline{m} \times \underline{n}$ as a partially ordered set, show that the canonical map $|\underline{m} \times \underline{n}| \to |\underline{m}| \times |\underline{n}|$ is a homeomorphism.*

**Example 2.5.** *Let $\mathcal{C}$ be the category $\mathsf{SS}$ of all functors $X : \mathsf{Ord}^{\mathrm{op}} \to \mathsf{Sets}$, where $\mathsf{Ord}^{\mathrm{op}}$ denotes the opposite category of $\mathsf{Ord}$. The arrows of the category are the natural transformations of functors. Let $\Delta_{\mathsf{SS}} : \mathsf{Ord} \to \mathsf{SS}$ be the functor that sends $\underline{n}$ to the functor it represents; i.e., $\Delta^n_{\mathsf{SS}} = (\underline{m} \mapsto \mathrm{Hom}_{\mathsf{Ord}}(\underline{m}, \underline{n}))$.*

*An object $X$ of $\mathsf{SS}$ is called a* simplicial set. *An element of $X(\underline{n})$ is called an $n$-simplex of $X$.*

*Colimits and limits exist in $\mathsf{SS}$. The geometric realization functor $\mathsf{SS} \to \mathsf{Top}$ commutes with colimits.*

*Using Yoneda's lemma, one may prove that $\Delta_{\mathsf{SS}}$ is a fully faithful embedding, and that every simplicial set is canonically a colimit of representable ones. This allows us to regard* $\mathsf{SS}$ *as obtained from* $\mathsf{Ord}$ *by adding colimits in a formal way, a remark I first heard from Voevodsky; the same remark applies when* $\mathsf{Ord}$ *is replaced by any small category.*

**Remark 2.6.** *For any $X \in \mathcal{C}$, the geometric realization $|X|$ can be realized in a natural way as the geometric realization of a simplicial set, as follows. We define the* nerve $\mathsf{N}X \in \mathsf{SS}$ *of $X$ by $\mathsf{N}X(\underline{n}) := \mathrm{Hom}_{\mathcal{C}}(\Delta^n_{\mathcal{C}}, X)$. One produces a natural homeomorphism $|\mathsf{N}X| \cong |X|$ by first using Yoneda's lemma to produce a natural equivalence $\mathrm{I}_{\mathsf{N}X} \cong \mathrm{I}_X$ of categories. Thus simplicial sets appear (via the nerve) as the universal intermediary between $X$ and its geometric realization $|X|$, rendering simplicial sets ubiquitous wherever geometric realizations are used.*

**Example 2.7.** *Let* $\mathsf{Ab}$ *be the category of abelian groups. Let* $\mathsf{SAb}$ *be the category of all functors $X : \mathsf{Ord}^{\mathrm{op}} \to \mathsf{Ab}$. An object of the category is called a* simplicial abelian group. *Let $\Delta_{\mathsf{SAb}} : \mathsf{Ord} \to \mathsf{SAb}$ be the functor that sends $\underline{n}$ to the simplicial abelian group that sends $\underline{m}$ to the free abelian group on the set $\mathrm{Hom}_{\mathsf{Ord}}(\underline{m}, \underline{n})$. In this case $\Delta_{\mathsf{SAb}}$ is not fully faithful. With respect to this system of simplices, the nerve $\mathsf{N}X$ is the simplicial set underlying $X$, obtained by forgetting the abelian group structure.*

**Remark 2.8.** *We may replace the topological simplices, as basic building blocks, by other simple things. For example, we may consider the functor $\mathsf{Ord} \times \mathsf{Ord} \to \mathsf{Top}$ that sends $(\underline{m}, \underline{n})$ to $\Delta^m_{\mathsf{Top}} \times \Delta^n_{\mathsf{Top}}$. That leads naturally to the notion of* bisimplicial set *and to* geometric realization of bisimplicial sets. *The spaces constructible in this way are, up to homeomorphism, the same ones as before.*

**Example 2.9.** *Let* $\mathsf{CAT}$ *be the category of small categories, and for each $n \geq 0$ let $\Delta^n_{\mathsf{CAT}}$ be the category whose objects are $0, 1, 2, \ldots, n$, and where there is an arrow $i \to j$ if an only if $i \leq j$, and when there is one, it's unique. The functor $\Delta_{\mathsf{CAT}}$ is the one that arises by identifying the objects of $\Delta^n_{\mathsf{CAT}}$ with the elements of $\underline{n}$. In this case, $\Delta_{\mathsf{CAT}}$ is fully faithful.*

*The geometric realization of the category $0 \rightrightarrows 1$ is homeomorphic to the circle, and is obtained by gluing together two 1-simplices along their vertices.*

*If $G$ is a group, then let $\underline{G}$ denote the category with one object and with $G$ as its set of arrows; the multiplication in $G$ gives the composition in $\underline{G}$. The geometric realization $|\underline{G}|$ is a model for the classifying space $BG$ of $G$.*

*A natural transformation between two functors $F, G : \mathcal{C} \to \mathcal{D}$ is conveniently encoded as a single functor $h : \mathcal{C} \times \Delta^1_{\mathsf{CAT}} \to \mathcal{D}$. Geometric realization yields a continuous map $|\mathcal{C}| \times \Delta^1_{\mathsf{Top}} \to |\mathcal{D}|$ and hence a homotopy between $|F|$ and $|G|$. This observation provides a convenient dictionary between the language of category theory and the language of homotopy theory.*

**Example 2.10.** *Let $\mathcal{C}$ be the category* $\mathsf{CC}$ *of positive chain complexes of abelian groups. Let $\Delta^n_{\mathsf{CC}}$ be the cellular chain complex of $\Delta^n_{\mathsf{Top}}$, described explicitly as follows. In degree $k$ it is the free abelian group $(\Delta^n_{\mathsf{Top}})_k$ whose basis consists of symbols $\langle A \rangle$ corresponding to the subsets $A \subseteq \underline{n}$ of cardinality $k+1$. The boundary* *map* $d : (\Delta^n_{\mathsf{Top}})_k \to (\Delta^n_{\mathsf{Top}})_{k-1}$ *sends* $\langle A \rangle$ *to* $\sum_{i=0}^{k} (-1)^i \langle A_i \rangle$ *where* $A_i$ *is obtained from* $A$ *by removing its* $i$*-th element. Let* $\Delta_{\mathsf{CC}} : \mathsf{Ord} \to \mathsf{CC}$ *be the functor that assigns to a map* $\varphi : \underline{n} \to \underline{m}$ *the map* $\Delta^n_{\mathsf{CC}} \to \Delta^m_{\mathsf{CC}}$ *that sends* $\langle A \rangle$ *to* $\langle \varphi(A) \rangle$ *if* $\varphi$ *is injective on* $A$*, and to* 0 *otherwise. In this case,* $\Delta_{\mathsf{CC}}$ *is not fully faithful.*

*If* $X \in \mathsf{CC}$*, then* $|X|$ *is a space whose homotopy groups are the homology groups of* $X$*, i.e.,* $\pi_n|X| \cong H_n(X)$*, for all* $n \in \mathbb{Z}$*; see the works of Dold* [12] *and of Kan* [23].

*If* $0 \to X \to Y \to Z \to 0$ *is a short exact sequence in* $\mathsf{CC}$*, then the long exact sequence* $\cdots \to H_i X \to H_i Y \to H_i Z \to H_{i-1} X \to \cdots$ *of homology groups can be rewritten as a long exact sequence* $\cdots \to \pi_i|X| \to \pi_i|Y| \to \pi_i|Z| \to \pi_{i-1}|X| \to \cdots$. *Much of algebraic* $K$*-theory is concerned with producing long exact sequences of homotopy groups from various sequences of spaces* $|X| \to |Y| \to |Z|$*. Such sequences of spaces are called fibration sequences, up to homotopy.*

*The main thing to remember about this work of Dold and Kan is that it makes homological algebra into a part of homotopy theory, and thus it provides a source of motivation for results in homotopy theory.*

**Remark 2.11.** *If* $\mathcal{C}$ *is a* $\mathbb{Z}$*-linear category, then for any* $X \in \mathcal{C}$*, the nerve can be regarded as a simplicial abelian group. Thus simplicial abelian groups appear (via the nerve) as the universal intermediary between* $X$ *and its geometric realization* $|X|$*, rendering simplicial abelian groups ubiquitous wherever geometric realizations are used in a* $\mathbb{Z}$*-linear category.*

*For* $X \in \mathsf{SS}$ *define* $\mathbb{Z}X \in \mathsf{SAb}$ *by replacing each set of* $X$ *by the free abelian group it generates. The map* $\pi_i|X| \to \pi_i|\mathbb{Z}X| \cong H_i|X|$ *is the Hurewicz map. If* $|X|$ *is a sphere of dimension* $q$*, then we see that the connected component of* $|\mathbb{Z}X|$ *containing the basepoint is an Eilenberg-MacLane space of type* $K(\mathbb{Z}, q)$*, and thus it represents the cohomology functor* $Y \mapsto H^i(Y, \mathbb{Z})$*. Thus spheres, made "additive" and reduced, represent cohomology.*

*The category* $\mathsf{CC}$ *of example 2.10 is a* $\mathbb{Z}$*-linear category. The resulting nerve functor* $\mathsf{N} : \mathsf{CC} \to \mathsf{SAb}$ *is an equivalence of categories. Hence chain complexes are as useful as simplicial abelian groups to encapsulate the information retrieved by geometric realizations in a* $\mathbb{Z}$*-linear category, and that explains why homological algebra, conceived as the study of chain complexes, is so important. Alternatively, this shows that geometric realization in* $\mathbb{Z}$*-linear categories is the natural realm of homological algebra.*

**Example 2.12.** *Let* $F$ *be a field and let* $\mathcal{C}$ *be the category* $\mathsf{Sm}_F$ *of smooth varieties over* $F$*. Put* $\Delta^n_{\mathsf{Sm}_F} = \operatorname{Spec}(F[T_0, \ldots, T_n]/(\sum T_i - 1))$*; the equations are the same as for* $\Delta^n_{\mathsf{Top}}$*, but the inequalities are gone. Maps from these simplices to a smooth variety* $X$ *often do not exist; for example, the Hurwitz formula tells that all such maps are constant for a curve* $X$ *of positive genus, and thus* $|X|$ *turns out to be a discrete space equivalent to the set of rational points of* $X$*. In section 5 we see how to repair this.*

**Example 2.13.** *Let* $Y$ *be a topological space. Let* $\mathcal{C} = \mathsf{Top}$ *equipped with the system of simplices defined by sending* $\underline{n}$ *to* $Y \times \Delta^n_{\mathsf{Top}}$*. The realization of any other topological space* $X$ *with respect to this system of simplices is a CW-approximation to the mapping space* $\operatorname{Map}(Y, X)$.

**Example 2.14.** *Suppose $\mathcal{C}$ is equipped with a system $\Delta_{\mathcal{C}}$ of simplices, suppose $Y \in \mathcal{C}$, and suppose $\mathcal{C}$ has products. As in example 2.13, the product $Y \times \Delta_{\mathcal{C}}$ is a new system of simplices. We may use $|Y, X|$ as notation for the geometric realization of $X \in \mathcal{C}$ with respect to it.*

# 3. Direct sum $K$-theory

Given a (small) additive category $\mathcal{M}$ the direct sum Grothendieck group $K_0^{\oplus}(\mathcal{M})$ is defined to be the abelian group with one generator $[M]$ for each object $M$, modulo the relations of the form $[M'] + [M''] = [M]$ whenever $M' \oplus M'' \cong M$.

For each $n \geq 0$, define a category $\mathrm{sub}\,\underline{\underline{n}}$ with objects $(i, j)$ for each $0 \leq i \leq j \leq n$, and with a unique arrow $(i, j) \to (r, s)$ for each $i \leq r \leq s \leq j$. In this way we obtain a functor $\mathrm{sub} : \mathsf{Ord} \to \mathsf{CAT}$.

Given an additive category $\mathcal{M}$ with a (choice of) zero object $0 \in \mathcal{M}$, a functor $M\colon \mathrm{sub}\,\underline{\underline{n}} \to \mathcal{M}$ is called *additive* if:

1. $M(i, i) = 0$ for all $i \in \underline{\underline{n}}$
2. $M(i, k) \to M(i, j) \oplus M(j, k)$ is an isomorphism for all $i \leq j \leq k \in \underline{\underline{n}}$

An additive functor should be thought of as a direct sum diagram in $\mathcal{M}$. For example, an additive functor $\mathrm{sub}\,\underline{\underline{2}} \to \mathcal{M}$ boils down to three objects $P = M(0, 1)$, $Q = M(1, 2)$, and $R = M(0, 2)$, of $\mathcal{M}$, together with an isomorphism $R \xrightarrow{\cong} P \oplus Q$.

**Definition 3.1.** *We let $S^{\oplus}\mathcal{M}$ denote the simplicial set that sends $\underline{\underline{n}}$ to the set of additive functors $M : \mathrm{sub}\,\underline{\underline{n}} \to \mathcal{M}$.*

Observe that $\pi_0|S^{\oplus}\mathcal{M}| = 0$ and $\pi_1|S^{\oplus}\mathcal{M}| \cong K_0^{\oplus}(\mathcal{M})$. We define the higher direct sum $K$-groups

$$K_n^{\oplus}\mathcal{M} := \pi_{n+1}|S^{\oplus}\mathcal{M}|.$$

Given an associative unital ring $R$ we introduce the additive category $\mathcal{P}_R$ of finitely generated projective left $R$-modules. We define the higher $K$-groups of $R$ by setting $K_n R := K_n^{\oplus}\mathcal{P}_R$.

The definition above is an example of the $K$-theory *gambit*:

> If you have a group $K$ defined by generators and relations (*or* as a quotient set of a set) then make a space $X$ with $\pi_1 X \cong K$ (*or* $\pi_0 X \cong K$, respectively) putting in higher dimensional simplices naturally. The space $X$ is likely to have an interesting homotopy type, and its homotopy groups are worthy of study.

Here are the classical definitions of $K_1$ and $K_2$, in terms of group homology of the infinite general linear group $GL(R)$ of $R$, and its subgroup $E(R)$ generated by the elementary matrices:

$$K_1(R) = H_1(BGL(R), \mathbb{Z}), \quad K_2(R) = H_2(BE(R), \mathbb{Z}).$$

In order to relate these to our definitions we need first to find a way to construct a combinatorial model of the loop space $\Omega|S^{\oplus}\mathcal{P}_R|$.

Given two objects $\underline{\underline{m}}$ and $\underline{\underline{n}}$ of $\mathsf{Ord}$, let $\underline{\underline{m}} * \underline{\underline{n}}$ denote the ordered set in $\mathsf{Ord}$ (uniquely) isomorphic to the ordered set obtained by concatenating $\underline{\underline{m}}$ and $\underline{\underline{n}}$ so that the elements of $\underline{\underline{m}}$ are less than the elements of $\underline{\underline{n}}$, when viewed in $\underline{\underline{m}} * \underline{\underline{n}}$. In this way we obtain a functor $* : \mathsf{Ord} \times \mathsf{Ord} \to \mathsf{Ord}$, as well as natural injections $\underline{\underline{m}} \to \underline{\underline{m}} * \underline{\underline{n}}$ and $\underline{\underline{n}} \to \underline{\underline{m}} * \underline{\underline{n}}$.

Given a simplicial set $Y$ with a base point $y_0$ we define a naive approximation $\omega Y$ to the loop space $\Omega|Y|$ of $Y$ as a simplicial set by setting

$$\omega Y(\underline{\underline{n}}) := \lim \left( \begin{array}{ccccccc} & & & & & & \{y_0\} \\ & & & & & & \downarrow \\ & & & & Y(\underline{\underline{0}} * \underline{\underline{n}}) & \longrightarrow & Y(\underline{\underline{0}}) \\ & & & & \downarrow & & \\ & & Y(\underline{\underline{0}} * \underline{\underline{n}}) & \longrightarrow & Y(\underline{\underline{n}}) & & \\ & & \downarrow & & & & \\ \{y_0\} & \longrightarrow & Y(\underline{\underline{0}}) & & & & \end{array} \right).$$

There is a natural map $|\omega Y| \to \Omega|Y|$, which can easily fail to be a homotopy equivalence.

The 0-simplices of $\omega S^{\oplus}\mathcal{M}$ are pairs $(M, N)$ of objects of $\mathcal{M}$. A 1-simplex $(M', N') \to (M, N)$ consists of an object $P \in \mathcal{M}$ and a pair of isomorphisms $M \to M' \oplus P$, $N \to N' \oplus P$. Taking the isomorphisms to be identities yields 1-simplices $(M', N') \to (M' \oplus P, N' \oplus P)$. Alternatively, taking $P = 0$ allows such a 1-simplex $(\alpha, \beta) : (M', N') \to (M, N)$ to be constructed from a pair of isomorphisms $\alpha : M' \xrightarrow{\cong} M$, $\beta : N' \xrightarrow{\cong} N$. An elementary isomorphism $\pi_0|\omega S^{\oplus}\mathcal{M}| \cong K_0\mathcal{M}$ arises by associating the path-connected component containing the vertex $(M, N)$ to the element $[M] - [N]$.

The following theorem is analogous to Quillen's original $+ = Q$ theorem [16], but since exact sequences are not involved, it is simpler, and is derived mainly from Waldhausen's discussion of his $S$-construction and of Segal's $\Gamma$-spaces in [59, 60].

**Theorem 3.2** ([19, section 3]). *There is a homotopy equivalence* $|\omega S^{\oplus}\mathcal{M}| \xrightarrow{\sim} \Omega|S^{\oplus}\mathcal{M}|$.

**Definition 3.3.** *We introduce the $K$-theory space of an additive category $\mathcal{M}$ by setting $K^{\oplus}\mathcal{M} := |\omega S^{\oplus}\mathcal{M}|$, and the $K$-theory space of a ring $R$ by setting $K^{\oplus}R := K^{\oplus}\mathcal{P}_R$.*

The theorem implies that $\pi_i K^{\oplus}\mathcal{M} \cong K_i\mathcal{M}$.

Quillen's original definition of $K$-theory for additive categories was the $S^{-1}S$-construction of [16]. It yields a category $S^{-1}S\mathcal{M}$ with $|S^{-1}S\mathcal{M}| \simeq |\omega S^{\oplus}\mathcal{M}|$. Indeed, $\omega S^{\oplus}\mathcal{M}$ can be viewed as a slight modification of $S^{-1}S\mathcal{M}$.

Let $\operatorname{Aut} M$ be the automorphism group of an object $M \in \mathcal{M}$. There is a map $\mathsf{N}\underline{\operatorname{Aut} M} \to \omega S^{\oplus}\mathcal{M}$ in $\mathsf{SS}$ that sends $\theta \in \operatorname{Aut} M$ to $(1_M, \theta)$. Thus we get a map $B\operatorname{Aut} M \to K^{\oplus}(\mathcal{M})$. Taking $\mathcal{M} = \mathcal{P}_R$ and $M = R^n$ we get a map $BGL_nR \to K^{\oplus}R$. The triangle

$$\begin{array}{ccc} BGL_n(R) & \longrightarrow & K^{\oplus}R \\ \downarrow & \nearrow & \\ BGL_{n+1}(R) & & \end{array}$$

commutes up to homotopy, and with some work this yields a homotopy class of maps $BGL(R) \to K^{\oplus}R$ whose image (since $BGL(R)$ is connected) is contained in the connected component of $K^{\oplus}R$ that contains the base point. We may call that connected component $BGL(R)^+$, although originally, Quillen constructed it a different way. Each of the path-connected components of $K^{\oplus}R$ is homotopy equivalent to $BGL(R)^+$, for we may use direct sum to translate from one component to another. *Warning:* avoid writing a homotopy equivalence $K^{\oplus}R \simeq K_0R \times BGL(R)^+$, for this equivalence is not canonical when $K_0R \not\cong \mathbb{Z}$, and it's not possible to make the equivalence respect the $H$-space structures.

**Theorem 3.4** (Quillen). *The map $BGL(R) \to BGL(R)^+$ is* acyclic[3], *in particular*

$$H_*\big(BGL(R), \mathbb{Z}\big) \xrightarrow{\cong} H_*\left(BGL(R)^+, \mathbb{Z}\right)$$

*is an isomorphism (even with local coefficients). On $\pi_1$ it is just abelianization.*

**Corollary 3.5.** *$K_1(R)$ and $K_2(R)$ defined here agree with Bass' and Milnor's definitions.*

By the way, there is a way to interpret each set $S^{\oplus}\mathcal{M}(\underline{n})$ as the objects of an additive category, allowing a new simplicial set $S^{\oplus}S^{\oplus}\mathcal{M}(\underline{n})$ to be introduced. The resulting object $S^{\oplus}S^{\oplus}\mathcal{M}$ is a functor $\mathsf{Ord}^{\mathrm{op}} \times \mathsf{Ord}^{\mathrm{op}} \to \mathsf{Sets}$, which has a suitable geometric realization. It turns out that there is a homotopy equivalence $|S^{\oplus}\mathcal{M}| \simeq \Omega|S^{\oplus}S^{\oplus}\mathcal{M}|$. This process can be iterated, showing that the $K$-theory space $K^{\oplus}\mathcal{M}$ appears as the 0-th space in a connective (symmetric) $\Omega$-spectrum (essentially because $\oplus$ is commutative).

**Theorem 3.6** ([36]).

$$K_i(\mathbb{F}_q) \cong \begin{cases} \mathbb{Z} & i = 0, \\ \mathbb{F}_q^{\times} & i = 1, \\ 0 & i = 2n,\ n > 0, \\ \mathbb{Z}/(q^r - 1) & i = 2r - 1,\ r \geq 1. \end{cases}$$

*Idea of proof.* Compare with $\mathbb{Z} \times BU = K^{\mathrm{top}}(\mathrm{pt})$. On $K(\mathbb{F}_q)$ the $q$-th Adams operation $\Psi_q$ satisfies $\Psi_q = 1$. Use the *Brauer lifting* $BGL(\mathbb{F}_q)^+ \to K^{\mathrm{top}}(\mathrm{pt})$ to get a fibration sequence

$$BGL(\mathbb{F}_q)^+ \to K^{\mathrm{top}}(\mathrm{pt}) \xrightarrow{\Psi_q - 1} K^{\mathrm{top}}(\mathrm{pt})$$

[3] A continuous map $f : X \to Y$ is acyclic if for every local coefficient system $\mathcal{L}$ on $Y$ and for all $n \in \mathbb{N}$, the map $H_n(X, f^*\mathcal{L}) \to H_n(Y, \mathcal{L})$ is an isomorphism.

and then use long exact sequences of homotopy groups. Compute the action of $\Psi_q$ on $K_{2r}^{\text{top}}(\text{pt})$ by observing that $K_{2r}^{\text{top}}(\text{pt})$ is an infinite cyclic group generated by $\beta^r$, where $\beta \in K_2(pt)$ is the Bott element. Since $\Psi_q\beta = q\beta$, it follows from multiplicativity that $\Psi_q(\beta^r) = (q\beta)^r = q^r\beta^r$, i.e., $\Psi_q$ is multiplication by $q^r$ on $K_{2r}^{\text{top}}(\text{pt}) \cong \mathbb{Z}$. Hence $K_{2r}(\mathbb{F}_q) = 0$ and $K_{2r-1}(\mathbb{F}_q) \cong \mathbb{Z}/(q^r - 1)$ for all $r > 0$. □

## 4. Exact sequence $K$-theory

In a category such as $\mathcal{P}_R$, every short exact sequence is split, and hence direct sum diagrams provide all the information about exact sequences. Other categories, such as the category $\mathcal{M}_R$ of finitely generated left $R$-modules, have exact sequences that don't split, and $K_0\mathcal{M}_R$ is not necessarily isomorphic to $K_0^{\oplus}\mathcal{M}_R$. For such a category $\mathcal{M}$, the Grothendieck group $K_0(\mathcal{M})$ is defined to be the abelian group with one generator $[M]$ for each object $M$, modulo the relations of the form $[M'] + [M''] = [M]$ whenever there is a short exact sequence $0 \to M' \to M \to M'' \to 0$.

The higher $K$-theory of such categories is handled as follows.

**Definition 4.1** ([38]). *An* exact category *is an additive category $\mathcal{M}$, together with a collection of sequences in $\mathcal{M}$ of the form $0 \to M' \to M \to M'' \to 0$ (called* short exact sequences*), such that there is a fully faithful embedding of $\mathcal{M}$ in an abelian category $\mathcal{A}$ so that (1) any short exact sequence $0 \to M' \to A \to M'' \to 0$ of $\mathcal{A}$ with $M' \in \mathcal{M}$ and $M'' \in \mathcal{M}$ has $A \in \mathcal{M}$, too; and (2) the short exact sequences of $\mathcal{M}$ are exactly those sequences that are sent to exact ones in $\mathcal{A}$.*

**Definition 4.2** ([38]). *An* admissible monomorphism *of an exact category $\mathcal{M}$ is a map $M' \rightarrowtail M$ that appears as the left hand map in a short exact sequence of $\mathcal{M}$. An* admissible epimorphism *of $\mathcal{M}$ is a map $M \twoheadrightarrow M''$ that appears as the right hand map in a short exact sequence of $\mathcal{M}$.*

In an exact category $0 \to M' \to M' \oplus M'' \to M'' \to 0$ is always a short exact sequence, so a short exact sequence is a generalization of a direct sum diagram. Generalizing the definition of $S^{\oplus}\mathcal{M}$ appropriately, one defines, as in [59], a simplicial set $S\mathcal{M}$ for any exact category $\mathcal{M}$ so that $K_0(\mathcal{M}) \cong \pi_1|S\mathcal{M}|$. The higher $K$-groups are defined as $K_i\mathcal{M} := \pi_{i+1}|S\mathcal{M}|$.

**Theorem 4.3** ([14, 15]). *There is a natural homotopy equivalence $|\omega S\mathcal{M}| \xrightarrow{\simeq} \Omega|S\mathcal{M}|$.*

We may then define $K\mathcal{M} := |\omega S\mathcal{M}|$, and it also appears as the first space in an $\Omega$-spectrum.

Quillen's original definition of $K$-theory was the $Q$-construction [38]. It yields a category $Q\mathcal{M}$ with $|Q\mathcal{M}| \simeq |S\mathcal{M}|$. The objects of $Q\mathcal{M}$ are the objects of $\mathcal{M}$, and an arrow from $M$ to $N$ in $Q\mathcal{M}$ is an isomorphism of $M$ with a subquotient object of $N$, where the subquotient objects under consideration are the ones that arise from admissible monomorphisms and admissible epimorphisms of $\mathcal{M}$. The proofs of some important theorems of Quillen in $K$-theory depend on the use of the $Q$-construction.

**Theorem 4.4** (Quillen's resolution theorem [38]). *If $\mathcal{P} \subseteq \mathcal{M}$ is a full exact subcategory closed under extensions and closed under kernels and such that for every $M$ in $\mathcal{M}$ there exists a short exact sequence $0 \to P' \to P \to M \to 0$ with $P$ and $P'$ in $\mathcal{P}$, then the map $K(\mathcal{P}) \to K(\mathcal{M})$ is a homotopy equivalence.*

**Fact 4.5.** *If $X$ and $Y$ are simplicial sets, then the projections induce a homeomorphism $|X \times Y| \xrightarrow{\cong} |X| \times |Y|$, provided the product is given the compactly generated topology, or if one of the factors is generated by a finite number of simplices. (Putting the compactly generated topology on a space does not change the homotopy groups.) Similarly for small categories.*

In particular, if $\mathcal{M}$ and $\mathcal{N}$ are exact categories then $K_*(\mathcal{M} \times \mathcal{N}) \cong K_*(\mathcal{M}) \times K_*(\mathcal{N})$. If $R$, $S$ are rings then $K_*(R \times S) \cong K_*(R) \times K_*(S)$. This gives the first part of the following theorem.

**Theorem 4.6** ([38]). *$K$-theory commutes with finite products and filtering colimits (of exact categories).*

If $F\colon \mathcal{A} \to \mathcal{C}$ is an exact functor between abelian categories then $\mathcal{B} = \ker F$ is abelian, too, and for all exact sequences $0 \to A' \to A \to A'' \to 0$ in $\mathcal{A}$ then $A \in \mathcal{B}$ if and only if $A', A'' \in \mathcal{B}$. A full subcategory $\mathcal{B}$ of an abelian category $\mathcal{A}$ satisfying this property is called a *Serre subcategory.*

**Fact 4.7.** *If $\mathcal{B} \subseteq \mathcal{A}$ is a Serre subcategory of an abelian category then there exists a* quotient *abelian category $\mathcal{A}/\mathcal{B}$ and an exact functor $G\colon \mathcal{A} \to \mathcal{A}/\mathcal{B}$ with $\ker G = \mathcal{B}$ such that for any exact functor $F\colon \mathcal{A} \to \mathcal{C}$ to any exact category $\mathcal{C}$ with $\mathcal{B} \subseteq \ker F$ there is an exact functor $\overline{F}\colon \mathcal{A}/\mathcal{B} \to \mathcal{C}$ (unique up to natural isomorphism) such that $\overline{F} \circ G \cong F$.*

**Theorem 4.8** (Quillen's localization theorem for exact categories [38]). *If $\mathcal{B} \subseteq \mathcal{A}$ is a Serre subcategory of an abelian category then there is a natural long exact sequence in $K$-theory*

$$\cdots \to K_n(\mathcal{B}) \to K_n(\mathcal{A}) \to K_n(\mathcal{A}/\mathcal{B}) \to K_{n-1}(\mathcal{B}) \to \cdots.$$

We apply this to understand $K_*\mathbb{Z} \to K_*\mathbb{Q}$. First of all, the resolution theorem shows that $K_*\mathbb{Z} = K_*\mathcal{P}_\mathbb{Z} \xrightarrow{\cong} K_*\mathcal{M}_\mathbb{Z}$, and similarly for $\mathbb{Q}$. Now look at the exact functor $F\colon \mathcal{M}_\mathbb{Z} \to \mathcal{M}_\mathbb{Q}$ that sends $M$ to $M \otimes \mathbb{Q}$. Then $\ker F$ is the abelian category $\mathcal{M}^1_\mathbb{Z}$ of torsion finitely generated $\mathbb{Z}$-modules, i.e., the category of finite abelian groups. One can check that the quotient Abelian category $\mathcal{M}_\mathbb{Z}/\mathcal{M}^1_\mathbb{Z}$ is equivalent to $\mathcal{M}_\mathbb{Q}$. Moreover,

$$\mathcal{M}^1_\mathbb{Z} \cong \operatorname*{colim}_{n \in \mathbb{N}} \mathcal{M}^1_\mathbb{Z}(n) \cong \operatorname*{colim}_{n \in \mathbb{N}} \mathcal{M}_{\mathbb{Z}/n\mathbb{Z}}$$

where $\mathcal{M}^1_\mathbb{Z}(n)$ is the full subcategory of $\mathcal{M}_\mathbb{Z}$ whose objects are the abelian groups annihilated by $n$.

Now what is $K_*\mathcal{M}_{\mathbb{Z}/n\mathbb{Z}}$? In terms of the prime factorization $n = p_1^{e_1} \cdots p_n^{e_n}$ the Chinese remainder theorem yields $\mathbb{Z}/n\mathbb{Z} \cong \mathbb{Z}/p_1^{e_1}\mathbb{Z} \times \cdots \times \mathbb{Z}/p_n^{e_n}\mathbb{Z}$, hence $\mathcal{M}_{\mathbb{Z}/n\mathbb{Z}} \cong \mathcal{M}_{\mathbb{Z}/p_1^{e_1}\mathbb{Z}} \times \cdots \times \mathcal{M}_{\mathbb{Z}/p_n^{e_n}\mathbb{Z}}$, and $K_n\mathcal{M}_{\mathbb{Z}/n\mathbb{Z}} \cong K_n\mathcal{M}_{\mathbb{Z}/p_1^{e_1}\mathbb{Z}} \times \cdots \times K_n\mathcal{M}_{\mathbb{Z}/p_n^{e_n}\mathbb{Z}}$ so we need understand only $K_n\mathcal{M}_{\mathbb{Z}/p^e\mathbb{Z}}$.

**Theorem 4.9** (Quillen's dévissage (untwisting) theorem [38]). *If $\mathcal{A}$ is an abelian category and $\mathcal{B}$ a full subcategory such that:*

(1) *for all $B \in \mathcal{B}$ and $A \in \mathcal{A}$, if $A \rightarrowtail B$ then $A \in \mathcal{B}$;*
(2) *for all $B \in \mathcal{B}$ and $A \in \mathcal{A}$, if $B \twoheadrightarrow A$ then $A \in \mathcal{B}$;*
(3) *for all $B', B'' \in \mathcal{B}$ then $B' \oplus B'' \in \mathcal{B}$;*
(4) *for all $A \in \mathcal{A}$ there is a filtration $A = A^0 \supseteq A^1 \supseteq \cdots \supseteq A^r = 0$ with $A^i/A^{i+1} \in \mathcal{B}$;*

*then the map $K_*\mathcal{B} \xrightarrow{\cong} K_*\mathcal{A}$ is an isomorphism.*

In particular $K_*\mathcal{M}_{\mathbb{Z}/p^e\mathbb{Z}} \xleftarrow{\cong} K_*\mathcal{M}_{\mathbb{Z}/p\mathbb{Z}} = K_*\mathcal{P}_{\mathbb{F}_p} = K_*\mathbb{F}_p$, and these groups are known, by Theorem 3.6. Then

$$\begin{aligned} K_*\mathcal{M}^1_{\mathbb{Z}} &\cong \operatorname*{colim}_n K_*\mathcal{M}_{\mathbb{Z}/n\mathbb{Z}} \cong \operatorname*{colim}_{p_1^{e_1},\dots,p_n^{e_n}} K_*(\mathcal{M}_{\mathbb{Z}/p_1^{e_1}\mathbb{Z}} \times \cdots \times \mathcal{M}_{\mathbb{Z}/p_n^{e_n}\mathbb{Z}}) \\ &\cong \operatorname*{colim}_{p_1,\dots,p_n} K_*\mathbb{F}_{p_1} \times \cdots \times K_*\mathbb{F}_{p_n} \cong \bigoplus_{\substack{p\in\mathbb{Z} \\ \text{prime}}} K_*\mathbb{F}_p \end{aligned}$$

All this is true more generally for the ring $\mathcal{O}_F$ of integers in a number field $F$ and yields the following long exact sequence (where the sums run over all prime ideals $\mathfrak{p}$ of $\mathcal{O}_F$).

$$\begin{aligned} \cdots \to \bigoplus_{\mathfrak{p}} K_{2n}\mathcal{O}_F/\mathfrak{p} \to K_{2n}\mathcal{O}_F \to K_{2n}F \\ \to \bigoplus_{\mathfrak{p}} K_{2n-1}\mathcal{O}_F/\mathfrak{p} \to K_{2n-1}\mathcal{O}_F \to K_{2n-1}F \to \cdots \\ \to K_0\mathcal{O}_F \to K_0F \to 0 \end{aligned}$$

It is known that for any field $F$, the rank homomorphism $K_0F \to \mathbb{Z}$ is an isomorphism, and the determinant homomorphism $K_1F \to F^\times$ is an isomorphism. For any Dedekind domain, such as $\mathcal{O}_F$, the rank and the determinant maps yield an isomorphism $K_0\mathcal{O}_F \xrightarrow{\cong} \mathbb{Z}\oplus\operatorname{Pic}\mathcal{O}_F$, where $\operatorname{Pic}\mathcal{O}_F$ is the Picard group of isomorphism classes of invertible modules. The $K$-groups of the finite fields $\mathcal{O}_F/\mathfrak{p}$ are known, see Theorem 3.6. Bass, Milnor, and Serre [2] proved that $K_1\mathcal{O}_F \to K_1F$ is injective, and hence that $K_1\mathcal{O}_F \cong \mathcal{O}_F^\times$. Soulé [40] has generalized that to show that $K_n\mathcal{O}_F \to K_nF$ is injective for $n > 0$. Quillen has shown [37] that $K_n\mathcal{O}_F$ is a finitely generated abelian group for all $n$. The Dirichlet unit theorem shows that the rank of $K_1\mathcal{O}_F$ is $r_1+r_2-1$, where the isomorphism of rings $F\otimes\mathbb{R} \cong \mathbb{R}^{r_1}\times\mathbb{C}^{r_2}$ defines the numbers $r_1$ and $r_2$. Moreover, Borel [10] has computed the ranks of the groups $K_n\mathcal{O}_F$ for $n \geq 2$, showing that the ranks are $0, r_1+r_2, 0, r_2$ provided $n$ is congruent to $0,1,2,3$ modulo 4, respectively.

Taking all that information into account, we see that the long exact sequence above splits into the folllowing simpler pieces, for $n > 0$.

$$0 \to K_{2n+1}\mathcal{O}_F \xrightarrow{\cong} K_{2n+1}F \to 0$$

$$0 \hookrightarrow K_{2n}\mathcal{O}_F \to K_{2n}F \to \bigoplus_{\mathfrak{p}} K_{2n-1}\mathcal{O}_F/\mathfrak{p} \to 0$$

$$0 \to \mathcal{O}_F^\times \to F^\times \to \bigoplus_{\mathfrak{p}} \mathbb{Z} \to \operatorname{Pic}\mathcal{O}_F \to 0$$

Specializing to the case $F = \mathbb{Q}$ one gets

$$0 \to K_2\mathbb{Z} \to K_2\mathbb{Q} \to \bigoplus_{p} \mathbb{F}_p^\times \to 0$$

where the sum runs over all prime numbers $p$. We know that $\mathbb{F}_p^\times \cong \mathbb{Z}/(p-1)\mathbb{Z}$, and by Milnor [33] we know that $K_2\mathbb{Z} \cong \mathbb{Z}/2\mathbb{Z}$. It follows (after showing that the sequence splits) that

$$K_2\mathbb{Q} \cong \mathbb{Z}/2\mathbb{Z} \oplus \bigoplus_{p} \mathbb{Z}/(p-1)\mathbb{Z}.$$

Hence $K_2\mathbb{Q}$ is not a finitely generated abelian group, because there are infinitely many primes.

# 5. Weight filtrations and motivic cohomology

In this section we present the connection between $K$-theory and motivic cohomology. Based on ideas of Beilinson, motivic cohomology was developed initially by Bloch and Suslin, and then was further developed by Friedlander, Geisser, Levine, and Lichtenbaum. It was molded into its final form by the Fields Medal winning work of Voevodsky, who proved the Bloch-Kato conjecture in characteristic 2 and the related conjecture of Milnor. Work of Rost on norm varieties and their motives, complemented by work of Haesemeyer, Joukhovitski, Suslin, and Weibel, recently seems to have settled the Bloch-Kato conjecture for fields of higher characteristic.

Recall first étale cohomology, i.e., an algebraic construction of $H^i(Y, \mathbb{Z}/n\mathbb{Z})$, where $Y$ is an algebraic variety. Let $\mathbb{G}_m = \operatorname{Spec}(\mathbb{C}[T, T^{-1}])$ be the multiplicative group. Its group of complex points $\mathbb{G}_m(\mathbb{C}) \cong \mathbb{C}^\times$, with the analytic topology, is not simply connected; its fundamental group is $\mathbb{Z}$, and the universal covering space $\mathbb{C} \to \mathbb{C}^\times$ is given by the exponential map, which is not a polynomial map, so algebra alone cannot see that the fundamental group is infinite. However, algebra can see the finite quotients $\mathbb{Z}/n\mathbb{Z}$ of the fundamental group, because the $n$-fold covering map $\mathbb{C}^\times \to \mathbb{C}^\times$ sending $z$ to $z^n$ is a polynomial map. The construction of étale cohomology and the fact that it works for finite coefficients is based on that observation.

The goal of motivic cohomology is to provide a purely algebraic construction of something analogous to the singular cohomology groups $H^i(Y, \mathbb{Z})$, where $Y$ is an algebraic variety. Following remarks 2.14 and 2.11, the idea is to work in an appropriate $\mathbb{Z}$-linear category, to look for "spheres" there, and to use them to represent cohomology.

We define a new $\mathbb{Z}$-linear category $\mathsf{KSm}_F$ with the same objects as $\mathsf{Sm}_F$, but whose morphisms are "$K_0$-correspondences from $X$ to $Y$". The morphisms from

$X = \operatorname{Spec} A$ to $Y = \operatorname{Spec} B$ will be the elements of $K_0(\mathcal{P}(A,B))$, where $\mathcal{P}(A,B)$ is the exact category of $A$-$B$-bimodules $M$ such that $M \in \mathcal{P}_A$. For non-affine varieties, a similar definition works.

Composition

$$K_0(\mathcal{P}(A,B)) \otimes K_0(\mathcal{P}(B,C)) \to K_0(\mathcal{P}(A,C))$$

is given by tensor product over $B$ of bimodules.

We get a functor $\mathsf{Sm}_F \to \mathsf{KSm}_F$ and hence a system of simplices $\Delta_{\text{alg}} \colon \mathsf{Ord} \to \mathsf{KSm}_F$, better than that defined in example 2.12, because correspondences from the affine line to curves of high genus exist in abundance.

Now, by analogy with remark 2.11, we seek a suitable "algebraic sphere" $S^q_{\text{alg}}$ in $\mathsf{KSm}_F$. Motivated by the observation that $\mathbb{G}_m(\mathbb{C})$, with the analytic topology, is homotopy equivalent to the circle $S^1_{\text{Top}}$, we define $S^1_{\text{alg}} = \mathbb{G}_m = \mathbb{A}^1 - \{0\}$ and denote by $\widetilde{S}^1_{\text{alg}}$ its reduced version. That means that, later, after applying a functor with values in $\mathsf{Ab}$ to it, we will remove the direct summand arising as the image of the projection operator $\widetilde{S}^1_{\text{alg}} \to * \to \widetilde{S}^1_{\text{alg}}$. Alternatively, think of it as the pair $(\widetilde{S}^1_{\text{alg}}, *)$. Then define $\widetilde{S}^q_{\text{alg}} = \widetilde{S}^1_{\text{alg}} \wedge \cdots \wedge \widetilde{S}^1_{\text{alg}}$, where $\wedge$ denotes the product of varieties, mixed with the product of pairs.

As in 2.14, for a variety $Y$ look at the system of simplices given by $Y \times \Delta_{\text{alg}}$ and the corresponding geometric realization $|Y, X|$. In topology, if we work in the $\mathbb{Z}$-linear category obtained from $\mathsf{Top}$ by replacing the Hom-sets by the free abelian groups they generate, $\pi_i|Y, \widetilde{S}^q_{\text{Top}}|$ is $H^{q-i}(Y,\mathbb{Z})$ if $i \leq q$ and is 0 if $i > q$. Thus we may expect (or hope) $\widetilde{S}^q_{\text{alg}}$ to represent cohomology in the algebraic context. We define motivic cohomology $H^i(Y,\mathbb{Z}(q))$ to be $\pi_{q-i}|Y, \widetilde{S}^q_{\text{alg}}|$, but only if $Y = \operatorname{Spec} A$ and $A$ is a local ring, such as a field. (When $Y$ is not local, one must use sheaf cohomology, too.) Unraveling the definitions, this is $\pi_{q-i}$ of the simplicial set whose $n$-simplices are the elements of the group

$$\widetilde{K}_0(\mathcal{P}(A[T_0,\ldots,T_n]/\sum T_i - 1, F[U_1,U_1^{-1},\ldots,U_q,U_q^{-1}])),$$

where $\widetilde{K}_0$ is obtained from $K_0$ as the quotient by the subgroup generated by the classes of all objects where one of the variables $U_i$ acts trivially.

**Theorem 5.1** ([19, 42, 20]). *Suppose $Y = \operatorname{Spec} A$ is smooth (local) over $F$. Then there is a filtration of the $K$-theory space of $Y$*

$$K(Y) = W^0 \leftarrow W^1 \leftarrow W^2 \leftarrow \cdots$$

*and there are fibration sequences*

$$W^{q+1} \to W^q \to \Omega^{-q}|Y, \widetilde{S}^q_{\text{alg}}|.$$

*Therefore there is a* motivic *spectral sequence*

$$E_2^{pq} = H^{p-q}(Y,\mathbb{Z}(-q)) \Rightarrow K_{-p-q}(Y)$$

*analogous to the Atiyah-Hirzebruch spectral sequence in topology.*

The spectral sequence of the theorem above was presented first by Bloch and Lichtenbaum in [9] in the case where $Y$ is the spectrum of a field, and extended to the global case by Friedlander and Suslin in [13]. Levine's approach to the spectral sequence in [30] manages to extract it from formal properties of $K$-theory in a clean way.

Taking into account remark 2.11, we can interpret $|Y, \widetilde{S}^q_{\mathrm{alg}}|$ as an explicit chain complex of abelian groups. Taking into account its contravariant dependence on $Y$ and sheafifying, we obtain a chain complex of sheaves, which is called simply $\mathbb{Z}(q)$, so that $H^i(Y, \mathbb{Z}(q))$ can be defined as its sheaf hypercohomology (for the Zariski topology on $Y$). This definition of motivic cohomology is not the one most often cited; we present it here because it is the one most closely related to algebraic $K$-theory.

**Properties 5.2.** *If $\ell$ is a prime number and char$F \nmid \ell$ and $Y \in \mathrm{Sm}_F$, then:*

(1) $\mathbb{Z}(0) \cong \mathbb{Z}$ *and* $\mathbb{Z}(1) \cong \mathbb{G}_m[-1]$*;*
(2) $\mathbb{Z}(q)$ *vanishes in degree greater than* $q$*;*
(3) *there are associative and commutative pairings* $\mathbb{Z}(q_1) \otimes \mathbb{Z}(q_2) \to \mathbb{Z}(q_1 + q_2)$*;*
(4) *after sheafification in the étale topology, there is an isomorphism* $\mu_\ell^{\otimes q} \cong \mathbb{Z}(q) \otimes \mathbb{Z}/\ell\mathbb{Z}$*, where* $\mu_\ell$ *denotes the étale sheaf of* $\ell$*-th roots of unity;*
(5) *the natural map* $K_q^M(F) \xrightarrow{\cong} H^q(\mathrm{Spec}\, F, \mathbb{Z}(q))$*, where* $K_q^M(F)$ *denotes the* $q$*-th Milnor* $K$*-group of* $F$*, is an isomorphism;*
(6) *there is a natural isomorphism* $H^{2q}(Y, \mathbb{Z}(q)) \cong CH^q(Y)$*, where* $CH^q(Y)$ *denotes the Chow group of algebraic cycles of codimension* $q$ *on* $Y$ *modulo rational equivalence;*
(7) localization: *if* $Z$ *is a smooth subvariety of pure codimension* $c$ *in* $Y$*, then there is a natural long exact sequence* $\cdots \to H^{i-1}(Y - Z, \mathbb{Z}(q)) \to H^{i-2c}(Z, \mathbb{Z}(q - c)) \to H^i(Y, \mathbb{Z}(q)) \to H^i(Y - Z, \mathbb{Z}(q)) \to \cdots$*;*
(8) Beilinson-Lichtenbaum conjecture: *The natural map* $H^i(Y, \mathbb{Z}(q) \otimes \mathbb{Z}/\ell\mathbb{Z}) \to H^i_{\text{ét}}(Y, \mathbb{Z}(q) \otimes \mathbb{Z}/\ell\mathbb{Z})$ *is an isomorphism if* $i \leq q$*;*
(9) Bloch-Kato conjecture: *the natural map*

$$K_q^M(F) \otimes \mathbb{Z}/\ell\mathbb{Z} \xrightarrow{\cong} H^q_{\mathrm{cont}}(\mathrm{Gal}(\overline{F}/F), \mu_\ell^{\otimes q})$$

*from Milnor* $K$*-theory to Galois cohomology is an isomorphism.*
(10) Beilinson-Soulé vanishing conjecture: $H^i(Y, \mathbb{Z}(q)) = 0$ *for* $i \leq 0$, $q > 0$.

The first seven properties are known.

The Bloch-Kato conjecture (9) is the Beilinson-Lichtenbaum conjecture (8) in the case where $Y = \mathrm{Spec}\, F$ and $i = q$, interpreted using (4) and (5). It is known [46] that the Bloch-Kato conjecture (9) implies the Beilinson-Lichtenbaum conjecture (9), rendering them equivalent. For the latest word on the proof of the Bloch-Kato conjecture (9), which seems to be completely written down now, see [21].

The Beilinson-Lichtenbaum conjecture (8) and the motivic spectral sequence, together with computations of étale cohomology, lead to computations [61] of the $K$-groups of rings of integers in number fields, some of which depend on the (unsettled) Vandiver conjecture of number theory.

The Beilinson-Soulé vanishing conjecture (10) is completely out of reach at the moment.

For introductions to motivic cohomology see [28, 58, 32, 51]. See also [25, 4, 8, 7, 35, 6, 27, 45, 48, 47, 29, 34, 57, 49, 56, 50, 52, 54, 44, 55, 53, 63] for the development of the theory and the proofs of the properties above.

# References

[1] M. F. Atiyah. *K-theory.* Advanced Book Classics. Addison-Wesley Publishing Company Advanced Book Program, Redwood City, CA, second edition, 1989. Notes by D. W. Anderson.

[2] H. Bass, J. Milnor, and J.-P. Serre. Solution of the congruence subgroup problem for $SL_n$ $(n \geq 3)$ and $Sp_{2n}$ $(n \geq 2)$. *Inst. Hautes Études Sci. Publ. Math.*, (33):59–137, 1967.

[3] Hyman Bass. *Algebraic K-theory.* W. A. Benjamin, Inc., New York-Amsterdam, 1968.

[4] Alexander Beilinson. Letter to Soulé, 1982. Preprint, June 12, 2004, K-theory Preprint Archives, http://www.math.uiuc.edu/K-theory/0694/.

[5] A. Jon Berrick. *An approach to algebraic K-theory*, volume 56 of *Research Notes in Mathematics.* Pitman (Advanced Publishing Program), Boston, Mass., 1982.

[6] S. Bloch. The moving lemma for higher Chow groups. *J. Algebraic Geom.*, 3(3):537–568, 1994.

[7] Spencer Bloch. Algebraic cycles and higher $K$-theory. *Adv. in Math.*, 61(3):267–304, 1986.

[8] Spencer Bloch and Kazuya Kato. $p$-adic étale cohomology. *Inst. Hautes Études Sci. Publ. Math.*, (63):107–152, 1986.

[9] Spencer Bloch and Steve Lichtenbaum. A Spectral Sequence for Motivic Cohomology. http://www.math.uiuc.edu/K-theory/0062/, March 3, 1995.

[10] Armand Borel. Stable real cohomology of arithmetic groups. *Ann. Sci. École Norm. Sup. (4)*, 7:235–272 (1975), 1974.

[11] Gunnar Carlsson. Deloopings in algebraic $K$-theory. In *Handbook of K-theory. Vol. 1, 2*, pages 3–37. Springer, Berlin, 2005.

[12] Albrecht Dold. Homology of symmetric products and other functors of complexes. *Ann. of Math. (2)*, 68:54–80, 1958.

[13] Eric M. Friedlander and Andrei Suslin. The spectral sequence relating algebraic $K$-theory to motivic cohomology. *Ann. Sci. École Norm. Sup. (4)*, 35(6):773–875, 2002.

[14] Henri Gillet and Daniel R. Grayson. The loop space of the $Q$-construction. *Illinois J. Math.*, 31(4):574–597, 1987.

[15] Henri Gillet and Daniel R. Grayson. Erratum to: "The loop space of the $Q$-construction" [Illinois J. Math. **31** (1987), no. 4, 574–597; MR0909784 (89h:18012)]. *Illinois J. Math.*, 47(3):745–748, 2003.

[16] Daniel Grayson. Higher algebraic $K$-theory. II (after Daniel Quillen). In *Algebraic K-theory (Proc. Conf., Northwestern Univ., Evanston, Ill., 1976)*, pages 217–240. Lecture Notes in Math., Vol. 551. Springer, Berlin, 1976.

[17] Daniel R. Grayson. On the $K$-theory of fields. In *Algebraic $K$-theory and algebraic number theory (Honolulu, HI, 1987)*, volume 83 of *Contemp. Math.*, pages 31–55. Amer. Math. Soc., Providence, RI, 1989.

[18] Daniel R. Grayson. Weight filtrations in algebraic $K$-theory. In *Motives (Seattle, WA, 1991)*, volume 55 of *Proc. Sympos. Pure Math.*, pages 207–237. Amer. Math. Soc., Providence, RI, 1994.

[19] Daniel R. Grayson. Weight filtrations via commuting automorphisms. *$K$-Theory*, 9(2):139–172, 1995.

[20] Daniel R. Grayson. The motivic spectral sequence. In *Handbook of $K$-theory. Vol. 1, 2*, pages 39–69. Springer, Berlin, 2005.

[21] Christian Haesemeyer and Charles A. Weibel. Norm Varieties and the Chain Lemma (after Markus Rost). Preprint, June 22, 2008, K-theory Preprint Archives, http://www.math.uiuc.edu/K-theory/0900/.

[22] Hvedri Inassaridze. *Algebraic $K$-theory*, volume 311 of *Mathematics and its Applications*. Kluwer Academic Publishers Group, Dordrecht, 1995.

[23] Daniel M. Kan. Functors involving c.s.s. complexes. *Trans. Amer. Math. Soc.*, 87:330–346, 1958.

[24] Max Karoubi. *$K$-theory*. Springer-Verlag, Berlin, 1978. An introduction, Grundlehren der Mathematischen Wissenschaften, Band 226.

[25] Kazuya Kato. A generalization of local class field theory by using $K$-groups. II. *J. Fac. Sci. Univ. Tokyo Sect. IA Math.*, 27(3):603–683, 1980.

[26] Aderemi Kuku. *Representation theory and higher algebraic $K$-theory*, volume 287 of *Pure and Applied Mathematics (Boca Raton)*. Chapman & Hall/CRC, Boca Raton, FL, 2007.

[27] Marc Levine. Bloch's higher Chow groups revisited. *Astérisque*, (226):10, 235–320, 1994. $K$-theory (Strasbourg, 1992).

[28] Marc Levine. *Mixed motives*, volume 57 of *Mathematical Surveys and Monographs*. American Mathematical Society, Providence, RI, 1998.

[29] Marc Levine. Techniques of localization in the theory of algebraic cycles. *J. Algebraic Geom.*, 10(2):299–363, 2001.

[30] Marc Levine. Chow's moving lemma and the homotopy coniveau tower. *$K$-Theory*, 37(1-2):129–209, 2006.

[31] E. Lluis-Puebla, J.-L. Loday, H. Gillet, C. Soulé, and V. Snaith. *Higher algebraic $K$-theory: an overview*, volume 1491 of *Lecture Notes in Mathematics*. Springer-Verlag, Berlin, 1992.

[32] Carlo Mazza, Vladimir Voevodsky, and Charles Weibel. *Lecture notes on motivic cohomology*, volume 2 of *Clay Mathematics Monographs*. American Mathematical Society, Providence, RI, 2006.

[33] John Milnor. *Introduction to algebraic $K$-theory*. Princeton University Press, Princeton, N.J., 1971. Annals of Mathematics Studies, No. 72.

[34] Fabien Morel and Vladimir Voevodsky. $\mathbf{A}^1$-homotopy theory of schemes. *Inst. Hautes Études Sci. Publ. Math.*, (90):45–143 (2001), 1999.

[35] Yu. P. Nesterenko and A. A. Suslin. Homology of the general linear group over a local ring, and Milnor's $K$-theory. *Izv. Akad. Nauk SSSR Ser. Mat.*, 53(1):121–146, 1989.

[36] Daniel Quillen. On the cohomology and $K$-theory of the general linear groups

over a finite field. *Ann. of Math. (2)*, 96:552–586, 1972.

[37] Daniel Quillen. Finite generation of the groups $K_i$ of rings of algebraic integers. In *Algebraic K-theory, I: Higher K-theories (Proc. Conf., Battelle Memorial Inst., Seattle, Wash., 1972)*, pages 179–198. Lecture Notes in Math., Vol. 341. Springer, Berlin, 1973.

[38] Daniel Quillen. Higher algebraic $K$-theory. I. In *Algebraic K-theory, I: Higher K-theories (Proc. Conf., Battelle Memorial Inst., Seattle, Wash., 1972)*, pages 85–147. Lecture Notes in Math., Vol. 341. Springer, Berlin, 1973.

[39] Jonathan Rosenberg. *Algebraic K-theory and its applications*, volume 147 of *Graduate Texts in Mathematics*. Springer-Verlag, New York, 1994.

[40] C. Soulé. $K$-théorie des anneaux d'entiers de corps de nombres et cohomologie étale. *Invent. Math.*, 55(3):251–295, 1979.

[41] V. Srinivas. *Algebraic K-theory*, volume 90 of *Progress in Mathematics*. Birkhäuser Boston Inc., Boston, MA, 1991.

[42] A. Suslin. On the Grayson spectral sequence. *Tr. Mat. Inst. Steklova*, 241(Teor. Chisel, Algebra i Algebr. Geom.):218–253, 2003.

[43] A. A. Suslin. Algebraic $K$-theory. In *Algebra. Topology. Geometry, Vol. 20*, Itogi Nauki i Tekhniki, pages 71–152, 196. Akad. Nauk SSSR, Vsesoyuz. Inst. Nauchn. i Tekhn. Informatsii, Moscow, 1982.

[44] Andrei Suslin and Seva Joukhovitski. Norm Varieties. Preprint, February 17, 2006, K-theory Preprint Archives, http://www.math.uiuc.edu/K-theory/0742/.

[45] Andrei Suslin and Vladimir Voevodsky. Singular homology of abstract algebraic varieties. *Invent. Math.*, 123(1):61–94, 1996.

[46] Andrei Suslin and Vladimir Voevodsky. Bloch-Kato conjecture and motivic cohomology with finite coefficients. In *The arithmetic and geometry of algebraic cycles (Banff, AB, 1998)*, volume 548 of *NATO Sci. Ser. C Math. Phys. Sci.*, pages 117–189. Kluwer Acad. Publ., Dordrecht, 2000.

[47] Andrei Suslin and Vladimir Voevodsky. Relative cycles and Chow sheaves. In *Cycles, transfers, and motivic homology theories*, volume 143 of *Ann. of Math. Stud.*, pages 10–86. Princeton Univ. Press, Princeton, NJ, 2000.

[48] V. Voevodsky. Homology of schemes. *Selecta Math. (N.S.)*, 2(1):111–153, 1996.

[49] V. Voevodsky. On the zero slice of the sphere spectrum. *Tr. Mat. Inst. Steklova*, 246(Algebr. Geom. Metody, Svyazi i Prilozh.):106–115, 2004.

[50] Vladimir Voevodsky. Cancellation theorem. Preprint, January 28, 2002, K-theory Preprint Archives, http://www.math.uiuc.edu/K-theory/0541/.

[51] Vladimir Voevodsky. Lectures on motivic cohomology 2000/2001 (written by Pierre Deligne). Preprint, November 19, 2001, K-theory Preprint Archives, http://www.math.uiuc.edu/K-theory/0527/.

[52] Vladimir Voevodsky. Motives over simplicial schemes. Preprint, June 16, 2003, K-theory Preprint Archives, http://www.math.uiuc.edu/K-theory/0638/.

[53] Vladimir Voevodsky. Motivic Eilenberg-MacLane spaces. Preprint, September 10, 2007, K-theory Preprint Archives, http://www.math.uiuc.edu/K-theory/0864/.

[54] Vladimir Voevodsky. On motivic cohomology with Z/l-coefficients. Preprint,

June 16, 2003, K-theory Preprint Archives, http://www.math.uiuc.edu/K-theory/0639/.

[55] Vladimir Voevodsky. Simplicial radditive functors. Preprint, September 10, 2007, K-theory Preprint Archives, http://www.math.uiuc.edu/K-theory/0863/.

[56] Vladimir Voevodsky. Unstable motivic homotopy categories in Nisnevich and cdh-topologies. Preprint, September 6, 2000, K-theory Preprint Archives, http://www.math.uiuc.edu/K-theory/0444/.

[57] Vladimir Voevodsky. Reduced power operations in motivic cohomology. *Publ. Math. Inst. Hautes Études Sci.*, (98):1–57, 2003.

[58] Vladimir Voevodsky, Andrei Suslin, and Eric M. Friedlander. *Cycles, transfers, and motivic homology theories*, volume 143 of *Annals of Mathematics Studies*. Princeton University Press, Princeton, NJ, 2000.

[59] Friedhelm Waldhausen. Algebraic $K$-theory of generalized free products. I, II. *Ann. of Math. (2)*, 108(1):135–204, 1978.

[60] Friedhelm Waldhausen. Algebraic $K$-theory of generalized free products. III, IV. *Ann. of Math. (2)*, 108(2):205–256, 1978.

[61] Charles Weibel. Algebraic $K$-theory of rings of integers in local and global fields. In *Handbook of $K$-theory. Vol. 1, 2*, pages 139–190. Springer, Berlin, 2005.

[62] Charles A. Weibel. The $K$-book: An introduction to algebraic $K$-theory. A graduate textbook in progress, http://www.math.rutgers.edu/~weibel/Kbook.html.

[63] Charles A. Weibel. Patching the Norm Residue Isomorphism Theorem. Preprint, May 23, 2007, K-theory Preprint Archives, http://www.math.uiuc.edu/K-theory/0844/.

*Cohomology of Groups and*
*Algebraic K-theory*
ALM 12, pp. 187–195



# The Braid Groups of $\mathbb{R}P^2$ Satisfy the Fibered Isomorphism Conjecture

Daniel Juan-Pineda* and Silvia Millan-López**

**Abstract**

In this article we prove that the braid groups of the 2-projective plane satisfy the Fibered Isomorphism Conjecture.

**2000 Mathematics Subject Classification:** 19B28; Secondary: 19A31

**Keywords and Phrases:** Braid group, fibered isomorphism conjecture

## 1. Introduction

Let $\Gamma$ be a discrete group and $\mathbb{Z}[\Gamma]$ its integral group ring. Let $K_i(\mathbb{Z}[\Gamma])$ denote the algebraic $K$-theory groups of $\mathbb{Z}[\Gamma]$. These groups are very difficult to compute, the Fibered Isomorphism Conjecture (FIC) has proven to be an effective way to approach $K_i(\mathbb{Z}[\Gamma])$, the input comes from the geometry and homological data of the group $\Gamma$. FIC has been verified for important classes of groups, see [2, 3, 4, 6]. In this paper we prove the following

**Theorem 1.** *The full and the pure braid groups on any number of strands of the 2-dimensional real projective space satisfy FIC.*

Braid groups of $\mathbb{R}P^2$ do not fall into the cases studied in [1]. One of the new features is the presence of torsion in these groups. A key concept that S. Aravinda, T. Farrell and K. Roushon introduced in [1] is that of a *strongly poly free* group, which we now recall:

**Definition 2.** *A discrete group $\Gamma$ is called strongly poly-free of rank $\leq n$, if there exists a finite filtration by subgroups $1 = \Gamma_0 \subseteq \Gamma_1 \subseteq \cdots \subseteq \Gamma_n = \Gamma$ such that the following conditions are satisfied:*

*Instituto de Matemáticas, Unidad Morelia, Universidad Nacional Autónoma de México, Campus Morelia, Apartado Postal 61-3 (Xangari), Morelia, Michoacán, MEXICO 58089. E-mail: daniel@matmor.unam.mx

**Department of Mathematics, Binghamton University, Binghamton, N. Y. 13902. E-mail: millan@math.binghamton.edu

(1) $\Gamma_i$ *is normal in* $\Gamma$ *for each* $i$.
(2) $\Gamma_{i+1}/\Gamma_i$ *is a finitely generated free group for all* $i$.
(3) *For each* $\gamma \in \Gamma$ *there is a compact surface* $F$ *and a diffeomorphism* $f : F \to F$ *such that the induced homomorphism* $f_\#$ *on* $\pi_1(F)$ *is equal to* $C_\gamma$ *in* $Out(\pi_1(F))$, *where* $C_\gamma$ *is the action of* $\gamma$ *on* $\Gamma_{i+1}/\Gamma_i$ *by conjugation and* $\pi_1(F)$ *is identified with* $\Gamma_{i+1}/\Gamma_i$ *via a suitable isomorphism, i.e.* $f_\# = \phi^{-1} \circ C_\gamma \circ \phi$ *from the following diagram:*

$$\begin{array}{ccc} \pi_1(F, f_0) & \xrightarrow{f_\#} & \pi_1(F, f_0) \\ \phi \downarrow & & \uparrow \phi^{-1} \\ \Gamma_{i+1}/\Gamma_i & \xrightarrow{C_\gamma} & \Gamma_{i+1}/\Gamma_i \end{array}$$

For the rest of this paper we will use $spf$ to denote a strongly poly-free group.

We dedicated this paper to Tom Farrell's 65th birthday.

## 2. The main results

In this section we define the full braid group and prove that this group satisfies the Fibered Isomorphism Conjecture. We start with some notation and definitions.

Let $Q_m = \{q_1, \ldots, q_m\}$ a fixed set of $m-$ distinct points of the 2-projective plane, $\mathbb{R}P^2$. Define

$$F(\mathbb{R}P^2, n) = \{(x_1, \ldots, x_n) \mid x_i \in \mathbb{R}P^2 - Q_m,\, x_i \neq x_j,\, if\, i \neq j\}.$$

Observe that when $m = 0$ in the above definition we have

$$F(\mathbb{R}P^2, n) = \{(x_1, \cdots, x_n) \mid x_i \in \mathbb{R}P^2,\, x_i \neq x_j,\, if\, i \neq j\}.$$

and that $S_n$ the symmetric group on $n$-symbols acts naturally on $F(\mathbb{R}P^2, n)$.

**Definition 3.** *The Pure Braid Group,* $PB_n(\mathbb{R}P^2)$, *of* $n-$ *strands of* $\mathbb{R}P^2$ *is the fundamental group of* $F(\mathbb{R}P^2, n)$.

**Definition 4.** *The Full Braid Group,* $B_n(\mathbb{R}P^2)$, *of* $n-$ *strands of* $\mathbb{R}P^2$ *is the fundamental group of* $F(\mathbb{R}P^2, n)/S_n$.

**Theorem A.** *The Pure Braid group on* $n$*-strands of the projective plane* $\mathbb{R}P^2$ *satisfies* $FIC$ *for all* $n > 0$.

*Proof.* First, we study the cases for $n = 1, 2$, and 3 which were studied by Van Buskirk in [13] and found the following:

For $n = 1$, $PB_1(\mathbb{R}P^2) = \pi_1(F(\mathbb{R}P^2, 1)) = \mathbb{Z}_2$, for $n = 2$, $PB_2(\mathbb{R}P^2) = \pi_1(F(\mathbb{R}P^2, 2)) \approx \mathbf{Q}_8$, where $\mathbf{Q}_8$ is the quaternion group, and if $n = 3$, $PB_3(\mathbb{R}P^2) = \pi_1(F(\mathbb{R}P^2, 3)) \cong F_2 \rtimes \mathbf{Q}_8$.

For $n = 1$ and 2 the FIC is true, since they are finite groups. To show that $F_2 \rtimes \mathbf{Q}_8$ satisfies the FIC, first we construct the short exact sequence

$$1 \longrightarrow F_2 \xrightarrow{i} F_2 \rtimes \mathbf{Q}_8 \xrightarrow{p} \mathbf{Q}_8 \longrightarrow 1$$

and use lemma 11 to obtain that

$$F_2 \rtimes \mathbf{Q}_8 \leqslant F_2^{\mathbf{Q}_8} \rtimes \mathbf{Q}_8.$$

Note that $F_2 = \pi_1(\mathbb{R}^2 - \mathbf{Q}_2)$ is the fundamental group of the surface $\mathbb{R}^2 - \mathbf{Q}_2$. Hence $F_2^{\mathbf{Q}_8} \rtimes \mathbf{Q}_8$ satisfies the FIC by applying lemma 9 to $F_2^{\mathbf{Q}_8} \rtimes \mathbf{Q}_8$.

Now we can conclude that $F_2 \rtimes \mathbf{Q}_8$ satisfies the FIC, since this is a subgroup of $F_2^{\mathbf{Q}_8} \rtimes \mathbf{Q}_8$.

For the case $n > 3$, we look at the following locally trivial fiber bundle

$$F(\mathbb{R}P^2, n) \xrightarrow{p} F(\mathbb{R}P^2, 2)$$
$$(x_1, ..., x_n) \mapsto (x_1, x_2)$$

with fiber $F(\mathbb{R}P^2 - Q_2, n-2)$.

Using the long exact sequence on this bundle we obtain the following:

$$\begin{aligned} ...\pi_2(F(\mathbb{R}P^2, 2)) &\longrightarrow \pi_1(F(\mathbb{R}P^2 - Q_2, n-2)) \longrightarrow \pi_1(F(\mathbb{R}P^2, n)) \\ &\longrightarrow \pi_1(F(\mathbb{R}P^2, 2)) \longrightarrow \pi_0(F(\mathbb{R}P^2 - Q_2, n-2)) \cdots . \end{aligned}$$

But, $\pi_2(F(\mathbb{R}P^2, 2)) \cong \pi_2(\mathbb{S}^3) = 1$, see [8], $\pi_1(F(\mathbb{R}P^2, 2)) \cong \mathbf{Q}_8$, see [13], and, $\pi_0(F(M - *, n-2)) = 1$, where $M$ is the Möbius band with boundary.

Thus, this long exact sequence leads us to the following short exact sequence:

$$1 \longrightarrow \pi_1(F(\mathbb{R}P^2 - Q_2, n-2)) \longrightarrow PB_n(\mathbb{R}P^2) \longrightarrow \mathbf{Q}_8 \longrightarrow 1.$$

Assuming the following claim is true, we use Lemma 8 to conclude that $PB_n(\mathbb{R}P^2)$ satisfies FIC.

**Claim 5.** *$\Gamma = \pi_1(F(\mathbb{R}P^2 - Q_2, n-2))$ is strongly poly free.*

*Proof of claim 5.* We start by observing that $\mathbb{R}P^2 - Q_2$ is homotopically equivalent to $M - *$. Hence $\pi_1(F(\mathbb{R}P^2 - Q_2, n-2)) \cong \pi_1(F(M - *, n-2))$ for all $n > 3$. For simplicity, let $\Gamma_i = \pi_1(F(M - *, i))$ for $i \geqslant 1$.

Now the problem is reduced to showing that $\Gamma_i = \pi_1(F(M - *, i))$ is *spf* for all $i \geqslant 1$. Aravinda, Farrell, and Roushon showed in [1] that $PB_{i+1}(M) = \pi_1(F(M, i+1))$, the pure braid group of $(i+1)$ – *strands* of the Möbius band with boundary, is an *spf* group with

$$\Gamma_k = \pi_1(F(M - Q_{i+1-k}, k)$$

a filtration for $PB_{i+1}(M)$. Note that,

$$\Gamma_i \leqslant \Gamma_{i+1} = PB_{i+1}(M)$$

is a subgroup of this filtration and so $\Gamma_i$ is also *spf*. Therefore $\Gamma = \pi_1(F(M - *, n-2))$ is strongly poly free for all $n > 3$.

**Theorem B.** *The Full Braid group of n-strands of the projective plane $\mathbb{R}P^2$, $B_n(\mathbb{R}P^2)$ satisfies FIC for all $n > 0$.*

*Proof.* The cases $n = 1$ and $n = 2$ follow immediately as $B_n(\mathbb{R}^2)$ are finite groups, see [6] and [13].

For $n > 2$ note that the projection $F(\mathbb{R}P^2, n) \xrightarrow{p} F(\mathbb{R}P^2, 2)$ to the first two coordinates is a locally trivial fiber bundle with fiber $F(\mathbb{R}P^2 - Q_2, n-2)$. Using the homotopy exact sequence on this fiber bundle it follows that the following

$$1 \longrightarrow \Gamma \xrightarrow{\theta} PB_n(\mathbb{R}P^2) \xrightarrow{\delta} \mathbf{Q}_8 \longrightarrow 1$$

is a short exact sequence, where $\mathbf{Q}_8$ is the quaternion group and $\Gamma = \pi_1(F(\mathbb{R}P^2 - Q_2, n-2))$ is an *spf* group, see claim 5.

Also, note that since $S_n$ acts freely on $F(\mathbb{R}P^2, n)$, the natural map

$$F(\mathbb{R}P^2, n) \xrightarrow{p} F(\mathbb{R}P^2, n)/S_n$$

is a covering map with fiber $S_n$ and since $F(\mathbb{R}P^2, n)$ is path connected

$$p_{\#}\pi_1(F(\mathbb{R}P^2, n)) \trianglelefteq \pi_1(F(\mathbb{R}P^2, n)/S_n)$$

and so,

$$1 \longrightarrow PB_n(\mathbb{R}P^2) \xrightarrow{\mu} B_n(\mathbb{R}P^2) \xrightarrow{\rho} S_n \longrightarrow 1$$

is also a short exact sequence. Let $H = (\mu \circ \theta)(\Gamma) < B_n(\mathbb{R}P^2)$. Since $\mu \circ \theta$ is an injective homomorphism, $H \cong \Gamma$ and so the index of $H$ in $B_n(\mathbb{R}P^2)$ is finite and $H$ is an *spf* group. Now define,

$$S = \bigcap_{g \in B_n(\mathbb{R}P^2)} gHg^{-1}$$

and recall from group theory the following

**Proposition 6.** *Let $H$ be a subgroup of $G$ such that the index of $H$ in $G$ is finite. Then $K = \bigcap_{g \in G} gHg^{-1} \trianglelefteq G$ and the index of $K$ in $G$ is finite.*

Applying this proposition to $S$, we have that $S \trianglelefteq B_n(\mathbb{R}P^2)$ and $[B_n(\mathbb{R}P^2) : S]$ is finite. Hence $S \trianglelefteq H$ and $[H : S]$ is also finite.

To continue with the proof we state the following result that will be proved in the next section.

**Theorem C.** *Let $\Gamma$ be a strongly poly-free group of rank $\leq n$ and $H$ a normal subgroup of finite index. Then $H$ is also a strongly poly-free group of rank $\leq n$.*

Assuming this theorem is true, we conclude that $S$ is an *spf* group and construct the following short exact sequence

$$1 \longrightarrow S \longrightarrow B_n(\mathbb{R}P^2) \longrightarrow B_n(\mathbb{R}P^2)/S \longrightarrow 1.$$

Now we apply Lemma 8 to conclude the desired result. Hence the Full Braid Group of the 2-projective plane satisfies the FIC for all $n > 0$. □

## 3. Strongly poly-free groups

In this section we provide the proof of Theorem C:

*Proof.* To show that $H$ is an $spf$, it is necessary to give a filtration of $H$ such that conditions (1)-(3) are satisfied. We define a filtration of $H$ as follows: Let $H_i = \Gamma_i \cap H$. From this definition, it is clear that $H_i \lhd H$ for all $i \geq 1$. So we have verified that a filtration of $H$ exists and satisfies condition (1) in the definition of $spf$.

Now we will show that $H_{i+1}/H_i$ is a finitely generated free group. But first we recall the following result:

If G is a finitely generated free group and $H$ is a subgroup of $G$ of finite index, then $H$ is finitely generated free group, [5]. Thus, we will show the following:

i) $H_{i+1}/H_i$ is a subgroup of $\Gamma_{i+1}/\Gamma_i$
ii) The index of $H_{i+1}/H_i$ in $\Gamma_{i+1}/\Gamma_i$ is finite.

*Proof of* i) Observe that $(\Gamma_{i+1} \cap H) \cdot \Gamma_i$ is a subgroup of $\Gamma_{i+1}$ and since $\Gamma_i, \Gamma_{i+1}$, and $\Gamma_{i+1}H$ are normal in $\Gamma$, we have that

$$(\Gamma_{i+1} \cap H) \cdot \Gamma_i \trianglelefteq \Gamma_{i+1}.$$

Thus, we can look at the following sequence of normal subgroups of $\Gamma$:

$$\Gamma_i \trianglelefteq (\Gamma_{i+1} \cap H) \cdot \Gamma_i \trianglelefteq \Gamma_{i+1} \trianglelefteq \Gamma.$$

Let $\phi : \Gamma \longrightarrow \Gamma/\Gamma_i$ be the natural map, then by the fourth isomorphism theorem of group theory we have the following two results:

**a)** $(\Gamma_{i+1} \cap H) \cdot \Gamma_i \trianglelefteq \Gamma_{i+1}$ if and only if $\frac{(\Gamma_{i+1} \cap H) \cdot \Gamma_i}{\Gamma_i} \trianglelefteq \Gamma_{i+1}/\Gamma_i$.

**b)** The index of $(\Gamma_{i+1} \cap H) \cdot \Gamma_i$ in $\Gamma_{i+1}$ is equal to the index of $\frac{(\Gamma_{i+1} \cap H) \cdot \Gamma_i}{\Gamma_i}$ in $\Gamma_{i+1}/\Gamma_i$.

Applying the second isomorphism theorem we have that

$$\frac{(\Gamma_{i+1} \cap H) \cdot \Gamma_i}{\Gamma_i} \cong \frac{\Gamma_{i+1} \cap H}{\Gamma_i \cap (\Gamma_{i+1} \cap H)} = H_{i+1}/H_i.$$

By identifying $H_{i+1}/H_i$ with its image in $\Gamma_{i+1}/\Gamma_i$ we can say that $H_{i+1}/H_i$ is a normal subgroup of $\Gamma_{i+1}/\Gamma_i$.

To show ii) we will show that the index of $\frac{(\Gamma_{i+1} \cap H) \cdot \Gamma_i}{\Gamma_i}$ in $\Gamma_{i+1}/\Gamma_i$ is finite. But first, we state and prove the following:

**Claim:** The index of $\Gamma_{i+1} \cap H$ in $\Gamma_{i+1}$ is finite.
*Proof of the claim*: by the second isomorphism theorem $\frac{\Gamma_{i+1}}{\Gamma_{i+1} \cap H}$ is isomorphic to $\frac{\Gamma_{i+1} \cdot H}{H}$, but $H \trianglelefteq \Gamma_{i+1} \cdot H \trianglelefteq \Gamma$ and the index of $H$ in $\Gamma$ is finite. Hence the index

of $\Gamma_{i+1} \cap H$ in $\Gamma_{i+1}$ is finite.

Now we can conclude the desired result by using the claim in the following relation of normal subgroups of $\Gamma$:

$$\Gamma_{i+1} \cap H \trianglelefteq (\Gamma_{i+1} \cap H) \cdot \Gamma_i \trianglelefteq \Gamma_{i+1}.$$

To check condition (3) in the definition of *spf* we recall the following

**Fact 7.** *Let $p : E \longrightarrow B$ be a covering map. If $B$ is compact and $p^{-1}(b)$ is finite for all $b \in B$, then $E$ is compact.*

*Proof of the third condition of spf.* Let $\gamma \in H \trianglelefteq \Gamma$. Since $\Gamma$ is *spf* there exists a compact surface $X$ and a diffeomorphism $f : X \longrightarrow X$ such that $\pi_1(X, x_0) \cong \Gamma_{i+1}/\Gamma_i$ and $f_\# : \pi_1(X, x_0) \longrightarrow \pi_1(X, x_0)$ is conjugation by $\gamma$ in the sense of property three of an *spf* group. Recall that:

$$H_{i+1}/H_i \overset{\rho}{\cong} (\Gamma_{i+1} \cap H) \cdot \Gamma_i/\Gamma_i \trianglelefteq \Gamma_{i+1}/\Gamma_i \overset{\phi}{\cong} \pi_1(X, x_0).$$

Let $K = (\phi \circ \rho)(H_{i+1}/H_i)$ and Let $p : (E, e_0) \longrightarrow (X, x_0)$ be a covering map such that $p_\#(\pi_1(E, e_0)) = K$. Since $p_\#$ is injective, $p_\#(\pi_1(E, e_0)) \cong \pi_1(E, e_0) \cong K$. Hence we have a surface $E$ such that $\pi_1(E, e_0) \cong H_{i+1}/H_i$. This surface is compact as $[\Gamma_{i+1}/\Gamma_i : H_{i+1}/H_i]$ is finite.

For the existence of the diffemorphism, first we will show that there exists $g : (E, e_0) \longrightarrow (E, e_0)$ such that the following diagram commutes.

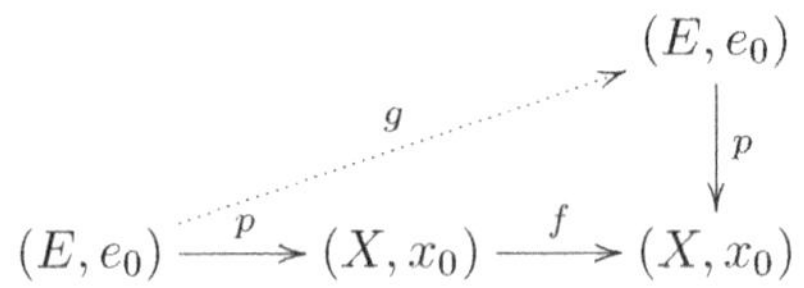


In other words we will show that $fop$ can be lifted.

Since $p$ is a covering map with $p(e_0) = x_0$ we can use the lifting theorem and show that,

$$(fop)_\#(\pi_1(E, e_0)) \subset p_\#(\pi_1(E, e_0))$$

equivalently,

$$f_\#(K) \subset K.$$

Recall that $f_\#$ is conjugation by $[\gamma]$ as stated in the definition of an *spf* group. The following diagram shows that $K \lhd \pi_1(X, x_0)$ which concludes the desired result.

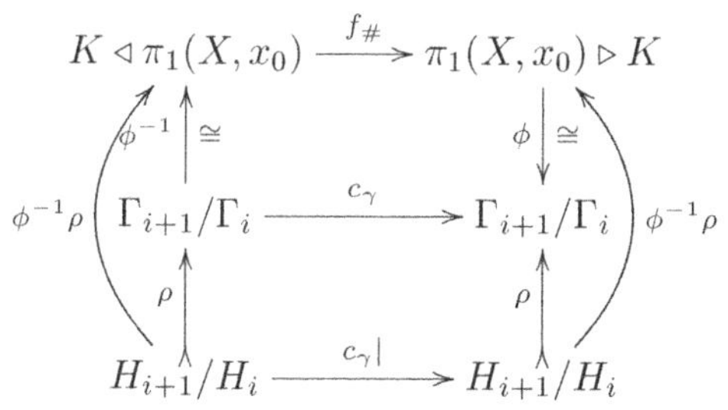


(1)

Thus, we have a map $g : (E, e_0) \longrightarrow (E, e_0)$ and we will show that this map is a homeomorphism. But before we do this, we can show in a similar way that there exists a map $h : (E, e_0) \longrightarrow (E, e_0)$ such that the following diagram commutes:

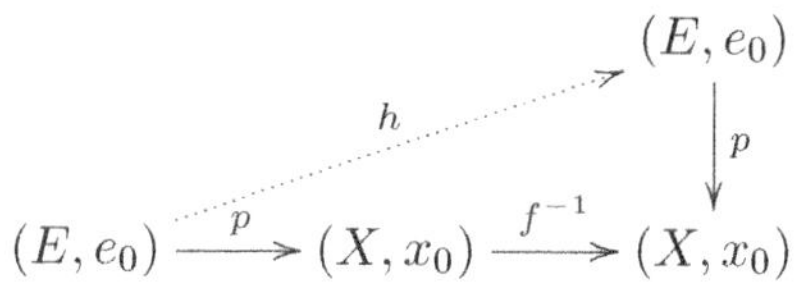


where $f_{\#}^{-1}$ is conjugation by $[\gamma^{-1}]$ in the sense of the definition of an *spf* group. Using these two diagrams we have that

$$p \circ h \circ g = f \circ f^{-1} \circ p = p,$$

hence $h \circ g$ and $id$ are two lifts for $p$. Thus, $h$ must be the inverse of $g$ and since these lifts are continuous then $g$ is a homeomorphism.

For the case of compact surfaces the third condition in the definition of an *spf* group can be modified and require only a homeomorphism instead of a diffeomorphism. But if one wants to be more careful, we can choose a differentiable covering map $p : (E, e_0) \longrightarrow (X, x_0)$ to adjust the proof.

It is left to show that $g_{\#}$ is conjugation by $[\gamma]$ as stated in the definition of an *spf* group. But as we can see in diagram 1

$$g_{\#} = f_{\#}|\pi_1(E, e_0) = C_{[\gamma]}. \qquad \square$$

# 4. FIC and useful lemmas

We recall the Fibered Isomorphism Conjecture formulated in [6]. Let $\mathcal{S} : TOP \to SPECTRA$ be a covariant homotopy functor. Let $\mathbf{B}$ be the category of continuous surjective maps: objects in $\mathbf{B}$ are continuous maps $p : E \to B$, where $E, B$ are objects in $TOP$, and morphisms between $p_1 : E_1 \to B_1$ and $p_2 : E_2 \to B_2$ consist of continuous maps $f : E_1 \to E_2$ and $g : B_1 \to B_2$ making the following diagram commute

$$\begin{array}{ccc} E_1 & \xrightarrow{f} & E_2 \\ {\scriptstyle p_1}\downarrow & & {\scriptstyle p_2}\downarrow \\ B_1 & \xrightarrow{g} & B_2. \end{array}$$

In this setup, Quinn [10] constructs a functor from $\mathbf{B}$ to $\Omega - SPECTRA$. The value of this spectrum at $p : E \to B$ is denoted by

$$\mathbb{H}(B; \mathcal{S}(p)),$$

and has the property that its value at the object $E \to *$ is $\mathcal{S}(E)$. The map of spectra $\mathbb{A}$ associated to

$$\begin{array}{ccc} E & \xrightarrow{id} & E \\ {\scriptstyle p}\downarrow & & {\scriptstyle p}\downarrow \\ B & \xrightarrow{g} & * \end{array}$$

is known as *Quinn* assembly map.

Given a discrete group $\Gamma$, let $\mathcal{E}$ be a universal $\Gamma$-space for the family of virtually cyclic subgroups of $\Gamma$ [6] and denote by $\mathcal{B}$ the orbit space $\mathcal{E}/\Gamma$. Let $X$ be any free and properly discontinuous $\Gamma$- space, and $p : \mathcal{E} \times_\Gamma X \to \mathcal{B}$ be the map determined by the projection onto $\mathcal{B}$. The *Fibered Isomorphism Conjecture* (FIC) for $\mathcal{S}$ and $\Gamma$ is the assertion that

$$\mathbb{A} : \mathbb{H}(\mathcal{B}; \mathcal{S}(p)) \to \mathcal{S}(X/\Gamma)$$

is a weak equivalence of spectra. This conjecture was made in [6] for the functors $\mathcal{S} = \mathcal{P}(), \mathcal{K}()$, and $\mathcal{L}^{-\infty}$, the pseudoisotopy, algebraic $K$-theory and $\mathcal{L}^{-\infty}$-theory functors. In this paper we mean FIC as FIC for the functor $\mathcal{S} = \mathcal{P}()$.

We now recall some useful lemmas from [7] that we use in this work.

**Lemma 8.** *Let $\Gamma$ be an extension of a finite group by an spf,* i.e.

$$1 \longrightarrow H \longrightarrow \Gamma \longrightarrow F \longrightarrow 1$$

*is a short exact sequence where $H$ is an spf group and $F$ is a finite group. Then $\Gamma$ satisfies the FIC.*

**Lemma 9.** *Let $\Gamma$ be the fundamental group of a surface, then the FIC is true for $\Gamma^G \rtimes G$ for a finite group $G$.*

**Lemma 10.** *If the FIC is true for $\Gamma$, then it is also true for every subgroup $S$ of $\Gamma$.*

**Lemma 11.** *Let*

$$1 \longrightarrow \Pi \longrightarrow \Gamma \longrightarrow G \longrightarrow 1$$

*be a group extension. Then there are injective homomorphisms $\Gamma \xrightarrow{\delta} \Pi^G \rtimes G$ and $\Pi \xrightarrow{\theta} \Pi^G$ which together with id: $G \longrightarrow G$ define a map to the group extension*

$$1 \longrightarrow \Pi^G \longrightarrow \Pi^G \rtimes G \longrightarrow G \longrightarrow 1.$$

## References


[1] C.S. Aravinda, F.T. Farrell, and S. K. Roushon, *Algebraic K-theory of pure braid groups*, Asian J. Math., 4(2): 337–343, 2000.

[2] A. C. Bartels, T. Farrell, L. Jones, and H. Reich, *On the Isomorphism Conjecture in algebraic K-Theory*, Topology, 43(1): 157–231, 2004.

[3] A. C. Bartels and H. Reich *On the Farrell-Jones conjecture for higher algebraic K-theory*, J. Amer. Math. Soc., 18(3): 501–545. 2005.

[4] A. C. Bartels, W. Lück, and H. Reich, *The K-theoretic Farrell-Jones Conjecture for hyperbolic groups*, Preprint 813 in http://www.math.uiuc.edu/K-theory/.

[5] Glen E. Bredon, *Topology and Geometry*, GTM 139, Springer-Verlag, 1997.

[6] F.T. Farrell and L. Jones, *Isomorphism Conjecture in algebraic K-theory*, J. Amer. Math. Soc., 6(2): 249–297, 1993.

[7] F.T. Farrell and Sayed K. Roushon, *The Whitehead Groups of Braid Groups Vanish*, Internat. Math. Res. Notices, (10): 515–526, 2000.

[8] D. Lima-Goncalves and J. Guaschi, *The braid groups of the projective plane*, Algebraic and Geometric Topology, 4: 757–780, 2004.

[9] S. J. Millán-Lopez, *The Whitehead group and the lower algebraic K-theory of braid groups*, thesis work in progress.

[10] F. Quinn, *Ends of maps II*, Invent. Math., 68(3): 353–424, 1982.

[11] J. Rotman, *An Introduction to the Theory of Groups*, Springer-Verlag, 1984.

[12] S. K. Roushon, *K-Theory of virtually poly-surface groups*, Algeb. Geom. Top., 3: 103–116, 2003.

[13] J. Van Buskirk, *Braid Groups of compact 2-manifolds with elements of finite order*, Trans. of AMS, 122: 81–87, 1966.

*Cohomology of Groups and*
*Algebraic K-theory*
ALM 12, pp. 197–215



# $K$-Theory, an Elementary Introduction

Max Karoubi*

**Abstract**

In this elementary paper, we try to give some basic definitions in $K$-theory which will be suitable for students with a reasonable interest to mathematics in general. Almost no proofs are given since many references are provided. On the other hand, nontrivial applications are given.

**Keywords and Phrases:** $K$-theory, homotopy groups, classical groups, projective modules, Bott periodicity, Banach algebras, Bernoulli numbers, Brauer group

The purpose of these notes is to give a feeling of "$K$-theory", a new interdisciplinary subject within Mathematics. This theory was invented by Alexander Grothendieck [1][BS] in the 50's in order to solve some difficult problems in Algebraic Geometry (the letter "$K$" comes from the German word "Klassen", the mother tongue of Grothendieck. This idea of $K$-theory has invaded other parts of Mathematics, for example Number Theory [Ga], Topology [Bott] (AH1] and Functional Analysis [Connes]. Among many successes of $K$-theory, one should mention the solution of classical topological problems ([K] chapter V), the Atiyah-Singer index theorem [AS], its wide generalization to the new subject called "Noncommutative geometry" [Connes] and many algebraic applications [MS] [Sr].

This paper is by no means complete and almost no proof is given but the basic ideas are quite simple and no special training is needed to start the subject. However, if one looks for applications in a specific field of Mathematics, some knowledge of this field is required (e.g. Algebraic Geometry, Number Theory, Functional Analysis, etc.). Our presentation in this paper is driven by one of the most beautiful aspects of the subject which is "Bott periodicity". From its various facets explained here, one guesses some of the main problems and conjectures of the theory.

The prerequisite to read the notes is some familiarity with Topology and elementary Algebra: see for instance the excellent book of Allen Hatcher [Hatcher]

*Clay Mathematics Academy.
Université Diderot /Paris 7, UFR de Mathématiques. Case 7012, 175, rue du chevaleret, 75205 Paris cedex 13, France.

[1]The symbol [ ] means a reference at the end of the paper.

or the references below. However, the basic definitions are given in the first section of this paper.

Here are some possible directions for advanced readings. More complete references are listed inside the paper and recalled at the end.

**Introduction to the subject and applications:** [At1], [Ba], [Ber] [Connes], [K1], [M], [R], [S]. One should not be discouraged by unavoidable technicalities and just browse through the main ideas, very often explained at the beginning of these books.
**Some introductory books in Algebraic Topology:**[Hatcher], [ES], [Hu].
**More advanced books and papers:** [FW], [S], [AH1], [AS], [BS], [KV], [L], [Q2], etc.

Finally, and not the least, I would like to thank David Alexandre Ellwood, Research Director of the Clay Mathematics Institute, for helpful comments after a first draft of this paper.

# 1. Preliminaries on homotopy theory. Classical Bott periodicity

Let $X$ and $Y$ be two "nice" topological spaces (for instance metric spaces). Two continuous maps $f_0$ and $f_1$ from $Y$ to $X$ are homotopic if we can find a continuous map[2]

$$F : Y \times [0,1] \longrightarrow X$$

such that $F(y,i) = f_i(x)$ for $i = 0, 1$. This is an equivalence relation and we denote by $[Y, X]$ the set of equivalence classes. This set is difficult to compute explicitly in general, even for spaces as simple as spheres (see below the partial computation of homotopy groups of spheres).

Two spaces $Y$ and $X$ are homotopically equivalent (or have the same **homotopy type**) if there exist two maps $f : Y \longrightarrow X$ and $g : X \longrightarrow Y$ such that $f \cdot g$ is homotopic to $Id_X$ and $g \cdot f$ is homotopic to $Id_Y$. The idea behind this definition is the notion of "shape" of space which just another word for "homotopy type". Roughly speaking, two spaces have the same shape if they can be deformed one into another. The mathematical meaning of "deformation" is the "homotopy type" which we just defined.

A variant of these considerations is to take spaces $X$ with a "base point", say $x_0$, and maps preserving base points. For instance, in the definition above of homotopic maps we assume moreover that $F(y_0, t) = x_0$. One reason for this variant is the definition of the **homotopy groups** of a space $X$ (with a fixed base point). These are defined when $Y$ is a sphere $S^p$ and play an important role in Algebraic Topology.

More precisely, the homotopy classes of base point preserving maps from the sphere[3] $S^p$ to $X$ is the set of path connected components if $p = 0$, form a group if

[2] From now on, we assume that all the maps considered in this paper are continuous.

[3] The sphere $S^p$ is the subset of $\mathbb{R}^{p+1}$ defined by the equation $(x_1)^2 + \cdots + (x_{p+1})^2 = 1$.

$p > 0$ which is abelian if $p > 1$. This group (or set if $p = 0$) is written $\pi_p(X, x_0)$ or simply $\pi_p(X)$ if the base point is implicit. If two points $x_0$ and $x_1$ are in the same path connected component of $X$, one can show that $\pi_p(X, x_0)$ and $\pi_p(X, x_1)$ are isomorphic: make a picture ! It is also not difficult to prove that two homotopically equivalent spaces have isomorphic homotopy groups.

One should add that $\pi_1(X)$ is called the "fundamental group" or "Poincaré group" of a space: Poincaré introduced this group in order to understand the behaviour of the solutions of differential equations around critical points.

The other homotopy groups $\pi_i(X)$ for $i > 1$ play also a central role in Differential Topology like for example the classification of manifolds. The "**Poincaré conjecture**" for instance is asking the following question. Let $M$ be a topological manifold of dimension 3 which has the homotopy type of $S^3$; is $M$ homeomorphic to $S^3$? This is one of the problems (with prizes !) raised by the Clay Mathematics Institute.

The homotopy groups of a space are quite difficult to compute. For example, $\pi_p(S^r)$ is not known in general if $r > 1$. We know however that this group is 0 if $p < r$ and isomorphic to $\mathbb{Z}$ if $r = p$. For $p > r$, some partial information, coming from the fundamental work of Serre, is the following: $\pi_p(S^r)$ is finite if $r$ is odd or if $r$ is even, $p \neq 2r - 1$. If $r$ is even and $p = 2r - 1$, $\pi_p(S^r)$ is the direct sum of $\mathbb{Z}$ and a finite group. For these results and more general ones, see for instance the basic book of S.-T. Hu on Homotopy Theory [Hu].

However, as it was pointed out by Bott, the homotopy groups of "Lie groups" might be easier to determine in certain circumstances. The Lie groups considered here are defined as the sets of invertible $n \times n$ matrices with coefficients in one of the fundamental fields $\mathbb{R}, \mathbb{C}$ or $\mathbb{H}$: real numbers, complex numbers or quaternions respectively. We denote these groups by $G_n = GL_n(\mathbb{R}), GL_n(\mathbb{C})$ or $GL_n(\mathbb{H})$ with the obvious topology (and metric) induced by the natural embedding in an Euclidean space of a suitable dimension.

One can see that $\pi_p(G_n)$ "stabilizes" as $n$ increases. More precisely, the obvious inclusion of $G_n$ in $G_{n+1}$ (adding 0's except for the element at the bottom right corner which is equal to 1) induces an isomorphism

$$\pi_p(G_n) \cong \pi_p(G_{n+1})$$

for sufficiently large $n$ (cf. [K] p. 13-14). More precisely, we have to take $n > p+1$ (resp. $n > p/2$, resp. $n > p/4 + 1/2$ for $GL_n(\mathbb{R})$ (resp. $GL_n(\mathbb{C})$, resp. $GL_n(\mathbb{H})$). We shall denote these stabilized homotopy groups by $\pi_p(GL(\mathbb{R}))$,$\pi_p(GL(\mathbb{C}))$ or $\pi_p(GL(\mathbb{H}))$ respectively. Bott's remarkable result states that:

**THEOREM .** [Bott] *These stabilized homotopy groups are periodic. More precisely*

$$\pi_p(GL(\mathbb{R})) \cong \pi_{p+4}(GL(\mathbb{H})) \cong \pi_{p+8}(GL(\mathbb{R})) \quad \text{and}$$
$$\pi_p(GL(\mathbb{C})) \cong \pi_{p+2}(GL(\mathbb{C}))$$

**Remark.** This theorem looks at the first glance quite technical. But, it appeared soon after Bott's discovery that it has a central role in many applications. Topological $K$-theory (see next sections of the paper) would not have existed without this theorem.

There are other ways to view this theorem and generalize it. If $X$ is a space with base point $x_0$, we call $\Omega X$ the space of base point preserving continuous maps $\sigma$ from the circle $S^1$ to $X$: this is the "loop space" of $X$. In the case of a metric space, the topology of $\Omega X$ is induced by the metric $d(\sigma,\tau) = \sup d_X(\sigma(t),\tau(t))$ where $t$ runs through all the points of the circle. The homotopy groups of $\Omega X$ are those of $X$ with a shift:$\pi_n(\Omega X) \cong \pi_{n+1}(X)$. Moreover, this construction may be iterated: we define for instance $\Omega^2 X$ as $\Omega(\Omega X)$, $\Omega^3 X$ as $\Omega(\Omega^2 X)$, etc. It is easy to show that $\pi_n(\Omega^r X) \cong \pi_{n+r}(X)$.

With this new definition, Bott periodicity can be stated in a more general form: we have the homotopy equivalences[4]

$$GL(\mathbb{R}) \approx \Omega^4(GL(\mathbb{H})) \approx \Omega^8(GL(\mathbb{R})) \quad \text{and}$$

$$GL(\mathbb{C}) \approx \Omega^2(GL(\mathbb{C}))$$

If $H$ is a closed subgroup of a Lie group $G$, the homogeneous space $G/H$ is the quotient of $G$ by the equivalence relation

$$g_1 \sim g_2 \Leftrightarrow (g_2^{-1})g_1 \in H$$

The "homogeneous space" $GL(\mathbb{C})/GL(\mathbb{R})$ is defined as some kind of limit of the spaces $GL_n(\mathbb{C})/GL_n(\mathbb{R})$ (any real matrix is a complex matrix); in the same way $GL(\mathbb{R})/GL(\mathbb{C})$ is the limit of $GL_{2n}(\mathbb{R})/GL_n(\mathbb{C})$, where $GL_n(\mathbb{C})$ is embedded in $GL_{2n}(\mathbb{R})$ via the classical map

$$A + iB \longmapsto \begin{pmatrix} A & -B \\ B & A \end{pmatrix}$$

It has also been shown in [Bott] that we have the following homotopy equivalences[5]

$$\Omega(\mathbb{Z} \times BGL(\mathbb{R})) \approx GL(\mathbb{R})$$

$$\Omega(GL(\mathbb{R})) \approx GL(\mathbb{R})/GL(\mathbb{C})$$

$$\Omega(GL(\mathbb{R})/GL(\mathbb{C})) \approx GL(\mathbb{C})/GL(\mathbb{H})$$

$$\Omega(GL(\mathbb{C})/GL(\mathbb{H})) \approx \mathbb{Z} \times BGL(\mathbb{H})$$

$$\Omega(\mathbb{Z} \times BGL(\mathbb{H})) \approx GL(\mathbb{H})$$

$$\Omega(GL(\mathbb{H})) \approx GL(\mathbb{H})/GL(\mathbb{C})$$

$$\Omega(GL(\mathbb{H})/GL(\mathbb{C})) \approx GL(\mathbb{C})/GL(\mathbb{R})$$

$$\Omega(GL(\mathbb{C})/GL(\mathbb{R})) \approx \mathbb{Z} \times BGL(\mathbb{R})$$

[see [K1] p. 148 for a proof and for a concrete definition of the homogeneous spaces $G/H$ in various situations].

[4]For a more precise definition of $G = GL(\mathbb{R}), GL(\mathbb{C})$ or $GL(\mathbb{H})$, see the end of this section.
[5]See again below a concrete definition of $BG$ when $G = GL(\mathbb{R}), GL(\mathbb{C})$ or $GL(\mathbb{H})$.

For $G = GL(\mathbb{R}), GL(\mathbb{C})$ or $GL(\mathbb{H})$, the "classifying space" $BG$ may be explicitly described (up to homotopy) as follows. It is the space consisting of linear continuous maps

$$Q : H \longrightarrow H$$

where $H$ is an infinite Hilbert space (over $\mathbb{R}$, $\mathbb{C}$ or $\mathbb{H}$), such that $Q^2 = Q$, the image of $Q$ being a finite dimensional vector space.

In the same way, $G$ has the homotopy type of the space of linear continuous maps

$$a : H \longrightarrow H$$

such that $\alpha - 1$ is a compact operator (i.e. a limit of operators of finite rank).

**Comments and references for further readings.** A nice account of Bott's work can be found in various textbooks and papers, according to one's own taste
1) The original paper of Bott [Bott], however difficult to grasp for a beginner (heavy use of "Morse theory", a deep subject in Riemannian Geometry). We recommend [M2] for a nice overview.
2) An elementary proof (in the complex case) by Atiyah and Bott [AB], also found in the fundamental book of Atiyah [At1]
3) A systematic use of real and complex $K$-theory (see the following section), where Bott periodicity appears as a consequence of the periodicity of Clifford algebras [K1].

Concerning Algebraic Topology, there are many good textbooks aimed at undergraduates: we recommend [Hatcher] as the most accessible reference.

## 2. The beginning of $K$-theory

The starting point of $K$-theory was a simple idea of Grothendieck [BS]. If $M$ is a semi-group, one can "symmetrize" it in the same way that we define $\mathbb{Z}$ from the semi-group $\mathbb{N}$ of natural integers. More precisely, we associate to $M$ a new group $S(M)$ defined as the quotient of $M \times M$ by the equivalence relation

$$\begin{gathered}(a, b) \sim (c, d) \Leftrightarrow \text{there exists } e \text{ such that} \\ a + d + e = b + c + e\end{gathered}$$

There is a map from $M$ to $S(M)$ sending $m$ to the class of $(m, 0)$ which is not injective in general. As a matter of fact, the semi-group $M$ is difficult to compute in the applications we have in mind (see the example $\phi(A)$ below), whereas $S(M)$ is reachable through standard techniques of Algebra and Topology.

Here is our basic example: let $A$ be a ring with unit and let $M = \phi(A)$ be the set of finitely generated projective (right) $A$-modules up to isomorphism. Concretely, such a module $E$ is defined as the image of an $A$-linear map

$$Q : A^n \longrightarrow A^n$$

such that $Q^2 = Q$. If $A$ is a field, $E$ is just a finite dimensional vector space.

As suggested by the notation, $M$ is a semi-group with the following composition law

$$\overline{E}+\overline{F}=\overline{E\oplus F}$$

where $\overline{E}$ denotes in general the isomorphism class of the module $E$. The group $S(M)$ obtained from $M$ by symmetrization is called the $K$-theory of $A$ and denoted by $K(A)$.

One should notice that $K(A)$ is a "functor": a ring homomorphism $f:A\longrightarrow B$ induces a group map $K(A)\longrightarrow K(B)$. On the level of modules, one associates to an $A$-module $E$ the $B$-module defined by the formula $F=E\otimes_A B$. If we write $E$ as the image of $Q$ as above, $F$ is the image of the projection operator $\widetilde{f}(Q)$ where $\widetilde{f}$ is the associated ring homomorphism $M_n(A)\longrightarrow M_n(B)$.

In particular, there is a unique ring map $\mathbb{Z}\longrightarrow A$. The cokernel of the natural associated map[6] $K(\mathbb{Z})\longrightarrow K(\mathbb{A})$ is the reduced $K$-theory of $A$ and denoted by $\widetilde{K}(A)$. If $A$ is commutative, $\widetilde{K}(A)$ is isomorphic (non naturally) to the direct sum $\mathbb{Z}\oplus\widetilde{K}(A)$. Therefore, we may view $\widetilde{K}(A)$ as the non trivial part of the $K$-theory of the ring $A$.

Despite the apparent simplicity of this definition, $K$-theory has been very useful in many parts of Mathematics. Two extreme examples will show the versatility of the notion.

The first example comes from Number Theory: $A$ is the quotient ring $\mathbb{Z}[x]/(1+x+\ldots+x^{p-1})$, where $p$ is an odd prime number. Then $\widetilde{K}(A)$ is a finite group studied in the $19^{th}$ century by Kummer (under a different name) and related to Fermat's last theorem: if this group has no $p$ torsion, Kummer proved that Fermat's equation $X^p+Y^p=Z^p$ has no integral solution (except if $X$ or $Y=0$ of course). See for instance [BorS] and [M1] where reduced $K$-theory is related to ideals in number fields (i.e. finite extensions of $\mathbb{Q}$).

The second example comes from Topology: $A$ is the ring of (real or complex) continuous functions on a compact space $X$. We shall denote this example by $C_{\mathbb{R}}(X)$ or $C_{\mathbb{C}}(X)$ in the real or complex case respectively. Then we have an isomorphism between $K(A)\otimes_{\mathbb{Z}}\mathbb{Q}$ and some part of the **cohomology**[7] of $X$ (see [K3] for a proof). More precisely:

$$K(C_{\mathbb{C}}(X))\otimes_{\mathbb{Z}}\mathbb{Q}\cong\oplus_k H^{2k}(X;\mathbb{Q})$$

$$K(C_{\mathbb{R}}(X))\otimes_{\mathbb{Z}}\mathbb{Q}\cong\oplus_k H^{4k}(X;\mathbb{Q})$$

One should also notice that $K$-theory of these rings is a "homotopy invariant": replacing $X$ by $X\times[0,1]$ does not change the $K$-theory.

The following table lists some basic examples (with comments) one can try to compute. Calculating "$K$-groups" is the best way to learn about the "theory" behind $K$-theory.

**Examples of computations (with comments)**

[6] It is relatively easy to show that $K(\mathbb{Z})$ is isomorphic to $\mathbb{Z}$; see the table of $K$-groups below.
[7] The "cohomology" is defined for instance in [Hatcher] or [ES].

| $A$ | $K(A)$ |
|---|---|
| $\mathbb{Z}$ | $\mathbb{Z}$<br>This means that every abelian group which is a direct summand of some $\mathbb{Z}^n$ is itself of the type $\mathbb{Z}^m$ for a certain $m$. |
| $\mathbb{Z}(\sqrt{-5})$ | $\mathbb{Z}/2 \oplus \mathbb{Z}$<br>The factor $\mathbb{Z}/2$ is related to the existence of a non principal ideal [Sa]. |
| $\mathbb{C}[G]$ | $\mathbb{Z}^N$<br>Here $G$ is a finite group, $N$ the number of its irreducible representations. This computation is the starting point of the beautiful theory of representations of finite groups [Se]. |
| $C_{\mathbb{C}}(S^1)$ | $\mathbb{Z}$<br>The geometric interpretation of this result is the following: any complex vector bundle over the circle is trivial ([K] p. 13). |
| $C_{\mathbb{R}}(S^1)$ | $\mathbb{Z}/2 \oplus \mathbb{Z}$<br>The factor $\mathbb{Z}/2$ is due to the existence of the *Mœbius* band ([K1] p. 13). |
| $C_{\mathbb{C}}(S^2)$ | $\mathbb{Z} \oplus \mathbb{Z}$<br>One of the factor $\mathbb{Z}$ is due to the existence of a "canonical" complex line bundle $L$ over the sphere of dimension 2 ([K1] p.13 or [K2] p.22). |
| $C_{\mathbb{R}}(S^2)$ | $\mathbb{Z}/2 \oplus \mathbb{Z}$<br>The underlying real bundle to $L$ - let us call it $L'$ - is not trivial and is responsible for the factor $\mathbb{Z}/2$ since $L' \oplus L'$ is a trivial real bundle (use [K1] p. 13 again). |
| $M_n(A)$ | $K(A)$<br>This means that $K$-theory does not change when $A$ is replaced by the ring of $n \times n$ matrices with coefficients in $A$ (Morita invariance). Many elementary proofs are possible; see for instance [K1] p. 138. |
| $\mathcal{K}$ | $\mathbb{Z}$<br>Here $\mathcal{K}$ denotes the algebra of compact operators in an infinite dimensional Hilbert space $H$. The proof requires some analysis. See for instance [K] p. 176 (exercise 7.9). |
| $\mathcal{B}$ | 0<br>$\mathcal{B}$ denotes the algebra of all bounded operators in $H$. The proof is quite elementary as soon we notice that the category of finitely generated projective modules over this algebra is equivalent to the category which objects are Hilbert spaces (direct summands of some $H^n$) and morphisms linear continuous maps between them. |

**Comments and recommendations for further readings**

There are many books on the elementary aspects of $K$-theory. The interested student should read for instance [At1] or the first two chapters of [K1] (try to solve the relevant exercises pp. 44-51 and pp. 105-111).

# 3. Relation between $K$-theory and Bott periodicity

We shall limit ourselves to the complex case for simplicity and describe an explicit map from $\pi_{p-1}(GL(\mathbb{C}))$ to $\widetilde{K}(A)$, where $A$ is $C_{\mathbb{C}}(S^p)$, the ring of complex continuous functions defined on the sphere $S^p$. For this purpose, we decompose the sphere $S^p$ into two "fat" hemispheres $S^p_+$ and $S^p_-$, each hemisphere being defined by $x_{p+1} > -1/2$ (resp $x_{p+1} < 1/2$).

Using the last coordinate $x_{p+1}$ as a parameter, we define two continous functions $\alpha_+$ and $\alpha_-$ from the sphere $S^p$ to the unit interval $[0,1]$ such[8] that $\alpha_+$ (resp. $\alpha_-$ ) is zero outside $S^p_+$ (resp. $S^p_+$) and such that $\alpha_+ + \alpha_- = 1$. We define

$$\beta = \frac{\alpha}{\sqrt{(\alpha_+)^2 + (\alpha_-)^2}}$$

where $\alpha = \alpha_+$ or $\alpha_-$ and $\beta = \beta_+$ or $\beta_-$ accordingly.

After these preliminaries, let us consider an element of $\pi_{p-1}(GL(\mathbb{C}))$ represented by a continuous map (for large enough $n$)

$$f : S^{p-1} \longrightarrow GL_n(\mathbb{C})$$

We extend this map to $S^p_+ \cap S^p_-$ by meridian projection on the equator

$$\widetilde{f} : S^p_+ \cap S^p_- \longrightarrow GL_n(\mathbb{C})$$

The matrix

$$Q = \begin{pmatrix} (\beta_+)^2 & \beta_+\beta_-\widetilde{f} \\ \beta_+\beta_-\widetilde{f}^{-1} & (\beta_-)^2 \end{pmatrix}$$

(where $(\beta_+)^2, (\beta_-)^2$ or $(\beta_+\beta_-)$ means this scalar times the identity matrix of order $n$) is a $2n \times 2n$ matrix whose entries are continuous functions on $S^p$ with values in $M_{2n}(\mathbb{C})$. It is easy to see that $Q^2 = Q$. Therefore, the image of $Q$ defines a projective module over the ring $A = C_{\mathbb{C}}(S^p)$, hence an element of $\widetilde{K}(A)$ which is independent of the choice of the partition of unity $(\beta_+, \beta_-)$ ( this can be proved by an homotopy argument).

**THEOREM .** *The correspondence $f \longmapsto Im(Q)$ defines an isomorphism*

$$\pi_{p-1}(GL(\mathbb{C})) \xrightarrow{\approx} \widetilde{K}(C_{\mathbb{C}}(S^p))$$

[for a proof, see for instance [K1] p. 36 and 55 §1.18].

The previous theorem gives us a translation of Bott periodicity in terms of $K$-theory: one simply has to show that $\widetilde{K}(C_{\mathbb{C}}(S^p))$ is periodic of period 2 with respect to $p$. Equivalently, using the same construction in the real case, one has to show that $\widetilde{K}(C_{\mathbb{R}}(S^p))$ is periodic of period 8 with respect to $p$.

---

[8]such a data is called a partition of unity associated to the covering of the sphere by its two hemispheres (see [Mu] p.222 in a much broader context).

# 4. $K$-theory as a homology theory on Banach algebras

So far, we have defined $K$-theory for unital rings only. A slight modification allows us to generalize this definition to non unital rings $A$ as follows. We assume that $A$ is a $k$-algebra where $k$ is any commutative ring with unit (for instance $\mathbb{Z}$). We define a new unital ring $A^+$ as $A \times k$ with the obvious addition and the following "twisted" multiplication

$$(a, \lambda)(a', \lambda') = (aa' + \lambda a' + \lambda' a, \lambda\lambda')$$

There is an obvious "augmentation" $A^+ \longrightarrow k$ and the $K$-theory of $A$ is then defined as the kernel of the induced map

$$K(A^+) \longrightarrow K(k)$$

It can be shown (not quite easily) that this definition is in fact independent of $k$.

An interesting and motivating example is the ring of $k$-valued continuous functions $f$ on a locally compact space $X(k = \mathbb{R}$ or $\mathbb{C})$ such that $f(x)$ goes to 0 when $x$ goes to $\infty$. Then $A^+$ is the ring of continuous functions on the "one point compactification" of $X$. For instance, if $X = \mathbb{R}^p$, $K(A)$ is isomorphic to $\widetilde{K}(C(S^p))$.

Also note that this method enables us to define a morphism $K(A) \longrightarrow K(B)$ each time we have a general ring map $A \longrightarrow B$ (we no longer assume that $f(1) = 1$, even if $A$ and $B$ are unital rings)

A sequence of rings and maps

$$0 \longrightarrow A' \longrightarrow A \longrightarrow A'' \longrightarrow 0$$

is called exact if the underlying sequence of abelian groups is exact. In other words $A'$ is a two sided ideal in $A$, whereas $A''$ may be identified with the quotient ring $A/A'$.

**THEOREM .** *The previous sequence induces an exact sequence of $K$-groups*

$$K(A') \xrightarrow{\alpha} K(A) \xrightarrow{\beta} K(A'')$$

*that is $Im(\alpha) = Ker(\beta)$*

[for a proof see for instance [M] or [KV]].

It is natural to ask what is $Ker(\alpha)$ and $Coker(\beta)$. If one is familiar with homological algebra, one should define **"derived functors"** $K_n$, $n \in \mathbb{Z}$, of the $K$-group in order to extend the previous exact sequence to the left and to the right (the group $K_0(A)$ being $K(A)$). A partial solution (to the left) is given by the theorem a few lines below.

There are at least two ways to solve this problem of derived functors. The first one is to put some topology on the rings involved (i.e. consider Banach algebras as explained below), The other one is to stay in pure algebra, which is paradoxically much harder. We shall begin with the first approach.

We recall that a **Banach algebra** (over $k = \mathbb{R}$ or $\mathbb{C}$) is a $k$-algebra $A$ (not necessarily unital) with the following properties:

1) A norm $a \mapsto \|a\|$ is defined on the vector space $A$ in such a way that $A$ is complete for the distance $d(a,b) = \|a - b\|$
2) We have the inequality $\|ab\| \leq \|a\| \cdot \|b\|$

A typical example is the ring of continuous functions $f$ on a locally compact space $X$ such that $f(x)$ goes to 0 when $x$ goes to $\infty$. The norm $\|f\|$ of $f$ is then the maximum of the values of $|f(x)|$ when $x \in X$.

More generally, if $A$ is a Banach algebra and $X$ a locally compact space, we define a new Banach algebra $A(X)$ as the ring of continuous functions $f$ on $X$ with values in $A$ with the same condition at infinity. For the definition of the norm, we just replace $|f(x)|$ in the previous example by $\|f(x)\|$. With obvious notations, we have the following isomorphism $A(X)(Y) \cong A(X \times Y)$.

**THEOREM .** *Up to isomorphism, there is a unique way to define functors $K_n(A)$, for a Banach algebra $A$ and $n \geq 0$, such that the following axioms are satisfied:*

**1) Exactness.** *Given an exact sequence of Banach algebras*

$$0 \longrightarrow A' \longrightarrow A \longrightarrow A'' \longrightarrow 0$$

*such that $A'$ has the induced norm and $A''$ the quotient norm, we have a long exact sequence (infinite to the left)*

$$\longrightarrow K_{n+1}(A'') \longrightarrow K_n(A') \longrightarrow K_n(A) \longrightarrow K_n(A'') \longrightarrow K_{n-1}(A') \longrightarrow$$

**2) Homotopy invariance.** *The canonical inclusion of $A$ into $B = A([0,1])$ induces an isomorphism*

$$K_n(A) \cong K_n(B)$$

**3) Normalization.**

$$K_0(A) \cong K(A)$$

[for a reference, see for instance [K1] p. 109, exercise 6.14].

It is not too difficult to deduce from these axioms the following isomorphism

$$K_n(A) \cong K(A(\mathbb{R}^n))$$

In particular, $\widetilde{K}(C_{\mathbb{R}}(S^n)) \approx K_n(\mathbb{R})$ and $\widetilde{K}(C_{\mathbb{C}}(S^n)) \approx K_n(\mathbb{C})$. Moreover, if $A$ has a unit, the group $K_n(A)$ may also be defined as a suitable homotopy group: for $n > 0$, this is just the homotopy group $\pi_{n-1}(GL(A))$ which is "limit" of the homotopy groups $\pi_{n-1}(GL_r(A))$ when $r$ goes to $\infty$.

The previous theorem enables us to define a "cohomology theory" on the category of compact spaces in the sense of Eilenberg and Steenrod [ES]: if $X$ is a compact space and $Y$ a closed subset, we define $K^{-n}(X,Y)$ as the $K_n$ group[9] of the Banach algebra $C(X-Y)$. There are in fact two theories involved according to whether we consider real or complex continuous functions on $X-Y$. These groups satisfy all the axioms of a cohomology theory [ES], except for the dimension axiom. This cohomology theory was defined initially by Atiyah and Hirzebruch [AH1] and was the starting point of "**topological $K$-theory**" for compact spaces and later on for Banach algebras [K]. This theory for Banach algebras (developped in the last 20 years) got an enormous importance in a new field of Mathematics called "Noncommutative Geometry" [Connes]. This shows the importance of the following theorem in this new field:

**THEOREM .** ([K1]§III)
*Let $A$ be a complex (resp. real) Banach algebra. Then we have an isomorphism*

$$K_n(A) \cong K_{n+2}(A)(\text{resp.} K_n(A) \cong K_{n+8}(A))$$

**Remark.** If we stay in the complex case for simplicity and if $n \in \mathbb{Z}$, we see that we can define formally $K_n(A)$ as $K_{n+2r}(A)$ for $r$ sufficiently large such that $n+2r>0$. In other words, we can define $K_{-1}(A)$ as $K_1(A)$, $K_{-2}(A)$ as $K_0(A)$, etc. Therefore, to an exact sequence of Banach algebras as above

$$0 \longrightarrow A' \longrightarrow A \longrightarrow A'' \longrightarrow 0$$

we can associate an "hexagonal" exact sequence

$$\begin{array}{ccccc} K_0(A') & \longrightarrow & K_0(A) & \longrightarrow & K_0(A'') \\ \uparrow & & & & \downarrow \\ K_1(A'') & \longleftarrow & K_1(A) & \longleftarrow & K_1(A') \end{array}$$

This solves the problem we had before about the kernel of the map $K_0(A') \longrightarrow K_0(A)$ and the cokernel of the map $K_0(A) \longrightarrow K_0(A'')$.

There are many proofs of Bott periodicity in the literature. They all require some sophistication. One conceptual proof is to introduce "cup-products" in $K$-theory. More precisely, a bilinear pairing between three unital rings $A$, $B$ and $C$ is given by a $\mathbb{Z}$-bilinear map

$$\varphi : A \times C \longrightarrow B$$

such that

$$\varphi(aa', cc') = \varphi(a,c)\varphi(a',c')$$
$$\varphi(1,1) = 1$$

It is easy to see that such a pairing induces a group homomorphism

$$K(A) \otimes K(C) \longrightarrow K(A \otimes_{\mathbb{Z}} C) \longrightarrow K(B)$$

[9]Note that $K^n(X,Y)$ has not yet been defined for $n>0$.

This definition may be generalized easily to non unital rings. Moreover, if we assume that $A$, $B$ and $C$ are Banach algebras and that $\|\varphi(a,c)\| \leq \|a\| \cdot \|c\|$, such a bilinear map $\varphi$ induces another one

$$A(\mathbb{R}^n) \times C(\mathbb{R}^p) \longrightarrow B(\mathbb{R}^{n+p})$$

and therefore a bilinear pairing (called the "cup-product")

$$K_n(A) \times K_p(C) \longrightarrow K_{n+p}(B)$$

Another version of Bott periodicity is then the following:

**THEOREM .** (see for instance [K4]§2)
a) *Let $C = \mathbb{C}$ and let $A = B$ be a complex Banach algebra. Then the group $K_2(\mathbb{C})$ is isomorphic to $\mathbb{Z}$ and the cup-product with a generator $u$ induces an isomorphism*

$$\beta : K_n(A) \longrightarrow K_{n+2}(A)$$

b) *Let $C = \mathbb{R}$ and let $A = B$ be a real Banach algebra. Then the group $K_8(\mathbb{R})$ is isomorphic to $\mathbb{Z}$ and the cup-product with a generator induces an isomorphism*

$$K_n(A) \longrightarrow K_{n+8}(A)$$

**Sketch of a proof of the theorem, part a).**

It relies on the construction of a full theory $K_n(A)$ defined[10] for all $n \in \mathbb{Z}$ and satisfying the formal properties of a cup-product (for instance associativity). We then construct an element $v$ in $K_{-2}(\mathbb{C})$ such that $uv = vu = 1$ in $K_0(\mathbb{C}) = \mathbb{Z}$. This element $v$ is used in order to construct a morphism "going backwards"

$$\beta' : K_{n+2}(A) \longrightarrow K_n(A)$$

It is simply defined by $\beta'(x) = x \cdot v$. The fact that $\beta$ and $\beta'$ are inverse to each other is a formal exercise. For instance, $(\beta\beta')(x) = \beta(x \cdot v) = (x \cdot v) \cdot u = x \cdot (v \cdot u) = x \cdot 1 = x$.

**Applications.** A variety of applications of Bott periodicity and $K$-theory may be found in various papers and books [At1], [AH2], [AS2] [Connes]. We just mention two topological applications, due to J.F. Adams (the proofs may be found in [K], chapter V).

1) The sphere $S^n$ is a topological group only if $n = 0, 1$, or 3.
2) What is the maximum number $\rho(t)$ of linearly independent tangent fields on the sphere $S^{t-1}$?
  a) According to an old theorem of Hopf, $\rho(t) = 0$ if $t$ is odd.
  b) If $t$ is even, we can write it as (odd number). $2^\beta$ and we divide $\beta$ by 4, say $\beta = \gamma + 4\delta$ where $0 \leq \gamma \leq 3$.
  Then $\rho(t) = 2^\gamma + 8\delta - 1$.

**Suggestions for advanced readings:** the books [At1] and [K], already quoted many times.

[10] In order to avoid a vicious circle, we DON'T assume periodicity to define $K_n$ for $n < 0$.

# 5. $K$-theory as a homology theory on discrete rings

For various reasons, especially the applications of $K$-theory to Algebraic Geometry and Number Theory, the definition of the functors $K_n$ we have just given is not very satisfactory. We would rather not use the topology of the ring $A$. For such a purpose, a definition has been proposed by Quillen in 1970 [Q2]. Unfortunately, Quillen's definition requires some sophistication in Algebraic Topology. Therefore, we shall use another one, introduced slightly before by Villamayor and the author [KV] which coincides with Quillen's definition in many favourable cases. It is much easier to define, in the same spirit as the higher $K$-groups for Banach algebras.

**Caution:** those **algebraic** $K$-groups will now be called $K_n(A)$. In order to avoid confusion, the topological $K$-groups for Banach algebras defined earlier will be denoted by $K_n^{top}(A)$.

As we have seen previously, homotopy groups play an important role in the definition and properties of the topological $K$-groups. Therefore, it is natural to look for an "algebraic" definition of them, especially for spaces like the general linear groups $GL_r(A)$ or their limit

$$GL(A) = \cup GL_r(A)$$

The first thing is to **define algebraically** $\pi_0(GL(A))$. In other words, one should say when two invertible matrices $\alpha_0$ and $\alpha_1$ are homotopic (or are in the same "algebraic" connected path component).

**DEFINITION .** *Two invertible matrices $\alpha_0$ and $\alpha_1$ are called **algebraically homotopic** if there exists an element $\alpha_x$ of $GL(A[x])$ such that $\alpha(0) = \alpha_0$ and $\alpha(1) = \alpha_1$. This is an equivalence relation. The algebraically connected components form an abelian group denoted by $K_1(A)$.*

**Examples.** If $F$ is a field or an Euclidean ring, it is easy to show (Gauss reduction) that $K_1(F) \cong F^*$, the group of invertible elements in $F$ (take the kernel of the determinant map). One should note however that $K_1^{top}(\mathbb{C}) = 0$ and $K_1^{top}(\mathbb{R}) = \mathbb{Z}/2$.

A non trivial example [BMS] is the ring $A$ of integers in a number field: one finds again $K_1(A) = A^*$. Note that $A^*$ has been determined by a theorem of Dirichlet (see [Sa] for a nice introduction to Number Theory).

For instance, if $A = \mathbb{Z}[x]/(1 + x + \cdots + x^{p-1})$, where $p$ is an odd prime number, as in §2, we have $K_1(A) \cong A^* \cong \mathbb{Z}^{(p-3)/2} \oplus \mathbb{Z}/p$. An example of a ring $A$ such that $K_1(A) \neq A^*$ is the following

$$A = \mathbb{Z}[x, y]/(x^2 + y^2 - 1)$$

One can show that the following matrix with coefficients in $A$:

$$\begin{pmatrix} x & -y \\ y & x \end{pmatrix}$$

is not algebraically homotopic to a $1 \times 1$ matrix, although its determinant is one (Hint: imbed $A$ in the ring $C_{\mathbb{R}}(S^1)$).

Here is a table of $K_1$-groups which repeats the rings listed in the §2

| $A$ | $K_1(A)$ |
|---|---|
| $\mathbb{Z}$ | $\mathbb{Z}/2$ The determinant of an invertible matrix with integral coefficients is $\pm 1$. Moreover, a matrix of determinant 1 is a product of elementary matrices (use the Euclid algorithm). |
| $\mathbb{Z}(\sqrt{-5})$ | $\mathbb{Z}/2$ This follows from a general theorem in Number Theory about the units in imaginary quadratic fields [Sa]. |
| $\mathbb{C}[G]$ | $(\mathbb{C}^*)^N$ Here $G$ is a finite group, $N$ the number of its irreducible representations. This computation follows from the fact that $\mathbb{C}[G]$ is a product of matrix algebras [Se]. |
| $C_{\mathbb{C}}(S^1)$ | $\mathrm{Map}(S^1, \mathbb{C}^*)$ In general, if $A$ is a commutative Banach algebra, one has $K_1(A)$ isomorphic to $\pi_0(SL(A)) \oplus A^*$ (Proof in [M]). |
| $C_{\mathbb{R}}(S^1)$ | $\mathrm{Map}(S^1, \mathbb{R}^*)$ Same comment. |
| $C_{\mathbb{C}}(S^2)$ | $\mathrm{Map}(S^2, \mathbb{C}^*)$ Same comment. |
| $C_{\mathbb{R}}(S^2)$ | $\mathrm{Map}(S^2, \mathbb{R}^*)$ Same comment. |
| $M_n(A)$ | $K(A)$ This means that $K_1$-theory does not change when $A$ is replaced by the ring of $n \times n$ matrices with coefficients in $A$ (Morita invariance). Essentially the same proof as for the $K_0$-group. |
| $\mathcal{B}$ | 0 Essentially the same proof as for the $K_0$-group. |

Coming back to the theory, it is easy to generalize this definition of $K_1(A)$ for rings without unit. If $A$ is a $k$-algebra, following the previous definition for $K(A)$, we define $K_1(A)$ as the kernel of the natural map

$$K_1(A^+) \longrightarrow K_1(k)$$

(it is easy to show that this definition is independent of $k$).

**Example.** An important example in Functional Analysis is the ring $\mathcal{K}$ of compact operators in an infinite dimensional Hilbert space. It is a challenging exercise to show that the $K_1$-group of this ring is reduced to 0.

We can push the analogy with topology further and introduce the "**loop ring**" $\Omega A$ of $A$ as the non unital ring $x(x-1)A[x]$, i.e. the two-sided ideal in $A[x]$ consisting of polynomials $P(x)$ such that $P(0) = P(1) = 0$. The topological analog of $\Omega A$ is $A(\mathbb{R}) = A(]0,1[)$ in the previous section if $A$ is a Banach algebra.

**DEFINITION .** *For $n \geq 1$, the* ***algebraic*** *groups $K_n(A)$ are defined inductively by the following formula*

$$K_{n+1}(A) = K_n(\Omega A) = \cdots = K_1(\Omega^n A)$$

**Caution.** This definition does not coincide with Quillen's definition, except if the ring $A$ is regular noetherian. This means that every finitely generated module admits a finite projective resolution. Many rings of geometric nature are of this type, for instance those associated to algebraic functions on affine algebraic varieties without singularities.

As for Banach algebras, a bilinear pairing

$$A \times C \longrightarrow B$$

induces a "**cup-product**"

$$K_n(A) \times K_p(C) \longrightarrow K_{n+p}(B)$$

An interesting example is $A = C = B$ a commutative field (denoted by $F$). The above cup-product gives a "symbol map"

$$S : K_1(F) \otimes_{\mathbb{Z}} K_1(F) = F^* \otimes_{\mathbb{Z}} F^* \longrightarrow K_2(F)$$

**THEOREM** (Matsumoto).
*The symbol map $S$ is surjective. Its kernel is generated by the tensor products of the type $x \otimes (1-x)$, where $x \in F/0, 1$*

[for a proof see [M1] ].

For example, in the case of the group $K_2(\mathbb{Q})$, where $\mathbb{Q}$ is the field of rational numbers, one finds

$$K_2(\mathbb{Q}) \cong \mathbb{Z}/2 \oplus [\oplus_p (\mathbb{Z}/p)^*]$$

where $p$ runs through the set of all odd prime numbers.

Let us now state a striking conjecture [Ga] linking the groups $K_n(\mathbb{Z})$ with the **Bernoulli numbers** $B_k$. These rational numbers $B_k$ appear in the series expansion of the function $x/(1-e^{-x})$. More precisely, one has

$$x/(1-e^{-x}) = 1 + x/2 + \sum_{k=1}^{\infty} (-1)^{k-1} \frac{B_k}{(2k)!} x^{2k} + \cdots$$

Let us express the number $B_k$ as an irreducible fraction $c_k/d_k$. The conjecture is then the following:

$$\begin{array}{l} K_{8n}(\mathbb{Z}) = 0 \text{ (for } n > 0) \\ K_{8n+1}(\mathbb{Z}) = \mathbb{Z} \oplus \mathbb{Z}/2 \text{ (for } n > 0) \\ K_{8n+2}(\mathbb{Z}) = \mathbb{Z}/2 \oplus \mathbb{Z}/c_k \text{ (for } k = 2n+1) \\ K_{8n+3}(\mathbb{Z}) = \mathbb{Z}/8k \cdot d_k \text{ (for } k = 2n+1) \\ K_{8n+4}(\mathbb{Z}) = 0 \\ K_{8n+5}(\mathbb{Z}) = \mathbb{Z} \\ K_{8n+6}(\mathbb{Z}) = \mathbb{Z}/c_k \text{ (for } k = 2n+2) \\ K_{8n+7}(\mathbb{Z}) = \mathbb{Z}/4k \cdot d_k \text{ (for } k = 2n+2) \end{array}$$

This conjecture has been verified for $K_i(\mathbb{Z})$ if $i \leq 6$. Otherwise, we know at least that the right hand side is a lower bound for the groups $K_i(\mathbb{Z})$.

For instance, the conjecture predicts that $K_{23}(\mathbb{Z})$ is a cyclic group of order 65520(!). It is striking to see some partial periodicity appearing in the table as for real Banach algebras. Unfortunately, we are still far from understanding the full story. We recommend [Ga] for an overview and the relation with classical (unsolved) problems in Number Theory, like the Vandiver conjecture.

On the positive side, the first computations of algebraic $K$-groups were made by Quillen [Q1]. For a finite field with $q$ elements (traditionally called $\mathbb{F}_q$), he proved that

$$K_{2i}(\mathbb{F}_q) = 0 \ for \ i > 0 \text{ and}$$

$$K_{2i-1}(\mathbb{F}_q) \cong \mathbb{Z}/(q^i - 1)\mathbb{Z}$$

From this result, we see that Bott periodicity cannot be true in a *naïve* sense. However, the situation is more favourable if one considers $K$-theory with finite coefficients $\mathbb{Z}/n$, say $K_i(A;\mathbb{Z}/n)$. Such a theory has been introduced by $W$. Browder and the author [Br] [So] in the 70's: it fits into an exact sequence

$$K_i(A) \xrightarrow{.n} K_i(A) \longrightarrow K_i(A;\mathbb{Z}/n) \longrightarrow K_{i-1}(A) \xrightarrow{.n} K_{i-1}(A)$$

If $A$ is a field $F$, one has for instance $K_1(F;\mathbb{Z}/n) \cong F^*/(F^*)^n$.

If $F$ is an algebraically closed field, let us write $\mu_n$ for the group of $n^{th}$ roots of unity in $F$. It is isomorphic to $\mathbb{Z}/n$, but not canonically. Note that $K_*(F;\mathbb{Z}/n)$ has a well defined ring structure in this case [Br] which is compatible with one shown before on $K_*(F)$.

**THEOREM .** [Su] *Let $F$ be an algebraically closed field with characteristic prime to $n$. Then*

$$K_{2i-1}(F;\mathbb{Z}/n) = 0$$

$$K_{2i}(F;\mathbb{Z}/n) \cong (\mu_n)^{\otimes i}$$

*Moreover, these isomorphisms are compatible with the ring structure. In other words, the cup-product*

$$K_{2i}(F;\mathbb{Z}/n) \otimes K_2(F;\mathbb{Z}/n) \longrightarrow K_{2i+1}(F;\mathbb{Z}/n)$$

*is an isomorphism.*

This result is the true analog of Bott periodicity, since it implies that the $K$-groups with coefficients are periodic. Note however the existence of a twist since $\mu_n$ is not isomorphic to $\mathbb{Z}/n$ canonically.

**THEOREM .** [Su] *The obvious maps*

$$K_i(\mathbb{C};\mathbb{Z}/n) \longrightarrow K_i^{top}(\mathbb{C};\mathbb{Z}/n)$$

$$K_i(\mathbb{R};\mathbb{Z}/n) \longrightarrow K_i^{top}(\mathbb{R};\mathbb{Z}/n)$$

*are isomorphisms. In particular, the former groups are periodic of period 2 and the latter are periodic of period 8.*

At this point, one may ask how to compute $K_i(F;\mathbb{Z}/n)$ for an arbitrary field (the groups $K_*(F)$ themselves seem untractable for the moment). The general feeling among specialists (not yet completely understood) is that we should calculate them from the groups $K_*(\overline{F};\mathbb{Z}/n)$, where $\overline{F}$ is the separable closure of $F$, taking into account the Galois action on them: this requires a lot of machinery from Algebra and Algebraic Topology.

**Applications.** One can find many applications of Algebraic $K$-theory in the references [R][Sr]. We just mention two of them.

The first one is relevant to Differential Topology. Let $M$ be a sphere $S^n$. The space $\mathcal{P}(M)$ of pseudo-isotopies of $M$ is the space of diffeomorphisms of the product $M \times [0,1]$ which induce the identity on $M \times \{0\}$. By a result of Waldhausen [L], the homotopy groups $\pi_i(\mathcal{P}(M)) \otimes_{\mathbb{Z}} \mathbb{Q}$ are isomorphic to $K_{i+2}(\mathbb{Z}) \otimes_{\mathbb{Z}} \mathbb{Q}$. In particular, according to Borel [Bo], these groups are periodic of period 4, isomorphic to $0, 0, 0, \mathbb{Q} \ldots$ (starting from $i = 0$).

Another spectacular application is due to Merkujev and Suslin [MS] and involves a well known algebraic invariant called the Brauer group of a field $F$. One considers central simple algebras $A$ over $F$. Two such algebras $A$ and $A'$ are called equivalent if $M_p(A)$ is isomorphic to $M_p(A')$ for some $p$ and $p'$. The Brauer group is the quotient of this set of algebras by the equivalence relation (it is a group by the law induced by the tensor product of algebras). If the field $F$ contains the $n^{th}$ roots of unity, then the $n$-torsion of the Brauer group is isomorphic to $K_2(F)/nK_2(F)$. For instance, if $n = 2$, this result implies that any element of the 2-torsion of the Brauer group can be represented by the tensor product of finitely many quaternion algebras.

# References

[At1] ATIYAH M.-F.: *K-theory.* Benjamin, New-York (1967).

[Ba] BASS H.: *Algebraic K-theory.* New York: Benjamin (1968).

[Ber] BERRICK J.: *An approach to Algebraic K-theory.* Pitman (1982).

[Connes] CONNES A.: *Noncommutative Geometry.* Academic Press (1994).

[FW] FRIEDLANDER E. and WEIBEL C.: *School on Algebraic K-theory and applications. An overview of the subject.* ICTP Trieste (1997).

[K1] KAROUBI M.: *K-theory, an introduction.* Grundlehren der Math. Wissen. 226, Springer, Berlin (1978).

[K2] KAROUBI M.: *Cyclic homology and K-theory.* Astérique 149. Société Mathématique de France (1987).

[M1] MILNOR J.: *Introduction to Algebraic K-theory.* Ann. of Math. Studies 197. Princeton, NJ. Princeton University Press (1974).

[R] ROSENBERG J.: *Algebraic K-theory and its applications.* Graduate Texts N147, Springer, Berlin (1994).

[S] SRINIVAS V.: *Algebraic K-theory.* Birkhäuser, Boston (1996).

**Books on Algebraic Topology**

[ES] EILENBERG S. and STEENROD N.: *Foundations of Algebraic Topology.* Princeton University Press (1952).

[Hatcher] HATCHER A.: *Algebraic Topology.* Cambridge University Press (2002).

[Hu] HU S-T.: *Homotopy theory.* Academic Press (1959).

[Mu] MUNKRES J.R.: *Topology, a first course.* Prentice Hall (1975).

**More books and papers**

[Ad] ADAMS J.F.: *Vector fields on the sphere.* Ann. Math. 75, 603-632 (1962).

[At2] ATIYAH M.F.: *Global theory of elliptic operators.* Proc. Internat. Conf. on Functional Analysis and related topics, p. 21-30. Univ. of Tokyo Press (1970).

[AB] ATIYAH M.F. and BOTT R.: *On the periodicity theorem for complex vector bundles.* Acta Math. 112, p. 229-247 (1964).

[AH1] ATIYAH M.F. and HIRZEBRUCH F.: *Vector bundles and homogeneous spaces.* Proc. Symposium in Pure Maths, Amer. Math. Soc. 3, 7-38 (1961).

[AH2] ATIYAH M.F. and HIRZEBRUCH F.: *Riemann-Roch theorems for differentiable manifolds.* Bull. Amer. Math. Soc. 65, 276-281 (1959).

[AS] ATIYAH M.F. and SINGER I.M.: *The index of elliptic operators.* Ann. of Math. 87, 484-530, 564-604 (1968), 93, 119-138 (1971).

[BMS] BASS H., MILNOR J. et SERRE J.-P.: *Solution of the congruence subgroup problem for $SL_n (n \geq 3)$ and $Sp_n (n \geq 2)$.* Publ. Math. IHES 33, p. 59-137 (1967).

[Bo] BOREL A.: *Stable real cohomology of arithmetic groups.* Ann. Sc. Ec. Norm. Sup. 7, 235-272 (1974).

[BS] BOREL A. et SERRE J.-P.: *Le théorème de Riemann-Roch (d'après Grothendieck).* Bull. Soc. Math. France 86, 97-136 (1958).

[BorS] BOREVICH Z.I. and SHAVAREVICH I.R.: *Number theory.* Academic Press (1966).

[Bott] BOTT R.: *The stable homotopy of the classical groups.* Ann. of Math. 70, 313-337 (1959).

[Br] BROWDER W.: *Algebraic K-theory with coefficients $\mathbb{Z}/p$.* Springer Lecture Notes in Math. Nr 657 (1978).

[Ga] GAJDA W.: *On $K_*(\mathbb{Z})$ and classical conjectures in the arithmetic of cyclotomic fields.* Contemporary Math. 346. American Math. Society (2005).

[K3] KAROUBI M.: *Les isomorphismes de Chern et de Thom-Gysin en K-théorie.* Séminaire Cartan-Schwartz 1963/64. Benjamin (1967).

[K4] KAROUBI M.: *Algèbres de Clifford et K-théorie.* Ann. Sci. Ec. Norm. Sup., p. 161-270 (1968).

[KV] KAROUBI M. et VILLAMAYOR O.: *Foncteurs $K^n$ en algèbre et en topologie, Comptes Rendus Acad.* Sci. Paris, t. 269, p. 416 (1969).

[L] LODAY J.-L.: *Homotopie des espaces de concordance.* Séminaire Bourbaki N 516 (1977/78).

[M2] MILNOR J.: *Morse theory.* Annals of Math Studies 51. Princeton University Press (1963).

[MS] MERKURJEV A.S. and SUSLIN A.A.: *K-cohomology of Severi-Brauer varieties and the norm residue homomorphism.* Math. USSR Izv. 21, 307-440 (1983).

[Q1] QUILLEN D.: *On the cohomology and K-theory of the general linear groups over a finite field.* Ann. of Math. 96, p. 552-586 (1972).

[Q2] QUILLEN D.: *Higher Algebraic K-theory.* Lecture Notes in Math 341, 85-147. Berlin-Heidelberg-New York: Springer (1973).

[Sa] SAMUEL P.: *Théorie algébrique des nombres.* Hermann (1962).

[Se] SERRE J.-P.: *Représentations linéaires des groupes finis.* Hermann (1978).

[So] SOULE C.: *K-théorie des anneaux d'entiers de corps de nombres et cohomologie étale.* Inventiones Math. 55, p. 251-295 (1979).

[Su] SUSLIN A.: *Algebraic K-theory of fields.* Proceedings of the International Congress of Mathematicians 1986. American Math. Society (1987).

*Cohomology of Groups and*
*Algebraic K-theory*
ALM 12, pp. 217–268



# Lectures on $K$-Theory*

Max Karoubi**

**Abstract**

This series of lectures gives an introduction to $K$-theory, including both topological $K$-theory and algebraic $K$-theory, and also discussing some aspects of Hermitian $K$-theory and cyclic homology. They also include a survey of several contemporary developments and applications.

**2000 Mathematics Subject Classification:**19-01; Secondary 19-02

**Keywords and Phrases:** $K$-group, $K$-theory

## Lecture 1 The beginning of $K$-theory: Grothendieck, Atiyah and Hirzebruch

In order to formalize his work on the Riemann-Roch theorem (in the spirit of Hirzebruch), Grothendieck introduced a new contravariant functor [BS] defined on the category of non singular algebraic varieties $X$. He named this functor $K(X)$, the "$K$-theory" of $X$. It seems that the terminology "$K$" came out from the German word "Klassen", since $K(X)$ may be thought of as a group of "classes" of vector bundles on $X$. Grothendieck could not use the terminology $C(X)$ since his thesis (in functional analysis) made an heavy use of the ring $C(X)$ of continuous functions on a space $X$.

In order to define $K(X)$, one considers first the free abelian group $L(X)$ generated by the isomorphism classes $[E]$ of vector bundles $E$ on $X$. The group $K(X)$ is then the quotient of $L(X)$ by the subgroup generated by the following relations

$$[E'] + [E''] = [E]$$

*Printed with permission from ICTP, Trieste. It was originally electronically published in *Contemporary developments in algebraic K-theory*, pp. 1–95, ICTP Lect. Notes, XV, Abdus Salam Int. Cent. Theoret. Phys., Trieste, 2004.

**Université Diderot/Paris 7,UFR de Mathématiques. Case 7012,175,rue du chevaleret,75205 Paris cedex 13,France.

when we have an exact sequence of vector bundles

$$0 \longrightarrow E' \longrightarrow E \longrightarrow E'' \longrightarrow 0$$

**Exercise:** compute $K(X)$ when $X$ is a Riemann surface in complex Algebraic Geometry.

This group $K(X)$ has nice algebraic properties, very much related to the usual ones in cohomology. As a matter of fact, the theory of characteristic classes, e.g. the Chern character denoted by $Ch$ - which we shall review later on - might be thought of as defining a functor from $K(X)$ to suitable cohomology theories. For instance, if $f : X \longrightarrow Y$ is a morphism of projective varieties, we have a "Gysin map" $f_*^H : K(X) \longrightarrow K(Y)$ which is related to the usual Gysin map $f_*^H$ in rational cohomology by the following formula

$$Ch(f_*^H(x)) = f_*^H(Ch(x).Todd(f))$$

where $x$ is an arbitrary element of $K(X)$ and $Todd(f)$ is the "Todd class" (see the third lecture for more details) of the normal bundle of $f$, as introduced by Hirzebruch [AH2]. More precisely, the Chern character $Ch$ is a natural transformation from $K(X)$ to the even Betti cohomology groups $H^{ev}(X;\mathbb{Q})$ (which depend only on the topology of $X$) and $Todd(f)$ is well defined in terms of the Chern classes of the tangent bundles associated to $X$ and $Y$. If $x \in K(X), Ch(x)$ is the sum of homogeneous elements $Ch_0(x), Ch_1(x), \ldots,$ where $Ch_n(x) \in H^{2n}(X;\mathbb{Q})$.

The formula above shows that $Todd(f)$ is the defect of commutativity of the following diagram

$$\begin{array}{ccc} K(X) & \xrightarrow{f_*^K} & K(Y) \\ {\scriptstyle Ch}\downarrow & & {\scriptstyle Ch}\downarrow \\ H^{ev}(X;\mathbb{Q}) & \xrightarrow{f_*^H} & H^{ev}(Y;\mathbb{Q}) \end{array}$$

The group $K(X)$ is hard to compute in general: it is just a nice functor used for a better understanding of characteristic classes. This "Algebraic $K$-theory" was investigated deeply by other mathematicians Bass, Milnor, Quillen... at the end of the 60's and will be studied in detail during the Lectures 4 and 5.

Just after the discovery of $K(X)$, Atiyah and Hirzebruch considered a topological analog named $K^{top}(X)$ - now for an arbitrary compact space $X$ - by considering topological vector bundles instead of algebraic vector bundles. The main difference between the two definitions lies in the fact that an exact sequence of topological vector bundles[1]

$$0 \longrightarrow E' \longrightarrow E \longrightarrow E'' \longrightarrow 0$$

always splits. Therefore, $K^{top}(X)$ is just the symmetrized group of the semi-group of isomorphism classes $[E]$ of vector bundles, with the addition rule $[E] + [F] =$

[1] *complex* vector bundles, in order to fix the ideas.

$[E \oplus F]$, where $E \oplus F$ is the "Whitney sum" of the vector bundles $E$ and $F$. It is clearly a contravariant functor of $X$. Moreover, every vector bundle is a direct summand in a trivial vector bundle (i.e. of the type $X \times \mathbb{C}^n$, with an easily guessed vector bundle structure). This fact enables us to compute $K^{top}(X)$ in purely homotopical terms. More precisely, let us define $\widetilde{K}^{top}(X)$, the reduced $K$-theory of $X$, as the cokernel of the obvious map $\mathbb{Z} \approx K^{top}(Point) \longrightarrow K^{top}(X)$ induced by the projection $X \longrightarrow Point$. Then $\widetilde{K}(X)$ may be identified with the group of *stable* isomorphism classes of vector bundles: two vector bundles $E$ and $E'$ are stably isomorphic if $E \oplus T \approx E' \oplus T'$ for some trivial bundles $T$ and $T'$. It follows that $K^{top}(X)$ is isomorphic to the set of homotopy classes of maps from $X$ to $\mathbb{Z} \times BU$, where $BU$ is the "infinite Grassmannian", i.e. the direct limit of the Grasmannians $G_n(\mathbb{C}^m)$ with obvious inclusions between them.

Here is another important difference between the two groups: $K^{top}(X)$ is much easier to compute that its algebraic counterpart $K(X)$. For instance, $K^{top}(X) \otimes \mathbb{Q}$ is isomorphic to $H^{ev}(X;\mathbb{Q})$, where $H^{ev}(X;\mathbb{Q})$ denotes the rational even Čech cohomology groups of $X$ via the topological Chern character. There is in fact an obvious factorization of the algebraic Chern character

$$K(X) \longrightarrow K^{top}(X) \longrightarrow H^{ev}(X;\mathbb{Q})$$

Therefore, if $X$ is a finite $CW$ complex, we find that $K^{top}(X)$ is isomorphic to $G \oplus \mathbb{Z}^d$, where $d$ is the dimension of the vector space $H^{ev}(X;\mathbb{Q})$ and $G$ is a finite group.

**Exercise**: compute $K^{top}(X)$ when $X$ is a sphere $S^n$ for $n \leq 4$.

A famous theorem of Bott which we shall see in the next lecture is the following: $\widetilde{K}^{top}(S^n)$ is isomorphic to $\mathbb{Z}$ if $n$ is even and $= 0$ if $n$ is odd. We shall see at the end how the Chern character enables us to construct a non trivial stable bundle over $S^n$ when $n$ is even. The idea of the proof of the isomorphism $K^{top}(X) \otimes \mathbb{Q} \approx H^{ev}(X;\mathbb{Q})$ is to notice that $K^{top}(X) \otimes \mathbb{Q}$ and $H^{ev}(X;\mathbb{Q})$ are half-exact functors[2] and that the Chern character induces an isomorphism when $X$ is a sphere. It is now a general statement that in such a situation the two half-exact functors are indeed isomorphic (for the details see for instance the Cartan-Schwartz seminar 1963/64, exposé 16).

In order to make the Chern character $K^{top}(X) \longrightarrow H^{2n}(X;\mathbb{Q})$ more explicit, let us describe it when $X$ is a $C^\infty$-manifold of finite dimension and let us imbed $\mathbb{Q}$ in $\mathbb{C}$ via the group homomorphism $\lambda \longrightarrow (2i\pi)^n\lambda$. The vector space $H^{2n}(X;\mathbb{Q})$ is then isomorphic to the de Rham cohomology of differential forms of degree $n$. On the other hand, it is not difficult to show that the classification of topological bundles on $X$ is equivalent to the classification of $C^\infty$ -bundles. Moreover, any $C^\infty$-bundle $E$ is a direct summand in a trivial bundle as in the topological situation: this means there is a $C^\infty$-map $p : X \longrightarrow M_n(\mathbb{C})$ such that for $x$ in $X$, the fiber $E_x$ is the image of $p(x)$, with $p(x)^2 = p(x)$.

[2]According to Dold, a homotopy (contravariant) functor on pointed spaces is called half-exact if we have an sequence $F(X/Y) \longrightarrow F(X) \longrightarrow F(Y)$ each time we have a closed subspace $Y$ of the compact space $X$.

Now, as popularized a few years ago in the framework of cyclic homology and non commutative de Rham homology, the Chern character of $E$, denoted by $Ch_n(E)$, is the cohomology class of the trace of the following product of matrices which entries are differential forms

$$Ch_n(E) = \frac{1}{n!(2i\pi)^n} Tr(p \cdot dp \cdots dp)$$

($2n$ factors $dp$). It is a nice exercise to show that $Ch_n(E)$ is a closed differential form and that its cohomology class is just a function of the class of $E$ in $K^{top}(X)$. This is the "modern" version of the Chern character which we shall see in Lecture 7.

In order to check the non triviality of the Chern character as defined above, let us choose the example of the Hopf bundle over the sphere $X = S^2$ which we write as the set of point $(x, y, z)$ such that $x^2 + y^2 + z^2 = 1$. We then put $p = (1 + J)/2$ , where $J$ is the involution of the trivial bundle $X \times \mathbb{C}^2$ defined over the point $(x, y, z)$ by the following matrix

$$\begin{pmatrix} z & x+iy \\ x-iy & -z \end{pmatrix}$$

The Hopf bundle is then the image of the projection operator $p$. The explicit computation of $Ch_1(E)$ gives the canonical volume form on the sphere $S^2$. Therefore the Chern character (as a cohomology class) is non trivial in general.

As it was pointed out by Hirzebruch, the Chern character can be caracterized by the following axioms (on the category of differentiable manifolds for instance)

1) Naturality: if $f : X \longrightarrow Y$ is a $C^\infty$-map and if $E$ is a vector bundle over $Y$, then $Ch_r(f^*(E)) = f^*(Ch_r(E))$.
2) Additivity: $Ch_r(E \oplus F) = Ch_r(E) + Ch_r(F)$
3) Normalisation: if $L$ is the canonical line bundle over $P^n(\mathbb{C})$ (which restricts to the Hopf bundle over the sphere $S^2 = P_1(\mathbb{C})$), one has

$$Ch_r(L) = x^r/r!$$

where $x$ is the unique element of $H^2(P^n(\mathbb{C}))$ which restricts to the volume form of $S^2$.

More generally, using the Chern character and Clifford algebras, we are going to show that $Ch_n$ is non trivial when $X$ is the sphere $S^{2n}$ (cf. [K2], chapter 1) by choosing an explicit vector bundle on $S^{2n}$ (as a matter of fact the generator of $\widetilde{K}(S^{2n}) \cong \mathbb{Z}$. More precisely, let $E$ be a complex vector space of finite dimension and let $e_1, ..., e_{2n+1}$ be automorphisms of $E$ such that

$(e_\alpha)^2 = 1$
$e_\alpha e_\beta = -e_\beta e_\alpha$ for $\alpha \neq \beta$
$e_1 ... e_{2n+1} = i^n$

An example of such a data is given by the exterior algebra $\Lambda\mathbb{C}^n = \Lambda\mathbb{C}\widehat{\otimes}\cdots\widehat{\otimes}\Lambda\mathbb{C}$

($n$ factors), where the couple $(e_{2r-1}, e_{2r})$ is acting on the $r^{th}$-factor $\Lambda\mathbb{C} = \mathbb{C}^2$ via the $(2 \times 2)$ matrix defined above. The last automorphism $e_{2n+1}$ is of course determined by the equation $e_1 ... e_{2n+1} = i^n$. It can be shown, using the theory of Clifford modules, that any such $E$ is isomorphic to a direct sum of copies of this example.

Let now $X$ be the sphere $S^{2n} = \{x_1, ..., x_{2n+1} | \sum(x_i)^2 = 1\}$ and let $V$ be the vector bundle which is the image of the projection operator $p = (1+J)/2$ on the trivial bundle $X \times E$, where

$$J = x_1 e_1 + \cdots + x_{2n+1} e_{2n+1}$$

Modulo an exact form, we have

$$n!(2i\pi)^n Ch_n(V) = 2^{-2n-1} Tr(J.(dJ)^{2n})$$

with $(dJ)^{2n} = (dx_1 e_1 + \cdots + dx_{2n+1} e_{2n+1})^{2n}$

Since the entries $(dx_\alpha e_\alpha)$ commute which each other, we may write the previous expression as the following sum

$$(2n)! \sum_\alpha (dx_1 e_1) \cdots \widehat{(dx_\alpha e_\alpha)} \cdots (dx_{2n+1} e_{2n+1})$$

where the symbol $\widehat{(dx_\alpha e_\alpha)}$ means that we omit $dx_\alpha e_\alpha$. If we expand it, we find

$$-i^n (2n)! \sum_\alpha (-1)^\alpha dx_1 \cdots \widehat{dx_\alpha} \cdots dx_{2n+1} e_\alpha$$

Therefore,

$$Trace[J.(dJ)^{2n}] = -i^n (2n)! \sum_\alpha (-1)^\alpha x_\alpha dx_1 \cdots \widehat{dx_\alpha} \cdots dx_{2n+1}$$

modulo an exact form. Since the expression $\sum_\alpha (-1)^\alpha x_\alpha dx_1 \cdots \widehat{dx_\alpha} \cdots dx_{2n+1}$ is the volume form of $S^{2n}$, which we denote by $Vol(S^{2n})$, we finally find that

$$Ch_n(V) = \frac{(2n)!}{2^{2n+1} n! (2\pi)^n} Vol(S^{2n}).dim(E)$$

In our example $E = \Lambda\mathbb{C}^n$ above, we have $dim(E) = 2^n$. Therefore, the integral of $Ch_n(V)$ on the sphere $S^{2n}$ is equal to 1 which implies that $Ch_n(V) \neq 0$ in the cohomology of $S^{2n}$ (and is in fact a generator of the integral class).

Let us come back now to the general theory $K(X)$ of Grothendieck where, for simplicity, we assume $X$ to be a complex projective variety. We may view $X$ as a subvariety of the complex projective space $P^n(\mathbb{C})$ for a certain $n$. As it is well known, $P^n(\mathbb{C})$ may be written as the quotient of the group $GL_{n+1}(\mathbb{C})$ by the subgroup consisting of matrices of the type

$$\begin{pmatrix} a & b \\ 0 & c \end{pmatrix}$$

where $a$ is an $(n \times n)$ matrix, $c$ is a $(1 \times 1)$ matrix and finally $b$ is an $(n \times 1)$ matrix.

Over $P^n(\mathbb{C})$, there is an affine variety $P = Proj_1(\mathbb{C}^{n+1})$ consisting of projection operators $p$ in $M_{n+1}(\mathbb{C})$ such that $Tr(p) = 1$, which may also be written as the homogeneous space $GL_{n+1}(\mathbb{C})/H$, where $H$ is the subgroup $GL_n(\mathbb{C}) \times GL_1(\mathbb{C})$. Note that $P$ is an affine variety in $M_{n+1}(\mathbb{C}) = \mathbb{C}^{(n+1)(n+1)}$ since it is defined by the following equations

$$\sum_j p_{ij}p_{jk} = p_{ik} \text{ and } \sum_j p_{jj} = 1$$

Let now $X'$ be the pull-back variety defined by the cartesian square

$$\begin{array}{ccc} X' & \longrightarrow & GL_{n+1}(\mathbb{C})/GL_n(\mathbb{C}) \times GL_1(\mathbb{C}) = P \\ \downarrow & & \downarrow \\ X & \longrightarrow & P_n(\mathbb{C}) \end{array}$$

[the second vertical map associates to a projection operator $p$ its image $Im(p)$].

Then $X'$ is called a "torsor" over $X$ (there is a vector bundle acting simply transitively on the fibers) and Grothendieck proved that $K(X) \simeq K(X')$. But now $X'$ is an affine variety as a closed subvariety of $P$ and Serre has shown that the category of algebraic vector bundles over $X'$ is equivalent to the category $\mathcal{P}(A)$ of finitely generated projective modules over the ring $A$ of coordinates of $X'$. For example, if $X = P_n(\mathbb{C})$ itself, $A$ is the algebra generated by the elements $p_{ij}$ with $1 \leq i, j \leq n+1$, subject to the relations $\sum p_{ij} \cdot p_{jk} = p_{ik}$ and $\sum p_{jj} = 1$. There is a topological analogy which we shall see in the next lecture, but already this point of view gives rise to several questions:

1) What will be the analog of the Chern character if we work with $A$-modules instead of vector bundles? As we shall see in Lecture 7, this opens a wide range of applications belonging to the new domain of "noncommutative geometry", linked with homology and cohomology theories for arbitrary rings.

2) Since the target of the Chern character is even cohomology, this suggests that there might be "derived functors" $K_n(A), n \in \mathbb{Z}$, for which $K_0(A)$ is just the first group. However, we shall see later on that this question is more complicated that it looks at the first glance. For instance, the topology of $A$ (if any) plays an important role: this is not seen of course in the definition of $K(A)$.

Finally, $K$-theory has nice cohomology operations coming from exterior powers of bundles or modules (more geometric in nature than their cohomology counterparts which are the Steenrod operations). These operations gave spectacular applications of $K$-theory in the 60's. For instance, J.-F. Adams was able to compute the maximum number of independent tangent vector fields on the sphere $S^n$ using these methods. Other applications were found in global analysis and in the theory of $C^*$−algebras. We shall review all these applications in Lectures 2 and 3.

At this point in history we have two theories, quite different in nature, coming from Algebraic Geometry and Algebraic Topology. We shall analyse the topological theory first in the spirit of noncommutative geometry. However, we shall see during this historical sketch that the algebraic and topological methods are in fact deeply linked one to another.

# Lecture 2 $K$-theory of Banach algebras. Bott Periodicity theorems

Let $E$ be a complex vector bundle over a compact space $X$ et let $A$ be the Banach algebra $C(X)$ of continuous functions $f : X \longrightarrow \mathbb{C}$ (with the Sup norm). If $M = \Gamma(X, E)$ denotes the vector space of continuous sections $s : X \longrightarrow E$ of the vector bundle $E$, $M$ is clearly a right $A$-module if we define $s.f$ to be the continuous section $x \longmapsto s(x)f(x)$.

As a matter of fact, since $X$ is compact, we may find another vector bundle $E'$ such that the Whitney sum $E \oplus E'$ is trivial, say $X \times \mathbb{C}^n$. Therefore, if we put $M' = \Gamma(X, E')$, we have $M \oplus M' \cong A^n$ as $A$-modules, which means that $M$ is a finitely generated projective $A$-module. The theorem of Serre and Swan [K1] says precisely that the correspondance $E \longmapsto M$ induces a functor from the category $\mathcal{E}(X)$ of vector bundles over $X$ to the category $\mathcal{P}(A)$ of finitely generated projective $A$-modules, which is in fact an equivalence of categories. In particular, isomorphism classes of vector bundles correspond bijectively to isomorphism classes of modules.

These considerations lead to the following definition of the $K$-theory of a ring with unit $A$: we just mimic the definition of $K^{top}(X)$ by replacing vector bundles by (projective finitely generated) $A$-modules. We call this group $K(A)$ by abuse of notation. It is clearly a covariant functor of the ring $A$ (by extension of scalars). We have of course $K^{top}(X) \approx K(A)$ , when $A = C(X)$.
As it was done in usual cohomology theory, one would like to "derive" this functor $K(A)$ and try to define $K_n(A), n \in \mathbb{Z}$, with some nice formal properties with $K_0(A) = K(A)$. This task is in fact more difficult than it looks for general rings $A$, as we shall see in Lectures 4 and 5.

On the other hand, if we avoid too much generality by working in the category of (complex) Banach algebras, there is essentially one way to do it (at least for $n \in \mathbb{N}$).

Firstly, we extend the definition of $K(A)$ to non necessarily unital algebras $A$ by adding a unit to $A$. For this purpose, we consider the vector space $\mathbb{C} \oplus A = \overline{A}$ provided with the "twisted" multiplication defined by

$$(\lambda, a).(\lambda', a') = (\lambda.\lambda', \lambda.a' + a.\lambda' + a.a')$$

The algebra has now a unit $= (1, 0)$. There is an obvious augmentation

$$\overline{A} \longrightarrow \mathbb{C}$$

and we define $K(A)$ as the kernel of the induced map $K(\overline{A}) \longrightarrow K(\mathbb{C}) = \mathbb{Z}$. It is easy to see that if $A$ already has a unit, we recover the previous definition $K(A)$.

**Exercise:** Let $\mathcal{K}$ be the ideal of compact operators in a Hilbert space $H$. Prove that $K(\mathcal{K}) \cong K(\mathbb{C}) \cong \mathbb{Z}$.

Secondly, for $n \in \mathbb{N}$, we define $K_n(A)$ as $K(A_n)$, where $A_n = A(\mathbb{R}^n)$ is the Banach algebra of continuous functions $f = f(x)$ from $\mathbb{R}^n$ to $A$ which vanish when $x$ goes to $\infty$.

**Exercise:** show that $K_1(A) \cong injlim\pi_0(GL_n(A)) \cong \pi_0(GL(A))$, where $GL(A)$ is the direct limit of the $GL_n(A)$ with respect to the obvious inclusions $GL_n(A) \subset GL_{n+1}(A)$.

**Exercise:** let $A = C(X)$, where $X$ is locally compact. Let $X^+$ be the one point compactification of $X$ ($X$ with one point added outside) and let $Y = S^n(X^+)$ , the $n^{th}$ suspension of $X^+$. Show that $K^{top}(Y)$ is isomorphic to $K_n(A) \oplus \mathbb{Z}$ (hint: notice that $C(Y)$ is isomorphic to $\overline{C}(X \times \mathbb{R}^n)$. In particular $K^{top}(S^n)$ is isomorphic to $K_n(\mathbb{C}) \oplus \mathbb{Z}$.

The following theorem is not too difficult to prove. It can be extracted easily from [K1].

**THEOREM .** *The functors $K_n(A), n \in \mathbb{N}$, are characterized by the following properties*

**1) Exactness:** *for any exact sequence of Banach algebras (where $A''$ has the quotient norm and $A'$ the induced norm)*

$$0 \longrightarrow A' \longrightarrow A \longrightarrow A'' \longrightarrow 0$$

*we have an exact sequence of $K$-groups*

$$K_{n+1}(A) \longrightarrow K_{n+1}(A'') \longrightarrow K_n(A') \longrightarrow K_n(A) \longrightarrow K_n(A'')$$

**2) Homotopy:** *$K_n(A(I)) \cong K_n(A)$, where $A(I)$ is the ring of continuous functions on the unit interval $I$ with values in $A$.*

**3) Normalization:** *$K_0(A)$ is the group $K(A)$ defined above.*

Another type of $K$-groups which will be quite useful later on, although more technical, is the <u>*relative*</u> $K$-group associated to a functor. If $\phi : \mathcal{P}(A) \longrightarrow \mathcal{P}(A'')$ is an additive functor (with some extra topological conditions), one can also define groups $K_n(\phi)$ which fit into exact sequences

$$K_{n+1}(A) \longrightarrow K_{n+1}(A'') \longrightarrow K_n(\phi) \longrightarrow K_n(A) \longrightarrow K_n(A'')$$

For instance, if $\phi$ is asssociated to an epimorphism of Banach algebras $A \longrightarrow A''$ as above, $K_n(\phi)$ may be identified with $K_n(A')$. Another case of interest is when $A$ (resp. $A''$) is the ring of continuous functions on a compact space $X$ (resp. a closed subspace $X''$). Then, the group $K_n(\phi)$ is also written $K_n(X, X'')$ or $K^{-n}(X, X'')$.

The functors $K_*(A)$ have some other nice properties like the following: a continuous bilinear pairing of Banach algebras

$$A \times C \longrightarrow B$$

induces a "cup-product"

$$K_i(A) \otimes K_j(C) \longrightarrow K_{i+j}(B)$$

In particular, if $C = \mathbb{C}$, the field of complex numbers and if $A = B$, we have a pairing

$$K_i(A) \otimes K_j(C) \longrightarrow K_{i+j}(A)$$

**Exercise:** compute this pairing when $A = \mathbb{C}$ with $i$ and $j \leq 2$.

We can now state the Bott periodicity theorem in the Banach algebras setting.

**THEOREM .** *The group $K_2(\mathbb{C})$ is isomorphic to $\mathbb{Z}$ and the cup-product with a generator $u_2$ induces an isomorphism $\beta : K_n(A) \longrightarrow K_{n+2}(A)$ for any Banach algebra $A$.*

This theorem has a spectacular application in homotopy: let $GL(A)$ be the infinite general linear group which is the direct limit of the $GL_r(A)$ with respect to the obvious inclusions. Then an analysis of the group $K(A(\mathbb{R}^n))$ shows that it is isomorphic to the homotopy group $\pi_{n-1}(GL(A))$. More precisely, if $A = \mathbb{C}$ for instance, the group $K(A(\mathbb{R}^n))$ is linked with the classification of stable complex vector bundles over the sphere $S^n$ which are determined by homotopy classes of "gluing functions" $f : S^{n-1} \longrightarrow GL(A)$. For a general Banach algebra $A$, one has to consider bundles over the sphere whose fibers are $A^r$ instead of $\mathbb{C}^r$. Then the previous theorem has the following corollary:

**COROLLARY .** *Let $A$ be any complex Banach algebra. Then we have isomorphisms*

$\pi_i(GL(A)) \cong \pi_{i+2}(GL(A)) \cong K_1^{top}(A)$ *if $i$ is even;*
$\pi_i(GL(A)) \cong \pi_{i+2}(GL(A)) \cong K(A)$ *if $i$ is odd;*
*In particular, we have $\pi_1(GL(A)) \cong K(A)$.*

As a matter of fact, the last isomorphism (inspired by Atiyah and Bott) is the basis of the proof of the theorem. The idea is to show that any loop in $GL(A)$ can be deformed into a loop of the following type

$$\theta \longmapsto pz + 1 - p$$

where $p$ is an idempotent matrix and $z = e^{i\theta}$ . Such an idempotent matrix $p$ is of course associated to the projective module $Im(p)$. A more conceptual proof will be sketched at the end of this lecture.

**Remark.** If $A = \mathbb{C}$, the groups $\pi_i(GL_r(\mathbb{C})) = \pi_i(U(r))$ stabilize and are equal to $K_{i+1}^{top}(\mathbb{C})$ if $r > i/2$.

**Exercise:** Prove the following facts:

1) $K^{top}(S^n)$ is isomorphic to $\mathbb{Z}$ if $n$ is odd and to $\mathbb{Z} \oplus \mathbb{Z}$ if $n$ is even.

2) $\pi_i(GL_r(\mathbb{C})) \cong \pi_i(U(r)) \cong \mathbb{Z}$ if $i$ is odd and $i > r/2$ and $= 0$ if $i$ is even and $i > r/2$.
3) $K^{top}(X) \oplus K^{top}(X) \cong K^{top}(X \times S^0) \cong K^{top}(X \times S^2) \cong K^{top}(X \times S^{2n})$ for any compact space $X$.

In the previous considerations, we were working with modules over *complex* Banach algebras and their topological counterparts which are complex vector bundles. We could as well consider *real* Banach algebras and real vector bundles. Most considerations are valid in the real case, with the notable exception of the last theorem: in this case, we have an eight-periodicity and not a two-periodicity. More precisely:

**THEOREM .** *The group $K_8(\mathbb{R})$ is isomorphic to $\mathbb{Z}$ and the cup-product with a generator induces an isomorphism between the groups $K_n(A)$ and $K_{n+8}(A)$ for any real Banach algebra $A$.*

For instance, the groups $K_n(\mathbb{R})$ are equal to $\mathbb{Z}, \mathbb{Z}/2, \mathbb{Z}/2, 0, \mathbb{Z}, 0, 0, 0$ respectively for $n = 0, 1, \cdots, 7$ mod. 8.

The original proof of this theorem used Morse theory [Bott]. Other elementary proofs (whom we mentioned already) were found by Atiyah and Bott, Wood and the author, the last one in the framework of Banach algebras and Banach categories. See [At], [K1] and [Wo] for references.

As a matter of fact, Bott proved other theorems in the real case using loop spaces of homogeneous spaces. He gave the following impressive list of homotopy equivalences:

$$\Omega(\mathbb{Z} \times BGL(\mathbb{R})) \approx GL(\mathbb{R})$$
$$\Omega(GL(\mathbb{R})) \approx GL(\mathbb{R})/GL(\mathbb{C})$$
$$\Omega(GL(\mathbb{R})/GL(\mathbb{C})) \approx GL(\mathbb{C})/GL(\mathbb{H})$$
$$\Omega(GL(\mathbb{C})/GL(\mathbb{H})) \approx \mathbb{Z} \times BGL(\mathbb{H})$$
$$\Omega(\mathbb{Z} \times BGL(\mathbb{H})) \approx GL(\mathbb{H})$$
$$\Omega(GL(\mathbb{H})) \approx GL(\mathbb{H})/GL(\mathbb{C})$$
$$\Omega(GL(\mathbb{H})/GL(\mathbb{C})) \approx GL(\mathbb{C})/GL(\mathbb{R})$$
$$\Omega(GL(\mathbb{C})/GL(\mathbb{R})) \approx \mathbb{Z} \times BGL(\mathbb{R})$$

One way to understand these eight homotopy equivalences is to use Clifford algebras as it was pointed out for the first time by Atiyah, Bott and Shapiro. As a matter of fact, if we denote by $C_n$ the Clifford algebra of $\mathbb{R}^n$ provided with the quadratic form $(x_1)^2 + \cdots + (x_n)^2$, there is a kind of "periodicity" of the $C_n$[3]: we have graded algebra isomorphisms

$$C_{n+8} \approx M_{16}(C_n)$$

[3]Here and until Lecture 4, we shall write $K$ instead of $K^{top}$.

On the other hand, the complexified Clifford algebras have a 2-periodicity

$$C_{n+2} \otimes \mathbb{C} \approx M_2(C_n) \otimes \mathbb{C}$$

These remarks gave rise to an elementary proof of the previous 8 homotopy equivalences and the fact that $\pi_7(GL(A)) \approx K_0(A)$ [K1] [Wo]. As a matter of fact, a uniform way to state these homotopy equivalences is to write (up to connected components)

$$GL(C_n)/GL(C_{n-1}) \approx \Omega[GL(C_{n+1})/GL(C_n)]$$

A second way to interpret Bott periodicity is to use Hermitian $K$-theory where one studies finitely generated projective modules provided with non degenerate Hermitian forms. This theory is associated to other classical Lie groups like the orthogonal group or the symplectic group. We shall investigate this theory in the sixth lecture and extend it to a discrete context (which is not possible with the Clifford algebra interpretation).

Before going back to classical Algebraic Topology, it might be interesting to give a conceptual proof of Bott periodicity (at least in the complex case).

Let us assume we have defined $K_n(A)$, not only for $n \in \mathbb{N}$, as we did before, but also for $n \in \mathbb{Z}$. Let us assume also that the pairing

$$K_i(A) \otimes K_j(C) \longrightarrow K_{i+j}(B)$$

roughly described above, extends to all values of $i$ and $j$ in $\mathbb{Z}$ and has some obvious "associative" properties. Finally, let us suppose the existence of a "negative" Bott element $u_{-2}$ in $K_{-2}(\mathbb{C})$ such that the cup-product with $u_2$ gives the unit element 1 in $K_0(\mathbb{C}) \cong \mathbb{Z}$. With these hypotheses, we may define an inverse

$$\beta' : K_{n+2}(A) \longrightarrow K_n(A)$$

of the Bott map

$$\beta : K_n(A) \longrightarrow K_{n+2}(A)$$

It is defined by the cup-product with $u_{-2}$. It is clear that the compositions of $\beta$ with $\beta'$ both ways are the identity.

The price to pay for this proof is of course the construction of this "negative" $K$-theory $K_n(A)$ for $n < 0$. This may be done, using the notion of "suspension" of a ring which is in some sense dual to the notion of the suspension of a space. More precisely, we define the "cone" $CA$ of a ring $A$ to be the set of all infinite matrices $M = (a_{ij}), i, j \in \mathbb{N}$, such that each row and each column only contains a finite number of non zero elements in $A$. This is clearly a ring for the usual rule of matrix multiplication. We make $CA$ into a Banach algebra by completing it with respect to the following norm

$$||M|| - \sup_j \sum_i ||a_{ij}||$$

Finally we define $\widetilde{A}$, the "stabilization" of $A$ as the closure of the set of finite matrices[4] in $CA$. It is a closed 2-sided ideal in $CA$ and the suspension of $A$ - denoted by $SA$ - is the quotient ring $CA/\widetilde{A}$.

[4]An infinite matrix is called finite if all its elements are 0, except a finite number.

**DEFINITION/THEOREM .** *Let $A$ be a Banach algebra. Let us define the groups $K_{-n}(A)$ to be $K(S^nA)$, where $S^nA$ is the $n^{th}$–suspension of $A$. Then $BGL(SA)$ is a "delooping" of $K_0(A) \times BGL(A)$, i.e. we have a homotopy equivalence*

$$\Omega(BGL(SA)) = K_0(A) \times BGL(A)$$

*Therefore, to any exact sequence of Banach algebras as above*

$$0 \longrightarrow A' \longrightarrow A \longrightarrow A'' \longrightarrow 0$$

*we can associate an exact sequence of $K$-groups*

$$K_{n+1}(A) \longrightarrow K_{n+1}(A'') \longrightarrow K_n(A') \longrightarrow K_n(A) \longrightarrow K_n(A'')$$

*for $n \in \mathbb{Z}$.*

For a better understanding of $SA$, it is interesting to notice that the ring of Laurent series $A < t, t^{-1} >$ is a good approximation of the suspension. Any element of $A < t, t^{-1} >$ is a series

$$S = \sum_{n \in Z} a_n t^n$$

such that $\sum_{n\in\mathbb{Z}} | a_n |< +\infty$ . We define a ring homomorphism

$$A < t, t^{-1} > \longrightarrow SA$$

which associates to the series above the class of the following infinite matrix

$$\begin{pmatrix} a_0 & a_1 & a_2 & \cdots\cdots \\ a_{-1} & a_0 & a_1 & \cdots\cdots \\ a_{-2} & a_{-1} & a_0 & \cdots\cdots \\ \cdots & \cdots & \cdots & \cdots\cdots \end{pmatrix}$$

For instance, in order to define the element $u_{-2}$ mentionned above, it is enough to define a projective module over the Banach algebra $\mathbb{C} < t, u, t^{-1}, u^{-1} >$, i.e. a non trivial complex vector bundle on the torus $S^1 \times S^1$, as we can see easily, using Fourier analysis.

Let us come back now to usual Algebraic Topology and define $\overline{K}^n(X), n \in \mathbb{Z}$, for a compact space with base point $*$ as $K_{-n}(A)$ where $A$ is the Banach algebra of continuous functions on $X$ which vanish on the base point. The (contravariant) functors $X \longmapsto \overline{K}^n(X)$ define a (reduced) cohomology theory for compact spaces. One may also define a relative $K$-theory by putting $K^n(X, Y) = \overline{K}^n(X/Y)$ which is also isomorphic to the group $K_{-n}(\phi)$ defined at the beginning, where the functor $\phi$ is associated to the surjection of rings

$$\theta : C(X) \longrightarrow C(Y)$$

To sum up, one has therefore an exact sequence

$$K^{n-1}(X) \longrightarrow K^{n-1}(Y) \longrightarrow K^n(X, Y) \longrightarrow K^n(X) \longrightarrow K^n(Y)$$

and an excision isomorphism

$$K^n(X, Y) \cong K^n(X/Y, *) \simeq K^n(X/Y)$$

Finally, it is easy to show that $K^0(X, \Phi) \cong K(X)$, the original $K$-theory of Grothendieck-Atiyah-Hirzebruch.

What we have defined above is a "cohomology theory" satisfying all the Eilenberg-Steenrod axioms, except the dimension axiom. This is just an example of what mathematicians called an "extraordinary" cohomology theory in the 60's (other famous examples are coming from cobordism). As we shall see in the next lecture, this theory has many interesting topological applications.

There are many variants of topological $K$-theory which were considered in the 60's. One of appealing interest is equivariant $K$-theory $K_G(X)$ where $G$ is a compact Lie group acting on $X$ which was introduced by Atiyah and Segal. It is defined as the $K$-theory of $G$-equivariant vector bundles on $X$. The analog of Bott periodicity in this context is the "Thom isomorphism": one considers a *complex* $G$-vector bundle $V$ on $X$ and we would like to compute the equivariant $K$-theory of $V$ (viewed as a locally compact space), i.e. $\overline{K}_G(V^+)$ , where $V^+$ is the one-point compactification of $V$. If we denote this group simply by $K_G(V)$, we have a Thom isomorphism (due to Atiyah)

$$K_G(X) \longrightarrow K_G(V)$$

More generally, if $V$ is a real vector bundle, we can define on $V$ a *positive* definite metric invariant under the action of $G$ and consider the associated Clifford bundle $C(V)$; the group $G$ also acts naturally on $C(V)$. We denote by $\mathcal{E}_G(X)$ the category of real vector bundles where $G$ and $C(V)$ act simultaneously; these two actions are linked together by the formula

$$g * (a.e) = (g * a).(g * e)$$

where the symbol $*$ (resp. .) denotes the action of $G$ (resp. of $C(V)$). For instance, if $G$ is finite, it is easy to show that $\mathcal{E}_G^V(X)$ is equivalent to the category of finitely generated projective modules over the crossed product algebra $G \ltimes \overline{C}(V)$, where $\overline{C}(V)$ is the algebra of continuous sections of the bundle $C(V)$. On the other hand, if "1" denotes the trivial bundle of rank one (with trivial action of $G$), we have a "restriction" functor

$$\varphi : \mathcal{E}_G^{V \oplus 1}(X) \longrightarrow \mathcal{E}_G^V(X)$$

A generalization of the theorem of Atiyah quoted above is the following: the relative group $K(\varphi)$ is isomorphic to the equivariant $K$-theory of the Thom space, i.e. $K_G(V)$, with the notations above.

We should notice that the last statement is true in both real and complex $K$-theory and implies Bott periodicity for the $K$-theory of vector bundles!

It is also worth mentioning that Atiyah's theorem and its generalization are not easy to prove even if $X$ is reduced to a point. A key ingredient in the proof is an index map which Atiyah constructed with suitable elliptic operators.

There is another generalization of topological $K$-theory which was considered by Donovan and the author at the beginning of the 70's and which became "fashionable" recently: this is "$K$-theory with twisted coefficients" $K^\alpha(X)$[5] defined for $\alpha \in H^3(X;\mathbb{Z})$ (we limit ourselves to the complex case). The simplest way to define it is to notice that the usual complex $K$-theory is also the $K$-theory of the Banach algebra of continuous functions

$$f : X \longrightarrow \mathcal{K}$$

where $\mathcal{K}$ is the ideal of compact operators in a Hilbert space $H$ (we just enlarge $\mathbb{C}$ to $\mathcal{K}$, which is quite usual in functional analysis). This is also the algebra of sections of the trivial bundle

$$\begin{array}{c} E = X \times \mathcal{K} \\ \downarrow \\ X \end{array}$$

viewed as a bundle of algebras. The idea is now to "twist" this algebra bundle by an automorphism. More precisely, using the fact that $Aut(H)$ is contractile by Kuiper's theorem, we have the following principal bundles with contractible total spaces (where $P(H) = Aut(H)/C^*$).

$$B\mathbb{C}^* \longrightarrow B(Aut(H)) \longrightarrow B(P(H))$$

$$K(\mathbb{Z},2) \longrightarrow B\mathbb{C} \longrightarrow B\mathbb{C}^*$$

This shows $BP(H)$ is the Eilenberg-Mac Lane space $K(\mathbb{Z},3)$. In other words, any element $\alpha$ of $H^3(X;\mathbb{Z})$ is associated to a principal bundle with fiber $P(H)$ which may be defined by transition functions $g_{ji}$ over an open cover $(U_i)$ of $X$. We can use this $P(H)$ bundle to twist the previous $\underline{algebra}$ bundle $E$ in the following way: the twisted algebra bundle $\overline{E}$ is obtained by gluing $U_i \times \mathcal{K}$ with $U_j \times \mathcal{K}$ over $U_i \cap U_j$: we identify $(x_i, k_i)$ with $(x_j, k_j)$ whenever $x = x_i = x_j \in U_i \cap U_j$ and $k_j = g_{ji}(x)k_i(g_{ji}(x))^{-1}$.

The space of sections of $\overline{E}$ is a Banach algebra whose $K$-theory is precisely the twisted $K$-theory of $X$ we wanted to define.

**Exercise:** Let $V$ be an oriented real vector bundle of even rank with second Stiefel-Whitney class $w_2 \in H^2(X;\mathbb{Z}/2)$ and a the Bockstein of $w_2$ in $H^3(X;\mathbb{Z})$. Show that $K^\alpha(X)$ is isomorphic to the $K$-theory of the Thom space of $V$.

Let us mention finally that in the classical case of untwisted $K$-theory $K^*(X)$, there is a spectral sequence due to Atiyah and Hirzebruch whose $E_2$ term is $H^p(X;K^q_{top}(*))$, where $*$ is a point. This spectral sequence converges to $K^{p+q}(X)$. There is a generalization of this spectral sequence in the case of twisted $K$-theory which will appear in a forthcoming paper by Atiyah and Segal.

[5] now called "twisted" $K$-theory.

# Lecture 3 Some applications of topological $K$-theory. The Atiyah-Singer index theorem

There are many applications of topological $K$-theory, most of them due to Atiyah and Adams. More recent applications in the $K$-theory of Banach algebras are also given.

### 1) Riemann-Roch theorems for differentiable manifolds [AH2][K1]

As it was explained in the first lecture, Grothendieck's original aim was to generalize the Hirzebruch-Riemann-Roch theorem in the context of projective algebraic smooth varieties. Soon after the discovery of topological $K$-theory, Atiyah and Hirzebruch achieved the same result in the category of differentiable manifolds.

The simplest case is the following: let us consider two $C^\infty-$ manifolds $X$ and $Y$, with an almost complex structure given on the tangent bundles, together with a smooth <u>proper</u> map

$$f : X \longrightarrow Y$$

Since $X$ and $Y$ are oriented, there is a Gysin map in cohomology

$$H^*(X) \longrightarrow H^*(Y)$$

which increases the degrees by the number $dim(Y) - dim(X)$. On the other hand, as briefly stated in the first lecture, any complex vector bundle $V$ has a "Todd class" $Todd(V)$ which can be formally defined as follows. We express the "total Chern class"

$$c(V) = 1 + c_1(V) + c_2(V) + \cdots + c_n(V) + \cdots$$

as a formal product $\prod(1 + x_i)$ , so that the $c_i(V)$ are the elementary symmetric functions of the $x_j$. Then we consider the formal product $\prod x_i/(1 - e^{-xi})$ which can be written as

$$Todd(V) = 1 + Todd_1(V) + Todd_2(V) + \cdots + Todd_n(V) + \cdots$$

For instance[6], one has the following formulas (with $c_i = c_i(V)$, and $T_i = Todd_i(V)$)

$T_1 = 1/2c_1$
$T_2 = 1/12(c_2 + (c_1)^2)$
$T_3 = 1/24c_2c_1$
$T_4 = 1/720[-c_4 + c_3c_1 + 3(c_2)^2 + 4c_2(c_1)^2 - (c_1)^4]$
............

[6]Note that this definition of $Todd(V)$ can be extended to elements $x$ of $K(X)$: if $x = V - W$, we put formally $Todd(x) = Todd(V)/Todd(W)$.

Using (complex) Bott periodicity and Thom isomorphism in $K$-theory, one may also define a Gysin map in $K$-theory

$$f_*^K : K^{top}(X) \longrightarrow K^{top}(Y)$$

The Atiyah-Hirzebruch theorem may then be stated as follows: for any element $x$ in $K(X)$, we have the formula

$$Ch(f_*^K(x)).Todd(TY) = f_*^H(Ch(x).Todd(TX))$$

If $Y$ is reduced to a point, $f_*^H$is given by integration over the compact manifold $X$ and we find that for any element $x$ in the $K$-theory of $X$, $Ch(x).Todd(TX)$ is always an *integral* class of degree $2n = dim(X)$ (a priori it is only a rational class). For instance, $Todd_n(TX)$ is an integral class when $X$ is a manifold of dimension $2n$ with an almost complex structure.

There are many interesting generalizations of this formula (all due to Atiyah and Hirzebruch) when $Tf$ (the tangent bundle of $f$ "along the fibers") is provided with a $^c$spinorial structure. There is an analog of the $Todd$ class, called the $\widehat{A}$-genus which can be expressed in terms of the Pontrjagin classes of $TX$. Here is the type of formula we get:

$$Ch(f_*^K(x)) = f_*^H(ch(x).\widehat{A}(Tf))$$

This relation implies other integrality theorems in terms of the Pontrjagin classes of the tangent bundles involved. Here is a typical example: if $X$ is a manifold of dimension 8 which is spinorial, the value of $7(p_1)^2 - 4p_2$ is divisible by 11720. Here $p_1$ and $p_2$ are the first two Pontjagin classes of $TX$ (see [K1] for more examples and details).

## 2) The Atiyah-Singer index theorem

There is an extensive literature about this famous theorem and the basic references are the series of papers of the authors themselves [AS1][AS2]. Therefore, we limit ourselves to the general ideas underlying this theorem. Let $X$ be a compact oriented manifold, $E$ and $F$ two $C^\infty$-vector bundles on $X$ and $D : \Gamma(X, E) \longrightarrow \Gamma(X, F)$ an elliptic differential operator. Then $Ker(D)$ and $Coker(D)$ are finite dimensional vector spaces and one would like to compute the index of $D$.

$$Index(D) = dim(KerD) - dim(CokerD)$$

in purely topological terms.

Since the operator $D$ is elliptic, its symbol may be viewed as a morphism of vector bundles

$$\sigma : \pi^*(E) \longrightarrow \pi^*(F),$$

where $\pi$ is the projection of the cotangent bundle $T^*X$ over $X$, which is an isomorphism outside the zero section of $T^*X$. If $B(X)$ (resp. $S(X)$) denotes the ball bundle (resp. the sphere bundle) of $T^*X$ with respect to a metric, the symbol gives rise to an element, called $[\sigma]$, of the relative group $K^0(B(X), S(X))$. Let us denote $C(\sigma)$ the image of $[\sigma]$ under the composition of the Chern character: $K^0(B(X), S(X)) \longrightarrow H^{ev}(B(X), S(X))$ and the Thom isomorphism $H^{ev}(B(X), S(X)) \longrightarrow H^*(X)$. The product of $C(\sigma)$ by the Todd class of the complexified bundle $T'(X)$ of $TX$ is a cohomology class $C(\sigma).Todd(T'(X))$ which we can evaluate on the fundamental class $[X]$ of the manifold. This evaluation is written classically as $C(\sigma).Todd(T'(X))[X]$. The Atiyah-Singer index theorem can be stated as follows:

$$Index(D) = C(\sigma).Todd(T'(X))[X]$$

In particular, $C(\sigma).Todd(T'(X))[X]$ is an integer[7]. On the other hand, various classical elliptic differential operators enable us to recover classical previous results: the Riemann-Roch theorem in the differentiable category and the signature theorem of Hirzebruch among many others [AS1][AS2]. We may also notice that the integrality theorems are a consequence of the index theorem. However, as it was shown in [AH2] (see also [K1]), these theorems are topological in nature and do not require the pseudo-differential machinery needed in the proof of the index theorem.

### 3) The Hopf invariant one problem

Before attacking this topological problem, we recall the operations $\lambda^k$ of Grothendieck, defined by the exterior powers of bundles. As Adams has pointed out, there exist more convenient operations, called $\psi^k$, which are ring endomorphisms of $K(X)$ (and also $K^{top}(X)$). Moreover, we have the remarkable relation $\psi^k\psi^1 = \psi^{k1}$. In order to define this operation $\psi^k$, we notice the universal formula expressing the "fundamental" symmetric functions

$$S_K = \sum (x_i)^k$$

in terms of the elementary symmetric functions

$$\sigma_k = \sum_{i_1<\ldots<i_k} x_{i_1}\ldots x_{i_k}$$

For instance, $S_1 = \sigma_1$, $S_2 = (\sigma_1)^2 - 2\sigma_2, S_3 = (\sigma_1)^3 - 3\sigma_1\sigma_2 + 3\sigma_3$, etc. In general $S_k = Q_k(\sigma_1, \ldots, \sigma_k)$ , where the $Q_k$ are the so-called Newton polynomials. Then, for any element $x$ of $K(X)$, we define $\varphi^k(x)$ as the expression $Q_k(\lambda^1(x), \ldots, \lambda^k(x))$.

The Hopf invariant one problem in Algebraic Topology is linked with another problem of more appealing interest: for which values of $n$ the sphere $S^{n-1}$ may be provided with an $H$-space structure ? It is not difficult to show (using Künneth theorem in rational cohomology for instance) that this is not possible if $n$ is odd $> 1$. On the other hand, $S^0, S^1, S^3, S^7$ are easily seen to have a structure of $H$-spaces, in fact topological groups, except for $S^7$. It is a remarkable theorem of Adams that these spheres are the only ones with an $H$-space structure. The $K$-theory proof (due to Adams and Atiyah) is as follows (see [K1] for more details).

[7]As a matter of fact, one has to be careful about signs, according to the chosen conventions.

Starting with the given product $S^{n-1}\times S^{n-1} \longrightarrow S^{n-1}$, with $n$ even, and following Hopf, we construct a map $f : S^{2n-1} \longrightarrow S^n$ to which we associate the "Puppe sequence"

$$S^{2n-1} \longrightarrow S^n \longrightarrow Cf \longrightarrow S^{2n} \longrightarrow S^{n+1}$$

More precisely, $Cf$, the cone of $f$, is a $CW$-complex with 2 cells of dimensions $n$ and $2n$ respectively: we attach a $2n$-cell to $S^n$ using the map $f$. Now the reduced $K$-theory of $Cf$ is easily computed: it is a free group with 2 generators $u$ and $v$ and we have $u^2 = hv$ for a certain scalar $h$ which is precisely $\pm 1$ (for a general map $f, h = h(f)$ is called the *Hopf invariant* of $f$).

We now follow the pattern given by Atiyah and Adams to show that $h(f)$ cannot be odd in general (except if $n = 2, 4$ *or* $8$). From the general properties of the Adams operations $\psi^k$, we have $\psi^k(v) = k^{2r}v$ and $\psi^k(v) = k^r u + \sigma(k)v$ , where $n = 2r$ and $\sigma(k) \in \mathbb{Z}$. On the other hand, since $\psi^2 = (\lambda^1)^2 - 2\lambda^2$, we have $\psi^2(u) = u^2$ *mod.* $2 = h(f)v$ *mod.* $2$. Hence $\sigma(2)$ must be odd, since $h(f)$ is odd by hypothesis. On the other hand, from the relation $\psi^k\psi^l = \psi^l\psi^k$, we deduce (for $l \neq 1$)

$$k^r(k^r - 1)\sigma(l) = l^r(l^r - 1)\sigma(k)$$

If we choose $l = 2$ and $k$ odd, we see that $2^r$ should divide $k^r - 1$ for all odd integers $k$, a property of $r$ which is only true if $r = 1, 2$ *or* $4$ (an easy exercise left to the reader).

## 4) The vector field problem on the sphere

Let $M$ be a compact connected oriented manifold. Here is a classical result due to Hopf: there exists a non zero tangent vector field on $M$ if and only if $\chi(M)$ - the Euler Poincaré characteristic of $M$ - vanishes. A much harder problem is to find the maximum number of linearly independent tangent vector fields on $M$, even if $M$ is the sphere $S^{t-1}$. The Gram-Schmidt orthonormalization procedure may be used to replace any field of $(n-1)-$ linearly independent tangent vectors by a field of $n-1$ tangent vectors of norm one which are orthogonal to each other. Therefore, if $O_{n,t}$ denotes the Stiefel manifold $O(t)/O(t-n)$, the existence of a field of $(n-1)$ linearly independent tangent vectors on $S^{t-1}$ is equivalent to the existence of a continuous section $\sigma : O_{1,t} = S^{t-1} \longrightarrow O_{n,t}$ of the natural projection

$$O(n,t) \longrightarrow O_{1,t}$$

Note that each element $a$ of $O(n,t)$ defines a linear map $\varphi_a : \mathbb{R}^n \longrightarrow \mathbb{R}^t$ which is injective and depends continuously on $a$.

Let us introduce now the (real) Hopf bundle $\zeta$ on $RP^{n-1} = S^{n-1}/\mathbb{Z}^2$ as the quotient of $S^{n-1} \times \mathbb{R}^t$ by the equivalence relation $(x, \lambda) \approx (-x, -\lambda_1, ..., -\lambda_t)$. In the same way, the Whitney sum of $t$ copies of $\zeta$ (denoted by $t\zeta$) is the quotient of $S^{n-1} \times \mathbb{R}^t$ by the identification

$$(x, \lambda_1, ..., \lambda_t) \approx (-x, -\lambda_1, ..., -\lambda_t)$$

If we write $S(t\zeta)$ (resp. $S(t\varepsilon)$) for the sphere bundle of $t\zeta$ (resp. $t\varepsilon$), the existence of $\sigma$ shows the existence of a commutative diagram

$$\begin{array}{ccc} S(t\varepsilon) & \longrightarrow & S(t\varepsilon) \\ & \searrow \quad \swarrow & \\ & X = RP^{n-1} & \end{array}$$

Here $\theta$ is induced by the map $(x, v) \longmapsto (x, \varphi_{\sigma(v)}(x))$ from $S^{n-1} \times S^{t-1}$ into itself, where $\varphi_a$ is the linear map $\mathbb{R}^n \longrightarrow \mathbb{R}^t$ defined above. Moreover, $\theta$ induces a homotopy equivalence on each fiber.

We are now in the general situation of two vector bundles $V$ and $W$ (here $t\varepsilon$ and $t\zeta$) such that the associated sphere bundles $S(V)$ and $S(W)$ have the same *fiber* homotopy type. The same property is true if we compactify the fibers instead of considering the spheres. Therefore, the map $\theta$ (extended to the vector bundles by radial extension) has the property that for each point $x$ in $X$ we have the commutative diagram

$$\begin{array}{ccc} K_{\mathbb{R}}(W) & \xrightarrow{\theta^*} & K_{\mathbb{R}}(V) \\ \downarrow & & \downarrow \\ K_{\mathbb{R}}(W_x) & \xrightarrow{\theta_x^*} & K_{\mathbb{R}}(V_x) \end{array}$$

where $K_{\mathbb{R}}$ means *real* $K$-theory. As it was shown by Adams, this condition implies the existence of an element $y$ of $\overline{K}_{\mathbb{R}}(X)$ such that for each $k$ we have the relation

$$\rho^k(V) = \rho^k(W)\frac{\psi^k(1+y)}{1+y}$$

under the condition that $V$ and $W$ are spinorial bundles of rank $\equiv 0$ mod. 8.

Here $\rho^k(V)$ is the characteristic class in $K$-theory[8] introduced by Bott: it is defined as the image of 1 under the composition of the following maps

$$K_{\mathbb{R}}(X) \xrightarrow{\varphi} K_{\mathbb{R}}(V) \xrightarrow{\psi^k} K_{\mathbb{R}}(V) \xrightarrow{\varphi^{-1}} K_{\mathbb{R}}(X)$$

where $\varphi$ is the Thom isomorphism in $K_{\mathbb{R}}$-theory. These remarks, together with the computation of the order of the group $\overline{K}(RP^{n-1})$ are the key ingredients for the solution in two steps of our original problem.

**PROPOSITION .** *Let $a_n$ be the order of the group $\overline{K}_{\mathbb{R}}(RP^{n-1})$, that is $a_n = 2^f$, where $f$ is the number of integers $i$ such that $0 < i < n$ and $i = 0, 1, 2$ or $4$ mod $8$. If $S^{t-1}$ admits $n-1$ linearly independent tangent vector fields, then $t$ is a multiple of $a_n$.*

From this proposition and the theory of Clifford algebras over the real numbers, we deduce the following fundamental theorem of Adams

[8] Note that this class is constructed in complete analogy with the Stiefel-Whitney classes in cohomology mod.2: one just has to replace the Adams operations by the Steenrod squares.

**THEOREM** . *Let us write each integer t in the form* $t = (2\alpha - 1).2^{\beta}$ *where* $\beta = \gamma + 4\delta$, *with* $0 \leq \delta \leq 3$. *Then the maximum number of linearly independent tangent vector fields on the sphere* $S^{t-1}$ *is* $\sigma(t) = 2^{\gamma} + 8\delta - 1$.

## 5) Applications to $C^*$-algebras

Here is a theorem due to Connes: if $A = C_r^*(G)$ is the reduced $C^*$-algebra of a free group on $n$ generators, then the only idempotents of $A$ are 0 and 1. The proof is too technical to give the details here (see [R] p. 358-361); $K$-theory comes into the picture since one has to consider in the proof extensions of $C^*$algebras of the type

$$0 \longrightarrow \mathcal{K} \longrightarrow D \longrightarrow A \longrightarrow 0$$

where $\mathcal{K}$ is the ideal of compact operators in a Hilbert space. A splitting of this extension induces a homomorphism $K(A) \longrightarrow K(\mathcal{K}) \approx \mathbb{Z}$ which Connes identifies with the usual positive normalized trace on $A$.

Another beautiful application in Connes's book [C1] is a description of the Penrose tiling in terms of inductive limits of finite dimensional algebras (so-called $AF$-algebras). Such an inductive limit $A$ (or rather its closure in the algebra of bounded operators in a Hilbert space) has of course a Grothendieck group $K(A)$. Moreover, $K(A)$ is an *ordered group* (an element of $K(A)$ is called $\geq 0$ if it comes from a genuine projective module). In the case of the Penrose tiling, one has $K(A) \cong \mathbb{Z}^2$ , the positive cone being given by the set of all $(m, n)$ such that $n(1 + \sqrt{5})/2 + m \geq 0$. As a matter of fact, all $AF$-algebras $A$ are classified by their ordered Grothendieck group $K(A)$.

In order to get more applications of the theory to $C^*$-algebras, a considerable generalization was made by Kasparov [Kas], in relation with Atiyah-Singer theorem. In Kasparov's theory, one associates to two $C^*$algebras $A$ and $B$ a new group called $KK(A, B)$ with remarkable formal properties. According to Higson [*H*] for instance, this new theory $KK(A, B)$ - at least for separable $C^*$-algebras, not necessarily unitary - is characterized by the following properties

1. Homotopy invariance: the group is unchanged (functorially) if we replace $A$ or $B$ by the ring of continuous function on $[0, 1]$ with values in $A$ or $B$.

2. The group is $C^*$-stable with respect to the second variable, i.e. we have the functorial isomorphism

$$KK(A, B) \approx KK(A, \mathcal{K} \otimes B),$$

where $\mathcal{K}$ is the $C^*$-algebra of compact operators in a separable Hilbert space.

3. If

$$0 \longrightarrow J \longrightarrow \mathcal{E} \longrightarrow B \longrightarrow 0$$

is a split exact sequence of $C^*$-algebras, we have $KK(A, \mathcal{E}) \approx KK(A, J) \oplus KK(A, B)$.

4. A "composition" (called Kasparov product)

$$KK(A,B) \times KK(B,C) \longrightarrow KK(A,C)$$

may be defined.

Therefore, the correspondence $(A,B) \longmapsto KK(A,B)$ defines a functor from the category of $C^*$-algebras couples $(A,B)$ to the the category of abelian groups, which is an extended definition of the usual $K$-theory in the following sense. We have $KK(\mathbb{C},A) \approx K(A)$ and every morphism $f : A \longrightarrow B$ gives rise to an element of $KK(A,B)$. By Kasparov's product, this induces a morphism from $K(A) = KK(\mathbb{C},A)$ to $KK(\mathbb{C},B) = K(B)$ which is the usual one: to every $A$-module $M$ we associate the $B$-module $f_*(M) = B \otimes_A M$.

A concrete definition of $KK(A,B)$ and of the Kasparov product has been given by Cuntz [Cu]. He describes $KK(A,B)$ as the set of homotopy classes of $*$-homomorphisms

$$qA \longrightarrow \mathcal{K} \otimes B$$

Here $qA$ is the ideal in the free product $A * A$ generated by the elements of type $x * 1 - 1 * x$. One can prove that this definition is equivalent to the first one of Kasparov and also to the definition of Connes-Skandalis ([C1] p. 428-436).

The relation with the Atiyah-Singer theorem is the following. Let $B = \mathbb{C}$ be the field of complex numbers and let $A$ be the algebra of continuous functions on a compact manifold of even dimension. Then, an elliptic operator $D$ on $X$ defines a "Fredholm module" on $A$ [At2] which gives rise to an element of $KK(A,\mathbb{C})$. The image of $1 \in K(A)$ by the associated morphism

$$K(A) \longrightarrow K(\mathbb{C}) = \mathbb{Z}$$

is the index of the operator $D$.

An open problem is to define a group $KK(A,B)$ for general algebras $A$ and $B$ and to find the relation with Algebraic $K$-theory on one side (Lecture 5) and cyclic homology on the other side (Lecture 7).

# Lecture 4 Algebraic $K$-theory of Bass and Milnor Applications

After the success of topological $K$-theory, the algebraists felt challenged to "derive" the $K_0$ group for ANY ring $A$, as it was done for Banach algebras and, hopefully, prove some kind of "Bott periodicity" for these hypothetic $K_n(A)$.

The first problem was of course to define the $K_1$ group in a purely algebraic context. Remember that for a Banach algebra $A$, $K_1^{top}(A)$ (called before $K_1(A)$) was defined either as $K(A(\mathbb{R}))$ or $\pi_0(GL(A)) = injlim_n \pi_0(GL_n(A))$. The first definition does not make any sense a priori for an abstract ring $A$. However, Bass managed to give a meaning to the second definition thanks to the notion of an elementary matrix. Such a matrix - called $e_{ij}^{\lambda}$ - has all entries $= 0$, except for the diagonal entries = 1 and the entry at the spot $(i,j), i \neq j$, which is equal to the scalar $\lambda$. These matrices satisfy the following identities

$$e_{ij}^{\lambda} e_{ij}^{\mu} = e_{ij}^{\lambda+\mu}$$
$$[e_{ij}^{\lambda}, e_{jk}^{\mu}] = e_{ik}^{\lambda\mu} \text{ if } i \neq k$$
$$[e_{ij}^{\lambda}, e_{kl}^{\mu}] = 1 \text{ if } j \neq k \text{ and } i \neq l$$

We call $E(A)$ the subgroup generated by these elementary matrices. One has to think of the elements of $E(A)$ as those "homotopic to the identity matrix".

**DEFINITION/THEOREM** . *The subgroup $G' = E(A)$ is equal to the commutator subgroup $[G,G]$ of $G = GL(A)$. Moreover it is perfect, i.e. $[G',G'] = G'$. We define $K_1(A)$ as the quotient group $G/G'$, in other words the group $G$ made abelian.*

This theorem is quite easy to prove, using the relations above between elementary matrices. One of the key point is to show that a commutator $ghg^{-1}h^{-1}$ is a product of elementary matrices if we stabilize 3 times the size of the matrix, say $n$. Indeed, let us write $Diag(u,v,w)$ for a matrix of size $3n$ with $u$, $v$ and $w$ as $n$-diagonal blocks. Then we have $Diag(ghg^{-1}h^{-1},1,1) = [u,v]$, where $u = Diag(g,1,g^{-1})$ and $v = Diag(h,h^{-1},1)$. On the other hand, a matrix of type

$$\begin{pmatrix} g & 0 \\ 0 & g^{-1} \end{pmatrix}$$

may be written as the product

$$\begin{pmatrix} 0 & 1 \\ -1 & 0 \end{pmatrix} \begin{pmatrix} 0 & -g^{-1} \\ g & 0 \end{pmatrix}$$

We notice finally that the matrix

$$\begin{pmatrix} 0 & -g^{-1} \\ g & 0 \end{pmatrix}$$

may be written as the product of 3 matrices, each of them being a product of elementary matrices, i.e.

$$\begin{pmatrix} 1 & -g^{-1} \\ 0 & 1 \end{pmatrix} \begin{pmatrix} 1 & 0 \\ g & 1 \end{pmatrix} \begin{pmatrix} 1 & -g^{-1} \\ 0 & 1 \end{pmatrix}$$

**Caution:** if $A$ is a Banach algebra, this group $K_1(A)$ does not coincide with its topological analog. For instance, if $A$ is the field of complex numbers $\mathbb{C}$, we have $K_1(\mathbb{C}) \approx \mathbb{C}^*$, whereas the topological analog $\pi_0(GL(\mathbb{C}))$ is equal to 0.

**THEOREM** . *Let us consider a cartesian square of rings with units*

$$\begin{array}{ccc} A & \longrightarrow & A_1 \\ \downarrow & & \downarrow \phi_1 \\ A_2 & \xrightarrow{\phi_2} & A' \end{array}$$

*with $\phi_1$ or $\phi_2$ surjective. Then we have a "Mayer-Vietoris exact sequence" of $K$-groups*

$$K_1(A) \longrightarrow K_1(A_1) \oplus K_1(A_2) \longrightarrow K_1(A') \longrightarrow K(A) \longrightarrow K(A_1) \oplus K(A_2) \longrightarrow K(A')$$

This theorem implies "excision" for the $K_0$-group; in other words, if $I$ is an ideal in a ring $A$, the $K$-group of the functor[9] $\mathcal{P}(A) \longrightarrow \mathcal{P}(A/I)$ only depends on the ideal $I$, i.e. is isomorphic to $Ker(K(\overline{I}) \longrightarrow K(\mathbb{Z}))$, where $\overline{I}$ denotes the $\mathbb{Z}$-algebra $I$ with a unit added.

In the case of a group ring $A = \mathbb{Z}[G]$, the group $K_1$ has applications in differential topology as it was shown in the 60's. A striking one is the $s$-cobordism theorem ([R] p. 86): if $M$ is a compact connected manifold of dimension $n \geq 5$, any $h-$cobordism built on $M$ is homeomorphic (rel. $\partial M$) to a product $M \times [0,1]$. There is just one algebraic condition for this: $Wh(G) = K_1(A)/G$ should be trivial. A spectacular application is the famous theorem of Smale which solves the Poincaré conjecture in dimension $\geq 5$: any manifold of dimension $n \geq 5$ homotopically equivalent to the sphere $S^n$ is indeed homeomorphic to $S^n$.

A more algebraic application is the famous "congruence subgroup problem": if $R$ is a commutative ring and $SL(R)$ the subgroup of $GL(R)$ consisting of matrices of determinant 1, is every normal subgroup of $SL(R)$ a congruence subgroup, i.e. the kernel of a group homomorphism $SL(R) \longrightarrow SL(R/I)$ for a certain ideal $I$ ? An analysis of a "relative" $K_1$ group shows this is the case for instance if $R = \mathbb{Z}$. In general one should have $SK_1(A, I) = 0$ for all ideals $I$([R] p.106).

Let us mention an important result with many applications, due to Bass, Milnor and Serre: if $A$ is the ring of integers in a number field $F$, then $K_1(A) \cong A^*$ by the determinant map. In other words, the map $K_1(A) \longrightarrow K_1(F) \cong F^*$ is injective.

One question we may ask is whether $K_1(A)$ is the $K_0$-group of another ring; this is inspired by Topology where we saw that $K^{-1}(X)$ is essentially $K^0$ of the suspension of $X$. There is indeed a dual notion of the suspension of a space which is the "loop ring" $\Omega A$ of $A$: it is the subring of $A[x]$, consisting of polynomials $P(x)$ such that $P(0) = P(1) = 0$. Under certain conditions (for instance if $A$ is regular noetherian), one can show that $K_1(A) \cong K_0(\Omega A)$. This is the starting point of the so-called "Karoubi-Villamayor" $K$-theory which will be sketched in the next lecture.

**Exercise:** Let $\Omega^+(A)$ be the $A-$algebra $\Omega A$ with a unit added. Show that $\Omega^+(A)$ may be identified with the coordinate ring of the following cubic

$$\Omega^+(A) \cong A[u,v]/(u-v)^3 - uv$$

**Exercise** (Milnor): Let $A$ be the Banach algebra of continuous functions on a

[9] Here we should be more precise about the definition of the Grothendieck group of an additive functor $\varphi : \mathcal{C} \longrightarrow \mathcal{C}$. It is the quotient of the free group generated by isomorphism classes of triples $(E, F, \alpha)$, where $E$ and $F$ are objects of $\mathcal{C}$ and $\alpha : \varphi(E) \longrightarrow \varphi(F)$ is an isomorphism, by the following relations:
$(E, F, \alpha) + (E', F', \alpha') = (E \oplus E', F \oplus F', \alpha \oplus \alpha')$
$(E, F, \alpha) + (F, G, \beta) = (E, G, \beta.\alpha)$.

compact space $X$ with complex values. Then show that $K_1(A) \cong A^* \oplus \pi_0(SL(A))$.

**Remark.** In contrast with the situation of the $K_0$-group, the definition of $K_1(I)$ when $I$ is a ring without unit is unclear. For instance, if we define $K_1(I)$ as $Ker[K_1(I^+) \longrightarrow K_1(k)]$ if $I$ is a $k-$algebra, this definition depends strongly on the choice of $k$.

For certain special rings $I$ however, there is a "good" definition of $A$, for instance the ideal $\mathcal{K}$ of compact operators in a Hilbert space, or the ring of continuous functions from a compact space $X$ to $\mathcal{K}$.

The next step in Algebraic $K-$theory was done by Milnor who gave an algebraic definition of $K_2(A)$. He first introduced the (infinite) Steinberg group $ST(A)$ which is generated by elements $x_{ij}^{\lambda}$, $i \neq j$, subject to the "universal" relations between elementary matrices, i.e.

$x_{ij}^{\lambda} x_{ij}^{\mu} = x_{ij}^{\lambda+\mu}$
$[x_{ij}^{\lambda}, x_{jk}^{\mu}] = x_{ik}^{\lambda\mu}$ if $i \neq k$
$[x_{ij}^{\lambda}, x_{kl}^{\mu}] = 1$ if $j \neq k$ and $i \neq l$

**Note:** for convenience, we shall sometimes write these symbols as $x_{ij}(\lambda)$.

There is an obvious epimorphism

$$ST(A) \longrightarrow E(A)$$

As we said before, the group $E(A)$ is perfect (i.e. is equal to its own commutator subgroup $[E(A), E(A)]$ ) and Milnor proved that the exact sequence

$$1 \longrightarrow K_2(A) \longrightarrow ST(A) \longrightarrow E(A) \longrightarrow 0$$

is in fact the universal central extension of the perfect group $E(A)$. In particular, $K_2(A)$ is abelian and may be identified with $H_2(E(A); \mathbb{Z})$, the second homology group of $E(A)$ with $\mathbb{Z}$ coefficients. Analogously, one may remark that $K_1(A)$ is the homology group $H_1(GL(A); \mathbb{Z})$.

An important part of the structure of $K_2$ is defined by Milnor's cup-product

$$K_1(A) \times K_1(A) \longrightarrow K_2(A)$$

when $A$ is commutative. In particular, if $u$ and $v \in A^* \subset K_1(A)$, the *Steinberg symbol* $\{u, v\}$ is the cup-product of $u$ and $v$ in $K_2(A)$. It satisfies the following relations

(a) $(u, 1-u) = 1$ if $u$ and $1-u$ belong to $A^*$
(b) $(u, -u) = 1$ if $u \in A^*$

If $A$ is a field, the second relation follows from the first.

**THEOREM (Matsumoto).** *Let $F$ be a (commutative) field. The Steinberg symbol identifies $K_2(F)$ with the quotient of $F^* \otimes_{\mathbb{Z}} F^*$ by the subgroup generated by the relations $u \otimes (1-u)$ for $u \neq 0, 1$.*

This theorem is the starting point for applications of $K_2$ in Algebra and Number Theory, in relation with the Brauer group and Galois cohomology. Let $\overline{F}$ be the separable closure of $F, G$ the Galois group of this extension and $\mu_n$ the multiplicative group consisting of $n$-roots of unity in $\overline{F}$. Then, the exact sequence

$$0 \longrightarrow \mu_n \longrightarrow \overline{F}^* \longrightarrow \overline{F}^* \longrightarrow 0$$

induces an isomorphism $H^1(G;\mu_n) \cong F^*/F^{*n}$ , since $H^1(G;\overline{F}^*) = 0$ by a famous theorem of Hilbert. By a classical argument, the composition

$$s : F^* \times F^* \longrightarrow F^*/F^{*n} \times F^*/F^{*n} \cong H^1(G;\mu_n) \times H^1(G;\mu_n) \longrightarrow H^2(G;(\mu_n)^{\otimes 2})$$

where the last map is the cup-product in Galois cohomology, is a Steinberg symbol (i.e. $s(u, 1-u) = 0$ if $u \neq 0, 1$). Therefore, $s$ induces a homomorphism from $K_2(F)/nK_2(F)$ to $H^2(G;(\mu_n)^{\otimes 2})$. The following classical theorem ([Sr] p. 149) is one of the nicest in the subject:

**THEOREM (Merkurjev-Suslin).** *The homomorphism*

$$K_2(F)/nK_2(F) \longrightarrow H^2(G;(\mu_n)^{\otimes 2})$$

*defined above is an isomorphism.*

As a concrete application let us consider the Brauer group of $F$, $Br(F)$, generated (multiplicatively) by central simple algebras $A$ with the relation $A \sim B$ iff $M_n(A) \cong M_s(B)$ for some numbers $n$ et $s$. It is well known that $Br(F)$ is canonically isomorphic to the Galois cohomology group $H^2(G, \overline{F}^*)$.

**COROLLARY .** *Let us assume $\mu_n \subset F$. Then $H^2(G;(\mu_n)^{\otimes 2}) \cong H^2(G;\mu_n) \cong Br(F)_n$, the $n$-torsion of the Brauer group. Therefore, we have an isomorphism*

$$K_2(F)/nK_2(F) \longrightarrow Br(F)_n$$

A striking example is the case when $n = 2$ and $F$ is of characteristic $\neq 2$. Then, the central simple algebra associated to the symbol $\{a, b\}$ is the quaternion algebra generated by the symbols $i$ and $j$ with the relations $i^2 = a, j^2 = b$ and $ij = -ji$. This corollary implies that any element of the 2-torsion in $Br(F)$ is stably isomorphic to a product of quaternion algebras, an open problem before Merkurjev and Suslin proved their theorem.

As another example, let us consider the case when $A$ is a Banach algebra. As we have seen in a previous lecture, there is a topologically $K_2^{top}(A)$ which is simply the fundamental group $\pi_1(GL(A))$. In order to see the relation with the group $K_2(A)$, we send a generator $x_{ij}^{\lambda}$ of the Steinberg group to the path $t \longmapsto x_{ij}^{\lambda t}$. The relations

$$x_{ij}^{\lambda}x_{ij}^{\mu} = x_{ij}^{\lambda+\mu}$$
$$[x_{ij}^{\lambda}, x_{kl}^{\mu}] = 1 \text{ if } j \neq k \text{ and } i \neq l$$

still hold if we replace $\lambda$ and $\mu$ by $\lambda t$ and $\mu t$. Now if we compute the commutator $[x_{ij}^{\lambda t}, x_{jk}^{\mu t}]$ for $i \neq k$, we find the path $x_{ij}^{\lambda\mu t^2}$ which is homotopic (end points fixed) to the path $x_{ij}^{\lambda\mu t}$. This shows that the previous correspondence induces a well defined homomorphism:

$$\varphi : ST(A) \longrightarrow E(GL(A))/\mathcal{R}$$

Here $EX$ denotes in general the path space of $X$, i.e. the set of continuous paths $f : [0,1] \longrightarrow X$ such that $f(0) = I$, the unit matrix, and $\mathcal{R}$ the equivalence relation defined by homotopy with fixed end points. Since $K_2(A)$ is the kernel of the homomorphism $ST(A) \longrightarrow GL(A)$, we see immediately that $\varphi(x)$ is a loop in $GL(A)$ if $x \in K_2(A)$ and that $\varphi$ induces a well defined homomorphism

$$\varphi : K_2(A) \longrightarrow \pi_1(GL(A)) = K_2^{top}(A)$$

In general, $\varphi$ is neither an injection nor a surjection. As a typical example, if we choose $A$ to be the field of complex numbers $\mathbb{C}$, we see that $\varphi$ is reduced to 0 since its image is contained in $\pi_1(SL(\mathbb{C})) = 0$. Therefore, the kernel of $\varphi$ is the full group $K_2(\mathbb{C})$, which is known to be an uncountable group ([M]).

There are however Banach algebras $A$ such that $\varphi$ is an isomorphism. A typical example is the algebra $\mathcal{K}$ of compact operators in a Hilbert space or, more generally, the algebra of continuous functions $X \longrightarrow \mathcal{K}$, where $X$ is a compact space. A technical problem appears here since these algebras have no unit. However, as it was proved in [SW][10], these algebras satisfy excision and we may choose to add a unit the way we want, as for the group $K_0$. For these algebras, Bott periodicity holds: $K_2(A) \cong K_0(A)$, which is quite exceptional in Algebraic $K$-theory.

The previous considerations suggest to replace topological homotopy by algebraic homotopy for the definition of the group $\pi_1(GL(A))$ (A being now a discrete ring). An "algebraic loop" is just an element $\sigma$ of $GL(A[t])$ such that $\sigma(0) = \sigma(1)$. Two algebraic loops $\sigma_0$ and $\sigma_1$ are called homotopic if there is an element $\sum$ of $GL(A[t,u])$ such that

$$\sum(t,0) = \sigma_0(t), \sum(t,1) = \sigma_1(t)$$

$$\sum(0,u) = \sum(1,u) = 1$$

This is an equivalence relation because $GL(A)$ is a group and we define the algebraic fundamental group $\pi_1^{alg}(GL(A))$ as the group of homotopy classes of

[10]See also M. Karoubi: Homologie des groups discrets associés à des algèbres d'opérateurs. Journal of Operator Theory. 15, p. 109-161(1986).

algebraic loops (like in Topology). As before, we have an homomorphism

$$K_2(A) \longrightarrow \pi_1^{alg}(GL(A))$$

which is an isomorphism if $A$ is a regular Noetherian ring. This gives an alternative definition of $K_2$, closer to Topology for this type of rings.

From another viewpoint, we also have a cup-product in Topological $K$-theory (for commutative Banach algebras)

$$K_1^{top}(A) \times K_1^{top}(A) \longrightarrow K_2^{top}(A)$$

The obvious diagram

$$\begin{array}{ccc} K_1(A) \times K_1(A) & \longrightarrow & K_2(A) \\ \downarrow & & \downarrow \\ K_1^{top}(A) \times K_1^{top}(A) & \longrightarrow & K_2^{top}(A) \end{array}$$

is then commutative. As a typical example, it is easy to show that $K_1(\mathbb{Z}) = \mathbb{Z}/2$ and the composition $K_1(\mathbb{Z}) \longrightarrow K_1(\mathbb{R}) \longrightarrow K_1^{top}(\mathbb{R})$is injective. It follows that the cup-product of the generator of $K_1(\mathbb{Z})$ by itself is a non trivial element of $K_2(\mathbb{Z})$, because the same is true for the topological $K$-theory of $\mathbb{R}$. As a matter of fact, as it was proved in Milnor's book, we have $K_2(\mathbb{Z}) = \mathbb{Z}/2$.

There are many other examples where we can compute the group $K_2(A)$. For finite fields $F_q$ for instance, Milnor showed that $K_2(F_q) = 0$. Another fundamental example is the case of a number field $F$ and its ring of integers $A$. We then have a "localization" exact sequence:

$$K_2(A) \longrightarrow K_2(F) \longrightarrow \oplus_{\mathcal{P}} K_1(A/\mathcal{P}) \longrightarrow K_1(A) \longrightarrow K_1(F)$$

where $\mathcal{P}$ runs over all the prime ideals of $A$. As a matter of fact, by a famous result of Bass, Milnor and Serre quoted in the previous lecture, the map $K_1(A) \longrightarrow K_1(F) \cong F^*$ is injective and we therefore have a well defined surjective map $K_2(F) \longrightarrow \oplus K_1(A/\mathcal{P}) \cong \oplus(A/\mathcal{P})^*$. Each component

$$K_2(F) \longrightarrow (A/\mathcal{P})^*$$

is called the tame symbol and may be defined explicitly, using the $\mathcal{P}$-adic valuation.

Finally, there is a "fundamental theorem" due to Bass, Heller and Swan which enables us to compute the Algebraic $K$-theory $K_i$ (for $i = 1$ and 2) of the ring of Laurent polynomials $A[t, t^{-i}]$ in terms of the groups $K_{i-1}$ of $A$. More precisely, we have an exact sequence

$$0 \longrightarrow K_i A \longrightarrow K_i(A[t]) \oplus K_i(A[t^{-i}]) \longrightarrow K_i(A[t, t^{-i}]) \longrightarrow K_{i-1}(A) \longrightarrow 0$$

If $A$ is a regular Noetherian ring there is a "homotopy invariance" (first proved by

Grothendieck) i.e. $K_i(A) \cong K_i(A[t])$ for $i = 0, 1, 2$. Therefore, the previous exact sequence reduces to the isomorphism

$$K_i(A[t, t^{-i}]) \cong K_i(A) \oplus K_{i-1}(A)$$

in this case. We shall come back to this matter for the definition of the negative $K$-groups in the next lecture.

As for the functor $K_1$ there is an application of $K_2$ in differential topology which is due to Hatcher and Wagoner. This application is related to pseudo-isotopy classes of diffeomorphisms on manifolds. As it is well known, isotopy classes of diffeomorphisms of the sphere $S^n$ for instance parametrize the set of differentiable structures on $S^{n+1}$ for $n \geq 5$. However, it is difficult to tell in practice when two diffeomorphisms $h_0$ and $h_1$ of a manifold $M$ are isotopic. There is a weaker notion of "pseudo-isotopy" which is a diffeomorphism of a cyclinder $M \times [0, 1]$, whose restrictions to $M \times \{0\}$ and $M \times \{1\}$ give $h_0$ and $h_1$ respectively. Let us call $\mathcal{P}(M)$ the space of pseudo-isotopies of $M$, i.e. diffeomorphisms of $M \times [0, 1]$ which restrict to the identity on $M \times \{0\}$.

Before stating the result about this space $\mathcal{P}(M)$, we need a preliminary definition: in the Steinberg group $St(A)$ of a group algebra $A = \mathbb{Z}[G]$, we define $w_{ij}(u) = x_{ij}(u).x_{ji}(-u^{-1}).x_{ij}(u)$ and $W_G$ the subgroup of $St(A)$ generated by the $w_{ij}(u)$. Finally, we define

$$Wh_2(G) = K_2(A)/K_2(A) \cap W_G$$

**THEOREM (Hatcher-Wagoner).** ([R] p. 242) *Let $M$ be a smooth compact connected manifold without boundary, of dimension $n \geq 5$ and with fundamental group $G$. Then there is a surjection $\pi_0(\mathcal{P}(M)) \longrightarrow Wh_2(G)$.*

In some cases, the kernel has been computed and identified with a group related to $Wh_1(G)$.

# Lecture 5 Higher Algebraic $K$-theory Some computations

After the definition of $K_1(A)$ and $K_2(A)$ by Bass and Milnor, the problem was open at the beginning of the 70's for a "good" definition of $K_n(A)$ for $n \geq 3$. It is now widely accepted that this definition was proposed by Quillen at the International Congress of Mathematicians in Nice (1970). However, other definitions were proposed at the same time and are in fact isomorphic to Quillen's definition in favourable cases. We shall discuss briefly one of them at the end of the lecture.

Let us point out first that there is no "axiomatic" definition of these groups $K_n$ in the spirit of what has been done for Banach algebra for instance (Lecture 2). More precisely, Swan has pointed out the following disturbing fact: we cannot have a natural Mayer-Vietoris exact sequence associated to a cartesian square of rings with units

$$\begin{array}{ccc} A & \longrightarrow & A_1 \\ \downarrow & & \downarrow \phi_1 \\ A_2 & \xrightarrow{\phi_2} & A' \end{array}$$

with $\phi_1$ and $\phi_2$ surjective; in other words, in exact

$$K_{n+1}(A) \to K_{n+1}(A_1) \oplus K_{n+1}(A_2) \to K_{n+1}(A') \\ \to K_n(A) \to K_n(A_1) \oplus K_n(A_2) \to K_n(A')$$

with $K_0$ isomorphic to the Grothendieck group.

This means that instead of waiting for a "good" axiomatic definition of $K_n(A)$, Quillen's proposal is justified by the nice theorems and applications proved with it (the Merkurjev-Suslin theorem in the previous lecture is for instance a good example of an application; it could not be proved without Quillen's theory).

The definition of $K_n(A)$ uses the Algebraic Topology machinery, as we have seen in Berrick's series of lectures: more precisely, we have $K_n(A) = \pi_n(BGL(A)^+)$, for $n > 0$, where $BGL(A)^+$ is a space obtained from the classifying space of the infinite general linear group $GL(A)$ by adding cells of dimension 2 and 3. For a change, we shall adopt a more geometric viewpoint, using the concept of "virtual" flat $A$–bundles.

Firstly, we define a flat $A$–bundle on a space $X$ as a covering $P \longrightarrow X$ such that each fiber is an f.g.p. (= finitely generated projective) $A$-module. Note that if $A = \mathbb{C}$, we recover the usual notion of flat complex vector bundle.

A *virtual* flat $A$–bundle on a space $X$ (assumed to be a finite $CW$-complex) is given by

1) an acyclic map $Y \longrightarrow X$ (i.e. a map inducing an isomorphism in homology with local coefficients or, equivalently, such that its homotopy fiber is an acyclic space).

2) a flat $A$–bundle $E$ on $Y$.

Two virtual $A$–bundles $E$ and $E'$ on $Y$ and $Y'$ respectively are called equivalent if we can find a commutative diagram (up to homotopy)

$$\begin{array}{ccc} Y & \longrightarrow & X \\ \downarrow & & \uparrow \\ Z & \longleftarrow & Y' \end{array}$$

with where $g : Z \longrightarrow X$ is an acyclic map, together with a flat $A$-bundle $F$ on $Z$ such that the pull-back of $F$ through the map $Y \longrightarrow Z$ (resp. $Y' \longrightarrow Z$) is isomorphic to $E$ (resp. $E'$). Two virtual flat $A$-bundles (for instance $E$ and $E'$ with the previous notations) may be added by taking the homotopy fiber product $T$ of $Y$ and $Y'$ over $X$ and adding $\gamma^*(E)$ and $\gamma'^*(E')$, where $\gamma$ and $\gamma'$ are the obvious projections of $T$ on $Y$ and $Y'$. This addition is compatible with the equivalence relation.

**DEFINITION/THEOREM .** [K2] *Let us call $K_A(X)$ the Grothendieck group of equivalence classes of virtual flat A-bundles on $X$. Then, as a functor of the finite CW-complex $X$, $K_A(X)$ is representable and we have the following isomorphism*

$$K_A(X) \cong [X, K_0(A) \times BGL(A)^+]$$

*In particular, $K_A(S^n) \cong K_0(A) \oplus K_n(A)$, where $K_n(A)$ are the $K$-groups of Quillen.*

**Remarks.** This theorem should be related to the analogous result in topological $K$-theory (for complex vector bundles): $K^{top}(S^n) \cong \mathbb{Z} \oplus K_n^{top}(\mathbb{C})$. Note also that this definition both includes $K_0$ and $K_n$ for $n > 0$.

As in topological $K$-theory, one can define tensor products of virtual flat $A$-bundles and construct this way a bilinear pairing

$$K_A(X) \times K_B(Z) \longrightarrow K_{A \otimes B}(X \times Z)$$

In particular, if $A$ is commutative $K_*(A)$ becomes a graded ring and one sees in more geometrical terms Milnor's cup-product

$$K_1(A) \times K_1(A) \longrightarrow K_2(A)$$

The $\lambda^k$ and $\psi^k$ operations also extend to the ring $K_A(X)$ (if $A$ is commutative).

The first $K$-groups were calculated by Quillen of course. If $A$ is a finite field with $q$ elements, Quillen proved that

$$K_{2n}(F_q) = 0 \text{ for } n > 0$$

$$K_{2n+1}(F_q) \cong \mathbb{Z}/(q^n - 1)\mathbb{Z}$$

On the other hand, as we have seen in Berrick's lectures, $K_*(A) \otimes \mathbb{Q}$ is the primitive part of the homology of $GL(A)$ (with rational coefficients). If $A$ is the ring of integers in a number field, Quillen has shown on one side that $K_i(A)$ is a finitely generated abelian group; on the other side Borel has computed the cohomology of $SL(A)$ with complex coefficients. Comparing these results, one finds the following isomorphisms with $i > 0$ (mod. finite abelian groups)

$$K_{2i}(A) = 0$$

$$K_{4i+1}(A) = \mathbb{Z}^{r_1 + r_2}$$

$$K_{4i-1}(A) = \mathbb{Z}^{r_2}$$

where $r_1$ and $r_2$ are determined by the isomorphism

$$A \otimes_{\mathbb{Z}} \mathbb{R} \cong \mathbb{R}^{r_1} \times \mathbb{C}^{r_2}$$

Following Grothendieck and Bass, Quillen has succeeded in generalizing to higher $K$-groups some theorems proved for $K_0$ and $K_1$. Let us mention a few of them.

**THEOREM (homotopy invariance).** *Let $A$ be a regular Noetherian ring. Then $K_n(A[t]) \cong K_n(A)$. Therefore, we also have $K_n(A[t_1, ..., t_r]) \cong K_n(A)$.*

Note that if $A$ is a field, it was an old question of Serre whether an f.g.p. module over $A[t_1, ..., t_r]$ is free. This question was answered in the affirmative by Quillen and Suslin in the 70's.

**THEOREM ("fundamental" theorem).** *Let $A$ be a regular Noetherian ring and let $A[t, t^{-1}]$ be the ring of Laurent polynomial with coefficients in $A$. Then we have the exact sequence*

$$0 \to K_n(A) \to K_n(A[t]) \oplus K_n(A[t^{-1}]) \to K_n(A[t, t^{-1}]) \to K_{n-1}(A) \to 0$$

**THEOREM (one form of the "dévissage" and localization theorems).** *Let $A$ be a regular Noetherian ring and $f$ a non zero divisor such that $A/f$ is regular Noetherian. Then we have an exact sequence*

$$K_{n+1}(A) \to K_{n+1}(A_f) \to K_n(A/f) \to K_n(A) \to K_n(A_f)$$

*where $A_f$ is the ring $A$ localised at $f$ (i.e. making $f$ invertible).*

Another variant of this theorem is the following: let $A$ be a Dedekind ring and $F$ its field of fractions. Then we have an exact sequence

$$K_{n+1}(A) \to K_{n+1}(F) \to \oplus_{\mathcal{P}} K_n(A/p) \to K_n(A) \to K_n(F)$$

where $\mathcal{P}$ runs through the set of all prime ideals in $A$.

**Remarks.** All these theorems may be paraphrased with a parameter space $X$ instead of the sphere $S^n$ by introducing a suitable fibration. For instance, the homotopy invariance may be restated as

$$K_{A[t]}(X) \cong K_A(X)$$

for any regular Noetherian ring $A$. We should also notice that Quillen was able to prove his theorems by using another construction in Algebraic $K$-theory, called the "$Q$-construction". This has been detailed in Berrick's lectures (see also [FW]).

Having stated all these theorems, one may ask what the definition of $K_n(A)$ for $n < 0$ should be, so that the previous theorems can be extended to all values of $n \in \mathbb{Z}$. This definition is due independently to Bass and the author of these notes. Since we are going to use the second definition in Hermitian $K$-theory, let us choose the latter. The key ingredient is the use of "infinite" matrices. More precisely, let us define the "cone" of $A$, called $CA$, as the set of infinite matrices such that in each row and each column we have a finite number of non zero elements chosen among a finite number of elements in $A$. Clearly $CA$ is a ring by matrix multiplication, containing the finite matrices (i.e. whose entries are 0, except for a finite number) as a 2-sided ideal. We define the suspension $SA$ of $A$ as the quotient ring. This definition may be iterated and $S^n(A)$ will denote the $n^{th}$ suspension of $A$.

**DEFINITION/THEOREM .** *The group $K_{-n}(A)$ is by definition $K(S^nA)$. All the theorems stated before are true for the functors $K_n$ , $n \in \mathbb{Z}$. Note however that $K_n(A) = 0$ for $n < 0$ if $A$ is regular Noetherian.*

A theorem which is missing in the picture is "Bott periodicity". Of course, it does not work in general: for finite fields for instance, we are far from getting the same answer for $K_n$ and $K_{n+\alpha}$. However, we can get a partial answer for other types of $K$-groups which we shall introduce now. As a matter of fact, since we know that Algebraic $K$-theory is represented by a spectrum $\mathcal{K}(A)$ defined through the $BGL(S^nA)^+$, there is a well known procedure in Algebraic Topology taking this spectrum *mod.* $n$ for any integer $n$. A more concrete way is to consider the Puppe sequence associated to a map of degree $n$ between spheres

$$S^r \longrightarrow S^r \longrightarrow M(n,r) = X \longrightarrow S^{r+1} \longrightarrow S^{r+1}$$

Therefore, $X$ (so-called Moore space) is a space with 2 cells of dimension $r$ and $r+1$ respectively, the second cell being attached to the first by a map of degree $n$. If $n = 2$ and $r = 1$ for instance, $X$ is just the real projective space of dimension 2.

We now define $K$-theory with coefficients in $\mathbb{Z}/n$, denoted by $K_{r+1}(A;\mathbb{Z}/n)$, as the quotient group $K_A(X)/K_0(A)$. From the Puppe sequence, we get an exact sequence

$$K_{r+1}(A) \longrightarrow K_{r+1}(A) \longrightarrow K_{r+1}(A;\mathbb{Z}/n) \longrightarrow K_r(A) \longrightarrow K_r(A)$$

where the arrows between the groups $K_i(A)$ are the multiplication by $n$.

Here is a fundamental theorem of Suslin which is the true analog of Bott periodicity.

**THEOREM .** *Let $F$ be an algebraically closed field and let $n$ be a number prime to the characteristic of $F$. Then there is a canonical isomorphism of graded rings $K_{2r}(F;\mathbb{Z}/n) \cong (\mu_n)^{\otimes r}$ and $K_{2r+1}(F;\mathbb{Z}/n) = 0$.*

A sketch of the proof of this theorem may be found (among other things) in [FW], Lecture 8.

With this theorem in mind, one may naturally ask if there is a way to compute $K_*(F;\mathbb{Z}/n)$ for an arbitrary field (not necessarily algebraically closed). If $F$ is the field of real numbers, and if we work in topological $K$-theory instead, we know that it is not an easy task since we get an 8-periodicity which looks mysterious compared to the 2-periodicity of complex $K$-theory. All these questions are in fact related to the so-called homotopy fixed point set relative to a group action. Roughly speaking, we know[11] that $\mathcal{K}(F)$ is the fixed point set of $\mathcal{K}(\overline{F})$, where $\overline{F}$ denotes the separable closure of $F$, with respect to the action of the Galois group $G$. We have a fundamental map

$$\phi : \mathcal{K}(F) = \mathcal{K}(\overline{F})^G \longrightarrow \mathcal{K}(\overline{F})^{hG}$$

[11] We now denote by $\mathcal{K}(F)$ the classifying space of Algebraic $K$-theory mod. $n$: its homotopy groups are the $K$-groups with coefficients in $\mathbb{Z}/n$.

where $\mathcal{K}(\overline{F})^{hG}$ is the "homotopy fixed point set" of $\mathcal{K}(\overline{F})$, i.e. the set of equivariant maps $EG \longrightarrow \mathcal{K}(\overline{F})$ where $EG$ is the "universal" principal $G$ bundle over $BG$. One version of the Lichtenbaum-Quillen conjecture is that $\phi$ induces an isomorphism on the homotopy groups $\pi_i$ for $i > n$-cohomological dimension of $G = d_n$. This implies the existence of a spectral sequence

$$E_r^{p,q} = H^p(G; \mu_n^{\otimes q}) \to K_{2q-p}(F; \mathbb{Z}/n)$$

if the characteristic of the field does not divide $n$ and if $2q - p > d_n$. For instance, $d_n = 1$ if $n$ is odd and if $F$ is a number field: then the spectral sequence degenerates and we get a direct link between Algebraic $K$-theory and Galois cohomology, quite interesting in Number Theory. By the recent work of Voevodsky, the conjecture is true for $n = 2$. We shall hear more about this in Morel's lectures.

We should notice that the topological analog of this conjecture is true: one should replace the classifying space of Algebraic $K$-theory by the classifying space of (complex) topological $K$-theory, where $\mathbb{Z}/2$ acts by complex conjugation. Then the fixed point set (i.e. the classifying space of *real* topological $K$-theory) has the homotopy type of the homotopy fixed point set.

Although the definition of the $K_n(A)$ is getting relatively old now, they are still quite difficult to compute, even for simple rings as the ring of integers $\mathbb{Z}$ (although we know these groups rationally by the work of Borel on the rational cohomology of arithmetic groups). Thanks to the remarkable work of Voevodsky followed by Bökstedt, Rognes and Weibel, we can now compute the 2 primary torsion of $K_n(\mathbb{Z})$ through the following homotopy cartesian square (where $\mathbb{Z}' = \mathbb{Z}[1/2]$)

$$\begin{array}{ccc} BGL(\mathbb{Z}')^+_{\#} & \longrightarrow & BGL(\mathbb{R})_{\#} \\ \downarrow & & \downarrow \\ BGL(\mathbb{F}_3)^+_{\#} & \longrightarrow & BGL(\mathbb{C})_{\#} \end{array}$$

where the symbol $\#$ means 2-adic completion and where $BGL(\mathbb{R})$ and $BGL(\mathbb{C})$ are the classifying spaces of the *topological* groups $GL(\mathbb{R})$ and $GL(\mathbb{C})$ respectively. From this homotopy cartesian square, Rognes and Weibel found the following results (mod. a finite odd torsion group and with $n > 0$ for the first 2 groups and $n \geq 0$ for the others)

$K_{8n}(\mathbb{Z}) = 0$
$K_{8n+1}(\mathbb{Z}) = \mathbb{Z} \oplus \mathbb{Z}/2$
$K_{8n+2}(\mathbb{Z}) = \mathbb{Z}/2$
$K_{8n+3}(\mathbb{Z}) = \mathbb{Z}/16$
$K_{8n+4}(\mathbb{Z}) = 0$
$K_{8n+5}(\mathbb{Z}) = \mathbb{Z}$
$K_{8n+6}(\mathbb{Z}) = 0$
$K_{8n+7}(\mathbb{Z}) = \mathbb{Z}/2^r$ where $2^r$ is the 2 primary component of the number $4n + 4$.

One should also mention (in reverse historical order) that Milnor gave another definition of the groups $K_n(F)$, $n > 2$, for a *field* $F$. This definition does not agree with Quillen's definition in general and we should denote it by $K_n^M(F)$: it is the quotient of the tensor product $F^* \otimes_{\mathbb{Z}} \cdots \otimes_{\mathbb{Z}} F^*$ ($n$ factors) by the subgroup generated by the tensors $x_1 \otimes \cdots \otimes x_n$ such that $x_{i+1} = 1 - x_i$ for a certain $i$. By the usual cup-product in Algebraic $K$-theory, there is an obvious map

$$K_n^M(F) \longrightarrow K_n(F)$$

which is an isomorphism if $n = 1$ or 2 according to Matsumoto's theorem. The next interesting case is $n = 3$. Then we have the following nice theorem of Suslin:

**THEOREM .** *The map $K_3^M(F) \longrightarrow K_3(F)$ is injective and the quotient $K_3(F)_{ind}$ fits into an exact sequence*

$$0 \longrightarrow \mu \longrightarrow K_3(F)_{ind} \longrightarrow B(F) \longrightarrow 0$$

In this sequence, $B(F)$ is the "Bloch group" i.e. the group which fits into the exact sequence

$$0 \longrightarrow B(F) \longrightarrow P(F) \xrightarrow{r} \Lambda^2(F^*) \longrightarrow K_2(F) \longrightarrow 0$$

Here $P(F)$ is the quotient of the free group generated by elements [x] in $F^*$ by the $\lambda$ group generated by the "cross-ratio relations":

$$[x] - [y] + [\frac{y}{x}] - [\frac{1-x^{-1}}{1-y^{-1}}] + [\frac{1-x}{1-y}] = 0$$

and $r([x] = x \wedge (1-x)$. Moreover, $\mu$ is a torsion group which is deduced from the group $\mu_F$ of roots of unity in $F$. If, following [FW], we write $\mu_F(2)$ as $Tor(\mu F, \mu F)$, $m$ is the group $\mu_F(2)$ if the characteristic is 2 and the non trivial extension of $\mathbb{Z}/2$ by $\mu_F(2)$ otherwise.

It is an open problem to describe the other $K_n$-groups of a field in such a nice algebraic way...

One of the most striking applications of Algebraic $K$-theory to Differential Topology is due to the hard work of Waldhausen which we have not touched in this report. For instance, in the Lecture 4 we introduced the space $\mathcal{P}(M)$ of pseudo-isotopies of $M$, i.e. diffeomorphisms $q$ of $M \times [0,1]$ which restrict to the identity on $M \times 0$. If $M$ has a boundary, we ask moreover that $q$ is the identity on $\partial M \times [0,1]$.

**THEOREM .** *Let $M$ be the ball $B^n$ of dimension $n$. Then, we have an isomorphism*

$$\pi_i(\mathcal{P}(M)) \otimes_{\mathbb{Z}} \mathbb{Q} \cong K_{i+2}(\mathbb{Z}) \otimes_{\mathbb{Z}} \mathbb{Q}$$

*for $0 \leq i \leq n/6 - 7$. Therefore, mod. torsion elements, and within this range of degrees, we have $\pi_i(\mathcal{P}(M)) = \mathbb{Z}$ for $i = 4\delta - 1$ and $\pi_i(\mathcal{P}(M)) = 0$ for $i \neq 4\delta - 1$.*

The same type of results applies to the group of diffeomorphisms of $B^n$ mod. its boundary. We get $\pi_i(Diff(B^n, \partial) \otimes_{\mathbb{Z}} \mathbb{Q} = K_{i+2}(\mathbb{Z}) \otimes_{\mathbb{Z}} \mathbb{Q}$ for the same range $0 \leq i \leq n/6 - 7$ and $n$ odd if $i = 4k - 1$.

For Banach algebras $A$, there are now two competing theories: topological $K$-theory on one side which we denoted by $K_n^{top}(A)$ previously and Algebraic $K$-theory $K_n(A)$. There is a canonical homomorphism

$$\phi_n : K_n(A) \longrightarrow K_n^{top}(A)$$

If $A$ is the field of complex numbers and if we take $K$-theory with finite coefficients, we know by Suslin's theorem that this map is an isomorphism. Other Banach algebras of interest in functional analysis are the so-called "stable" $C^*$-algebras; this means that $A \otimes \mathcal{K}$ is isomorphic to $A$ (for a suitable notion of tensor product), where $\mathcal{K}$ is the ring of compact operators in a Hilbert space. A surprising result proved by the author of these notes at the end of the 80's is that $\phi_n$ is an isomorphism for $n \leq 2$ ($n$ may be negative). The question arose then whether $\phi_n$ is also an isomorphism for $n > 2$. This question was answered positively by Suslin and Wodzicki [SW]. In fact, they proved excision for this kind of rings $A$ (without unit): $K_n(A)$ does not depend of the ambiant ring where the ideal $A$ lives. The idea of the proof (assuming excision) is quite simple and is based on cup-products with Bott elements as it was explained in the second lecture.

Let us conclude this lecture by trying to answer a frustration the algebraists might feel at this point. Is there a way to define $K_n(A)$ which avoids the Algebraic Topology machinery? For *regular noetherian rings*, this definition was indeed proposed by Villamayor and myself [KV] and is based on polynomial homotopies (much related to the $A^1$-homotopy in Morel's lectures). The definition is the following. Let us consider the ring

$$A_n = A[x_0, x_1, ..., x_n]/(x_0 + x_1 + \cdots + x_n - 1)$$

and the subgroup[12] $G_n$ of $GL(A_n)$ consisting of matrices which are equal to the identity matrix if one of the variables $x_i = 0$ (for $i < n$). We define a homomorphism $d_n : G_n \longrightarrow G_{n-1}$ by setting $x_n = 0$ (if $n < 0$, $G_n = \{1\}$ by convention). Then $Kerd_n/Imd_{n+1}$ is naturally isomorphic to $K_n(A)$ for $n > 0$.

**Exercise:** check this for $n = 1$.

We can put these last considerations in a more sophisticated framework by considering $A_n$ , $n \in \mathbb{N}$, as a simplicial ring and constructing its classifying space $BGL(A_*)$ as it is usual in simplicial topology ($A$ is now an arbitrary ring). If we view $A$ as a "constant" simplicial ring, we have a map

$$\theta : BGL(A) \longrightarrow BGL(A_*)$$

Now $BGL(A_*)$ is a connected space with a fundamental group easily seen to be $GL(A)/U(A)$, where $U(A)$ is the group generated by unipotent matrices. Since we have the inclusions

$$[GL(A), GL(A)] = E(A) \subset U(A) \subset GL(A)$$

[12] We should notice that $G*$ is a simplicial group. In modern language, we just say that $K_{n+1}(A)$ is the $n$-homotopy group of this simplicial group.

the quotient $GL(A)/U(A)$ is an abelian group and therefore $\theta$ leads to a map

$$BGL(A)^+ \longrightarrow BGL(A_*)$$

If we follow [FW] by putting $KV_n(A) = \pi_n(BGL(A_*))$, we therefore have a homomorphism

$$\gamma : K_n(A) \longrightarrow KV_n(A)$$

Now the groups $KV_n(A)$ are just the groups described above by polynomial homotopies and another way of stating the previous result (due essentially to Quillen) is the following fact: $\gamma$ is an isomorphism if $A$ is a regular Noetherian ring.

# Lecture 6 Hermitian $K$-theory

As we know, usual $K$-theory is deeply linked with the general linear group and there is no a priori reason why we should not consider the other classical groups on an equal footing. As we shall see, it is not only desirable, but this setting appears to be suitable for a generalization of Bott periodicity and the computation of the homology of classical groups in terms of classical Witt groups.

The starting point is a ring with antiinvolution $a \mapsto \overline{a}$ together with an element $\varepsilon$ of the center of $A$ such that $\varepsilon\overline{\varepsilon} = 1$. In most examples, $\varepsilon = \pm 1$. For sake of simplicity, we also assume the existence of an element $\lambda$ in the center of $A$ such that $\lambda + \overline{\lambda} = 1$ (if 2 is invertible in $A$, we might choose $\lambda = 1/2$). We refer to the book of A. Bak (Annals of Math Studies) for a more refined notion of hermitian forms using the so called "form-parameters" (when there is no $\lambda$ satisfying this property).

If $M$ is a *right* f.g.p. module over $A$, we define its dual $M^*$ to be the group of $\mathbb{Z}$ -linear maps $f : M \longrightarrow A$ such that $f(m.a) = \overline{a}.f(m)$ for $m \in M$ and $a \in A$. It is again a right f.g.p. $A$-module if we put $(f.b)(m) = f(m).b$ for $b \in A$. An $\varepsilon$-hermitian form on $M$ is roughly speaking a $A$-linear map $M \longrightarrow M^*$ satisfying some conditions of $\varepsilon$-symmetry. More precisely, it is given by a $\mathbb{Z}$-bilinear map

$$\phi : M \times M \longrightarrow A$$

such that

$$\phi(ma, m'b) = \overline{a}\phi(m, m')b$$

$$\phi(m', m) = \varepsilon\overline{\phi(m, m')}$$

with obvious notations. Such a $(M, \phi)$ is called an $\varepsilon$-*hermitian module* . The correspondence

$$\overline{\phi} : m' \mapsto [m \mapsto \phi(m, m')]$$

defines a morphism from $M$ to $M^*$ and we say that $\phi$ is non-degenerate if $\overline{\phi}$ is an isomorphism.

**Fundamental example (the hyperbolic module).** Let $N$ be an f.g.p. module and $M = N \oplus N^*$. Then a non degenerate $\varepsilon$-hermitian form on $M$ is given by the formula

$$\phi((x,f),(x',f')) = \overline{f(x')} + \varepsilon f'(x)$$

We denote this module by $H(N)$. If $N = A^n$, we may identify $N$ with its dual via the map $y \mapsto f_y$ defined by $f_y(x) = \overline{x}y$ . The hermitian form on $A^n \oplus A^n$ may then be written as

$$\phi((x,y),(x',y')) = \overline{y}x' + \varepsilon\overline{x}y'$$

There is an obvious definition of the direct sum of $\varepsilon$-hermitian modules and of an isomorphism between them. We write ${}_{\varepsilon}L(A)$ for the Grothendieck group constructed from such modules[13].

**Exercise:** compute ${}_{\varepsilon}L(A)$ for $A = \mathbb{R}$, $\mathbb{C}$ or $\mathbb{H}$ and all possible antiinvolutions and $\varepsilon$.

**Exercise:** if $(M,\phi)$ is an $\varepsilon$-hermitian module, prove that $(M,\phi) \oplus (M,-\phi)$ is isomorphic to $H(M)$ (one has to use the existence of the $\lambda$ above).

**Exercise:** let $A$ be the ring of continuous functions on a compact space with *complex* values. If $A$ is provided with the trivial involution and if we take $\varepsilon = 1$, prove that ${}_{\varepsilon}L(A)$ is isomorphic to the *real* topological $K$-theory of $X$.

The analog of the general linear group is the $\varepsilon$-orthogonal group which is the group of automorphisms of $H(A^n)$. We denote this group by ${}_{\varepsilon}O_{2n}(A)$: it can be described concretely by matrices in $n$-blocks

$$M = \begin{pmatrix} a & b \\ c & d \end{pmatrix}$$

such that $M^*M = MM^* = I$ where

$$M^* = \begin{pmatrix} {}^t\overline{d} & \varepsilon{}^t\overline{b} \\ \overline{\varepsilon}{}^t\overline{c} & {}^t\overline{a} \end{pmatrix}$$

**Example:** if $A =$ the field of real numbers $\mathbb{R}$, ${}_1O_{2n}(A)$ is the classical group $O(n,n)$ which has the homotopy type of $O(n) \times O(n)$. On the other hand, ${}_{-1}O_{2n}(A)$ is the classical group $Sp(2n,\mathbb{R})$ which has the homotopy type of the unitary group $U(n)$.

The infinite orthogonal group

$${}_{\varepsilon}O(A) = lim_{\varepsilon}O_{2n}(A)$$

[13]We use here the letter $L$ which is quite convenient, but the reader should not be confused with the definition of surgery groups, also denoted by the letter $L$.

has the same formal property as the infinite general linear group. In particular, its commutator subgroup is perfect. Therefore, we can perform the + construction (or, more geometrically, consider virtual flat $A$-bundles with $\varepsilon$-hermitian forms as we did in Lecture 5).

**DEFINITION .** *The higher Hermitian $K$-theory of a ring $A$ (for $n > 0$) is defined as*

$$_{\varepsilon}L_n(A) = \pi_n(B_{\varepsilon}O(A)^+)$$

**Example:** Let $F$ be a field of characteristic different from 2 provided with the trivial involution. Then $_{\varepsilon}L_1(F) = 0$ if $\varepsilon = -1$ and $_{\varepsilon}L_1(F) = \mathbb{Z}/2 \times F^*/F^{*^2}$ if $\varepsilon = +1$.

**Notation.** We write

$$\mathcal{K}(A) = K(A) \times BGL(A)^+$$

for the classifiying space of Algebraic $K$-theory and

$$_{\varepsilon}\mathcal{L}(A) =_{\varepsilon} L(A) \times B_{\varepsilon}O(A)^+$$

for the classifying space of Hermitian $K$-theory.

There are two interesting functors between Hermitian $K$-theory and Algebraic $K$-theory. One of them is the forgetful functor from modules with hermitian forms to modules (with no forms) and the other one from modules to modules with forms, sending $N$ to $H(N)$, the hyperbolic module associated to $N$. These functors induce two maps

$$F :_{\varepsilon} \mathcal{L}(A) \longrightarrow \mathcal{K}(A) \quad \text{and} \quad \mathrm{H} : \mathcal{K}(\mathrm{A}) \longrightarrow_{\varepsilon} \mathcal{L}(\mathrm{A})$$

We define $_{\varepsilon}\mathcal{V}(A)$ as the homotopy fiber of $F$ and $_{\varepsilon}\mathcal{U}(A)$ as the homotopy fiber of $H$. We define this way two "relative" theories:

$$_{\varepsilon}V_nA = \pi_n(_{\varepsilon}\mathcal{V}(A)) \text{ and } _{\varepsilon}\mathrm{U_nA} = \pi_{\mathrm{n}}(_{\varepsilon}\mathcal{U}(\mathrm{A}))$$

**THEOREM (the fundamental theorem of Hermitian $K$-theory).** *There is a natural homotopy equivalence between $_{\varepsilon}\mathcal{V}(A)$ and the loop space of $_{-\varepsilon}\mathcal{U}(A)$. In particular,*

$$_{\varepsilon}V_n(A) \cong {}_{-\varepsilon}U_{n+1}(A)$$

*Moreover, if we work within the framework of Banach algebras, the same statement is valid for the topological analogs (i.e. replacing $BGL(A)^+$ by $BGL(A)^{top}$ and $B_{\varepsilon}O(A)^+$ by $B_{\varepsilon}O(A)^{top}$).*

In order to get a feeling for this theorem, it is worthwhile to work out the classical topological examples $A = \mathbb{R}$, $\mathbb{C}$ or $\mathbb{H}$ , with various antiinvolutions (and $\varepsilon = \pm 1$). Note in general that the connected component of $_{\varepsilon}\mathcal{V}(A)$ (resp.

${}_{\varepsilon}\mathcal{U}(A)$) is the connected component of the homogeneous space $GL(A)/{}_{\varepsilon}O(A)$ (resp. ${}_{-\varepsilon}O(A)/GL(A)$). If for instance $A = \mathbb{R}$, $\varepsilon = -1$, we get the spaces $GL(\mathbb{R})/{}_{-1}O(\mathbb{R})$ (which has the homotopy type of $O/U$) and ${}_{1}O(R)/GL(R)$ (which has the homotopy type of $O$, the infinite orthogonal group). Therefore, the previous theorem implies that $O/U$ has the homotopy type of the loop space $\Omega O$, one of the eight homotopy equivalences of Bott (see the end of Lecture 2, where $O$ has the homotopy type of $GL(\mathbb{R})$ and $U$ the homotopy type of $GL(\mathbb{C})$). It is a pleasant exercise to recover the seven other homotopy equivalences by dealing with other classical groups. Since the list of classical groups is finite, it is "reasonable" to expect some periodicity...

There are two remarkable involutions[14] on the $H$-spaces $\mathcal{K}(A)$ and ${}_{\varepsilon}\mathcal{L}(A)$ and it is better to describe them in the context of classical groups. On $GL(A)$, one takes the contragredient

$$M \mapsto \overline{M}^{-1}$$

On ${}_{\varepsilon}O(A)$ the involution is more delicate. The idea is to take the functor which associates to any hermitian module $(M, \phi)$ the "opposite" module $(M, -\phi)$. On the level of groups, this means that we have to identify the hyperbolic module $H(A^n) = A^n \oplus A^n$ with its opposite Hermitian form. In terms of the orthogonal group, this involution sends the matrix

$$M = \begin{pmatrix} a & b \\ c & d \end{pmatrix}$$

to the matrix

$$\begin{pmatrix} 1 & 0 \\ 0 & -1 \end{pmatrix} \begin{pmatrix} a & b \\ c & d \end{pmatrix} \begin{pmatrix} 1 & 0 \\ 0 & -1 \end{pmatrix} = \begin{pmatrix} a & -b \\ -c & d \end{pmatrix}$$

Therefore, if we localize the spaces $\mathcal{K}(A)$ and ${}_{\varepsilon}\mathcal{L}(A)$ away from 2, we obtain spaces $\mathcal{K}(A)'$ and ${}_{\varepsilon}\mathcal{L}(A)'$, together with splittings according to these involutions

$$\mathcal{K}(A)' \approx \mathcal{K}(A)'_+ \times \mathcal{K}(A)'_- \text{ and } {}_{\varepsilon}\mathcal{L}(\mathrm{A})' \approx_{\varepsilon} \mathcal{L}(\mathrm{A})'_+ \times_{\varepsilon} \mathcal{L}(\mathrm{A})'_-$$

From to this decompositon, the hyperbolic functor and the forgetful functor are both represented by the matrix

$$\begin{pmatrix} 1 & 0 \\ 0 & 0 \end{pmatrix}$$

If follows that, after localisation which we denote by the symbol $a'$, the space ${}_{\varepsilon}\mathcal{V}(A)'$ has the homotopy type of ${}_{\varepsilon}\mathcal{L}(A)'_- \times \Omega\mathcal{K}(A)'_-$ , whereas the space ${}_{-\varepsilon}\mathcal{U}(A)'$ has the homotopy type of $\mathcal{K}(A)'_- \times \Omega_{-\varepsilon}\mathcal{L}(A)'_-$. Therefore, the fundamental theorem of Hermitian $K$-theory is equivalent to saying that ${}_{\varepsilon}\mathcal{L}(A)'_-$ has the homotopy type of $\Omega^2_{-\varepsilon}\mathcal{L}(A)'_-$.

[14] One has to be careful about the precise definition of these spaces when a group is acting. For instance, the decomposition $\mathcal{K}(A) = K(A) \times BGL(A)^+$ is NON canionical, A correct definition (among others) is to choose $\Omega(BGL(SA))^+$.

The preceding considerations give us an idea of how to prove the fundamental theorem in Hermitian $K$-theory after localisation (a more delicate proof, taking into account the 2-torsion is presented in [K4]).We define the <u>*Witt groups*</u> ${}_{\varepsilon}W_n(A)$ as Coker $(K_n(A) \longrightarrow {}_{\varepsilon}L_n(A))$. Then the fundamental theorem amounts to saying that

$$ {}_{\varepsilon}W_n(A) \cong_{-\varepsilon} W_{n+2}(A) \text{ mod.2 torsion} $$

since these Witt groups are essentially the homotopy groups of the space ${}_{\varepsilon}\mathcal{L}(A)'_{-}$. In order to prove such a statement, one may follow the pattern used to prove complex Bott periodicity in the second lecture. The base ring which plays the role of $\mathbb{C}$ is the ring $\mathbb{Z}[x]$ with the involution $x \mapsto 1-x$. Then the scheme is the following

1) Construct non connected deloopings of Hermitian $K$-theory with a good notion of cup-product. More precisely if we have a pairing of rings with involution

$$ A \times B \longrightarrow C $$

we should have a pairing between the representing spaces of Hermitian $K$-theory, taking into account the sign of symmetry:

$$ {}_{\varepsilon}\mathcal{L}(A) \times_{\eta} \mathcal{L}(B) \longrightarrow_{\varepsilon\eta} \mathcal{L}(C) $$

This is done as in usual $K$-theory, using the suspension of a ring. One can proceed in the same way for Banach algebras by completing the cone and the suspension.

2) It is a well known theorem that ${}_{1}W(\mathbb{Z}[x]) \cong \mathbb{Z} \approx_{1} W(\mathbb{R})$. The key point is now to construct elements ${}_{-1}W_{-2}(\mathbb{Z}[x])$ and $v \in_{-1} W_{-2}(\mathbb{Z}[x])$ such that their cup-product is 4 times a generator of ${}_{1}W(\mathbb{Z}[x])$. The construction of $u$ and $v$ is done very explicitly in [K4]. To check that their cup-product is non trivial, one has to send $\mathbb{Z}[x]$ into $\mathbb{R}$, the image of $x$ being $1/2$ and use topological $K$-theory of Banach algebras. This is a nice point where Algebraic and Topological $K$-theory are interacting.

3) Using $u$ and $v$, we define morphisms

$$ {}_{\varepsilon}W_n(A) \longrightarrow_{-\varepsilon} W_{n+2}(A) \text{ and } {}_{-\varepsilon}W_{n+2}(A) \longrightarrow_{\varepsilon} W_n(A) $$

such that the two compositions are 4 times the identity. This essentially concludes the proof.

Another theorem of the same spirit is the following.

**THEOREM .** *There exist $u' \in_{-1} W_2(\mathbb{Z})$ and $v' \in_{-1} W_2(\mathbb{Z})$ such that their cup-product is an element of ${}_{1}W(\mathbb{Z})$ which image in ${}_{1}W(\mathbb{R}) \cong \mathbb{Z}$ is 32 times the canonical generator. Now let A be ANY ring with involution (we no longer assume the existence of $\lambda$ such that $\lambda + \overrightarrow{\lambda} = 1$). Then, using $u'$ and $v'$, we define as before two morphisms between ${}_{\varepsilon}W_n(A)$ and ${}_{-\varepsilon}W_{n+2}(A)$ such that their composite is 32 times the identity. In particular,*

$$ {}_{\varepsilon}W_n(A) \otimes \mathbb{Z}[1/2] \cong_{-\varepsilon} W_{n+2}(A) \otimes \mathbb{Z}[1/2] $$

*for ANY ring A.*

The fundamental theorem and the slight generalization above have nice applications for the computation of the homology of the general $\varepsilon$-orthogonal group. The philosophy is that this homology splits into two parts, one coming from the general linear group which is unknown in general and the other coming from more accessible invariants which are ${}_{\pm\varepsilon}W_n(A)$ for $n = 0$ and 1 only (since we have periodicity). As a typical example, let us consider the homology of the symplectic group $Sp(\mathbb{Z}) = lim Sp_{2n}(\mathbb{Z})$ and the orthogonal group ${}_1O(\mathbb{Z}) = \lim_1 O_{2n}(\mathbb{Z})$ .

**THEOREM .** *Let $\mathbb{Z}' = \mathbb{Z}[1/2]$. Then for the homology with coefficients in $\mathbb{Z}'$, we have the following decomposition of the homology of $H_*(BSp(\mathbb{Z}))$ :*

$$H_*(BSp(\mathbb{Z})) \cong \mathbb{Z}'[x_2, x_6, ...] \otimes M_*$$

*where $M_0 = \mathbb{Z}'$ and $M_i$ is a finite group for $i > 0$.*

The theorem follows from the decomposition of $BSp(\mathbb{Z})^+$ after localisation

$$BSp(\mathbb{Z})^+ \approx_{-1} \mathcal{L}(\mathbb{Z})'_+ \times_{-1} \mathcal{L}(\mathbb{Z})'_-$$

where $\pi_i({}_{-1}\mathcal{L}(\mathbb{Z})'_-) = \mathbb{Z}'$ for i $\equiv 2$ mod. 4 and 0 otherwise and $\pi_i({}_{-1}\mathcal{L}(\mathbb{Z})'_+))$ is the symmetric part of $K_i(\mathbb{Z})$ after localisation which is 0 mod. torsion since the regulator map sends symmetric elements to 0.

**THEOREM .** *Let $\mathbb{Z}' = \mathbb{Z}[1/2]$. Then for the homology with coefficients in $\mathbb{Z}'$, we have the following decomposition of the homology of $H_*({}_1O(\mathbb{Z}))$:*

$$H_*({}_1O(\mathbb{Z})) \cong \mathbb{Z}'[x_4, x_8, ...] \otimes M_*$$

*where $M_0 = \mathbb{Z}'$ and $M_i$ is a finite group for $i > 0$.*

For both groups, it is important to notice that the elements $x_{4i+2}$ in the symplectic case are determined by embedding $Sp(\mathbb{Z})$ in the topological group $Sp(\mathbb{R})$ which has the homotopy type of the infinite (topological) unitary group. In the same way, the elements $x_{4i}$ in the orthogonal case are determined by embedding ${}_1O(\mathbb{Z})$ in the topological group ${}_1O(\mathbb{R})$ which has the homotopy type of the product of two copies of the classical (topological) infinite orthogonal group. The situation there is in sharp contrast with what happens for the general linear group: by Chern-Weil theory, we know that the homomorphism

$$K_n(R) \longrightarrow K_n^{top}(\mathbb{R})$$

has a finite cokernel. This is not true for the homomorphism

$${}_1L_n(\mathbb{R}) \longrightarrow_1 L_n^{top}(\mathbb{R}) \text{ (resp.} {}_{-1}\mathrm{L}_n(\mathbb{R}) \longrightarrow_{-1} \mathrm{L}_n^{top}(\mathbb{R}))$$

for $n \equiv 0$ mod. 4 (resp. $n \equiv 2$ mod. 4).

Let us now turn to some computations, starting with finite fields. It has been proved by Quillen that $BGL(\mathbb{F}_q)^+$ has the homotopy fiber of the map $\psi^q - 1$, where

$$\psi^q : BGL(\mathbb{C}) \longrightarrow BGL(\mathbb{C})$$

is the map induced by the Adams operation. In the same way, for $q$ odd, it has been proved by Friedlander that $B_1O(F_q)$ [resp.$B_{-1}O(F_q)$] is the homotopy fiber of the map $\psi^q - 1$, where

$$\psi^q : B_1O(\mathbb{C}) \longrightarrow B_1O(\mathbb{C}) \text{ [resp. } \psi^q : \mathrm{B}_{-1}\mathrm{O}(\mathbb{C}) \longrightarrow \mathrm{B}_{-1}\mathrm{O}(\mathbb{C})]$$

Note that ${}_1O(\mathbb{C})$ has the homotopy type of the usual infinite orthogonal group $O = injlimO(n)$ and ${}_{-1}O(\mathbb{C})$ has the homotopy type of $Sp = injlimSp(n)$. This leads to the following computations of ${}_1L_i(F_q)$ for i > 0 and $i$ mod.8.

| i (mod. 8) | ${}_1L_i(F_q)$ | ${}_{-1}L_i(F_q)$ |
|---|---|---|
| i=0 | $\mathbb{Z}/2$ | 0 |
| i = 1 | $\mathbb{Z}/2 \oplus \mathbb{Z}/2$ | 0 |
| i = 2 | $\mathbb{Z}/2$ | 0 |
| i = 3 | $\mathbb{Z}/(q^{(i+1)/2} - 1)$ | $\mathbb{Z}/(q^{(i+1)/2} - 1)$ |
| i = 4 | 0 | 0 |
| i = 5 | 0 | $\mathbb{Z}/2 \oplus \mathbb{Z}/2$ |
| i = 6 | 0 | $\mathbb{Z}\ /2$ |
| i = 7 | $\mathbb{Z}/(q^{(i+1)/2} - 1)$ | $\mathbb{Z}/(q^{(i+1)/2} - 1)$ |

Friedlander has also noticed that the Witt groups ${}_1W_i(F_q)$ and ${}_{-1}W_i(F_q)$ are periodic of period 8 with respect to $i$. More precisely, we have ${}_1W_i(F_q) \cong_{-1} W_{i+4}(F_q)$ and these are isomorphic to the following list, starting for $i \equiv 0$ mod.8

$$\mathbb{Z}/2, \mathbb{Z}/2, 0, 0, 0, 0, \mathbb{Z}/2, \mathbb{Z}/2$$

Another case of interest is when our basic ring is $\mathbb{Z}' = \mathbb{Z}[1/2]$. Some computations have been done recently by A.J. Berrick and the author. They will be published shortly. Here are the results for $i > 0$ and mod. an odd torsion finite group

| i (mod. 8) | ${}_1L_i(Z')$ | ${}_{-1}L_i(Z')$ |
|---|---|---|
| i = 0 | $\mathbb{Z} \oplus \mathbb{Z}/2$ | 0 |
| i = 1 | $\mathbb{Z}/2 \oplus \mathbb{Z}/2 \oplus \mathbb{Z}/2$ | 0 |
| i = 2 | $\mathbb{Z} \oplus \mathbb{Z}/2$ | $\mathbb{Z}$ |
| i = 3 | $\mathbb{Z}\ /8$ | $\mathbb{Z}\ /16$ |
| i = 4 | $\mathbb{Z}$ | $\mathbb{Z}\ /2$ |
| i = 5 | 0 | $\mathbb{Z}\ /2$ |
| i = 6 | 0 | $\mathbb{Z}$ |
| i = 7 | $\mathbb{Z}/2^{t+1}$ | $\mathbb{Z}/2^{t+1}$ |

where $2^t$ is the 2-primary component of $i + 1$.

# Lecture 7 Cyclic homology and $K$-theory

Let me say a few historical words on the subject before explaining more recent achievements. Cyclic homology (and cohomology) grew from at least 3 sources:

1) In his attempt to understand the Atiyah-Singer index theorem in noncommutative geometry, Connes was forced to find the analog of the de Rham complex for suitable dense subalgebras of $C^*$-algebras (typical examples are the Schatten ideals introduced in Higson's lectures). Connes explained his ideas in the $K$-theory seminar in Paris in 1981 and with more details in his course at the College de France the following years. He called his groups "cyclic homology" and denoted them by $HC_*(A)$. These groups are very much related to Hochschild homology $HH_*(A)$, introduced 50 years ago. As a matter of fact, there is a well defined homomorphism

$$B : HC_*(A) \longrightarrow HH_{*+1}(A)$$

which plays a central role in the theory.

2) From a quite different perspective, Tsygan was trying to compute the Lie algebra homology of infinite matrices over a ring $A$ of characteristic 0. He proved that this homology is primitively generated (as a Hopf algebra) by precisely the cyclic homology groups of $A$, introduced by Connes independently. This result was proved also by Quillen and Loday in 1983, using invariant theory.

3) Finally, from my work in $K$-theory during the late 70's, I was looking for a kind of Chern character for general rings. The target was then a new homology theory of rings called "noncommutative de Rham homology" (see [K2] § 1) and denoted $H_*^{dR}(A)$, but the relaton with cyclic homology was unclear at the first glance. However, as it was realized in 1983 by Connes and myself, this theory $H_*^{dR}(A)$ happens to be just the kernel of the map $B : HC_*(A) \longrightarrow HH_{*+1}(A)$ mentioned above. From this clarified viewpoint, $HC_*(A)$ and $H_*^{dR}(A)$ are like looking very much as Deligne cohomology in a noncommutative context.

The purpose of this lecture is to fill the gap between $K$-theory and cyclic homology, via various types of "Chern characters". In order not to overlap with the lectures of Berrick and Higson, we shall take here the viewpoint of functional analysis and deal essentially with Algebraic and Topological $K$-theory of Fréchet algebras (which are more general than Banach algebras), comparing them with the corresponding cyclic homology groups.

More precisely, let $A$ be a unitary, but not necessarily commutative Fréchet algebra. As we have seen in previous lectures, we can define Algebraic and Topological $K$-theory denoted respectively by $K_n(A)$ and $K_n^{top}(A)$. For $n > 0$, we have $K_n(A) = \pi_n(BGL(A)^+)$; here $BGL(A)^+$ denotes the $+$ construction of Quillen applied to the classifying space $BGL(A)^\delta$, with $GL(A)^\delta = GL(A)$, viewed as a <u>*discrete*</u> group. On the other hand, $K_n^{top}(A)$ is defined as the homotopy group $\pi_n(BGL(A))$, where $BGL(A)$ denotes the classifying space of the <u>*topological*</u> group

$GL(A)$ (at least if $A$ is a Banach algebra; the general case needs simplicial methods). The obvious map $BGL(A)^+ \longrightarrow BGL(A)$ induces a natural homomorphism

$$K_n(A) \longrightarrow K_n^{top}(A)$$

We shall use cyclic homology as a tool to construct an intermediary $K$-theory, called the "multiplicative $K$-theory of $A$" and denoted by $\mathcal{K}_n(A)$, which is more accessible than Algebraic $K$-theory. The previous homomorphism can be factored as

$$K_n(A) \longrightarrow \mathcal{K}_n(A) \longrightarrow K_n^{top}(A)$$

The homomorphism $K_n(A) \longrightarrow \mathcal{K}_n(A)$ contains various interesting invariants of the Algebraic $K$-theory of the Fréchet algebra $A$. For instance $\mathcal{K}_n(A) \cong \mathbb{C}^*$ when $A = \mathbb{C}$ and $n$ odd. The map $K_{2n-1}(\mathbb{C}) \longrightarrow \mathbb{C}^*$ is essentially the regulator map. We also have $\mathcal{K}_2(C^\infty(S^1)) \cong \mathbb{C}^*$ and the map $K_2(C^\infty(S^1)) \longrightarrow \mathbb{C}^*$ gives rise to the well-known Kac-Moody extension of $SL(C^\infty(S^1))$ by $\mathbb{C}^*$ ([CK]).

In order to put all these informations together, we shall construct a commutative diagram with 16 entries (when $n \geq 1$) and exact horizontal sequences:

$$\begin{array}{ccccccc}
K_n^{rel}(A) & \longrightarrow & K_n(A) & \longrightarrow & K_n^{top}(A) & \longrightarrow & K_{n-1}^{rel}(A) \\
\downarrow & & \downarrow & & \| & & \downarrow \\
HC_{n-1}(A) & \longrightarrow & \mathcal{K}_n(A) & \longrightarrow & K_n^{top}(A) & \longrightarrow & HC_{n-2}(A) \\
\| & & \downarrow & & \downarrow & & \| \\
HC_{n-1}(A) & \longrightarrow & HC_n(A) & \longrightarrow & HC_n^{top}(A) & \longrightarrow & HC_{n-2}(A) \\
\| & & \downarrow & & \downarrow & & \| \\
HC_{n-1}(A) & \longrightarrow & HH_n(A) & \longrightarrow & HC_n(A) & \longrightarrow & HC_{n-2}(A)
\end{array}$$

Most of the work describing this diagram is included in a paper by Connes and myself [CK] and in a paper of Weibel[15]. In the first row, the group $K_n^{rel}(A)$ is defined as the $n^{th}$ homotopy group of the homotopy fiber of the map $BGL(A)^+ \longrightarrow BGL(A)$. Therefore, the first exact sequence is tautological. The other terms and arrows of this diagram will be detailed now. At the same time we shall explain what the Chern character should be in this context.

Cyclic homology has already been defined in previous lectures. Therefore, we shall innovate slightly by taking cyclic homology of Fréchet algebras[16] instead, since it is our viewpoint here. If $\otimes_\pi$ denotes the completed tensor product of Grothendieck, we define a double complex by the following formulas

$$C_{p,q}(A) = A \otimes_\pi A \otimes_\pi \cdots \otimes_\pi A(q+1 \text{ times})$$

for $p \in \mathbb{Z}$ and $q \in \mathbb{N}$. The first differential $b : C_{p,q}(A) \longrightarrow C_{p,q-1}(A)$ is the usual Hochschild boundary

$$b(a_0, ..., a_q) = \sum_{i=0}^{q-1}(-1)^i(a_0, ..., a_i a_{i+1}, ..., a_q) + (-1)^q(a_q a_0, a_1, ..., a_{q-1})$$

[15] C.Weibel. Nil $K$-theory maps to cyclic homology. Trans. AMS 303(1987), 541-558.

[16] For instance, a Banach algebra or the algebra of $C^\infty$ functions on a manifold.

The second differential $\partial : C_{p,q}(A) \longrightarrow C_{p-1,q}(A)$ is defined by $1 - t$ if $p$ is even and $-N$ if $p$ is odd. Here the operators $t$ and $N : C_{p,q}(A) \longrightarrow C_{p,q}(A)$ are defined by

$$t(a_0, ..., a_q) = (-1)^q (a_q, a_0, ..., a_{q-1})$$
$$N = 1 + t + \cdots + t^q$$

We therefore get the following picture:

$$\begin{array}{ccccccccccc}
\downarrow & & \downarrow & & \downarrow & & \downarrow & & \downarrow & & \downarrow \\
\longleftarrow & C_{-2,2} & \longleftarrow & C_{-1,2} & \longleftarrow & C_{0,2} & \longleftarrow & C_{1,2} & \longleftarrow & C_{1,2} & \longleftarrow \\
\downarrow & & \downarrow & & \downarrow & & \downarrow & & \downarrow & & \downarrow \\
\longleftarrow & C_{-2,1} & \longleftarrow & C_{-1,1} & \longleftarrow & C_{0,1} & \longleftarrow & C_{1,1} & \longleftarrow & C_{1,1} & \longleftarrow \\
\downarrow & & \downarrow & & \downarrow & & \downarrow & & \downarrow & & \downarrow \\
\longleftarrow & C_{-2,0} & \longleftarrow & C_{-1,0} & \longleftarrow & C_{0,0} & \longleftarrow & C_{1,0} & \longleftarrow & C_{1,0} & \longleftarrow
\end{array}$$

The "total" complex of this diagram is defined by a <u>*product*</u> formula:

$$C_n^{per}(A) = \prod_{p+q=n} C_{p,q}(A)$$

We denote by $HC_*^{per}(A)$ the (periodic cyclic) homology of this complex. The restriction of this double complex to the first quadrant defines a complex whose homology $HC_*(A)$ is by definition the cyclic homology of the Fréchet algebra $A$ (taking the topology into account ). The restriction to the second quadrant (denoted $C_n^-(A)$ ) of the double complex gives the definition of the negative cyclic homology $HC_*^-(A)$. We have an exact sequence (see [L1] p. 160 for instance):

$$0 \longrightarrow \underline{\lim}_r^1 HC_{n+2r+1}(A) \longrightarrow HC_n^{per}(A) \longrightarrow \underline{\lim}_r HC_{n+2r}(A) \longrightarrow 0$$

Here the projective limits lim and $\lim^1$ are computed using the periodicity operator S of Connes [C1]. We should notice that the homology spaces obtained this way are not Hausdorff in general.

For technical reasons which will appear later on, we shall truncate the complexes $E_n = C_n^{per}$ or $C_n^-$ by 0 if $n < 0$ and $Ker(E_0 \longrightarrow E_{-1})$ if $n = 0$, the other $E_n$ staying unchanged. For $n \geq 0$, the groups $HC_n^{per}$ and $HC_n^-$ are still the same as before. Finally, we notice the following diagram with two exact sequences

$$\begin{array}{ccccccc}
HC_{n-1}(A) & \longrightarrow & HC_n^-(A) & \longrightarrow & HC_n^{per}(A) & \longrightarrow & HC_{n-2}(A) \\
\| & & \downarrow & & \downarrow & & \| \\
HC_{n-1}(A) & \longrightarrow & HH_n(A) & \longrightarrow & HC_n(A) & \longrightarrow & HC_{n-2}(A)
\end{array}$$

Another way to introduce cyclic homology is to consider the double complex $(B, b)$ of Connes (cf. [L1] p. 56 for instance). It can be written over 3 quadrants in the following way:

$$\begin{array}{ccccccccc}
 & \longleftarrow & & \longleftarrow & D_2 & \longleftarrow & D_1 & \overset{B}{\longleftarrow} & D_0 \\
\downarrow & & \downarrow & & \downarrow & & \downarrow{\scriptstyle b} & & \\
 & \longleftarrow & D_2 & \longleftarrow & D_1 & \longleftarrow & D_0 & & \\
\downarrow & & \downarrow & & \downarrow & & & & \\
D_2 & \longleftarrow & D_1 & \longleftarrow & D_0 & & & &
\end{array}$$

where we put in general $D_q = C_{p,q} = A^{\otimes_\pi^{q+1}}$. If we truncate as before, we can restrict it to the first two quadrants. We denote by $B$ (resp. $b$) the horizontal arrows (resp. the vertical ones). The bidegrees are obvious (the $D_0$ of the second line has bidegree $(0,0)$).

We may normalize this double complex by replacing $D_q$ by $\Omega_q = \Omega_q(A) = A \otimes_\pi \overline{A} \otimes_\pi ... \otimes_\pi \overline{A}$ ($q$ copies of$\overline{A} = A/C.1$). The bicomplex

$$\begin{array}{ccccccccc}
 & \longleftarrow & & \longleftarrow & \Omega_2 & \longleftarrow & \Omega_1 & \overset{B}{\longleftarrow} & \Omega_0 \\
\downarrow & & \downarrow & & \downarrow & & \downarrow{\scriptstyle b} & & \\
 & \longleftarrow & \Omega_2 & \longleftarrow & \Omega_1 & \longleftarrow & \Omega_0 & & \\
\downarrow & & \downarrow & & \downarrow & & & & \\
\Omega_2 & \longleftarrow & \Omega_1 & \longleftarrow & \Omega_0 & & & &
\end{array}$$

also enables us to compute cyclic homology and its variants $(H\mathbb{C}_*^{per}, H\mathbb{C}_*^-, ...)$.

Finally, there is a last variant of cyclic homology, called non commutative de Rham homology, mentionned at the beginning, which we shall define now. The direct sum of the $\Omega_n(A)$ above is a differential graded algebra: we write formally an element of $A \otimes_\pi \overline{A} \otimes_\pi \cdots \otimes_\pi \overline{A}$ as a linear combination (infinite since we are dealing with the topological case) of elements of the type

$$a_0 da_1 ... da_n$$

and we multiply them formally using the Leibniz rule: for instance, $(x.dy).(z.dt) = x.d(yz).dt - xy.dz.dt$. The differential is defined simply by $d(a_0 da_1 ... da_n) = 1.da_0 da_1 ... da_n$. Let us now consider the algebra $\overline{A}$ obtained by adding a unit to $A$ and the quotient $\overline{\Omega}_*(\overline{A})$ of $\Omega_*(\overline{A})$ by the $\mathbb{C}$-module generated by commutators $\lambda\omega - \omega\lambda, \lambda \in \overline{A}$, and their differentials. This $\mathbb{C}$-module is easily seen to be the quotient of the algebra $\overline{\Omega}_*(\overline{A})$ by the module generated by graded commutators.

**THEOREM** ([C1][K2]). *We have a natural isomorphism*

$$H_n^{dR}(\overline{A}) \cong Ker[B : HC_n(A) \longrightarrow HH_{n+1}(A)]$$

From the classical proof of the Poincaré lemma in de Rham cohomology, we deduce that $H_n^{dR}$ is a homotopy invariant. More precisely, we have $H_n^{dR}(\overline{A}) \cong H_n^{dR}(\overline{A} \otimes C^\infty(\mathbb{R}))$. As a consequence, we deduce from the previous theorem that $Ker B$ is also a homotopy invariant which in turn implies the following theorem:

**THEOREM** . *Let $A_r$ be the algebra of $C^\infty$ functions on the canonical simplex*[17] *$\Delta_r$ with values in $A$, interpreted as the completed tensor product $C^\infty(\Delta_r) \otimes_\pi A$. Then, the obvious map $A \longrightarrow A_r$ induces an isomorphism $HC_n^{per}(A) \approx HC_n^{per}(A_r)$.*

In order to go further, we need a "differential form" version of the Dold-Kan correspondance. If we denote by $\mathcal{C}$ the category of chain complexes (positively graded and assumed to be $\mathbb{C}$-vector spaces, which is our framework) and by $\mathcal{G}$ the category of simplicial $\mathbb{C}$-vector spaces, they are equivalent. In particular, we can associate to a chain complex $(C_n)$ a space such that its homotopy groups are the homology groups of the chain complex. There are many choices of spaces[18] which are equivalent up to homotopy. The one which is the most suitable for our purpose is the following. Let $\Omega^*(\Delta_r)$ be the DGA of usual $C^\infty$ differential forms on the $r$-simplex. We associate now to $(C_n)$ a double simplicial complex defined on the second quadrant by

$$D_{p,q}(\Delta_r) = \Omega^{-p}(\Delta_r) \otimes_{\mathbb{C}} C_q$$

The two differentials (which decrease the degrees $p$ and $q$ by 1) are induced by the differentials on $\Omega^{-*}$ and $C_*$. To this double complex we associate the simplicial complex defined by

$$C_n(\Delta_r) = \bigoplus_{p+q=n} D_{p,q}(\Delta_r) = \bigoplus_{q \geq n} \Omega^{q-n}(\Delta_r) \otimes_{\mathbb{C}} C_q$$

where $n \in \mathbb{Z}$. There is a remarkable simplicial subgroup of $C_0(\Delta_*)$: it consists of closed chains of total degree 0: it is a subgroup of the direct sum $\Omega^q(\Delta_*) \otimes_{\mathbb{C}} C_q$. The face and degeneracy operators are induced by those on $\Omega^q(\Delta_*)$. We call it $DK(C_*)$.

**THEOREM** . *The functor $DK : C_* \mapsto DK(C_*)$ from $\mathcal{C}$ to $\mathcal{G}$ is inverse up to homotopy to the well-known category equivalence $F : \mathcal{G} \longrightarrow \mathcal{C}$ . We shall simply write $|C_*|$ for $DK(C_*)$ and we call it the "simplicial realization" of the chain complex $C_*$.*

Let $X_*$ be a simplicial abelian group and let $C_* = F(X_*)$ be the associated chain complex. According to the previous theorem, there should be an explicit homotopy equivalence

$$\phi : X_* \longrightarrow |C_*|$$

This may be applied to the particular case when $X_*$ is the $\mathbb{C}$-vector space with basis the simplicial set $BG$ or $EG$, whose $G$ is a discrete group. In order to be

[17] We may avoid "angle problem" by defining $\Delta_r$ as the *affine space* of points with coordinates $(t_0, t_1, ..., t_r)$ in $\mathbb{R}^{r+1}$ such that $t_0 + t_1 + \cdots + t_r = 1$ (we do not assume that $t_i$ are $\geq 0$.) The face and degeneracy operators enable us to define a simplicial ring from the $C^\infty(\Delta_r)$. If $X$ is a manifold and $A$ a Banach algebra, it is a well know fact that $C^\infty(X) \otimes_\pi A$ may be identified with the algebra of $C^\infty$ functions on $X$ with values in $A$.

[18] M. karoubi. Formes différentielles et théotème de Dold-Kan. Journal of Algebra 198, p. 618-626 (1997).

more explicit, let $C_* = C_*(EG)$ be the "homogeneous" Eilenberg-Mac Lane chain complex defining the homology of $BG = G \setminus EG$. We define a simplicial map

$$\Phi : BG \longrightarrow |C_*|$$

by the following formula for $g = (g_0, g_1, ..., g_r)$ of degree $r$ in $BG = G \setminus EG$:

$$\Phi(g) = \sum_i x_i \otimes g_j + \sum_{(i,j)} x_i dx_j \otimes \{g_i, g_j\} + \sum_{(i,j,k)} x_i dx_j \wedge dx_k \otimes \{g_i, g_j, g_k\} + \cdots$$

where$\{g_i, g_j\}$, $\{g_i, g_j, g_k\}$, etc... are the corresponding classes in $C_*(BG)$. Here $x_i dx_j$, $x_i dx_j \wedge dx_k$, etc... are the usual differential forms on the standard $r$-simplex $\Delta_r$ with the $x_i$ as barycentric coordinates. It is easy to check that $\phi$ is the expected map.

With these definitions, let us now describe the Chern character in Algebraic $K$-theory. For this purpose we consider the double complex

$$\overline{C}_{**}(G)$$

defined in [K2] and $C_{**} = G/\overline{C}_{**}(G)$ the quotient complex by the left action of $G$. The definition of this double complex follows the same pattern as $C_{**}$, using the obvious cyclic module associated to the bar construction of $EG$. The double indices in $C_{p,q}(G)$ or $\overline{C}_{p,q}(G)$ are in the same range as before: $p \in \mathbb{Z}, q \in \mathbb{N}$. In the same way, we put

$$C_n^{per}(G) = \prod_{p+q=n} C_{p,q}(G) \ for\ n > 0 \text{ and}$$

$$C_0^{per}(G) = Ker[\prod_{p+q=0} C_{p,q}(G) \longrightarrow \prod_{p+q=-1} C_{p,q}(G)],$$

and analogous definitions for $C_n^-(G)$ if $n \geq 0$. In particular, we define $HC_*(G)$ as the homology of the total complex associated to the previous double complex restricted to the first quadrant. With the method described in [K2], it is easy to show, using classical homological algebra (see [K2] or the book of Weibel) that

$$HC_n(G) = H_n(G) \oplus H_{n-2}(G) \oplus \cdots$$

$$HC_n^-(G) = \prod_{k \geq 0} H_{n+2k}(G)$$

$$HC_n^{per}(G) = \prod_k H_{n+2k}(G)$$

According to the previous considerations, we therefore have well defined maps of simplicial complexes

$$BG \longrightarrow |C_*(G)| \longrightarrow |C_*^-(G)|$$

In particular, if $G = GL(A)$, we get a morphism $|C_*(G)| \longrightarrow |C_*^-(\overline{A})|$. If we apply the + construction, we obtain a well defined map up to homotopy $BGL(A)^+ \longrightarrow |C_*^-(\overline{A})| \longrightarrow |C_*^-(A)|$ and therefore a homomorphism

$$K_n(A) \longrightarrow HC_n^-(A)$$

defined by Hood and Jones[19], generalizing the results obtained in [K2], with the same method.

In order to define the same type of invariants in topological $K$-theory, we just apply the same ideas, but with $G = GL(A)$ replaced by the simplicial group $G_* = GL(A_*)$. We write the composition

$$BG_* \longrightarrow |C_*(G_*)| \longrightarrow |C_*^-(G_*)| \longrightarrow |C_*^-(A_*)| \longrightarrow |C_*^{per}(A_*)|$$

Now comes the crucial fact: since periodic cyclic homology is homotopy invariant, the space $|C_*^{per}(A_*)|$ is homotopically equivalent to the space $|C_*^{per}(A)|$ and finally we get the Chern character in topological $K$-theory defined as a map

$$BGL(A) \longrightarrow |C_*^{per}(A)|$$

By taking homotopy groups, one finally gets a homomorphism

$$K_n^{top}(A) \longrightarrow HC_n^{per}(A)$$

**DEFINITION/THEOREM .** *Let $G_*$ be the simplicial group $GL(A_*)$ and let $\mathcal{K}(A)$ be the homotopy fiber of the map*

$$BG_* \longrightarrow |C_*^{per}(A)| \longrightarrow |C_*^{per}(A)|/|C_*^-(A)|$$

*For $n > 0$, the multiplicative $K$-theory groups $\mathcal{K}_n(A)$ are then defined as the homotopy groups of the space $\mathcal{K}(A)$. We have an exact sequence*

$$K_{n+1}^{top}(A) \longrightarrow HC_{n-1}(A) \longrightarrow \mathcal{K}_n(A) \longrightarrow K_n^{top}(A) \longrightarrow HC_{n-2}(A)$$

It is easy to see that the composite map

$$BGL(A)^\delta \longrightarrow BGL(A_*) \longrightarrow |C_*^{per}(A_*)| \approx |C_*^{per}(A)|$$

can be factored through $|C_*^-(A)|$. We deduce from this observation a <u>canonical</u> map from $BGL(A)^\delta$ to $\mathcal{K}(A)$. In the same way, if we denote by $\mathcal{G}$ the homotopy fiber of the map

$$BGL(A) \longrightarrow BGL(A_*)$$

we have a canonical map from $\mathcal{G}$ to $|C_{*-1}(A)|$. Thanks to the $+$ construction, we can write the following diagram of homotopy fibrations (with $\mathcal{F} = \mathcal{G}^+$):

$$\begin{array}{ccccccc}
\Omega BGL(A_*) & \longrightarrow & \mathcal{F} & \longrightarrow & BGL(A)^+ & \longrightarrow & BGL(A_*) \\
\downarrow & & \downarrow & & \downarrow & & \downarrow \\
\Omega BGL(A_*) & \longrightarrow & \Omega|C_{*-2}(A)| & \longrightarrow & \mathcal{K}(A) & \longrightarrow & BGL(A_*)
\end{array}$$

[19] C. HOOD and J.D.S. JONES. Some algebraic properties of cyclic homology groups. $K$-theory 1(1987),361-384.

Applying the functor $\pi_n$, we find the commutative diagram

$$\begin{array}{ccccccc} K_n^{rel}(A) & \longrightarrow & K_n(A) & \longrightarrow & K_n^{top}(A) & \longrightarrow & K_{n-1}^{rel}(A) \\ \downarrow & & \downarrow & & \| & & \downarrow \\ HC_{n-1}(A) & \longrightarrow & \mathcal{K}_n(A) & \longrightarrow & K_n^{top}(A) & \longrightarrow & HC_{n-2}(A) \end{array}$$

The commutativity of the diagram

$$\begin{array}{ccccccc} HC_{n-1}(A) & \longrightarrow & \mathcal{K}_n(A) & \longrightarrow & K_n^{top}(A) & \longrightarrow & HC_{n-2}(A) \\ \| & & \downarrow & & \downarrow & & \| \\ HC_{n-1}(A) & \longrightarrow & HC_n^-(A) & \longrightarrow & HC_n^{per}(A) & \longrightarrow & HC_{n-2}(A) \end{array}$$

is a consequence of the homotopy fibrations:

$$\begin{array}{ccccc} \mathcal{K}_n(A) & \longrightarrow & BGL(A_*) & \longrightarrow & |C_*^{per}(A)|/|C_*^-(A)| \\ \downarrow & & \downarrow & & \downarrow \\ |C_*^-(A)| & \longrightarrow & |C_*^{per}(A)| & \longrightarrow & |C_*^{per}(A)|/|C_*^-(A)| \end{array}$$

Finally, the commutativity of the diagram

$$\begin{array}{ccccccc} HC_{n-1}(A) & \longrightarrow & HC_n^-(A) & \longrightarrow & HC_n^{per}(A) & \longrightarrow & HC_{n-2}(A) \\ \| & & \downarrow & & \downarrow & & \| \\ HC_{n-1}(A) & \longrightarrow & HH_n(A) & \longrightarrow & HC_n(A) & \longrightarrow & HC_{n-2}(A) \end{array}$$

follows from the classical definitions (cf. [L], p. 158 for instance).

# References


[At1] ATIYAH M.-F.: *K-theory.* Benjamin, New-York (1967).

[Ba] BASS H.: *Algebraic K-theory.* Benjamin, New York. (1968).

[Ber] BERRICK J.: *An approach to Algebraic K-theory.* Pitman (1982).

[Bla] BLACKLADAR: *K-theory of operator algebras.* Springer (1986).

[C1] CONNES A.: *Noncommutative Geometry.* Academic Press, San Diego (1994).

[FW] FRIEDLANDER E. and WEIBEL C.: *School on Algebraic K-theory and applications.* An overview of the subject. ICTP Trieste (1997).

[K1] KAROUBI M.: *K-theory, an introduction.* Grundlehren der Math. Wissen. 226, Springer, Berlin (1978).

[K2] KAROUBI M.: *Homologie cyclique et K-théorie.* Astérisque 149. Société Mathématique de France (1987).

[L1] LODAY J.-L.: *Cyclic homology.* Grundelehren der Math. Wissen. 301, Springer, Berlin (1992).

[M] MILNOR J.: *Introduction to Algebraic K-theory.* Ann. of Math. Studies 197. Princeton, NJ. Princeton University Press (1974).

[R] ROSENBERG J.: *Algebraic $K$-theory and its applications.* Graduate Texts *No* 147, Springer, Berlin (1994).
[Sr] SRINIVAS V.: *Algebraic $K$-theory.* Birkhauser, Boston (1996).
[SLN] Springer Lecture Notes *Nr* 341, 342 and 342 (1973).

**Papers**

[Ad] ADAMS J.-F.: *Vector fields on the sphere.* Ann. of Math. 75, 603-632 (1962).
[At2] ATIYAH M.-F.: *Global theory of elliptic operators.* Proc. Internat. Conf. on Functional Analysis and related topics, p. 21-30. Univ. of Tokyo Press (1970).
[AH1] ATIYAH M.-F. and HIRZEBRUCH F.: *Vector bundles and homogeneous spaces.* Proc. Symposium in Pure Maths, Amer. Math. Soc. 3, 7-38 (1961).
[AH2] ATIYAH M.-F. and HIRZEBRUCH F.: *Riemann-Roch theorems for differentiable manifolds.* Bull. Amer. Math. Soc. 65, 276-281 (1959).
[AS1] ATIYAH M.-F. and SINGER I.M.: *The index of elliptic operators on compact manifolds.* Bull. Amer. Math. Soc. 69, 422-433 (1963).
[AS2] ATIYAH M.-F. and SINGER I.M.: *The index of elliptic operators.* Ann. of Maths. 87, 484-530, 564-604 (1968), 93, 119-138 (1971).
[Be] BEILINSON A.A.: *Higher regulators and values of L-functions.* J. Soviet. Math. 30, 2036-2070 (1985).
[BHM] BOKSTEDT M., HSIANG W.-C. and MADSEN I.: *The cyclotomic trace and algebraic $K$-theory of spaces.* Invent. Math. 111, 465-540 (1993).
[Bor] BOREL A.: *Stable real cohomology of arithmetic groups.* Ann. Sc. Ec. Norm. Sup. 7, 235-272 (1974).
[BS] BOREL A. et SERRE J.-P.: *Le théorème de Riemann-Roch (d'après Grothendieck).* Bull. Soc. Math. France 86, 97-136 (1958).
[Bott] BOTT R.: *The stable homotopy of the classical groups.* Ann. of Math. 70, 313-337 (1959).
[Br] BROWDER W.: *Surgery on simply connected manifolds.* Ergebnisse der Mathematik, Springer (1970).
[Bu] BURGHELEA D.: *Cyclic homology and the algebraic $K$-theory of spaces I.* Contemporary Mathematics, 55, 89-115 (1986).
[BF] BURGHELEA D. and FIEDOROWICZ Z.: *Cyclic homology and the algebraic $K$-theory of spaces II.* Topology 25, 303-317 (1986).
[C2] CONNES A.: *Non commutative differential geometry.* Publ. Institut Hautes Etudes Scientifiques N 62, 257-360 (1985).
[CK] CONNES A. et KAROUBI M.: *Caractère multiplicatif d'un module de Fredholm.* $K$-theory 2, 431-463 (1988).
[Cu] CUNTZ J.: *A new look at $KK$-theory.* $K$-theory 1, 31-51 (1987).
[HH] HAUSMANN J.-C. and HUSEMOLLER D.: *Acyclic maps.* L'Enseignement mathématique, XXV, 53-75 (1979).
[H] HIGSON N. *A characterization of $KK$-theory.* Pacific J. Math. 126, 253-276 (1987).
[HJ] HOOD C., JONES J.D.S.: *Some algebraic properties of cyclic homology*

*groups.* $K$-theory 1, 361-384 (1987).

[Gra] GRAYSON D.: *Higher Algebraic $K$-theory II (after Daniel Quillen).* Lecture Notes in Math. 551 (1976).

[K3] KAROUBI M.: *Quatre Notes sur l'homologie cyclique.* C.R. Acad. Sci. Paris 297, 381-384, 447-450, 513-516 et 557-560 (1983).

[K4] KAROUBI M.: *Le théorème fondamental de la $K$-théorie hermitienne.* Annals of Math. 112, 259-282 (1980).

[KV] KAROUBI M. et VILLAMAYOR O.: *Foncteurs $K^n$ en algèbre et en topologie, Comptes Rendus Acad.* Sci. Paris, t. 269, p. 416 (1969).

[KL] KAROUBI M. et LAMBRE T.: *Quelques classes caractéristiques en théorie des nombres (avec T. Lambre).* J. Reine angew. Math. 543, p. 169-186 (2002)

[Kas] KASPAROV G.: *The operator $K$-functor and extension of $C^*$-algebras.* Izv. Akad. Nauk. SSSR, Ser. Math. 44, 571-636 (1980).

[L2] LODAY J.-L.: *Homotopie des espaces de concordance.* Séminaire Bourbaki N 516 (1977/78).

[LQ] LODAY J.-L. and QUILLEN D.: *Cyclic homology and the Lie algebra homology of matrices.* Comm. Math. Helvetici 59, 565-591 (1984).

[MS] MERKURJEV A.S. and SUSLIN A.A.: *$K$-cohomology of Severi-Brauer varieties and the norm residue homomorphism.* Math. USSR Izv. 21, 307-440 (1983).

[NT] NEST R. and TSYGAN B.: *Algebraic Index Theorem.* Comm. Math. Phys. 172, 223-262 (1995).

[Q] QUILLEN D.: *Higher Algebraic $K$-theory.* Lecture Notes in Maths 341, 85-147. Berlin-Heidelberg-New York: Springer (1973).

[So] SOULE C.: *$K$-théorie des anneaux d'entiers de corps de nombres.* Inventiones math. 55, 251-295 (1979).

[Su] SUSLIN A.: *On the $K$-theory of local fields.* J. Pure and Applied Algebra 34, 301-318 (1984).

[SW] SUSLIN A. and WODZICKI M.: *Excision in Algebraic $K$-theory.* Ann. of Maths 136, 51-122 (1992).

[Th] THOMASON R.: *Algebraic $K$-theory and etale cohomology.* Ann. Sci. Ec. Norm. Supér. 13, 437-552 (1985).

[Ts] TSYGAN B.L.: *The homology of matrix Lie algebras over rings and the Hochschild homology.* Russian Math. Survey 38 *N* 2, 198-199 (1983).

[Vo] VOEVODSKY V. (to appear).

[Wald] WALDHAUSEN W.: *Algebraic $K$-theory of topological spaces I.* Proc. Symp. Pure Math. 32 AMS, 35-60 (1978), II Springer Lecture Notes in Math. 1051, 173-196 (1984).

[Wall] WALL C.T.C.: *Surgery on compact manifolds.* Academic Press (1967).

[Wo] WOOD R.: *Banach algebras and Bott periodicity.* Topology 4, 371-389 (1966).

*Cohomology of Groups and*
*Algebraic K-theory*
ALM 12, pp. 269–341



# On the Farrell-Jones and Related Conjectures

Wolfgang Lück*

**Abstract**

These extended notes are based on a series of six lectures presented at the summer school "Cohomology of groups and algebraic $K$-theory" which took place in Hangzhou, China from July 1 until July 12 in 2007. They give an introduction to the *Farrell-Jones* and the *Baum-Connes Conjecture.*

**Mathematics Subject Classification 2000:** 19A31, 19B28, 19D99, 19G24, 19K99, 46L80, 57R67

**Key words and Phrases:** $K$- and $L$-groups of group rings and group $C^*$-algebras, Farrell-Jones Conjecture, Baum-Connes Conjecture

## 0. Introduction

These extended notes are based on a series of six lectures presented at the summer school "Cohomology of groups and algebraic $K$-theory" which took place in Hangzhou, China from July 1 until July 12 in 2007. They contain an introduction to the *Farrell-Jones* and the *Baum-Connes Conjecture.*

Given a group $G$, the Farrell-Jones Conjecture and the Baum-Connes Conjecture respectively predict the values of the algebraic $K$- and $L$-theory of the group ring $RG$ and of the topological $K$-theory of the reduced group $C^*$-algebra respectively. These are very hard to compute directly. These conjectures identify them via assembly maps to much easier to handle equivariant homology groups of certain classifying spaces. This is the computational aspect of these conjectures.

But also the structural aspect is very important. The assembly maps have geometric or analytic interpretations. Hence the Farrell-Jones Conjecture and the Baum-Connes Conjecture imply many very well-known conjectures such as the

*Mathematisches Institut, Universität Münster, D-48149 Münster, Germany. Url: http://www.math.uni-muenster.de/u/lueck E-mail: lueck@math.uni-muenster.de

ones due to Bass, Borel, Kaplansky, and Novikov. The point is that the Farrell-Jones Conjecture and the Baum-Connes Conjecture have been proven for many groups for which the other conjectures were a priori not known.

The prerequisites consist of a solid knowledge of homology theory and $CW$-complexes and of a basic knowledge of rings, modules, homological algebra, groups, group homology, finite dimensional representation theory of finite groups, group actions, categories, homotopy groups, and manifolds. The challenge for the reader but also the beauty, impact and fascination of these conjectures come from the broad scope of mathematics which they address and which is needed for proofs and applications.

For a more advanced survey on the Farrell-Jones and the Baum-Cones Conjecture we refer to Lück-Reich [126]. There more details are given and more aspects are discussed but it is addressed to a more advanced reader and requires much more previous knowledge.

We fix some notation. Ring will always mean associative ring with unit (which is not necessarily commutative). Examples are the ring of integers $\mathbb{Z}$, the fields of rational numbers $\mathbb{Q}$, of real numbers $\mathbb{R}$ and of complex numbers $\mathbb{C}$, the finite field $\mathbb{F}_p$ of $p$ elements, and the group ring $RG$ for a ring $R$ and a group $G$. Ring homomorphisms are unital. Modules are understood to be left modules unless explicitly stated differently. Groups are understood to be discrete unless explicitly stated differently.

The author wants to express his deep gratitude to all the organizers of the summer school for their excellent work, support and hospitality.

# 1. The role of lower and middle $K$-theory in topology

The outline of this section is:

- Introduce the projective class group $K_0(R)$.
- Discuss its algebraic and topological significance (e.g., finiteness obstruction).
- Introduce $K_1(R)$ and the Whitehead group $\mathrm{Wh}(G)$.
- Discuss its algebraic and topological significance (e.g., $s$-cobordism theorem).
- Introduce negative $K$-theory and the Bass-Heller-Swan decomposition.

**Definition 1.1** (Projective $R$-module). An $R$-module $P$ is called *projective* if it satisfies one of the following equivalent conditions:

(1) $P$ is a direct summand in a free $R$-module;
(2) The following lifting problem has always a solution

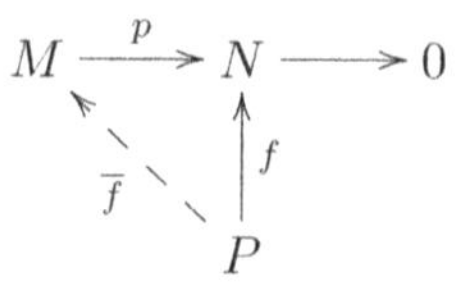

(3) If $0 \to M_0 \to M_1 \to M_2 \to 0$ is an exact sequence of $R$-modules, then $0 \to \hom_R(P, M_0) \to \hom_R(P, M_1) \to \hom_R(P, M_2) \to 0$ is exact.

**Example 1.2** (Principal ideal domains). Over a field or, more generally, over a principal ideal domain every projective module is free. If $R$ is a principal ideal domain, then a finitely generated $R$-module is projective (and hence free) if and only if it is torsionfree. For instance $\mathbb{Z}/n$ is for $n \geq 2$ never projective as $\mathbb{Z}$-module.

**Example 1.3** (Product of rings). Let $R$ and $S$ be rings and $R \times S$ be their product. Then $R \times \{0\}$ is a finitely generated projective $R \times S$-module which is not free.

**Example 1.4** (Trivial representation of a finite group). Let $F$ be a field of characteristic $p$ for $p$ a prime number or $p = 0$. Let $G$ be a finite group. Then $F$ with the trivial $G$-action is a projective $FG$-module if and only if $p = 0$ or $p$ does not divide the order of $G$. It is a free $FG$-module only if $G$ is trivial.

**Definition 1.5** (Projective class group). Let $R$ be a ring. Define its *projective class group* $K_0(R)$ to be the abelian group whose generators are isomorphism classes $[P]$ of finitely generated projective $R$-modules $P$ and whose relations are $[P_0] + [P_2] = [P_1]$ for every exact sequence $0 \to P_0 \to P_1 \to P_2 \to 0$ of finitely generated projective $R$-modules.

The projective class group $K_0(R)$ is the same as the Grothendieck construction applied to the abelian monoid of isomorphism classes of finitely generated projective $R$-modules under direct sum. There do exists rings $R$ with $K_0(R) = 0$, e.g., $R = \text{end}(F)$ for a field $F$.

**Definition 1.6** (Reduced projective class group). The *reduced projective class group* $\widetilde{K}_0(R)$ is the quotient of $K_0(R)$ by the subgroup generated by the classes of finitely generated free $R$ modules, or, equivalently, the cokernel of $K_0(\mathbb{Z}) \to K_0(R)$.

**Remark 1.7** (Stably finitely generated free modules). Let $P$ be a finitely generated projective $R$-module. It is stably finitely generated free, i.e., $P \oplus R^m \cong R^n$ for appropriate $m, n \in \mathbb{Z}$, if and only if $[P] = 0$ in $\widetilde{K}_0(R)$. Hence $\widetilde{K}_0(R)$ measures the deviation of finitely generated projective $R$-modules from being stably finitely generated free.

There exists finitely generated projective $R$-modules which are stably finitely generated free but not finitely generated free. An example is $R = C(S^2)$ and $P = C(TS^2)$, where $C(S^2)$ is the ring of continuous functions $S^2 \to \mathbb{R}$ and $C(TS^2)$ is the $C(S^2)$-module of sections of the tangent bundle of $S^2$. However, in most of the applications the relevant question is whether a finitely generated projective $R$-module is stable finitely generated free and not whether it is finitely generated free.

**Remark 1.8** (Universal dimension function). The assignment $P \mapsto [P] \in K_0(R)$ is the *universal additive invariant* or *dimension function* for finitely generated projective $R$-modules in the following sense. Given an abelian group $A$ and an assignment associating to a finitely generated projective $R$-module $P$ an element

$a(P) \in A$ such that $a(P_0) - a(P_1) + a(P_2) = 0$ holds for any exact sequence of finitely generated projective $R$-modules $0 \to P_0 \to P_1 \to P_2 \to 0$, there exists precisely one homomorphism of abelian groups $\phi\colon K_0(R) \to A$ satisfying $\phi([P]) = a(P)$ for every finitely generated projective $R$-module $P$.

**Remark 1.9** (Induction). Let $f\colon R \to S$ be a ring homomorphism. Consider $S$ as a $S$-$R$-bimodule via $f$. Given an $R$-module $M$, let $f_*M$ be the $S$-module $S \otimes_R M$. We obtain a homomorphism of abelian groups

$$f_*\colon K_0(R) \to K_0(S), \quad [P] \mapsto [f_*P]$$

called *induction* or *change of rings homomorphism*. Thus $K_0$ becomes a covariant functor from the category of rings to the category of abelian algebras.

**Lemma 1.10** ($K_0$ and products). *Let $R$ and $S$ be rings. Then the two projections from $R \times S$ to $R$ and $S$ induce isomorphisms*

$$K_0(R \times S) \xrightarrow{\cong} K_0(R) \times K_0(S).$$

**Theorem 1.11** (Morita equivalence). *Let $R$ be a ring and $M_n(R)$ be the ring of $(n,n)$-matrices over $R$. We can consider $R^n$ as a $M_n(R)$-$R$-bimodule and as a $R$-$M_n(R)$-bimodule by scalar and matrix multiplication. Tensoring with these yields mutually inverse isomorphisms*

$$\begin{aligned} K_0(R) &\xrightarrow{\cong} K_0(M_n(R)), & [P] &\mapsto [{}_{M_n(R)}R^n{}_R \otimes_R P]; \\ K_0(M_n(R)) &\xrightarrow{\cong} K_0(R), & [Q] &\mapsto [{}_R R^n{}_{M_n(R)} \otimes_{M_n(R)} Q]. \end{aligned}$$

**Example 1.12** (Principal ideal domains). Let $R$ be a principal ideal domain. Let $F$ be its quotient field. Then we obtain mutually inverse isomorphisms

$$\begin{aligned} \mathbb{Z} &\xrightarrow{\cong} K_0(R), & n &\mapsto n \cdot [R]; \\ K_0(R) &\xrightarrow{\cong} \mathbb{Z}, & [P] &\mapsto \dim_F(F \otimes_R P). \end{aligned}$$

**Example 1.13** (Representation ring). Let $G$ be a finite group and let $F$ be a field of characteristic zero. Then the representation ring $R_F(G)$ is the same as $K_0(FG)$. Taking the character of a representation yields an isomorphism

$$R_{\mathbb{C}}(G) \otimes_{\mathbb{Z}} \mathbb{C} = K_0(\mathbb{C}G) \otimes_{\mathbb{Z}} \mathbb{C} \xrightarrow{\cong} \mathrm{class}(G, \mathbb{C}),$$

where $\mathrm{class}(G; \mathbb{C})$ is the complex vector space of class functions $G \to \mathbb{C}$, i.e., functions, which are constant on conjugacy classes. We refer for instance to the book of Serre [167] for more information about the representation theory of finite groups.

**Definition 1.14** (Dedekind domain). A commutative ring $R$ is called *Dedekind domain* if it is an integral domain, i.e., contains no non-trivial zero-divisors, and for every pair of ideals $I \subseteq J$ of $R$ there exists an ideal $K \subseteq R$ with $I = JK$.

A ring is called *hereditary* if every ideal is projective, or, equivalently, if every submodule of a projective $R$-module is projective.

**Theorem 1.15** (Characterization of Dedekind domains). *The following assertions are equivalent for a commutative integral domain with quotient field $F$:*

(1) *$R$ is a Dedekind domain;*
(2) *$R$ is hereditary;*
(2) *Every finitely generated torsionfree $R$-module is projective;*
(4) *$R$ is Noetherian and integrally closed in its quotient field $F$ and every non-zero prime ideal is maximal.*

*Proof.* This follows from [49, Proposition 4.3 on page 76 and Proposition 4.6 on page 77] and the fact that a finitely generated torsionfree module over an integral domain $R$ can be embedded into $R^n$ for some integer $n \geq 0$ (see Auslander-Buchsbaum [8, Proposition 3.3 in Chapter 9 on page 321]). □

**Example 1.16** (Ring of integers). Recall that an *algebraic number field* is a finite algebraic extension of $\mathbb{Q}$ and the *ring of integers* in $F$ is the integral closure of $\mathbb{Z}$ in $F$. The ring of integers in an algebraic number field is a Dedekind domain. (see [154, Theorem 1.4.18 on page 22]).

**Example 1.17** (Dedekind domains). Let $R$ be a Dedekind domain. We call two ideals $I$ and $J$ in $R$ equivalent if there exists non-zero elements $r$ and $s$ in $R$ with $rI = sJ$. The *ideal class group* $C(R)$ is the abelian group of equivalence classes of ideals under multiplication of ideals. Then $C(R)$ is finite and we obtain an isomorphism

$$C(R) \xrightarrow{\cong} \widetilde{K}_0(R), \quad [I] \mapsto [I].$$

A proof of the claim above can be found for instance in [132, Corollary 11 on page 14] and [154, Theorem 1.4.12 on page 20 and and Theorem 1.4.19 on page 23].

The structure of the finite abelian group

$$C(\mathbb{Z}[\exp(2\pi i/p)]) \cong \widetilde{K}_0(\mathbb{Z}[\exp(2\pi i/p)]) \simeq \widetilde{K}_0(\mathbb{Z}[\mathbb{Z}/p])$$

is only known for small prime numbers $p$ (see [132, Remark 3.4 on page 30]).

**Theorem 1.18** (Swan (1960)). *If $G$ is finite, then $\widetilde{K}_0(\mathbb{Z}G)$ is finite.*

*Proof.* See [175, Theorem 8.1 and Proposition 9.1]. □

Let $X$ be a compact space. Let $K^0(X)$ be the Grothendieck group of isomorphism classes of finite-dimensional complex vector bundles over $X$. This is the zero-th term of a generalized cohomology theory $K^*(X)$ called topological $K$-theory. It is 2-periodic, i.e., $K^n(X) = K^{n+2}(X)$, and satisfies $K^0(\{\bullet\}) = \mathbb{Z}$ and $K^1(\{\bullet\}) = \{0\}$, where $\{\bullet\}$ is the space consisting of one point.

Let $C(X)$ be the ring of continuous functions from $X$ to $\mathbb{C}$.

**Theorem 1.19** (Swan (1962)). *If $X$ is a compact space, then there is an isomorphism*

$$K^0(X) \xrightarrow{\cong} K_0(C(X)).$$

*Proof.* See [176]. □

**Definition 1.20** (Finitely dominated). A $CW$-complex $X$ is called *finitely dominated* if there exists a finite $CW$-complex $Y$ together with maps $i\colon X \to Y$ and $r\colon Y \to X$ satisfying $r \circ i \simeq \mathrm{id}_X$.

Obviously a finite $CW$-complex is finitely dominated.

**Problem 1.21.** Is a given finitely dominated $CW$-complex homotopy equivalent to a finite $CW$-complex?

A finitely dominated $CW$-complex $X$ defines an element

$$o(X) \in K_0(\mathbb{Z}[\pi_1(X)])$$

called its *finiteness obstruction* as follows. Let $\widetilde{X}$ be the universal covering. The fundamental group $\pi = \pi_1(X)$ acts freely on $\widetilde{X}$. Let $C_*(\widetilde{X})$ be the cellular chain complex. It is a free $\mathbb{Z}\pi$-chain complex. Since $X$ is finitely dominated, there exists a finite projective $\mathbb{Z}\pi$-chain complex $P_*$ with $P_* \simeq_{\mathbb{Z}\pi} C_*(\widetilde{X})$. Finite projective means that every $P_i$ is finitely generated projective and $P_i \neq 0$ holds only for finitely many element $i \in \mathbb{Z}$.

**Definition 1.22** (Wall's finiteness obstruction). Define

$$o(X) := \sum_n (-1)^n \cdot [P_n] \in K_0(\mathbb{Z}\pi).$$

This definition is indeed independent of the choice of $P_*$.

**Theorem 1.23** (Wall (1965)). *A finitely dominated $CW$-complex $X$ is homotopy equivalent to a finite $CW$-complex if and only if its reduced finiteness obstruction $\widetilde{o}(X) \in \widetilde{K}_0(\mathbb{Z}[\pi_1(X)])$ vanishes.*

*Given a finitely presented group $G$ and $\xi \in K_0(\mathbb{Z}G)$, there exists a finitely dominated $CW$-complex $X$ with $\pi_1(X) \cong G$ and $o(X) = \xi$.*

*Proof.* See [187] and [188]. □

A finitely dominated simply connected $CW$-complex is always homotopy equivalent to a finite $CW$-complex since $\widetilde{K}_0(\mathbb{Z}) = \{0\}$.

**Corollary 1.24** (Geometric characterization of $\widetilde{K}_0(\mathbb{Z}G) = \{0\}$). *The following statements are equivalent for a finitely presented group $G$:*

(1) *Every finite dominated $CW$-complex with $G \cong \pi_1(X)$ is homotopy equivalent to a finite $CW$-complex;*
(2) $\widetilde{K}_0(\mathbb{Z}G) = \{0\}$.

**Conjecture 1.25** (Vanishing of $\widetilde{K}_0(\mathbb{Z}G)$ for torsionfree $G$). *If $G$ is torsionfree, then*

$$\widetilde{K}_0(\mathbb{Z}G) = \{0\}.$$

For more information about the finiteness obstruction we refer for instance to [75], [76], [116], [134], [150], [156], [182], [187] and [188].

**Definition 1.26** ($K_1$-group)**.** Define the $K_1(R)$ to be the abelian group whose generators are conjugacy classes $[f]$ of automorphisms $f\colon P \to P$ of finitely generated projective $R$-modules with the following relations:

(1) Given an exact sequence $0 \to (P_0, f_0) \to (P_1, f_1) \to (P_2, f_2) \to 0$ of automorphisms of finitely generated projective $R$-modules, we get $[f_0] + [f_2] = [f_1]$;
(2) $[g \circ f] = [f] + [g]$.

**Theorem 1.27** ($K_1(R)$ and matrices)**.** *There is a natural isomorphism*

$$K_1(R) \cong GL(R)/[GL(R), GL(R)],$$

*where the target is the abelianization of the general linear group* $GL(R) = \bigcup_{n \geq 1} GL_n(R)$.

*Proof.* See [154, Theorem 3.1.7 on page 113]. □

**Remark 1.28** ($K_1(R)$ and row and column operations)**.** An invertible matrix $A \in GL(R)$ can be reduced by elementary row and column operations and (de-)stabilization to the empty matrix if and only if $[A] = 0$ holds in the *reduced* $K_1$*-group*

$$\widetilde{K}_1(R) := K_1(R)/\{\pm 1\} = \operatorname{cok}\left(K_1(\mathbb{Z}) \to K_1(R)\right).$$

**Remark 1.29** ($K_1(R)$ and determinants)**.** If $R$ is commutative, the determinant induces an epimorphism

$$\det\colon K_1(R) \to R^{\times},$$

which in general is not bijective.

The assignment $A \mapsto [A] \in K_1(R)$ can be thought of as the universal determinant for $R$, where $R$ is not necessarily commutative. Namely, given any abelian group $A$ together with an assignment which associates to an $R$-automorphism $f\colon P \to P$ of a finitely generated projective $R$-module an element $[f]$ such that the obvious analogues of the relations appearing in Definition 1.26 hold, there exists precisely one homomorphism of abelian groups $\phi\colon K_1(R) \to A$ sending $[f]$ to $a(f)$ for every $R$-automorphism $f$ of a finitely generated projective $R$-module.

There do exists rings $R$ with $K_1(R) = 0$, e.g. $R = \operatorname{end}(F)$ for a field $F$.

**Remark 1.30** ($K_1(R)$ of principal ideal domains)**.** There exists principal ideal domains $R$ such that $\det\colon K_1(R) \to R^{\times}$ is not bijective. For instance Grayson [81] gives such an example, namely, take $\mathbb{Z}[x]$ and invert $x$ and all polynomials of the shape $x^m - 1$ for $m \geq 1$. Other examples can be found in Ischebeck [94].

**Theorem 1.31** ($K_1$ of ring of integers, Bass-Milnor-Serre (1967))**.** *Let $R$ be the ring of integers in an algebraic number field. Then the determinant induces an isomorphism*

$$\det\colon K_1(R) \xrightarrow{\cong} R^{\times}.$$

*Proof.* See [23, 4.3]. □

**Definition 1.32** (Whitehead group)**.** The *Whitehead group* of a group $G$ is defined to be

$$\mathrm{Wh}(G) = K_1(\mathbb{Z}G)/\{\pm g \mid g \in G\}.$$

**Lemma 1.33.** *We have* $\mathrm{Wh}(\{1\}) = \{0\}$.

*Proof.* The ring $\mathbb{Z}$ possesses an Euclidean algorithm. Hence every invertible matrix over $\mathbb{Z}$ can be reduced via elementary row and column operations and destabilization to a $(1,1)$-matrix $(\pm 1)$. For every ring such operations do not change the class of a matrix in $K_1(R)$. □

Let $G$ be a finite group. Let $F$ be $\mathbb{Q}$, $\mathbb{R}$ or $\mathbb{C}$. Define $r_F(G)$ to be the number of irreducible $F$-representations of $G$. This is the same as the number of $F$-conjugacy classes of elements of $G$. Here $g_1 \sim_{\mathbb{C}} g_2$ if and only if $g_1 \sim g_2$, i.e., $gg_1g^{-1} = g_2$ for some $g \in G$. We have $g_1 \sim_{\mathbb{R}} g_2$ if and only if $g_1 \sim g_2$ or $g_1 \sim g_2^{-1}$ holds. We have $g_1 \sim_{\mathbb{Q}} g_2$ if and only if $\langle g_1 \rangle$ and $\langle g_1 \rangle$ are conjugated as subgroups of $G$.

**Theorem 1.34** ($\mathrm{Wh}(G)$ for finite groups $G$)**.**

(1) *The Whitehead group* $\mathrm{Wh}(G)$ *is a finitely generated abelian group;*
(2) *Its rank is* $r_{\mathbb{R}}(G) - r_{\mathbb{Q}}(G)$.
(3) *The torsion subgroup of* $\mathrm{Wh}(G)$ *is the kernel of the map* $K_1(\mathbb{Z}G) \to K_1(\mathbb{Q}G)$.

In contrast to $\widetilde{K}_0(\mathbb{Z}G)$ the Whitehead group $\mathrm{Wh}(G)$ is computable (see Oliver [138]).

**Definition 1.35** ($h$-cobordism)**.** An *$h$-cobordism* over a closed manifold $M_0$ is a compact manifold $W$ whose boundary is the disjoint union $M_0 \amalg M_1$ such that both inclusions $M_0 \to W$ and $M_1 \to W$ are homotopy equivalences.

**Theorem 1.36** ($s$-Cobordism Theorem, Barden, Mazur, Stallings, Kirby-Siebenmann)**.**

(1) *Let* $M_0$ *be a closed (smooth) manifold of dimension* $\geq 5$. *Let* $(W; M_0, M_1)$ *be an $h$-cobordism over* $M_0$.
*Then* $W$ *is homeomorphic (diffeomorpic) to* $M_0 \times [0,1]$ *relative* $M_0$ *if and only if its Whitehead torsion*

$$\tau(W, M_0) \in \mathrm{Wh}(\pi_1(M_0)).$$

*vanishes;*
(2) *Let* $G$ *be a finitely presented group* $G$, $n$ *an integer* $n \geq 5$ *and* $x$ *an element in* $\mathrm{Wh}(G)$. *Then there exists an $n$-dimensional $h$-cobordism* $(W; M_0, M_1)$ *over* $M_0$ *with* $\tau(W, M_0) = x$.

**Corollary 1.37** (Geometric characterization of $\mathrm{Wh}(G) = \{0\}$)**.** *The following statements are equivalent for a finitely presented group* $G$ *and a fixed integer* $n \geq 6$

(1) *Every compact $n$-dimensional $h$-cobordism* $W$ *with* $G \cong \pi_1(W)$ *is trivial;*
(2) $\mathrm{Wh}(G) = \{0\}$.

**Conjecture 1.38** (Vanishing of $\mathrm{Wh}(G)$ for torsionfree $G$). *If $G$ is torsionfree, then*

$$\mathrm{Wh}(G) = \{0\}.$$

**Conjecture 1.39** (Poincaré Conjecture). *Let $M$ be an $n$-dimensional topological manifold which is a homotopy sphere, i.e., homotopy equivalent to $S^n$.*

*Then $M$ is homeomorphic to $S^n$.*

**Theorem 1.40.** *For $n \geq 5$ the Poincaré Conjecture is true.*

*Proof.* We sketch the proof for $n \geq 6$. Let $M$ be an $n$-dimensional homotopy sphere. Let $W$ be obtained from $M$ by deleting the interior of two disjoint embedded disks $D_1^n$ and $D_2^n$. Then $W$ is a simply connected $h$-cobordism. Since $\mathrm{Wh}(\{1\})$ is trivial, we can find a homeomorphism $f\colon W \xrightarrow{\cong} \partial D_1^n \times [0,1]$ which is the identity on $\partial D_1^n = \partial D_1^n \times \{0\}$. By the Alexander trick, i.e., by coning the homeomorphism of $\partial D^n$ to the cone of $\partial D^n$ which is $D^n$, we can extend the homeomorphism $f|_{\partial D_2^n}\colon \partial D_2^n \xrightarrow{\cong} \partial D_1^n = \partial D_1^n \times \{1\}$ to a homeomorphism $g\colon D_2^n \to D_1^n$. The three homeomorphisms $id_{D_1^n}$, $f$ and $g$ fit together to a homeomorphism $h\colon M \to D_1^n \cup_{\partial D_1^n \times \{0\}} \partial D_1^n \times [0,1] \cup_{\partial D_1^n \times \{1\}} D_1^n$. The target is obviously homeomorphic to $S^n$. □

**Remark 1.41** (Exotic spheres). The argument above does not imply that for a smooth manifold $M$ we obtain a diffeomorphism $g\colon M \to S^n$. The problem is that the Alexander trick does not work smoothly. Indeed, there exists so called exotic spheres, i.e., closed smooth manifolds which are homeomorphic but not diffeomorphic to $S^n$. For more information about exotic spheres we refer for instance to [101], [110], [113] and [119, Chapter 6].

**Remark 1.42** (The Poincaré Conjecture and the $s$-cobordism theorem in low dimensions). The Poincaré Conjecture has been proved in dimension 4 by Freedman [79] and in dimension 3 by Perelman (see [144], [145] and for more details for instance [103], [137]). It is true in dimensions 1 and 2 for elementary reasons.

The $s$-cobordism theorem is known to be false (smoothly) for $n = \dim(M_0) = 4$ in general, by the work of Donaldson [55], but it is true for $n = \dim(M_0) = 4$ for so called "good" fundamental groups in the topological category by results of Freedman [79], [80]. The trivial group is an example of a "good" fundamental group. Counterexamples in the case $n = \dim(M_0) = 3$ are constructed by Cappell-Shaneson [41].

**Remark 1.43** (Surgery program). The $s$-cobordism theorem is a key ingredient in the surgery program for the classification of closed manifolds due to Browder, Novikov, Sullivan and Wall. For more information about surgery theory we refer for instance to [33], [38], [39], [73], [74], [98], [104], [148], [173], [172], and [189].

More information about Whitehead torsion and the $s$-cobordism theorem can be found for instance in [47], [100], [119, Chapter 1], [130], [131], [157, page 87-90].

**Definition 1.44** (Bass-Nil-groups). Define for $n = 0, 1$

$$\mathrm{NK}_n(R) := \mathrm{coker}\left(K_n(R) \to K_n(R[t])\right).$$

**Theorem 1.45** (Bass-Heller-Swan decomposition for $K_1$, Bass-Heller-Swan(1964)). *There is an isomorphism, natural in R,*

$$K_0(R) \oplus K_1(R) \oplus \mathrm{NK}_1(R) \oplus \mathrm{NK}_1(R) \xrightarrow{\cong} K_1(R[t,t^{-1}]) = K_1(R[\mathbb{Z}]).$$

*Proof.* See for instance [22] (for regular rings), [19, Chapter XII], [154, Theorem 3.2.22 on page 149]. □

Notice that the Bass-Heller-Swan decomposition for $K_1$ gives the possibility to define $K_0(R)$ in terms of $K_1$. This motivates the following definition.

**Definition 1.46** (Negative $K$-theory). Define inductively for $n = -1, -2, \ldots$

$$K_n(R) := \operatorname{coker}\left(K_{n+1}(R[t]) \oplus K_{n+1}(R[t^{-1}]) \to K_{n+1}(R[t,t^{-1}])\right).$$

Define for $n = -1, -2, \ldots$

$$\mathrm{NK}_n(R) := \operatorname{coker}\left(K_n(R) \to K_n(R[t])\right).$$

**Theorem 1.47** (Bass-Heller-Swan decomposition for negative $K$-theory). *For $n \leq 1$ there is an isomorphism, natural in R,*

$$K_{n-1}(R) \oplus K_n(R) \oplus \mathrm{NK}_n(R) \oplus \mathrm{NK}_n(R) \xrightarrow{\cong} K_n(R[t,t^{-1}]) = K_n(R[\mathbb{Z}]).$$

**Definition 1.48** (Regular ring). A ring $R$ is called *regular* if it is Noetherian and every finitely generated $R$-module possesses a finite projective resolution.

Principal ideal domains are regular. In particular $\mathbb{Z}$ and any field are regular. If $R$ is regular, then $R[t]$ and $R[t,t^{-1}] = R[\mathbb{Z}]$ are regular. If $R$ is Noetherian, then $RG$ is not in general Noetherian. Theorem 1.47 implies

**Theorem 1.49** (Bass-Heller-Swan decomposition for regular rings). *Suppose that R is regular. Then*

$$\begin{aligned} K_n(R) &= 0 \quad \textit{for } n \leq -1; \\ \mathrm{NK}_n(R) &= 0 \quad \textit{for } n \leq 1, \end{aligned}$$

*and the Bass-Heller-Swan decomposition reduces for $n \leq 1$ to the natural isomorphism*

$$K_{n-1}(R) \oplus K_n(R) \xrightarrow{\cong} K_n(R[t,t^{-1}]) = K_n(R[\mathbb{Z}]).$$

There are also higher algebraic $K$-groups $K_n(R)$ for $n \geq 2$ due to Quillen (1973). They are defined as homotopy groups of certain spaces or spectra. We refer to the lectures of Grayson. Most of the well known features of $K_0(R)$ and $K_1(R)$ extend to both negative and higher algebraic $K$-theory. For instance the Bass-Heller-Swan decomposition holds also for higher algebraic $K$-theory.

**Remark 1.50** (Similarity between $K$-theory and group homology). Notice the following formulas for a regular ring $R$ and a generalized homology theory $\mathcal{H}_*$, which look similar:

$$\begin{aligned} K_n(R[\mathbb{Z}]) &\cong K_n(R) \oplus K_{n-1}(R); \\ \mathcal{H}_n(B\mathbb{Z}) &\cong \mathcal{H}_n(\{\bullet\}) \oplus \mathcal{H}_{n-1}(\{\bullet\}). \end{aligned}$$

If $G$ and $K$ are groups, then we have the following formulas, which look similar:

$$\begin{aligned} \widetilde{K}_n(\mathbb{Z}[G * K]) &\cong \widetilde{K}_n(\mathbb{Z}G) \oplus \widetilde{K}_n(\mathbb{Z}K); \\ \widetilde{\mathcal{H}}_n(B(G * K)) &\cong \widetilde{\mathcal{H}}_n(BG) \oplus \widetilde{\mathcal{H}}_n(BK). \end{aligned}$$

**Question 1.51** ($K$-theory of group rings and group homology). Is there a relation between $K_n(RG)$ and group homology of $G$?

# 2. The Isomorphism Conjectures in the torsionfree case

The outline of this section is:

- We introduce spectra and how they yield homology theories.
- We state the Farrell-Jones-Conjecture and the Baum-Connes Conjecture for torsionfree groups.
- We discuss applications of these conjectures such as the Kaplansky Conjecture and the Borel Conjecture.
- We explain that the formulations for torsionfree groups cannot extend to arbitrary groups.

Given two pointed spaces $X = (X, x_0)$ and $Y = (Y, y_0)$, their *one-point union* and their *smash product* are defined to be the pointed spaces

$$\begin{aligned} X \vee Y &:= \{(x, y_0) \mid x \in X\} \cup \{(x_0, y) \mid y \in Y\} \subseteq X \times Y; \\ X \wedge Y &:= (X \times Y)/(X \vee Y). \end{aligned}$$

We have $S^{n+1} \cong S^n \wedge S^1$.

**Definition 2.1** (Spectrum). A *spectrum*

$$\mathbf{E} = \{(E(n), \sigma(n)) \mid n \in \mathbb{Z}\}$$

is a sequence of pointed spaces $\{E(n) \mid n \in \mathbb{Z}\}$ together with pointed maps called *structure maps*

$$\sigma(n) \colon E(n) \wedge S^1 \longrightarrow E(n+1).$$

A *map of spectra*

$$\mathbf{f} \colon \mathbf{E} \to \mathbf{E}'$$

is a sequence of maps $f(n) \colon E(n) \to E'(n)$ which are compatible with the structure maps $\sigma(n)$, i.e., $f(n+1) \circ \sigma(n) = \sigma'(n) \circ (f(n) \wedge \mathrm{id}_{S^1})$ holds for all $n \in \mathbb{Z}$.

**Example 2.2** (Sphere spectrum). The *sphere spectrum* $\mathbf{S}$ has as $n$-th space $S^n$ and as $n$-th structure map the homeomorphism $S^n \wedge S^1 \xrightarrow{\cong} S^{n+1}$.

**Example 2.3** (Suspension spectrum). Let $X$ be a pointed space. Its *suspension spectrum* $\Sigma^\infty X$ is given by the sequence of spaces $\{X \wedge S^n \mid n \geq 0\}$ with the homeomorphisms $(X \wedge S^n) \wedge S^1 \cong X \wedge S^{n+1}$ as structure maps. We have $\mathbf{S} = \Sigma^\infty S^0$.

**Definition 2.4** ($\Omega$-spectrum). Given a spectrum $\mathbf{E}$, we can consider instead of the structure map $\sigma(n) \colon E(n) \wedge S^1 \to E(n+1)$ its adjoint

$$\sigma'(n) \colon E(n) \to \Omega E(n+1) = \operatorname{map}(S^1, E(n+1)).$$

We call $\mathbf{E}$ an *$\Omega$-spectrum* if each map $\sigma'(n)$ is a weak homotopy equivalence.

**Definition 2.5** (Homotopy groups of a spectrum). Given a spectrum $\mathbf{E}$, define for $n \in \mathbb{Z}$ its *$n$-th homotopy group*

$$\pi_n(\mathbf{E}) := \operatorname*{colim}_{k \to \infty} \pi_{k+n}(E(k))$$

to be the abelian group which is given by the colimit over the directed system indexed by $\mathbb{Z}$ with $k$-th structure map

$$\pi_{k+n}(E(k)) \xrightarrow{\sigma'(k)} \pi_{k+n}(\Omega E(k+1)) = \pi_{k+n+1}(E(k+1)).$$

Notice that a spectrum can have in contrast to a space non-trivial negative homotopy groups. If $\mathbf{E}$ is an $\Omega$-spectrum, then $\pi_n(\mathbf{E}) = \pi_n(E(0))$ for all $n \geq 0$.

**Example 2.6** (Eilenberg-MacLane spectrum). Let $A$ be an abelian group. The $n$-th *Eilenberg-MacLane space* $K(A,n)$ associated to $A$ for $n \geq 0$ is a $CW$-complex with $\pi_m(K(A,n)) = A$ for $m = n$ and $\pi_m(K(A,n)) = \{0\}$ for $m \neq n$.

The associated *Eilenberg-MacLane spectrum* $\mathbf{H}(A)$ has as $n$-th space $K(A,n)$ and as $n$-th structure map a homotopy equivalence $K(A,n) \to \Omega K(A,n+1)$.

**Example 2.7** (Algebraic $K$-theory spectrum). For a ring $R$ there is the *algebraic $K$-theory spectrum* $\mathbf{K}(R)$ with the property

$$\pi_n(\mathbf{K}(R)) = K_n(R) \quad \text{for } n \in \mathbb{Z}.$$

For its definition see [42], [115], and [143].

Next we state the $L$-theoretic version. Since we will not focus on $L$-theory in these lectures, we will use $L$-theory as a black box and will later explain its relevance when we discuss applications. At least we mention that $L$-theory may be thought of a kind of $K$-theory not for finitely generated projective modules and their automorphisms but for quadratic forms over finitely generated projective modules and their automorphisms modulo hyperbolic forms.

**Example 2.8** (Algebraic $L$-theory spectrum). For a ring with involution $R$ there is the *algebraic $L$-theory spectrum* $\mathbf{L}^{\langle -\infty \rangle}(R)$ with the property

$$\pi_n(\mathbf{L}^{\langle -\infty \rangle}(R)) = L_n^{\langle -\infty \rangle}(R) \quad \text{for } n \in \mathbb{Z}.$$

For its construction we refer for instance to Quinn [147] and Ranicki [151].

**Example 2.9** (Topological $K$-theory spectrum). By *Bott periodicity* there is a homotopy equivalence

$$\beta\colon BU \times \mathbb{Z} \xrightarrow{\simeq} \Omega^2(BU \times \mathbb{Z}).$$

The *topological $K$-theory spectrum* $\mathbf{K}^{\mathrm{top}}$ has in even degrees $BU \times \mathbb{Z}$ and in odd degrees $\Omega(BU \times \mathbb{Z})$. The structure maps are given in even degrees by the map $\beta$ and in odd degrees by the identity $\mathrm{id}\colon \Omega(BU \times \mathbb{Z}) \to \Omega(BU \times \mathbb{Z})$.

**Definition 2.10** (Homology theory). Let $\Lambda$ be a commutative ring, for instance $\mathbb{Z}$ or $\mathbb{Q}$. A *homology theory* $\mathcal{H}_*$ with values in $\Lambda$-modules is a covariant functor from the category of $CW$-pairs to the category of $\mathbb{Z}$-graded $\Lambda$-modules together with natural transformations

$$\partial_n(X,A)\colon \mathcal{H}_n(X,A) \to \mathcal{H}_{n-1}(A)$$

for $n \in \mathbb{Z}$ satisfying the following axioms:

- Homotopy invariance
- Long exact sequence of a pair
- Excision

  If $(X,A)$ is a $CW$-pair and $f\colon A \to B$ is a cellular map , then

$$\mathcal{H}_n(X,A) \xrightarrow{\cong} \mathcal{H}_n(X \cup_f B, B).$$

- Disjoint union axiom

$$\bigoplus_{i \in I} \mathcal{H}_n(X_i) \xrightarrow{\cong} \mathcal{H}_n\left(\coprod_{i \in I} X_i\right).$$

**Definition 2.11** (Smash product). Let $\mathbf{E}$ be a spectrum and $X$ be a pointed space. Define the *smash product* $X \wedge \mathbf{E}$ to be the spectrum whose $n$-th space is $X \wedge E(n)$ and whose $n$-th structure map is

$$X \wedge E(n) \wedge S^1 \xrightarrow{\mathrm{id}_X \wedge \sigma(n)} X \wedge E(n+1).$$

**Theorem 2.12** (Homology theories and spectra). *Let $\mathbf{E}$ be a spectrum. Then we obtain a homology theory $H_*(-;\mathbf{E})$ by*

$$H_n(X,A;\mathbf{E}) := \pi_n\left((X \cup_A \mathrm{cone}(A)) \wedge \mathbf{E}\right).$$

*It satisfies*

$$H_n(\{\bullet\};\mathbf{E}) = \pi_n(\mathbf{E}).$$

**Example 2.13** (Stable homotopy theory). The homology theory associated to the sphere spectrum $\mathbf{S}$ is *stable homotopy* $\pi_*^s(X)$. The groups $\pi_n^s(\{\bullet\})$ are finite abelian groups for $n \neq 0$ by a result of Serre (1953). Their structure is only known for small $n$.

**Example 2.14** (Singular homology theory with coefficients). The homology theory associated to the Eilenberg-MacLane spectrum $\mathbf{H}(A)$ is singular homology with coefficients in $A$.

**Example 2.15** (Topological $K$-homology). The homology theory associated to the topological $K$-theory spectrum $\mathbf{K}^{\mathrm{top}}$ is *$K$-homology* $K_*(X)$. We have

$$K_n(\{\bullet\}) \cong \begin{cases} \mathbb{Z} & n \text{ even;} \\ \{0\} & n \text{ odd.} \end{cases}$$

Next we give the formulation of the Farrell-Jones Conjecture for $K$- and $L$-theory and the Baum-Connes Conjecture in the case of a torsionfree group. The general formulations for arbitrary groups will require more prerequisites and will be presented later. We begin with the $K$-theoretic version. Recall:

- $K_n(RG)$ is the algebraic $K$-theory of the group ring $RG$;
- $\mathbf{K}(R)$ is the (non-connective) algebraic $K$-theory spectrum of $R$;
- $H_n(\{\bullet\}; \mathbf{K}(R)) \cong \pi_n(\mathbf{K}(R)) \cong K_n(R)$ for $n \in \mathbb{Z}$.
- BG is the classifying space of the group $G$, i.e., the base space of the universal $G$-principal $G$-bundle $G \to EG \to BG$. Equivalently, $BG = K(G, 1)$. The space $BG$ is unique up to homotopy.

**Conjecture 2.16** ($K$-theoretic Farrell-Jones Conjecture for torsionfree groups). *The* $K$-theoretic Farrell-Jones Conjecture *with coefficients in the regular ring $R$ for the torsionfree group $G$ predicts that the assembly map*

$$H_n(BG; \mathbf{K}(R)) \to K_n(RG)$$

*is bijective for all $n \in \mathbb{Z}$.*

Recall:

- $L_n^{\langle -\infty \rangle}(RG)$ is the algebraic $L$-theory of $RG$ with decoration $\langle -\infty \rangle$;
- $\mathbf{L}^{\langle -\infty \rangle}(R)$ is the algebraic $L$-theory spectrum of $R$ with decoration $\langle -\infty \rangle$;
- $H_n(\{\bullet\}; \mathbf{L}^{\langle -\infty \rangle}(R)) \cong \pi_n(\mathbf{L}^{\langle -\infty \rangle}(R)) \cong L_n^{\langle -\infty \rangle}(R)$ for $n \in \mathbb{Z}$.

**Conjecture 2.17** ($L$-theoretic Farrell-Jones Conjecture for torsionfree groups). *The* $L$-theoretic Farrell-Jones Conjecture *with coefficients in the ring with involution $R$ for the torsionfree group $G$ predicts that the assembly map*

$$H_n(BG; \mathbf{L}^{\langle -\infty \rangle}(R)) \to L_n^{\langle -\infty \rangle}(RG)$$

*is bijective for all $n \in \mathbb{Z}$.*

Recall:

- $K_n(BG)$ is the topological $K$-homology of $BG$, where $K_*(-) = H_*(-; \mathbf{K}^{\mathrm{top}})$ for $\mathbf{K}^{\mathrm{top}}$ the topological $K$-theory spectrum.
- $K_n(C_r^*(G))$ is the topological $K$-theory of the reduced complex group $C^*$-algebra $C_r^*(G)$ of $G$ which is the closure in the norm topology of $\mathbb{C}G$ considered as subalgebra of $\mathcal{B}(l^2(G))$.

**Conjecture 2.18** (Baum-Connes Conjecture for torsionfree groups). *The* Baum-Connes Conjecture *for the torsionfree group $G$ predicts that the assembly map*

$$K_n(BG) \to K_n(C_r^*(G))$$

*is bijective for all $n \in \mathbb{Z}$.*

*There is also a real version of the Baum-Connes Conjecture*

$$KO_n(BG) \to K_n(C_r^*(G; \mathbb{R})).$$

In order to illustrate the depth of the Farrell-Jones Conjecture and the Baum-Connes Conjecture, we present some conclusions which are interesting in their own right.

**Notation 2.19.** Let $\mathcal{FJ}_K(R)$ and $\mathcal{FJ}_L(R)$ respectively be the class of groups which satisfy the $K$-theoretic and $L$-theoretic respectively Farrell-Jones Conjecture for the coefficient ring (with involution) $R$.

Let $\mathcal{BC}$ be the class of groups which satisfy the Baum-Connes Conjecture.

**Theorem 2.20** (Lower and middle $K$-theory of group rings in the torsionfree case). *Suppose that $G$ is torsionfree.*

(1) *If $R$ is regular and $G \in \mathcal{FJ}_K(R)$, then*
   (a) *$K_n(RG) = 0$ for $n \leq -1$;*
   (b) *The change of rings map $K_0(R) \to K_0(RG)$ is bijective;*
   (c) *In particular $\widetilde{K}_0(RG)$ is trivial if and only if $\widetilde{K}_0(R)$ is trivial.*

(2) *If $G \in \mathcal{FJ}_K(\mathbb{Z})$, then the Whitehead group $\mathrm{Wh}(G)$ is trivial.*

*Proof.* The idea of the proof is to study the Atiyah-Hirzebruch spectral sequence. It converges to $H_n(BG; \mathbf{K}(R))$ which is isomorphic to $K_n(RG)$ by the assumption that $G$ satisfies the Farrell-Jones Conjecture. The $E^2$-term is given by

$$E^2_{p,q} = H_p(BG, K_q(R)).$$

(1) Since $R$ is regular by assumption, we get $K_q(R) = 0$ for $q \leq -1$. Hence the spectral sequence is a first quadrant spectral sequence. This implies $K_n(RG) \cong H_n(BG; \mathbf{K}(R)) = 0$ for $n \leq -1$ and that the edge homomorphism yields an isomorphism

$$K_0(R) = H_0(\{\bullet\}, K_0(R)) \xrightarrow{\cong} H_0(BG; \mathbf{K}(R)) \cong K_0(RG).$$

(2) We have $K_0(\mathbb{Z}) = \mathbb{Z}$ and $K_1(\mathbb{Z}) = \{\pm 1\}$. We get an exact sequence

$$0 \to H_0(BG; K_1(\mathbb{Z})) = \{\pm 1\} \to H_1(BG; \mathbf{K}(\mathbb{Z})) \cong K_1(\mathbb{Z}G) \\ \to H_1(BG; K_0(\mathbb{Z})) = G/[G,G] \to 0.$$

This implies $\mathrm{Wh}(G) := K_1(\mathbb{Z}G)/\{\pm g \mid g \in G\} = 0$. □

We summarize that we get for a torsionfree group $G \in \mathcal{FJ}_K(\mathbb{Z})$:

(1) $K_n(\mathbb{Z}G) = 0$ for $n \leq -1$;
(2) $\widetilde{K}_0(\mathbb{Z}G) = 0$;
(3) $\mathrm{Wh}(G) = 0$;
(4) Every finitely dominated $CW$-complex $X$ with $G = \pi_1(X)$ is homotopy equivalent to a finite $CW$-complex;
(5) Every compact $h$-cobordism $W$ of dimension $\geq 6$ with $\pi_1(W) \cong G$ is trivial;
(6) If $G$ belongs to $\mathcal{FJ}_K(\mathbb{Z})$, then it is of type FF if and only if it is of type FP (Serre's problem).

**Conjecture 2.21** (Kaplansky Conjecture). *The Kaplansky Conjecture says for a torsionfree group $G$ and an integral domain $R$ that $0$ and $1$ are the only idempotents in $RG$.*

In the next theorem we will use the notion of a *sofic group* that was introduced by Gromov and originally called *subamenable group*. Every residually amenable group is sofic but the converse is not true. The class of sofic groups is closed under taking subgroups, direct products, free amalgamated products, colimits and inverse limits, and, if $H$ is a sofic normal subgroup of $G$ with amenable quotient $G/H$, then $G$ is sofic. This is a very general notion, e.g., no group is known which is not sofic. For more information about the notion of a sofic group we refer to [60]. The next result is taken from [15, Theorem 0.12].

**Theorem 2.22** (The Farrell-Jones Conjecture and the Kaplansky Conjecture, Bartels-Lück-Reich(2007)). *Let $F$ be a skew-field and let $G$ be a group with $G \in \mathcal{FJ}_K(F)$. Suppose that one of the following conditions is satisfied:*

(1) *$F$ is commutative and has characteristic zero and $G$ is torsionfree;*
(2) *$G$ is torsionfree and sofic;*
(3) *The characteristic of $F$ is $p$, all finite subgroups of $G$ are $p$-groups and $G$ is sofic.*

*Then $0$ and $1$ are the only idempotents in $FG$.*

*Proof.* Let $p$ be an idempotent in $FG$. We want to show $p \in \{0,1\}$. Denote by $\epsilon\colon FG \to F$ the augmentation homomorphism sending $\sum_{g\in G} r_g \cdot g$ to $\sum_{g \in G} r_g$. Obviously $\epsilon(p) \in F$ is 0 or 1. Hence it suffices to show $p = 0$ under the assumption that $\epsilon(p) = 0$.

Let $(p) \subseteq FG$ be the ideal generated by $p$ which is a finitely generated projective $FG$-module. Since $G \in \mathcal{FJ}_K(F)$, we can conclude that

$$i_*\colon K_0(F) \otimes_{\mathbb{Z}} \mathbb{Q} \to K_0(FG) \otimes_{\mathbb{Z}} \mathbb{Q}$$

is surjective. Hence we can find a finitely generated projective $F$-module $P$ and integers $k, m, n \geq 0$ satisfying

$$(p)^k \oplus FG^m \cong_{FG} i_*(P) \oplus FG^n.$$

If we now apply $i_* \circ \epsilon_*$ and use $\epsilon \circ i = \mathrm{id}$, $i_* \circ \epsilon_*(FG^l) \cong FG^l$ and $\epsilon(p) = 0$ we obtain

$$FG^m \cong i_*(P) \oplus FG^n.$$

Inserting this in the first equation yields

$$(p)^k \oplus FG^m \cong FG^m.$$

Our assumptions on $F$ and $G$ imply that $FG$ is stably finite, i.e., if $A$ and $B$ are square matrices over $FG$ with $AB = I$, then $BA = I$. This implies $(p)^k = 0$ and hence $p = 0$. □

**Theorem 2.23** (The Baum-Connes Conjecture and the Kaplansky Conjecture). *Let $G$ be a torsionfree group with $G \in \mathcal{BC}$. Then 0 and 1 are the only idempotents in $C_r^*(G)$ and in particular in $\mathbb{C}G$.*

*Proof.* There is a trace map

$$\mathrm{tr}\colon C_r^*(G) \to \mathbb{C}$$

which sends $f \in C_r^*(G) \subseteq \mathcal{B}(l^2(G))$ to $\langle f(e), e\rangle_{l^2(G)}$. The $L^2$-index theorem due to Atiyah (1976) (see [6]) shows that the composite

$$K_0(BG) \to K_0(C_r^*(G)) \xrightarrow{\mathrm{tr}} \mathbb{C}$$

coincides with

$$K_0(BG) \xrightarrow{K_0(\mathrm{pr})} K_0(\{\bullet\}) = \mathbb{Z} \to \mathbb{C}.$$

Hence $G \in \mathcal{BC}$ implies $\mathrm{tr}(p) \in \mathbb{Z}$. Since $\mathrm{tr}(1) = 1$, $\mathrm{tr}(0) = 0$, $0 \le p \le 1$ and $p^2 = p$, we get $\mathrm{tr}(p) \in \mathbb{R}$ and $0 \le \mathrm{tr}(p) \le 1$. We conclude $\mathrm{tr}(0) = \mathrm{tr}(p)$ or $\mathrm{tr}(1) = \mathrm{tr}(p)$. Since the trace tr is faithful, this implies already $p = 0$ or $p = 1$. □

The next conjecture is one of the basic conjectures about the classification of topological manifolds.

**Conjecture 2.24** (Borel Conjecture). *The* Borel Conjecture for *$G$ predicts for two closed aspherical manifolds $M$ and $N$ with $\pi_1(M) \cong \pi_1(N) \cong G$ that any homotopy equivalence $M \to N$ is homotopic to a homeomorphism and in particular that $M$ and $N$ are homeomorphic.*

**Remark 2.25** (Borel versus Mostow). The Borel Conjecture can be viewed as the topological version of Mostow rigidity. A special case of Mostow rigidity says that any homotopy equivalence between closed hyperbolic manifolds of dimension $\geq 3$ is homotopic to an isometric diffeomorphism.

**Remark 2.26** (The Borel Conjecture fails in the smooth category). The Borel Conjecture is not true in the smooth category by results of Farrell-Jones [63], i.e., there exists aspherical closed manifolds which are homeomorphic but not diffeomorphic.

**Remark 2.27** (Topological rigidity for non-aspherical manifolds). There are also non-aspherical manifolds which are topologically rigid in the sense of the Borel Conjecture (see Kreck-Lück [106]).

**Theorem 2.28** (The Farrell-Jones Conjecture and the Borel Conjecture). *If the $K$- and $L$-theoretic Farrell-Jones Conjecture hold for $G$ in the case $R = \mathbb{Z}$, then the Borel Conjecture is true in dimension $\geq 5$ and in dimension 4 if $G$ is good in the sense of Freedman.*

**Remark 2.29** (The Borel Conjecture in dimension $\leq 3$). Thurston's Geometrization Conjecture implies the Borel Conjecture in dimension three. The Borel Conjecture in dimension one and two is obviously true.

Next we give some explanations about the proof of Theorem 2.28.

**Definition 2.30** (Structure set). The *structure set* $S^{top}(M)$ of a manifold $M$ consists of equivalence classes of orientation preserving homotopy equivalences $N \to M$ with a manifold $N$ as source.

Two such homotopy equivalences $f_0 \colon N_0 \to M$ and $f_1 \colon N_1 \to M$ are equivalent if there exists a homeomorphism $g \colon N_0 \to N_1$ with $f_1 \circ g \simeq f_0$.

The next result follows directly from the definitions.

**Theorem 2.31.** *The Borel Conjecture holds for a closed manifold $M$ if and only if $\mathcal{S}^{\mathrm{top}}(M)$ consists of one element.*

Let $\mathbf{L}\langle 1\rangle$ be the 1-*connective cover* of the $L$-theory spectrum $\mathbf{L}$. It is characterized by the following property. There is a natural map of spectra $\mathbf{L}\langle 1\rangle \to \mathbf{L}$ which induces an isomorphism on the homotopy groups in dimensions $n \geq 1$ and the homotopy groups of $\mathbf{L}\langle 1\rangle$ vanish in dimensions $n \leq 0$.

**Theorem 2.32** (Ranicki (1992)). *There is an exact sequence of abelian groups, called algebraic surgery exact sequence, for an $n$-dimensional closed manifold $M$*

$$\begin{aligned} \ldots \xrightarrow{\sigma_{n+1}} H_{n+1}(M;\mathbf{L}\langle 1\rangle) \xrightarrow{A_{n+1}} L_{n+1}(\mathbb{Z}\pi_1(M)) \xrightarrow{\partial_{n+1}} \mathcal{S}^{\mathrm{top}}(M) \\ \xrightarrow{\sigma_n} H_n(M;\mathbf{L}\langle 1\rangle) \xrightarrow{A_n} L_n(\mathbb{Z}\pi_1(M)) \xrightarrow{\partial_n} \ldots \end{aligned}$$

*It can be identified with the classical geometric surgery sequence due to Sullivan and Wall in high dimensions.*

*Proof.* See [151, Definition 15.19 on page 169 and Theorem 18.5 on page 198]. □

The $K$-theoretic version of the Farrell-Jones Conjecture ensures that we do not have to deal with decorations, e.g., it does not matter if we consider $\mathbf{L}$ or $\mathbf{L}^{\langle -\infty\rangle}$. (This follows from the so called *Rothenberg sequences*). The $L$-theoretic version of the Farrell-Jones Conjecture implies that $H_n(M;\mathbf{L}) \to L_n(\mathbb{Z}\pi_1(M))$ is bijective for all $n \in \mathbb{Z}$. An easy spectral sequence argument shows that $H_k(M;\mathbf{L}\langle 1\rangle) \to H_k(M;\mathbf{L})$ is bijective for $k \geq n+1$ and injective for $k = n$. For $k = n$ and $k = n+1$ the map $A_k$ is the composite of the map $H_k(M;\mathbf{L}\langle 1\rangle) \to H_k(M;\mathbf{L})$ with the map $H_k(M;\mathbf{L}) \to L_k(\mathbb{Z}\pi_1(M))$. Hence $A_{n+1}$ is surjective and $A_n$ is injective. Theorem 2.32 implies that $\mathcal{S}^{\mathrm{top}}(M)$ consist of one element. Now Theorem 2.28 follows from Theorem 2.31.

More information on the Borel Conjecture can be found for instance in [62], [63], [64], [67], [68], [72] [77], [105], [119], [126].

Next we explain that the versions of the Farrell-Jones and the Baum-Connes Conjecture above cannot be true if we drop the assumption that $G$ is torsionfree or that $R$ is regular.

**Example 2.33** (The condition torsionfree is essential). The versions of the Farrell-Jones Conjecture and the Baum-Connes Conjecture above become false for finite groups unless the group is trivial. For instance the version of the Baum-Connes Conjecture above would predict for a finite group $G$

$$K_0(BG) \cong K_0(C_r^*(G)) \cong R_{\mathbb{C}}(G).$$

However, $K_0(BG) \otimes_{\mathbb{Z}} \mathbb{Q} \cong_{\mathbb{Q}} K_0(\{\bullet\}) \otimes_{\mathbb{Z}} \mathbb{Q} \cong_{\mathbb{Q}} \mathbb{Q}$ and $R_{\mathbb{C}}(G) \otimes_{\mathbb{Z}} \mathbb{Q} \cong_{\mathbb{Q}} \mathbb{Q}$ holds if and only if $G$ is trivial.

**Example 2.34** (The condition regular is essential). If $G$ is torsionfree, then the version of the $K$-theoretic Farrell-Jones Conjecture predicts

$$\begin{aligned} H_n(B\mathbb{Z};\mathbf{K}(R)) &= H_n(S^1;\mathbf{K}(R)) = H_n(\{\bullet\};\mathbf{K}(R)) \oplus H_{n-1}(\{\bullet\};\mathbf{K}(R)) \\ &= K_n(R) \oplus K_{n-1}(R) \cong K_n(R\mathbb{Z}). \end{aligned}$$

In view of the Bass-Heller-Swan decomposition this is only possible if $\mathrm{NK}_n(R)$ vanishes which is true for regular rings $R$ but not for general rings $R$.

Next we want to discuss what we may have to take into account if we want to give a formulation of the Farrell-Jones and the Baum-Connes Conjecture which may have a chance to be true for all groups.

**Remark 2.35** (Assembly). For a field $F$ of characteristic zero and some groups $G$ one knows that there is an isomorphism

$$\operatorname*{colim}_{\substack{H \subseteq G \\ |H| < \infty}} K_0(FH) \xrightarrow{\cong} K_0(FG).$$

This indicates that one has at least to take into account the values for all finite subgroups to assemble $K_n(FG)$.

**Remark 2.36** (Degree Mixing). The Bass-Heller-Swan decomposition shows that the $K$-theory of finite subgroups in degree $m \leq n$ can affect the $K$-theory in degree $n$ and that at least in the Farrell-Jones setting finite subgroups are not enough.

**Remark 2.37** (No Nil-phenomena occur in the Baum-Connes setting). In the Baum-Connes setting Nil-phenomena do not appear. Namely, a special case of a result due to Pimsner-Voiculescu [146] says

$$K_n(C_r^*(G \times \mathbb{Z})) \cong K_n(C_r^*(G)) \oplus K_{n-1}(C_r^*(G)).$$

**Remark 2.38** (Homological behavior)**.** There is still a lot of homological behavior known for $K_*(C_r^*(G))$. For instance there exists a long exact Mayer-Vietoris sequence associated to amalgamated products $G_1 *_{G_0} G_2$ by Pimsner-Voiculescu [146].

$$\begin{aligned} \ldots \to K_n(C_r^*(G_0)) \to K_n(C_r^*(G_1)) \oplus K_n(C_r^*(G_2)) \to K_n(C_r^*(G)) \\ \to K_{n-1}(C_r^*(G_0)) \to K_{n-1}(C_r^*(G_1)) \oplus K_{n-1}(C_r^*(G_2)) \to \cdots \end{aligned}$$

This is very similar to the corresponding Mayer-Vietoris sequence in group homology theory

$$\begin{aligned} \ldots \to H_n(G_0) \to H_n(G_1)) \oplus H_n(G_2) \to H_n(G) \\ \to H_{n-1}(G_0) \to H_{n-1}(G_1) \oplus H_{n-1}G_2) \to \cdots \end{aligned}$$

It comes from the fact that there is a model for $BG$ which contains $BG_0$, $BG_1$ and $BG_2$ as $CW$-subcomplexes such that $BG = BG_1 \cup BG_2$ and $BG_0 = BG_1 \cap BG_2$.

An analogous similarity exists for the Wang-sequence associated to a semi-direct product $G \rtimes \mathbb{Z}$

Similar versions of the Mayer-Vietoris sequence and the Wang sequence in algebraic $K$-and $L$-theory of group rings are due to Cappell (1974) and Waldhausen (1978) provided one makes certain assumptions on $R$ or ignores certain Nil-phenomena.

**Question 2.39** (Classifying spaces for families)**.** Is there a version $E_{\mathcal{F}}(G)$ of the classifying space $EG$ which takes the structure of the family of finite subgroups or other families $\mathcal{F}$ of subgroups into account and can be used for a general formulation of the Farrell-Jones Conjecture?

**Question 2.40** (Equivariant homology theories)**.** Can one define appropriate $G$-homology theories $\mathcal{H}_*^G$ that are in some sense computable and yield when applied to $E_{\mathcal{F}}(G)$ a term which potentially is isomorphic to the groups $K_n(RG)$, $L_n^{-\langle\infty\rangle}(RG)$ or $K_n(C_r^*(G))$?

In the torsionfree case they should reduce to $H_n(BG; \mathbf{K}(R))$, $H_n(BG; \mathbf{L}^{-\langle\infty\rangle})$ and $K_n(BG)$.

# 3. Classifying spaces for families

The outline of this section is:

- We introduce the notion of the classifying space of a family $\mathcal{F}$ of subgroups $E_{\mathcal{F}}(G)$ and $J_{\mathcal{F}}(G)$.
- In the case, where $\mathcal{F}$ is the family $\mathcal{COM}$ of compact subgroups, we present some nice geometric models for $E_{\mathcal{F}}(G)$ and explain $E_{\mathcal{F}}(G) \simeq J_{\mathcal{F}}(G)$.
- We discuss finiteness properties of these classifying spaces.

The material of this section is an extract of the survey article by Lück [125], where more information and proofs of the results stated below are given.

In this section group means locally compact Hausdorff topological group with a countable basis for its topology, unless explicitly stated differently.

**Definition 3.1** ($G$-$CW$-complex)**.** A *$G$-$CW$-complex* $X$ is a $G$-space together with a $G$-invariant filtration

$$\emptyset = X_{-1} \subseteq X_0 \subseteq \ldots \subseteq X_n \subseteq \ldots \subseteq \bigcup_{n\geq 0} X_n = X$$

such that $X$ carries the colimit topology with respect to this filtration, and $X_n$ is obtained from $X_{n-1}$ for each $n \geq 0$ by attaching equivariant $n$-dimensional cells, i.e., there exists a $G$-pushout

$$\begin{array}{ccc} \coprod_{i\in I_n} G/H_i \times S^{n-1} & \xrightarrow{\coprod_{i\in I_n} q_i^n} & X_{n-1} \\ \downarrow & & \downarrow \\ \coprod_{i\in I_n} G/H_i \times D^n & \xrightarrow{\coprod_{i\in I_n} Q_i^n} & X_n \end{array}$$

**Example 3.2** (Simplicial actions)**.** Let $X$ be a (geometric) simplicial complex. Suppose that $G$ acts simplicially on $X$. Then $G$ acts simplicially also on the barycentric subdivision $X'$, and all isotropy groups are open and closed. The $G$-space $X'$ inherits the structure of a $G$-$CW$-complex.

**Definition 3.3** (Proper $G$-action)**.** A $G$-space $X$ is called *proper* if for each pair of points $x$ and $y$ in $X$ there are open neighborhoods $V_x$ of $x$ and $W_y$ of $y$ in $X$ such that the closure of the subset $\{g \in G \mid gV_x \cap W_y \neq \emptyset\}$ of $G$ is compact.

**Lemma 3.4.** (1) *A proper $G$-space has always compact isotropy groups.*
(2) *A $G$-$CW$-complex $X$ is proper if and only if all its isotropy groups are compact.*

*Proof.* See [117, Theorem 1.23 on page 19]. □

**Example 3.5** (Smooth actions)**.** Let $G$ be a Lie group acting properly and smoothly on a smooth manifold $M$. Then $M$ inherits the structure of $G$-$CW$-complex (see Illman [93]).

**Definition 3.6** (Family of subgroups)**.** A *family $\mathcal{F}$ of subgroups* of $G$ is a set of (closed) subgroups of $G$ which is closed under conjugation and finite intersections.

Examples for $\mathcal{F}$ are:

| | | |
|---|---|---|
| $\mathcal{TR}$ | = | {trivial subgroup}; |
| $\mathcal{FIN}$ | = | {finite subgroups}; |
| $\mathcal{VCYC}$ | = | {virtually cyclic subgroups}; |
| $\mathcal{COM}$ | = | {compact subgroups}; |
| $\mathcal{COMOP}$ | = | {compact open subgroups}; |
| $\mathcal{ALL}$ | = | {all subgroups}. |

**Definition 3.7** (Classifying $G$-$CW$-complex for a family of subgroups)**.** Let $\mathcal{F}$ be a family of subgroups of $G$. A model for the *classifying $G$-$CW$-complex for the family $\mathcal{F}$* is a $G$-$CW$-complex $E_{\mathcal{F}}(G)$ which has the following properties:

(1) All isotropy groups of $E_{\mathcal{F}}(G)$ belong to $\mathcal{F}$;
(2) For any $G$-$CW$-complex $Y$, whose isotropy groups belong to $\mathcal{F}$, there is up to $G$-homotopy precisely one $G$-map $Y \to E_{\mathcal{F}}(G)$.

We abbreviate $\underline{E}G := E_{\mathcal{COM}}(G)$ and call it the *universal $G$-$CW$-complex for proper $G$-actions*. We also write $EG = E_{\mathcal{TR}}(G)$.

**Theorem 3.8** (Homotopy characterization of $E_{\mathcal{F}}(G)$). *Let $\mathcal{F}$ be a family of subgroups.*

(1) *There exists a model for $E_{\mathcal{F}}(G)$ for any family $\mathcal{F}$;*
(2) *A $G$-$CW$-complex $X$ is a model for $E_{\mathcal{F}}(G)$ if and only if all its isotropy groups belong to $\mathcal{F}$ and for each $H \in \mathcal{F}$ the $H$-fixed point set $X^H$ is weakly contractible.*

**Example 3.9** ($E_{\mathcal{ALL}}(G)$). A model for $E_{\mathcal{ALL}}(G)$ is $G/G$;

**Example 3.10** (Universal principal $G$-bundle). The projection $EG \to BG := G\backslash EG$ is the universal $G$-principal bundle for $G$-$CW$-complexes.

**Example 3.11** (Infinite dihedral group). Let $D_\infty = \mathbb{Z} \rtimes \mathbb{Z}/2 = \mathbb{Z}/2 * \mathbb{Z}/2$ be the infinite dihedral group. A model for $ED_\infty$ is the universal covering of $\mathbb{RP}^\infty \vee \mathbb{RP}^\infty$. A model for $\underline{E}D_\infty$ is $\mathbb{R}$ with the obvious $D_\infty$-action. Notice that every model for $ED_\infty$ or $BD_\infty$ must be infinite-dimensional, whereas there exists a cocompact 1-dimensional model for $\underline{E}D_\infty$.

**Lemma 3.12.** *If $G$ is totally disconnected, then $E_{\mathcal{COMOP}}(G) = \underline{E}G$.*

**Definition 3.13** ($\mathcal{F}$-numerable $G$-space). An *$\mathcal{F}$-numerable $G$-space* is a $G$-space, for which there exists an open covering $\{U_i \mid i \in I\}$ by $G$-subspaces satisfying:

(1) For each $i \in I$ there exists a $G$-map $U_i \to G/G_i$ for some $G_i \in \mathcal{F}$;
(2) There is a locally finite partition of unity $\{e_i \mid i \in I\}$ subordinate to $\{U_i \mid i \in I\}$ by $G$-invariant functions $e_i \colon X \to [0,1]$.

Notice that we do not demand that the isotropy groups of a $\mathcal{F}$-numerable $G$-space belong to $\mathcal{F}$.

If $f\colon X \to Y$ is a $G$-map and $Y$ is $\mathcal{F}$-numerable, then $X$ is also $\mathcal{F}$-numerable.

**Lemma 3.14.** *A $G$-$CW$-complex is $\mathcal{F}$-numerable if and only if each isotropy group appears as a subgroup of an element in $\mathcal{F}$.*

**Definition 3.15** (Classifying numerable $G$-space for a family of subgroups). Let $\mathcal{F}$ be a family of subgroups of $G$. A model $J_{\mathcal{F}}(G)$ for the *classifying numerable $G$-space for the family of subgroups $\mathcal{F}$* is a $G$-space which has the following properties:

(1) $J_{\mathcal{F}}(G)$ is $\mathcal{F}$-numerable;
(2) For any $\mathcal{F}$-numerable $G$-space $X$ there is up to $G$-homotopy precisely one $G$-map $X \to J_{\mathcal{F}}(G)$.

We abbreviate $\underline{J}G := J_{\mathcal{COM}}(G)$ and call it the *universal numerable $G$-space for proper $G$-actions* or briefly *the universal space for proper $G$-actions*. We also write $JG = J_{\mathcal{TR}}(G)$.

**Theorem 3.16** (Homotopy characterization of $J_{\mathcal{F}}(G)$). *Let $\mathcal{F}$ be a family of subgroups.*

(1) *For any family $\mathcal{F}$ there exists a model for $J_{\mathcal{F}}(G)$ whose isotropy groups belong to $\mathcal{F}$;*
(2) *Let $X$ be an $\mathcal{F}$-numerable $G$-space. Equip $X \times X$ with the diagonal action and let $\mathrm{pr}_i \colon X \times X \to X$ be the projection onto the $i$-th factor for $i = 1, 2$. Then $X$ is a model for $J_{\mathcal{F}}(G)$ if and only if for each $H \in \mathcal{F}$ there is $x \in X$ with $H \subseteq G_x$ and $\mathrm{pr}_1$ and $\mathrm{pr}_2$ are $G$-homotopic.*
(3) *For $H \in \mathcal{F}$ the $H$-fixed point set $J_{\mathcal{F}}(G)^H$ is contractible.*

*Proof.* See [125, Theorem 2.5]. □

**Example 3.17** (Universal $G$-principal bundle). The projection $JG \to G\backslash JG$ is the universal $G$-principal bundle for numerable free proper $G$-spaces.

**Theorem 3.18** (Comparison of $E_{\mathcal{F}}(G)$ and $J_{\mathcal{F}}(G)$, Lück (2005)).

(1) *There is up to $G$-homotopy precisely one $G$-map*

$$\phi \colon E_{\mathcal{F}}(G) \to J_{\mathcal{F}}(G);$$

(2) *It is a $G$-homotopy equivalence if one of the following conditions is satisfied:*
  (a) *Each element in $\mathcal{F}$ is open and closed;*
  (b) *$G$ is discrete;*
  (c) *$\mathcal{F}$ is $\mathcal{COM}$;*
(3) *Let $G$ be totally disconnected. Then $EG \to JG$ is a $G$-homotopy equivalence if and only if $G$ is discrete.*

*Proof.* See [125, Theorem 3.7]. □

Next we want to illustrate that the space $\underline{E}G = \underline{J}G$ often has very nice geometric models and appear naturally in many interesting situations.

Let $C_0(G)$ be the Banach space of complex valued functions of $G$ vanishing at infinity with the supremum-norm. The group $G$ acts isometrically on $C_0(G)$ by $(g \cdot f)(x) := f(g^{-1}x)$ for $f \in C_0(G)$ and $g, x \in G$. Let $PC_0(G)$ be the subspace of $C_0(G)$ consisting of functions $f$ such that $f$ is not identically zero and has non-negative real numbers as values.

**Theorem 3.19** (Operator theoretic model, Abels (1978)). *The $G$-space $PC_0(G)$ is a model for $\underline{J}G$.*

*Proof.* See [1, Theorem 2.4]. □

**Theorem 3.20.** *Let $G$ be discrete. A model for $\underline{J}G$ is the space*

$$X_G = \left\{ f \colon G \to [0,1] \,\middle|\, f \text{ has finite support}, \sum_{g \in G} f(g) = 1 \right\}$$

*with the topology coming from the supremum norm.*

**Theorem 3.21** (Simplicial Model). *Let $G$ be discrete. Let $P_\infty(G)$ be the geometric realization of the simplicial set whose $k$-simplices consist of $(k+1)$-tupels $(g_0, g_1, \ldots, g_k)$ of elements $g_i$ in $G$.*

*Then $P_\infty(G)$ is a model for $\underline{E}G$.*

**Remark 3.22** (Comparison of $X_G$ and $P_\infty(G)$). The spaces $X_G$ and $P_\infty(G)$ have the same underlying sets but in general they have different topologies. The identity map induces a $G$-map $P_\infty(G) \to X_G$ which is a $G$-homotopy equivalence, but in general not a $G$-homeomorphism.

**Theorem 3.23** (Almost connected groups, Abels (1978)). *Suppose that $G$ is almost connected, i.e., the group $G/G^0$ is compact for $G^0$ the component of the identity element.*

*Then $G$ contains a maximal compact subgroup $K$ which is unique up to conjugation, and the $G$-space $G/K$ is a model for $\underline{J}G$.*

*Proof.* See [1, Corollary 4.14]. □

As a special case we get:

**Theorem 3.24** (Discrete subgroups of almost connected Lie groups). *Let $L$ be a Lie group with finitely many path components.*

*Then $L$ contains a maximal compact subgroup $K$ which is unique up to conjugation, and the $L$-space $L/K$ is a model for $\underline{E}L$.*

*If $G \subseteq L$ is a discrete subgroup of $L$, then $L/K$ with the obvious left $G$-action is a finite dimensional $G$-$CW$-model for $\underline{E}G$.*

**Theorem 3.25** (Actions on CAT(0)-spaces). *Let $G$ be a (locally compact second countable Hausdorff) topological group. Let $X$ be a proper $G$-$CW$-complex. Suppose that $X$ has the structure of a complete simply connected* CAT(0)*-space for which $G$ acts by isometries.*

*Then $X$ is a model for $\underline{E}G$.*

*Proof.* By [31, Corollary II.2.8 on page 179] the $K$-fixed point set of $X$ is a non-empty convex subset of $X$ and hence contractible for any compact subgroup $K \subset G$. □

**Remark 3.26.** The result above contains as special case isometric $G$ actions on simply-connected complete Riemannian manifolds with non-positive sectional curvature and $G$-actions on trees.

Let $\Sigma$ be an *affine building* sometimes also called *Euclidean building*. This is a simplicial complex together with a system of subcomplexes called *apartments* satisfying the following axioms:

(1) Each apartment is isomorphic to an affine Coxeter complex;
(2) Any two simplices of $\Sigma$ are contained in some common apartment;
(3) If two apartments both contain two simplices $A$ and $B$ of $\Sigma$, then there is an isomorphism of one apartment onto the other which fixes the two simplices $A$ and $B$ pointwise.

The precise definition of an *affine Coxeter complex*, which is sometimes called also *Euclidean Coxeter complex*, can be found in [35, Section 2 in Chapter VI], where also more information about affine buildings is given. An affine building comes with metric $d\colon \Sigma \times \Sigma \to [0, \infty)$ which is non-positively curved and complete. The building with this metric is a CAT(0)-space. A simplicial automorphism of $\Sigma$ is always an isometry with respect to $d$. For two points $x, y$ in the affine building there is a unique line segment $[x, y]$ joining $x$ and $y$. It is the set of points $\{z \in \Sigma \mid d(x, y) = d(x, z) + d(z, y)\}$. For $x, y \in \Sigma$ and $t \in [0, 1]$ let $tx + (1-t)y$ be the point $z \in \Sigma$ uniquely determined by the property that $d(x, z) = td(x, y)$ and $d(z, y) = (1-t)d(x, y)$. Then the map

$$r\colon \Sigma \times \Sigma \times [0, 1] \to \Sigma, \qquad (x, y, t) \mapsto tx + (1-t)y$$

is continuous. This implies that $\Sigma$ is contractible. All these facts are taken from [35, Section 3 in Chapter VI] and [31, Theorem 10A.4 on page 344].

Suppose that the group $G$ acts on $\Sigma$ by isometries. If $G$ maps a non-empty bounded subset $A$ of $\Sigma$ to itself, then the $G$-action has a fixed point (see [35, Theorem 1 in Section 4 in Chapter VI on page 157]). Moreover the $G$-fixed point set must be contractible since for two points $x, y \in \Sigma^G$ also the segment $[x, y]$ must lie in $\Sigma^G$ and hence the map $r$ above induces a continuous map $\Sigma^G \times \Sigma^G \times [0, 1] \to \Sigma^G$. This implies together with Example 3.2, Theorem 3.8 (2), Lemma 3.12 and Theorem 3.18

**Theorem 3.27** (Affine buildings). *Let $G$ be a topological (locally compact second countable Hausdorff) group. Suppose that $G$ acts on the affine building by simplicial automorphisms such that each isotropy group is compact. Then each isotropy group is compact open, $\Sigma$ is a model for $J_{\mathcal{COMOP}}(G)$ and the barycentric subdivision $\Sigma'$ is a model for both $J_{\mathcal{COMOP}}(G)$ and $E_{\mathcal{COMOP}}(G)$. If we additionally assume that $G$ is totally disconnected, then $\Sigma$ is a model for both $\underline{J}G$ and $\underline{E}G$.*

**Example 3.28** (Bruhat-Tits building). An important example is the case of a reductive $p$-adic algebraic group $G$ and its associated affine Bruhat-Tits building $\beta(G)$ (see [179], [180]). Then $\beta(G)$ is a model for $\underline{J}G$ and $\beta(G)'$ is a model for $\underline{E}G$ by Theorem 3.27.

For more information about buildings we refer to the lectures of Abramenko.

The Rips complex $P_d(G, S)$ of a group $G$ with a symmetric finite set $S$ of generators for a natural number $d$ is the geometric realization of the simplicial set whose set of $k$-simplices consists of $(k+1)$-tuples $(g_0, g_1, \ldots g_k)$ of pairwise distinct elements $g_i \in G$ satisfying $d_S(g_i, g_j) \leq d$ for all $i, j \in \{0, 1, \ldots, k\}$.

The obvious $G$-action by simplicial automorphisms on $P_d(G, S)$ induces a $G$-action by simplicial automorphisms on the barycentric subdivision $P_d(G, S)'$.

**Theorem 3.29** (Rips complex, Meintrup-Schick (2002)). *Let $G$ be a discrete group with a finite symmetric set of generators. Suppose that $(G, S)$ is $\delta$-hyperbolic for the real number $\delta \geq 0$. Let $d$ be a natural number with $d \geq 16\delta + 8$.*

*Then the barycentric subdivision of the Rips complex $P_d(G, S)'$ is a finite $G$-CW-model for $\underline{E}G$.*

*Proof.* See [129]. □

Arithmetic groups in a semisimple connected linear $\mathbb{Q}$-algebraic group possess finite models for $\underline{E}G$. Namely, let $G(\mathbb{R})$ be the $\mathbb{R}$-points of a semisimple $\mathbb{Q}$-group $G(\mathbb{Q})$ and let $K \subseteq G(\mathbb{R})$ be a maximal compact subgroup. If $A \subseteq G(\mathbb{Q})$ is an arithmetic group, then $G(\mathbb{R})/K$ with the left $A$-action is a model for $\underline{E}A$ as already explained above. However, the $A$-space $G(\mathbb{R})/K$ is not necessarily cocompact. But there is a finite model for $\underline{E}A$ by the following result.

**Theorem 3.30** (Borel-Serre compactification). *The Borel-Serre compactification (see [29], [168]) of $G(\mathbb{R})/K$ is a finite $A$-CW-model for $\underline{E}A$.*

*Proof.* This is pointed out in Adem-Ruan [2, Remark 5.8], where a private communication with Borel and Prasad is mentioned. A detailed proof is given by Ji [97]. □

For more information about arithmetic groups we refer to the lectures of Abramenko.

Let $\Gamma^s_{g,r}$ be the mapping class group of an orientable compact surface $F$ of genus $g$ with $s$ punctures and $r$ boundary components. We will always assume that $2g+s+r > 2$, or, equivalently, that the Euler characteristic of the punctured surface $F$ is negative. It is well-known that the associated Teichmüller space $\mathcal{T}^s_{g,r}$ is a contractible space on which $\Gamma^s_{g,r}$ acts properly.

We could not find a clear reference in the literature for the to experts known statement that there exist a finite $\Gamma^s_{g,r}$-$CW$-model for $\underline{E}\Gamma^s_{g,r}$. The work of Harer [85] on the existence of a spine and the construction of the spaces $T_S(\epsilon)^H$ due to Ivanov [95, Theorem 5.4.A] seem to lead to such models. However, a detailed proof can be found in a manuscript by Mislin [135].

**Theorem 3.31** (Teichmüller space). *The $\Gamma^s_{g,r}$-space $\mathcal{T}^s_{g,r}$ is a model for $\underline{E}\Gamma^s_{g,r}$.*

Let $F_n$ be the free group of rank $n$. Denote by $\operatorname{Out}(F_n)$ the group of outer automorphisms of $F_n$, i.e., the quotient of the group of all automorphisms of $F_n$ by the normal subgroup of inner automorphisms. Culler-Vogtmann (see [48], [183]) have constructed a space $X_n$ called *outer space* on which $\operatorname{Out}(F_n)$ acts with finite isotropy groups. It is analogous to the Teichmüller space of a surface with the action of the mapping class group of the surface. Fix a graph $R_n$ with one vertex $v$ and $n$-edges and identify $F_n$ with $\pi_1(R_n, v)$. A *marked metric graph* $(g, \Gamma)$ consists of a graph $\Gamma$ with all vertices of valence at least three, a homotopy equivalence $g\colon R_n \to \Gamma$ called marking and to every edge of $\Gamma$ there is assigned a positive length which makes $\Gamma$ into a metric space by the path metric. We call two marked metric graphs $(g, \Gamma)$ and $(g', \Gamma')$ equivalent if there is a homothety $h\colon \Gamma \to \Gamma'$ such that $g \circ h$ and $h'$ are homotopic. Homothety means that there is a constant $\lambda > 0$ with $d(h(x), h(y)) = \lambda \cdot d(x, y)$ for all $x, y$. Elements in outer space $X_n$ are equivalence classes of marked graphs. The main result in [48] is that $X$ is contractible. Actually, for each finite subgroup $H \subseteq \operatorname{Out}(F_n)$ the $H$-fixed point set $X_n^H$ is contractible (see [107, Proposition 3.3 and Theorem 8.1], [190, Theorem 5.1]).

The space $X_n$ contains a *spine* $K_n$ which is an $\mathrm{Out}(F_n)$-equivariant deformation retraction. This space $K_n$ is a simplicial complex of dimension $(2n-3)$ on which the $\mathrm{Out}(F_n)$-action is by simplicial automorphisms and cocompact. Actually the group of simplicial automorphisms of $K_n$ is $\mathrm{Out}(F_n)$ (see Bridson-Vogtmann [32]). We conclude

**Theorem 3.32** (Spine of outer space). *The barycentric subdivision $K_n'$ is a finite $(2n-3)$-dimensional model of $\underline{E}\,\mathrm{Out}(F_n)$.*

**Example 3.33** ($SL_2(\mathbb{R})$ and $SL_2(\mathbb{Z})$). In order to illustrate some of the general statements above we consider the special example $SL_2(\mathbb{R})$ and $SL_2(\mathbb{Z})$.

Let $\mathbb{H}^2$ be the 2-dimensional hyperbolic space. We will use either the upper half-plane model or the Poincaré disk model. The group $SL_2(\mathbb{R})$ acts by isometric diffeomorphisms on the upper half-plane by Moebius transformations, i.e., a matrix $\begin{pmatrix} a & b \\ c & d \end{pmatrix}$ acts by sending a complex number $z$ with positive imaginary part to $\frac{az+b}{cz+d}$. This action is proper and transitive. The isotropy group of $z = i$ is $SO(2)$. Since $\mathbb{H}^2$ is a simply-connected Riemannian manifold, whose sectional curvature is constant $-1$, the $SL_2(\mathbb{R})$-space $\mathbb{H}^2$ is a model for $\underline{E}SL_2(\mathbb{R})$ by Remark 3.26.

One easily checks that $SL_2(\mathbb{R})$ is a connected Lie group and $SO(2) \subseteq SL_2(\mathbb{R})$ is a maximal compact subgroup. Hence $SL_2(\mathbb{R})/SO(2)$ is a model for $\underline{E}SL_2(\mathbb{R})$ by Theorem 3.24. Since the $SL_2(\mathbb{R})$-action on $\mathbb{H}^2$ is transitive and $SO(2)$ is the isotropy group at $i \in \mathbb{H}^2$, we see that the $SL_2(\mathbb{R})$-manifolds $SL_2(\mathbb{R})/SO(2)$ and $\mathbb{H}^2$ are $SL_2(\mathbb{R})$-diffeomorphic.

Since $SL_2(\mathbb{Z})$ is a discrete subgroup of $SL_2(\mathbb{R})$, the space $\mathbb{H}^2$ with the obvious $SL_2(\mathbb{Z})$-action is a model for $\underline{E}SL_2(\mathbb{Z})$ (see Theorem 3.24).

The group $SL_2(\mathbb{Z})$ is isomorphic to the amalgamated product $\mathbb{Z}/4 *_{\mathbb{Z}/2} \mathbb{Z}/6$. This implies that $SL_2(\mathbb{Z})$ acts on a tree $T$ which consists of two 0-dimensional equivariant cells with isotropy groups $\mathbb{Z}/4$ and $\mathbb{Z}/6$ and one 1-dimensional equivariant cell with isotropy group $\mathbb{Z}/2$. From Remark 3.26 we conclude that a model for $\underline{E}SL_2(\mathbb{Z})$ is given by the following $SL_2(\mathbb{Z})$-pushout

$$\begin{array}{ccc} SL_2(\mathbb{Z})/(\mathbb{Z}/2) \times \{-1,1\} & \xrightarrow{F_{-1} \coprod F_1} & SL_2(\mathbb{Z})/(\mathbb{Z}/4) \coprod SL_2(\mathbb{Z})/(\mathbb{Z}/6) \\ \downarrow & & \downarrow \\ SL_2(\mathbb{Z})/(\mathbb{Z}/2) \times [-1,1] & \longrightarrow & T = \underline{E}SL_2(\mathbb{Z}) \end{array}$$

where $F_{-1}$ and $F_1$ are the obvious projections. This model for $\underline{E}SL_2(\mathbb{Z})$ is a tree, which has alternately two and three edges emanating from each vertex. The other model $\mathbb{H}^2$ is a manifold. These two models must be $SL_2(\mathbb{Z})$ homotopy equivalent. They can explicitly be related by the following construction.

Divide the Poincaré disk into fundamental domains for the $SL_2(\mathbb{Z})$-action. Each fundamental domain is a geodesic triangle with one vertex at infinity, i.e., a vertex on the boundary sphere, and two vertices in the interior. Then the union of the edges, whose end points lie in the interior of the Poincaré disk, is a tree $T$ with $SL_2(\mathbb{Z})$-action. This is the tree model above. The tree is a $SL_2(\mathbb{Z})$-equivariant deformation retraction of the Poincaré disk. A retraction is given by moving a point $p$ in the Poincaré disk along a geodesic starting at the vertex at infinity, which belongs to the triangle containing $p$, through $p$ to the first intersection point of this geodesic with $T$.

The tree $T$ above can be identified with the Bruhat-Tits building of $SL_2(\widehat{\mathbb{Q}_p})$ and hence is a model for $\underline{E}SL_2(\widehat{\mathbb{Q}_p})$ (see [35, page 134]). Since $SL_2(\mathbb{Z})$ is a discrete subgroup of $SL_2(\widehat{\mathbb{Q}_p})$, we get another reason why this tree is a model for $\underline{E}SL_2(\mathbb{Z})$.

**Definition 3.34** (Cohomological dimension). Let $\Lambda$ be a commutative ring. The cohomological dimension $\mathrm{cd}_\Lambda(G)$ of a group $G$ over $\Lambda$ is defined to be the infimum over all integers $d$ for which there exist a $d$-dimensional projective $\Lambda G$-resolution of the trivial $\Lambda G$-module $\Lambda$. If $\Lambda = \mathbb{Z}$, we abbreviate $\mathrm{cd}(G) = \mathrm{cd}_\mathbb{Z}(G)$.

By definition $\mathrm{cd}_\Lambda(G) = \infty$ if there is no finite-dimensional projective $\Lambda G$-resolution of the trivial $\Lambda G$-module $\Lambda$.

**Example 3.35.** If $G$ is a non-trivial finite group, then $\mathrm{cd}(G) = \infty$ and $\mathrm{cd}_\mathbb{Q}(G) = 0$. We conclude that a group $G$ with $\mathrm{cd}(G) < \infty$ must be torsionfree.

**Definition 3.36** (Virtual cohomological dimension). A group $G$ is called virtually torsionfree if it contains a torsionfree subgroup $\Delta \subset G$ with finite index $[G:\Delta]$.

Let $\Lambda$ be a commutative ring. Define the virtual cohomological dimension of a virtually torsionfree group $G$ over $\Lambda$ by

$$\mathrm{vcd}_\Lambda(G) = \mathrm{cd}_\Lambda(\Delta)$$

for any torsionfree subgroup $\Delta \subset G$ with finite index $[G:\Delta]$.

If $\Lambda = \mathbb{Z}$, we abbreviate $\mathrm{vcd}(G) = \mathrm{vcd}_\mathbb{Z}(G)$.

This definition is indeed independent of the choice of $\Delta \subseteq G$.

Next we investigate the relation between the minimal dimension of a model $\underline{E}G$ with the virtual cohomological dimension provided that $G$ is virtually torsionfree.

**Theorem 3.37** (Discrete subgroups of Lie groups). *Let $L$ be a Lie group with finitely many path components. Let $K \subseteq L$ be a maximal compact subgroup $K$. Let $G \subseteq L$ be a discrete subgroup of $L$. Then $L/K$ with the left $G$-action is a model for $\underline{E}G$.*

*Suppose additionally that $G$ is virtually torsionfree, i.e., contains a torsionfree subgroup $\Delta \subseteq G$ of finite index.*

*Then we have for its virtual cohomological dimension*

$$\mathrm{vcd}(G) \leq \dim(L/K).$$

*Equality holds if and only if $G\backslash L$ is compact.*

*Proof.* We have already mentioned in Theorem 3.24 that $L/K$ is a model for $\underline{E}G$. The restriction of $\underline{E}G$ to $\Delta$ is a $\Delta$-$CW$-model for $\underline{E}\Delta$ and hence $\Delta\backslash\underline{E}G$ is a $CW$-model for $B\Delta$. This implies $\mathrm{vcd}(G) := \mathrm{cd}(\Delta) \leq \dim(L/K)$. Obviously $\Delta\backslash L/K$ is a manifold without boundary. Suppose that $\Delta\backslash L/K$ is compact. Then $\Delta\backslash L/K$ is a closed manifold and hence its homology with $\mathbb{Z}/2$-coefficients in the top dimension is non-trivial. This implies $\mathrm{cd}(\Delta) \geq \dim(\Delta\backslash L/K)$ and hence $\mathrm{vcd}(G) = \dim(L/K)$. If $\Delta\backslash L/K$ is not compact, it contains a $CW$-complex $X \subseteq \Delta\backslash L/K$ of dimension smaller than $\Delta\backslash L/K$ such that the inclusion of $X$ into $\Delta\backslash L/K$ is a homotopy equivalence. Hence $X$ is another model for $B\Delta$. This implies $\mathrm{cd}(\Delta) < \dim(L/K)$ and hence $\mathrm{vcd}(G) < \dim(L/K)$. □

**Theorem 3.38** (A criterion for 1-dimensional models for $BG$, Stallings (1968), Swan (1969)). *Let $G$ be a discrete group.*

*The following statements are equivalent:*

- *There exists a 1-dimensional model for $EG$;*
- *There exists a 1-dimensional model for $BG$;*
- *The cohomological dimension of $G$ is less or equal to one;*
- *$G$ is a free group.*

*Proof.* See [171] and [177]. □

**Theorem 3.39** (A criterion for 1-dimensional models for $\underline{E}G$, Dunwoody (1979)). *Let $G$ be a discrete group. Then there exists a 1-dimensional model for $\underline{E}G$ if and only if the cohomological dimension of $G$ over the rationals $\mathbb{Q}$ is less or equal to one.*

*Proof.* See Dunwoody [56, Theorem 1.1]. □

**Theorem 3.40** (Virtual cohomological dimension and $\dim(\underline{E}G)$, Lück (2000)). *Let $G$ be a discrete group which is virtually torsionfree.*

(1) *Then*
$$\mathrm{vcd}(G) \leq \dim(\underline{E}G)$$
*for any model for $\underline{E}G$.*

(2) *Let $l \geq 0$ be an integer such that for any chain of finite subgroups $H_0 \subsetneq H_1 \subsetneq \ldots \subsetneq H_r$ we have $r \leq l$.*
*Then there exists a model for $\underline{E}G$ whose dimension is*
$$\max\{3, \mathrm{vcd}(G)\} + l.$$

*Proof.* See Lück [118, Theorem 6.4]. □

The following problem has been stated by Brown [34, page 32] and has created a lot of activities.

**Problem 3.41.** For which discrete groups $G$, which are virtually torsionfree, does there exist a $G$ $CW$-model for $\underline{E}G$ of dimension $\mathrm{vcd}(G)$?

**Remark 3.42.** The results above give some evidence for the hope that the problem above has a positive answer for every discrete group. However, Leary-Nucinkis [112] have constructed virtually torsionfree groups $G$ for which the answer is negative, i.e., for which the dimension of any model for $\underline{E}G$ is different from $\mathrm{vcd}(G)$.

The following result shows that in general one can say nothing about the quotient $G\backslash\underline{E}G$ although in many interesting cases there do exist small models for it.

**Theorem 3.43** (Leary-Nucinkis (2001)). *Let $X$ be a CW-complex. Then there exists a group $G$ with $X \simeq G\backslash\underline{E}G$.*

*Proof.* See [111]. □

**Question 3.44** (Homological Computations based on nice models for $\underline{E}G$). Can nice geometric models for $\underline{E}G$ be used to compute the group homology and more general homology and cohomology theories of a group $G$?

**Question 3.45** ($K$-theory of group rings and group homology). Is there a relation between $K_n(RG)$ and the group homology of $G$?

**Question 3.46** (Isomorphism Conjectures and classifying spaces of families). Can classifying spaces of families be used to formulate a version of the Farrell-Jones Conjecture and the Baum-Connes Conjecture which may hold for all group $G$ and all rings?

# 4. Equivariant homology theories

The outline of this section is:

- We introduce the notion of an equivariant homology theory.
- We present the general formulation of the Farrell-Jones Conjecture and the Baum-Connes Conjecture.
- We discuss equivariant Chern characters.
- We present some explicit computations of equivariant topological $K$-groups and of homology groups associated to classifying spaces of groups.

**Definition 4.1** ($G$-homology theory). A *$G$-homology theory $\mathcal{H}_*^G$* is a covariant functor from the category of $G$-$CW$-pairs to the category of $\mathbb{Z}$-graded $\Lambda$-modules together with natural transformations

$$\partial_n^G(X,A)\colon \mathcal{H}_n^G(X,A) \to \mathcal{H}_{n-1}^G(A)$$

for $n \in \mathbb{Z}$ satisfying the following axioms:

- $G$-homotopy invariance;
- Long exact sequence of a pair;
- Excision;
- Disjoint union axiom.

The following definition is taken from [120, Section 1].

**Definition 4.2** (Equivariant homology theory). An *equivariant homology theory* $\mathcal{H}_*^?$ assigns to every group $G$ a $G$-homology theory $\mathcal{H}_*^G$. These are linked together with the following so called *induction structure*: given a group homomorphism

$\alpha\colon H \to G$ and a $H$-$CW$-pair $(X, A)$, there are for all $n \in \mathbb{Z}$ natural homomorphisms

$$\mathrm{ind}_\alpha\colon \mathcal{H}_n^H(X, A) \to \mathcal{H}_n^G(\mathrm{ind}_\alpha(X, A))$$

satisfying

- Bijectivity
  If $\ker(\alpha)$ acts freely on $X$, then $\mathrm{ind}_\alpha$ is a bijection;
- Compatibility with the boundary homomorphisms;
- Functoriality in $\alpha$;
- Compatibility with conjugation.

We have the following examples of equivariant homology theories.

**Example 4.3** (Borel homology). Given a non-equivariant homology theory $\mathcal{K}_*$, put

$$\mathcal{H}_*^G(X) := \mathcal{K}_*(X/G);$$
$$\mathcal{H}_*^G(X) := \mathcal{K}_*(EG \times_G X) \quad \text{(Borel homology)}.$$

**Example 4.4** (Equivariant bordism). Equivariant bordism $\Omega_*^?(X)$ based on proper cocompact equivariant smooth manifolds with reference map to the $G$-space $X$.

**Example 4.5** (Equivariant topological $K$-theory). Equivariant topological $K$-theory $K_*^?(X)$ defined for proper equivariant $CW$-complexes has the property that for any finite subgroup $H \subseteq G$ we get

$$K_n^H(\{\bullet\}) \cong K_n^G(G/H) \cong \begin{cases} R_{\mathbb{C}}(H) & n \text{ even;} \\ 0 & n \text{ odd.} \end{cases}$$

**Theorem 4.6** (Lück-Reich (2005)). *Given a functor* $\mathbf{E}\colon \mathsf{Groupoids} \to \mathsf{Spectra}$ *sending equivalences to weak equivalences, there exists an equivariant homology theory* $\mathcal{H}_*^?(-;\mathbf{E})$ *satisfying*

$$\mathcal{H}_n^H(\{\bullet\}) \cong \mathcal{H}_n^G(G/H) \cong \pi_n(\mathbf{E}(H)).$$

*Proof.* See [126, Proposition 6.4 on page 738]. □

**Theorem 4.7** (Equivariant homology theories associated to $K$ and $L$-theory, Davis-Lück (1998)). *Let $R$ be a ring (with involution). There exist covariant functors*

$$\mathbf{K}_R\colon \mathsf{Groupoids} \to \mathsf{Spectra};$$
$$\mathbf{L}_R^{\langle\infty\rangle}\colon \mathsf{Groupoids} \to \mathsf{Spectra};$$
$$\mathbf{K}^{\mathrm{top}}\colon \mathsf{Groupoids}^{\mathrm{inj}} \to \mathsf{Spectra}$$

*with the following properties:*

- *They send equivalences of groupoids to weak equivalences of spectra;*
- *For every group $G$ and all $n \in \mathbb{Z}$ we have*

$$\pi_n(\mathbf{K}_R(G)) \cong K_n(RG);$$
$$\pi_n(\mathbf{L}_R^{\langle -\infty\rangle}(G)) \cong L_n^{\langle -\infty\rangle}(RG);$$
$$\pi_n(\mathbf{K}^{\mathrm{top}}(G)) \cong K_n(C_r^*(G)).$$

*Proof.* See [52, Section 2]. □

Combining the last two theorems we get

**Example 4.8** (Equivariant homology theories associated to $K$ and $L$-theory). We get equivariant homology theories

$$H_*^?(-;\mathbf{K}_R);$$
$$H_*^?(-;\mathbf{L}_R^{\langle -\infty\rangle});$$
$$H_*^?(-;\mathbf{K}^{\mathrm{top}}),$$

satisfying for $H \subseteq G$

$$\begin{array}{lll} H_n^G(G/H;\mathbf{K}_R) & \cong H_n^H(\{\bullet\};\mathbf{K}_R) & \cong K_n(RH); \\ H_n^G(G/H;\mathbf{L}_R^{\langle -\infty\rangle}) & \cong H_n^H(\{\bullet\};\mathbf{L}_R^{\langle -\infty\rangle}) & \cong L_n^{\langle -\infty\rangle}(RH); \\ H_n^G(G/H;\mathbf{K}^{\mathrm{top}}) & \cong H_n^H(\{\bullet\};\mathbf{K}^{\mathrm{top}}) & \cong K_n(C_r^*(H)). \end{array}$$

Now we are ready to give the general formulation of the Farrell-Jones and the Baum-Connes Conjecture.

**Conjecture 4.9** ($K$-theoretic Farrell-Jones-Conjecture). *The* $K$-theoretic Farrell-Jones Conjecture *with coefficients in $R$ for the group $G$ predicts that the assembly map*

$$H_n^G(E_{\mathcal{VCYC}}(G),\mathbf{K}_R) \to H_n^G(\{\bullet\},\mathbf{K}_R) = K_n(RG),$$

*which is the map induced by the projection $E_{\mathcal{VCYC}}(G) \to \{\bullet\}$, is bijective for all $n \in \mathbb{Z}$.*

**Conjecture 4.10** ($L$-theoretic Farrell-Jones-Conjecture). *The* $L$-theoretic Farrell-Jones Conjecture *with coefficients in $R$ for the group $G$ predicts that the assembly map*

$$H_n^G(E_{\mathcal{VCYC}}(G),\mathbf{L}_R^{\langle -\infty\rangle}) \to H_n^G(\{\bullet\},\mathbf{L}_R^{\langle -\infty\rangle}) = L_n^{\langle -\infty\rangle}(RG),$$

*which is the map induced by the projection $E_{\mathcal{VCYC}}(G) \to \{\bullet\}$, is bijective for all $n \in \mathbb{Z}$.*

**Conjecture 4.11** (Baum-Connes Conjecture). *The* Baum-Connes Conjecture *predicts that the assembly map*

$$K_n^G(\underline{E}G) = H_n^G(E_{\mathcal{FIN}}(G),\mathbf{K}^{\mathrm{top}}) \to H_n^G(\{\bullet\},\mathbf{K}^{\mathrm{top}}) = K_n(C_r^*(G))$$

*which is the map induced by the projection $E_{\mathcal{FIN}}(G) \to \{\bullet\}$, is bijective for all $n \in \mathbb{Z}$.*

**Remark 4.12** (Original sources for the Farrell-Jones and the Baum-Connes Conjecture). These conjectures were stated in Farrell-Jones [66, 1.6 on page 257] and Baum-Connes-Higson [24, Conjecture 3.15 on page 254]. Our formulations differ from the original ones, but are equivalent (see [10, Section 6], [52, Section 6], and [84]). In the case of the Farrell-Jones Conjecture we slightly generalize the original conjecture by allowing arbitrary coefficient rings instead of $\mathbb{Z}$.

We will discuss these conjectures and their applications in the next section. We will now continue with equivariant homology theories.

Let $\mathcal{H}_*$ be a (non-equivariant) homology theory. There is the Atiyah-Hirzebruch spectral sequence which converges to $\mathcal{H}_{p+q}(X)$ and has as $E^2$-term

$$E^2_{p,q} = H_p(X; \mathcal{H}_q(\{\bullet\})).$$

Rationally it collapses completely by the following result.

**Theorem 4.13** (Non-equivariant Chern character, Dold (1962)). *Let $\mathcal{H}_*$ be a homology theory with values in $\Lambda$-modules for $\mathbb{Q} \subseteq \Lambda$.*

*Then there exists for every $n \in \mathbb{Z}$ and every $CW$-complex $X$ a natural isomorphism*

$$\bigoplus_{p+q=n} H_p(X; \Lambda) \otimes_\Lambda \mathcal{H}_q(\{\bullet\}) \xrightarrow{\cong} \mathcal{H}_n(X),$$

*where $H_p(X; \Lambda)$ is the singular or cellular homology of $X$ with coefficients in $\Lambda$.*

*Proof.* At least we give the definition of Dold's Chern character for a $CW$-complex $X$, for more details we refer to Dold [54]. It is given by the following composite:

$$\mathrm{ch}_n\colon \bigoplus_{p+q=n} H_p(X; \mathcal{H}_q(*)) \xrightarrow{\alpha^{-1}} \bigoplus_{p+q=n} H_p(X; \mathbb{Z}) \otimes_{\mathbb{Z}} \mathcal{H}_q(*)$$

$$\xrightarrow{\bigoplus_{p+q=n} (\mathrm{hur} \otimes \mathrm{id})^{-1}} \bigoplus_{p+q=n} \pi_p^s(X_+, *) \otimes_{\mathbb{Z}} \mathcal{H}_q(*) \xrightarrow{\bigoplus_{p+q=n} D_{p,q}} \mathcal{H}_n(X),$$

Here the canonical map $\alpha$ is bijective, since any $\Lambda$-module is flat over $\mathbb{Z}$ because of the assumption $\mathbb{Q} \subset \Lambda$, The map hur is the Hurewicz homomorphism which is bijective because of Serre's Theorem (see [166], [102]) which says

$$\pi_m^s \otimes \mathbb{Q} \cong \begin{cases} \mathbb{Q} \text{ for } m = 0, \\ 0 \text{ else.} \end{cases}$$

The map $D_{p,q}$ sends $[f\colon (S^{p+k}, \{\bullet\}) \to (S^k \wedge X_+, \{\bullet\})] \otimes \eta$ to the image of $\eta$ under the composite

$$\mathcal{H}_q(*) \cong \mathcal{H}_{p+k+q}(S^{p+k}, \{\bullet\}) \xrightarrow{\mathcal{H}_{p+k+q}(f)} \mathcal{H}_{p+k+q}(S^k \wedge X_+, \{\bullet\}) \cong \mathcal{H}_{p+q}(X).$$

□

We want to extend this to the equivariant setting. This requires an extra structure on the coefficients of an equivariant homology theory $\mathcal{H}_*^?$.

We define a covariant functor called induction

$$\operatorname{ind}\colon \mathcal{FGI} \to \Lambda\text{-}\operatorname{Mod}$$

from the category $\mathcal{FGI}$ of finite groups with injective group homomorphisms as morphisms to the category of $\Lambda$-modules as follows. It sends $G$ to $\mathcal{H}_n^G(\{\bullet\})$ and an injection of finite groups $\alpha\colon H \to G$ to the morphism given by the induction structure

$$\mathcal{H}_n^H(\{\bullet\}) \xrightarrow{\operatorname{ind}_\alpha} \mathcal{H}_n^G(\operatorname{ind}_\alpha\{\bullet\}) \xrightarrow{\mathcal{H}_n^G(\operatorname{pr})} \mathcal{H}_n^G(\{\bullet\}).$$

**Definition 4.14** (Mackey extension). We say that $\mathcal{H}_*^?$ has a Mackey extension if for every $n \in \mathbb{Z}$ there is a contravariant functor called restriction

$$\operatorname{res}\colon \mathcal{FGI} \to \Lambda\text{-}\operatorname{Mod}$$

such that the two functors ind and res agree on objects and satisfy the double coset formula, i.e., we have for two subgroups $H, K \subset G$ of the finite group $G$

$$\operatorname{res}_G^K \circ \operatorname{ind}_H^G = \sum_{KgH \in K\backslash G/H} \operatorname{ind}_{c(g)\colon H\cap g^{-1}Kg \to K} \circ \operatorname{res}_H^{H\cap g^{-1}Kg},$$

where $c(g)$ is conjugation with $g$, i.e., $c(g)(h) = ghg^{-1}$.

**Remark 4.15** (Existence of Mackey extensions). In every case we will consider such a Mackey extension does exist and is given by an actual restriction. For instance for $H_0^?(-;\mathbf{K}^{\mathrm{top}})$ induction is the functor complex representation ring $R_{\mathbb{C}}$ with respect to induction of representations. The restriction part is given by the restriction of representations.

We need some notation. Consider a subgroup $H \subseteq G$. Denote by $C_G H$ the centralizer and by $N_G H$ the normalizer of $H \subseteq G$. Put

$$W_G H := N_G H / H \cdot C_G H.$$

This is always a finite group. Define for an equivariant homology theory $\mathcal{H}_*^?$

$$S_H\left(\mathcal{H}_q^H(*)\right) := \operatorname{cok}\left(\bigoplus_{\substack{K \subset H\\ K \neq H}} \operatorname{ind}_K^H \colon \bigoplus_{\substack{K \subset H\\ K \neq H}} \mathcal{H}_q^K(*) \to \mathcal{H}_q^H(*)\right).$$

**Theorem 4.16** (Equivariant Chern character, Lück (2002)). *Let $\mathcal{H}_*^?$ be an equivariant homology theory with values in $\Lambda$-modules for $\mathbb{Q} \subseteq \Lambda$. Suppose that $\mathcal{H}_*^?$ has a Mackey extension. Let $I$ be the set of conjugacy classes $(H)$ of finite subgroups $H$ of $G$.*

*Then there is for every group $G$, every proper $G$-CW-complex $X$ and every $n \in \mathbb{Z}$ a natural isomorphism called equivariant Chern character*

$$\mathrm{ch}_n^G\colon \bigoplus_{p+q=n} \bigoplus_{(H)\in I} H_p(C_G H\backslash X^H;\Lambda) \otimes_{\Lambda[W_G H]} S_H\left(\mathcal{H}_q^H(*)\right) \xrightarrow{\cong} \mathcal{H}_n^G(X).$$

*Actually* $\mathrm{ch}_*^?$ *is an equivalence of equivariant homology theories.*

*Proof.* See [120, Theorem 0.2] □

Recall the following basic result from complex representation theory of finite groups.

**Theorem 4.17** (Artin's Theorem). *Let $G$ be finite. Then the map*

$$\bigoplus_{C\subset G} \mathrm{ind}_C^G\colon \bigoplus_{C\subset G} R_{\mathbb{C}}(C) \to R_{\mathbb{C}}(G)$$

*is surjective after inverting $|G|$, where $C \subset G$ runs through the cyclic subgroups of $G$.*

*Proof.* See for instance [167, Theorem 17 in 9.2 on page 70]. □

Let $C$ be a finite cyclic group. The Artin defect is the cokernel of the map

$$\bigoplus_{D\subset C, D\neq C} \mathrm{ind}_D^C\colon \bigoplus_{D\subset C, D\neq C} R_{\mathbb{C}}(D) \to R_{\mathbb{C}}(C).$$

For an appropriate idempotent $\theta_C \in R_{\mathbb{Q}}(C) \otimes_{\mathbb{Z}} \mathbb{Z}\left[\frac{1}{|C|}\right]$ the Artin defect is after inverting the order of $|C|$ canonically isomorphic to

$$\theta_C \cdot R_{\mathbb{C}}(C) \otimes_{\mathbb{Z}} \mathbb{Z}\left[\frac{1}{|C|}\right]$$

by [120, Lemma 7.4].

**Example 4.18** (An improvement of Artin's Theorem). Let $K_*^G = H_*^?(-;\mathbf{K}^{\mathrm{top}})$ be equivariant topological $K$-theory. We get for a finite subgroup $H \subseteq G$

$$K_n^G(G/H) = K_n^H(\{\bullet\}) = \begin{cases} R_{\mathbb{C}}(H) & \text{if } n \text{ is even;} \\ \{0\} & \text{if } n \text{ is odd.} \end{cases}$$

Hence $S_H\left(K_q^H(*)\right) \otimes_{\mathbb{Z}} \mathbb{Q} = 0$, if $H$ is not cyclic and $q$ is even or if $q$ is odd, and we have $S_C\left(K_q^C(*)\right) \otimes_{\mathbb{Z}} \mathbb{Q} = \theta_C \cdot R_{\mathbb{C}}(C) \otimes_{\mathbb{Z}} \mathbb{Q}$, if $C$ is finite cyclic and $q$ is even.

Let $G$ be finite, $X = \{*\}$ and $\mathcal{H}_*^? = K_*^?$. In this very special case Theorem 4.16 yields already something new, namely, an improvement of Artin's theorem, i.e., the equivariant Chern character induces an isomorphism

$$\mathrm{ch}_0^G(\{\bullet\})\colon \bigoplus_{(C)} \mathbb{Z} \otimes_{\mathbb{Z}[W_G C]} \theta_C \cdot R_{\mathbb{C}}(C) \otimes_{\mathbb{Z}} \mathbb{Z}\left[\frac{1}{|G|}\right] \xrightarrow{\cong} R_{\mathbb{C}}(G) \otimes_{\mathbb{Z}} \mathbb{Z}\left[\frac{1}{|G|}\right]$$

where $(C)$ runs over the conjugacy classes of finite cyclic subgroups. (Theorem 4.16 yields only a statement after applying $-\otimes_{\mathbb{Z}}\mathbb{Q}$ but the statement above, where we only invert the order of the group $G$ is proved in [122, Theorem 0.7]).

**Theorem 4.19** (Rational computation of $K_*^G(\underline{E}G)$). *For every group G and every $n\in\mathbb{Z}$ we obtain an isomorphism*

$$\bigoplus_{(C)}\bigoplus_{k} H_{p+2k}(BC_GC)\otimes_{\mathbb{Z}[W_GC]}\theta_C\cdot R_{\mathbb{C}}(C)\otimes_{\mathbb{Z}}\mathbb{Q}\xrightarrow{\cong} K_n^G(\underline{E}G)\otimes_{\mathbb{Z}}\mathbb{Q}.$$

*Proof.* This follows from Theorem 4.16 applied to the case $X=\underline{E}G$ and $\mathcal{H}_*^?=K_*^?$ using the following facts.

- $\underline{E}G^C$ is a contractible proper $C_GC$- space. Hence the canonical map $BC_GC\to C_GC\backslash\underline{E}G^C$ induces an isomorphism

$$H_p(BC_GC)\otimes_{\mathbb{Z}}\mathbb{Q}\xrightarrow{\cong}H_p(C_GC\backslash\underline{E}G^C)\otimes_{\mathbb{Z}}\mathbb{Q}.$$

- $S_H\left(K_q^H(*)\right)\otimes_{\mathbb{Z}}\mathbb{Q}=0$ if $H$ is not cyclic and $q$ is even or if $q$ is odd.
- $S_C\left(K_q^C(*)\right)\otimes_{\mathbb{Z}}\mathbb{Q}=\theta_C\cdot R_{\mathbb{C}}(C)\otimes_{\mathbb{Z}}\mathbb{Q}$ if $C$ is finite cyclic and $q$ is even. □

**Remark 4.20** (Rational computation of $K_*(C_r^*(G))$). If the Baum-Connes Conjecture holds for $G$, Theorem 4.19 yields an isomorphism

$$\bigoplus_{(C)}\bigoplus_{k} H_{p+2k}(BC_GC)\otimes_{\mathbb{Z}[W_GC]}\theta_C\cdot R_{\mathbb{C}}(C)\otimes_{\mathbb{Z}}\mathbb{Q}\xrightarrow{\cong} K_n(C_r^*(G))\otimes_{\mathbb{Z}}\mathbb{Q}.$$

Next we introduce some notation. For a prime $p$ denote by $r(p)=|\operatorname{con}_p(G)|$ the number of conjugacy classes $(g)$ of elements $g\neq 1$ in $G$ of $p$-power order. Let $\mathbb{I}_G$ is the augmentation ideal of $R_{\mathbb{C}}(G)$. Denote by $\mathbb{I}_p(G)$ the image of the restriction homomorphism $\mathbb{I}(G)\to\mathbb{I}(G_p)$ for the inclusion of the $p$-Sylow subgroup $G_p\to G$.

**Theorem 4.21** (Completion Theorem, Atiyah-Segal (1969)). *Let G be a finite group. Then there are isomorphisms of abelian groups*

$$\begin{aligned}K^0(BG)&\cong R_{\mathbb{C}}(G)\widehat{_{\mathbb{I}_G}}\\&\cong\mathbb{Z}\times\prod_{p\ prime}\mathbb{I}_p(G)\otimes_{\mathbb{Z}}\widehat{\mathbb{Z}_p}\cong\mathbb{Z}\times\prod_{p\ prime}(\widehat{\mathbb{Z}_p})^{r(p)};\\K^1(BG)&\cong 0.\end{aligned}$$

*Proof.* See [7] and for the explicit formula for instance [96, page 125] or [124, Theorem 3.5]. □

**Theorem 4.22** (Lück (2005)). *Let G be a discrete group. Denote by $K^*(BG)$ the topological (complex) K-theory of its classifying space BG. Suppose that there is a cocompact G-CW-model for the classifying space $\underline{E}G$ for proper G-actions.*

*Then there is a* $\mathbb{Q}$*-isomorphism*

$$\overline{\mathrm{ch}}_G^n\colon K^n(BG)\otimes_{\mathbb{Z}}\mathbb{Q}\xrightarrow{\cong}$$
$$\left(\prod_{i\in\mathbb{Z}} H^{2i+n}(BG;\mathbb{Q})\right)\times\prod_{p\ prime}\ \prod_{(g)\in\mathrm{con}_p(G)}\left(\prod_{i\in\mathbb{Z}} H^{2i+n}(BC_G\langle g\rangle;\widehat{\mathbb{Q}_p})\right).$$

*Proof.* See [124]. □

**Remark 4.23** (Multiplicative structure)**.** The multiplicative structure is also determined in [124].

**Remark 4.24** (Finiteness condition about $\underline{E}G$)**.** We have presented in the previous section many groups for which a cocompact $G$-$CW$-model for $\underline{E}G$ exists, e.g., hyperbolic groups. Notice that this condition appears in Theorem 4.22 although the conclusion in Theorem 4.22 is about $BG$ and not about $\underline{E}G$ or $G\backslash\underline{E}G$.

**Example 4.25** ($SL_3(\mathbb{Z})$)**.** It is well-known that its rational cohomology satisfies $\widetilde{H}^n(BSL_3(\mathbb{Z});\mathbb{Q})=0$ for all $n\in\mathbb{Z}$. Actually, by a result of Soulé [170, Corollary on page 8] the quotient space $SL_3(\mathbb{Z})\backslash\underline{E}SL_3(\mathbb{Z})$ is contractible and compact. From the classification of finite subgroups of $SL_3(\mathbb{Z})$ we see that $SL_3(\mathbb{Z})$ contains up to conjugacy two elements of order 2, two elements of order 4 and two elements of order 3 and no further conjugacy classes of non-trivial elements of prime power order. The rational homology of each of the centralizers of elements in $\mathrm{con}_2(G)$ and $\mathrm{con}_3(G)$ agrees with the one of the trivial group. Hence we get

$$K^0(BSL_3(\mathbb{Z}))\otimes_{\mathbb{Z}}\mathbb{Q}\cong\mathbb{Q}\times(\widehat{\mathbb{Q}_2})^4\times(\widehat{\mathbb{Q}_3})^2;$$
$$K^1(BSL_3(\mathbb{Z}))\otimes_{\mathbb{Z}}\mathbb{Q}\cong 0.$$

The identification of $K^0(BSL_3(\mathbb{Z}))\otimes_{\mathbb{Z}}\mathbb{Q}$ above is compatible with the multiplicative structures.

Actually the computation using Brown-Petersen cohomology and the Conner-Floyd relation by Tezuka-Yagita [178] gives the integral computation

$$K^0(BSL_3(\mathbb{Z}))\cong\mathbb{Z}\times(\widehat{\mathbb{Z}_2})^4\times(\widehat{\mathbb{Z}_3})^2;$$
$$K^1(BSL_3(\mathbb{Z}))\cong 0.$$

Soulé [170] has computed the integral cohomology of $SL_3(\mathbb{Z})$.

Let $G$ be a discrete group. Let $\mathcal{MFIN}$ be the subset of $\mathcal{FIN}$ consisting of elements in $\mathcal{FIN}$ which are maximal in $\mathcal{FIN}$. Consider the following conditions about $G$:

(M) Every non-trivial finite subgroup of $G$ is contained in a unique maximal finite subgroup;

(NM) If $M\in\mathcal{MFIN}, M\neq\{1\}$, then $N_GM=M$.

**Example 4.26** (Groups satisfying (M) and (NM))**.** By Davis-Lück [53, page 101-102] the following groups satisfy conditions (M) and (NM):

- Extensions $1 \to \mathbb{Z}^n \to G \to F \to 1$ for finite $F$ such that the conjugation action of $F$ on $\mathbb{Z}^n$ is free outside $0 \in \mathbb{Z}^n$;
- Fuchsian groups;
- One-relator groups $G$.

For such a group there is a nice model for $\underline{E}G$ with as few non-free cells as possible. Let

$$\{(M_i) \mid i \in I\}$$

be the set of conjugacy classes of maximal finite subgroups of $M_i \subseteq G$. By attaching free $G$-cells we get an inclusion of $G$-$CW$-complexes $j_1 \colon \coprod_{i\in I} G \times_{M_i} EM_i \to EG$. Define the $G$-$CW$-complex $X$ as the $G$-pushout

$$\begin{array}{ccc} \coprod_{i\in I} G \times_{M_i} EM_i & \xrightarrow{j_1} & EG \\ \downarrow{\scriptstyle u_1} & & \downarrow{\scriptstyle f_1} \\ \coprod_{i\in I} G/M_i & \xrightarrow{k_1} & X \end{array} \tag{4.27}$$

where $u_1$ is the obvious $G$-map obtained by collapsing each $EM_i$ to a point.

**Theorem 4.28** (Model for $\underline{E}G$ for groups satisfying (M) and (NM))**.** *Let $G$ be a group satisfying conditions* (M) *and* (NM)*. Then the $G$-$CW$-complex $X$ defined by the $G$-pushout* (4.27) *is a model for $\underline{E}G$.*

*Proof.* The isotropy groups of $X$ are all finite. We have to show for $H \subseteq G$ finite that $X^H$ contractible. We begin with the case $H \neq \{1\}$. Because of conditions (M) and (NM) there is precisely one index $i_0 \in I$ such that $H$ is subconjugated to $M_{i_0}$ and is not subconjugated to $M_i$ for $i \neq i_0$. We get

$$\left(\coprod_{i\in I} G/M_i\right)^H = (G/M_{i_0})^H = \{\bullet\}.$$

Hence $X^H = \{\bullet\}$. It remains to treat $H = \{1\}$. Since $u_1$ is a non-equivariant homotopy equivalence and $j_1$ is a cofibration, $f_1$ is a non-equivariant homotopy equivalence. Hence $X$ is contractible. □

**Example 4.29** (The homology of groups satisfying (M) and (NM))**.** Let $G$ be a group satisfying conditions (M) and (NM). Because of Theorem 4.28 we obtain the following pushout by taking the $G$-quotient of the $G$-pushout (4.27)

$$\begin{array}{ccc} \coprod_{i\in I} BM_i & \longrightarrow & BG \\ \downarrow & & \downarrow \\ \coprod_{i\in I} \{\bullet\} & \longrightarrow & G\backslash\underline{E}G \end{array}$$

The associated long exact Mayer-Vietoris sequence yields

$$\cdots \to \widetilde{H}_{n+1}(G\backslash\underline{E}G) \to \bigoplus_{i\in I} \widetilde{H}_n(BM_i) \to \widetilde{H}_n(BG) \to \widetilde{H}_n(G\backslash\underline{E}G) \to \cdots.$$

In particular we obtain an isomorphism for $n \geq \dim(\underline{E}G) + 2$

$$\bigoplus_{i \in I} H_n(BM_i) \xrightarrow{\cong} H_n(BG).$$

So we get an explicit computation of $H_n(BG)$ for large $n$ and it is obvious why it is useful to have models for $\underline{E}G$ of as small as possible dimension. Computations for low values of $n$ can sometimes be carried out by spectral sequence arguments or specific arguments.

The following identifications follow from the definition of the *Whitehead groups* $\mathrm{Wh}_n(G)$ for $n \geq 0$ due to Waldhausen [184, Definition 15.6 on page 228 and Proposition 15.7 on page 229] which also makes sense for all $n \in \mathbb{Z}$ if we use the non-connective $K$-theory spectrum

$$\begin{aligned}
\mathrm{Wh}(G) &= \mathrm{Wh}_1(G); \\
\widetilde{K}_0(\mathbb{Z}G) &= \mathrm{Wh}_0(G); \\
K_n(\mathbb{Z}G) &= \mathrm{Wh}_n(G) \quad \text{for } n \leq -1.
\end{aligned}$$

For a finite group $H$ define $\widetilde{R}_{\mathbb{C}}(H)$ as the kernel of the ring homomorphism $R_{\mathbb{C}}(H) \to \mathbb{Z}$ sending $[V]$ to $\dim_{\mathbb{C}}(V)$.

**Theorem 4.30** (Davis-Lück (2003)). *Let $G$ be a discrete group which satisfies the conditions* (M) *and* (NM) *above.*

(1) *Then there is an isomorphism*

$$K_1^G(\underline{E}G) \xrightarrow{\cong} K_1(G \backslash \underline{E}G),$$

*and a short exact sequence*

$$0 \to \bigoplus_{i \in I} \widetilde{R}_{\mathbb{C}}(M_i) \to K_0(\underline{E}G) \to K_0(G \backslash \underline{E}G) \to 0.$$

(2) *The short exact sequence above splits if we invert the orders of all finite subgroups of $G$;*

(3) *Suppose that $G$ belongs to $\mathcal{BC}$. (This is the case for the groups appearing in Example 4.26). Then*

$$K_n(C_r^*(G)) \cong K_n^G(\underline{E}G);$$

(4) *Suppose that $G$ belongs to $\mathcal{FJ}_K(\mathbb{Z})$. Then there is for $n \in \mathbb{Z}$ an isomorphism of Whitehead groups*

$$\bigoplus_{i \in I} \mathrm{Wh}_n(M_i) \xrightarrow{\cong} \mathrm{Wh}_n(G),$$

*where $\mathrm{Wh}_n(M_i) \to \mathrm{Wh}_n(G)$ is induced by the inclusion $M_i \to G$.*

*Proof.* See [53, Theorem 5.1]. □

**Remark 4.31** (Small models for $\underline{E}G$ and computations). We see that for computations of group homology or of $K$- and $L$-groups of group rings and group $C^*$-algebras it is important to understand the spaces $G\backslash\underline{E}G$. Often geometric input ensures that $G\backslash\underline{E}G$ is a fairly small $CW$-complex.

On the other hand recall from Theorem 3.43 that for any $CW$-complex $X$ there exists a group $G$ with $X \simeq G\backslash\underline{E}G$.

**Question 4.32** (Consequences). What are the consequences of the Farrell-Jones Conjecture and the Baum-Connes Conjecture?

# 5. The Isomorphism Conjectures for arbitrary groups

The outline of this section is:

- We discuss the difference between the families $\mathcal{FIN}$ and $\mathcal{VCYC}$.
- We discuss consequences of the Farrell-Jones and the Baum-Connes Conjecture.

Throughout this section $G$ will always be a discrete group. We have introduced the following notions and conjectures:

- Families of subgroups, e.g., the families $\mathcal{FIN}$ and $\mathcal{VCYC}$ of finite and virtually cyclic subgroups (see Definition 3.6);
- Classifying $G$-$CW$-complex $E_{\mathcal{F}}(G)$ for a family of subgroups $\mathcal{F}$ (see Definition 3.7).
- Equivariant homology theories $\mathcal{H}_*^?$ (see Definition 4.2);
- Specific examples of equivariant homology theories associated to $K$- and $L$-theory (see Example 4.8)

$$H_*^?(-;\mathbf{K}_R);$$
$$H_*^?(-;\mathbf{L}_R^{\langle -\infty\rangle});$$
$$H_*^?(-;\mathbf{K}^{\mathrm{top}}),$$

satisfying for $H \subseteq G$

$$\begin{aligned} H_n^G(G/H;\mathbf{K}_R) &\cong H_n^H(\{\bullet\};\mathbf{K}_R) &&\cong K_n(RH); \\ H_n^G(G/H;\mathbf{L}_R^{\langle -\infty\rangle}) &\cong H_n^H(\{\bullet\};\mathbf{L}_R^{\langle -\infty\rangle}) &&\cong L_n^{\langle -\infty\rangle}(RH); \\ H_n^G(G/H;\mathbf{K}^{\mathrm{top}}) &\cong H_n^H(\{\bullet\};\mathbf{K}^{\mathrm{top}}) &&\cong K_n(C_r^*(H)); \end{aligned}$$

- The Farrell-Jones Conjecture for algebraic $K$-theory which predicts the bijectivity of the assembly map induced by the projection $E_{\mathcal{VCYC}}(G) \to \{\bullet\}$

$$H_n^G(E_{\mathcal{VCYC}}(G),\mathbf{K}_R) \to H_n^G(\{\bullet\},\mathbf{K}_R) = K_n(RG)$$

for all $n \in \mathbb{Z}$ (see Conjecture 4.9);

- The Farrell-Jones Conjecture for algebraic $L$-theory which predicts the bijectivity of the assembly map induced by the projection $E_{\mathcal{VCYC}}(G) \to \{\bullet\}$

$$H_n^G(E_{\mathcal{VCYC}}(G), \mathbf{L}_R^{\langle -\infty \rangle}) \to H_n^G(\{\bullet\}, \mathbf{L}_R^{\langle -\infty \rangle}) = L_n^{\langle -\infty \rangle}(RG)$$

for all $n \in \mathbb{Z}$ (see Conjecture 4.10);
- The Baum-Cones Conjecture which predicts the bijectivity of the assembly map induced by the projection $E_{\mathcal{FIN}}(G) = \underline{E}G \to \{\bullet\}$

$$K_n^G(\underline{E}G) = H_n^G(E_{\mathcal{FIN}}(G), \mathbf{K}^{\text{top}}) \to H_n^G(\{\bullet\}, \mathbf{K}^{\text{top}}) = K_n(C_r^*(G))$$

for all $n \in \mathbb{Z}$ (see Conjecture 4.11).

**Remark 5.1** (The Isomorphism Conjectures interpreted as induction theorems). These Conjecture can be thought of a kind of generalized induction theorem. They allow to compute the value of a functor such as $K_n(RG)$ on $G$ in terms of its values $K_m(RH)$ for all $m \leq n$ and all virtually cyclic subgroups subgroups $H$ of $G$.

Next we want to investigate, whether one can pass to smaller or larger families in the formulations of the Conjectures. The point is to find the family as small as possible.

**Theorem 5.2** (Transitivity Principle). *Let $\mathcal{F} \subseteq \mathcal{G}$ be two families of subgroups of $G$. Let $\mathcal{H}_*^?$ be an equivariant homology theory. Assume that for every element $H \in \mathcal{G}$ and $n \in \mathbb{Z}$ the assembly map*

$$\mathcal{H}_n^H(E_{\mathcal{F}|_H}(H)) \to \mathcal{H}_n^H(\{\bullet\})$$

*is bijective, where $\mathcal{F}|_H = \{K \cap H \mid K \in \mathcal{F}\}$.*

*Then the relative assembly map induced by the up to $G$-homotopy unique $G$-map $E_{\mathcal{F}}(G) \to E_{\mathcal{G}}(G)$*

$$\mathcal{H}_n^G(E_{\mathcal{F}}(G)) \to \mathcal{H}_n^G(E_{\mathcal{G}}(G))$$

*is bijective for all $n \in \mathbb{Z}$.*

*Proof.* See [12, Theorem 1.4]. □

**Example 5.3** (Passage from $\mathcal{FIN}$ to $\mathcal{VCYC}$ for the Baum-Connes Conjecture). The Baum-Connes Conjecture 4.11 is known to be true for virtually cyclic groups. The Transitivity Principle 5.2 implies that the relative assembly

$$K_n^G(\underline{E}G) \xrightarrow{\cong} K_n^G(E_{\mathcal{VCYC}}(G))$$

is bijective for all $n \in \mathbb{Z}$.

Hence it does not matter in the context of the Baum-Connes Conjecture whether we consider the family $\mathcal{FIN}$ or $\mathcal{VCYC}$.

**Example 5.4** (Passage from $\mathcal{FIN}$ to $\mathcal{VCYC}$ for the Farrell-Jones Conjecture). In general the relative assembly maps

$$H_n^G(\underline{E}G; \mathbf{K}_R) \to H_n^G(E_{\mathcal{VCYC}}(G); \mathbf{K}_R);$$
$$H_n^G(\underline{E}G; \mathbf{L}_R^{\langle -\infty \rangle}) \to H_n^G(E_{\mathcal{VCYC}}(G); \mathbf{L}_R^{\langle -\infty \rangle}),$$

are not bijective. Hence in the Farrell-Jones setting one has to pass to $\mathcal{VCYC}$ and cannot use the easier to handle family $\mathcal{FIN}$.

**Example 5.5** (The Farrell-Jones Conjecture for algebraic $K$-theory for the group $\mathbb{Z}$). The Farrell-Jones Conjecture 4.9 for algebraic $K$-theory for the group $\mathbb{Z}$ is true for trivial reasons since $\mathbb{Z}$ is virtually cyclic and hence the projection $E_{\mathcal{VCYC}}(\mathbb{Z}) \to \{\bullet\}$ is a homotopy equivalence.

**Example 5.6** (The Farrell-Jones Conjecture for algebraic $K$-theory for the group $\mathbb{Z}$ and the family $\mathcal{FIN}$). One may wonder what happens if we insert the family of finite subgroups, i.e., whether the map induced by the projection $E_{\mathcal{FIN}}(\mathbb{Z}) = \underline{E}\mathbb{Z} \to \{\bullet\}$

$$H_n^{\mathbb{Z}}(\underline{E}\mathbb{Z}, \mathbf{K}_R) \to H_n^{\mathbb{Z}}(\{\bullet\}, \mathbf{K}_R) = K_n(R[\mathbb{Z}]) \tag{5.7}$$

is bijective. Since $\mathbb{Z}$ is torsionfree, $\underline{E}\mathbb{Z}$ is the same as $E\mathbb{Z}$ and the induction structure yields an isomorphism

$$H_n^{\mathbb{Z}}(\underline{E}\mathbb{Z}, \mathbf{K}_R) = H_n(BZ, \mathbf{K}_R) = H_n(S^1, \mathbf{K}_R) = K_n(R[\mathbb{Z}]) \oplus K_{n-1}(R[\mathbb{Z}]).$$

Hence the map (5.7) can be identified with the map

$$K_n(R) \oplus K_{n-1}(R) \to K_n(R[\mathbb{Z}]).$$

However, by the Bass-Heller Swan decomposition we have the isomorphism

$$K_n(R) \oplus K_{n-1}(R) \oplus \mathrm{NK}_n(R) \oplus \mathrm{NK}_n(R)) \xrightarrow{\cong} K_n(R[t, t^{-1}]) \cong K_n(R[\mathbb{Z}]).$$

Hence the map (5.7) is bijective if and only if $\mathrm{NK}_n(R) = 0$. We have $\mathrm{NK}_n(R) = 0$ under the assumption that $R$ is regular. This is the reason why we have required $R$ to be regular in the version of the Farrell-Jones Conjecture for torsionfree groups 2.16.

**Definition 5.8** (Types of virtually cyclic groups). An infinite virtually cyclic group $G$ is called of type I if it admits an epimorphism onto $\mathbb{Z}$ and of type II if and only if admits an epimorphism onto $D_\infty$. Let $\mathcal{VCYC}_I$ be the family of virtually cyclic subgroups which are either finite or of type I.

An infinite virtually cyclic group is either of type I or of type II. An infinite subgroups of a virtually cyclic subgroup of type I is again of type I.

**Theorem 5.9** (Lück (2004), Quinn (2007), Reich (2007)). *The following maps are bijective for all $n \in \mathbb{Z}$*

$$H_n^G(E_{\mathcal{VCYC}_I}(G); \mathbf{K}_R) \to H_n^G(E_{\mathcal{VCYC}}(G); \mathbf{K}_R);$$
$$H_n^G(\underline{E}G; \mathbf{L}_R^{\langle -\infty \rangle}) \to H_n^G(E_{\mathcal{VCYC}_I}(G); \mathbf{L}_R^{\langle -\infty \rangle}).$$

*Proof.* See [123, Lemma 4.2] and [153]. □

**Theorem 5.10** (Cappell (1973), Grunewald (2005), Waldhausen (1978)).

(1) *The following maps are bijective for all $n \in \mathbb{Z}$.*

$$H_n^G(\underline{E}G; \mathbf{K}_{\mathbb{Z}}) \otimes_{\mathbb{Z}} \mathbb{Q} \to H_n^G(E_{\mathcal{VCYC}}(G); \mathbf{K}_{\mathbb{Z}}) \otimes_{\mathbb{Z}} \mathbb{Q};$$
$$H_n^G(\underline{E}G; \mathbf{L}_R^{\langle -\infty \rangle})\left[\frac{1}{2}\right] \to H_n^G(E_{\mathcal{VCYC}}(G); \mathbf{L}_R^{\langle -\infty \rangle})\left[\frac{1}{2}\right];$$

(2) *If $R$ is regular and $\mathbb{Q} \subseteq R$, then for all $n \in \mathbb{Z}$ we get a bijection*

$$H_n^G(\underline{E}G; \mathbf{K}_R) \to H_n^G(E_{\mathcal{VCYC}}(G); \mathbf{K}_R).$$

*Proof.* See [40], [82, Theorem 5.6], [126, Proposition 2.6 on page 686, Proposition 2.9 and Proposition 2.10 on page 688]. □

**Theorem 5.11** (Bartels (2003)). *For every $n \in \mathbb{Z}$ the two maps*

$$H_n^G(\underline{E}G; \mathbf{K}_R) \to H_n^G(E_{\mathcal{VCYC}}(G); \mathbf{K}_R);$$
$$H_n^G(\underline{E}G; \mathbf{L}_R^{\langle -\infty \rangle}) \to H_n^G(E_{\mathcal{VCYC}}(G); \mathbf{L}_R^{\langle -\infty \rangle}),$$

*are split injective.*

*Proof.* See [18]. □

Hence we get (natural) isomorphisms

$$(5.12)\quad H_n^G(E_{\mathcal{VCYC}}(G); \mathbf{K}_R) \cong H_n^G(\underline{E}G; \mathbf{K}_R) \oplus H_n^G(E_{\mathcal{VCYC}}(G), \underline{E}G; \mathbf{K}_R);$$

and

$$H_n^G(E_{\mathcal{VCYC}}(G); \mathbf{L}_R^{\langle -\infty \rangle}) \cong H_n^G(\underline{E}G; \mathbf{L}_R^{\langle -\infty \rangle}) \oplus H_n^G(E_{\mathcal{VCYC}}(G), \underline{E}G; \mathbf{L}_R^{\langle -\infty \rangle}).$$

The analysis of the terms $H_n^G(E_{\mathcal{VCYC}}(G), \underline{E}G; \mathbf{K}_R)$ and $H_n^G(E_{\mathcal{VCYC}}(G), \underline{E}G; \mathbf{L}_R^{\langle -\infty \rangle})$ boils down to investigating Nil-terms and UNil-terms in the sense of Waldhausen and Cappell. The analysis of the terms $H_n^G(\underline{E}G; \mathbf{K}_R)$ and $H_n^G(\underline{E}G; \mathbf{L}_R^{\langle -\infty \rangle})$ is using the methods of the previous lecture (e.g., equivariant Chern characters).

**Remark 5.13** (Relating the torsionfree versions to the general versions). Obviously the general version of the Baum-Connes Conjecture 4.11 reduces in the torsionfree case to the version 2.18 since for torsionfree $G$ we have $EG = \underline{E}G$.

The general version of the Farrell-Jones Conjecture 4.9 for $K$-theory reduces in the torsionfree case to the version 2.16 because of the Transitivity Principal 5.2 since a torsionfree virtually cyclic group is isomorphic to $\mathbb{Z}$ and for a regular ring $R$ the Bass-Heller-Swan decomposition shows that the map $H_n^{\mathbb{Z}}(\underline{E}\mathbb{Z}; \mathbf{K}_R) \to K_n(R[\mathbb{Z}])$ is bijective (as explained in Example 5.6).

The general version of the Farrell-Jones Conjecture 4.10 for $L$-theory reduces in the torsionfree case to the version 2.17 because of the Transitivity Principal 5.2 since a torsionfree virtually cyclic group is isomorphic to $\mathbb{Z}$ and the map $H_n^{\mathbb{Z}}(\underline{E}\mathbb{Z}; \mathbf{L}_R^{\langle -\infty \rangle}) \to L_n^{\langle -\infty \rangle}(R[\mathbb{Z}])$ is bijective by Theorem 5.9.

Next we explain in the case $G = SL_2(\mathbb{Z})$ how computations are made possible by the Farrell-Jones and the Baum-Connes Conjecture.

**Example 5.14** ($K$-theory of $C_r^*(SL_2(\mathbb{Z}))$ and of $\mathbb{Z}[SL_2(\mathbb{Z})]$). From Example 3.33 we obtain a $SL_2(\mathbb{Z})$-pushout

$$\begin{array}{ccc} SL_2(\mathbb{Z})/(\mathbb{Z}/2) \times \{-1,1\} & \xrightarrow{F_{-1} \coprod F_1} & SL_2(\mathbb{Z})/(\mathbb{Z}/4) \coprod SL_2(\mathbb{Z})/(\mathbb{Z}/6) \\ \downarrow & & \downarrow \\ SL_2(\mathbb{Z})/(\mathbb{Z}/2) \times [-1,1] & \longrightarrow & T = \underline{E}SL_2(\mathbb{Z}) \end{array}$$

Let $\mathcal{H}_*^?$ be an equivariant homology theory. Then the Mayer-Vietoris sequence applied to the $SL_2(\mathbb{Z})$-pushout above together with the induction structure yields a long exact sequence

$$(5.15) \quad \cdots \to \mathcal{H}_n^{\mathbb{Z}/2}(\{\bullet\}) \to \mathcal{H}_n^{\mathbb{Z}/4}(\{\bullet\}) \oplus \mathcal{H}_n^{\mathbb{Z}/6}(\{\bullet\}) \to \mathcal{H}_n^{SL_2(\mathbb{Z})}(\underline{E}SL_2(\mathbb{Z})) \\ \to \mathcal{H}_{n-1}^{\mathbb{Z}/2}(\{\bullet\}) \to \mathcal{H}_{n-1}^{\mathbb{Z}/4}(\{\bullet\}) \oplus \mathcal{H}_{n-1}^{\mathbb{Z}/6}(\{\bullet\}) \to \cdots.$$

The Baum-Connes Conjecture 4.11 is known to be true for $SL_2(\mathbb{Z})$ (see for instance [89]). Hence in the case, where $\mathcal{H}_*^?$ is equivariant topological $K$-theory, the long exact sequence (5.15) reduces to the exact sequences

$$0 \to K_1(C_r^*(SL_2(\mathbb{Z}))) \to R_{\mathbb{C}}(\mathbb{Z}/2) \to R_{\mathbb{C}}(\mathbb{Z}/4) \oplus R_{\mathbb{C}}(\mathbb{Z}/6) \\ \to K_0(C_r^*(SL_2(\mathbb{Z}))) \to 0,$$

where the map between the representation rings are induced by the obvious inclusions of groups. Since the inclusion $\mathbb{Z}/2 \to \mathbb{Z}/6$ is split injective and $R_{\mathbb{C}}(\mathbb{Z}/2) \cong \mathbb{Z}^2$, $R_{\mathbb{C}}(\mathbb{Z}/4) \cong \mathbb{Z}^4$ and $R_{\mathbb{C}}(\mathbb{Z}/6) \cong \mathbb{Z}^6$, we conclude

$$K_n(C_r^*(SL_2(\mathbb{Z}))) \cong \begin{cases} \mathbb{Z}^8 & n \text{ even;} \\ 0 & n \text{ odd.} \end{cases}$$

The Farrell-Jones Conjecture for $K$-theory is known to be true for $SL_2(\mathbb{Z})$ for any coefficient ring $R$ by [14] since it contains a finitely generated free subgroup of finite index and is hence a hyperbolic group. Because of Theorem 5.9 and Theorem 5.11 we obtain an isomorphism

$$(5.16) \quad K_n(R[SL_2(\mathbb{Z})]) \cong H_n^{SL_2(\mathbb{Z})}(E_{\mathcal{FIN}}(SL_2(\mathbb{Z})); \mathbf{K}_R) \\ \oplus H_n^{SL_2(\mathbb{Z})}(E_{\mathcal{VCY}_I}(SL_2(\mathbb{Z})), E_{\mathcal{FIN}}(SL_2(\mathbb{Z})); \mathbf{K}_R).$$

Let $V \subseteq SL_2(\mathbb{Z})$ be a virtually cyclic subgroup of type I, i.e., there is an exact sequence $1\colon F \to V \to \mathbb{Z} \to 1$ for a finite subgroup $F \subseteq V$. Since $SL_2(\mathbb{Z}) \cong \mathbb{Z}/4 *_{\mathbb{Z}/2} \mathbb{Z}/6$, $F$ is conjugated to $\mathbb{Z}/4$, $\mathbb{Z}/6$ or the subgroup $\mathbb{Z}/2$. Since the normalizers of $\mathbb{Z}/6$ and $\mathbb{Z}/4$ are finite, $F$ must be subconjugated to $\mathbb{Z}/2$. Since $\mathbb{Z}/2$ is the center of $SL_2(\mathbb{Z})$, the group $V$ is isomorphic to $\mathbb{Z}/2 \times \mathbb{Z}$. The group $\mathrm{NK}_n(\mathbb{Z}[\mathbb{Z}/2])$ vanishes for $n \leq 1$. Using the Bass-Heller-Swan decomposition we see that $H_n^V(E_{\mathcal{VCYC}_I}(V), E_{\mathcal{FIN}}(V); \mathbf{K}_{\mathbb{Z}}) = 0$ for $n \leq 1$. An obvious modification of the Transitivity Principal 5.2 (see [12, Theorem 1.4]) implies that for $n \leq 1$ the group $H_n^{SL_2(\mathbb{Z})}(E_{\mathcal{VCYC}_I}(SL_2(\mathbb{Z})), E_{\mathcal{FIN}}(SL_2(\mathbb{Z})); \mathbf{K}_{\mathbb{Z}})$ vanishes. Thus from (5.16) we obtain an isomorphism for $n \leq 1$.

$$K_n(\mathbb{Z}[SL_2(\mathbb{Z})]) \cong H_n^{SL_2(\mathbb{Z})}(E_{\mathcal{FIN}}(SL_2(\mathbb{Z})); \mathbf{K}_{\mathbb{Z}}).$$

Hence the long exact sequence (5.15) yields the long exact sequence

$$\begin{aligned} K_1(\mathbb{Z}[\mathbb{Z}/2]) &\to K_1(\mathbb{Z}[\mathbb{Z}/4]) \oplus K_1(\mathbb{Z}[\mathbb{Z}/6]) \to K_1(\mathbb{Z}[SL_2(\mathbb{Z})]) \to K_0(\mathbb{Z}[\mathbb{Z}/2]) \\ &\to K_0(\mathbb{Z}[\mathbb{Z}/4]) \oplus K_0(\mathbb{Z}[\mathbb{Z}/6]) \to K_0(\mathbb{Z}[SL_2(\mathbb{Z})]) \to K_{-1}(\mathbb{Z}[\mathbb{Z}/2]) \\ &\to K_{-1}(\mathbb{Z}[\mathbb{Z}/4]) \oplus K_{-1}(\mathbb{Z}[\mathbb{Z}/6]) \to K_{-1}(\mathbb{Z}[SL_2(\mathbb{Z})]) \to K_{-2}(\mathbb{Z}[\mathbb{Z}/2]) \\ &\to K_{-2}(\mathbb{Z}[\mathbb{Z}/4]) \oplus K_{-2}(\mathbb{Z}[\mathbb{Z}/6]) \to K_{-2}(\mathbb{Z}[SL_2(\mathbb{Z})]) \to \cdots . \end{aligned}$$

It induces the long exact sequence

$$\begin{aligned} \mathrm{Wh}(\mathbb{Z}/2) &\to \mathrm{Wh}(\mathbb{Z}/4) \oplus \mathrm{Wh}(\mathbb{Z}/6) \to \mathrm{Wh}(SL_2(\mathbb{Z})) \to \widetilde{K}_0(\mathbb{Z}[\mathbb{Z}/2]) \\ &\to \widetilde{K}_0(\mathbb{Z}[\mathbb{Z}/4]) \oplus \widetilde{K}_0(\mathbb{Z}[\mathbb{Z}/6]) \to \widetilde{K}_0(\mathbb{Z}[SL_2(\mathbb{Z})]) \to K_{-1}(\mathbb{Z}[\mathbb{Z}/2]) \\ &\to K_{-1}(\mathbb{Z}[\mathbb{Z}/4]) \oplus K_{-1}(\mathbb{Z}[\mathbb{Z}/6]) \to K_{-1}(\mathbb{Z}[SL_2(\mathbb{Z})]) \to K_{-2}(\mathbb{Z}[\mathbb{Z}/2]) \\ &\to K_{-2}(\mathbb{Z}[\mathbb{Z}/4]) \oplus K_{-2}(\mathbb{Z}[\mathbb{Z}/6]) \to K_{-2}(\mathbb{Z}[SL_2(\mathbb{Z})]) \to \cdots . \end{aligned}$$

The groups $\mathrm{Wh}(\mathbb{Z}/2)$, $\mathrm{Wh}(\mathbb{Z}/4)$, $\mathrm{Wh}(\mathbb{Z}/6)$, $\widetilde{K}_0(\mathbb{Z}[\mathbb{Z}/2])$, $\widetilde{K}_0(\mathbb{Z}[\mathbb{Z}/4])$, $\widetilde{K}_0(\mathbb{Z}[\mathbb{Z}/6])$, $\widetilde{K}_{-1}(\mathbb{Z}[\mathbb{Z}/2])$, $\widetilde{K}_{-1}(\mathbb{Z}[\mathbb{Z}/4])$ vanish, whereas $\widetilde{K}_{-1}(\mathbb{Z}[\mathbb{Z}/6]) \cong \mathbb{Z}$ (see Bass [19, Theorem 10.6 on page 695], Carter [44], Cassou-Noguès [45], Curtis-Rainer [50, Corollary 50.17 on page 253]), Oliver [138, Theorem 14.1 on page 328]. The groups $K_n(\mathbb{Z}[H])$ vanish for all $n \geq -2$ and all finite groups $H$ (see Carter [44]). Hence we get

$$\begin{array}{ll} \mathrm{Wh}(SL_2(\mathbb{Z})) & \cong 0; \\ \widetilde{K}_0(\mathbb{Z}[SL_2(\mathbb{Z})]) & \cong 0; \\ K_{-1}(\mathbb{Z}[SL_2(\mathbb{Z})]) & \cong \mathbb{Z}: \\ K_n(\mathbb{Z}[SL_2(\mathbb{Z})]) & \cong 0 \quad \text{for } n \leq -2. \end{array}$$

Next we show that the Farrell-Jones Conjecture and the Baum-Conjecture imply certain other well-known conjectures.

**Conjecture 5.17** (Novikov Conjecture). *The* Novikov Conjecture for $G$ *predicts for a closed oriented manifold $M$ together with a map $f\colon M \to BG$ that for any $x \in H^*(BG)$ the higher signature*

$$\mathrm{sign}_x(M, f) := \langle \mathcal{L}(M) \cup f^*x, [M] \rangle$$

*is an oriented homotopy invariant of $(M, f)$, i.e., for every orientation preserving homotopy equivalence of closed oriented manifolds $g\colon M_0 \to M_1$ and homotopy equivalence $f_i\colon M_0 \to M_1$ with $f_1 \circ g \simeq f_2$ we have*

$$\mathrm{sign}_x(M_0, f_0) = \mathrm{sign}_x(M_1, f_1).$$

**Theorem 5.18** (The Farrell-Jones, the Baum-Connes and the Novikov Conjecture). *Suppose that one of the following assembly maps*

$$H_n^G(E_{\mathcal{VCYC}}(G), \mathbf{L}_R^{\langle -\infty \rangle}) \to H_n^G(\{\bullet\}, \mathbf{L}_R^{\langle -\infty \rangle}) = L_n^{\langle -\infty \rangle}(RG);$$
$$K_n^G(\underline{E}G) = H_n^G(E_{\mathcal{FIN}}(G), \mathbf{K}^{\mathrm{top}}) \to H_n^G(\{\bullet\}, \mathbf{K}^{\mathrm{top}}) = K_n(C_r^*(G)),$$

*is rationally injective.*

*Then the Novikov Conjecture holds for the group $G$.*

*Proof.* See for instance [126, Proposition 3.1 and Proposition 3.5 on page 699] and [152, Proposition 6.6 on page 300]. □

For more information about the Novikov Conjecture we refer for instance to [42], [43], [51], [72], [77], [105], [151] and [155].

**Theorem 5.19** (Induction from finite subgroups, Bartels-Lück-Reich (2007)).

(1) *Let $R$ be a regular ring such that the order of any finite subgroup of $G$ is invertible in $R$. Suppose $G \in \mathcal{FJ}_K(R)$. Then the map given by induction from finite subgroups of $G$*

$$\operatorname*{colim}_{\mathrm{Or}_{\mathcal{FIN}}(G)} K_0(RH) \to K_0(RG)$$

*is bijective;*

(2) *Let $F$ be a field of characteristic $p$ for a prime number $p$. Suppose that $G \in \mathcal{FJ}_K(F)$. Then the map*

$$\operatorname*{colim}_{\mathrm{Or}_{\mathcal{FIN}}(G)} K_0(FH)[1/p] \to K_0(FG)[1/p]$$

*is bijective;*

(3) *If $G \in \mathcal{F}_K(\mathbb{Z})$, then the canonical map*

$$\operatorname*{colim}_{\mathrm{Or}_{\mathcal{FIN}}(G)} K_{-1}(\mathbb{Z}H) \to K_{-1}(\mathbb{Z}G)$$

*is bijective;*

(4) *If $G \in \mathcal{F}_K(\mathbb{Z})$, then*

$$K_n(\mathbb{Z}G) = 0 \text{ for } n \leq -2.$$

*Proof.* See Bartels-Lück-Reich [15, Theorem 0.5], [66, 1.65 on page 260], and Lück-Reich [126, Section 3.1.1 on page 690]. □

**Theorem 5.20** (Permutation Modules, Bartels-Lück-Reich (2007)). *Suppose that $G \in \mathcal{FJ}_K(\mathbb{Q})$. Then for every finitely generated projective $\mathbb{Q}[G]$-module $P$ there exists integers $k \geq 1$ and $l \geq 0$ and finitely many finite subgroups $H_1$, $H_2$, ..., $H_r$ such that*

$$P^k \oplus \mathbb{Q}[G]^l \cong_{\mathbb{Q}[G]} \mathbb{Q}[G/H_1] \oplus \mathbb{Q}[G/H_2] \oplus \cdots \oplus \mathbb{Q}[G/H_r].$$

*Proof.* Because of [15, Lemma 4.3 and Lemma 4.4] it suffices to prove the claim in the case, where $G$ is finite cyclic. This special case follows from Segal [165]. □

Next we introduce some notation. Let $R$ be commutative ring and let $G$ be a group. Let $\operatorname{class}(G,R)$ be the $R$-module of class functions $G \to R$, i.e., functions $G \to R$ which are constant on conjugacy classes. Let $\operatorname{tr}_{RG}\colon RG \to \operatorname{class}(G,R)$ be the $R$-homomorphism which sends $g \in G$ to the class function which takes the value one on the conjugacy class of $g$ and the value zero otherwise. It extends to a map

$$\operatorname{tr}_{RG}\colon M_n(RG) \to \operatorname{class}(G,R)$$

by taking the sums of the values of the diagonal entries.

Let $P$ be a finitely generated $RG$-module. Choose a finitely generated projective $RG$-module $Q$ and an isomorphism $\phi\colon RG^n \xrightarrow{\cong} P \oplus Q$. Let $A \in M_n(RG)$ be a matrix such that $\phi^{-1} \circ (\operatorname{id}_P \oplus 0) \circ \phi\colon RG^n \to RG^n$ is given by $A$.

**Definition 5.21** (Hattori-Stallings rank)**.** Define the Hattori-Stallings rank of $P$ to be the class function

$$\operatorname{HS}_{RG}(P) := \operatorname{tr}_{RG}(A).$$

This definition is independent of the choice of $Q$ and $\phi$. Let $G$ be a finite group and let $F$ be a field of characteristic zero. Then a finitely generated $RG$-module $P$ is the same as a finite dimensional $G$-representation over $F$ and the Hattori-Stallings rank can be identified with the character of the $G$-representation (see (5.27)).

**Conjecture 5.22** (Bass Conjecture)**.** *Let $R$ be a commutative integral domain and let $G$ be a group. Let $g \neq 1$ be an element in $G$. Suppose that either the order $|g|$ is infinite or that the order $|g|$ is finite and not invertible in $R$.*

*Then the* Bass Conjecture *predicts that for every finitely generated projective $RG$-module $P$ the value of its* Hattori-Stallings rank $\operatorname{HS}_{RG}(P)$ *at $(g)$ is trivial.*

If $G$ is finite, the Bass Conjecture 5.22 reduces to a theorem of Swan (1960) (see [175, Theorem 8.1], [21, Corollary 4.2]).

The next results follows from the argument in[70, Section 5].

**Theorem 5.23** (Linnell-Farrell (2003))**.** *Let $G$ be a group. Suppose that*

$$\operatorname*{colim}_{\operatorname{Or}_{\mathcal{FIN}}(G)} K_0(FH) \otimes_{\mathbb{Z}} \mathbb{Q} \to K_0(FG) \otimes_{\mathbb{Z}} \mathbb{Q}$$

*is surjective for all fields $F$ of prime characteristic. (This is true if $G \in \mathcal{FJ}_K(F)$ for every field $F$ of prime characteristic).*

*Then the Bass Conjecture is satisfied for every integral domain $R$.*

**Remark 5.24** (Geometric interpretation of the Bass Conjecture)**.** The Bass Conjecture 5.22 can be interpreted topologically. Namely, the Bass Conjecture 5.22 is true for a finitely presented group $G$ in the case $R = \mathbb{Z}$ if and only if every homotopy idempotent selfmap of an oriented smooth closed manifold whose dimension is greater than 2 and whose fundamental group is isomorphic to $G$ is homotopic to a selfmap which has precisely one fixed point (see Berrick-Chatterji-Mislin [28]).

The Bass Conjecture 5.22 for $G$ in the case $R = \mathbb{Z}$ (or $R = \mathbb{C}$) also implies for a finitely dominated $CW$-complex with fundamental group $G$ that its Euler characteristic agrees with the $L^2$-Euler characteristic of its universal covering (see Eckmann [59]).

Next we present another version of the Bass Conjecture. Let $F$ be a field of characteristic zero. Fix an integer $m \geq 1$. Let $F(\zeta_m) \supset F$ be the Galois extension given by adjoining the primitive $m$-th root of unity $\zeta_m$ to $F$. Denote by $\Gamma(m, F)$ the Galois group of this extension of fields, i.e., the group of automorphisms $\sigma\colon F(\zeta_m) \to F(\zeta_m)$ which induce the identity on $F$. It can be identified with a subgroup of $\mathbb{Z}/m^*$ by sending $\sigma$ to the unique element $u(\sigma) \in \mathbb{Z}/m^*$ for which $\sigma(\zeta_m) = \zeta_m^{u(\sigma)}$ holds. Let $g_1$ and $g_2$ be two elements of $G$ of finite order. We call them *$F$-conjugate* if for some (and hence all) positive integers $m$ with $g_1^m = g_2^m = 1$ there exists an element $\sigma$ in the Galois group $\Gamma(m, F)$ with the property that $g_1^{u(\sigma)}$ and $g_2$ are conjugate. Two elements $g_1$ and $g_2$ are $F$-conjugate for $F = \mathbb{Q}$, $\mathbb{R}$ or $\mathbb{C}$ respectively if the cyclic subgroups $\langle g_1 \rangle$ and $\langle g_2 \rangle$ are conjugate, if $g_1$ and $g_2$ or $g_1$ and $g_2^{-1}$ are conjugate, or if $g_1$ and $g_2$ are conjugate respectively.

Denote by $\mathrm{con}_F(G)_f$ the set of $F$-conjugacy classes $(g)_F$ of elements $g \in G$ of finite order. Let $\mathrm{class}_F(G)_f$ be the $F$-vector space with the set $\mathrm{con}_F(G)_f$ as basis, or, equivalently, the $F$-vector space of functions $\mathrm{con}_F(G)_f \to F$ with finite support.

**Conjecture 5.25** (Bass Conjecture for fields of characteristic zero as coefficients). *Let $F$ be a field of characteristic zero and let $G$ be a group. The Hattori-Stallings (see Definition 5.21) induces an isomorphism*

$$\mathrm{HS}_{FG}\colon K_0(FG) \otimes_{\mathbb{Z}} F \to \mathrm{class}_F(G)_f.$$

**Lemma 5.26.** *Suppose that $F$ is a field of characteristic zero and $G$ is a finite group. Then Conjecture 5.25 is true.*

*Proof.* Since $G$ is finite, an $FG$-module is a finitely generated projective $FG$-module if and only if it is a (finite-dimensional) $G$-representation with coefficients in $F$ and $K_0(FG)$ is the same as the representation ring $R_F(G)$. The Hattori-Stallings rank $\mathrm{HS}_{FH}(V)$ and the character $\chi_V$ of a $G$-representation $V$ with coefficients in $F$ are related by the formula

$$\chi_V(g) = |Z_G\langle g \rangle| \cdot \mathrm{HS}_{FG}(V)(g) \tag{5.27}$$

for $g \in G$, where $Z_G\langle g \rangle$ is the centralizer of $g$ in $G$. Hence Lemma 5.26 follows from representation theory, see for instance [169, Corollary 1 on page 96]. □

Here is a conjecture related to the Bass Conjecture

**Conjecture 5.28.** *Let $R$ be an integral domain with quotient field $F$. Suppose that no prime divisor of the order of a finite subgroup of $G$ is a unit in $R$. Then the change of rings homomorphism*

$$K_0(RG) \otimes_{\mathbb{Z}} \mathbb{Q} \to K_0(FG) \otimes_{\mathbb{Z}} \mathbb{Q}$$

*factorizes as*

$$K_0(RG) \otimes_{\mathbb{Z}} \mathbb{Q} \to K_0(R) \otimes_{\mathbb{Z}} \mathbb{Q} \to K_0(F) \otimes_{\mathbb{Z}} \mathbb{Q} \to K_0(FG) \otimes_{\mathbb{Z}} \mathbb{Q}.$$

**Theorem 5.29** (Bartels-Lück-Reich (2007)). *Let $R$ be an integral domain with quotient field $F$. Suppose that no prime divisor of the order of a finite subgroup of $G$ is a unit in $R$. Suppose that $G$ belongs to $\mathcal{F}_K(R)$.*

*Then Conjecture 5.28 is true for $G$ and $R$.*

*Proof.* See [15, Theorem 0.10]. □

More information and further references about the Bass Conjecture can be found for instance in [20], [27, Section 7], [36], [58], [59], [70], [114] [121, Subsection 9.5.2], and [136, page 66ff].

**Conjecture 5.30** (Vanishing of Bass-Nil-groups). *Let $R$ be a regular ring with $\mathbb{Q} \subseteq R$. Then we get for all groups $G$ and all $n \in \mathbb{Z}$ that*

$$\mathrm{NK}_n(RG) = 0.$$

The relation of this conjecture to the Farrell-Jones Conjecture is discussed in [15, Section 6.3].

Next we discuss some connections of the Farrell-Jones Conjecture to $L^2$-invariants. For more information and some explanations about $L^2$-invariants we refer for instance to Lück [121].

The $L^2$-torsion of a closed Riemannian manifold $M$ is defined in terms of the heat kernel on the universal covering. If $M$ is hyperbolic and has odd dimension, its $L^2$-torsion is up to a (non-vanishing) dimension constant its volume (see [87]).

**Conjecture 5.31** (Homotopy invariance of $L^2$-torsion). *Let $X$ and $Y$ be det-$L^2$-acyclic finite $G$-$CW$-complexes, which are $G$-homotopy equivalent.*

*Then their* $L^2$-torsion *agree:*

$$\rho^{(2)}(X; \mathcal{N}(G)) = \rho^{(2)}(Y; \mathcal{N}(G)).$$

The conjecture above allows to extend the notion of volume to hyperbolic groups whose $L^2$-Betti numbers all vanish.

**Theorem 5.32** (Lück (2002)). *Suppose that $G \in \mathcal{FJ}_K(\mathbb{Z})$. Then $G$ satisfies the Conjecture above.*

*Proof.* See [15, Theorem 0.14]. □

**Remark 5.33** ($p$-adic Fuglede-Kadison determinant). Deninger defines a $p$ adic Fuglede-Kadison determinant for a group $G$ and relates it to $p$-adic entropy provided that $\mathrm{Wh}^{\mathbb{F}_p}(G) \otimes_{\mathbb{Z}} \mathbb{Q}$ is trivial.

**Remark 5.34** (Atiyah Conjecture). The surjectivity of the map

$$\operatorname*{colim}_{\mathrm{Or}_{\mathcal{FIN}}(G)} K_0(\mathbb{C}H) \to K_0(\mathbb{C}G)$$

plays a role (33 %) in a program to prove the Atiyah Conjecture. It says that for a closed Riemannian manifold with torsionfree fundamental group the $L^2$-Betti numbers of its universal covering are all integers.

The Atiyah Conjecture is rather surprising in view of the analytic definition of the $L^2$-Betti numbers by

$$b_p^{(2)}(M) := \lim_{t\to\infty} \int_F e^{-t\widetilde{\Delta}_p}(\widetilde{x},\widetilde{x})\,dvol_{\widetilde{M}},$$

where $F$ is a fundamental domain for the $\pi_1(M)$-action on $\widetilde{M}$.

Next we explain the relation of the Baum-Connes Conjecture to the Gromov-Lawson-Rosenberg Conjecture.

**Definition 5.35** (Bott manifold). A *Bott manifold* is any simply connected closed Spin-manifold $B$ of dimension 8 whose $\widehat{A}$-genus $\widehat{A}(B)$ is 8.

We fix a choice of a Bott manifold. (The particular choice does not matter.) Notice that the index defined in terms of the Dirac operator $\mathrm{ind}_{C_r^*(\{1\};\mathbb{R})}(B) \in KO_8(\mathbb{R}) \cong \mathbb{Z}$ is a generator and the product with this element induces the Bott periodicity isomorphisms $KO_n(C_r^*(G;\mathbb{R})) \xrightarrow{\cong} KO_{n+8}(C_r^*(G;\mathbb{R}))$. In particular

$$\mathrm{ind}_{C_r^*(\pi_1(M);\mathbb{R})}(M) = \mathrm{ind}_{C_r^*(\pi_1(M\times B);\mathbb{R})}(M\times B),$$

if we identify $KO_n(C_r^*(\pi_1(M);\mathbb{R})) = KO_{n+8}(C_r^*(\pi_1(M);\mathbb{R}))$ via Bott periodicity. If $M$ carries a Riemannian metric with positive scalar curvature, then the index

$$\mathrm{ind}_{C_r^*(\pi_1(M);\mathbb{R})}(M) \in KO_n(C_r^*(\pi_1(M);\mathbb{R})),$$

which is defined in terms of the Dirac operator on the universal covering, must vanish by the Bochner-Lichnerowicz formula.

**Conjecture 5.36** ((Stable) Gromov-Lawson-Rosenberg Conjecture). *Let $M$ be a closed connected* Spin*-manifold of dimension $n \geq 5$.*

*Then $M \times B^k$ carries for some integer $k \geq 0$ a Riemannian metric with positive scalar curvature if and only if*

$$\mathrm{ind}_{C_r^*(\pi_1(M);\mathbb{R})}(M) = 0 \qquad \in KO_n(C_r^*(\pi_1(M);\mathbb{R})).$$

**Theorem 5.37** (Stolz (2002)). *Suppose that the assembly map for the real version of the Baum-Connes Conjecture*

$$H_n^G(\underline{E}G;\mathbf{KO}^{\mathrm{top}}) \to KO_n(C_r^*(G;\mathbb{R}))$$

*is injective for the group $G$.*

*Then the Stable Gromov-Lawson-Rosenberg Conjecture is true for all closed* Spin*-manifolds of dimension $\geq 5$ with $\pi_1(M) \cong G$.*

*Proof.* See [174]. □

The requirement $\dim(M) \geq 5$ in Theorem 5.37 is essential in the Stable Gromov-Lawson-Rosenberg Conjecture, since in dimension four new obstructions, the Seiberg-Witten invariants, occur.

Since the Baum-Connes Conjecture is true for finite groups (for the trivial reason that $\underline{E}G = \{\bullet\}$ for finite groups $G$), the Stable Gromov-Lawson Conjecture holds for finite fundamental groups by Theorem 5.37.

**Remark 5.38** (The unstable version of the Gromov-Lawson-Rosenberg Conjecture)**.** The unstable version of the Gromov-Lawson-Rosenberg Conjecture says that $M$ carries a Riemannian metric with positive scalar curvature if and only if the index $\operatorname{ind}_{C_r^*(\pi_1(M);\mathbb{R})}(M)$ vanishes. Schick(1998) [163] has constructed counterexamples to the unstable version using minimal hypersurface methods due to Schoen and Yau. It is not known whether the unstable version is true or false for finite fundamental groups.

**Question 5.39** (Status)**.** For which groups are the Farrell-Jones Conjecture and the Baum-Connes Conjecture known to be true?

What are open interesting cases?

**Question 5.40** (Methods of proof)**.** What are the methods of proof?

**Question 5.41** (Relations)**.** What are the relations between the Farrell-Jones Conjecture and the Baum-Connes Conjecture?

# 6. Summary, status and outlook

The outline of this section is:

- We present other versions of the Isomorphism Conjecture;
- We give a summary about the status of the Farrell-Jones and the Baum-Connes Conjecture:
- We discuss open questions and problems.

**Conjecture 6.1** (Isomorphism Conjecture)**.** *Let $\mathcal{H}_*^?$ be an equivariant homology theory. It satisfies the* Isomorphism Conjecture *for the group $G$ and the family $\mathcal{F}$ if the projection $E_{\mathcal{F}}(G) \to \{\bullet\}$ induces for all $n \in \mathbb{Z}$ a bijection*

$$\mathcal{H}_n^G(E_{\mathcal{F}}(G)) \to \mathcal{H}_n^G(\{\bullet\}).$$

**Example 6.2** (The Farrell-Jones and the Baum-Connes Conjecture as special cases of the Isomorphism Conjecture)**.** The Farrell-Jones Conjecture for $K$-theory or $L$-theory respectively with coefficients in $R$ (see 4.9 and 4.10 respectively) is the Isomorphism Conjecture 6.1 for $\mathcal{H}_*^? = H_*^?(-;\mathbf{K}_R)$ or $\mathcal{H}_*^? = H_*^?(-;\mathbf{L}_R^{\langle -\infty \rangle})$ respectively and $\mathcal{F} = \mathcal{VCYC}$.

The Baum-Connes Conjecture 4.11 is the Isomorphism Conjecture 6.1 for $\mathcal{H}_*^? = K_*^? = H_*^?(-;\mathbf{K}^{\mathrm{top}})$ and $\mathcal{F} = \mathcal{FIN}$.

There are functors $\mathcal{P}$ and $A$ which assign to a space $X$ the space of pseudo-isotopies and its $A$-theory. Composing it with the functor sending a groupoid to its classifying space yields functors $\mathbf{P}$ and $\mathbf{A}$ from $\mathsf{Groupoids}$ to $\mathsf{Spectra}$. Thus we obtain equivariant homology theories $H_*^?(-;\mathbf{P})$ and $H_*^?(-;\mathbf{A})$. They satisfy $H_n^G(G/H;\mathbf{P}) = \pi_n(\mathcal{P}(BH))$ and $H_n^G(G/H;\mathbf{A}) = \pi_n(A(BH))$.

Pseudo-isotopy and $A$-theory are important theories. In particular they are closely related to the space of selfhomeomorphisms and the space of selfdiffeomorphisms of closed manifolds. For more information about $A$-theory and pseudoisotopy we refer for instance to [37], [57, Section 9]), [86], [92], [185], [186].

**Conjecture 6.3** (The Farrell-Jones Conjecture for pseudo-isotopies and $A$-theory)**.** The Farrell-Jones Conjecture for pseudo-isotopies and $A$-theory *respectively is the Isomorphism Conjecture for* $H_*^?(-;\mathbf{P})$ *and* $H_*^?(-;\mathbf{A})$ *respectively for the family* $\mathcal{VCYC}$.

**Theorem 6.4** (Relating pseudo-isotopy and $K$-theory)**.** *The rational versions of the $K$-theoretic Farrell-Jones Conjecture for coefficients in $\mathbb{Z}$ and of the Farrell-Jones Conjecture for Pseudoisotopies are equivalent.*

*In degree $n \le 1$ this is even true integrally.*

*Proof.* See [66, 1.6.7 on page 261]. □

There are functors **THH** and **TC** which assign to a ring (or more generally to an $\mathbf{S}$-algebra) a spectrum describing its topological Hochschild homology and its topological cyclic homology. These functors play an important role in $K$-theoretic computations. Composing it with the functor sending a groupoid to a kind of group ring yields functors $\mathbf{THH}_R$ and $\mathbf{TC}_R$ from $\mathsf{Groupoids}$ to $\mathsf{Spectra}$. Thus we obtain equivariant homology theories $H_*^?(-;\mathbf{THH}_R)$ and $H_*^?(-;\mathbf{TC}_R)$. They satisfy $H_n^G(G/H;\mathbf{THH}_R) = \mathbf{THH}_n(RH)$ and $H_n^G(G/H;\mathbf{TC}_R) = \mathbf{TC}_n(RH)$.

**Conjecture 6.5** (The Farrell-Jones Conjecture for topological Hochschild homology and cyclic homology)**.** The Farrell-Jones Conjecture for topological Hochschild homology and for topological cyclic homology *respectively is the Isomorphism Conjecture for* $H_*^?(-;\mathbf{THH})$ *and* $H_*^?(-;\mathbf{TC})$ *respectively for the family* $\mathcal{CYC}$ *of cyclic subgroups.*

We can apply the functor topological $K$-theory also to Banach algebras such that $l^1(G)$. Let $\mathbf{K}_{l^1}^{\mathrm{top}}$ be the functor from $\mathsf{Groupoids}$ to $\mathsf{Spectra}$ which assign to a groupoid the topological $K$-theory spectrum of its $l^1$-algebra. We obtain an equivariant homology theory $H_*^?(-;\mathbf{K}_{l^1}^{\mathrm{top}})$. It satisfies $H_n^G(G/H,\mathbf{K}_{l^1}^{\mathrm{top}}) = K_n(l^1(H))$.

**Conjecture 6.6** (Bost Conjecture)**.** *The* Bost Conjecture *is the Isomorphism Conjecture for* $H_*^?(-;\mathbf{K}_{l^1}^{\mathrm{top}})$ *and the family* $\mathcal{FIN}$.

**Remark 6.7** (Relating the Baum-Connes Conjecture and the Bost Conjecture)**.** The assembly map appearing in the Bost Conjecture 6.6

$$H_n^G(\underline{E}G;\mathbf{K}_{l^1}^{\mathrm{top}}) \to H_n^G(\{\bullet\};\mathbf{K}_{l^1}^{\mathrm{top}}) = K_n(l^1(G))$$

composed with the change of algebras homomorphism

$$K_n(l^1(G)) \to K_n(C_r^*(G))$$

is precisely the assembly map appearing in the Baum-Connes Conjecture 4.11

$$H_n^G(\underline{E}G; \mathbf{K}^{\mathrm{top}}) = H_n^G(\underline{E}G; \mathbf{K}_{l^1}^{\mathrm{top}}) \to H_n^G(\{\bullet\}; \mathbf{K}^{\mathrm{top}}) = K_n(C_r^*(G)).$$

**Remark 6.8** (Relating the Farrell-Jones Conjecture for $L$-theory and the Baum-Connes Conjecture)**.** We discuss the relation between the Farrell-Jones Conjecture for $L$-theory and the Baum-Connes Conjecture. Mainly these come from the sequence of inclusions of rings

$$\mathbb{Z}G \to \mathbb{R}G \to C_r^*(G;\mathbb{R}) \to C_r^*(G)$$

and the change of theories from algebraic to topological $K$-theory and from algebraic $L$-theory to topological $K$-theory for $C^*$-algebras. Namely, we obtain the following commutative diagram

$$\begin{array}{ccc}
H_n^G(E_{\mathcal{FIN}}(G); \mathbf{L}_{\mathbb{Z}}^p)[1/2] & \longrightarrow & L_n^p(\mathbb{Z}G)[1/2] \\
\downarrow \cong & & \downarrow \cong \\
H_n^G(E_{\mathcal{FIN}}(G); \mathbf{L}_{\mathbb{Q}}^p)[1/2] & \longrightarrow & L_n^p(\mathbb{Q}G)[1/2] \\
\downarrow \cong & & \downarrow \\
H_n^G(E_{\mathcal{FIN}}(G); \mathbf{L}_{\mathbb{R}}^p)[1/2] & \longrightarrow & L_n^p(\mathbb{R}G)[1/2] \\
\downarrow \cong & & \downarrow \\
H_n^G(E_{\mathcal{FIN}}(G); \mathbf{L}_{C_r^*(?;\mathbb{R})}^p)[1/2] & \longrightarrow & L_n^p(C_r^*(G;\mathbb{R}))[1/2] \\
\downarrow \cong & & \downarrow \cong \\
H_n^G(E_{\mathcal{FIN}}(G); \mathbf{K}_{\mathbb{R}}^{\mathrm{top}})[1/2] & \longrightarrow & K_n(C_r^*(G;\mathbb{R}))[1/2] \\
\downarrow & & \downarrow \\
H_n^G(E_{\mathcal{FIN}}(G); \mathbf{K}^{\mathrm{top}})[1/2] & \longrightarrow & K_n(C_r^*(G))[1/2]
\end{array}$$

The arrows marked with $\cong$ are known to be bijective (see [149, page 376], [151, Proposition 22.34 on page 252], [155]). If $G$ satisfies the Farrell-Jones Conjecture 4.10 for $L$-theory for $R = \mathbb{Z}$ and $R = \mathbb{R}$ and the Baum-Connes Conjecture 4.11 for both the real and the complex case, then all horizontal arrows are bijective and hence all arrows except the two lowest vertical ones are isomorphisms. The Baum-Connes Conjecture for the complex case does imply the Baum-Connes Conjecture for the real case (see Baum-Karoubi [25]).

**Theorem 6.9** (Rational computations of $K$-groups, Lück (2002))**.** *Let $G$ be a group. Let $T$ be the set of conjugacy classes $(g)$ of elements $g \in G$ of finite order.*

*Then there is a commutative diagram*

$$\begin{array}{ccc} \bigoplus_{p+q=n} \bigoplus_{(g)\in T} H_p(BC_G\langle g\rangle;\mathbb{C}) \otimes_{\mathbb{Z}} K_q(\mathbb{C}) & \longrightarrow & K_n(\mathbb{C}G) \otimes_{\mathbb{Z}} \mathbb{C} \\ \downarrow & & \downarrow \\ \bigoplus_{p+q=n} \bigoplus_{(g)\in T} H_p(BC_G\langle g\rangle;\mathbb{C}) \otimes_{\mathbb{Z}} K_q^{\mathrm{top}}(\mathbb{C}) & \longrightarrow & K_n^{\mathrm{top}}(C_r^*(G)) \otimes_{\mathbb{Z}} \mathbb{C} \end{array}$$

*Proof.* See [120, Theorem 0.5]. □

The horizontal arrows can be identified with the assembly maps occurring in the Farrell-Jones Conjecture 4.9 and the Baum-Connes Conjecture 4.11 by the equivariant Chern character. In particular they are isomorphisms if these conjecture hold for $G$.

**Remark 6.10** (Splitting principle.). The calculation of the relevant $K$-and $L$-groups often split into a universal group homology part which is independent of the theory, and a second part which essentially depends on the theory in question and the coefficients.

**Remark 6.11** (Integral Computations). In contrast to general rational computations such as the one appearing in Theorem 6.9, complete integral computations of $K_n(\mathbb{Z}G)$, $L_n(\mathbb{Z}G)$ or $K_n(C_r^*(G))$ seem to be possible only in special cases. Here are some examples, where some of these groups are computed. They are always based on the assumption that the Farrell-Jones Conjecture for algebraic $K$ or $L$-theory or the Baum-Connes Conjecture is true what is in most cases known to be true.

| | |
|---|---|
| Three-dimensional Heisenberg group and finite extensions | Lück [123] |
| 2-dimensional crystallographic groups and more general cocompact NEC-groups | Lück-Stamm [128], Pearson [142]) |
| Three-dimensional crystallographic groups | Alves-Ontaneda [3] |
| Fuchsian groups | Berkhove-Juan-Pineda-Pearson [26], Davis-Lück [53], Lück-Stamm [128] |
| Extensions $1 \to \mathbb{Z}^n \to G \to F \to 1$ for finite $F$ and free conjugation action of $F$ in $\mathbb{Z}^n$ | Davis-Lück [53], Lück-Stamm [128] |
| One relator groups | Davis-Lück [53] |
| $SL_3(\mathbb{Z})$ | Sanchez-Garcia [161], Upadhyay [181] |
| (Pure) braid groups | Aravinda-Farrell-Roushon [5], Farrell-Roushon [71] |
| fundamental groups of knot and link complements | Aravinda-Farrell-Roushon [4] |
| Certain Coxeter groups | Lafont-Ortiz [109], Sanchez-Garcia [162] |

Next we discuss the status of the various Conjectures such as the Farrell-Jones Conjecture and the Baum-Connes Conjecture.

**Theorem 6.12** (Bartels-Lück-Reich (2007)). *Let $R$ be a ring. Then every subgroup of a hyperbolic group belongs to $\mathcal{FJ}_K(R)$.*

*Proof.* See [14]. □

**Theorem 6.13** (Bartels-Lück). *Let $R$ be a ring with involution. Then every subgroup of a hyperbolic group belongs to $\mathcal{FJ}_L(R)$.*

*Proof.* The proof will appear in the paper [13] which is in preparation. □

**Theorem 6.14** (Bartels-Echterhoff-Reich (2007)). *Let $R$ be a ring (with involution). Let $\{G_i \mid i \in I\}$ be a directed system of groups (with not necessarily injective structure maps). Let $G$ be a subgroup of the colimit $\operatorname{colim}_{i\in I} G_i$.*

(1) *Suppose that for all $i \in I$ and every subgroup $H \subseteq G_i$ we have $H \in \mathcal{F}_K(R)$. Then $G \in \mathcal{FJ}_K(R)$;*
(2) *Suppose that for all $i \in I$ and every subgroup $H \subseteq G_i$ we have $H \in \mathcal{F}_L(R)$. Then $G \in \mathcal{FJ}_L(R)$;*
(3) *Suppose that for all $i \in I$ and every subgroup $H \subseteq G_i$ the Bost Conjecture holds for $H$. Then the Bost Conjecture holds for $G$.*

*Proof.* See [9, Theorem 0.8]. □

**Corollary 6.15.** *Let $\{G_i \mid i \in I\}$ be a directed system of hyperbolic groups (with not necessarily injective structure maps). Let $G$ be the colimit $\operatorname{colim}_{i\in I} G_i$. Let $H \subseteq G$ be any subgroup of $G$. Let $R$ be a ring (with involution).*

*Then $H \in \mathcal{FJ}_K(R)$, $H \in \mathcal{FJ}_L(R)$ and $H$ satisfies the Bost Conjecture 6.6.*

*Proof.* In the case $\mathcal{FJ}_K(R)$ and $\mathcal{FJ}_L(R)$ one has just to combine Theorem 6.12, Theorem 6.13, and Theorem 6.14. In the case of the Bost Conjecture one has to use Theorem 6.14 and the results of Lafforgue [108]). Details can be found in [9, Theorem 0.9]. □

**Example 6.16** (Groups satisfying the hypothesis of Corollary 6.15). The groups appearing in Theorem 6.12 are certainly wild in Bridson's universe of groups (see [30]). Many recent constructions of groups with exotic properties are given by colimits of directed systems of hyperbolic groups. Examples are

- groups with expanders in the sense of Gromov;
- Lacunary hyperbolic groups in the sense of Olshanskii-Osin-Sapir [139];
- Tarski monsters, i.e., infinite groups whose proper subgroups are all finite cyclic of $p$-power order for a given prime $p$;

Notice that Gromov's groups with expanders belong to $\mathcal{FJ}_K(R)$ for all $R$, whereas the Baum-Connes Conjecture with coefficients is not true for them by Higson-Lafforgue-Skandalis [90].

**Remark 6.17** (Twisted coefficients). The results above do extend to the more general case, where one allows twisted group rings or more general crossed product rings $R * G$ in the setting of the Farrell-Jones Conjecture and coefficients in a $G$-$C^*$-algebra in the setting of the Bost Conjecture. There are also so called fibered versions of the Farrell-Jones Conjecture and of an Isomorphism Conjecture in general (see for instance [12, Definition 1.2], [11, Definition 1.1], [66, 1.7]. In these more advanced settings with coefficients or the fibered setting one has that the class of groups for which the conjectures with coefficients or the fibered version are true is closed under taking finite direct products and taking subgroups.

Proofs of these claims can be found in [12, Lemma 1.3], [11, Lemma 1.2], [46, Theorem 2.5 and Theorem 3.17], [66, Theorem A.8 on page 289], [126, 5.5.4], [140, Corollary 7.12].

**Example 6.18** (Torsionfree hyperbolic groups)**.** If $G$ is a torsionfree hyperbolic group and $R$ any ring, then we get from Theorem 6.12 as explained in [14, page 2] an isomorphism

$$H_n(BG;\mathbf{K}(R)) \oplus \left( \bigoplus_{\substack{(C),C\subseteq G,C\neq 1 \\ C \text{ maximal cyclic}}} \mathrm{NK}_n(R) \right) \xrightarrow{\cong} K_n(RG).$$

**Remark 6.19** (Program for CAT(0)-groups)**.** Bartels and Lück have a program to prove $G \in \mathcal{FJ}_K(R)$ and $G \in \mathcal{FJ}_L(R)$ if $G$ acts properly and cocompactly on a simply connected CAT(0)-space. This would imply $G \in \mathcal{FJ}_K(R)$ and $G \in \mathcal{FJ}_L(R)$ for all subgroups $G$ of cocompact lattices in almost connected Lie groups and for all limit groups $G$.

**Theorem 6.20** (Mineyev-Yu (2002))**.** *Every subgroup of a hyperbolic group belongs to $\mathcal{BC}$.*

*Proof.* See [133]. □

**Definition 6.21** (a-T-menable group)**.** A group $G$ is *a-T-menable*, or, equivalently, has the *Haagerup property* if $G$ admits a metrically proper isometric action on some affine Hilbert space.

The class of a-T-menable groups is closed under taking subgroups, under extensions with finite quotients and under finite products. It is not closed under semi-direct products. Examples of a-T-menable groups are:

- countable amenable groups;
- countable free groups;
- discrete subgroups of $SO(n,1)$ and $SU(n,1)$;
- Coxeter groups;
- countable groups acting properly on trees, products of trees, or simply connected CAT(0) cubical complexes.

A group $G$ has *Kazhdan's property (T)* if, whenever it acts isometrically on some affine Hilbert space, it has a fixed point. An infinite a-T-menable group does not have property (T). Since $SL(n,\mathbb{Z})$ for $n \geq 3$ has property (T), it cannot be a-T-menable.

**Theorem 6.22** (Higson-Kasparov(2001))**.** *A group $G$ which is a-T-menable satisfies the Baum Connes Conjecture (with coefficients).*

*Proof.* See [89]. □

**Theorem 6.23** (Farrell-Jones (1993)). *Let $G$ be a subgroup of a cocompact lattice in an almost connected Lie group. Then the Farrell-Jones Conjecture for pseudo-isotopy is true for $G$.*

*Proof.* See [66, Theorem 2.1 on page 263]. □

**Theorem 6.24** (Lück-Reich-Rognes-Varisco (2007)). *The Farrell-Jones Conjecture for topological Hochschild homology is true for all groups.*

*Proof.* See [127]. □

For more information about the theorems above and further results we refer to the talks by Bartels, Rosenthal and Varisco.

**Remark 6.25** (Borel-Conjecture). Recall that the Borel Conjecture 2.24 is true for a closed $n$-dimensional manifold $M$ with fundamental group $G$ if $G$ belongs to both $\mathcal{FJ}_K(\mathbb{Z})$ and $\mathcal{FJ}_K(\mathbb{Z})$ and $n \geq 5$ (see Theorem 2.28). Recall from Corollary 6.15 that any subgroup of a colimit over a directed system of hyperbolic groups (with not necessarily injective structure maps) satisfy this assumption and that very exotic groups occur in this way (see Example 6.16).

Here are other groups for which the Borel Conjecture has been proved.

**Theorem 6.26** (Farrell-Jones). *The Borel Conjecture and the L-theoretic Farrell-Jones Conjecture with coefficients in $\mathbb{Z}$ are true for a group $G$ if one of the following conditions are satisfied:*

- *$G$ is the fundamental group of a closed Riemannian manifold with non-positive curvature;*
- *$G$ is the fundamental group of a complete Riemannian manifold with pinched negative curvature;*
- *$G$ is a torsionfree subgroup of $GL(n, \mathbb{R})$.*

*Proof.* See [67], [68]. □

For more information we refer to [126, Section 5].

In the following table we list prominent classes of groups and state whether they are known to satisfy the Farrell-Jones Conjectures 4.9 and 4.10 and the Baum-Connes Conjecture 4.11 or versions of them. Some of the classes are redundant. A question mark means that the author does not know about a corresponding result. A phrase like injectivity or after inverting 2 is true means that the corresponding assembly map is injective or is bijective after inverting 2. The reader should keep in mind that there may exist results of which the authors are not aware. The following table is an updated version of the one appearing in [126, 5.3].

| type of group | Baum-Connes Conjecture 4.11 | Farrell-Jones Conjecture 4.9 for $K$-theory | Farrell-Jones Conjecture 4.10 for $L$-theory |
|---|---|---|---|
| a-T-menable groups | true with coefficients (see Theorem 6.22) | ? | injectivity is true after inverting 2 for $R = \mathbb{Z}$ (see Remark 6.8) |
| amenable groups | true with coefficients (see Theorem 6.22) | ? | injectivity is true after inverting 2 for $R = \mathbb{Z}$ (see Remark 6.8) |
| elementary amenable groups | true with coefficients (see Theorem 6.22) | true fibered for a ring with finite characteristic $N$ after inverting $N$ (see [15, Theorem 0.3]) | true fibered after inverting 2 for $R = \mathbb{Z}$ (see [15, Lemma 1.12 and Lemma 7.1] or see [69, Theorem 5.2]) |
| virtually poly-cyclic | true with coefficients (see Theorem 6.22) | true rationally for $R = \mathbb{Z}$, true fibered for $R = \mathbb{Z}$ in the range $n \leq 1$ (see [126, Remark 5.3] | true fibered after inverting 2 for $R = \mathbb{Z}$ (see [15, Lemma 1.12 and Lemma 7.1] or see [69, Theorem 5.2]) |
| torsion free virtually solvable subgroups of $GL(n, \mathbb{C})$ | true with coefficients (see Theorem 6.22) | true in the range $\leq 1$ [69, Theorem 1.1] | true fibered after inverting 2 [69, Theorem 5.2] |
| discrete subgroups of Lie groups with finitely many path components | injectivity true (see [126, Theorem 5.9 and Remark 5.11 on page 718]) | injectivity is true for the family $\mathcal{FIN}$ and all rings $R$ (see [17]) | injectivity is true for the family $\mathcal{FIN}$ and rings $R$ with vanishing $K_n(RH)$ for $n \leq -2$ and $H \subseteq G$ finite (see [17]) |
| subgroups of groups which are discrete cocompact subgroups of Lie groups with finitely many path components | injectivity is true (see [126, Theorem 5.9 and Remark 5.11 on page 718]) | true rationally, true fibered in the range $n \leq 1$ (see [66, 1.6.7 on page 261 and Theorem 2.1 on page 263].) | injectivity is true for the family $\mathcal{FIN}$ and rings $R$ with vanishing $K_n(RH)$ for $n \leq -2$ and $H \subseteq G$ finite (see [17]) |

Continued

| type of group | Baum-Connes Conjecture 4.11 | Farrell-Jones Conjecture 4.9 for $K$-theory | Farrell-Jones Conjecture 4.10 for $L$-theory |
|---|---|---|---|
| linear groups | injectivity is true (see [83]) | ? | injectivity is true after inverting 2 for $R = \mathbb{Z}$ (see Remark 6.8) |
| finitely generated subgroup of $GL_n(k)$ for a global field $k$ | injectivity is true (see [83]) | injectivity is true for $R = \mathbb{Z}$ (see [97]) | injectivity is true for $R = \mathbb{Z}$ (see [97]) |
| torsion free discrete subgroups of $GL(n, \mathbb{R})$ | injectivity is true (see [83]) | true in the range $n \leq 1$ (see [68] and also [126, Theorem 5.5 on page 722]) | true for $R = \mathbb{Z}$ (see [68] and also [126, Theorem 5.5 on page 722]) |
| Groups with finite $\underline{E}G$ and finite asymptotic dimension | injectivity is true with coefficients (see [88, Theorem 1.1], [91, Theorem 1,1 and Lemma 4.3], | injectivity is true for the family $\mathcal{FIN}$ and all rings $R$ (see [17]) | injectivity is true for the family $\mathcal{FIN}$ and rings $R$ with vanishing $K_n(RH)$ for $n \leq -2$ and $H \subseteq G$ finite (see [17]) |
| $G$ acts properly and isometrically on a complete Riemannian manifold $M$ with non-positive sectional curvature | rational injectivity is true (see [99]) | ? | injectivity is true after inverting 2 for $R = \mathbb{Z}$ (see Remark 6.8) |
| $\pi_1(M)$ for a complete Riemannian manifold $M$ with non-positive sectional curvature | rational injectivity is true (see [99]) | ? | injectivity is true for $R = \mathbb{Z}$ (see [78, Corollary 2.3] |
| $\pi_1(M)$ for a complete Riemannian manifold $M$ with non-positive sectional curvature which is A-regular | rational injectivity is true (see [99]) | true in the range $n \leq 1$ for $R = \mathbb{Z}$ (see [68, Proposition 0.10 and Lemma 0.12]) | true for $R = \mathbb{Z}$ (see [68]) |

Continued

| type of group | Baum-Connes Conjecture 4.11 | Farrell-Jones Conjecture 4.9 for $K$-theory | Farrell-Jones Conjecture 4.10 for $L$-theory |
|---|---|---|---|
| $\pi_1(M)$ for a complete Riemannian manifold $M$ with pinched negative sectional curvature | rational injectivity is true (see [99]) | true in the range $n \leq 1$ and true rationally for $= \mathbb{Z}$ (see [68, Proposition 0.10 and Lemma 0.12 and page 216]) | true for $R = \mathbb{Z}$ (see and also [126, Theorem 5.5 on page 722]) |
| $\pi_1(M)$ for a closed Riemannian manifold $M$ with non-positive sectional curvature | rational injectivity is true (see [99]) | true fibered in the range $n \leq 1$, true rationally for $R = \mathbb{Z}$ (see [65]). | true for $R = \mathbb{Z}$ (see [68] and also [126, Theorem 5.5 on page 722]) |
| $\pi_1(M)$ for a closed Riemannian manifold $M$ with negative sectional curvature | true for all subgroups (see [133]) | true for all coefficients $R$ (see [16]) | true for $R = \mathbb{Z}$ (see [68] and also [126, Theorem 5.5 on page 722]) |
| subgroups of directed colimits of word hyperbolic groups | ? | true for all $R$ (see [14] and [9, Theorem 0.9]) | true for all $R$ (see [13] and [9, Theorem 0.9]) |
| subgroups of word hyperbolic groups | true (see [133]) | true for all $R$ (see [14]) | true for all $R$ (see [13]) |
| one-relator groups | true with coefficients (see [141]) | rational injectivity is true for $R = \mathbb{Z}$ or for regular $R$ with $\mathbb{Q} \subseteq R$ (see [11]) | true after inverting 2 for all $R$ (see [11, Proposition 0.9 and Theorem 0.13]), true after inverting 2 for $R = \mathbb{Z}$ fibered (see [160]) |
| torsion free one-relator groups | true with coefficients (see [141]) | true for $R$ regular [184, Theorem 19.4 on page 249 and Theorem 19.5 on page 250] | true after inverting 2 for all $R$ (see [40, Corollary 8], [11, Proposition 0.9 and Theorem 0.13]), true after inverting 2 for $R = \mathbb{Z}$ fibered (see [160]) |

Continued

| type of group | Baum-Connes Conjecture 4.11 | Farrell-Jones Conjecture 4.9 for $K$-theory | Farrell-Jones Conjecture 4.10 for $L$-theory |
|---|---|---|---|
| 3-manifold groups | ? | true fibered for $R = \mathbb{Z}$ in the range $n \leq 1$ (see [158, Corollary 4.2] and [159, Corollary 1.1.5]) | ? |
| Haken 3-manifold groups (in particular knot groups) | true with coefficients (see [136, Theorem 5.23]) | true for $R$ regular (see [184, Theorem 19.4 on page 249 and Theorem 19.5 on page 250]) | true after inverting 2 for all $R$ (see [40, Corollary 8]) |
| $SL(n, \mathbb{Z}), n \geq 3$ | injectivity is true (see [83]) | injecivity is true for the family $\mathcal{FIN}$ and $R = \mathbb{Z}$ (see [97]) | injecivity is true for the family $\mathcal{FIN}$ and $R = \mathbb{Z}$ (see [97]) |
| Artin's braid group $B_n$ | true with coefficients (see [136, Theorem 5.25], [164]) | true for $R = \mathbb{Z}$ fibered in the range $n \leq 1$, true for $R = \mathbb{Z}$ rationally (see [71]) | injectivity is true after inverting 2 for $R = \mathbb{Z}$ (see Remark 6.8) |
| pure braid group $C_n$ | true with coefficients (see [136, Theorem 5.25], [164]) | true for $R = \mathbb{Z}$ in the range $n \leq 1$ (see [5]) | true after inverting 2 for all $R$ (see [11, Proposition 0.9 and Theorem 0.13]) |
| Thompson's group $F$ | true with coefficients [61] | ? | injectivity is true after inverting 2 for $R = \mathbb{Z}$ (see Remark 6.8) |

**Remark 6.27** (Open cases). We mention some interesting groups or classes of groups for which the Conjectures are still open.

- The Farrell-Jones Conjecture for $K$-theory 4.9 and for $L$-theory 4.10 and the Baum-Connes Conjecture 4.11 are to the authors's knowledge open for $SL_n(\mathbb{Z})$ for $n \geq 3$, mapping class groups and $\mathrm{Out}(F_n)$;
- The Farrell-Jones Conjecture for $K$-theory 4.9 and for $L$-theory 4.10 are to the author's knowledge open for solvable groups and one-relator groups, whereas the Baum-Connes Conjecture 4.11 is known for these groups.
- There are certain groups with expanders for which the Baum-Connes Conjecture 4.11 is to the author's knowledge open and the version with coefficients is

actually false (see Higson-Lafforgue-Skandalis [90]). The Farrell-Jones Conjecture for $K$-theory 4.9 and for $L$-theory 4.10 are known for these groups since they are examples of directed colimits of hyperbolic groups.

**Remark 6.28** (Possible candidates for counterexamples)**.** It is not known whether there are counterexamples to the Farrell-Jones Conjecture or the Baum-Connes Conjecture. There seems to be no promising candidate of a group $G$ which is a potential counterexample to the $K$- or $L$-theoretic Farrell-Jones Conjecture or the Bost Conjecture. We cannot name a property or a lack of a certain property of a group which may be a reason for this group to be counterexample. There are many groups with rather exotic properties for which these Conjectures are known to be true.

**Remark 6.29** (The suspicious Baum-Connes Conjecture)**.** The Baum-Connes Conjecture is the one for which it is most likely that there may exist a counterexample. One reason is the existence of counterexamples to the version with coefficients (see Higson-Lafforgue-Skandalis [90]). Another reason is that $K_n(C_r^*(G))$ has certain failures concerning functoriality which do not occur for $K_n^G(\underline{E}G)$. For instance $K_n(C_r^*(G))$ is not known to be functorial for arbitrary group homomorphisms since the reduced group $C^*$-algebra is not functorial for arbitrary group homomorphisms. These failures are not present for the Farrell-Jones and the Bost Conjecture, i.e., for $K_n(RG)$, $L^{\langle -\infty \rangle}(RG)$ and $K_n(l^1(G))$.

**Remark 6.30** (Methods of proof)**.** Most of the proofs of the Farrell-Jones Conjecture use methods from controlled topology. Roughly speaking, controlled topology means that one considers free modules with a basis and thinks of these basis elements as sitting in a metric space. Then a map between such modules can be visualized by arrows between these basis elements. Control means that these arrows are small. Our homological approach to the assembly map is good for structural investigations but not for proofs. For proofs of the Farrell-Jones Conjecture or the Baum-Connes Conjecture it is often helpful to get some geometric input. In the Farrell-Jones setting the door to geometry is opened by interpreting the assembly map as a forget control map. The task to show for instance surjectivity is to manipulate a representative of the $K$-or $L$-theory class such that its class is unchanged but one has gained control. This is done by geometric constructions which yield contracting maps. These constructions are possible if some geometry connected to the group is around, such as negative curvature. We refer to the lectures of Bartels for such controlled methods.

The approach using topological cyclic homology goes back to Böckstedt-Hsiang-Madsen. It is of homotopy theoretic nature. We refer to the lecture of Varisco for more information about that approach.

The methods of proof for the Baum-Connes Conjecture are of analytic nature. The most prominent one is the Dirac-Dual-Dirac method based on $KK$-theory due to Kasparov. $KK$-theory is a bivariant theory together with a product. The assembly map is given by multiplying with a certain element in a certain $KK$-group. The essential idea is to construct another element in a dual $KK$-group which implements the inverse of the assembly map.

The analytic methods for the proof of the Baum-Connes Conjecture do not seem to be applicable to the Farrell-Jones setting. One would hope for a transfer of methods from the Farrell-Jones setting to the Baum-Connes Conjecture. So far not much has happened in this direction.

# References


[1] H. Abels. A universal proper $G$-space. *Math. Z.*, 159(2):143–158, 1978.

[2] A. Adem and Y. Ruan. Twisted orbifold $K$-theory. *Comm. Math. Phys.*, 237(3):533–556, 2003.

[3] A. Alves and P. Ontaneda. A formula for the Whitehead group of a three-dimensional crystallographic group. *Topology*, 45(1):1–25, 2006.

[4] C. S. Aravinda, F. T. Farrell, and S. K. Roushon. Surgery groups of knot and link complements. *Bull. London Math. Soc.*, 29(4):400–406, 1997.

[5] C. S. Aravinda, F. T. Farrell, and S. K. Roushon. Algebraic $K$-theory of pure braid groups. *Asian J. Math.*, 4(2):337–343, 2000.

[6] M. F. Atiyah. Elliptic operators, discrete groups and von Neumann algebras. *Astérisque*, 32-33:43–72, 1976.

[7] M. F. Atiyah and G. B. Segal. Equivariant $K$-theory and completion. *J. Differential Geometry*, 3:1–18, 1969.

[8] M. Auslander and D. A. Buchsbaum. *Groups, rings, modules*. Harper & Row Publishers, New York, 1974. Harper's Series in Modern Mathematics.

[9] A. Bartels, S. Echterhoff, and W. Lück. Inheritance of isomorphism conjectures under colimits. Preprintreihe SFB 478 — Geometrische Strukturen in der Mathematik, Heft 452, Münster, arXiv:math.KT/0702460v2, 2007.

[10] A. Bartels, T. Farrell, L. Jones, and H. Reich. On the isomorphism conjecture in algebraic $K$-theory. *Topology*, 43(1):157–213, 2004.

[11] A. Bartels and W. Lück. Isomorphism conjecture for homotopy $K$-theory and groups acting on trees. *J. Pure Appl. Algebra*, 205(3):660–696, 2006.

[12] A. Bartels and W. Lück. Induction theorems and isomorphism conjectures for $K$- and $L$-theory. *Forum Math.*, 19:379–406, 2007.

[13] A. Bartels and W. Lück. The Borel conjecture for hyperbolic groups. in preparation, 2008.

[14] A. Bartels, W. Lück, and H. Reich. The $K$-theoretic Farrell-Jones Conjecture for hyperbolic groups. Preprintreihe SFB 478 — Geometrische Strukturen in der Mathematik, Heft 434, Münster, arXiv:math.GT/0609685, to appear in Inventiones, 2007.

[15] A. Bartels, W. Lück, and H. Reich. On the Farrell-Jones Conjecture and its applications. Preprintreihe SFB 478 — Geometrische Strukturen in der Mathematik, Heft 457, Münster, arXiv:math.KT/0703548, to appear in Journal of Topology, 2007.

[16] A. Bartels and H. Reich. On the Farrell-Jones conjecture for higher algebraic $K$-theory. *J. Amer. Math. Soc.*, 18(3):501–545 (electronic), 2005.

[17] A. Bartels and D. Rosenthal. On the $K$-theory of groups with finite asymptotic dimension. Preprintreihe SFB 478 — Geometrische Strukturen in

der Mathematik, Heft 425, Münster, arXiv:math.KT/0605088, to appear in Crelle, 2006.

[18] A. C. Bartels. On the domain of the assembly map in algebraic $K$-theory. *Algebr. Geom. Topol.*, 3:1037–1050 (electronic), 2003.

[19] H. Bass. *Algebraic $K$-theory.* W. A. Benjamin, Inc., New York-Amsterdam, 1968.

[20] H. Bass. Euler characteristics and characters of discrete groups. *Invent. Math.*, 35:155–196, 1976.

[21] H. Bass. Traces and Euler characteristics. In *Homological group theory (Proc. Sympos., Durham, 1977)*, volume 36 of *London Math. Soc. Lecture Note Ser.*, pages 1–26. Cambridge Univ. Press, Cambridge, 1979.

[22] H. Bass, A. Heller, and R. G. Swan. The Whitehead group of a polynomial extension. *Inst. Hautes Études Sci. Publ. Math.*, 22:61–79, 1964.

[23] H. Bass, J. Milnor, and J.-P. Serre. Solution of the congruence subgroup problem for $\mathrm{SL}_n\,(n \geq 3)$ and $\mathrm{Sp}_{2n}\,(n \geq 2)$. *Inst. Hautes Études Sci. Publ. Math.*, 33:59–137, 1967.

[24] P. Baum, A. Connes, and N. Higson. Classifying space for proper actions and $K$-theory of group $C^*$-algebras. In *$C^*$-algebras: 1943–1993 (San Antonio, TX, 1993)*, pages 240–291. Amer. Math. Soc., Providence, RI, 1994.

[25] P. Baum and M. Karoubi. On the Baum-Connes conjecture in the real case. *Q. J. Math.*, 55(3):231–235, 2004.

[26] E. Berkove, D. Juan-Pineda, and K. Pearson. The lower algebraic $K$-theory of Fuchsian groups. *Comment. Math. Helv.*, 76(2):339–352, 2001.

[27] A. J. Berrick, I. Chatterji, and G. Mislin. From acyclic groups to the Bass conjecture for amenable groups. *Math. Ann.*, 329(4):597–621, 2004.

[28] A. J. Berrick, I. Chatterji, and G. Mislin. Homotopy idempotents on manifolds and Bass' conjectures. *Geometry and Topology Monographs*, 10:41–62, 2007.

[29] A. Borel and J.-P. Serre. Corners and arithmetic groups. *Comment. Math. Helv.*, 48:436–491, 1973. Avec un appendice: Arrondissement des variétés à coins, par A. Douady et L. Hérault.

[30] M. R. Bridson. Non-positive curvature and complexity for finitely presented groups. In *International Congress of Mathematicians. Vol. II*, pages 961–987. Eur. Math. Soc., Zürich, 2006.

[31] M. R. Bridson and A. Haefliger. *Metric spaces of non-positive curvature.* Springer-Verlag, Berlin, 1999. Die Grundlehren der mathematischen Wissenschaften, Band 319.

[32] M. R. Bridson and K. Vogtmann. The symmetries of outer space. *Duke Math. J.*, 106(2):391–409, 2001.

[33] W. Browder. *Surgery on simply-connected manifolds.* Springer-Verlag, New York, 1972. Ergebnisse der Mathematik und ihrer Grenzgebiete, Band 65.

[34] K. Brown. Groups of virtually finite dimension. In *Proceedings* "Homological group theory", *editor: Wall, C.T.C., LMS Lecture Notes Series 36*, pages 27–70. Cambridge University Press, 1979.

[35] K. S. Brown. *Buildings.* Springer-Verlag, New York, 1998. Reprint of the 1989 original.

[36] M. Burger and A. Valette. Idempotents in complex group rings: theorems of Zalesskii and Bass revisited. *J. Lie Theory*, 8(2):219–228, 1998.
[37] D. Burghelea and R. Lashof. Stability of concordances and the suspension homomorphism. *Ann. of Math. (2)*, 105(3):449–472, 1977.
[38] S. Cappell, A. Ranicki, and J. Rosenberg, editors. *Surveys on surgery theory. Vol. 1.* Princeton University Press, Princeton, NJ, 2000. Papers dedicated to C. T. C. Wall.
[39] S. Cappell, A. Ranicki, and J. Rosenberg, editors. *Surveys on surgery theory. Vol. 2.* Princeton University Press, Princeton, NJ, 2001. Papers dedicated to C. T. C. Wall.
[40] S. E. Cappell. Unitary nilpotent groups and Hermitian $K$-theory. I. *Bull. Amer. Math. Soc.*, 80:1117–1122, 1974.
[41] S. E. Cappell and J. L. Shaneson. On 4-dimensional $s$-cobordisms. *J. Differential Geom.*, 22(1):97–115, 1985.
[42] G. Carlsson. Deloopings in algebraic $K$-theory. In this Handbook.
[43] G. Carlsson and E. K. Pedersen. Controlled algebra and the Novikov conjectures for $K$- and $L$-theory. *Topology*, 34(3):731–758, 1995.
[44] D. W. Carter. Lower $K$-theory of finite groups. *Comm. Algebra*, 8(20):1927–1937, 1980.
[45] P. Cassou-Noguès. Classes d'idéaux de l'algèbre d'un groupe abélien. *C. R. Acad. Sci. Paris Sér. A-B*, 276:A973–A975, 1973.
[46] J. Chabert and S. Echterhoff. Permanence properties of the Baum-Connes conjecture. *Doc. Math.*, 6:127–183 (electronic), 2001.
[47] M. M. Cohen. *A course in simple-homotopy theory.* Springer-Verlag, New York, 1973. Graduate Texts in Mathematics, Vol. 10.
[48] M. Culler and K. Vogtmann. Moduli of graphs and automorphisms of free groups. *Invent. Math.*, 84(1):91–119, 1986.
[49] C. W. Curtis and I. Reiner. *Methods of representation theory. Vol. I.* John Wiley & Sons Inc., New York, 1981. With applications to finite groups and orders, Pure and Applied Mathematics, A Wiley-Interscience Publication.
[50] C. W. Curtis and I. Reiner. *Methods of representation theory. Vol. II.* John Wiley & Sons Inc., New York, 1987. With applications to finite groups and orders, A Wiley-Interscience Publication.
[51] J. F. Davis. Manifold aspects of the Novikov conjecture. In *Surveys on surgery theory, Vol. 1*, pages 195–224. Princeton Univ. Press, Princeton, NJ, 2000.
[52] J. F. Davis and W. Lück. Spaces over a category and assembly maps in isomorphism conjectures in $K$- and $L$-theory. *$K$-Theory*, 15(3):201–252, 1998.
[53] J. F. Davis and W. Lück. The $p$-chain spectral sequence. *$K$-Theory*, 30(1):71–104, 2003. Special issue in honor of Hyman Bass on his seventieth birthday. Part I.
[54] A. Dold. Relations between ordinary and extraordinary homology. Colloq. alg. topology, Aarhus 1962, 2-9, 1962.
[55] S. K. Donaldson. Irrationality and the $h$-cobordism conjecture. *J. Differential Geom.*, 26(1):141–168, 1987.

[56] M. J. Dunwoody. Accessibility and groups of cohomological dimension one. *Proc. London Math. Soc. (3)*, 38(2):193–215, 1979.
[57] W. Dwyer, M. Weiss, and B. Williams. A parametrized index theorem for the algebraic $K$-theory Euler class. *Acta Math.*, 190(1):1–104, 2003.
[58] B. Eckmann. Cyclic homology of groups and the Bass conjecture. *Comment. Math. Helv.*, 61(2):193–202, 1986.
[59] B. Eckmann. Projective and Hilbert modules over group algebras, and finitely dominated spaces. *Comment. Math. Helv.*, 71(3):453–462, 1996.
[60] G. Elek and E. Szabó. On sofic groups. *J. Group Theory*, 9(2):161–171, 2006.
[61] D. S. Farley. Proper isometric actions of Thompson's groups on Hilbert space. *Int. Math. Res. Not.*, (45):2409–2414, 2003.
[62] F. T. Farrell. *Lectures on surgical methods in rigidity.* Published for the Tata Institute of Fundamental Research, Bombay, 1996.
[63] F. T. Farrell and L. E. Jones. A topological analogue of Mostow's rigidity theorem. *J. Amer. Math. Soc.*, 2(2):257–370, 1989.
[64] F. T. Farrell and L. E. Jones. Rigidity and other topological aspects of compact nonpositively curved manifolds. *Bull. Amer. Math. Soc. (N.S.)*, 22(1):59–64, 1990.
[65] F. T. Farrell and L. E. Jones. Stable pseudoisotopy spaces of compact non-positively curved manifolds. *J. Differential Geom.*, 34(3):769–834, 1991.
[66] F. T. Farrell and L. E. Jones. Isomorphism conjectures in algebraic $K$-theory. *J. Amer. Math. Soc.*, 6(2):249–297, 1993.
[67] F. T. Farrell and L. E. Jones. Topological rigidity for compact non-positively curved manifolds. In *Differential geometry: Riemannian geometry (Los Angeles, CA, 1990)*, pages 229–274. Amer. Math. Soc., Providence, RI, 1993.
[68] F. T. Farrell and L. E. Jones. Rigidity for aspherical manifolds with $\pi_1 \subset GL_m(\mathbb{R})$. *Asian J. Math.*, 2(2):215–262, 1998.
[69] F. T. Farrell and P. A. Linnell. $K$-theory of solvable groups. *Proc. London Math. Soc. (3)*, 87(2):309–336, 2003.
[70] F. T. Farrell and P. A. Linnell. Whitehead groups and the Bass conjecture. *Math. Ann.*, 326(4):723–757, 2003.
[71] F. T. Farrell and S. K. Roushon. The Whitehead groups of braid groups vanish. *Internat. Math. Res. Notices*, 10:515–526, 2000.
[72] T. Farrell. The Borel conjecture. In T. Farrell, L. Göttsche, and W. Lück, editors, *High dimensional manifold theory*, number 9 in ICTP Lecture Notes, pages 225–298. Abdus Salam International Centre for Theoretical Physics, Trieste, 2002. Proceedings of the summer school "High dimensional manifold theory" in Trieste May/June 2001, Number 1. http://www.ictp.trieste.it/~pub_off/lectures/vol9.html.
[73] T. Farrell, L. Göttsche, and W. Lück, editors. *High dimensional manifold theory.* Number 9 in ICTP Lecture Notes. Abdus Salam International Centre for Theoretical Physics, Trieste, 2002. Proceedings of the summer school "High dimensional manifold theory" in Trieste May/June 2001, Number 1. http://www.ictp.trieste.it/~pub_off/lectures/vol9.html.
[74] T. Farrell, L. Göttsche, and W. Lück, editors. *High dimensional manifold*

*theory*. Number 9 in ICTP Lecture Notes. Abdus Salam International Centre for Theoretical Physics, Trieste, 2002. Proceedings of the summer school "High dimensional manifold theory" in Trieste May/June 2001, Number 2. http://www.ictp.trieste.it/~pub_off/lectures/vol9.html.

[75] S. Ferry. A simple-homotopy approach to the finiteness obstruction. In *Shape theory and geometric topology (Dubrovnik, 1981)*, pages 73–81. Springer-Verlag, Berlin, 1981.

[76] S. Ferry and A. Ranicki. A survey of Wall's finiteness obstruction. In *Surveys on surgery theory, Vol. 2*, volume 149 of *Ann. of Math. Stud.*, pages 63–79. Princeton Univ. Press, Princeton, NJ, 2001.

[77] S. C. Ferry, A. A. Ranicki, and J. Rosenberg. A history and survey of the Novikov conjecture. In *Novikov conjectures, index theorems and rigidity, Vol. 1 (Oberwolfach, 1993)*, pages 7–66. Cambridge Univ. Press, Cambridge, 1995.

[78] S. C. Ferry and S. Weinberger. Curvature, tangentiality, and controlled topology. *Invent. Math.*, 105(2):401–414, 1991.

[79] M. H. Freedman. The topology of four-dimensional manifolds. *J. Differential Geom.*, 17(3):357–453, 1982.

[80] M. H. Freedman. The disk theorem for four-dimensional manifolds. In *Proceedings of the International Congress of Mathematicians, Vol. 1, 2 (Warsaw, 1983)*, pages 647–663, Warsaw, 1984. PWN.

[81] D. R. Grayson. $SK_1$ of an interesting principal ideal domain. *J. Pure Appl. Algebra*, 20(2):157–163, 1981.

[82] J. Grunewald. The behaviour of nil-groups under localization. arXiv:math. KT/0005194, 2006.

[83] E. Guentner, N. Higson, and S. Weinberger. The Novikov conjecture for linear groups. *Publ. Math. Inst. Hautes Études Sci.*, 101:243–268, 2005.

[84] I. Hambleton and E. K. Pedersen. Identifying assembly maps in $K$- and $L$-theory. *Math. Ann.*, 328(1-2).27–57, 2004.

[85] J. L. Harer. The virtual cohomological dimension of the mapping class group of an orientable surface. *Invent. Math.*, 84(1):157–176, 1986.

[86] A. E. Hatcher. Concordance spaces, higher simple-homotopy theory, and applications. In *Algebraic and geometric topology (Proc. Sympos. Pure Math., Stanford Univ., Stanford, Calif., 1976), Part 1*, pages 3–21. Amer. Math. Soc., Providence, R.I., 1978.

[87] E. Hess and T. Schick. $L^2$-torsion of hyperbolic manifolds. *Manuscripta Math.*, 97(3):329–334, 1998.

[88] N. Higson. Bivariant $K$-theory and the Novikov conjecture. *Geom. Funct. Anal.*, 10(3):563–581, 2000.

[89] N. Higson and G. Kasparov. $E$-theory and $KK$-theory for groups which act properly and isometrically on Hilbert space. *Invent. Math.*, 144(1):23–74, 2001.

[90] N. Higson, V. Lafforgue, and G. Skandalis. Counterexamples to the Baum-Connes conjecture. *Geom. Funct. Anal.*, 12(2):330–354, 2002.

[91] N. Higson and J. Roe. Amenable group actions and the Novikov conjecture. *J. Reine Angew. Math.*, 519:143–153, 2000.

[92] K. Igusa. The stability theorem for smooth pseudoisotopies. *$K$-Theory*, 2(1-2):vi+355, 1988.
[93] S. Illman. Existence and uniqueness of equivariant triangulations of smooth proper $G$-manifolds with some applications to equivariant Whitehead torsion. *J. Reine Angew. Math.*, 524:129–183, 2000.
[94] F. Ischebeck. Hauptidealringe mit nichttrivialer $SK_1$-Gruppe. *Arch. Math. (Basel)*, 35(1-2):138–139, 1980.
[95] N. V. Ivanov. Mapping class groups. In *Handbook of geometric topology*, pages 523–633. North-Holland, Amsterdam, 2002.
[96] S. Jackowski and B. Oliver. Vector bundles over classifying spaces of compact Lie groups. *Acta Math.*, 176(1):109–143, 1996.
[97] L. Ji. Integral Novikov conjectures and arithmetic groups cointaning torsion elements. preprint, 2006.
[98] M. Karoubi. Bott periodicity, generalizations and variants. In this Handbook.
[99] G. G. Kasparov. Equivariant $KK$-theory and the Novikov conjecture. *Invent. Math.*, 91(1):147–201, 1988.
[100] M. A. Kervaire. Le théorème de Barden-Mazur-Stallings. *Comment. Math. Helv.*, 40:31–42, 1965.
[101] M. A. Kervaire and J. Milnor. Groups of homotopy spheres. I. *Ann. of Math. (2)*, 77:504–537, 1963.
[102] S. Klaus and M. Kreck. A quick proof of the rational Hurewicz theorem and a computation of the rational homotopy groups of spheres. *Math. Proc. Cambridge Philos. Soc.*, 136(3):617–623, 2004.
[103] B. Kleiner and J. Lott. Notes on Perelman's papers. preprint, arXiv:math.DG/0605667v2, 2006.
[104] M. Kreck. Surgery and duality. *Ann. of Math. (2)*, 149(3):707–754, 1999.
[105] M. Kreck and W. Lück. *The Novikov conjecture*, volume 33 of *Oberwolfach Seminars*. Birkhäuser Verlag, Basel, 2005. Geometry and algebra.
[106] M. Kreck and W. Lück. Topological rigidiyt for non-aspherical manifolds. preprint, arXiv:math.GT/0509238v1, 2005.
[107] S. Krstić and K. Vogtmann. Equivariant outer space and automorphisms of free-by-finite groups. *Comment. Math. Helv.*, 68(2):216–262, 1993.
[108] V. Lafforgue. $K$-théorie bivariante pour les algèbres de Banach et conjecture de Baum-Connes. *Invent. Math.*, 149(1):1–95, 2002.
[109] J.-F. Lafont and I. Ortiz. Lower algebraic K-theory of hyperbolic 3-simplex reflection groups. preprint, arXiv:math.KT/0705.0844, to appear in Comm. Helv., 2007.
[110] T. Lance. Differentiable structures on manifolds. In *Surveys on surgery theory, Vol. 1*, pages 73–104. Princeton Univ. Press, Princeton, NJ, 2000.
[111] I. J. Leary and B. E. A. Nucinkis. Every CW-complex is a classifying space for proper bundles. *Topology*, 40(3):539–550, 2001.
[112] I. J. Leary and B. E. A. Nucinkis. Some groups of type $VF$. *Invent. Math.*, 151(1):135–165, 2003.
[113] J. P. Levine. Lectures on groups of homotopy spheres. In *Algebraic and geometric topology (New Brunswick, N.J., 1983)*, pages 62–95. Springer, Berlin,

1985.

[114] P. A. Linnell. Decomposition of augmentation ideals and relation modules. *Proc. London Math. Soc. (3)*, 47(1):83–127, 1983.

[115] J.-L. Loday. $K$-théorie algébrique et représentations de groupes. *Ann. Sci. École Norm. Sup. (4)*, 9(3):309–377, 1976.

[116] W. Lück. The geometric finiteness obstruction. *Proc. London Math. Soc. (3)*, 54(2):367–384, 1987.

[117] W. Lück. *Transformation groups and algebraic $K$-theory*. Springer-Verlag, Berlin, 1989.

[118] W. Lück. The type of the classifying space for a family of subgroups. *J. Pure Appl. Algebra*, 149(2):177–203, 2000.

[119] W. Lück. A basic introduction to surgery theory. In F. T. Farrell, L. Göttsche, and W. Lück, editors, *High dimensional manifold theory*, number 9 in ICTP Lecture Notes, pages 1–224. Abdus Salam International Centre for Theoretical Physics, Trieste, 2002. Proceedings of the summer school "High dimensional manifold theory" in Trieste May/June 2001, Number 1. http://www.ictp.trieste.it/~pub_off/lectures/vol9.html.

[120] W. Lück. Chern characters for proper equivariant homology theories and applications to $K$- and $L$-theory. *J. Reine Angew. Math.*, 543:193–234, 2002.

[121] W. Lück. *$L^2$-invariants: theory and applications to geometry and $K$-theory*, volume 44 of *Ergebnisse der Mathematik und ihrer Grenzgebiete. 3. Folge. A Series of Modern Surveys in Mathematics [Results in Mathematics and Related Areas. 3rd Series. A Series of Modern Surveys in Mathematics]*. Springer-Verlag, Berlin, 2002.

[122] W. Lück. The relation between the Baum-Connes conjecture and the trace conjecture. *Invent. Math.*, 149(1):123–152, 2002.

[123] W. Lück. $K$- and $L$-theory of the semi-direct product of the discrete 3-dimensional Heisenberg group by $\mathbb{Z}/4$. *Geom. Topol.*, 9:1639–1676 (electronic), 2005.

[124] W. Lück. Rational computations of the topological $K$-theory of classifying spaces of discrete groups. Preprintreihe SFB 478 — Geometrische Strukturen in der Mathematik, 391, Münster, arXiv:math.GT/0507237, to appear in Journal für die Reine und Angewandte Mathematik, 2005.

[125] W. Lück. Survey on classifying spaces for families of subgroups. In *Infinite groups: geometric, combinatorial and dynamical aspects*, volume 248 of *Progr. Math.*, pages 269–322. Birkhäuser, Basel, 2005.

[126] W. Lück and H. Reich. The Baum-Connes and the Farrell-Jones conjectures in $K$- and $L$-theory. In *Handbook of $K$-theory. Vol. 1, 2*, pages 703–842. Springer, Berlin, 2005.

[127] W. Lück, H. Reich, J. Rognes, and M. Varisco. Algebraic $K$-theory of integral group rings and topological cyclic homology. in preparation, 2007.

[128] W. Lück and R. Stamm. Computations of $K$- and $L$-theory of cocompact planar groups. *$K$-Theory*, 21(3):249–292, 2000.

[129] D. Meintrup and T. Schick. A model for the universal space for proper actions of a hyperbolic group. *New York J. Math.*, 8:1–7 (electronic), 2002.

[130] J. Milnor. *Lectures on the $h$-cobordism theorem*. Princeton University Press,

Princeton, N.J., 1965.
[131] J. Milnor. Whitehead torsion. *Bull. Amer. Math. Soc.*, 72:358–426, 1966.
[132] J. Milnor. *Introduction to algebraic $K$-theory*. Princeton University Press, Princeton, N.J., 1971. Annals of Mathematics Studies, No. 72.
[133] I. Mineyev and G. Yu. The Baum-Connes conjecture for hyperbolic groups. *Invent. Math.*, 149(1):97–122, 2002.
[134] G. Mislin. Wall's finiteness obstruction. In *Handbook of algebraic topology*, pages 1259–1291. North-Holland, Amsterdam, 1995.
[135] G. Mislin. Mapping class groups. Unpublished preprint of a talk on the conference "Classifying spaces for families in Münster in June 2004, 2004.
[136] G. Mislin and A. Valette. *Proper group actions and the Baum-Connes conjecture*. Advanced Courses in Mathematics. CRM Barcelona. Birkhäuser Verlag, Basel, 2003.
[137] J. Morgan and G. Tian. Ricci flow and the Poincaé conjecture. preprint, arXiv:math.DG/0607607v2, 2006.
[138] R. Oliver. *Whitehead groups of finite groups*. Cambridge University Press, Cambridge, 1988.
[139] A. Ol'shankskii, D. Osin, and M. Sapir. lacunary hyperbolic groups. Arxiv:math.GR/0701365v1, 2007.
[140] H. Oyono-Oyono. Baum-Connes Conjecture and extensions. *J. Reine Angew. Math.*, 532:133–149, 2001.
[141] H. Oyono-Oyono. Baum-Connes conjecture and group actions on trees. *$K$-Theory*, 24(2):115–134, 2001.
[142] K. Pearson. Algebraic $K$-theory of two-dimensional crystallographic groups. *$K$-Theory*, 14(3):265–280, 1998.
[143] E. K. Pedersen and C. A. Weibel. A non-connective delooping of algebraic $K$-theory. In *Algebraic and Geometric Topology; proc. conf. Rutgers Uni., New Brunswick 1983*, volume 1126 of *Lecture notes in mathematics*, pages 166–181. Springer, 1985.
[144] G. Perelman. The entropy formula for the Ricci flow and its geometric applications. preprint, arXiv:math.DG/0211159, 2002.
[145] G. Perelman. Finite extinction time for the solutions to the Ricci flow on certain three-manifolds. preprint, arXiv:math.DG/0307245, 2003.
[146] M. Pimsner and D. Voiculescu. $K$-groups of reduced crossed products by free groups. *J. Operator Theory*, 8(1):131–156, 1982.
[147] F. Quinn. A geometric formulation of surgery. In *Topology of Manifolds (Proc. Inst., Univ. of Georgia, Athens, Ga., 1969)*, pages 500–511. Markham, Chicago, Ill., 1970.
[148] A. Ranicki. Foundations of algebraic surgery. In T. Farrell, L. Göttsche, and W. Lück, editors, *High dimensional manifold theory*, number 9 in ICTP Lecture Notes, pages 491–514. Abdus Salam International Centre for Theoretical Physics, Trieste, 2002. Proceedings of the summer school "High dimensional manifold theory" in Trieste May/June 2001, Number 2. http://www.ictp.trieste.it/~pub_off/lectures/vol9.html.
[149] A. A. Ranicki. *Exact sequences in the algebraic theory of surgery*. Princeton University Press, Princeton, N.J., 1981.

[150] A. A. Ranicki. The algebraic theory of finiteness obstruction. *Math. Scand.*, 57(1):105–126, 1985.
[151] A. A. Ranicki. *Algebraic L-theory and topological manifolds.* Cambridge University Press, Cambridge, 1992.
[152] A. A. Ranicki. On the Novikov conjecture. In *Novikov conjectures, index theorems and rigidity, Vol. 1 (Oberwolfach, 1993)*, pages 272–337. Cambridge Univ. Press, Cambridge, 1995.
[153] H. Reich. On Quinns improvement of the Farrell-Jones conjecture. preprint, 2007.
[154] J. Rosenberg. *Algebraic K-theory and its applications.* Springer-Verlag, New York, 1994.
[155] J. Rosenberg. Analytic Novikov for topologists. In *Novikov conjectures, index theorems and rigidity, Vol. 1 (Oberwolfach, 1993)*, pages 338–372. Cambridge Univ. Press, Cambridge, 1995.
[156] J. Rosenberg. Comparison between algebraic and topological $K$-theory for Banach algebras and $C^*$-algebras. In *Handbook of K-theory. Vol. 1, 2*, pages 843–874. Springer, Berlin, 2005.
[157] C. P. Rourke and B. J. Sanderson. *Introduction to piecewise-linear topology.* Springer-Verlag, Berlin, 1982. Reprint.
[158] S. Roushon. The Farrell-Jones isomorphism conjecture for 3-manifold groups. arXiv:math.KT/0405211v4, 2006.
[159] S. Roushon. The isomorphism conjecture for 3-manifold groups and $K$-theory of virtually poly-surface groups. arXiv:math.KT/0408243v4, 2006.
[160] S. Roushon. The isomorphism conjecture in $L$-theory: poly-free groups and one-relator groups. arXiv:math.KT/0703879v3, 2007.
[161] R. Sánchez-García. Equivariant $K$-homology of $SL_3(\mathbb{Z})$. preprint, arXiv:math.KT/0601586, 2006.
[162] R. Sánchez-García. Equivariant $K$-homology of some Coxeter groups. preprint, arXiv:math.KT/0604402, 2006.
[163] T. Schick. A counterexample to the (unstable) Gromov-Lawson-Rosenberg conjecture. *Topology*, 37(6):1165–1168, 1998.
[164] T. Schick. Finite group extensions and the Baum-Connes conjecture. preprint, arXiv:math.KT/0209165v3, to appear in GT, 2007.
[165] G. Segal. Permutation representations of finite $p$-groups. *Quart. J. Math. Oxford Ser. (2)*, 23:375–381, 1972.
[166] J.-P. Serre. Groupes d'homotopie et classes de groupes abéliens. *Ann. of Math. (2)*, 58:258–294, 1953.
[167] J.-P. Serre. *Linear representations of finite groups.* Springer-Verlag, New York, 1977. Translated from the second French edition by Leonard L. Scott, Graduate Texts in Mathematics, Vol. 42.
[168] J.-P. Serre. Arithmetic groups. In *Homological group theory (Proc. Sympos., Durham, 1977)*, volume 36 of *London Math. Soc. Lecture Note Ser.*, pages 105–136. Cambridge Univ. Press, Cambridge, 1979.
[169] J.-P. Serre. *Galois cohomology.* Springer-Verlag, Berlin, 1997. Translated from the French by Patrick Ion and revised by the author.
[170] C. Soulé. The cohomology of $\mathrm{SL}_3(\mathbf{Z})$. *Topology*, 17(1):1–22, 1978.

[171] J. R. Stallings. On torsion-free groups with infinitely many ends. *Ann. of Math. (2)*, 88:312–334, 1968.

[172] C. Stark. Topological rigidity theorems. In R. Daverman and R. Sher, editors, *Handbook of Geometric Topology*, chapter 20. Elsevier, 2002.

[173] C. W. Stark. Surgery theory and infinite fundamental groups. In *Surveys on surgery theory, Vol. 1*, volume 145 of *Ann. of Math. Stud.*, pages 275–305. Princeton Univ. Press, Princeton, NJ, 2000.

[174] S. Stolz. Manifolds of positive scalar curvature. In *Topology of high-dimensional manifolds, No. 1, 2 (Trieste, 2001)*, volume 9 of *ICTP Lect. Notes*, pages 661–709. Abdus Salam Int. Cent. Theoret. Phys., Trieste, 2002.

[175] R. G. Swan. Induced representations and projective modules. *Ann. of Math. (2)*, 71:552–578, 1960.

[176] R. G. Swan. Vector bundles and projective modules. *Trans. Amer. Math. Soc.*, 105:264–277, 1962.

[177] R. G. Swan. Groups of cohomological dimension one. *J. Algebra*, 12:585–610, 1969.

[178] M. Tezuka and N. Yagita. Complex $K$-theory of $B\mathrm{sl}_3(\mathbf{z})$. *$K$-Theory*, 6(1):87–95, 1992.

[179] J. Tits. On buildings and their applications. In *Proceedings of the International Congress of Mathematicians (Vancouver, B. C., 1974), Vol. 1*, pages 209–220. Canad. Math. Congress, Montreal, Que., 1975.

[180] J. Tits. Reductive groups over local fields. In *Automorphic forms, representations and L-functions (Proc. Sympos. Pure Math., Oregon State Univ., Corvallis, Ore., 1977), Part 1*, Proc. Sympos. Pure Math., XXXIII, pages 29–69. Amer. Math. Soc., Providence, R.I., 1979.

[181] S. Upadhyay. Controlled algebraic $K$-theory of integral group ring of $\mathrm{SL}(3, \mathbf{Z})$. *$K$-Theory*, 10(4):413–418, 1996.

[182] K. Varadarajan. *The finiteness obstruction of C. T. C. Wall.* Canadian Mathematical Society Series of Monographs and Advanced Texts. John Wiley & Sons Inc., New York, 1989. A Wiley-Interscience Publication.

[183] K. Vogtmann. Automorphisms of free groups and outer space. To appear in the special issue of Geometriae Dedicata for the June, 2000 Haifa conference, 2003.

[184] F. Waldhausen. Algebraic $K$-theory of generalized free products. I, II. *Ann. of Math. (2)*, 108(1):135–204, 1978.

[185] F. Waldhausen. Algebraic $K$-theory of topological spaces. I. In *Algebraic and geometric topology (Proc. Sympos. Pure Math., Stanford Univ., Stanford, Calif., 1976), Part 1*, Proc. Sympos. Pure Math., XXXII, pages 35–60. Amer. Math. Soc., Providence, R.I., 1978.

[186] F. Waldhausen. Algebraic $K$-theory of spaces. In *Algebraic and geometric topology (New Brunswick, N.J., 1983)*, pages 318–419. Springer-Verlag, Berlin, 1985.

[187] C. T. C. Wall. Finiteness conditions for $CW$-complexes. *Ann. of Math. (2)*, 81:56–69, 1965.

[188] C. T. C. Wall. Finiteness conditions for $CW$ complexes. II. *Proc. Roy. Soc. Ser. A*, 295:129–139, 1966.

[189] C. T. C. Wall. *Surgery on compact manifolds.* American Mathematical Society, Providence, RI, second edition, 1999. Edited and with a foreword by A. A. Ranicki.

[190] T. White. Fixed points of finite groups of free group automorphisms. *Proc. Amer. Math. Soc.*, 118(3):681–688, 1993.

*Cohomology of Groups and*
*Algebraic K-theory*
ALM 12, pp. 343–385



# Introduction to Controlled Topology and Its Applications

Stratos Prassidis*

*To Tom Farrell and Lowell Jones*

**Abstract**

Controlled topology is an important tool in geometric topology. We present the basic definitions, results and applications of controlled topology, including the sucking principle and the $\alpha$-approximation theorem. Also, we give an interpretation of splitting theorems in $K$-theory for spaces over the circle and the interval using controlled topology. In particular, we identify the 'non-controlled' part as the exotic Nil-groups. Furthermore, we give a formulations of the Farrell–Jones Isomorphism Conjecture in the language of controlled topology. As an application we give an alternate proof of the Isomorphism Conjecture in $K$-theory for free groups.

**2000 Mathematics Subject Classification:** 57M60, 20F65, 57Q99, 57R19

**Key words and Phrases:** Controlled homotopy equivalence, approximate fibration, $K$-theory, Nil-groups, homotopy colimits, Isomorphism Conjecture

## 1. Introduction

The idea of using estimates in geometric topology appeared in the early developments of the field. The work that used controlled methods to approach classical problems in geometric topology was done by E. H. Connell and J. Hollingsworth ([30]). In this paper, the authors introduce estimates in computing algebraic obstructions. Later, controlled topology was developed by T. Chapman, S. Ferry and F. Quinn and it became an important tool in the topological classification of topological manifolds. The basic idea in controlled topology is to put an estimate on geometric obstructions. Typical theorems in controlled topology state that if the obstructions are small enough, in an appropriate sense, then they vanish. Another feature of controlled topology is that it captures the homological properties of the obstruction theory. That makes controlled obstructions easier to compute.

*Department of Mathematics Canisius College, Buffalo, NY 14208, U.S.A. E-mail: prasside@canisius.edu

Essentially, the controlled components reflect the smooth and cellular nature of the obstruction and their complements, the non-controlled components reflect purely topological phenomena. Beyond classification and rigidity theorems controlled topology was applied in the study of group actions on manifolds, equivariant and stratified phenomena.

In the first sections, we present the basic definitions and properties of lower algebraic $K$-theory. We connect the algebraic definitions with the geometric ones and we present the basic topological applications. The obstructions that appear in topology split into two pieces, the non-controlled and the controlled piece. We continue with the basic constructions of controlled $K$-theory and its applications. At the same time, we present the non-controlled part of $K$-theory, namely the Nil-groups. Usually the Nil-groups encode splitting obstructions that correspond to obstructions of reducing the problem to a problem in controlled topology.

There are two applications that will be discussed. The first is the controlled approach to the classical Fundamental Theorem of Algebraic $K$-theory. The theorem states that

$$Wh(X\times S^1) \cong Wh(X)\oplus\tilde{K}_0(X)\oplus\widetilde{\mathrm{Nil}}(X)\oplus\widetilde{\mathrm{Nil}}(X).$$

Thus, the theorem calculates the $K$-theory of $X\times S^1$ in terms of the $K$-theory of $X$. Using controlled topology, the theorem can be expressed as follows:

$$Wh(X\times S^1) \cong Wh(X\times S^1, p)_c\oplus\widetilde{\mathrm{Nil}}(X)\oplus\widetilde{\mathrm{Nil}}(X)$$

where the $p : X\times S^1 \to S^1$ is the projection map. The group $Wh(X\times S^1, p)_c$ is the controlled Whitehead group and it has the following homological property:

$$Wh(X\times S^1, p)_c \cong Wh(X)\oplus\tilde{K}_0(X)$$

the proof presented here uses geometric methods that were based on the controlled topology of Hilbert cube manifolds ([62]).

The Fundamental Theorem of $K$-theory calculates the $K$-theory of spaces that admit a natural map to $S^1$. Waldhausen's Splitting Theorem calculates the $K$-theory of spaces that admit a natural map to the unit interval. We interpret Waldhausen's result using controlled topology. Again, the $K$-theory of such spaces splits into two summands, the one that its the controlled $K$-theory over the interval and the other that is Waldhausen's Nil-groups.

The second result that is presented here is a controlled approach to Farrell–Jones Isomorphism Conjecture. The conjecture states that certain obstruction groups are isomorphic to homology groups. In other words, the obstruction groups have some homological properties. That means they can be calculated using induction techniques, which in this case are encoded using spectral sequences. More precisely, the conjecture in $K$-theory predicts that the $K$-theory of any group can be calculated from the knowledge of the $K$-theory of its virtually cyclic subgroups (finite subgroups and finite group extensions by the infinite cyclic group) and their lattice structure induced by inclusion and conjugation. The homological part of the Isomorphism Conjecture can be interpreted as a controlled $K$-group. We show that in the case that the space admits a natural map to the circle or the interval, then the controlled part of the $K$-theory can be described always as a homology theory with the appropriate control. As an application, we give an alternate proof of the fact that the free groups satisfy the $K$-theory Isomorphism Conjecture. It should be noted that the existing proofs of the Isomorphism Conjecture use controlled techniques. From the classical proof in [43] to more recent results like in [12], first a theorem in controlled topology is proved and then it is used to prove special cases of the Isomorphism Conjecture.

The author would like to thank Bruce Hughes, Daniel Juan Pineda, and Jean François Lafont for their useful comments. Also, the author would like to extend his gratitude to the referee whose comments made the paper more coherent. Furthermore, the author would like to thank the Mathematics Department at the University of the Aegean for their hospitality during the completion of the paper.

## 2. Very Short Introduction to Algebraic $K$-Theory

Let $R$ be a ring. To avoid degenerate situations, we assume that $R$ is a ring with identity and that finitely generated free modules have well-defined rank i.e.,

$$R^n \cong R^m \Longleftrightarrow n = m.$$

*Remark* 2.1. Classical examples of such rings are commutative rings. The proof is very short and elegant: Let $R$ be a commutative ring and $M$ a maximal ideal in $R$. Then $R/M$ is a field. Assume that $R^n \cong R^m$, as $R$-modules, which implies $MR^n \cong MR^m$. Then:

$$(R/M)^n = (R/MR)^n \cong R^n/MR^n \cong R^m/MR^m \cong (R/MR)^m = (R/M)^m.$$

But $(R/M)^n$ and $(R/M)^m$ are vector spaces over the field $R/M$. Since they are isomorphic, they must have the same dimension. Thus $n = m$.

For a group $G$ and a ring $R$, $RG$ denotes the group ring of $G$ with coefficients in $R$:

$$RG = \left\{ \sum_{i=1}^{n} r_i g_i \, : \, r_i \in R, \; g_i \in G, \text{ for } i = 1, 2, \ldots, n \right\}.$$

For example,

(1) Let $G = \mathbb{Z}^n$ be the free abelian group generated by $t_i$, $i = 1, \ldots, n$. Then $RG = R[t_1^{\pm 1}, \ldots, t_n^{\pm 1}]$, the ring of Laurent polynomials on $n$ variables with coefficients in $R$. In particular, if $G = C_\infty$, the infinite cyclic group generated by $t$,

$$RC_\infty = R[t, t^{-1}] = \left\{ \sum_{i=-m}^{n} r_i t^i \, : \, r_i \in R, \; m, n \in \mathbb{N} \right\}.$$

(2) If $G = C_n$, the cyclic group of order $n$, generated by $t$ then $RC_n$ is a quotient of the polynomial ring on one variable:

$$RC_n \cong R[t]/(t^n - 1).$$

(3) Let $G = F_n$ be the free group generated by $x_i$, $i = 1, \ldots, n$, then $RF_n$ is the Laurent ring on $n$ non-commuting variables with coefficients in $R$:

$$RF_n = R\{x_1^{\pm 1}, \ldots, x_n^{\pm 1}\}.$$

(4) Let $G = H \rtimes_\alpha \mathbb{Z}$ be the semi-direct product of $H$ be $\mathbb{Z}$ with $\alpha$ the action of the generator of $\mathbb{Z}$ on $H$. Then $RG \cong RH_\alpha[t, t^{-1}]$, the twisted Laurent polynomial ring of $RH$. More precisely, for $r, s \in RH$,

$$(rt^n)(st^m) = r\alpha^{-n}(s)t^{n+m}.$$

*Remark* 2.2. If $R$ has well-defined rank for finitely generated free modules, so does $RG$, for any group $G$. The proof is the same as in Remark 2.1 where the maximal ideal $M$ is replaced by the augmentation ideal $I$:

$$I = \mathrm{Ker}(\varepsilon : RG \to R)$$

where $\varepsilon$ is the augmentation homomorphism.

## 2.1. $K_0$, $K_1$ and Nil-groups of rings

By a module over a ring $R$ we will mean a **left module**. Everything works the same when right modules are chosen instead. The basic definitions are included in [6] and [78].

**Definition 2.3.** The group $K_0(R)$ is defined as $\mathcal{F}_R/N_R$ where:

- $\mathcal{F}_R$ is the free abelian group generated by the isomorphism classes of finitely generated projective $R$-modules.
- $\mathcal{N}_R$ is the subgroup of $\mathcal{F}_R$ generated by elements of the form $[P_0 \oplus P_1] - [P_0] - [P_1]$.

**Examples.** Some very basic examples:

(1) If $F$ is a field, then $K_0(F) \cong \mathbb{Z}$, and the isomorphism is given by the dimension.
(2) For $G = \mathbb{Z}^n$, $K_0(\mathbb{Z}G) \cong \mathbb{Z}$ (Bass–Heller–Swan, [7]).
(3) If $G = F_n$, $K_0(\mathbb{Z}F_n) \cong \mathbb{Z}$ (Gersten, [51]).
(4) If $R$ is the ring of integers in a number field, then $K_0(R)$ is isomorphic to the class group of $R$ ([78]).
(5) Let $X$ be a compact Hausdorff topological space and $C(X)$ be the ring of continuous functions from $X$ to $\mathbb{R}$. Then $K_0(C(X)) \cong K^0(X)$ where $K^0(X)$ is Atiyah's geometric $K$-group of real vectors bundles over $X$ (Swan, [86], [78]).

For the definition of $K_1(R)$, let $GL(n, R)$ be the group of invertible $n \times n$-matrices with entries in $R$. Let $E(n, R)$ be the subgroup generated by elementary matrices i.e., matrices that are formed from the identity by adding the multiple of a row (or column) to another row (or column). More precisely, $E(n, R)$ is generated by matrices $e_{i,j}(r)$, with $1 \leq i, j \leq n$, $i \neq j$, $r \in R$, of the form:

$$e_{i,j}(r) = \begin{pmatrix} 1 & 0 & \dots & \dots & \dots & \dots & \dots & 0 \\ 0 & 1 & \dots & \dots & \dots & \dots & \dots & 0 \\ \dots & \dots & \dots & \dots & \dots & \dots & \dots & \dots \\ \dots & \dots & \dots & 1 & \dots & r & \dots & \dots \\ \dots & \dots & \dots & \dots & \dots & \dots & \dots & \dots \\ 0 & 0 & \dots & \dots & \dots & \dots & \dots & 1 \end{pmatrix}$$

where $r$ is the $(i,j)$-position. There are monomorphisms:

$$GL(n, R) \to GL(n+1, R), \quad E(n, R) \to E(n+1, R)$$

induced by the map

$$A \mapsto \begin{pmatrix} A & 0 \\ 0 & 1 \end{pmatrix}.$$

Taking direct limits:

$$GL(R) = \varinjlim GL(n, R) = \bigcup_{n=1}^{\infty} GL(n, R), \quad E(R) = \varinjlim E(n, R) = \bigcup_{n=1}^{\infty} E(n, R).$$

**Lemma 2.4** (Whitehead's Lemma). $[GL(R), GL(R)] = E(R)$.

*Proof.* The following holds in $GL(2n, R)$ for each invertible $n \times n$-matrix $u$:

$$\begin{pmatrix} u & 0 \\ 0 & u^{-1} \end{pmatrix} = \begin{pmatrix} 1 & 0 \\ u^{-1} & 1 \end{pmatrix} \begin{pmatrix} 1 & 1-u \\ 0 & 1 \end{pmatrix} \begin{pmatrix} 1 & 0 \\ -1 & 1 \end{pmatrix} \begin{pmatrix} 1 & 1-u^{-1} \\ 0 & 1 \end{pmatrix}$$

So that the matrix on the left hand side of the equation is elementary. For the commutators, we see that

$$[g, h] = \begin{pmatrix} g & 0 \\ 0 & g^{-1} \end{pmatrix} \begin{pmatrix} h & 0 \\ 0 & h^{-1} \end{pmatrix} \begin{pmatrix} (hg)^{-1} & 0 \\ 0 & hg \end{pmatrix}$$

which implies that the commutators belong to $E(R)$. □

**Definition 2.5.** For a ring $R$ define $K_1(R) = GL(R)/E(R)$.

*Remark* 2.6.

(1) Whitehead's Lemma implies that $K_1(R)$ is abelian.
(2) $K_1(R)$ can be thought as the obstruction for solving a system of linear equations with coefficients in $R$ using Gauss operations of the third type (adding a multiple of an equation to another equation).

**Examples**

(1) If $F$ is a field, then the determinant map induces and isomorphism $K_1(F) \cong F \setminus \{0\}$.

(2) In general, if $R$ is a commutative ring then the determinant induces a homomorphism:

$$\det : K_1(R) \to U(R),$$

where $U(R)$ are the units of $R$. The map det is a split surjection (Proposition 2.2.1 [78]). We write $SK_1(R) = \ker(\det)$. When $R = \mathbb{Z}$, the determinant map induces an isomorphism $K_1(\mathbb{Z}) \cong \mathbb{Z}/2\mathbb{Z}$.

(3) If $R$ is the ring of integers in a number field, $SK_1(R)$ vanishes and $K_1(R)$ is the group of units of $R$ ([78]).

(4) For a compact Hausdorff topological space $X$ and $C(X)$ be the ring of continuous functions from $X$ to $\mathbb{R}$, then $K_1(C(X)) \cong K^1(X)$ Atiyah's geometric $K$-group of real vectors bundles over $X$ ([78]). Actually, the Serre–Swan's Theorem states that the category of real vector bundles over $X$ is equivalent to the category of finitely generated projective modules over $C(X)$.

Let $\alpha : R \to R$ be a ring automorphism. Let $M$ be an $R$-module. A map $f : M \to M$ is called $\alpha$-linear if:

- $f(m+n) = f(m) + f(n)$, for all $m$, $n$ in $M$.
- $f(rm) = \alpha(r)f(m)$, for all $r \in R$, $m \in M$.

Consider pairs $(P, \nu)$ where $P$ is a finitely generated projective $R$-module and $\nu$ is an $\alpha$-linear nilpotent endomorphism of $P$. We call such pairs *nil pairs*. A morphism of nil pairs

$$f : (P, \nu) \to (Q, \mu)$$

is an $R$-linear homomorphism $f : P \to Q$ making the diagram commutative:

$$\begin{array}{ccc} P & \xrightarrow{f} & Q \\ {\scriptstyle\nu}\downarrow & & \downarrow{\scriptstyle\mu} \\ P & \xrightarrow{f} & Q \end{array}$$

Exact sequences of nil pairs are given by:

$$0 \to (P_0, \nu_0) \xrightarrow{f} (P, \nu) \xrightarrow{g} (P_1, \nu_1) \to 0 \quad (E)$$

where the following diagram of exact sequences commutes:

$$\begin{array}{ccccccccc} 0 & \longrightarrow & P_0 & \xrightarrow{f} & P & \xrightarrow{g} & P_1 & \longrightarrow & 0 \\ & & {\scriptstyle\nu_0}\downarrow & & {\scriptstyle\nu}\downarrow & & {\scriptstyle\nu_1}\downarrow & & \\ 0 & \longrightarrow & P_0 & \xrightarrow{f} & P & \xrightarrow{g} & P_1 & \longrightarrow & 0 \end{array}$$

Notice that, in general, the exact sequence (E) does not split i.e., $(P, \nu)$ is not the direct sum of $(P_0, \nu_0)$ and $(P_1, \nu_1)$. Actually, if we identify $P \cong P_0 \oplus P_1$ then:

$$\nu = \begin{pmatrix} \nu_0 & \chi \\ 0 & \nu_1 \end{pmatrix} : P_0 \oplus P_1 \to P_0 \oplus P_1,$$

where $\chi : P_1 \to P_0$ is any $\alpha$-linear homomorphism.

**Definition 2.7.** The group $\mathrm{Nil}(R, \alpha)$ is defined as $\mathcal{F}N_R/\mathcal{N}N_R$ where

- $\mathcal{F}N_R$ is the free abelian group generated by the isomorphism classes of nil objects $(P, \nu)$.
- $\mathcal{N}N_R$ is the subgroup generated by:
  - Elements of the form
  $$[P, \nu] - [P_0, \nu_0] - [P_1, \nu_1]$$
  for each exact sequence (E).
  - Elements $[F, 0]$ where $F$ is a finitely generated free $R$-module.

The reduced Nil-group $\widetilde{\mathrm{Nil}}(R, \alpha)$ is defined to be the subgroup of $\mathrm{Nil}(R, \alpha)$ generated by elements of the form $[F, \nu]$ where $F$ is a finitely generated free $R$-module. We write $\widetilde{\mathrm{Nil}}(R)$ for $\widetilde{\mathrm{Nil}}(R, \mathrm{id}_R)$.

*Remark* 2.8.

(1) Suppose that $\alpha$ has finite order. Then if $\widetilde{\mathrm{Nil}}(R, \alpha) \neq 0$ it is infinitely generated (Farrell for $\alpha = 1$ ([39]), Ramos ([76]) and Grunewald ([54], [55]) for the general case). The question of finiteness or not of Nil-groups is open when $\alpha$ has infinite order.

(2) Remember that a ring $R$ is called *regular coherent* if every finitely presented $R$-module $M$ admits a resolution by finitely generated projective modules. The ring is called *regular Noetherian* is if in addition any finitely generated $R$-module is finitely presented. If $R$ is a regular coherent ring then $\widetilde{\mathrm{Nil}}(R, \alpha)$ vanishes (Bass ([6], [7]) and Waldhausen ([90])).

(3) Group rings of finitely generated free groups are regular coherent. Group rings of finitely generated free abelian groups are regular Noetherian. Even though twisted polynomial rings of Noetherian rings are Noetherian, the same is not true for coherent rings ([52]).

The remarks leave open two conjectures.

**Conjecture 1.** *Let $R$ be any ring and $\alpha : R \to R$ a ring automorphism of infinite order. Then $\widetilde{Nil}(R, \alpha)$ is either zero or infinitely generated.*

**Conjecture 2.** *Let $R$ be a ring of finite cohomological dimension (i.e., there is a number $n$ so that every module has a resolution of finite length at most $n$). Then $\widetilde{Nil}(R, \alpha) = 0$ for each ring automorphism $\alpha$.*

For the second conjecture, recent developments are presented in [14] and [15].

*Remark* 2.9. The following question has been asked by Tom Farrell: Given a ring $R$ and a ring automorphism $\alpha$ of $R$, is there a ring $S$ so that $\widetilde{\mathrm{Nil}}(R, \alpha) \cong \widetilde{\mathrm{Nil}}(S)$? For partial results the reader should check [5]. Related to the above question is the following: Let $R$ be a ring so that $\widetilde{\mathrm{Nil}}(R) = 0$. Is it true that $\widetilde{\mathrm{Nil}}(R, \alpha) = 0$ for any ring automorphism $\alpha$ of $R$?

*Remark* 2.10. There is a common pattern in the definition of the $K$-groups given above. The $K_0$-groups should be considered as the Grothedieck groups of matrices that are projections, the $K_1$-groups of invertible matrices and the Nil-groups of nilpotent matrices.

## 2.2. Algebraic $K$-theory in Topology

We present some of the basic applications of algebraic $K$-theory to topology. For the topological applications we need certain quotients of the algebraic $K$-groups.

**Definition 2.11.** Let $R$ be a ring with 1. Let $\mathbb{Z} \to R$ be the ring homomorphism induced by mapping 1 to the identity. Define the reduced $K_0$-group

$$\tilde{K}_0(R) = \mathrm{Coker}(K_0(\mathbb{Z}) \to K_0(R)).$$

Actually,

$$\tilde{K}_0(R) = K_0(R)/\langle [R] \rangle$$

where $\langle [R] \rangle$ is the infinite cyclic subgroup generated by the free $R$-module $[R]$ of rank 1.

*Remark* 2.12.

(1) An element $[P]$ in $\tilde{K}_0(R)$ is zero iff $P$ is finitely generated *stably free module* i.e., there is a finitely generated free module $F$ such that $P \oplus F$ is free.

(2) For any module $P$ there is $Q$ such that $P \oplus Q$ which is free. Then there is an isomorphism

$$P \oplus (Q \oplus P \oplus Q \oplus \cdots) \cong P \oplus Q \oplus P \oplus Q \oplus \cdots \Longrightarrow P \oplus F \cong F,$$

where $F$ is a countably generated free module. Thus if the finite generation condition is omitted in (1), every projective module is stably free (Eilenberg swindle).

**Definition 2.13.** A space $Y$ is called finitely dominated space iff there is a finite complex $K$ and maps

$$Y \xrightarrow{d} K \xrightarrow{u} Y,$$

such that $u \circ d \simeq \mathrm{id}_Y$.

*Remark* 2.14. Let $Y$ be a finitely dominated space as above. Then the map

$$e : K \xrightarrow{u} Y \xrightarrow{d} K,$$

satisfies

$$e^2 = dudu = d(ud)u \simeq d(\mathrm{id}_Y)u = du = e.$$

Thus $e$ is an idempotent up to homotopy (for the general theory of homotopy idempotents see [37], [57]).

Let $Y$ be a finitely dominated connected with $\pi_1(Y) = \pi$ space and $\tilde{Y}$ its universal cover. By adding a finite number of cells we arrange that the finite complex $K$ that dominates $Y$ has $\pi_1(K) = \pi$ ([91]). Then $C_*(\tilde{Y})$ is a $\mathbb{Z}\pi$-chain complex which is dominated by a finite complex of free modules namely $C_*(\tilde{K})$. That means there are chain maps

$$C_*(\tilde{Y}) \xrightarrow{u} C_*(\tilde{K}) \xrightarrow{d} C_*(\tilde{Y})$$

and a chain homotopy $d \circ u \simeq \mathrm{id}_{C_*(\tilde{Y})}$. Then $C_*(\tilde{Y})$ is chain homotopy equivalent to a chain complex $P_*$ of finitely generated projective modules (Wall ([91], [92])):

$$P_* : \cdots 0 \to P_n \to P_{n-1} \to \cdots P_1 \to P_0.$$

Thus a finitely dominated space determines an element

$$\sigma(Y) = \sum_{i=0}^{n} (-1)^i [P_i] \in \tilde{K}_0(\mathbb{Z}\pi).$$

**Theorem 2.15** (Wall's Finiteness Obstruction). *A finitely dominated space $Y$ determines an element in $\sigma(Y) \in \tilde{K}_0(\mathbb{Z}\pi)$ which vanishes iff $Y$ is homotopy equivalent to a finite complex.*

*Remark* 2.16. Using Wall's finiteness obstruction, we can interpret finitely dominated spaces as projective modules, up to homotopy, (they determine homotopy idempotents an in Remark 2.14) and the finite complexes as free modules.

A more geometric application of the finiteness obstruction comes from Siebenmann's Thesis ([79]). The question that this work addresses is when an end in a manifold can be completed i.e., it can be realized as the interior of a compact manifold with boundary. More precisely, the compact manifold with boundary $\bar{M}$ is a completion of a manifold $M$ if $M = \bar{M} - \partial\bar{M}$.

**Definition 2.17.** An end $\varepsilon$ in a topological space $X$ is called tame if:

(1) The end $\varepsilon$ is stable: There is a sequence of connected open neighborhoods of $\varepsilon$ such that there is a cofinal subsequence of the fundamental groups

$$G_1 \xleftarrow{f_1} G_2 \xleftarrow{f_2} \cdots G_n \xleftarrow{f_n} \cdots$$

which induces a sequence of isomorphism:

$$\mathrm{Im}(f_1) \xleftarrow{\cong} \mathrm{Im}(f_2) \xleftarrow{\cong} \cdots \xleftarrow{\cong} \mathrm{Im}(f_n) \xleftarrow{\cong} \cdots$$

(2) There are arbitrarily small neighborhoods of $\varepsilon$ that are finitely dominated.

For a complete exposition of the algebraic theory of ends a reference is Hughes–Ranicki ([63]).

**Theorem 2.18** (Siebenmann's Thesis). *Let $M^n$ be a manifold with $n \geq 6$ and $\varepsilon$ a tame end of $M$. Then $\varepsilon$ can be completed iff an obstruction $\sigma(\varepsilon) \in \tilde{K}_0(\mathbb{Z}\pi_1(\varepsilon))$ vanishes, where $\pi_1(\varepsilon)$ is the fundamental group of the end.*

Now we present some geometric applications of $K_1$. Again, for geometric applications we need a quotient of the algebraic $K_1$.

**Definition 2.19.** Let $G$ be a group. The Whitehead group of $G$ is defined as

$$\mathrm{Wh}(G) = K_1(\mathbb{Z}G)/\{\pm g : g \in G\}.$$

Here $\pm g \in GL(1, \mathbb{Z}G)$, for all $g \in G$.

The geometric applications of $\mathrm{Wh}(G)$ are derived from the fact that a strong deformation retraction:

$$f : K \to L$$

of connected finite CW-complexes determines an element in $\mathrm{Wh}(\pi_1(L))$ as follows ([29], [77]): Let $\pi = \pi_1(L)$. Choose a CW-decomposition of $K$ and $L$ so that $L$ is a subcomplex of $K$. Then the relative chain complex $C_* = C_*(\tilde{K}, \tilde{L})$ is a $\mathbb{Z}\pi$-chain complex, where $\tilde{K}$ (resp. $\tilde{L}$) is the universal cover of $K$ (resp. $L$). Liftings of the cells to the universal covers determines a $\mathbb{Z}\pi$-basis for $C_*$. The ambiguity of the choice of liftings is given by elements $\pm g$, $g \in \pi$ (the group element for the particular choice of the lift and the sign for the choice of the orientation of the cell). That is why we consider the quotient of $K_1(\mathbb{Z}\pi)$. Then $C_*$ is a based, finite, contractile chain complex. Thus there is an isomorphism ([77]):

$$\delta + \partial : C_{\text{even}} \to C_{\text{odd}}$$

where $\partial$ is the boundary map, $\delta$ is the chain contraction and $C_{\text{even}}$ (resp. $C_{\text{odd}}$) is the direct sum of the even (resp. odd) part of $C_*$. Using the bases determined by the cells, $\delta + \partial$ determines an invertible matrix with entries in $\mathbb{Z}\pi$. We define the *torsion* of $f$ to be that matrix: $\tau(f) = [\delta + \partial] \in \mathrm{Wh}(\pi)$. In general, if $f : K \to L$ is a homotopy equivalence then $\tau(f) = f_*(\tau(M(f), K))$, where $M(f)$ is the mapping cylinder of $f$.

**Definition 2.20.** Let $L$ be a space. An elementary expansion $K$ of $L$ corresponds to the attaching of a ball to $L$ along a face of the ball. More precisely, $K$ is formed as the push-out

$$\begin{array}{ccc} S^{n-1} & \xrightarrow{q} & L \\ \downarrow & & \downarrow \\ D^n & \xrightarrow{Q} & K \end{array}$$

In this case, we say that $L$ is formed by an elementary collapse of $K$. A simple homotopy equivalence $f : K \to L$ is a homotopy equivalence that can be described as a finite sequence of elementary expansions and collapses.

*Remark* 2.21. Simple homotopy equivalences were introduced by J. H. C. Whitehead ([93], [94]). His goal was to study homotopy equivalences combinatorially.

**Theorem 2.22.** *A homotopy equivalence $f : K \to L$ between finite CW-complexes is simple iff $\tau(f) = 0 \in Wh(\pi_1(L))$.*

The importance of the Whitehead torsion comes with its applications to the topology of manifolds.

**Definition 2.23.** A cobordism $(W; M, M')$ is a manifold $W$ such that $\partial W = M \coprod M'$. An $h$-cobordism is a cobordism such that the inclusion maps

$$M \hookrightarrow W, \;\; M' \hookrightarrow W$$

are homotopy equivalences.

The following works in all categories of manifolds (Diff, PL or Top).

**Theorem 2.24** (s-Cobordism Theorem). *Let dim(M) $\geq 5$. Then there is a bijection:*

$$\{h - \textit{cobordisms over } M\} \longrightarrow Wh(\pi_1(M)), \quad (W; M, M') \mapsto \tau(W, M).$$

*In particular* $\tau(W, M) = 0 \in Wh(\pi_1(M))$ *iff* $W \cong M \times [0,1]$ *which implies that* $M \cong M'$.

*Remark* 2.25.

(1) The equivalences in the s-cobordism theorem are equivalences in the given category (diffeomorphisms, PL-homeomorphisms, homeomorphisms respectively).
(2) The h-cobordism theorem (which is the s-cobordism theorem when $M$ is simply connected) was the main ingredient in the proof of the Poincaré Conjecture for $n \geq 5$ by Smale ([81]).

For the calculations of the reduced $K_0$ and the Whitehead group, there is the following conjecture.

**Conjecture 3.** *Let $G$ be a torsion free group. Then*

$$\tilde{K}_0(\mathbb{Z}G) = Wh(G) = 0.$$

Algebraically, Conjecture 3 is related to the following conjecture.

**Conjecture 4.** *Let $G$ be a torsion free group. Then*

$$(\mathbb{Z}G)^{\times} = \pm G$$

*i.e., the units of $\mathbb{Z}G$ are exactly the elements of $G$ and their negatives. Also, $\mathbb{Z}G$ contains no non-trivial nilpotent elements.*

*Remark* 2.26. Notice that if $x \in \mathbb{Z}G$ is nilpotent then $1 - x$ is a unit.

The geometric definition of Nil-groups is given in [62] and [72]. Let $X$ be a finite CW-complex. A *nil-pair* over $X$, $(Y, f)$ consists of a finite CW-complex $Y$ containing $X$ and $f : Y \to Y$ such that:

(1) $f_X = \mathrm{id}_X$.
(2) There is $s > 0$ such that $f^s \simeq r$ rel$X$, where $r$ is a retraction to $X$ (homotopy nilpotent condition).

In [72] it was shown that, if (1) holds, the second condition is equivalent to

(2*) The inclusion map $X \times S^1 \hookrightarrow T(f)$ is a homotopy equivalence.

Define $(Y, f) \sim (Y' f')$ if there is a homotopy equivalence $f : Y \to Y'$, rel$X$ such that $Ff \simeq f'$ rel$X$. Let $\mathcal{F}(X)$ be the free group on the equivalence classes of nil-pairs over $X$. Let $\mathcal{N}(X)$ be the subgroup generated by:

(1) For two nil-pairs $(Y_i, f_i)$, $i = 1, 2$,

$$[Y_1 \cup_X Y_2, f_1 \cup f_2] - [Y_1, f_1] - [Y_2, f_2]$$

(2) The nil-pairs $[Y, r]$, with $r : Y \to X$ a retraction.

Then $\widetilde{\mathrm{Nil}}(X) = \mathcal{F}(X)/\mathcal{N}(X)$.

The isomorphism between the algebraic and geometric Nil-groups is given:

$$\widetilde{\mathrm{Nil}}(X) \to \widetilde{\mathrm{Nil}}(\pi_1(X)), \quad (Y, f) \mapsto (C_*(\bar{Y}, \tilde{X}), f_*)$$

where $\tilde{X}$ is the universal cover of $X$, $\bar{Y}$ is the pull-back of the universal cover of $X$ under the retraction $r : Y \to X$, $C_*(\bar{Y}, \tilde{X})$ is a finite $\mathbb{Z}\pi_1(X)$-complex of free modules. Notice that $f_*$ is a chain homotopy nilpotent map. For the precise way that a pair $(C_*(\bar{Y}, \tilde{X}), f_*)$ determines an element of $\widetilde{\mathrm{Nil}}(\mathbb{Z}\pi_1(X))$ see [77].

Nil-groups appear as obstructions to splitting problems. We will present a brief summary of the applications after introducing the Fundamental Theorem of Algebraic $K$-theory.

## 2.3. The Fundamental Theorem of Algebraic $K$-theory

For a ring $R$, the Fundamental Theorem of Algebraic $K$-theory computes the $K$-theory of $R[t, t^{-1}]$ in terms of the $K$-theory of $R$.

**Theorem 2.27** (Fundamental Theorem of Algebraic $K$-theory). *For a ring $R$*

$$K_1(R[t, t^{-1}]) \cong K_1(R) \oplus K_0(R) \oplus \widetilde{Nil}(R) \oplus \widetilde{Nil}(R).$$

*For the geometrically significant groups,*

$$Wh(G \times \mathbb{Z}) \cong Wh(G) \oplus \tilde{K}_0(\mathbb{Z}G) \oplus \widetilde{Nil}(\mathbb{Z}G) \oplus \widetilde{Nil}(\mathbb{Z}G).$$

*Remark* 2.28.

(1) There is an analogue of the Fundamental Theorem for higher $K$-theory ([65], [82]).

(2) For twisted Laurent rings the Fundamental Theorem of Algebraic $K$-theory has the form ([42]):

$$K_1(R) \xrightarrow{1-\alpha_*} K_1(R) \to K_1(R_\alpha[t, t^{-1}]) / \left(\widetilde{\mathrm{Nil}}(R, \alpha) \oplus \widetilde{\mathrm{Nil}}(R, \alpha^{-1})\right)$$
$$\to K_0(R) \xrightarrow{1-\alpha_*} K_0(R)$$

(3) The Fundamental Theorem of Algebraic $K$-theory measures the failure of $K$-theory to be a homology theory. In particular, the Nil-terms are exactly the measure of that failure.

Let $X$ be a finite CW-complex (or a compact ANR ([25])) with $\pi_1(X) = G$. Then there is a transfer map

$$tr^n : Wh(G \times \mathbb{Z}) \to Wh(G \times \mathbb{Z})$$

defined geometrically as follows: Let $\tau = (f : Y \to X)$ be represented by a strong deformation retraction. Form the pull-back diagram

$$\begin{array}{ccc} \bar{Y} & \xrightarrow{\bar{f}} & X\times S^1 \\ \downarrow & & \downarrow \mathrm{id}_X\times n \\ Y & \xrightarrow{f} & X\times S^1 \end{array}$$

where $n$ is the map that sends an element of $S^1$ (considered as the unit circle in the complex plane) to its $n$-th power. Define $tr^n(\tau) = \tau(\bar{f})$. The summands of the splitting in the fundamental theorem behave as follows ([72], [77]):

(1) If $x \in Wh(G)$, then $tr^n(x) = nx$.
(2) If $x \in \tilde{K}_0(\mathbb{Z}G)$, $tr^n(x) = x$. Thus the elements of $\tilde{K}_0(\mathbb{Z}G)$ are exactly the elements of $Wh(G)$ that are transfer invariant.
(3) If $x \in \widetilde{\mathrm{Nil}}(\mathbb{Z}G) \oplus \widetilde{\mathrm{Nil}}(\mathbb{Z}G)$, then there is an $s > 0$ such that $tr^n(x) = 0$ for all $n > s$.

The Fundamental Theorem can be used for extending the definition of $K_i$-groups with $i < 0$. More precisely, inductively, $K_i(R)$ is defined to be the subgroup of the transfer invariant elements of $K_{i+1}(R)$ for $i \leq 1$. The Fundamental Theorem of Algebraic $K$-theory extends to the negative $K$-groups. The formal details for the delooping of the algebraic $K$-theory spectrum are given in [71].

## 2.4. Waldhausen's Splitting Theorem

Let $A_i$, $i = 0, 1$ be two rings containing $R$ as a subring. Assume that $A_i = R \oplus B_i$ as $R$-bimodules. Let $\Lambda$ be the push out in the category of $R$-algebras:

$$\begin{array}{ccc} R & \longrightarrow & A_1 \\ \downarrow & & \downarrow \\ A_2 & \longrightarrow & \Lambda \end{array}$$

As an $R$-bimodule $\Lambda$ splits as:

$$\Lambda = R \oplus B_1 \oplus B_2 \oplus B_1 \otimes B_2 \oplus B_2 \otimes B_1 \oplus \cdots$$

where $\otimes$ means $\otimes_R$. The multiplication is given by concatenation and using the ring structure of $A_i$.

**Theorem 2.29** (Waldhausen ([90])). *With the above notation, there is an exact sequence:*

$$\cdots \to K_1(R) \to K_1(A_1) \oplus K_1(A_2) \to K_1(\Lambda)/N_1 \to K_0(R) \to \cdots$$

*where* $N_1 = \widetilde{Nil}_0(R; B_0, B_1)$ *is Waldhausen's Nil-group.*

*Remark* 2.30. For more general splitting theorems, see [83] and [84].

We will give the definition of Waldhausen's Nil-groups. For the triple $(R; B_0, B_1)$ with $R$ a ring and $B_i$, $i = 0, 1$, two $R$-bimodules, consider quadruples $(P, Q, p, q)$ where $P$ and $Q$ are finitely generated projective $R$-modules and

$$p : P \to Q \otimes B_0, \quad q : Q \to P \otimes B_1$$

(here $\otimes$ means $\otimes_R$) are two $R$-maps such that the compositions

$$\begin{array}{l} P \xrightarrow{p} Q \otimes B_0 \xrightarrow{q \otimes \mathrm{id}} P \otimes B_1 \otimes B_0 \cdots \\ Q \xrightarrow{q} P \otimes B_1 \xrightarrow{p \otimes \mathrm{id}} P \otimes B_0 \otimes B_1 \cdots \end{array}$$

are eventually 0. Exact sequences have the obvious meaning. As before, define $\mathrm{Nil}_0(R; B_0, B_1)$ to be the free abelian group generated by isomorphism classes of quadruples as above modulo the additive relation resulting from the exact sequences. Notice that there are homomorphisms:

$$\begin{array}{c} K_0(R) \times K_0(R) \to \mathrm{Nil}_0(R; B_0, B_1),\ (P, Q) \mapsto (P, Q, 0, 0) \\ \mathrm{Nil}_0(R; B_0, B_1) \to K_0(R) \times K_0(R),\ (P, Q, p, q) \mapsto (P, Q) \end{array}$$

which make $K_0(R) \times K_0(R)$ a summand of $\mathrm{Nil}_0(R; B_0, B_1)$. Define

$$\widetilde{\mathrm{Nil}}_0(R; B_0, B_1) = \mathrm{Nil}_0(R; B_0, B_1) / K_0(R) \times K_0(R).$$

*Remark* 2.31.

(1) The quadruples defined above form an exact category $\mathcal{N}il(R; B_0, B_1)$. In general, define

$$\widetilde{\mathrm{Nil}}_i(R; B_0, B_1) = K_i(\mathcal{N}il(R; B_0, B_1)) / K_i(R) \times K_i(R).$$

The splitting theorem states that $\widetilde{\mathrm{Nil}}_{i-1}(R; B_0, B_1)$ is a summand of $K_i(\Lambda)$. Thus there is a dimension shift. Waldhausen's splitting theorem holds for all $i \in \mathbb{Z}$.

(2) For more information on the structure of lower Waldhausen Nil-groups see [66], [67].

# 3. Controlled Topology

Let $B$ be a metric space and $p : X \to B$ be a continuous map and $\varepsilon > 0$. A homotopy

$$F : Y \times [0, 1] \to X$$

is called a $p^{-1}(\varepsilon)$ homotopy if

$$\sup\{\mathrm{diam}(pF(\{y\} \times [0, 1])) : y \in Y\} < \varepsilon.$$

That means that the paths determined by the homotopy have diameter less that $\varepsilon$ when projected to $B$. In this case we write $F_0 \simeq_{p^{-1}(\varepsilon)} F_1$.

*Remark* 3.1. More generally, the assumption that $B$ is metric space is not needed. For a general topological space $B$ and an open coven $\mathcal{U}$ of $B$, the homotopy $F$ is called $p^{-1}(\mathcal{U})$-homotopy if, for each $y \in Y$, there is $U_y \in \mathcal{U}$ such that

$$pF(\{y\}\times[0,1]) \subset U_y.$$

**Definition 3.2.** Let $p : X \to B$ be as above and $\varepsilon > 0$. A map $f : Y \to X$ is called a $p^{-1}(\varepsilon)$-homotopy equivalence if there is $g : X \to Y$ such that

$$g \circ f \simeq_{(pf)^{-1}(\varepsilon)} \mathrm{id}_Y, \quad f \circ g \simeq_{p^{-1}(\varepsilon)} \mathrm{id}_X.$$

A continuous map $f : Y \to X$ is called a controlled homotopy equivalence (with control in $B$) if for each $\delta > 0$, $f$ is a $p^{-1}(\delta)$-homotopy equivalence.

**Definition 3.3.** A proper map $f : Y \to X$ is called cell-like (CE-map) if for each $x \in X$ and each neighborhood $U$ of $f^{-1}(x)$, the inclusion map $f^{-1}(x) \hookrightarrow U$ is null-homotopic.

*Remark* 3.4.

(1) If $f^{-1}(x)$ is an ANR for each $x \in X$, then a proper map $f$ is a CE-map iff $f^{-1}(x)$ is contractible for all $x \in X$. In particular, a proper cellular map $f$ between CW-complexes is a CE-map iff $f^{-1}(x)$ is contractible for all $x \in X$.
(2) Cell-like maps are the homotopy analogues of homeomorphisms. In a homeomorphism we require that the inverse image of a point is a point. In a CE-map we require that the inverse image of a point has the homotopy type of a point.
(3) If $f : Y \to X$ is a CE-map between locally compact ANR's then $f$ is a controlled homotopy equivalence with control in $X$.
(4) The basic properties of CE-maps are presented in [68].

Controlled homotopy equivalences and CE-maps are closely related to the controlled theory of fibrations.

**Definition 3.5.** Let $p : X \to B$ be map to a metric space $B$ and $\varepsilon > 0$. A map $q : E \to X$ is called a $p^{-1}(\varepsilon)$-fibration if the lifting problem

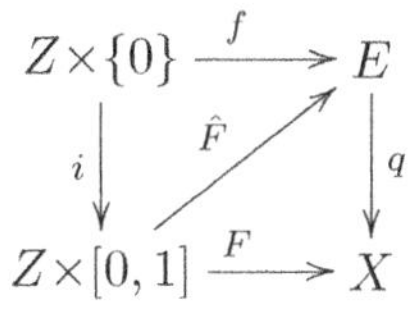


has a solution $\hat{F}$ such that:

- The top triangle commutes: $\hat{F} \circ i = f$.
- The bottom diagram $p^{-1}(\varepsilon)$-commutes: $d(pq\hat{F}(z,t), pF(z,t)) < \varepsilon$, for all $(z,t) \in Z\times[0,1]$.

The map $q$ is called an approximate fibration if it is a $\delta$-fibration for each $\delta$.

The properties of approximate fibrations are proved in [32] and [33]. The following result follows from the definitions ([68]).

**Theorem 3.6.** *A map $f : Y \to X$ between locally compact ANR's is a CE-map iff it is a homotopy equivalence and an approximate fibration.*

The following constructions are from [20]. Let $X$ be a compact ANR and $p : X \to B$ be a control map to metric space $B$ and $\varepsilon > 0$. Two $p^{-1}(\varepsilon)$-strong deformation retractions

$$f_i : Y_i \to X, \quad i = 1, 2,$$

with $Y_i$ compact ANR's, are called equivalent, denoted $f_1 \sim f_2$, if there is a compact ANR $Z$ such that the diagram:

$$\begin{array}{ccc} Z & \xrightarrow{\alpha} & Y_1 \\ {\scriptstyle\beta}\downarrow & & \downarrow{\scriptstyle f_1} \\ Y_2 & \xrightarrow{f_2} & X \end{array}$$

commutes up to $p^{-1}(\varepsilon)$-homotopy rel$X$. The above relation generates an equivalence relation on the set of $p^{-1}(\varepsilon)$-strong deformation retractions to $X$.

The $\varepsilon$-controlled Whitehead group of $X$, $Wh(X,p)_\varepsilon$ is defined as the group with elements equivalence classes of $p^{-1}(\varepsilon)$-sdr's whose inverses in the classical Whitehead group are $p^{-1}(\varepsilon)$-sdr's. The group operation is given by push-outs:

$$(f_1 : Y_1 \to X) + (f_2 : Y_2 \to X) = (f_1 \cup f_2 : Y_1 \cup_X Y_2 \to X).$$

The *controlled Whitehead group* is defined as the inverse limit

$$Wh(X,p)_c = \varprojlim Wh(X,p)_\varepsilon.$$

**Theorem 3.7** ([19]). *Let $p : X \to B$ be an approximate fibration with $X$ a compact ANR and $B$ a locally compact ANR. Then a strong deformation retraction $f : Y \to B$ represents an element in $Wh(X,p)_c$ iff $p \circ f$ is an approximate fibration.*

*Remark* 3.8.

(1) There is a more algebraic approach to controlled topology given in [74] and [75]. That approach is useful for complete calculations of the controlled groups.
(2) There are analogues of the $h$-cobordism theorem and the finite domination theorem in the controlled setting ([19], [74], [75]).

The controlled groups capture the homological properties of $K$-theory. In special cases, the homological properties of the controlled groups imply the vanishing of certain obstructions.

**Theorem 3.9** (The sucking principle or squeezing ([22], [24], [59])). *Let $X$ be a compact ANR. For each $\varepsilon > 0$, there is $\delta > 0$ such that every $\varepsilon$-fibration, with control in $X$, $p : E \to X$, with $E$ a compact ANR, is $\delta$-homotopic to an approximate fibration.*

**Theorem 3.10** (The thin $h$-cobordism theorem ([19], [74], [75]). *Let $M^n$ be a compact manifold with $n \geq 5$. Then for each $\varepsilon > 0$ there is $\delta > 0$ such that every $\varepsilon$-$h$-cobordism $(W; M, M')$ (i.e., such that the strong deformations retractions $W \to M$ and $W \to M'$ are $\varepsilon$-homotopy equivalences with control in $M$) is a $\delta$-product (i.e., there is a homeomorphism $h : W \to M \times [0,1]$ such that $p \circ h$ is $\delta$-close to the retraction of $W$ to $M$, here $p : M \times [0,1] \to M$ is the projection map).*

**Theorem 3.11** (The $\alpha$-approximation theorem ([22], [28], [48])). *Let $M^n$ be a compact manifold with $n \geq 5$. Then for each $\varepsilon > 0$ there is $\delta > 0$ such that every $\varepsilon$-homotopy equivalence $f : N^n \to M^n$, with control in $M^n$, and $N^n$ a compact manifold is $\delta$-homotopic to a homeomorphism.*

There are versions of the approximation theorems where the control map is not the identity ([41], [75]). In these cases, the map is required to be at least a stratified fibration and that the problem can be solved in the fibers.

## 3.1. The Fundamental Theorem of Algebraic $K$-theory Revisited

Using controlled topology, the Fundamental Theorem can be written as follows ([62]):

**Theorem 3.12.** *For a compact ANR $X$, there is a splitting:*

$$Wh(X \times S^1) \cong Wh(X \times S^1, p)_c \oplus \widetilde{Nil}(X) \oplus \widetilde{Nil}(X),$$

*where $p : X \times S^1 \to S^1$ is the projection map. Furthermore* ([64]),

$$Wh(X \times S^1, p)_c \cong Wh(X) \oplus \tilde{K}_0(X).$$

We outline the geometric proof of the Fundamental Theorem of algebraic $K$-theory. We will use Hilbert cube manifolds ([26]). Remember that the Hilbert cube is defined as

$$Q = \prod_{n=1}^{\infty} \left[ -\frac{1}{n}, \frac{1}{n} \right],$$

with the induced metric. Hilbert cube manifolds are separable metric spaces that have the local structure of $Q$. Some basic facts on $Q$-manifolds ([26]):

(1) If $X$ is a locally compact ANR, then $X \times Q$ is a $Q$-manifold.
(2) Any map $f : X \to Y$ with $X$ a locally compact ANR and $Y$ a locally compact $Q$-manifold, can be approximated by an embedding.
(3) Any homotopy $h : X \times I \to Y$, with $X$ and $Y$ as above, can be approximated by an isotopy.
(4) The $\alpha$-approximation theorem holds for locally compact $Q$-manifolds ([24], [21], [47]).

The main reason that $Q$-manifolds were introduced to geometric topology is the following theorem:

**Theorem 3.13** ([23], [27]). *A homotopy equivalence $f : X \to Y$ between compact ANR's is simple (i.e., $[f] = 0 \in Wh(X)$) iff $f\times id_Q$ is homotopic to a homeomorphism.*

This theorem gives a positive answer to Whitehead's Homeomorphism Conjecture.

**Theorem 3.14** (The topological invariance of the Whitehead torsion). *Let $f : K \to L$ be a homeomorphism between compact polyhedra. Then $[f] = 0 \in Wh(L)$.*

Also, in [26] there the following triangulation result for $Q$-manifolds.

**Theorem 3.15** (Triangulation of $Q$-manifolds). *Let $M$ be a locally compact $Q$-manifold. Then there is a locally compact polyhedron $K$ such that $M$ is homeomorphic to $L\times Q$.*

Combining the above theorem with the basic facts on $Q$-manifolds we get a positive answer to Borsuk's question.

**Theorem 3.16** ([23], [95]). *Let $X$ be a compact ANR. Then for each $\varepsilon > 0$ there is a finite CW-complex $K_\varepsilon$ that it is $\varepsilon$-homotopy equivalent to $X$ with control in $X$.*

Notice that in general given $X$ and $\varepsilon > 0$ a construction of a finite CW-complex that $\varepsilon$-dominates $X$ is not hard. The complex is constructed using nerves of open covers of $X$.

Now we can go back to the proof of the Fundamental Theorem of $K$-theory. The proof presented is based on the work in [62] and [72]. We can assume that $X$ is a compact $Q$-manifold and an element of $Wh(X\times S^1)$ is represented by a strong deformation retraction $f : Y \to X\times S^1$ with $Y$ a compact $Q$-manifold. Form the pull-back:

$$\begin{array}{ccc} \bar{Y} & \xrightarrow{\bar{f}} & X\times\mathbb{R} \\ \downarrow & & \downarrow \\ Y & \xrightarrow{f} & X\times S^1 \end{array}$$

Denote by $\zeta$ the covering translation on $\bar{Y}$ that corresponds to $+1$ action on the reals. For a subset $A \subset \mathbb{R}$ we write $Y_A = \bar{f}^{-1}(X\times A)$.

Let $K_\pm$ be compact sub-ANR's that dominate $Y_\pm$, rel$Y_0$, where $Y_- = \bar{Y}_{(-\infty,0]}$, $Y_+ = \bar{Y}_{[0,\infty)}$. The existence of such complexes follows from the fact that $\bar{f}$ is a bounded (over $\mathbb{R}$) strong deformation retraction ([20], [49]). Write

$$Y_\pm \xrightarrow{d_\pm} K_\pm \xrightarrow{u_\pm} Y_\pm$$

for the dominations. Write $K = K_- \cup_{Y_0} K_+$, and get the maps

$$\bar{Y} \xrightarrow{d} K \xrightarrow{u} \bar{Y},$$

induced from the maps of the dominations. Each of the spaces $K_\pm$, $K$, $Y$ and $Y_\pm$ contains a copy of $X$ times an interval containing 0. Let $L_\pm$, $L$, $Z$ and $Z_\pm$ be the spaces formed from the ones above by squeezing $X\times$interval to a single copy of $X$. Then the domination maps induce maps, denoted the same, on the new spaces. Set

- $z : L \xrightarrow{u} Z \xrightarrow{\zeta^{-1}} Z \xrightarrow{d} L$..
- $f_+ : L_+ \xrightarrow{u} Z_+ \xrightarrow{\zeta} Z_+ \xrightarrow{d} L_+$..
- $f_- : L_- \xrightarrow{u} Z_- \xrightarrow{\zeta^{-1}} Z_- \xrightarrow{d} L_-$..

#### 3.1.1. Relaxation

We start by recalling the construction of the relaxation of $f$ ([22], [24], [59], [62], [63], [80]). This construction is the map

$$Wh(X\times S^1) \to Wh(X\times S^1, p)_c$$

that splits the forget control map.

Let $q : X\times\mathbb{R} \to \mathbb{R}$ be the projection. Remember that the strong deformation retraction $\bar{f}$ is bounded over $\mathbb{R}$ i.e., it is a $q^{-1}(k)$-homotopy equivalence, for some $k > 0$. The number $k$ depends on how many times the image of $f$ wraps around $S^1$. Let

$$\phi_d : X\times\mathbb{R} \to X\times\mathbb{R}, \quad \phi_d(x,r) = (x, r/d).$$

be the homeomorphism that shrinks distances in the $\mathbb{R}$-direction. Then the composition $\phi_d \circ \bar{f}$ is a $q^{-1}(k/d)$-strong deformation retraction over $\mathbb{R}$. By choosing large enough $d$ and the Squeezing Theorem, we can assume that $\bar{f}$ is a controlled (over $\mathbb{R}$) strong deformation retraction. Thus, the composition

$$\bar{Y} \xrightarrow{\bar{f}} X\times\mathbb{R} \to \mathbb{R}$$

is an approximate fibration (Theorem 3.7).

Now we need the approximate isotopy lifting property for approximate fibrations. Remember that an isotopy of $X$ is a homeomorphism $g : B\times[0,1] \to B\times[0,1]$ which preserves the second coordinate. If $p : E \to B$ is an approximate fibration, the approximate isotopy lifting property for $p$ means that, given an isotopy $g$ of the base space and a $\varepsilon > 0$ there is an isotopy $G$ of $E$ making the diagram $\varepsilon$-commutative:

$$\begin{array}{ccc} E\times[0,1] & \xrightarrow{G} & E\times[0,1] \\ {\scriptstyle p\times \mathrm{id}_{[0,1]}}\downarrow & & \downarrow{\scriptstyle p\times \mathrm{id}_{[0,1]}} \\ B\times[0,1] & \xrightarrow{g} & B\times[0,1] \end{array}$$

More precisely,

$$d(p\times \mathrm{id}_{[0,1]}(G(e,s)), g(p(e),t)) < \varepsilon, \quad \text{for all } (e,t) \in E\times[0,1].$$

If both $X$ and $E$ are finite dimensional or $Q$-manifolds, then approximate fibrations have the approximate isotopy lifting property ([60], [61], [64], [63]).

In our setting, let $g : \mathbb{R}\times[0,1] \to \mathbb{R}\times[0,1]$ be the PL isotopy (i.e., a homeomorphism that preserves the second coordinate) such that

(1) $g$ is supported on $[-1,3]\times[0,1]$, that is $g$ is that identity outside the given set, and
(2) for each $s \in [0,1]$, $g_s$ takes $[-1,-0.5]$ linearly onto $[-1, s-0.5]$, takes $[-0.5,1.5]$ linearly onto $[s-0.5, s+1.5]$ and takes $[1.5,3]$ linearly onto $[s+1.5,3]$.

In particular, $g_1|[-0.5,1.5]$ has the form $t \mapsto t+1$. By choosing a small $\varepsilon > 0$ and applying the approximate isotopy lifting property, we get an isotopy:

$$h : \bar{Y}\times[0,1] \to \bar{Y}\times[0,1],$$

with compact support, from the identity such that

$$h_1(Y_{(-\infty,0)}) \supset Y_{(-\infty,1]}.$$

Define $\hat{Y} = \zeta h_1(Y_{(-\infty,0]}) \setminus Y_{(-\infty,0)}/\sim$, where a point $x \in Y_0$ is identified with $\zeta h_1(x) \in \zeta h_1(Y_0)$. There is a strong deformation retraction:

$$\hat{f} : \hat{Y} \to X\times S^1$$

Then $\hat{f} \in Wh(X\times S^1, p)_c$. The reason is that in the infinite cyclic cover induced by $\hat{f}$ the fundamental domain is 'larger' and that makes $\hat{f}$ more controlled over $S^1$. The fact that it is controlled follows from the Sucking Principle.

### 3.1.2. Maps to Nil-groups

We describe the Nil-groups as summands of the Whitehead group:

(1) $j_\pm : \widetilde{\mathrm{Nil}}(X) \to Wh(X\times S^1)$. The two injections are given by taking mapping tori:
  (a) $j_+(Y,g) = [T'(g), X\times S^1]$, where $T'(g) = Y\times[0,1]/(g(y),1) \sim (y,0)$.
  (b) $j_-(Y,f) = [T(g), X\times S^1]$, where $T(g) = Y\times[0,1]/(g(y),0) \sim (y,1)$.
(2) The projections are constructed by looking at the translation map on $\bar{Y}$ and $Z$. More precisely,

$$p_\pm : Wh(X\times S^1) \to \widetilde{\mathrm{Nil}}(X), \;\; p_\pm(f) = [L_\pm, f_\pm].$$

Since $K$ retracts to $X$, $L$ also retracts to $X$ and the pair $(T(z), X\times S^1) = 0 \in Wh(X\times S^1)$. In the mapping torus $T(z)$ (Figure 1):

(1) There is an obvious embedding of $Y$ if we follow the restriction of the mapping torus to $Y_0$.
(2) A variation of the previous argument is used to construct an embedding of $\hat{Y}$, disjoint from $Y$ into the part of $T(z)$ in $L_+$.
(3) The restriction, $N_-$, of the mapping torus to $L_-$ represents the image of $f$ under the composition:

$$Wh(X\times S^1) \xrightarrow{p_-} \widetilde{\mathrm{Nil}}(X) \xrightarrow{j_-} Wh(X\times S^1).$$

More precise, there is a strong deformation retraction $f_- : N_- \to X\times S^1$ such that $j_- \circ p_-([f]) = [f_-]$.

(4) Let $N_+$ be the region in $T(z)$ between $Y$ and $\bar{Y}$, considered "backwards" i.e., with the reverse orientation of that given by the mapping torus. Then $N_+$ represents the image of the element $[f]$ under the map.

$$Wh(X \times S^1) \xrightarrow{p_+} \widetilde{\mathrm{Nil}}(X) \xrightarrow{j_+} Wh(X \times S^1).$$

(5) The rest of the region of $T(z)$, $W$ has the property that it is homeomorphic to the mapping torus of a retraction and thus, there is a strong deformation retraction $r : W \to X \times S^1$ which has zero torsion in $Wh(X \times S^1)$.

Combining the above observations and the sum theorem for the Whitehead torsion we get that, in $Wh(X \times S^1)$:

$$0 = (T(z), X \times S^1) = [f_-] - [f] + [f_+] - [\bar{f}] + [r]$$

and thus

$$[f] = j_- \circ p_-([f]) + j_+ \circ p_+([f]) - [\bar{f}].$$

The last equation provides the splitting of $[f]$ into the various components. Figure 1 gives a presentation of the spaces used in the proof. It should be considered as the geometric analogue of Ranicki's pentagon ([77], p. 95, p. 99).

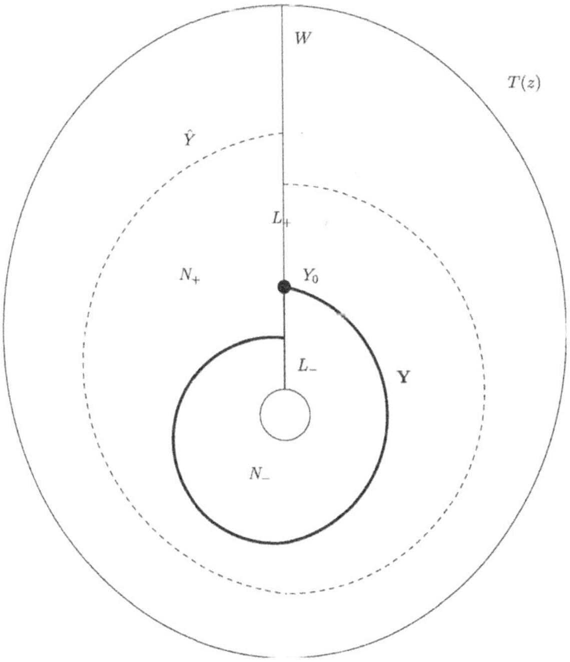


Figure 1 The mapping Torus of the Translation

## 3.2. Waldhausen's Splitting Theorem Revisited

There is a controlled interpretation of Waldhausen's Theorem. Let $H$ be a subgroup of $G_0 \cap G_1$ and $\Gamma = G_0 *_H G_1$. Set

$$M = M(\iota_0) \cup_{BH} M(\iota_1)$$

where $\iota_i : BH \to BG_i$ be the inclusion map of classifying spaces. Van Kampen's Theorem implies that $\pi_1(M) = \Gamma$.

**Theorem 3.17** (Waldhausen's Theorem). *There is a splitting*

$$Wh(M) = Wh(M,p)_c \oplus \widetilde{Nil}(\mathbb{Z}H; \mathbb{Z}[G_0 \backslash H], \mathbb{Z}[G_1 \backslash H]),$$

*where $p : M \to [-1,1]$ is the natural projection map to the second coordinate of the mapping cylinders. Furthermore, there is an exact sequence*

$$\begin{aligned} Wh(H) \to Wh(G_0) \oplus Wh(G_1) \to Wh(M,p) \to \tilde{K}_0(\mathbb{Z}H) \\ \to \tilde{K}_0(\mathbb{Z}G_0) \oplus \tilde{K}_0(\mathbb{Z}G_1) \to \cdots \end{aligned}$$

For this interpretation of Waldhausen's Theorem, see [70].

**Question**. Is there a geometric map that describes the splitting of the forget control map? More precisely, is there a geometric relaxation map:

$$Wh(M) \to Wh(M,p)_c$$

as in the case over the circle?

*Remark* 3.18. There is the analogue of the "infinite cyclic" cover of $M$ is this case. The natural action on the infinite cyclic cover is of $D_\infty$, the infinite dihedral group. The answer to the question seems to involve a version of $\mathbb{Z}/2\mathbb{Z}$-equivariant relaxation construction

## 3.3. Geometric Applications of Nil-groups

Nil-groups appear as obstructions to splitting theorems. Farrell's fibering theorem is such a result.

**Theorem 3.19** (Farrell ([38], Siebenmann [80])). *Let $M^n$, $n \geq 6$, be a closed connected manifold and $p : M \to S^1$ a map such that:*

(1) *$p_* : \pi_1(M) \to \pi_1(S^1)$ is an epimorphism with kernel $G$. Thus $\pi_1(M) = G \rtimes_\alpha \mathbb{Z}$.*
(2) *The infinite cyclic cover of $M$ induced by $p$ is finitely dominated.*

*Then*

(1) *There is an obstruction $n(f) \in \widetilde{Nil}(\mathbb{Z}G, \alpha)$ which vanishes iff $f$ is homotopic to an approximate fibration.*
(2) *If (1) is satisfied, then there is an obstruction $\tau(f) \in Wh(M,p)_c$ which vanishes iff $f$ is homotopic to a fiber bundle.*

*Remark* 3.20.

(1) Farrell's obstructions are encoded as a total obstruction geometrically in [80] and algebraically in [77]. Essentially, the obstruction is described as follows: Let $\bar{M}$ be the infinite cyclic cover of $M$ and $\zeta$ a generating cover translation. Then, the total obstructions is defined as:

$$\Phi(f) = \tau(T(\zeta) \to M) \in Wh(G \rtimes_\alpha \mathbb{Z}).$$

Because $T(\zeta)$ is not a finite complex, for the exact meaning of the torsion see [77].

(2) Originally, Farrell's theorem was not stated using controlled topology. To describe the classical setting, using transversality, first homotope $p$ so that the fundamental domain of the infinite cyclic cover is a cobordism $W$. The first obstruction was

$$\mathbf{n}(f) = n(f) + k(f) \in \widetilde{\mathrm{Nil}}(\mathbb{Z}G, \alpha) \oplus \tilde{K}_0(\mathbb{Z}G)$$

that vanishes iff $W$ is an $h$-cobordism. Once $\mathbf{n}(f)$ vanishes, the second obstruction

$$\tau(f) = \tau(W) \in Wh(G)_{\alpha_*} = Wh(G)/\{x - \alpha_*(x) : x \in Wh(G)\}$$

is the torsion of the $h$-cobordism $W$. Then $\tau(f)$ vanishes iff $p$ is homotopic to a fiber bundle.

(3) Notice that $Wh(M,p)_c$ fits into an exact sequence ([70]):

$$Wh(G) \xrightarrow{1-\alpha_*} Wh(G) \to Wh(M,p)_c \to \tilde{K}_0(\mathbb{Z}G) \xrightarrow{1-\alpha_*} \tilde{K}_0(\mathbb{Z}G).$$

(4) The controlled version of Part (1) of the Theorem is stated implicitly in different papers ([22], [62], [63]).

A more general splitting theorem is given in [40]. Let $M^n$, $n \geq 6$, be a closed connected manifold with $\pi_1(M) = G \rtimes_\alpha \mathbb{Z}$ ([42] for the algebraic version). Let $f : M \to M'$ be a homotopy equivalence. Also, let $N'$ be a codimension 1 submanifold of $M'$ such that $\pi_1(N') = G$. The map $f$ is called *splittable* along $N'$ if $f$ is homotopic to a map $g$ which is a homotopy equivalence of pairs:

$$g : (M, N) \to (M', N'),$$

where $N$ is a codimansion 1 submanifold of $M$.

**Theorem 3.21** ([40]). *Under the above notation, $f$ is splittable along $N'$ iff*

$$\mathbf{n}(f) = p(\tau(f)) \in \widetilde{Nil}(\mathbb{Z}G, \alpha) \oplus \tilde{K}_0(\mathbb{Z}G)$$

*vanishes.*

For the Waldhausen's splitting, let $M^n$, $n \geq 6$, be a closed connected manifold and $N \subset M$ a codimension 1 connected submanifold with trivial normal bundle such that $M \setminus N$ has two components. Thus $\pi_1(M) = G_0 *_H G_1$ where $\pi_1(N) = H$.

Assume that $H$ is *square root closed* in $G_i$ (i.e., if $g^2 \in H$ then $g \in H$). Let $f: W^n \to M^n$ be a homotopy equivalence. Then $f$ is called *splittable* along $N$ if $f$ is homotopic to $g$ so that there is a codimension 1 submanifold $V = g^{-1}(N)$ so that the restrictions $g$ is a homotopy equivalence of pairs

$$g : (W, V) \to (M, N).$$

**Theorem 3.22** ([17]). *In the above setting $f$ is splittable along $N$ if the component*

$$\begin{aligned}\mathbf{n}(f) = p(\tau(f)) \in \widetilde{Nil}(\mathbb{Z}H, \mathbb{Z}[G_0 \setminus H], \mathbb{Z}[G_1 \setminus H]) \oplus ker(\tilde{K}_0(\mathbb{Z}H) \\ \to \tilde{K}_0(\mathbb{Z}G_1) \oplus \tilde{K}_0(\mathbb{Z}G_2))\end{aligned}$$

*vanishes.*

*Remark* 3.23.

(1) The condition that $H$ is square root closed in both groups is needed for the vanishing of the UNil-groups, the surgery theory Nil-groups ([16]).
(2) In [17], a splitting theorem is given when $M \setminus N$ is a single component i.e., when $\pi_1(M)$ is an HNN-extension of $H$.

## 3.4. Remarks on Nil-groups

From the above discussion, it follows that the various Nil-groups are the obstructions for $K$-theory to be a homology theory. They measure the failure of $K$-theory to satisfy excision. The component of $K$-theory that satisfies homological properties, like excision, is given by the controlled component. In other words, algebraically, the controlled $K$-groups can be defined as the part of $K$-theory that satisfies excision. We will make two more remarks emphasizing these properties of the Nil-groups.

### 3.4.1. Categorical Double Mapping Cylinders

First, we specialize to amalgamated free products. $K$-theory can be defined for symmetric monoidal categories in general. Then $K_*(R) = K_*(\mathcal{F}_R^{Iso})$ where $\mathcal{F}_R^{Iso}$ is the category of finitely generated free (left) $R$-modules with morphisms $R$-isomorphisms. Let

$$\begin{array}{ccc} R & \longrightarrow & A_0 \\ \downarrow & & \downarrow \\ A_1 & \longrightarrow & \Lambda \end{array}$$

be a push out diagram of rings. Let $\mathcal{P}$ be the double mapping cylinder (the homotopy push-out) in the category of symmetric monoidal categories

$$\begin{array}{ccc} \mathcal{F}_R^{Iso} & \longrightarrow & \mathcal{F}_{A_0}^{Iso} \\ \downarrow & & \downarrow \\ \mathcal{F}_{A_1}^{Iso} & \longrightarrow & \mathcal{P} \end{array}$$

defined in [88], Construction 5.1. The above diagram induces a homotopy cartesian diagram of spectra ([88]):

$$\begin{array}{ccc} \mathbb{K}.(R) & \longrightarrow & \mathbb{K}.(A_0) \\ \downarrow & & \downarrow \\ \mathbb{K}.(A_1) & \longrightarrow & \mathbb{K}.(\mathcal{P}) \end{array}$$

Thus the $K$-theory of $\mathcal{P}$ is the 'controlled $K$-theory with control over the interval' (Section 3.2). Also, the universal properties of the double mapping cylinder imply that there is a natural functor

$$F : \mathcal{P} \to \mathcal{F}_{\Lambda}^{Iso}.$$

Waldhausen's Splitting Theorem implies that:

$$\widetilde{\mathrm{Nil}}_{i-1}(R; B_0, B_1) \cong \mathrm{Coker}(F_* : K_i(\mathcal{P}) \to K_i(\Lambda)).$$

where $A_i = R \oplus B_i$ as $R$-bimodules, for $i = 0, 1$.

### 3.4.2. Continuous Control $K$-theory

There is a connection between Nil-groups and continuously controlled $K$-groups. The definitions and the main calculations for the controlled $K$-groups are given in [1].

There are two cases to be consider. First we assume that $p : X \times S^1 \to S^1$ is the projection map and $X$ is a finite CW-complex (or a compact ANR). Let $M(p)$ denote the mapping cylinder of $p$. Then there is an exact sequence ([2]):

$$Wh(X \times S^1, p)_c \to Wh(X \times S^1) \to Wh(M(p), S^1)_{cc} \to \tilde{K}_0(X \times S^1, p)_c \to \tilde{K}_0(X \times S^1)$$

where $Wh(M(p), S^1)_{cc}$ is the continuously controlled Whitehead group. Thus, if we combine this with this exact sequence with the Fundamental Theorem of Algebraic $K$ theory, we get ([70]):

$$Wh(M(p), S^1)_{cc} \cong \widetilde{Nil}(X) \oplus \widetilde{Nil}(X).$$

Let $\Gamma = G_1 *_{G_0} G_2$ be an amalgamated free product. Consider the induced maps $\phi_i : BG_0 \to BG_i$, $i = 1, 2$. Let $X$ be the double mapping cylinder:

$$X = M(\phi_1) \cup_{BG_0} M(\phi_1).$$

Then $\pi(X) = \Gamma$ and $X$ comes with a natural map $\rho : X \to I$ to the unit interval. Then, as before we have an exact sequence:

$$Wh(X, \rho)_c \to Wh(X) \to Wh(M(\rho), I)_{cc} \to \tilde{K}_0(X, \rho)_c \to \tilde{K}_0(X).$$

Using Waldhausen's splitting theorem we get ([70]):

$$Wh(M(\rho), I)_{cc} \cong \widetilde{Nil}(\mathbb{Z}G_0, \mathbb{Z}[G_1 \backslash G_0], \mathbb{Z}[G_1 \backslash G_0]).$$

*Remark* 3.24. In [2], the continuously controlled groups are defined for holink triples. Our interpretation is equivalent to that one in the above cases and it is taken from [1].

# 4. Controlled Topology and the Isomorphism Conjecture

In this section, we describe the controlled aspects of the Farrell–Jones Isomorphism Conjecture ([43]). Most of the material is contained in [34]. Let $\Gamma$ be a discrete group and $\mathcal{C}_\Gamma$ a class of subgroups of $\Gamma$ (i.e. a collection of subgroups of $\Gamma$ closed under taking conjugates and subgroups). The *classifying space* of the class $\mathcal{C}_\Gamma$, $\mathcal{E}\mathcal{C}_\Gamma$ is the $\Gamma$-complex whose isotropy groups are in $\mathcal{C}_\Gamma$ and its non-empty fixed point sets are contractible. A model for this space, reminiscent of the bar construction ([43]), is given as follows: It is the realization of a semi-simplicial complex with $n$-simplex given by a sequence

$$\sigma = \gamma_0\Gamma_0(\gamma_1\Gamma_1, \gamma_2\Gamma_2, \dots, \gamma_n\Gamma_n)$$

with $\Gamma_i \in \mathcal{C}$ such that $\gamma_i^{-1}\Gamma_{i-1}\gamma_i \subset \Gamma_i$ for $i = 1, \dots, n$. The face operator is given by

$$\partial_i\sigma = \begin{cases} \gamma_0\gamma_1\Gamma_1(\gamma_2\Gamma_1, \dots, \gamma_n\Gamma_n), & i = 0, \\ \gamma_0\Gamma_0(\gamma_1\Gamma_1, \dots, \gamma_{i-1}\Gamma_{i-1}, \gamma_i\gamma_{i+1}\Gamma_{i+1}, \gamma_{i+1}\Gamma_{i+1}, \dots, \gamma_n\Gamma_n), & 0 < i < n, \\ \gamma_0\Gamma_0(\gamma_1\Gamma_1, \dots, \gamma_{n-1}\Gamma_{n-1}), & i = n. \end{cases}$$

The group $\Gamma$ acts on $E\mathcal{C}_\Gamma$ by

$$\gamma\sigma = (\gamma\gamma_0)\Gamma_0(\gamma_1\Gamma_1, \gamma_2\Gamma_2, \dots, \gamma_n\Gamma_n), \text{ for } \gamma \in \Gamma.$$

We write $\mathcal{B}\mathcal{C}_\Gamma$ for the orbit space $\mathcal{E}\mathcal{C}_\Gamma/\Gamma$. To ensure that $\mathcal{B}\mathcal{C}_\Gamma$ is a simplicial complex, we subdivide $\mathcal{E}\mathcal{C}_\Gamma$ twice, i.e. $\mathcal{B}\mathcal{C}_\Gamma = (\mathcal{E}\mathcal{C}_\Gamma)''/\Gamma$.

The construction of the classifying space is functorial with respect to group homomorphisms. Let $\mathcal{C}_\Gamma$ be a class of subgroups of $\Gamma$. For if $\rho : \Gamma \to G$ be a group homomorphism then $\rho$ induces a $\rho$-equivariant map

$$\begin{aligned} \bar{\rho} : \mathcal{E}\mathcal{C}_\Gamma \to \mathcal{E}\mathcal{C}_G,\ & \gamma_0\Gamma_0(\gamma_1\Gamma_1, \gamma_2\Gamma_2, \dots, \gamma_n\Gamma_n) \\ & \mapsto \rho(\gamma_0)\rho(\Gamma_0)(\rho(\gamma_1)\rho(\Gamma_1), \rho(\gamma_2)\rho(\Gamma_2), \dots, \rho(\gamma_n)\rho(\Gamma_n)) \end{aligned}$$

where $\mathcal{C}_G$ is a class of subgroups of $G$ that contains the images, under $\rho$, of the elements of $\mathcal{C}_\Gamma$. The map $\bar{\rho}$ induces a map $\rho'$ to the quotient spaces.

For each simplicial complex $K$, we write $\mathrm{cat}(K)$ for the category of simplices of $K$, viewed as a partially ordered set. Thus, objects are the simplices of $K$ and there is a single morphism from $\sigma$ to $\tau$ whenever $\sigma \leq \tau$.

**Definition 4.1** ([3]). Let $p : E \to B$ be a map with $B = |K|$, the geometric realization of a simplicial complex. The map is said to have a *homotopy colimit structure* if there is a functor

$$F : \mathrm{cat}(K)^{op} \to \mathbf{Top}$$

such that:

- $E = \mathrm{hocolim}_{\mathrm{cat}(K)^{op}}(F)$.

- $p = \text{hocolim}_{\text{cat}(K)^{op}}(\nu)$, where $\nu$ is the natural transformation from $F$ to the constant point functor.

*Remark* 4.2. For a simplicial complex $K$ and a functor $F : \text{cat}(K)^{op} \to \mathbf{Top}$, the homotopy colimit is defined as:

$$\text{hocolim}_{\text{cat}(K)^{op}}(F) = \coprod_{\sigma \in \text{cat}(K)^{op}} F(\sigma) \times |\sigma| / \sim$$

where for simplices $\tau \geq \sigma$, $x \in F(\tau)$, $t \in |\sigma|$,

$$(x, t) \sim (F(\tau \geq \sigma)(x), t).$$

Notice that $(x, t) \in F(\tau) \times |\sigma| \subset F(\tau) \times |\tau|$ and $F(\tau \geq \sigma) : F(\tau) \to F(\sigma)$.

Notation. Let $p : E \to B$ be a map, $B = |K|$ and $\text{cat}(K)^{op}$ the category of the simplicial complex $K$. In our setting, there is the *barycentre functor*

$$\text{bar}(p) : \text{cat}(K)^{op} \to \mathbf{Top},\ \sigma \mapsto p^{-1}(\hat{\sigma})$$

where $\hat{\sigma}$ is the barycentre of $\sigma$. More precisely, for maps that we will use, for each $\tau \geq \sigma$, there is a continuous map $p^{-1}(\hat{\tau}) \to p^{-1}(\hat{\sigma})$.

*Remark* 4.3. We give basic examples of maps that admit a homotopy colimit structure. The proofs follow from direct calculations (also [87]).

(1) Let $\Gamma$ be a discrete group, $E\Gamma$ a free contractible $\Gamma$-complex and $\mathcal{E}\mathcal{C}_\Gamma$ the classifying complex for the family of subgroups of $\Gamma$. Let

$$p_\Gamma : E\Gamma \times_\Gamma \mathcal{E}\mathcal{C}_\Gamma \to \mathcal{E}\mathcal{C}_\Gamma / \Gamma = \mathcal{B}\mathcal{C}_\Gamma$$

be the projection map to the second coordinate. Then $p_\Gamma$ has a homotopy colimit structure with respect to the barycentor functor $\text{bar}(p_\Gamma)$. Notice that, in this case, $p_\Gamma^{-1}(\hat{\sigma})$ is a space of type $B\Gamma_\sigma$, where $\Gamma_\sigma$ is the isotropy group of $\sigma$, an element in the class $\mathcal{C}_\Gamma$.

(2) Let $\rho : \Gamma \to G$ be a group epimorphism. Then the map

$$q : E\Gamma \times_\Gamma \mathcal{E}\mathcal{C}_\Gamma \xrightarrow{p_\Gamma} \mathcal{B}\mathcal{C}_\Gamma \xrightarrow{\rho'} \mathcal{B}\mathcal{C}_G$$

has a homotopy colimit structure with respect to the functor $\text{bar}(q)$, where $\mathcal{C}_G = \rho(\mathcal{C}_\Gamma)$, the class of subgroups of $G$ consisting of the images of elements of $\mathcal{C}_\Gamma$.

Let $F : \mathcal{C} \to \mathcal{D}$ and $X : \mathcal{C} \to \mathbf{Top}$ be two functors. Then *Segal's Pushdown Construction* (see for example [58]) defines a functor $F_*X : \mathcal{D} \to \mathbf{Top}$ such that

$$\text{hocolim}_\mathcal{C} X \simeq \text{hocolim}_\mathcal{D} F_* X.$$

We will explicitly describe the construction to the case of Part(2) in Remark 4.3. In this case, we start with a map:

$$q : E\Gamma \times_\Gamma \mathcal{E}\mathcal{C}_\Gamma \xrightarrow{p_\Gamma} \mathcal{B}\mathcal{C}_\Gamma \xrightarrow{\rho'} \mathcal{B}\mathcal{C}_G.$$

The map $\rho'$ induces a functor

$$P : \mathrm{cat}(\mathcal{BC}_\Gamma)^{op} \to \mathrm{cat}(\mathcal{BC}_G)^{op}.$$

We will describe the functor

$$P_* \mathrm{bar}(p) : \mathrm{cat}(\mathcal{BC}_G)^{op} \to \mathbf{Top}.$$

For each simplex $\sigma$ of $\mathcal{BC}_G$, let $P{\downarrow}\sigma$ be the over category. In this case, the objects of $P{\downarrow}\sigma$ are simplices $\tau$ of $\mathcal{BC}_\Gamma$ such that $\rho'(\tau)$ contains $\sigma$ as a face. Set

$$p_\sigma = p| : p^{-1}(|P{\downarrow}\sigma|) \to |P{\downarrow}\sigma|.$$

Then $P_*\mathrm{bar}(p)(\sigma) = \mathrm{hocolim}_{P\downarrow\sigma}\mathrm{bar}(p_\sigma)$. Summarizing:

**Proposition 4.4.** *There is a homotopy equivalence:*

$$hocolim_{cat(\mathcal{BC}_\Gamma)^{op}}(bar(p_\Gamma)) \simeq hocolim_{cat(\mathcal{BC}_G)^{op}}(P_* bar(p_\Gamma)).$$

*Remark* 4.5. Segal's pushdown construction it is the categorical analogue of the 'change of control' map. Essentially, for a map $p : X \to Y$ it shows how control over $X$ can be computed by computing the control over $Y$ with 'coefficients' the contribution of the fibers of $p$.

The class of subgroups of $\Gamma$ of interest in the Isomorphism Conjecture is the class of *virtually cyclic subgroups*, denoted $\mathcal{G}_\Gamma$. They split into two categories:

- Finite subgroups of $\Gamma$.
- Virtually infinite cyclic subgroups of $\Gamma$ i.e., subgroups which contain an infinite cyclic subgroup of finite index.

The subgroups of the second type are two-ended subgroups of $\Gamma$ ([36]) and they split into two types:

- Groups $H$ that admit an epimorphism to $\mathbb{Z}$ with finite kernel i.e.

  $$H \cong K \rtimes \mathbb{Z},$$

  with $K$ finite.
- Groups $H$ that admit an epimorphism to the infinite dihedral subgroup $D_\infty$ with finite kernel, i.e.

  $$H \cong A *_B C, \quad [B : 1] < \infty,\ [A : B] = [C : B] = 2.$$

Let $\mathbb{S}pectra$ denote the category of spectra with morphisms *strict maps* i.e., maps between spectra that commute with the bonding maps. Let

$$\mathbb{S} : \mathbf{Top} \to \mathbb{S}pectra$$

is a *homotopy invariant functor* i.e., a functor that maps homotopy equivalent spaces to homotopy equivalent spectra. Let $\Gamma$ be a discrete group As before, let

$\mathcal{G}_\Gamma$ be the class of virtually cyclic (finite or infinite) subgroups of $\Gamma$. Let $\mathcal{E}\mathcal{G}_\Gamma$ be the classifying $\Gamma$-complex for the class $\mathcal{G}_\Gamma$ and

$$p_\Gamma : E\Gamma \times_\Gamma \mathcal{E}\mathcal{G}_\Gamma \to \mathcal{E}\mathcal{G}_\Gamma / \Gamma = \mathcal{B}\mathcal{G}_\Gamma$$

be the projection map. Let $\mathrm{cat}(\mathcal{B}\mathcal{G}_\Gamma)^{op}$ be the category corresponding to the partially ordered set of simplices of $\mathcal{B}\mathcal{G}_\Gamma$. Let $r : Y \to B\Gamma$ be a bundle. Form the pull-back:

$$\begin{array}{ccc}
\bar{Y} & \longrightarrow & Y \\
{\scriptstyle \rho}\big\downarrow & & \big\downarrow{\scriptstyle r} \\
E\Gamma \times_\Gamma \mathcal{E}\mathcal{G}_\Gamma & \longrightarrow & B\Gamma \\
{\scriptstyle p_\Gamma}\big\downarrow & & \big\downarrow \\
\mathcal{B}\mathcal{G}_\Gamma & \longrightarrow & *
\end{array}$$

Notice that

$$\mathrm{hocolim}_{\mathrm{cat}(\mathcal{B}\mathcal{G}_\Gamma)^{op}} \mathrm{bar}(p_\Gamma \circ \rho) \cong \bar{Y} \simeq Y$$

Let $*$ denote the category with a single object and a single morphism. Let

$$F_\Gamma : * \to \mathbf{Top}, \quad F_\Gamma(*) = Y$$

<u>*The $\mathbb{S}$-Bundle Isomorphism Conjecture.*</u> With the notation above, the functor from $\mathrm{cat}(\mathcal{B}\mathcal{G}_\Gamma)^{op}$ to $*$, induces a homotopy equivalence of spectra:

$$\mathrm{hocolim}_{\mathrm{cat}(\mathcal{B}\mathcal{G}_\Gamma)^{op}} \mathbb{S} \circ \mathrm{bar}(p_\Gamma \circ \rho) \to \mathrm{hocolim}_{\mathrm{cat}(*)} \mathbb{S} \circ \mathrm{bar}(F_\Gamma) \simeq \mathbb{S}(Y).$$

*Remark* 4.6.

(1) In the classical setting the homotopy colimit was written as homology:

$$\mathrm{hocolim}_{\mathrm{cat}(\mathcal{B}\mathcal{G}_\Gamma)^{op}} \mathbb{S} \circ \mathrm{bar}(p_\Gamma \circ \rho) = \mathbb{H}.(\mathcal{B}\mathcal{G}_\Gamma, \mathbb{S}(p_\Gamma \circ \rho)).$$

The right hand side is Quinn homology. We will use both notations.

(2) Homotopy colimits were used in the formulation of the assembly map in [35] and [56]. Actually in [56], all the known constructions of the assembly are compared.

(3) The original fiber isomorphism conjecture was stated under the assumption that $r$ is a fibration ([43]). In that more general setting, Quinn homology is used in the formulation.

(4) If a torsion free group satisfies the $\mathbb{K}$-IC then it satisfies the Vanishing Conjectures for the Whitehead group. or a complete account of the connections between the Isomorphism Conjectures, Conjectures 3 and 4 and other conjectures in algebra is given in [11].

## 4.1. Spaces over the Circle

Let $\Gamma = G\rtimes_{\alpha}\mathbb{Z}$. Here $\alpha$ is the automorphism of $G$ induced by the action of the generator of $\mathbb{Z}$. The automorphism $\alpha$ is well-defined up to inner automorphisms. By choosing a suitable right $G$-space for $EG$, there is an $\alpha$-equivariant homeomorphism

$$\psi : EG \to EG$$

i.e., $\psi(xg) = \psi(x)\alpha(g)$. By taking quotients, we see that there is a homeomorphism

$$\phi : BG \to BG$$

that induces the map $\alpha$ in the fundamental group, again up to inner automorphisms. Choose as a model for $B\Gamma$ the mapping torus of $\phi$:

$$B\Gamma = BG\times[0,1]/\sim, \ (\phi(x), 0) \sim (x, 1).$$

Then a model for $E\Gamma$ is the infinite mapping telescope of $\psi$:

$$E\Gamma = EG\times[0,1]\times\mathbb{Z}/\sim, \quad (x, 1, n) \sim (\psi(x), 0, n+1).$$

The right action of $\Gamma$ on $E\Gamma$ is given by:

$$(x, t, n)(g, m) = (x\alpha^n(g), t, n+m).$$

The map

$$\Phi : BG\times[0,1] \to B\Gamma, \ \Phi(x,t) = [x,t]$$

has the property that:

$$\Phi(x,0) = [x,0], \quad \Phi(x,1) = [x,1] = [\phi(x),0]$$

i.e., it defines a homotopy between the identity map and the map $\phi$ on $BG$ inside $B\Gamma$. Also, the natural projection map to the second coordinate $\rho : B\Gamma \to S^1$ is a bundle. Consider the commutative diagram

$$\begin{array}{ccc} E\Gamma\times_{\Gamma}\mathcal{E}\mathcal{G}_{\Gamma} & \longrightarrow & E\Gamma\times_{\Gamma}\mathcal{E}\mathcal{G}_{\Gamma} \\ {\scriptstyle q_\Gamma}\downarrow & & \downarrow{\scriptstyle p_\Gamma} \\ S^1\times\mathcal{B}\mathcal{G}_{\Gamma} & \longrightarrow & \mathcal{B}\mathcal{G}_{\Gamma} \end{array}$$

where $q_\Gamma$ is the composite $E\Gamma \to B\Gamma \xrightarrow{\rho} S^1$ is the first coordinate and the projection $\mathcal{E}\mathcal{G}_\Gamma \to \mathcal{B}\mathcal{G}_\Gamma$ in the second. In the applications the spectrum $\mathbb{S}$ has an infinite loop space structure. Let $\mathcal{CF}$ be the homotopy cofiber:

$$\mathbb{S}(BG) \xrightarrow{1-\phi_*} \mathbb{S}(BG) \to \mathcal{CF}$$

The projection map:

$$BG\times[0,1] \to T(\phi) = B\Gamma$$

induces a homotopy between two maps from $BG$ to $B\Gamma$. The first map is

$$BG \to B\Gamma, \quad x \mapsto (x, 0).$$

The other map is

$$BG \to B\Gamma, \quad x \mapsto (x, 1) = (\phi(x), 0).$$

Thus the maps $\mathrm{id}_{BG}$ and $\phi$ are homotopic in $B\Gamma$. That mean the map $1 - \phi_*$ is null homotopic. Therefore, there is an induced map:

$$f : \mathcal{CF} \to \mathbb{S}(B\Gamma).$$

This is the map that 'forgets the control' over $S^1$. We will give a homological description of $\mathcal{CF}$.

We will study the homotopy colimit structure of the quotient map $q_\Gamma$. Equip $S^1$ with the structure of a simplicial complex with three 0-simplices $v_i$, $i = 0, 1, 3$ and three 1-simplices $e_i = \{v_i, v_{i+1}\}$, where $i$ is taken mod 3. The projection map induces a functor:

$$\rho : \mathrm{cat}(S^1 \times B\mathcal{G}_\Gamma)^{op} \to \mathrm{cat}(S^1)^{op}.$$

Using Segal's Pushdown Construction we get a homotopy equivalence:

$$\mathrm{hocolim}_{\mathrm{cat}(S^1 \times B\mathcal{G}_\Gamma)^{op}}(\mathbb{S}\mathrm{obar}(q_\Gamma)) \simeq \mathrm{hocolim}_{\mathrm{cat}(S^1)^{op}}(P_*\rho).$$

We will describe explicitly the functor $P_*\rho$ on $\mathrm{cat}(S^1)^{op}$:
For each simplex $t$ of $S^1$, let

$$\mathbb{S}\mathrm{obar}(q_\Gamma)(t, -) : \mathrm{cat}(B\mathcal{G}_\Gamma)^{op} \to \mathit{Spectra}, \ [\sigma]_\Gamma \mapsto \mathbb{S}\mathrm{obar}(q_\Gamma)(t, [\sigma]_\Gamma)$$

and thus

$$P_*\rho(t) = \mathrm{hocolim}_{\mathrm{cat}(B\mathcal{G}_\Gamma)^{op}}(\mathbb{S}\mathrm{obar}(q_\Gamma)(t, -)).$$

Now if we fix $t$, the functor $\mathrm{bar}(q_\Gamma)$ associates to $[\sigma]_\Gamma$, the space

$$q_\Gamma^{-1}(\hat{t} \times \hat{\sigma}) = (EG \times \{\hat{t}\} \times \mathbb{Z}) \times_\Gamma \Gamma\hat{\sigma} \cong (EG \times \{\hat{t}\} \times \mathbb{Z}) \times_{\Gamma_\sigma} \hat{\sigma}.$$

The inclusion map $G \to \Gamma$ induces a commutative diagram:

$$\begin{array}{ccc} EG \times_G \mathcal{E}\mathcal{G}_\Gamma & \xrightarrow{\bar{u}} & (EG \times \{\hat{t}\} \times \mathbb{Z}) \times_\Gamma \mathcal{E}\mathcal{G}_\Gamma \\ {\scriptstyle p_G}\downarrow & & \downarrow{\scriptstyle q_\Gamma} \\ \mathcal{E}\mathcal{G}_\Gamma / G = \mathcal{B}\mathcal{G}_G & \xrightarrow{u} & \hat{t} \times \mathcal{B}\mathcal{G}_\Gamma = \hat{t} \times \mathcal{E}\mathcal{G}_\Gamma / \Gamma \end{array}$$

where $u$ is just the quotient map. Notice that $\mathcal{E}\mathcal{G}_\Gamma$ is also a model for the space of type $\mathcal{E}\mathcal{G}_G$. Then $u$ induces a functor:

$$U : \mathrm{cat}(\mathcal{B}\mathcal{G}_G)^{op} \to \mathrm{cat}(\mathcal{B}\mathcal{G}_\Gamma)^{op}.$$

Start with the restriction:

$$p_G| : (p_G)^{-1}(|U{\downarrow}[\sigma]_\Gamma|) \to |U{\downarrow}[\sigma]_\Gamma|.$$

Define a functor

$$h : \mathrm{cat}(\mathcal{BG}_\Gamma)^{op} \to \mathbb{S}pectra, \quad h([\sigma]_\Gamma) = \mathrm{hocolim}_{\mathrm{cat}(|U\downarrow[\sigma]_\Gamma|)^{op}}(\mathbb{S}\mathrm{obar}(p_G|)).$$

**Claim.** $h([\sigma]_\Gamma \simeq \mathbb{S}\mathrm{obar}(q_\Gamma)(t, [\sigma]_\Gamma)$ and the homotopy equivalence is natural in $[\sigma]_\Gamma$.

**Proof.** For the over category, we have:

$$U\downarrow[\sigma]_\Gamma = \{[\gamma\tau]_G : \tau \geq \sigma, \gamma \in \Gamma\} = \{[(1,m)\tau]_G : \tau \geq \sigma, m \in \mathbb{Z}\},$$

where $[-]_G$ denotes the $G$-orbit of the simplex. To describe a splitting of $p_G^{-1}(|U\downarrow[\sigma]_\Gamma|)$ we need to start with an equivalence relation on $\mathbb{Z}$. Every simplex $\sigma$ of $\mathcal{EG}_\Gamma$ defines an equivalence relation on $\mathbb{Z}$,

$$m_1 \sim_\sigma m_2 \iff \Gamma_\sigma \cap (1, m_1)G(1, -m_2) \neq \emptyset, \quad \text{for each } m_1, m_2 \in \mathbb{Z}.$$

Also, for each $t \in S^1$, $m \in \mathbb{Z}$, we define a map

$$i_m : (EG \times \{t\} \times \{m\}) \times_{\Gamma_\sigma \cap G} \hat{\sigma} \to (EG \times \{t\} \times \mathbb{Z}) \times_\Gamma \Gamma\hat{\sigma}, \; [(x,t,m),\hat{\sigma}] \mapsto [(x,t,m),\hat{\sigma}]$$

(here $EG \times \{t\} \times \mathbb{Z} \subset E\Gamma$). Notice that the image of $i_m$ depends on the choice of $t \in S^1$. Direct calculations show that:

$$\begin{aligned}
|U\downarrow[\sigma]_\Gamma| &= \coprod_{[m_k]} |\mathrm{Id}_{\mathrm{cat}(\mathcal{BG}_G)^{op}}\downarrow[(1,m_k)\sigma]_G| \\
p_G^{-1}(|U\downarrow[\sigma]_\Gamma|) &= \coprod_{[m_k]} p_G^{-1}(|\mathrm{Id}_{\mathrm{cat}(\mathcal{BG}_G)^{op}}\downarrow[(1,m_k)\sigma]_G|) \\
&= \coprod_{[m_k]} EG \times_G G(|\mathrm{Id}_{\mathrm{cat}(\mathcal{EG}_\Gamma)^{op}}\downarrow(1,m_k)\sigma|) \\
&\cong \coprod_{[m_k]} EG \times_{\Gamma_\sigma \cap G} |\mathrm{Id}_{\mathrm{cat}(\mathcal{EG}_\Gamma)^{op}}\downarrow(1,m_k)\sigma|
\end{aligned}$$

where $m_k$ runs over a complete set of representatives of $\sim_\sigma$. Thus

$$\begin{aligned}
h([\sigma]_\Gamma) &= \mathrm{hocolim}_{\mathrm{cat}(|U\downarrow[\sigma]_\Gamma|)^{op}}(\mathbb{S}\mathrm{obar}(p_G|)) \\
&\simeq \bigvee_{[m_k]} \mathrm{hocolim}_{\mathrm{cat}(|\mathrm{Id}_{\mathcal{BG}_G}\downarrow[(1,m_k)\sigma]_G|)^{op}}(\mathbb{S}\mathrm{obar}(p_G|)).
\end{aligned}$$

But $[(1,m_k)\sigma]_G$ is a terminal object in the category $\mathrm{Id}_{\mathcal{BG}_G}\downarrow[(1,m_k)\sigma]_G$. Therefore,

$$h([\sigma]_\Gamma) \simeq \bigvee_{[m_k]} \mathbb{S}(p_G^{-1}([(1,m_k)\hat{\sigma}]_G))$$

Continuing the calculations,

$$\begin{aligned}
h([\sigma]_\Gamma) &\simeq \mathbb{S}\left(\coprod_{[m_k]} p_G^{-1}([(1,m_k)\hat{\sigma}]_G)\right) = \mathbb{S}\left(\coprod_{[m_k]} EG \times_G G(1,m_k)\hat{\sigma}\right) \\
&\simeq \mathbb{S}\left(\coprod_{[m_k]} \mathrm{Im}(i_k)\right) \simeq \mathbb{S}((EG \times \{\hat{t}\} \times \mathbb{Z}) \times_\Gamma \Gamma\hat{\sigma}) = \mathbb{S}\mathrm{obar}(q_\Gamma)(t, [\sigma]_\Gamma).
\end{aligned}$$

The naturality follows from the construction.

Now Segal's Theorem applied to $U$ implies:

$$\text{hocolim}_{\text{cat}(\mathcal{BG}_G)^{op}}(\mathbb{S}\circ\text{bar}(p_G)) \simeq \text{hocolim}_{\text{cat}(\mathcal{BG}_\Gamma)^{op}}(h) \simeq P_*\rho(t).$$

Thus, if we assume that $G$ **satisfies the $\mathbb{S}$-IC**, for each $t \in \text{cat}(S^1)^{op}$,

$$\mathbb{S}(BG) \simeq \text{hocolim}_{\text{cat}(\mathcal{BG}_G)^{op}}(\mathbb{S}\circ\text{bar}(p_G)) \simeq P_*\rho(t).$$

Since

$$\text{hocolim}_{\text{cat}(S^1\times\mathcal{BG}_\Gamma)^{op}}(\mathbb{S}\circ\text{bar}(q_\Gamma)) \simeq \text{hocolim}_{\text{cat}(S^1)^{op}}(P_*\rho) \simeq \text{hocolim}_{\text{cat}(S^1)^{op}}(BG),$$

the homological properties of homotopy colimits imply that (Section 3 in [70]) there is a long exact sequence

$$\cdots \to \pi_i(\mathbb{S}(BG)) \xrightarrow{1-\phi_*} \pi_i(\mathbb{S}(BG)) \to H_i(S^1\times B\mathcal{G}_\Gamma, \mathbb{S}(q_\Gamma)) \to \pi_{i-1}(\mathbb{S}(BG)) \to \cdots$$

Therefore we get that $\mathcal{CF} \simeq \mathbb{H}.(S^1\times B\mathcal{G}_\Gamma, \mathbb{S}(q_\Gamma))$ Thus $\mathbb{H}.(S^1\times B\mathcal{G}_\Gamma, \mathbb{S}(q_\Gamma))$ represents the controlled (over $S^1$) $\mathbb{S}$-theory.

The commutative diagram

$$\begin{array}{ccccc}
E\Gamma\times_\Gamma\mathcal{EG}_\Gamma & \longrightarrow & E\Gamma\times_\Gamma\mathcal{EG}_\Gamma & \longrightarrow & B\Gamma \\
{\scriptstyle q_\Gamma}\downarrow & & \downarrow{\scriptstyle p_\Gamma} & & \downarrow \\
S^1\times\mathcal{BG}_\Gamma & \longrightarrow & \mathcal{BG}_\Gamma & \longrightarrow & *
\end{array}$$

induces a map of spectra $f : \mathbb{H}.(S^1\times B\mathcal{G}_\Gamma, \mathbb{S}(q_\Gamma)) \to \mathbb{S}(B\Gamma)$. This map is the 'forget control' map. The inclusion induced map $\mathbb{S}(BG) \to \mathbb{S}(B\Gamma)$ factors through the spectrum $\mathbb{H}.(S^1\times B\mathcal{G}_\Gamma, \mathbb{S}(q_\Gamma))$ because of the commutative diagram

$$\begin{array}{ccccccccc}
BG & \longleftarrow & EG\times_G\mathcal{EG}_\Gamma & \longrightarrow & E\Gamma\times_\Gamma\mathcal{EG}_\Gamma & \longrightarrow & E\Gamma\times_\Gamma\mathcal{EG}_\Gamma & \longrightarrow & B\Gamma \\
\downarrow & & {\scriptstyle p_G}\downarrow & & {\scriptstyle q_\Gamma}\downarrow & & \downarrow{\scriptstyle p_\Gamma} & & \downarrow \\
* & \longleftarrow & \mathcal{BG}_G & \longrightarrow & S^1\times\mathcal{BG}_\Gamma & \longrightarrow & \mathcal{BG}_\Gamma & \longrightarrow & *
\end{array}$$

Notice that the map in the second square is not natural because it depends on the choice of an element of $S^1$ but the composition induced by the third square is not affected by that choice. Thus the following diagram commutes, up to homotopy:

$$\begin{array}{ccc}
\mathbb{S}(BG) & \longrightarrow & \mathbb{H}.(S^1\times B\mathcal{G}_\Gamma, \mathbb{S}(q_\Gamma)) \\
{\scriptstyle \text{id}}\downarrow & & \downarrow{\scriptstyle f} \\
\mathbb{S}(BG) & \longrightarrow & \mathbb{S}(B\Gamma)
\end{array}$$

For the Bundle version, start with a commutative diagram:

$$\begin{array}{ccccccc}
\bar{\bar{\bar{Y}}} & \longrightarrow & \bar{\bar{Y}} & \longrightarrow & \bar{Y} & \longrightarrow & Y \\
\downarrow{\scriptstyle \bar{\bar{\rho}}} & & \downarrow{\scriptstyle \bar{\bar{\rho}}} & & \downarrow{\scriptstyle \bar{\rho}} & & \downarrow{\scriptstyle \rho} \\
EG\times_G\mathcal{E}\mathcal{G}_\Gamma & \longrightarrow & EG\times_G\mathcal{E}\mathcal{G}_\Gamma & \longrightarrow & E\Gamma\times_\Gamma\mathcal{E}\mathcal{G}_\Gamma & \longrightarrow & B\Gamma \\
\downarrow{\scriptstyle q_G} & & \downarrow{\scriptstyle p_G} & & \downarrow{\scriptstyle p_\Gamma} & & \downarrow \\
S^1\times\mathcal{B}\mathcal{G}_G & \longrightarrow & \mathcal{B}\mathcal{G}_G & \longrightarrow & \mathcal{B}\mathcal{G}_\Gamma & \longrightarrow & *
\end{array}$$

where $\rho$ is a bundle and the top diagrams are pull-back diagrams. As before, we have the following.

**Theorem 4.7.** *Assume that the Bundle $\mathbb{S}$-IC holds for $G$. Then there is an exact sequence:*

$$\cdots \to \pi_i(\mathbb{S}(\bar{\bar{Y}})) \xrightarrow{1-\phi_*} \pi_i(\mathbb{S}(\bar{\bar{Y}})) \to H_i(S^1\times B\mathcal{G}_\Gamma, \mathbb{S}(\bar{\rho}\circ q_\Gamma)) \to \pi_{i-1}(\mathbb{S}(\bar{\bar{Y}})) \to \cdots$$

*where $\phi : \bar{\bar{Y}} \to \bar{\bar{Y}}$ is the homeomorphism induced by $\alpha$.*

*Remark* 4.8. The groups $H_*(S^1\times B\mathcal{G}_\Gamma, \mathbb{S}(\bar{\rho}\circ q_\Gamma))$ are the analogues of the controlled groups over $S^1$. Theorem 4.7 states the homological properties of the controlled groups.

We specialize to the case $\mathbb{S} = \mathbb{K}_R$, the $K$-theory spectrum with coefficients in a ring $R$. Then there is a commutative diagram of exact sequences for each $i$:

$$\begin{array}{ccccccc}
 & & & H_i(S^1\times\mathcal{B}\mathcal{G}_\Gamma, \mathbb{K}_R(q_\Gamma)) & & & \\
 & & \nearrow & \downarrow{\scriptstyle f} & \searrow & & \\
K_i(RG) & \xrightarrow{1-\phi_*} & K_i(RG) & & K_{i-1}(RG) & \xrightarrow{1-\phi_*} & K_{i-1}(RG) \\
 & & \searrow & \downarrow & \nearrow & & \\
 & & & K_i(R\Gamma) & & &
\end{array}$$

If the Nil-groups of $RG$ vanish, (for example, if $RG$ is regular coherent ring), the bottom sequence is exact for the Bass–Heller–Swan formula ([89], [90]). Thus, in this case,

$$f : H_i(S^1\times\mathcal{B}\mathcal{G}_\Gamma, \mathbb{K}_R(q_\Gamma)) \to K_i(R\Gamma)$$

is an isomorphism for all $i \in \mathbb{Z}$. From the definition of $f$ we have that it factors as:

$$f : H_i(S^1\times\mathcal{B}\mathcal{G}_\Gamma, \mathbb{K}_R(q_\Gamma)) \xrightarrow{p} H_i(\mathcal{B}\mathcal{G}_\Gamma, \mathbb{K}_R(p_\Gamma)) \xrightarrow{A} K_i(R\Gamma)$$

where $p$ is induced by the projection to the second coordinate and $A$ is the assembly map that appears in the Isomorphism Conjecture.

**Theorem 4.9.** *Assume that the Nil-groups of $RG$ vanish. Then*

$$A : H_i(\mathcal{B}\mathcal{G}_\Gamma, \mathbb{K}_R(p_\Gamma)) \to K_i(R\Gamma)$$

*is an epimorphism.*

## 4.2. Controlled Groups over the Interval

Let $\Gamma = G_1 *_{G_0} G_2$, where $G_0$ is a subgroup of $G_1 \cap G_2$. Let $BG_i$, $i = 0, 1, 2$, be classifying spaces for the corresponding groups with $BG_0$ a subcomplex of $BG_1 \cap BG_2$. Choose $B\Gamma$ to be the double mapping cylinder of the inclusion maps. Then there is a natural map $\rho : B\Gamma \to I$, where $I$ is the unit interval. Let $E\Gamma$ be the universal cover of $B\Gamma$. Let

$$E\Gamma \times_\Gamma \mathcal{E}\mathcal{G}_\Gamma \xrightarrow{q_\Gamma} I \times \mathcal{B}\mathcal{G}_\Gamma$$

be maps induced by the natural projection. As before, we have the commutative diagram

$$\begin{array}{ccc} E\Gamma \times_\Gamma \mathcal{E}\mathcal{G}_\Gamma & \longrightarrow & E\Gamma \times_\Gamma \mathcal{E}\mathcal{G}_\Gamma \\ {\scriptstyle q_\Gamma}\downarrow & & \downarrow{\scriptstyle p_\Gamma} \\ I \times \mathcal{B}\mathcal{G}_\Gamma & \longrightarrow & \mathcal{B}\mathcal{G}_\Gamma \end{array}$$

We will show that $\mathbb{H}.(I \times \mathcal{B}\mathcal{G}_\Gamma, \mathbb{S}(q_\Gamma))$ satisfies a Mayer–Vietoris type property.

We work as in Section 4.1. Equip $I$ with the structure of a simplicial complex with one 1-simplex $I$, and two 0-simplices $0, 1$. The projection map induces a functor:

$$\rho : \text{cat}(I \times \mathcal{B}\mathcal{G}_\Gamma)^{op} \to \text{cat}(I)^{op}$$

We will use Segal's Pushdown Construction. We start by giving a description of the functor $P_*\rho$ on $\text{cat}(I)^{op}$. For each simplex $t$ of $I$, define a functor:

$$\mathbb{S}\text{obar}(q_\Gamma)(t, -) : \text{cat}(\mathcal{B}\mathcal{G}_\Gamma)^{op} \to \mathbb{S}pectra, \; [\sigma]_\Gamma \mapsto \mathbb{S}\text{obar}(q_\Gamma)(t, [\sigma]_\Gamma).$$

Then the functor $P_*\rho$ is defined as:

$$P_*\rho : \text{cat}(I)^{op} \to \mathbb{S}pectra, \; P_*\rho(t) = \text{hocolim}_{\text{cat}(\mathcal{B}\mathcal{G}_\Gamma)^{op}}(\mathbb{S}\text{obar}(q_\Gamma)).$$

So if we fix $t$, the functor $\text{bar}(q_\Gamma)$ associates to $[\sigma]_\Gamma$, the space

$$q_\Gamma^{-1}(\hat{t} \times \hat{\sigma}) = (EG_i)\Gamma \times_\Gamma \Gamma\hat{\sigma} \cong (EG_i)\Gamma \times_{\Gamma_\sigma} \hat{\sigma},$$

after subdividing, where $i = 0$ if $t$ is an interior point and $G_i = G_{t+1}$ when $t \in \partial I$. Actually, the inverse image for $t \in \text{Int}(I)$, is

$$q_\Gamma^{-1}(\hat{t} \times \hat{\sigma}) = (EG_i \times \{t\})\Gamma \times_\Gamma \Gamma\hat{\sigma}.$$

The inclusion map $G_i \to \Gamma$ induces a commutative diagram:

$$\begin{array}{ccc} EG_i \times_{G_i} \mathcal{E}\mathcal{G}_\Gamma & \xrightarrow{\bar{u}} & EG_i\Gamma \times_\Gamma \mathcal{E}\mathcal{G}_\Gamma \\ {\scriptstyle p_{G_i}}\downarrow & & \downarrow{\scriptstyle q_\Gamma} \\ \mathcal{E}\mathcal{G}_\Gamma / G_i = \mathcal{B}\mathcal{G}_{G_i} & \xrightarrow{u} & \hat{t} \times \mathcal{B}\mathcal{G}_\Gamma = \hat{t} \times \mathcal{E}\mathcal{G}_\Gamma / \Gamma \end{array}$$

where $u$ is just the quotient map. As in the last section, $u$ induces a functor

$$U : \text{cat}(\mathcal{B}\mathcal{G}_{G_i})^{op} \to \text{cat}(\mathcal{B}\mathcal{G}_\Gamma)^{op}.$$

We consider the restriction

$$p_{G_i}| : (p_{G_i})^{-1}(|U{\downarrow}[\sigma]_\Gamma|) \to |U{\downarrow}[\sigma]_\Gamma|$$

and define a functor

$$h : \mathrm{cat}(\mathcal{B}\mathcal{G}_\Gamma)^{op} \to \mathbb{S}pectra, \;\; h([\sigma]_\Gamma) = \mathrm{hocolim}_{\mathrm{cat}(|U\downarrow[\sigma]_\Gamma|)^{op}}(\mathbb{S}\mathrm{obar}(p_{G_i}|)).$$

As in the last section, we get that there is a natural homotopy equivalence:

$$h([\sigma_\Gamma]) \simeq \mathbb{S}\mathrm{obar}(q_\Gamma)(t, [\sigma_\Gamma]).$$

We will indicate the modifications we need to get the result. Choose coset representatives:

$$\Delta_i = \{\gamma_{i,j} : \gamma_{i,j} \in \mathcal{A}_i\}, \; i = 0, 1, 2.$$

In other words,

$$\Gamma = \coprod_{j \in \mathcal{A}_i} G_i \gamma_{i,j}, \; i = 0, 1, 2.$$

With the above notation:

- Every simplex $[\sigma]_\Gamma$ of $\mathcal{E}\mathcal{G}_\Gamma$ defines an equivalence relation on $\Delta_i$,
$$\gamma_{i,j} \sim_\sigma \gamma'_{i,j} \iff \Gamma_\sigma \cap \gamma_{i,j} G_i (\gamma'_{i,j})^{-1} \neq \emptyset.$$
- Let $t \in I$. Set $i = 0$ if $t \in \mathrm{Int}(I)$ and $i = t$ if $t \in \partial I$. For each $\gamma_{i,j} \in \Delta_i$, a map
$$\iota_{\gamma_{i,j}} : EG_i \times_{\Gamma_\sigma \cap G_i} g_{i,j}\hat{\sigma} \to EG_i\Gamma \times_\Gamma \Gamma\hat{\sigma}, \; [x, \gamma_{i,j}\hat{\sigma}] \mapsto [x, \gamma_{i,j}\hat{\sigma}]$$
here $EG_i\Gamma \subset E\Gamma$. Notice that the image of $\iota_{\gamma_{i,j}}$ depends on the choice of $t \in I$.
- There is a homeomorphism:
$$\chi_i : EG_i \times_{G_i} G_i \gamma_{i,j} \hat{\sigma} \to E\Gamma \times_\Gamma \mathcal{E}\mathcal{G}_\Gamma$$
onto $\mathrm{Im}(\iota_{\gamma_{i,j}})$.

Then

$$\begin{aligned} |U{\downarrow}[\sigma_\Gamma]| &= \coprod_{[\gamma_{i,j}]} |\mathrm{Id}_{\mathrm{cat}(\mathcal{B}\mathcal{G}_{G_i})^{op}} {\downarrow} [\gamma_{i,j}\sigma]_{G_i}| \\ (p_{G_i})^{-1}(|U{\downarrow}[\sigma_\Gamma]|) &= \coprod_{[\gamma_{i,j}]} EG_i \times_{\Gamma_\sigma \cap G_i} |\mathrm{Id}_{\mathrm{cat}(\mathcal{B}\mathcal{G}_\Gamma)^{op}} {\downarrow} \gamma_{i,j}\sigma|. \end{aligned}$$

Segal's Theorem implies that

$$\mathrm{hocolim}_{\mathrm{cat}(\mathcal{B}\mathcal{G}_{G_i})^{op}}(\mathbb{S}\mathrm{obar}(p_{G_i})) \simeq \mathrm{hocolim}_{\mathrm{cat}(\mathcal{B}\mathcal{G}_\Gamma)^{op}}(h) \simeq P_*\rho(t).$$

Thus, if the $\mathbb{S}$-IC holds for $G_i$, $i = 0, 1, 2$, there is a natural homotopy equivalence:

$$P_*\rho(t) = \mathrm{hocolim}_{\mathrm{cat}(\mathcal{B}\mathcal{G}_\Gamma)^{op}}(\mathbb{S}\mathrm{obar}(q_\Gamma)(t, -)) \simeq \begin{cases} \mathbb{S}(BG_0), & \text{if } t \in \mathrm{Int}(I) \\ \mathbb{S}(BG_{t+1}), & \text{if } t \in \partial I. \end{cases}$$

Again, using the homological properties of homotopy colimits, we can show that the following is a homotopy cartesian diagram:

$$\begin{array}{ccc} \mathbb{S}(BG_0) & \longrightarrow & \mathbb{S}(BG_1) \\ \downarrow & & \downarrow \\ \mathbb{S}(BG_1) & \longrightarrow & \mathbb{H}.(I \times \mathcal{B}\mathcal{G}_\Gamma, \mathbb{S}(q_\Gamma)) \end{array}$$

In analogy with the circle, for the case $\mathbb{S} = \mathbb{K}_R$ we have that:

**Theorem 4.10.** *Let $R$ be a ring such that the Waldhausen Nil-groups of $RG_0$ vanish. Then*

(1) *the forgetful map*

$$f : \mathbb{H}.(I \times \mathcal{B}\mathcal{G}_\Gamma : \mathbb{K}_R(q_\Gamma)) \to \mathbb{K}_R(B\Gamma)$$

*is a homotopy equivalence. This map is induced by the universal properties of the homotopy cartesian square.*

(2) *the assembly map*

$$A : \mathbb{H}.(\mathcal{B}\mathcal{G}_\Gamma, \mathbb{K}_R(p_\Gamma)) \to \mathbb{K}_R(B\Gamma)$$

*induces an epimorphism on homotopy.*

# 5. Applications

We will apply the results to prove that $\mathbb{K}_R$-IC is true for certain types of groups. Most of the results in this section were obtained, with different methods, in [10]. Before we start we state the Novikov Conjecture for a discrete group $\Gamma$. Let $\mathcal{C}_\Gamma$ be the classifying space for the class of finite subgroups of $\Gamma$. Again, there is a commutative diagram:

$$\begin{array}{ccc} E\Gamma \times_\Gamma \mathcal{E}\mathcal{C}_\Gamma & \longrightarrow & B\Gamma \\ {\scriptstyle p}\downarrow & & \downarrow \\ \mathcal{B}\mathcal{C}_\Gamma & \longrightarrow & * \end{array}$$

A group $\Gamma$ satisfies the **integral $\mathbb{S}$-Novikov Conjecture** if the assembly map

$$A_\mathcal{C} : \mathbb{H}.(\mathcal{B}\mathcal{C}_\Gamma, \mathbb{S}(p)) \to \mathbb{S}(B\Gamma)$$

induces a split injection on the homotopy groups. It is an open question if all the torsion free groups satisfy the integral $\mathbb{K}_R$-Novikov Conjecture. It was proved that groups of finite cohomological dimension that also have finite asymptotic dimension satisfy the integral $\mathbb{K}_R$-Novikov Conjecture ([8]), generalizing the calculations in [18].

*Remark* 5.1. The geometric methods used in proving the Novikov Conjecture depend on controlling how 'large' compact subsets of $E\Gamma$ become as they are translated to infinity through the $\Gamma$-action. The assumption on the asymptotic dimension guarantees such control.

Let $R$ be a regular Noetherian ring. Let $G$ be a torsion free group that satisfies the integral $\mathbb{K}_R$-Novikov Conjecture. The commutative diagram

$$\begin{array}{ccccc} EG\times_G\mathcal{E}\mathcal{C}_G & \longrightarrow & EG\times_G\mathcal{E}\mathcal{G}_G & \longrightarrow & BG \\ {\scriptstyle p}\downarrow & & \downarrow{\scriptstyle p_G} & & \downarrow \\ \mathcal{B}\mathcal{C}_G & \longrightarrow & \mathcal{B}\mathcal{G}_G & \longrightarrow & * \end{array}$$

provides a splitting of the assembly map $A_{\mathcal{C}} = A \circ A_{\mathcal{C},\mathcal{G}}$:

$$\begin{array}{ccc} \mathbb{H}.(\mathcal{B}\mathcal{C}_G, \mathbb{K}_R(p)) & \xrightarrow{A_{\mathcal{C}}} & \mathbb{K}_R(BG) \\ {\scriptstyle A_{\mathcal{C},\mathcal{G}}}\downarrow & \nearrow{\scriptstyle A} & \\ \mathbb{H}.(\mathcal{B}\mathcal{G}_G, \mathbb{K}_R(p_G)) & & \end{array}$$

The relative assembly map is the map induced by the first commutative square. For every virtually cyclic group $S$ of $G$, let $\mathcal{C}_S$ be the class of finite subgroups of $S$. Theorem A.10 in [43] states that $A_{\mathcal{C},\mathcal{G}}$ is an equivalence if the assembly map $A_{\mathcal{C}_S}$ is an equivalence for all virtually cyclic subgroups $S$ of $G$ (a similar calculation appears in [31]). Since $G$ is torsion free, the only finite subgroup of $G$ is the trivial group and the only virtually cyclic subgroups are infinite cyclic subgroups. Thus $A_{\mathcal{C}_S}$ is the assembly map in the integral $\mathbb{K}_R$-Novikov conjecture for $S \cong \mathbb{Z}$. Since $R$ is regular Noetherian, this map is an equivalence ([43], Remark A.11). Therefore, if $A_{\mathcal{C}}$ induces a monomorphism on homotopy groups, so does $A$. Combining with the results in Theorems 4.9 and 4.10, we get the following:

**Theorem 5.2.** *Let $G$ be a torsion free group and $R$ a regular Noetherian ring. Assume that $G$ satisfies the $\mathbb{K}_R$-IC, and that $RG$ is a regular Noetherian ring. Let $\Gamma$ be a torsion free group defined by:*

(i) *$\Gamma = G \rtimes \mathbb{Z}$, or*
(ii) *$\Gamma = G_1 *_G G_2$, such that $G_i$, $i = 1, 2$, satisfy the $\mathbb{K}_R$-IC,*

*such that $\Gamma$ satisfies the integral $\mathbb{K}_R$-Novikov conjecture. Then $\Gamma$ satisfies the $\mathbb{K}_R$-IC.*

*Remark* 5.3. We assume that the groups involved satisfy the $\mathbb{K}_R$-IC as in the Theorem 5.2. More generally, it is enough if we assume that $R$ is regular Noetherian ring, and the twisted Nil-groups of $RG$ vanish in (i) and the Waldhausen Nil-groups of the triple $(RG, R[G_1\backslash G], R[G_2\backslash G])$ vanish in (ii).

Now we give applications of the Theorem.

**Corollary 5.4.** *Let $G$ be a torsion free group and $R$ a regular Noetherian ring. Assume that:*

(1) *$G$ satisfies the $\mathbb{K}_R$-IC,*
(2) *$G$ has finite asymptotic dimension,*
(3) *$RG$ is a regular coherent ring.*

*Let $\Gamma$ be a torsion free group defined by:*

(i) $\Gamma = G \rtimes \mathbb{Z}$, *or*
(ii) $\Gamma = G_1 *_G G_2$, *such that* $G_i$, $i = 1, 2$, *satisfy the* $\mathbb{K}_R$*-IC and they have finite cohomological and asymptotic dimensions.*

*Then* $\Gamma$ *satisfies the* $\mathbb{K}_R$*-IC.*

*Proof.* The assumptions on the groups imply that $\Gamma$ has finite asymptotic dimension ([13]). Also, $\Gamma$ has finite cohomological dimension and thus it admits a finite dimensional $B\Gamma$. By [8], $\Gamma$ satisfies the integral $\mathbb{K}_R$-Novikov conjecture. The result follows from Theorem 5.2. □

**Corollary 5.5.** *Let* $F$ *be a finitely generated free group and* $R$ *a regular Noetherian ring. Then*

(1) $F$ *satisfies the* $\mathbb{K}_R$*-IC.*
(2) $F \rtimes \mathbb{Z}$ *satisfies the* $\mathbb{K}_R$*-IC.*

*Proof.* The first statement follows by induction on the number $k$ of generators of $F$: If $k = 2$, then $F = \mathbb{Z} * \mathbb{Z}$ and the result follows from 5.4. For $k > 2$, $F = F_{k-1} * \mathbb{Z}$, where $F_{k-1}$ is the free group on $(k-1)$ generators. All the assumptions of Corollary 5.4 are satisfied and thus $F$ satisfies the $\mathbb{K}_R$-IC.

For the second statement, we use again Corollary 5.4. Part (1) implies that assumption (1) is satisfied. Also, the free group has finite asymptotic dimension and thus assumption (2) is satisfied. Assumption (3) follows because $R$ is regular Noetherian ring. □

*Remark* 5.6. In [4] and [46], there was a special assumption for groups of type $F \rtimes \mathbb{Z}$ to satisfy the fibered pseudoisotopy IC. Essentially the assumption was that the action of $\mathbb{Z}$ on $F$ has certain geometric properties. Corollary 5.5 is more general because such an assumption is not needed but it only gives the $\mathbb{K}_R$-IC and not the fibered version. A general fibered version could not follow along the same lines because the assumptions in Corollary 5.4 guarantee that all the Nil-groups that appear vanish.

# References


[1] D. R. Anderson, F. X. Connolly, H. J. Munkholm, *A comparison of continuously controlled and controlled K-theory*, Topology Appl. **71** (1996), 9–46.
[2] D. R. Anderson, H. J. Munkholm, *Continuously controlled K-theory with variable coefficients*, J. Pure Appl. Algebra **145** (2000), 215–266.
[3] D. R. Anderson, H. J. Munkholm, *Geometric modules and Quinn homology theory*, K-theory **7**, (1993) 443–475.
[4] C. S. Aravinda, F. T. Farrell, S. K. Roushon, *Algebraic K-theory of pure braid groups*, Asian J. Math. **4**, (2000) 337–343.
[5] B. Badzioch, *$K_1$ of twisted rings of polynomials*, $K$-Theory **16** (1999), 29–34.
[6] H. Bass, *Algebraic K-theory*, Benjamin (1968).
[7] H. Bass, A. Heller, R. G. Swan, *The Whitehead group of a polynomial extension*, Inst. Hautes Études Sci. Publ. Math. **22**, (1964) 61–79.

[8] A. C. Bartels, *Squeezing and Higher Algebraic K-Theory*, $K$-theory **28**, (2003) 19–37.
[9] A. C. Bartels, T. Farrell, L. Jones, H. Reich, *On the isomorphism conjecture in algebraic K-theory*, Topology **43**, (2004) 157–213.
[10] A. Bartels, W. Lück, *Isomorphism conjecture for homotopy K-theory and groups acting on trees*, J. Pure Appl. Algebra 205 (2006), no. 3, 660–696.
[11] A. Bartels, W. Lück, H. Reich, *On the Farrell–Jones Conjecture and its applications*, math.KT/0703548.
[12] A. Bartels, W. Lück, H. Reich, *The K-theoretic Farrell-Jones Conjecture for hyperbolic groups*, math.GT/0701434.
[13] G. Bell, A. Dranishnikov, *On asymptotic dimensions of groups*, Algebraic Geom. Topol. **1** (electronic), (2001) 57–71.
[14] F. Bihler, *Vanishing of the KNil groups: localization methods*, math.KT/0702320v1.
[15] F. Bihler, *Vogel's notion of regularity for non-coherent rings*, math.KT/0612569v1.
[16] S. E. Cappell, *Unitary nilpotent groups and Hermitian K-theory*, I. Bull. Amer. Math. Soc. **80** (1974), 1117–1122.
[17] S. E. Cappell, *A splitting theorem for manifolds*, Invent. Math. **33** (1976), 69–170.
[18] G. Carlsson, E. K. Pedersen, *Controlled algebra and the Novikov conjectures for K- and L-theory*, Topology **34**, (1995) 731–758.
[19] T. A. Chapman, *Controlled boundary and h-cobordism theorem*, Trans. Amer. Math. Soc. **280** (1983), 73–95.
[20] T. A. Chapman, *Controlled simple homotopy theory and applications*, Lecture Notes in Mathematics, 1009, Springer-Verlag, Berlin, 1983.
[21] T. A. Chapman, *A general approximation theorem for Hilbert cube manifolds* Compositio Math. **48** (1983), 373–407.
[22] T. A. Chapman, *Approximation results in topological manifolds*, Mem. Amer. Math. Soc. **34** (1981), no. 251.
[23] T. A. Chapman, *Invariance of torsion and the Borsuk conjecture*, Canad. J. Math. **32** (1980), 1333–1341.
[24] T. A. Chapman, *Approximation results in Hilbert cube manifolds*, Trans. Amer. Math. Soc. **262** (1980), 303–334.
[25] T. A. Chapman, *Simple homotopy theory for ANR's*, NSF-CBMS Regional Conference in Topology (Univ. Tennessee, Knoxville, Tenn., 1974), General Topology and Appl. **7** (1977), 165–174.
[26] T. A. Chapman, *Lectures on Hilbert cube manifolds*, Expository lectures from the CBMS Regional Conference held at Guilford College, October 11-15, 1975. Regional Conference Series in Mathematics, No. 28. American Mathematical Society, Providence, R. I., 1976.
[27] T. A. Chapman, Topological invariance of Whitehead torsion, Amer. J. Math. **96** (1974), 488–497.
[28] T. A. Chapman, S. Ferry, *Approximating homotopy equivalences by homeomorphisms*, Amer. J. Math. **101** (1979), 583–607.
[29] M. M. Cohen, *A course in simple-homotopy theory*, Graduate Texts in Math-

ematics, Vol. 10. Springer-Verlag, New York-Berlin, 1973.

[30] E. H. Connell, J. Hollingsworth, *Geometric groups and Whitehead torsion*, Trans. Amer. Math. Soc. **140** (1969), 161–181.

[31] F. X. Connolly, S. Prassidis, *On the exponent of the cokernel of the forget-control map on $K_0$-groups*, Fund. Math. **172** (2002), 201–216.

[32] D. Coram, P. Duvall, *Approximate fibrations and a movability condition for maps*, Pacific J. Math. **72** (1977), 41–56.

[33] D. S. Coram, P. F. Jr. Duvall, *Approximate fibrations*, Rocky Mountain J. Math. **7** (1977), 275–288.

[34] D. Juan-Pineda, S. Prassidis, *A Controlled Approach to the Isomorphism Conjecture*, math.KT/0404513.

[35] J. F. Davis, W. Lück, *Spaces over a category and assembly maps in isomorphism conjectures in $K$- and $L$-theory*, $K$-Theory **15** (1998) 201–252.

[36] W. Dicks, M. J. Dunwoody, *Groups acting on graphs*, Cambridge Studies in Advanced Mathematics vol. **17**, Cambridge University Press 1989.

[37] J. Dydak, J. Segal, *Shape theory. An introduction*, Lecture Notes in Mathematics, **688**. Springer, Berlin, 1978.

[38] F. T. Farrell, *The obstruction to fibering a manifold over a circle* Indiana Univ. Math. J. **21**, (1971/1972) 315–346.

[39] F. T. Farrell, *The nonfiniteness of Nil*, Proc. Amer. Math. Soc. **65** (1977), 215–216.

[40] F. T. Farrell, W. C. Hsiang, *Manifolds with $\pi_i = G \times \alpha T$*, Amer. J. Math. **95** (1973), 813–848.

[41] F. T. Farrell, W. C. Hsiang, *A structure set analogue of Chapman-Ferry-Quinn theory* Continua, decompositions, manifolds (Austin, Tex., 1980), 172–190, Univ. Texas Press, Austin, Tex., 1983.

[42] F. T. Farrell, W. C. Hsiang, *A formula for $K_1 R_\alpha[T]$*, Proc. Symp. Pure Math. vol. **17**, *Applications of Categorical Algebra*, Amer. Math. Soc., 192–218, (1970).

[43] F. T. Farrell, L. E. Jones, *Isomorphism conjectures in algebraic K-theory*, J. of AMS **6** (1993), 249–298.

[44] F. T. Farrell, L. E. Jones, *Rigidity for aspherical manifolds with $\pi_1 \subset$* $\mathrm{GL}_m(\mathbf{R})$, Asian J. Math., **2** (1998), 215–262.

[45] F. T. Farrell, P. Linnell, *$K$-theory of solvable groups*, Proc. London Math. Soc. **87**, (2003) 309–336.

[46] F. T. Farrell, S. K. Roushon, *The Whitehead groups of braid groups vanish*, Internat. Math. Res. Notices **10**, (2000) 515–526.

[47] S. Ferry, *The homeomorphism group of a compact Hilbert cube manifold is an* ANR, Ann. Math. **106** (1977), 101–119.

[48] S. Ferry, *Homotoping $\varepsilon$-maps to homeomorphisms*, Amer. J. Math. **101** (1979), 567–582.

[49] S. Ferry, *A simple-homotopy approach to the finiteness obstruction*, Shape theory and geometric topology (Dubrovnik, 1981), pp. 73–81, Lecture Notes in Math., 870, Springer, Berlin-New York, 1981.

[50] S. Gersten, *Whitehead groups of free associative algebras*, Bull. Amer. Math. Soc. **71**, (1965) 157–159.

[51] S. M. Gersten, *On class groups of free products*, Ann. of Math. **87**, (1968) 392–398.
[52] S. Glaz, *Commutative coherent rings*, Lecture Notes in Mathematics, 1371. Springer-Verlag, Berlin, 1989.
[53] D. Grayson, *Higher algebraic $K$-theory. II (after Daniel Quillen)*, Algebraic $K$-theory (Proc. Conf., Northwestern Univ., Evanston, Ill., 1976), pp. 217–240. Lecture Notes in Math., Vol. 551, Springer, Berlin, 1976
[54] J. Grunewald, *The behavior of Nil-groups under localization and the relative assembly map*, math.KT/0605194v2.
[55] J. Grunewald, *Non-finiteness results for Nil-groups*, preprint 2006.
[56] I. Hambleton, E. K. Pedersen, *Identifying assembly maps in $K$- and $L$-theory* Math. Ann. **328** (2004) 27–57.
[57] H. M. Hastings, A. Heller, *Homotopy idempotents on finite-dimensional complexes split*, Proc. Amer. Math. Soc., **85** (1982), 619–622.
[58] J. Hollender, R. M. Vogt, *Modules of topological spaces, applications to homotopy colimits and $E_\infty$ structures*, Arch. Math., **59** (1992), 115–129.
[59] C. B. Hughes, *Bounded homotopy equivalences of Hilbert cube manifolds*, Trans. Amer. Math. Soc. **287** (1985), 621–643.
[60] C. B. Hughes, *Approximate fibrations on topological manifolds*, Michigan Math. J. **32** (1985), 167–183.
[61] C. B. Hughes, *Approximate fibrations and bundle maps on Hilbert cube manifolds*, Topology Appl. **15** (1983), 159–172.
[62] C. B. Hughes, S. Prassidis, *Relaxation and control over the circle*, Memoirs of Amer. Math. Soc. **145** (2000), no. 691.
[63] B. Hughes, A. Ranicki, *Ends of complexes*, Cambridge Tracts in Mathematics, 123. Cambridge University Press, Cambridge, 1996.
[64] C. B. Hughes, L. R. Taylor, E. B. Williams, *Bundle theories for topological manifolds*, Trans. Amer. Math. Soc. **319** (1990), 1–65.
[65] H. Inassaridze, *Algebraic $K$-theory. Mathematics and its Applications*, 311. Kluwer Academic Publishers Group, Dordrecht, 1995.
[66] D. Juan-Pineda, S. Prassidis, *On the nil groups of Waldhausen nils*, Topology Appl. **146/147** (2005), 489–499.
[67] D. Juan-Pineda, S. Prassidis, *On the lower Nil-groups of Waldhausen*, Forum Math. **13** (2001), 261–285.
[68] R. C. Lacher, *Cell-like mappings and their generalizations*, Bull. Amer. Math. Soc. **83**, (1977) 495–552.
[69] W. Lück, R. Stamm, *Computations of $K$- and $L$-theory of cocompact planar groups*, $K$-Theory **21** (2000) 249–292.
[70] H. J. Munkholm, S. Prassidis, *Waldhausen's Nil-groups and continuously controlled $K$-theory*, Fund. Math. **161** 217-224 (1999).
[71] E. K. Pedersen, C. A. Weibel, *$K$-theory homology of spaces*, Algebraic topology (Arcata, CA, 1986), 346–361, Lecture Notes in Math., 1370, Springer, Berlin, 1989.
[72] S. Prassidis, *The Bass-Heller-Swan formula for the equivariant topological Whitehead group*, $K$-Theory **5** (1991/92), 395–448.
[73] D. Quillen, *Higher algebraic $K$-theory*, Lecture Notes in Math., vol 341,

Springer-Verlag, 1973, 85–147.
[74] F. Quinn, *Ends of maps. I*, Ann. of Math. **110** (1979), 275–331.
[75] F. Quinn, *Ends of maps. II*, Invent. Math. **68** (1982), 353–424.
[76] R. Ramos, *non-finiteness os twisted nils*, preprint 2006.
[77] A. Ranicki, *Lower K- and L-Theory*, Cambridge University Press, Cambridge 1992.
[78] J. Rosenberg, *Algebraic K-theory and its applications*, Graduate Texts in Mathematics, 147. Springer-Verlag, New York, 1994.
[79] L. Siebenmann, *The obstruction of finding the boundary of an open manifold*, Ph. D. thesis (1965).
[80] L. Siebenmann, *A total Whitehead torsion obstruction to fibering over the circle*, Comm. Math. Helv. **45**, (1972) 1–48.
[81] S. Smale, *Generalized Poincar's conjecture in dimensions greater than four*, Ann. of Math. **74**, 1961 391–406.
[82] V. Srinivas, *Algebraic K-theory*, Second edition. Progress in Mathematics, 90. Birkhäuser Boston, Inc., Boston, MA, 1996.
[83] R. Schwänzl, R. E. Staffeldt, *The approximation theorem and the K-theory of generalized free products*, Trans. Amer. Math. Soc. **347** (1995), 3319–3345.
[84] R. Schwänzl, R. Staffeldt, *The Seifert-van Kampen theorem and generalized free products of S-algebras*, Proc. Amer. Math. Soc. **130** (2002), 3193–3208 (electronic).
[85] J. Stallings, *Whitehead torsion of free products*, Ann. of Math. **82**, 354 – 363 (1965).
[86] R. G. Swan, *Vector bundles and projective modules*, Trans. Amer. Math. Soc. **105**, (1962) 264–277.
[87] R. Talbert, *An isomorphism between Bredon and Quinn homology*, Forum Math. **11**, 591–616 (1999).
[88] R. W. Thomason, *First quadrant spectral sequences in algebraic K-theory via homotopy colimits*, Comm. Algebra **10**, 1589–1668 (1982).
[89] F. Waldhausen, *Whitehead groups of generalized free products*, Proceedings of the Conference in Algebraic K-theory, Battelle 1972, Springer Lecture Notes in Mathematics v. 342 (1973).
[90] F. Waldhausen, *Algebraic K-theory of generalized free products*, Ann. of Math. **108**, 135–256 (1978).
[91] C. T. C. Wall, *Finiteness conditions for CW-complexes*, Ann. of Math. **81**, (1965) 56–69.
[92] C. T. C. Wall, *Finiteness conditions for CW complexes II*, Proc. Roy. Soc. Ser. A **295**, (1966) 129–139.
[93] J. H. C. Whitehead, *On incidence matrices, nuclei and homotopy types*, Ann. of Math. **42**, (1941) 1197–1239.
[94] J. H. C. Whitehead, *Simple homotopy types*, Amer. J. Math. **72**, (1950) 1–57.
[95] J. E. West, *Mapping Hilbert cube manifolds to ANR's: a solution of a conjecture of Borsuk*, Ann. Math. **106** (1977) 1–18.

*Cohomology of Groups and*
*Algebraic K-theory*
ALM 12, pp. 387–411



# Lecture Notes on $K$-Theory*

Hourong Qin**

**Abstract**

These notes give an elementary introduction to Milnor's $K_2$ functor and a brief explanation on the Milnor $K$-Theory, and also discuss some results finite order elements in $K_2\mathbb{Q}$ and the structure $K_2O_F$ for quadratic fields $F$.

**2000 Mathematics Subject Classification:** 18F25; Secondary: 12A65

**Keywords and Phrases:** Milnor $K$-group, algebraic $K$-theory

## 1. Introduction

These are the notes from lectures given for Summer School on Cohomology of Groups and Algebraic $K$-Theory (Hangzhou 2007). They provide an elementary introduction to Milnor's $K_2$ functor and a brief explanation on the Milnor $K$-Theory, including the Milnor Conjecture and the Bloch-Kato Conjecture. I would like to thank Professor Lizhen Ji and Professor Fang Li for inviting me to give lectures.

Nothing is original in these notes. Most of them are taken from Milnor's book "Introduction to Algebraic $K$-Theory" and Tate's paper "Relations between $K_2$ and Galois Cohomology". I will also mention some of our work on the finite elements on $K_2\mathbb{Q}$ and the structure $K_2O_F$ for quadratic fields $F$. I have to say that I have to leave many important results in this area unmentioned, because of time being very limited.

## 2. The Steinberg Group and the Functor $K_2$

Let $e_{ij}^{\lambda}$ (with $i \neq j$) $\in GL(n, \Lambda)$ denote the elementary matrix with entry $\lambda$ in the $(i, j)$-th place.

Clearly, $e_{ij}^{\lambda} e_{ij}^{u} = e_{ij}^{\lambda+u}$, in particular, $(e_{ij}^{\lambda})^{-1} = e_{ij}^{-\lambda}$.

*Supported by NSFC.

**Department of Mathematics, Nanjing University, China.

We have:

$$[e_{ij}^{\lambda}, e_{kl}^{u}] = \begin{cases} 1 & if\ j \neq k, i \neq l; \\ e_{il}^{\lambda u} & if\ j = k, i \neq l; \\ e_{kj}^{-u\lambda} & if\ j \neq k, i = l. \end{cases}$$

$([a,b] = aba^{-1}b^{-1}(\text{commutator}))$.

**Remark.** There is no simple formula for $[e_{ij}^{\lambda}, e_{ji}^{u}]$. For example, if $x, y \in \Lambda$ with $xy = yx$, then

$$e_{13}^{x} \cdot e_{31}^{y} \cdot e_{13}^{-x} \cdot e_{31}^{-y} = \begin{pmatrix} x^2y^2 + yx + 1 & 0 & -yx^2 \\ 0 & 1 & 0 \\ xy^2 & 0 & -xy + 1 \end{pmatrix}.$$

**Definition 2.1.** *Let $\Lambda$ be a ring. For $n \geqslant 3$, the Steinberg group $St(n, \Lambda)$ of $\Lambda$ is the group defined by generators $x_{ij}^{\lambda}$ subject to the relations*

$$\left.\begin{array}{l} (1)\ x_{ij}^{\lambda} x_{ij}^{u} = x_{ij}^{\lambda+u}; \\ (2)\ [x_{ij}^{\lambda}, x_{jl}^{u}] = x_{il}^{\lambda u}\ for\ i \neq l; \\ (3)\ [x_{ij}^{\lambda}, x_{kl}^{u}] = 1\ for\ j \neq k, i \neq l. \end{array}\right\}\ \textit{the Steinberg relations}$$

**Remark.** The assumption that $n \geqslant 3$ is needed to avoid technical complications.

Define the homomorphism (canonical)

$$\phi : St(n, \Lambda) \longrightarrow GL(n, \Lambda)$$

by the formula

$$\phi(x_{ij}^{\lambda}) = e_{ij}^{\lambda}.$$

Clearly, $\phi(St(n, \Lambda)) = E(n, \Lambda)$, the subgroup of $GL(n, \Lambda)$ generated by all elementary matrices.

Passing to the direct limit as $n \to \infty$, we obtain

$$\phi : St(\Lambda) \longrightarrow GL(\Lambda).$$

Of course, $\phi(St(\Lambda)) = E(\Lambda)$ is equal to the commutator subgroup of $GL(\Lambda)$.

**Definition 2.2.** *We call the kernel of the homomorphism $\phi : St(\Lambda) \to GL(\Lambda)$ the Milnor $K$-group of $\Lambda$ and denote it by $K_2\Lambda$.*

**Theorem 1.** The group $K_2\Lambda$ is precisely the center of the Steinberg group $St(\Lambda)$.
**Notation.** For a group $G$, we will write $C(G)$ for the center of $G$.

*Proof.* First, we have that $C(E(\Lambda))$ is trivial.

In fact, if $c \in C(E(\Lambda))$, by abuse of notation, we may assume $c \in E_n(\Lambda)$ for some $n$. Then the fact $C(E_n(\Lambda))$ is trivial implies that $c = I$.

Now if $c$ is an element in $C(St(\Lambda))$, then $\phi(c)$ belongs $C(E(\Lambda)) = \phi(St(\Lambda))$, hence $\phi(c) = I$.

Next we show that if $\phi(c) = I$ for some $c \in St(\Lambda)$, then $c$ is in $C(St(\Lambda))$.

Note that $St(\Lambda) =< x_{ij}^{\lambda} >$. Choose $n$ large enough so that $c$ can be expressed as a word in the generators $x_{ij}^{\lambda}$ with $i, j < n$.

Let $P_n$ denote the subgroup of $St(\Lambda)$ generated by the elements

$$x_{1n}^{\mu_1}, x_{2n}^{\mu_2}, \cdots, x_{n-1\;n}^{\mu_{n-1}},$$

where $\mu_1, \cdots, \mu_{n-1}$ range over $\Lambda$.

It is easy to see that $P_n$ is commutative (the Steinberg relation (3)). So each element of $P_n$ can be written as a product

$$x_{1n}^{\mu_1} x_{2n}^{\mu_2} \cdots x_{n-1\;n}^{\mu_{n-1}}.$$

Note that

$$e_{1n}^{\mu_1} e_{2n}^{\mu_2} \cdots e_{n-1\;n}^{\mu_{n-1}} = \begin{pmatrix} 1 & & & & \mu_1 \\ & 1 & & & \mu_2 \\ & & \ddots & & \vdots \\ & & & 1 & \mu_{n-1} \\ & & & & 1 \end{pmatrix}.$$

Hence $\mathrm{Ker}\phi|_{P_n} = \{I\}$, where $\phi$ is the canonical homomorphism.

If $i, j < n$, then

$$x_{ij}^{\lambda} x_{kn}^{\mu} x_{ij}^{-\lambda} = \begin{cases} x_{kn}^{\mu} & \text{if } j \neq k, \\ x_{in}^{\lambda\mu} x_{kn}^{\mu} & \text{if } j = k. \end{cases}$$

So $cP_nc^{-1} \subset P_n$.

$\phi(c) = I \Longrightarrow \forall P \in P_n, \phi(cPc^{-1}) = \phi(c)\phi(P)\phi(c)^{-1} = \phi(P)$.

$\mathrm{Ker}\phi|_{P_n} = I \Longrightarrow cPc^{-1} = P \Longrightarrow cP = Pc$. Hence $c$ commutes with every generator $x_{kn}^{\mu}, k < n$.

But a completely analogous argument shows that $c$ also commutes with every $x_{nl}^{\mu}, l < n$, so $c$ commutes with the commutators

$$[x_{kn}^{\mu}, x_{nl}^{1}] = x_{kl}^{\mu} \quad (k \neq l, k, l < n).$$

Since $n$ can be arbitrary large, the result follows. □

# 3. Universal Central Extensions

**Definition 3.1.** *Let $G$ be a group and $A$ an abelian group. A central extension of $G$ by $A$ is a short exact sequence of groups $1 \to A \to X \xrightarrow{\pi} G \to 1$ such that $A \subseteq C(X)$.*

*A central extension is called split if it is isomorphic to an extension of the form $1 \to A \to A \times G \xrightarrow{p_2} G \to 1$, where $p_2(a, g) = g$.*

Two extensions $X$ and $Y$ of $G$ by $A$ are said to be equivalent if there is an isomorphism $f : X \to Y$ which is the identity on $A$ and which induces the identity map on $G$.

$$\begin{array}{ccccccccc} 1 & \longrightarrow & A & \longrightarrow & X & \longrightarrow & G & \longrightarrow & 1 \\ & & \downarrow{\scriptstyle id} & & \cong\downarrow{\scriptstyle f} & & \downarrow{\scriptstyle id} & & \\ 1 & \longrightarrow & A & \longrightarrow & Y & \longrightarrow & G & \longrightarrow & 1. \end{array}$$

**FACT**: The equivalence classes of central extensions of $G$ by a fixed group $A$ are in 1-1 correspondence with the elements of the cohomology group $H^2(G, A)$.

**Definition 3.2.** *By a homomorphism over $G$ from $X \xrightarrow{\pi} G$ to another central extension $1 \to B \to Y \xrightarrow{\tau} G \to 1$, we mean a group homomorphism $f : X \to Y$ such that $\pi = \tau f$.*

$$\begin{array}{ccccccccc} & & & & X & \xrightarrow{\pi} & G & \longrightarrow & 1 \\ & & & & \big\downarrow{\scriptstyle f} & & \big\downarrow{\scriptstyle id} & & \\ 1 & \longrightarrow & B & \longrightarrow & Y & \xrightarrow{\tau} & G & \longrightarrow & 1. \end{array}$$

**Definition 3.3.** *A universal central extension of $G$ is a central extension $X \xrightarrow{\pi} G$ such that for every other central extension $Y \xrightarrow{\tau} G$ there is a unique homomorphism $f$ over $G$ from $X$ to $Y$.*

Recall that a group $G$ is perfect if $G = [G, G]$.

**Lemma 2.** If $G$ has a universal central extension $X$, then both $G$ and $X$ are perfect.

*Proof.* Let $\psi : X \longrightarrow B = X/[X, X]$ be the natural quotient homomorphism. If $B = X/[X, X]$ is nontrivial, then we would have two homomorphisms over $G$ from $X$ to the central extension $1 \to B \to B \times G \to G \to 1$ :

$$\begin{array}{ccccccccc} 1 & \longrightarrow & \mathrm{Ker} & \longrightarrow & X & \xrightarrow{\pi} & G & \longrightarrow & 1 \\ & & & & \big\downarrow{\scriptstyle 0\,|\,\pi} & & \big\| & & \\ 1 & \longrightarrow & B & \longrightarrow & B \times G & \longrightarrow & G & \longrightarrow & 1 \end{array}$$

and

$$\begin{array}{ccccccccc} 1 & \longrightarrow & \mathrm{Ker} & \longrightarrow & X & \xrightarrow{\pi} & G & \longrightarrow & 1 \\ & & & & \big\downarrow{\scriptstyle \psi\,|\,\pi} & & \big\| & & \\ 1 & \longrightarrow & B & \longrightarrow & B \times G & \longrightarrow & G & \longrightarrow & 1. \end{array}$$

Hence $B$ is trivial and so $X = [X, X]$. Therefore, $G = [G, G]$. □

**Lemma 3.** If $X$ and $Y$ are central extensions of $G$ and $X$ is a perfect group, then there is at most one homomorphism over $G$ from $X$ to $Y$.

*Proof.* Suppose $f_1$ and $f_2$ are two homomorphisms from $X$ to $Y$ over $G$.

$\forall x, x' \in X$, we have $f_1(x) = f_2(x)c$, $f_1(x') = f_2(x')c'$, with $c, c' \in C(Y)$. So

$$\begin{aligned} f_1(xx'x^{-1}(x')^{-1}) &= f_1(x)f_1(x')f_1(x)^{-1}f_1(x')^{-1} \\ &= f_2(x)cf_2(x')c'f_2(x)^{-1}c^{-1}f_2(x')^{-1}c'^{-1} \\ &= f_2(x)f_2(x')f_2(x)^{-1}f_2(x')^{-1} \\ &= f_2(xx'x^{-1}x'^{-1}). \end{aligned}$$

Note that $X$ can be generated by $xx'x^{-1}x'^{-1}$($X$ is perfect). So $f_1 = f_2$. □

**Definition 3.4.** *A presentation of $G$ is an exact sequence*

$$1 \longrightarrow R \longrightarrow F \longrightarrow G \longrightarrow 1,$$

*where $F$ is a free group (free group $\mathscr{F}(X)$ based on $X$, with generating set $X$ and an empty set of defining relations).*

**Lemma 4.** Every presentation of $G$ gives rise to two natural central extensions:

$$1 \longrightarrow R/[R,F] \longrightarrow F/[R,F] \longrightarrow G \longrightarrow 1, \tag{1}$$

$$1 \longrightarrow (R \cap [F,F])/[R,F] \longrightarrow [F,F]/[R,F] \longrightarrow [G,G] \longrightarrow 1. \tag{2}$$

The following fact is well known:

**FACT(Hopf)**: $(R \cap [F,F])/[R,F] \cong H_2(G,\mathbb{Z})$.

*Proof of Lemma* 4. If $r \in R$ and $f \in F$, then $rfr^{-1}f^{-1} = r(fr^{-1}f^{-1}) = rr_1 \in R$, where $r_1 = fr^{-1}f^{-1} \in R$ since $R$ is normal in $F$. Hence, $[R,F] \subset R$.

Note that

$$r(xyx^{-1}y^{-1})r^{-1} = (rxr^{-1})(ryr^{-1})(rxr^{-1})^{-1}(ryr^{-1})^{-1}.$$

On the other hand, $[R,F] \subset R, F/R \cong G$, hence $F/[R,F] \overset{\varphi}{\twoheadrightarrow} G$. We see that $\mathrm{Ker}\varphi = R/[R,F]$, i.e.,

$$1 \longrightarrow R/[R,F] \longrightarrow F/[R,F] \longrightarrow G \longrightarrow 1$$

is exact.

$F \twoheadrightarrow G \Longrightarrow [F,F] \twoheadrightarrow [G,G] \Longrightarrow\ [F,F]/[R,F] \twoheadrightarrow [G,G]$ and

$$1 \longrightarrow (R \cap [F,F])/[R,F] \overset{\psi}{\longrightarrow} [F,F]/[R,F] \overset{\varphi}{\longrightarrow} [G,G] \longrightarrow 1$$

is a complex ($\varphi\psi = 0$).

Clearly, $\psi$ is injective.

If $y \in [F,F]/[R,F]$ with $\varphi(y) = 1$, then by the exact sequence (1), we see that $y \in R/[R,F]$. Hence $y \in R \cap [F,F]/[R,F]$. Therefore (2) is exact. □

If $G = [G,G]$, then both (1) and (2) are central extensions of $G$.

**Lemma 5.** If $(X,\phi)$ is a central extension of a perfect group $G$, then the commutator subgroup $X' = [X,X]$ is perfect and maps onto $G$. Thus $(X', \phi|_{X'})$ is a perfect central extension of $G$.

*Proof.* Since $G$ is generated by commutators and $X$ maps onto $G$, $\phi$ maps $X'$ onto $P$.

Therefore, $\forall x \in X, x = x'c$ with $x' \in X'$ and $c$ central. The fact that diagram

$$\begin{array}{ccccc} X' & \longrightarrow & G & \longrightarrow & 1 \\ \downarrow & & \| & & \\ X & \longrightarrow & G & \longrightarrow & 1 \end{array}$$

is commutative implies that any generator $[x_1, x_2]$ of $X'$ is equal to $[x_1'c_1, x_2'c_2] = [x_1', x_2']$ for some $x_1', x_2' \in X'$, hence $X' = [X', X']$. □

**Theorem 6.** Every perfect group $G$ has a universal central extension:

$$1 \longrightarrow H_2(G, \mathbb{Z}) \longrightarrow [F,F]/[R,F] \longrightarrow G \longrightarrow 1.$$

*Proof.* For any central extension $X$ of $G$, the map $F \longrightarrow G$ lifts to a homomorphism $h : F \longrightarrow X$ since $F$ is free.

$$\begin{array}{ccccc} & {}^{h}\swarrow & F & & \\ & & \downarrow & & \\ X & \xrightarrow{\pi} & G & \longrightarrow & 1, \end{array} \qquad (*)$$

$$\begin{array}{ccccccccc} 1 & \longrightarrow & R & \longrightarrow & F & \longrightarrow & G & \longrightarrow & 1 \\ & & & & \downarrow h & & \| & & \\ 1 & \longrightarrow & Ker\pi & \longrightarrow & X & \xrightarrow{\pi} & G & \longrightarrow & 1. \end{array}$$

$\pi(h(R)) = 1 \in G \quad (*$ commutes),
$\Longrightarrow h(R) \subseteq X$,
$\Longrightarrow h([R, F]) = 1$.
Hence $h$ induces a homomorphism

$$[F,F]/[R,F] \longrightarrow X \text{ over } G.$$

However, $(F/[R,F])' = [F,F]/[R,F]$. By (1) of Lemma 4 and Lemma 5, we see that $[F,F]/[R,F]$ is perfect.

By Lemma 3, this homomorphism is unique, and so we are done. □

**FACT**: Lemma 2 implies that: If $Y$ is a central extension of $G$ which is not perfect, then for a suitable $(X, \phi)$, there exist more than one homomorphism from $Y$ to $X$ over $G$.

**Lemma 7.** Let $U$ be any central extension of $G$. The following are equivalent:
(1) $U$ is a universal central extension of $G$;
(2) $U$ is perfect and every central extension of $U$ splits.

*Proof.* Suppose every central extension of $U$ splits. Given a central extension $(X, \phi)$ of $G$, a homomorphism from $U$ to $X$ over $G$ can be constructed as follows:

Let $U \underset{G}{\times} X$ be the fiber product of $U$ and $X$ over $G$, i.e.,

$$U \underset{G}{\times} X = \{(u, x) \mid u \in U, x \in X,\ v(u) = \phi(x)\ \},$$

where $v$ is a homomorphism from $U$ to $G$.

Let $\pi$ be the projection $U \underset{G}{\times} X$ to $U$, i.e., $\pi(u, x) = u$.

Claim: $(U \underset{G}{\times} X, \pi)$ is a central extension of $U$. In fact,

(1) $\pi$ surjective: For any $u \in U$, since $\phi$ is surjective $\exists x \in X, s.t.\ \phi(x) = v(u)$, i.e., $\pi$ surjective.

(2) $\mathrm{Ker}\pi = \{(1, x) | 1 = v(1) = \phi(x)\}$.

The fact that $\mathrm{Ker}\phi$ is in the center of $X$ implies that $\mathrm{Ker}\pi$ is in the center of $U \underset{G}{\times} X$.

Hence we have a section $s : U \longrightarrow U \underset{G}{\times} X$. Write $s(u) = (u, h(u))$.

Then $h : U \longrightarrow X$ is over $G$, i.e., $\phi h = v$ $((u, h(u)) \in U \underset{G}{\times} X \Longleftrightarrow v(u) = \phi(h(u)))$.

$$\begin{array}{ccc} U & \xrightarrow{v} & G \\ {\scriptstyle h}\big\downarrow & & \big\| \\ X & \xrightarrow{\phi} & G \end{array}$$

Suppose $U$ is perfect. Then $h$ is unique.

This shows (2) $\Longrightarrow$ (1).

(1) $\Longrightarrow$ (2) :

Suppose $(U, v)$ is a universal central extension of $G$. Then $U$ is perfect.

It suffices to show that every central extension of $U$ splits.

Suppose $(X, \phi)$ is a central extension of $U$. Then the composition

$$X \xrightarrow{\phi} U \xrightarrow{v} G$$

is a central extension of $G$.

In fact, if $x_0 \in \mathrm{Ker}(v\phi)$, then $\phi(x_0) \in U$, so the correspondence

$$x \longmapsto x_0 x x_0^{-1}$$

defines a homomorphism from $X$ to $X$ over $U$. Restricting to the commutator subgroup $X'$, we see from Lemma 3 and Lemma 5 that the resulting homomorphism $X' \longrightarrow X'$ over $U$ is the identity.

Hence $x_0$ commutes with all elements of $X'$.

But $X$ is generated by $X'$ together with the central subgroup $\mathrm{Ker}\phi$, so it follows that $x_0$ commutes with all elements of $X$.

Therefore $(X, v \circ \phi)$ is a central extension of $G$. $(U, v)$ is universal implies that there exists a homomorphism $s : U \longrightarrow X$ over $G$. The composition $\phi \circ s$ is a homomorphism from $U$ to $U$ over $G$, so equals the identity. Hence $s$ is a section of $(X, \phi)$. Therefore every central extension of $U$ splits. □

**Theorem 8(Kervaire).** The Steinberg group $St(\Lambda)$ is the universal central extension of $E(\Lambda)$. So $K_2\Lambda \cong H_2(E(\Lambda), \mathbb{Z})$. □

This theorem follows from Theorem 6 and the following Lemma.

**Lemma 9.** For $n \geqslant 5$, every central extension $Y \xrightarrow{\pi} St(n, \Lambda)$ is split. Hence $St(n, \Lambda)$ is the universal central extension of $E(n, \Lambda)$.

*Proof.* We will construct a section $Y \xrightarrow{\pi} St(n, \Lambda)$, i.e., a homomorphism $St(n, \Lambda) \longrightarrow Y$.

(1) If $j \neq k$ and $l \neq i$, then every two elements $y, z \in Y$ with $\pi(y) = x^{\lambda}_{ij}, \pi(z) = x^{\mu}_{kl}$ must commute in $Y$.

Let $t \in \mathbb{N}$ with $t \notin \{i, j, k, l\}$. Choose $y', y'' \in Y$ with $\pi(y') = x^1_{it}, \pi(y'') = x^{\lambda}_{tj}$. Then

$$\pi([y', z]) = [x^1_{it}, x^{\mu}_{kl}] = 1 (\because i \neq l, t \neq k),$$

$$\pi([y'', z]) = [x^{\lambda}_{tj}, x^{\mu}_{kl}] = 1 (\because t \neq l, j \neq k).$$

$\Longrightarrow [y', z]$ and $[y'', z]$ are in $C(Y)$.

$\Longrightarrow zy' = c'y'z$, for some $c' \in C(Y)$.

Similarly, $zy'' = c''y''z$, for some $c'' \in C(Y)$.

$z[y', y''] = zy'y''y'^{-1}y''^{-1} = c'y'zy''y'^{-1}y''^{-1} = \cdots = [y', y'']z$.

Hence $\pi([y', y'']) = \pi(y)$

$\Longrightarrow c[y', y''] = y, c \in C(Y)$.

$\Longrightarrow zy = yz$.

(2) Some identities between commutators (for any group $G$), $u, v, w \in G$.

$$[u, v] = [v, u]^{-1},$$

$$[u, v][u, w] = [u, vw][v, [w, u]], \tag{i}$$

$$[u, [v, w]][v, [w, u]][w, [u, v]] \equiv 1 \bmod G'' \text{ (Jacobi identity)}, \tag{ii}$$

where $G'' = [[G, G], [G, G]]$ (The second commutator subgroup).

(i): $[u, vw][v, [w, u]]$

$= uvwu^{-1}w^{-1}v^{-1} \cdot v \cdot wuw^{-1}u^{-1} \cdot v^{-1}[u, w]$

$= [u, v] \cdot vuu^{-1}v^{-1}[u, w]$

$= [u, v][u, w]$.

(ii): (i)$\Longrightarrow [v, [w, u]] = [vw, u][u, v][u, w]$.

The left of (ii)

$= [uv, w][w, u][w, v] \cdot [vw, u][u, v][u, w] \cdot [wu, v][v, w][v, u]$

$\equiv [uv, w][vw, u][wu, v] \bmod G''$

$\equiv [uv, w][wu, v][vw, u] \bmod G''$

$= 1$.

If $uw = wu$, then

$$[[u, v], w] \equiv [u, [v, w]] \bmod G''. \tag{iii}$$

(3) We choose distinct indices $i, j, k, l$ and elements $u, v, w \in Y$ with

$$\pi(u) = x^1_{ij}, \pi(v) = x^{\lambda}_{jk}, \pi(w) = x^{\mu}_{kl}.$$

We have $[u, w] = 1$.

Let $G$ denote the subgroup of $Y$ generated by $u, v$ and $w$.

Then the commutator subgroup $G'$ is generated by elements lying in $\pi^{-1}(x_{hj}^{\lambda})$, $\pi^{-1}(x_{ik}^{\lambda\mu}), \pi^{-1}(x_{hk}^{\lambda\mu})$, so $G'' = 1$.

$([u, w] \longrightarrow [x_{hi}^{1}, x_{ij}^{\lambda}] = x_{hj}^{\lambda}$,
$[v, w] \longrightarrow [x_{ij}^{\lambda}, x_{jk}^{\mu}] = x_{ik}^{\lambda\mu}$,
$[[u, v], w] \longrightarrow [x_{hj}^{\lambda}, x_{jk}^{\mu}] = x_{hk}^{\lambda\mu}$,
$[u, w] = 1$,
$[x_{hj}^{*}, x_{ih}^{*}] = 1, [x_{hj}^{*}, x_{hk}^{*}] = 1, [x_{ik}^{*}, x_{hk}^{*}] = 1$,
$\Longrightarrow G'' = 1.)$

Hence $[[u, v], w] = [u, [v, w]]$. In other words,

$$[\pi^{-1}(x_{hj}^{\lambda}), \pi^{-1}(x_{jk}^{\mu})] = [\pi^{-1}(x_{hi}^{1}), \pi^{-1}(x_{ik}^{\lambda\mu})]. \tag{iv}$$

Taking $\lambda = 1$, we see that

$$s_{hk}^{\mu} \in [\pi^{-1}(x_{hj}^{1}), \pi^{-1}(x_{jk}^{\mu})]$$

does not depend on the choice of $j$.

Note that

$$s_{hk}^{\mu} \in \pi^{-1}(x_{hk}^{\mu}).$$

Rewrite (iv) as

$$[s_{hj}^{\lambda}, s_{jk}^{\mu}] = s_{hk}^{\lambda\mu} \text{ for } h, j, k \text{ distinct.}$$

This is the Steinberg relation (2).

We are going to prove the Steinberg relation (1):

Apply $[u, v][u, w] = [u, vw][v, [w, u]]$ to elements $u \in \pi^{-1}(x_{hj}^{1}), v \in \pi^{-1}(x_{jk}^{\lambda})$, $w \in \pi^{-1}(x_{jk}^{\mu})$.

Then $[u, v] = s_{hk}^{\lambda}, [u, w] = s_{hk}^{\mu}, [v, w] = s_{hk}^{\lambda+\mu}$ and $[v, [w, u]] \in [\pi_{-1}(x_{jk}^{\lambda})$, $\pi^{-1}(x_{hk}^{-\mu})] = 1$.

Hence

$$s_{hk}^{\lambda} \cdot s_{hk}^{\mu} = s_{hk}^{\lambda+\mu}.$$

The Steinberg relation (3) is a consequence of (i), since

$$[\pi^{-1}(x_{ij}^{\lambda}), \pi^{-1}(x_{kl}^{\mu})] = 1. \qquad \square$$

## 4. The $K_2$ of a field

**Definition 4.1.** *A Steinberg symbol on a field $F$ with values in a multiplicative abelian group $A$ is a bilinear map:*

$$c\colon F^{\cdot} \times F^{\cdot} \longrightarrow A$$

*satisfying $c(r, 1 - r) = 1$.*

**Examples.** (a) The Hilbert symbol of order 2. Let $F$ be a local field with characteristic $\neq 2$. The (quadratic residue) Hilbert symbol $(a,b)$ on $F^{\cdot} \times F^{\cdot}$ equals to $+1$ or $-1$, depending on whether or not the equation $ax^2 + by^2 = 1$ has a solution in $F$. In particular, $(a,b)_{\infty}$(on the field $\mathbb{R}$)$= -1$ if and only if both $a$ and $b$ are negative.

(b) The Hilbert symbol of order $m$. Let again $F$ be a local field with $\mu_m \subset F$ and characteristic $\neq m$. There is a finite "Kummer" extension $K$ containing the $m^{th}$ root of every element of $F$.

$$\begin{array}{lll} G_F = \mathrm{Gal}(K/F) & \overset{can.}{\cong} & \mathrm{Hom}(F^{\cdot}, \mu_m) \\ g & \longmapsto & \tilde{g} \quad (\tilde{g} : F^{\cdot} \longrightarrow \mu_m, \tilde{g}(a) = \frac{g(\sqrt[m]{a})}{\sqrt[m]{a}}). \end{array}$$

The cokernel of the norm map $K^{\cdot} \overset{N}{\longmapsto} F^{\cdot}$ is isomorphic to $G_F$ by the "norm residue" isomorphism of local class field theory. The composite

$$F^{\cdot} \rightarrow F^{\cdot}/NK^{\cdot} \cong G_F \cong \mathrm{Hom}(F^{\cdot}, \mu_m)$$

written as $x \longmapsto (x, -)_F$ is adjoint to a nondegenerate bilinear map $(,)_F$, the Hilbert symbol of order $m$: $F^{\cdot} \otimes F^{\cdot} \longrightarrow \mu_m$.

(c) Tame symbols. A discrete valuation on a field $F$ determines a Steinberg symbol on $F$ with values in $F$, the multiplicative group of the residue field $k$ of the valuation. This symbol is called the tame symbol and is defined by

$$\tau_{\mathfrak{p}}\{x, y\} = (-1)^{v_{\mathfrak{p}}(x)v_{\mathfrak{p}}(y)} x^{v_{\mathfrak{p}}(y)} y^{-v_{\mathfrak{p}}(x)} \pmod{\mathfrak{p}},$$

where $\mathfrak{p}$ is the valuation ideal.

(d) Let $F$ be a field, $\mu_n \subset F$. Let $\zeta_n \in \mu_n$ be a primitive $n$-th root of unity. Given elements $\alpha$ and $\beta$ in $F^{\cdot}$, let

$$A = A_{\zeta_n}(\alpha, \beta)$$

be the associative algebra with unit, of dimension $n^2$ over $F$, which is generated by two elements $x$ and $y$ subject to the relations:

$$x^n = \alpha \cdot 1_A, \; y^n = \beta_n \cdot 1_A, \; yx = \zeta_n xy$$

($x^i y^j, 0 \leqslant i < n, 0 \leqslant j < n$ are to form a basis for $A$ over $F$).

An algebra $A$ over $F$ (always associative with 1) is simple if it has no two-sided ideals other than 0 and $A$; and central if its center is equal to $F \cdot 1 \cong F$.

**Wedderburn's Theorem:** Every finite dimension simple algebra over $F \cong M_k(D)$ for some division algebra $D \supset F$ and $k \geq 1$.

The isomorphism class of $D$ are uniquely determined by $A$.

$A \sim A'$ (similar)$\Longleftrightarrow A = M_k(D), A' = M_{k'}(D'), D \overset{F}{\cong} D'$.

We use $[A]$ for the similarity class of $A$.

Define $[A] \cdot [B] = [A \otimes B]$ and $Br(F) := \{[A] \mid$ A finite dimensional central simple algebra$\}$.

Then $F^{\cdot} \times F^{\cdot} \longrightarrow Br(F)$,

$$(\alpha, \beta) \longrightarrow [A_{\zeta_n}(\alpha, \beta)]$$

is a Steinberg symbol, which depends on the choice of $\zeta_n$.

But

$$(\alpha, \beta) \longrightarrow \zeta_n \otimes [A_{\zeta_n}(\alpha, \beta)] \in \mu_n \otimes \; {}_nBr(F)$$

is independent of this choice.

In the local fields case, this symbol is equivalent to the Hilbert symbol on $F$ of order $n$.

$$\begin{array}{ccc} & \overset{(\frac{\cdot}{F})}{\nearrow} & \mu_n \\ F^{\cdot} \times F^{\cdot} & & \downarrow \cong \\ & \underset{\zeta_n \otimes [,]}{\searrow} & \mu_n \otimes \; {}_nBr(F) \end{array}$$

We now turn to the discussion of the $K_2$ of a field. We need some notation.

Let $\Lambda$ be a ring and $u \in \Lambda$ a unit.

We put:

$$w_{ij}(u) = x_{ij}^{u} x_{ji}^{-u^{-1}} x_{ij}^{u}, \; h_{ij}(u) = w_{ij}(u) w_{ij}(-1).$$

Then

$$w_{ij}(u) w_{ij}(-u) = 1, h_{ij}(1) = 1.$$

Let $W \subset St(n, \Lambda)$ denote the subgroup generated by all the $w_{ij}(u)$.

We have the following facts:

• The kernel $C_n$ of the homomorphism $\phi|_W : W \longrightarrow GL(n, \Lambda)$ lies in $C(St(n, \Lambda))$.

If $\Lambda$ is commutative, then the symbol

$$\{u, v\} = [h_{ij}(u), h_{ik}(v)]$$

is skew-symmetric and bi-multiplicative with values in the commutative group $C_n$. We have

$$C_n \longrightarrow \lim C_n \subset K_2\Lambda.$$

• If both $u$ and $1 - u$ are units, then $\{u, 1 - u\} = 1$. Furthermore, for any unit $u$ the identity $\{u, 1 - u\} = 1$ is valid.

• If $\Lambda$ is a field or skew-field, then the kernel of $\phi : St(n, \Lambda) \longrightarrow GL(n, \Lambda)$ is the $C_n$. Therefore the Steinberg group $St(n, \Lambda)$ is a central extension of $E(n, \Lambda)$ for every $n \geqslant 3$.

• If $\Lambda$ is a field, then $K_2\Lambda$ is generated by the symbols $\{u, v\}$.

**Theorem 10(Matsumoto).** The abelian group $K_2F$ has a presentation, in terms of generators $\{x, y\}$, with $x, y \in F^{\bullet}$, and relations as follows:

(1) $\{x, 1-x\} = 1$,
(2) $\{x_1x_2, y\} = \{x_1, y\}\{x_2, y\}$,
and (3) $\{x, y_1y_2\} = \{x, y_1\}\{x, y_2\}$.

*Proof.* Let $F^\bullet \times F^\bullet \longrightarrow A$ be the universal Steinberg symbol on the field $F$. In other words, let $A$ be the abelian group which is defined by generators $c(u, v)$ subject only to the relations $c(u_1u_2, v) = c(u_1, v)c(u_2, v), c(u, v_1v_2) = c(u, v_1)c(u, v_2)$ and $c(u, 1-u) = 1$ and to the consequences of these relations.

Let $\{x, y\} \in K_2F$ be the symbol as above. Since $\{u, v\}$ is bimultiplicative and satisfies $\{u, 1-u\} = 1$, there is a unique homomorphism

$$\eta : A \longrightarrow K_2F,$$

which carries $c(u, v)$ to $\{u, v\}$ for all $u$ and $v$ in $F$. We want to prove $\eta$ is an isomorphism.

We will use without proof the following assertion.

There is a group $G$ such that

$$1 \longrightarrow A \longrightarrow G \longrightarrow SL(F) \longrightarrow 1,$$

is a central extension of $SL(F)$ (see Milnor [20], §12).

But

$$1 \longrightarrow K_2F \longrightarrow St(F) \longrightarrow SL(F) \longrightarrow 1$$

is the universal central extension of $SL(F)$.

So there is a unique homomorphism $\xi : St(F) \longrightarrow G$ such that the following diagram is commutative:

$$\begin{array}{ccccccccc} 1 & \longrightarrow & A & \longrightarrow & G & \longrightarrow & SL(F) & \longrightarrow & 1 \\ & & & & \uparrow \exists!\ \xi & & \| & & \\ 1 & \longrightarrow & K_2F & \longrightarrow & St(F) & \longrightarrow & SL(F) & \longrightarrow & 1. \end{array}$$

Clearly, $\xi$ maps $K_2F$ into $A$.

Again from §12 in Milnor's book, we see that $\xi$ carries $\{u, v\}$ to $c(u, v)$ for all $u$ and $v$. But $\eta$ carries $c(u, v)$ to $\{u, v\}$ for all $u$ and $v$. Since $K_2F$ is generated by the symbols $c(u, v)$, this completes the proof that $K_2F \cong A$. □

**Example.** Let $F$ be a finite field. Then $K_2F = 0$.

In fact, $F^*$, the multiplicative group of $F$, is a cyclic group. Write $F^* = \langle g \rangle$. $\forall u, v \in F^*, u = g^n, v = g^m$. Hence,

$$\{u, v\} = \{g^m, g^n\} = \{g, g\}^{mn}.$$

It is clear that

$$\{g, g\} = \{-1, g\}.$$

If $F$ has characteristic 2, then $-1 = 1$, so $\{g, g\} = 1$. Therefore $K_2F = 0$.

If $F$ has odd characteristic, then consider the bijection

$$\begin{array}{ccc} F^* - \{1\} & \longrightarrow & F^* - \{1\}, \\ v & \longrightarrow & 1 - v \,. \end{array}$$

$F^* - \{1\}$ has $\frac{q-1}{2}(q = \#F)$ non-squares, $\frac{q-3}{2} = \frac{q-1}{2} - 1$ squares.
So there exists $v$ such that both $v$ and $1 - v$ are non-squares.
Hence $v = g^i, 1 - v = g^j$, for some $i, j$ odd.
$1 = \{v, 1 - v\} = \{g^i, g^j\} = \{g, g\}^{ij} = \{-1, g\}^{ij}$. But $\{-1, g\}^2 = 1$.
Hence $\{-1, g\} = 1$ and so $K_2F = 0$.

# 5. K-Theory and Galois Cohomology

## 5.1 Galois Cohomology

Let $G$ be a topological group and $M$ a topological $G$-module. Put

$$C^n(G, M) = \{f : G^n \longrightarrow M \mid f \text{ continuous}\}.$$

Homomorphisms $d_n : C^n(G, M) \longrightarrow C^{n+1}(G, M)$ are defined as usual:
$(d_0 f)(s) = sf(\cdot) - f(\cdot)(\cdot$ the unique element of $G^0)$,
$(d_1 f)(s, t) = sf(t) - f(st) + f(s)$,
$(d_2 f)(s, t, u) = sf(t, u) - f(st, u) + f(s, tu) - f(s, t)$, etc.

$$C^{\cdot}(G, M): \quad 0 \longrightarrow C^0(G, M) \xrightarrow{d_0} C^1(G, M) \xrightarrow{d_1} C^2(G, M) \xrightarrow{d_2} \cdots$$

is a complex.

Define

$$H^i(G, M) = \mathrm{Ker} d_i / \mathrm{Im} d_{i-1}.$$

In particular,

$$H^0(G, M) = M^G,$$

$$H^1(G, M) = \tilde{Z}^1(G, M)/\tilde{B}^1(G, M),$$

$$H^2(G, M) = \tilde{Z}^2(G, M)/\tilde{B}^2(G, M).$$

where

$\tilde{Z}^1(G, M) = \{f : G \longrightarrow M \text{ continuous}, \sigma f(\tau) - f(\sigma\tau) + f(\sigma) = 0, \forall\, \sigma, \tau \in G\}$,
$\tilde{B}^1(G, M) = \{f_a \mid f_a(\sigma) = \sigma a - a, \sigma \in G\}$;
$\tilde{Z}^2(G, M) = \{f : G^2 \longrightarrow M \text{ continuous}, \forall\, \sigma, \tau, \rho \in G,\ \sigma f(\tau, \rho) - f(\sigma\tau, \rho) + f(\sigma, \tau\rho) - f(\sigma, \tau) = 0\}$,
$\tilde{B}^2(G, M) = \{f : G \longrightarrow M,\ \partial f : G^2 \longrightarrow M \text{ continuous}, \forall\, \sigma, \tau \in G,\ \partial f(\sigma, \tau) = \sigma f(\tau) - f(\sigma\tau) + f(\sigma)\}$.

1. Exact cohomology sequence

Suppose

$$0 \longrightarrow L \longrightarrow M \longrightarrow N \longrightarrow 0$$

is an exact sequence of topological $G$-modules such that the topology of $L$ is induced by that of $M$, and such that the map $M \longrightarrow N$ has a continuous section (not necessarily a homomorphism).

Then we have the sequence of complexes

$$0 \longrightarrow C^{\cdot}(G, L) \longrightarrow C^{\cdot}(G, M) \longrightarrow C^{\cdot}(G, N) \longrightarrow 0,$$

which induces a long exact cohomology sequence

$$\cdots \longrightarrow H^r(G, M) \longrightarrow H^r(G, N) \xrightarrow{\delta} H^{r+1}(G, L) \longrightarrow H^{r+1}(G, M) \longrightarrow \cdots.$$

2. Cup product

Let $B : M_1 \times M_2 \longrightarrow M$ be continuous $G$-pairing, *i.e.*, continuous biadditive map such that

$$(sx_1) \cdot (sx_2) = s(x_1 \cdot x_2), x_i \in M_i, s \in G.$$

Then $B$ induces biadditive maps: $C^m(G, M_1) \times C^n(G, M_2) \longrightarrow C^{m+n}(G, M)$ via the formulas:

$(f \cup g)(s_1, \cdots, s_{m+n}) = f(s_1, \cdots, s_m)s_1 \cdots s_m g(s_{m+1}, \cdots, s_{m+n})$, where $f : G^m \longrightarrow M_1, g : G^n \longrightarrow M_2$.

We have

$$\delta(f \cup g) = (\delta f) \cup g + (-1)^m f \cup (\delta g),$$

which induces pairings

$$H^m(G, M_1) \times H^n(G, M_2) \longrightarrow H^{m+n}(G, M).$$

3. Restriction and transfer

Let $H$ be a subgroup of $G$ and $M$ a topological $G$-module.

The embedding $H \hookrightarrow G$ gives a homomorphism of complexes

$$\text{res} : H^n(G, M) \longrightarrow H^n(H, M).$$

Assume moreover that $H$ is open with finite index in $G$.

Let $R$ be a set of representative of the right cosets of $H$ in $G$, i.e., $G = \cup_{r \in R} Hr$, disjoint union.

Define:

$$(t_R f)(s_1, \cdots, s_n) = \sum_{r \in R} r_0^{-1} f(r_0 s_1 r_1^{-1}, r_1 s_2 r_2^{-1}, \cdots, r_{n-1} s_n r_n^{-1}),$$

where $r_i \in R$ are determined inductively as functions of $r : r_0 = r, r_i \in Hr_{i-1}s_i$, $1 \leqslant i \leqslant r$. This gives a homomorphism of complexes

$$t_R : C(H, M) \longrightarrow C(G, M).$$

The homotopy class of $t_R$ is independent of $R$, hence, we have a homomorphism:

$$\text{tr} : H^n(H, M) \longrightarrow H^n(G, M).$$

When $n = 0$, tr is the trace: $M^H \longrightarrow M^G : x \longrightarrow \sum_{r \in R} r^{-1}x$.

We have:

$$\operatorname{tr}(x \cup \operatorname{res} y) = (\operatorname{tr} x) \cup y, x \in H^m(H, M_1), y \in H^n(G, M_2).$$

For a field $F$, let $F_s$ be a separable algebraic closure of $F$. Let $G_F = \operatorname{Gal}(F_s/F)$. Write $H^r(F, M)$ for $H^r(G_F, M)$.

**Lemma 11.** Let $E/F$ be a finite Galois extension of degree $n$. Then the composition

$$H^i(F, M) \xrightarrow{\text{res}} H^i(E, M) \xrightarrow{\text{tr}} H^i(F, M)$$

satisfies

$$\operatorname{tr} \circ \operatorname{res} = n.$$

In the opposits direction, we have

$$\operatorname{res} \circ \operatorname{tr}(y) = \sum_{s \in G} s(y).$$

The same are true for $K_2F \longrightarrow K_2E$. □

Some useful $G_F$ modules $M$ :

1. $\mu_n$.
2. $\mu_n^{\otimes m} : \sigma \in G_F, \sigma(\zeta \otimes \zeta \otimes \cdots \otimes \zeta) = \sigma(\zeta) \otimes \sigma(\zeta) \otimes \cdots \otimes \sigma(\zeta)$.
3. $\mathbb{Z}_l(1) = \varprojlim(\mu_{l^i}), (\cdots \longrightarrow \mu_{l^{i+1}} \xrightarrow{*^l} \mu_{l^i} \longrightarrow \mu_{l^{i-1}} \longrightarrow \cdots)$.

So $\mathbb{Z}_l(1)$ is a free $\mathbb{Z}_l$-module of rank 1.
Define $G_F$-modules $\mathbb{Z}_l(m)$ inductively by
$\mathbb{Z}_l(0) = \mathbb{Z}_l, \mathbb{Z}_l(m+1) = \mathbb{Z}_l(m) \otimes_{\mathbb{Z}_l} \mathbb{Z}_l(1), m \geq 0;$
$\mathbb{Z}_l(m-1) = \operatorname{Hom}(\mathbb{Z}_l(1), \mathbb{Z}_l(m)), m \leq 0.$
For $(\mathbb{Z}_l, G_F)$-module $M$, define $M(m) = M \otimes_{\mathbb{Z}_l} \mathbb{Z}_l(m)$. We have $M(m)(n) \cong M(m+n)$.

4. $\mathbb{Q}_l/\mathbb{Z}_l$, $\mathbb{Q}_l/\mathbb{Z}_l(n)$.
5. $F_s^{\cdot}$.
6. $W = \{\zeta_{l^\alpha} \mid \alpha \geq 0\}$.

Let $F$ be a field. Consider the Kummer sequence of $G_F$-modules,

$$0 \longrightarrow \mu_n \longrightarrow \overline{F_s^*} \xrightarrow{\cdot n} \overline{F_s^*} \longrightarrow 0.$$

We have:

$$H^0(F, \overline{F_s}^*) \xrightarrow{\cdot n} H^0(F, \overline{F_s}^*) \longrightarrow H^1(F, \mu_n) \longrightarrow H^1(F, \overline{F_s}^*) = 0$$
(Hilbert Theorem 90),

i.e., $F^* \xrightarrow{\cdot n} F^* \longrightarrow H^1(F, \mu_n) \longrightarrow 0$ is exact, which gives an isomorphism

$$\chi_n : F^* \otimes_{\mathbb{Z}} \mathbb{Z}/n\mathbb{Z} \xrightarrow{\cong} H^1(F, \mu_n).$$

**Lemma 12.** If $a \in F^*, a \neq 1$, then

$$\chi_n(a) \cup \chi_n(1-a) = 0. \qquad \square$$

Consider the following maps:

$$\begin{aligned} F^* \underset{\mathbb{Z}}{\otimes} F^* &\longrightarrow (F^* \otimes F^*) \underset{\mathbb{Z}}{\otimes} \mathbb{Z}/n\mathbb{Z} \\ &\longrightarrow H^1(F, \mu_n) \otimes H^1(F, \mu_n) \\ &\longrightarrow H^2(F, \mu_n^{\otimes 2}). \end{aligned}$$

**Corollary 13.** The map

$$F^{\cdot} \otimes F^{\cdot} \longrightarrow H^2(F, \mu_n^{\otimes 2}) \qquad (*)$$

given by $a \otimes b \longmapsto \chi_n(a) \cup \chi_n(b)$ is a Steinberg symbol. $\square$

Let

$$R_{n,F} : K_2F \underset{\mathbb{Z}}{\otimes} \mathbb{Z}/n\mathbb{Z} \longrightarrow H^2(F, \mu_n^{\otimes 2})$$

be the homomorphism given by $(*)$. It is called the norm residue homomorphism or the Galois symbol.

**Theorem 14 (Merkurjev-Suslin).** Let $F$ be a field and $n$ a positive integer not divisible by $\mathrm{char}F$. Then

$$K_2F \underset{\mathbb{Z}}{\otimes} \mathbb{Z}/n\mathbb{Z} \overset{R_{n,F}}{\cong} H^2(F, \mu_n^{\otimes 2}) \cong {}_nBr(F).$$

$(Br(F) = H^2(F, \overline{F_s^{\cdot}}))$ $\square$

Recall that for a finite extension of fields $E/F$, the transfer:

$$\mathrm{tr}_{E/F} : K_2E \longrightarrow K_2F$$

is a group homomorphism. We have

(1) $x \in F, y \in E, \mathrm{tr}_{E/F}\{x, y\} = \{x, N_{E/F}y\}$;

(2) The composition

$$K_2F \xrightarrow{j} K_2E \xrightarrow{\mathrm{tr}_{E/F}} K_2F$$

is a multiplication by the degree of $E/F$ ($j$ : induced by $F \subset E$).

**Theorem 15 (Merkurjev-Suslin)(Hilbert 90 for $K_2$).** For $E/F$ a cyclic Galois extension with Galois group $\Gamma$ generated by $\sigma$, the following sequence is exact:

$$K_2E \xrightarrow{1-\sigma} K_2E \xrightarrow{\mathrm{tr}_{E/F}} K_2F.$$

So we have $(K_2E)_\Gamma \hookrightarrow K_2F$. $\square$

## 5.2 Torsion Elements in $K_2F$

It is interesting to investigate the torsion part of $K_2$ of a field $F$. In fact, for a global field $F$, $K_2F$ is a torsion group.
**Theorem 16.** If $\text{char}(F) = p \neq 0$, then the group $K_2F$ has no $p$-torsion.

The following theorem is first proved by Tate for global fields and then by Suslin for general.
**Theorem 17.** If $\zeta_n \in F$, then every element of order $n$ has the form $\{\zeta_n, x\}$ for some $x \in F^{\cdot}$.

Let $\Phi_n(x)$ be the $n$th cyclotomic polynomial, and let $G_n(F) = \{\{a, \Phi_n(a)\} | \, a, \Phi_n(a) \in F^*\}$, where $F$ is any field. For an abelian group $H$, and $n \in \mathbb{N}$, let ${}_nH = \{x \in H \mid x^n = 1\}$. J.Browkin[3] proved that

(1) For every field $F$ and every $n \in \mathbb{N}, \alpha^n = 1$ if $\alpha \in G_n(F)$;

(2) For every field $F \neq \mathbf{F}_2$ and $n = 1, 2, 3, 4$ or $6$, $G_n(F)$ is a subgroup of $K_2(F)$;

(3) ${}_3(K_2\mathbb{Q}) = G_3(\mathbb{Q})$;

(4) In $K_2\mathbb{Q}$ every element of order 4 has the form of $\{a, a^2 + 1\}v$, where $a \in \mathbb{Q}^*$ and $v \in K_2\mathbb{Q}, v^2 = 1$.

Moreover, he conjectured the following

(A) (2) does not hold for any other values of $n$ (and all fields);

(B) (3) and (4) hold for all fields.

J. Urbanowicz[43] proved that (3) is true for all fields and I[26] proved that (4) is true, so we have (B).

The author[32] confirmed (A) for $n = 5$ and 7, i.e., neither $G_5(\mathbb{Q})$ nor $G_7(\mathbb{Q})$ is a group. The same method is applied to prove $G_{11}(\mathbb{Q})$ is not a group, see [52]. Some other results in this direction can be found in [5,11,51] and among others. However,

$$G_p(\mathbb{Q}) \text{ is not a group for prime } \; p > 11,$$

is still open.

On the other hand, the following problem seems difficult to answer:

Let $p > 5$ be a prime.

$${}_p(K_2\mathbb{Q}) = < G_p(\mathbb{Q}) > ?$$

i.e., ${}_p(K_2(\mathbb{Q}))$ should be generated by $G_p(\mathbb{Q})$. Lenstra (A letter from Lenstra to Browkin, May 19, 1981) once proposed a computational proof for ${}_5(K_2\mathbb{Q}) = < G_5(\mathbb{Q}) >$ . An alternative proof has been given by Xu[50].

Browkin points out that ${}_n(K_2\mathbb{Q}) = < G_n(\mathbb{Q}) >$ does not hold for all $n$. However, we expect that the following should be true for all positive integers $n$ :

$${}_n(K_2\mathbb{Q}) = < G_m(\mathbb{Q}) | \;\; m \mid n > .$$

Of course, we can ask the same questions for all fields $F$. In particular, we may expect that

(i) For any global field $F, G_p(F)$ is not a group for prime $p > 3$;

(ii) For any field $F$ and any prime $p$, ${}_p(K_2F) = < G_p(F) >$;

(iii) For any field $F$ and any positive integer $n$, ${}_n(K_2F) = < G_m(F) \mid m \mid n >$ .

### 5.3 Milnor conjecture and Bloch-Kato Conjecture

Let $F$ be a field and $\ell$ a prime, with $\ell \nmid \mathrm{char} F$.

$$K_n^M(F) := F^{\cdot} \otimes \cdots \otimes F^{\cdot} / \langle a_1 \otimes \cdots \otimes a_n \mid a_i \in F^{\cdot}, \exists\, i, j,\ a_i + a_j = 1 > .$$

The isomorphism

$$F^{\cdot}/(F^{\cdot})^{\ell} \cong H^1(F, \mu_\ell)$$

gives the norm residue homomorphism

$$K_n^M(F)/\ell K_n^M(F) \longrightarrow H^n(F, \mu_\ell^{\otimes n}).$$

**Milnor Conjecture ($\ell = 2$):**

$$K_n^M(F)/2K_n^M(F) \cong H^n(F, \mu_2^{\otimes n}).$$

This conjecture was proved by Voevodsky[45].

**Bloch-Kato Conjecture:**

$$K_n^M(F)/\ell K_n^M(F) \cong H^n(F, \mu_\ell^{\otimes n}), \ell \text{ odd}.$$

This was also proved by Voevodsky[46]-modulo some assertions (e.g., the Rost variety has certain properties,...). See [48] for more in detail. One of the reductions of Voevodsky's proof is the following:

**Theorem 18(Voevodsky).** If $K_n^M(F)/\ell K_n^M(F) \cong H^n(F, \mu_\ell^{\otimes n})$ for all fields of characteristic=0, then it is an isomorphism for all fields of characteristic$\neq \ell$.

## 6. $K_n(O_F)$ for the Ring of Integers in an Algebraic Number Field

Let $F$ be a number field and $O_F$ the ring of its integers. It is well known that

$$K_0(O_F) \cong \mathbb{Z} \oplus Cl(O_F)$$

and by the Bass-Milnor-Serre theorem and the Dirichlet unit theorem,

$$K_1(O_F) \cong O_F^* \cong W \oplus \mathbb{Z}^{r_1+r_2-1},$$

where $Cl(O_F)$ is the ideal class group of $F$, $W$ is the group of roots of unity of $F$, and $r_1, 2r_2$ are the numbers of the real and complex embeddings of $F$ respectively.

By Garland [10], $K_2(O_F)$ is a finite (abelian) group. The following exact sequence is well-known:

$$0 \to K_2 O_F \to K_2(F) \xrightarrow{\tau} \oplus_{\mathfrak{p}\ \text{finite}} k_{\mathfrak{p}}^* \to 0,$$

where $k_{\mathfrak{p}}$ is the residue class field of $F$ at the finite prime $\mathfrak{p}$, the sum is extended over all finite places $\mathfrak{p}$ of the number field $F$, and the homomorphism $\tau = (\tau_{\mathfrak{p}})$ is the tame symbol.

Concerning $K_2(F)$, we have another famous exact sequence, which is due to Moore:

$$0 \to W(F) \to K_2(F) \xrightarrow{\eta} \oplus_{v\, non\, c.}\mu_v \xrightarrow{d} \mu \to 0,$$

where the sum is extended over all non-complex places $v$ of $F$, $\mu$ (resp. $\mu_v$) is the group of roots of unity in $F$ (resp. in the completion $F_v$), the homomorphism $\eta$ is induced by the Hilbert symbols and the homomorphism $d$ is given by $(\xi_v) \mapsto \prod_{v\, non\, c.} \xi_v^{m_v/m}$, where $m = |\mu|$ and $m_v = |\mu_v|$.

Let $\zeta_F(s)$ be the Dedekind zeta-function of a number field $F$.

**Birch-Tate Conjecture.** For any totally real number field $F$,

$$\#K_2O_F = |w_2(F) \cdot \zeta_F(-1)|,$$

where

$$w_2(F) = |H^0(F, \mathbb{Q}/\mathbb{Z}(n))|.$$

A general conjecture is proposed by Lichtenbaum:

**Lichtenbaum Conjecture.** *Let $F$ be a number field. For all $n \geq 2$*

$$\zeta_F^*(1-n) = \pm\frac{|K_{2n-2}(O_F)|}{K_{2n-1}(O_F)_{tors}} \cdot R_n^B(F)$$

*up to powers of* 2, *where $\zeta_F^*(1-n)$ is the special value of the Dedekind zeta-function $\zeta_F(s)$ at $s = 1-n$, and $R_n^B(F)$ is the Borel regulator.*

Both conjectures are known to be true when $F$ is abelian over $\mathbb{Q}$.

# 7. The 2-Sylow subgroup of $K_2O_F$ for quadratic number fields $F$

It is convenient for us to use the following notation: Let $F$ be a number field with $O_F$ the ring of integers in $F$. We will write $\bigtriangledown$ for $K_2O_F$. So for any natural number $n$, $\bigtriangledown^n = \{\alpha \in K_2O_F \mid \alpha = \beta^n \text{ for some } \beta \in K_2O_F\}$. For an integer $d \neq 0$, the set $S(d)$ is defined to be $\{\pm1, \pm2\}$ if $d > 0$ or $\{1, 2\}$ if $d < 0$. For any abelian group $A$, let ${}_2A = \{\alpha \in A \mid \alpha^2 = 1\}$. We will use the $2^n$-rank of $A$ for the number of cyclic factors of $A$ of order divisible by $2^n$. Let $p$ be a prime and $Q_p$ the field of $p$-adic numbers. As usual, $\left(\frac{a, b}{p}\right)$ will stand for the Hilbert symbol of order 2 over $Q_p$. $(a, b) \stackrel{2}{=} 1$ means that the integers $a$ and $b$ have no common odd divisor.

Let $F = \mathbb{Q}(\sqrt{d}), d \in \mathbb{Z}$ square-free, be a quadratic field. Recall from [5] that every element in ${}_2K_2O_F$ can be written as the form $\{-1, m\}, m \mid d$; together with $\{-1, m(u_i + \sqrt{d})\}$, if $\{-1, \pm2\} \bigcap NF \neq \emptyset$, where $u_i \in \mathbb{Z}$ such that $u_i^2 - d = c_i w_i^2$ for some $w_i \in \mathbb{Z}$ and $c_i \in \{-1, \pm2\} \bigcap NF$.

The following Theorem determines the 4-rank of $K_2O_F$ for all quadratic number fields $F$.

**Theorem 19.** $\{-1, m\} \in \nabla^2$ if and only if one can find an $\varepsilon \in S(d)$ such that for any odd prime $p \mid d$,

$$\left(\frac{-d, m}{p}\right) = \left(\frac{\varepsilon}{p}\right),$$

and $\{-1, m(u + \sqrt{d})\} \in \nabla^2$ if and only if one can find a $\delta \in S(d)$ such that for any odd prime $p \mid d$,

$$\left(\frac{-d, m}{p}\right) = \left(\frac{\delta(u + w)}{p}\right).$$

Let $F = \mathbb{Q}(\sqrt{d}), d \in \mathbb{Z}$ square-free, be a quadratic field. When $d$ has at most three odd prime divisors, we have known the 4-rank of $K_2O_F$ in every case, see [30,31]. Let $F = \mathbb{Q}(\sqrt{d})$ be a real quadratic field. We use $NF$ for the set of norms from $F$ over $\mathbb{Q}$. Let $d \in \mathbb{N}$ and let $d = 2^\sigma p_1 \ldots p_n$ be the prime factorization, where $\sigma = 0$ or 1. For $j = 1, 3, 5, 7$ we let $m_j$ denote the number of $p_i$'s which are $\equiv j \mod 8$, and we call $2^\sigma(m_1, m_3, m_5, m_7)$ the type of $d$. It is proved by B. Brauckmann[2] that if $d$ has a positive divisor $\equiv 7 \pmod 8$, then 4-rank of $K_2O_F \geq 1$.

We now give all real quadratic fields for which we have 4-rank of $K_2O_F \geq 1$. More precisely, we have the following theorem:

**Theorem 20.** Let $F = \mathbb{Q}(\sqrt{d})$ be a real quadratic field. Then 4-rank of $K_2O_F \geq 1$ in the following cases:

(1) $d$ has a positive divisor $\equiv 7 \pmod 8$;

(2) $d$ is odd and of type $(m_1, 0, m_5, 0)$ with $m_5 \geq 2$ even;

(3) $d = 2d'$ is even and $d'$ is of type $(m_1, m_3, 0, 0)$ with $m_3 \geq 2$ even.

Let $p \equiv 1 \pmod 8$ be a prime. It is known (Conner-Hurrelbrink[8]) that for $F = \mathbb{Q}(\sqrt{p})$ and also for $F = \mathbb{Q}(\sqrt{2p})$ both sets of primes

$$\{p \equiv 1 \pmod 8 \text{ and 4-rank of} K_2O_F = 0\}$$

and

$$\{p \equiv 1 \pmod 8 \text{ and 4-rank of} K_2O_F = 1\}$$

have density 1/2 in the set of all primes $p \equiv 1 \pmod 8$. The following theorem generalizes this phenomenon to more than one prime factor:

**Theorem 21.** Let $n \geq 2$. Except for the cases (1) – (3) listed above, there is a set of positive density consisting of square-free integers $d > 0$ of given parity and given type $2^\sigma(m_1, m_3, m_5, m_7)$, such that for $F = \mathbb{Q}(\sqrt{d})$ the 4-rank of $K_2O_F$ is 0, and there is a set of positive density consisting of square-free integers $d > 0$ of given parity and given type $2^\sigma(m_1, m_3, m_5, m_7)$, such that for $F = \mathbb{Q}(\sqrt{d})$ the 4-rank of $K_2O_F$ is $\geq 1$, where $\sigma = 0$ or 1.

Note that (2) and (3) are equivalent to the following (2′) and (3′) respectively.

(2′) $d \equiv 1 \pmod 8$ with $-1 \in NF$ and $-2 \notin NF$;
(3′) $d = 2d', d' \equiv 1 \pmod 8$ with $-1 \notin NF$ and $-2 \in NF$.

Therefore, in view of Wiles's result on the Birch-Tate conjecture, we know some information on the even part of $w_2(F)\zeta_F(-1)$.

As a consequence of our results, we obtain necessary and sufficient conditions, which are related only to some Legendre symbols, for the 2-Sylow subgroup of $K_2O_F$ to be elementary abelian, i.e., 4-rank of $K_2O_F = 0$. This turns out to be the same as to give necessary and sufficient conditions for 4-rank of $K_2O_F > 0$.

We have the similar results for imaginary quadratic fields.

**Theorem 22.** Let $F = \mathbb{Q}(\sqrt{-d}), d \in \mathbb{N}$ square-free, be an imaginary quadratic field. If $d \equiv 1 \pmod 8$ with $d \neq 1$, then 4-rank of $K_2O_F \geq 1$.

Unfortunately, for $F$ imaginary the corresponding statement of Theorem 21 is not valid in general. But we have

**Theorem 23.** Let $F = \mathbb{Q}(\sqrt{-d}), d \in \mathbb{N}$ square-free, be an imaginary quadratic field. If $d \not\equiv 1 \pmod 8$ has at least four odd prime divisors, then there is a set of positive density consisting of square-free integers $d' > 0$ with the same parity and the same type as $d$, such that for $F = \mathbb{Q}(\sqrt{-d'})$ the 4-rank of $K_2O_F$ is 0; and there exist infinitely many square-free integers $d' > 0$ with the same parity and the same type as $d$ such that 4-rank of $K_2O_F > 0$. With some exceptional cases, the same is true for $d$ having at most three odd prime divisors.

Finally we state our result on the 8-rank of $K_2O_F$.

It is easy to see that if $u_i^2 - d = -w_i^2$ or $-2w_i^2$, then $\{-1, m(u_i + \sqrt{d})\} \notin \nabla^2$. On the other hand, we know that a necessary condition for $\{-1, m\} \in \nabla^4$ is that there is an $\epsilon \in \{1, 2\}$ such that

$$\epsilon m Z^2 = X^2 + dY^2 \tag{E1}$$

is solvable. And a necessary condition for $\{-1, m(u + \sqrt{d})\} \in \nabla^4$ is that

$$m(u + w)Z^2 = X^2 + dY^2 \tag{E2}$$

is solvable.

For a square-free integer $d$ and $i = 1, 3, 5, 7$, denote by $d_i$ the product of all prime divisors of $d$ which are $\equiv i \pmod 8$ ($d_i = 1$ if $d$ has no prime divisor $\equiv i \pmod 8$).

We let $\sigma(l) = 1$ or 0 according to $l \mid m_5$ or not.

**Theorem 24.** Let $d$ be a square-free integer and $F = \mathbb{Q}(\sqrt{d})$, and let $m \mid d$. Write $m = \pm m_1 m_3 m_5 m_7$ with $m_i \mid d_i$ for $i = 1, 3, 5, 7$. Assume that (E1) is solvable and let $X_m, Y_m, Z_m \in \mathbb{N}$ with $(X_m, Y_m) = 1$ and $(Z_m, d) \overset{2}{=} 1$ is a solution of (E1).

(A) Suppose that $2 \notin NF$. Then $\{-1, m\} \in \nabla^4$ if and only if for $i = 1, 3, 5, 7$, there are $h_i \mid d_i$, in particular, $h_i = 1$ is permitted, and $\varepsilon \in \{\pm 1, \pm 2\}$ such that for any odd prime $l \mid d$,

$$\left(\frac{d, m_3 h_1 h_5}{l}\right)\left(\frac{-2^{\sigma(l)} d, m_5 h_3 h_7}{l}\right) = \left(\frac{\varepsilon Z_m}{l}\right).$$

(B) Suppose that $2 \in NF$.

(i) Then $\{-1, m\} \in \nabla^4$ if and only if for $i = 1, 7$, there are $h_i \mid d_i$ ($h_i = 1$ is permitted) and $\delta \in \{\pm 1\}$ such that for any odd prime $l \mid d$,

$$\left(\frac{d, h_1}{l}\right)\left(\frac{-d, h_7}{l}\right) = \left(\frac{\delta Z_m}{l}\right),$$

or

$$\left(\frac{d, h_1}{l}\right)\left(\frac{-d, h_7}{l}\right) = \left(\frac{\delta(u+w)Z_m}{l}\right).$$

(ii) Assume that (E2) is solvable and let $X_{mv}, Y_{mv}, Z_{mv} \in \mathbb{N}$ with $h = Y_{mv}, g = \dfrac{X_{mv} - wY_{mv}}{u+w} \in \mathbb{Z}$, $(g, h) = 1$ and $(Z_{mv}, dw) \overset{2}{=} 1$ be a solution. Then $\{-1, m(u + \sqrt{d})\} \in \nabla^4$ if and only if for $i = 1, 7$, there are $h_i \mid d_i$ ($h_i = 1$ is permitted) and $\eta \in \{\pm 1\}$ such that for any odd prime $l \mid d$,

$$\left(\frac{d, h_1}{l}\right)\left(\frac{-d, h_7}{l}\right) = \left(\frac{\eta u Z_{mv}}{l}\right),$$

or

$$\left(\frac{d, h_1}{l}\right)\left(\frac{-d, h_7}{l}\right) = \left(\frac{\eta(u+w)uZ_{mv}}{l}\right).$$

**Remark.** If (E1) or (E2) is solvable, then it must have a solution satisfying our assumption.

**Examples.** Let $F = \mathbb{Q}(\sqrt{89 \cdot 1481})$ or $F = \mathbb{Q}(\sqrt{-97 \cdot 241})$. Then 4-rank of $K_2O_F$= 8-rank of $K_2O_F = 2$.

Let $F = \mathbb{Q}(\sqrt{73 \cdot 89 \cdot 97 \cdot 353 \cdot 7481})$ (see [44]). Then we have 4-rank of $K_2O_F = 4$ and 8-rank of $K_2O_F$= 1. Moreover, with help of computer, we have $|w_2(F)\zeta_F(-1)| = 2^{13} \cdot 3^2 \cdot 3277923844289$. Applying the Birch-Tate conjecture, one gets the 16-rank of $K_2O_F = 1$ and the 32-rank of $K_2O_F = 0$.

It seems that there exists no the upper bound for the 8-rank of $K_2O_F$ for all quadratic number fields $F$.

## References

[1] A. Borel, Cojomologic réelle stable des groupes S-arithmétiques classiques, *C. R. Acad. Sci. Paris*, **7**(1974), 235-272.

[2] B. Brauckmann, The 2-Sylow subgroup of $K_2O_F$, *Canad. Journ. Math.*, **43**(1991), 255-264.

[3] J. Browkin, Elements of small order in $K_2(F)$, Lecture Notes in Math, Vol **966**, Springer-Verlag, 1982, 1-6.

[4] J. Browkin, Cyclotomic elements in $K_2(F)$, (preprint).

[5] J. Browkin, A. Schinzel, On Sylow 2-subgroups of $K_2O_F$ for quadratic number fields $F$, *J. Reine Angew. Math.*, **331**(1982), 104-113.

[6] J. Coates, On $K_2$ and some classical conjectures in algebraic number theory, *Ann. Math.*, **95**(1972), 99-116.

[7] P. E. Conner, J. Hurrelbrink, On elementary abelian 2-sylow $K_2$ of rings of integers of certain quadratic number fields, *Acta Arith.*, **73**(1995), 59-65.

[8] P. E. Conner, J. Hurrelbrink, On the 4-rank of the tame kernel $K_2(O)$ in positive definite terms, *J. Number Theory*, **88**(2001), 263-282.

[9] E. Friedlander, C. Weibel, An overview of algebraic $K$-theory, Algebraic $K$-theory and its application, Singapore: World Scientific, 1999, 1-119.

[10] H. Garland, A finiteness theorem for $K_2$ of a number field, *Ann. of Math.*, **94**(1971), 534-548.

[11] X. J. Guo, The torsion elements in $K_2$ of some local fields, *Acta Arith.*, **127**(2007), 97-102.

[12] X. J. Guo, H. R. Qin, The generalized Moore's reciprocity uniqueness theorem at prime 2 for real number fields, to appear in *J. K-theory*.

[13] J. Hurrelbrink, M. Kolster, Tame kernels under relative quadratic extensions and Hilbert symbols, *J. Reine Angew. Math.*, **499**(1998), 145-188.

[14] F. Keune, On the structure of the $K_2$ of the ring of integers in a number field, *K-theory*, **2**(1989), 625-645.

[15] S. Lichtenbaum, On the values of zata and $L$-functions, étale cohomology, and algebraic $K$-theory, Lecture Notes in Math., **342**, Springer-Verlag, 1973, 489-501.

[16] H. Matsumoto, Sur les sous-ǵroupes arithmétiques des ǵroupes semi-simples déployés, *Ann. Sci. Éc. Norm. Sup.* 4$^e$ *serie*, **2**(1969), 1-62.

[17] B. Mazur, A. Wiles, Class fields of abelian extensions of $\mathbb{Q}$, *Invent. Math.*, **76**(1984), 179-330.

[18] A. S. Merkurjev, A. A. Suslin, On the $K_3$ of a field, LOMI Preprints E-2-87, Leningrad, 1987.

[19] J. Milnor, Algebraic $K$-theory and quadratic forms, *Invent. Math.*, **9**(1970), 318-344.

[20] J. Milnor, Introduction to Algebraic $K$-Theory, Annals of Math., Studies **72**, Princeton Univ. Press, 1971.

[21] C. Moore, Group extensions of $p$-adic and adelic linear ǵroups, *Publ. Math. I.H.E.S.*, **35**(1969), 5-74.

[22] J. Neukirch, Class Field Theory, Springer, Berlin, 1986.

[23] J. Neukirch, A. Schmidt, K. Wingberg, Cohomology of Number Fields, Grundlehren Math. Wiss. 323, Springer, Berlin, 2000.

[24] R. Osburn, Densities of 4-ranks of $K_2(O)$, *Acta Arith.*, **69**(2002), 153-169.

[25] R. Osburn, B. Murray, Tame kernels and further 4-rank densities, *J. Number Theory*, **98**(2003), 390-406.

[26] H. R. Qin, Elements of finite order in $K_2F$ of fields, *Chin. Sci. Bull.*, **38**(24)(1994), 22-27.

[27] H. R. Qin, Computation of $K_2\mathbb{Z}[\sqrt{-6}]$, *J. Pure App. Algebra*, **96**(1994), 133-146.

[28] H. R. Qin, Computation of $K_2\mathbb{Z}[\frac{1+\sqrt{-35}}{2}]$, *Chin. Ann. of Math.(B)*, **17**(1)(1996), 63-72.

[29] H. R. Qin, 2-sylow subgroup of $K_2O_F$ for real quadratic fields $F$(in Chinese),

*Science in China*, **37**(1994), 1302-1313.

[30] H. R. Qin, The 2-sylow subgroups of the tame kernel of imaginary quadratic fields, *Acta Arith.*, **69**(1995), 153-169.

[31] H. R. Qin, The 4-rank of $K_2O_F$ for real quadratic fields, *Acta Arith.*, **72**(1995), 323-333.

[32] H. R. Qin, The subgroups of finite order in $K_2\mathbb{Q}$, Algebraic $K$-theory and its application, Singapore: World Scientific, 1999, 600-607.

[33] H. R. Qin, Tame kernels and Tate kernels of quadratic number fields, *J. Reine Angew. Math.*, **530**(2001), 105-144.

[34] H. R. Qin, The structure of the tame kernels of quadratic number fields (I), *Acta Arith.*, **113**(2004), 203-240.

[35] H. R. Qin, The 2-Sylow subgroup of $K_2O_F$ for number fields $F$, *J. Algebra*, **284**(2005), 494-519.

[36] H. R. Qin, H. Y. Zhou, The 3-Sylow subgroup of the tame kernel of real number fields, *J. Pure App. Algebra*, **209**(2007), 245-253.

[37] D. Quillen, Higher $K$-theory for categories with exact sequences, Proc. of the Symp. "New developments in topology", Oxford 1972, 95-103.

[38] D. Quillen, Finite Generation of the Groups $K_i$ of Rings of Algebraic Integers, Lectures Notes in Math. **341**, Springer-Verlag, 1973, 179-198.

[39] J. Rognes, C. Weibel, Two-primary algebraic $K$-theory of rings of integers in number fields, Appendix A by Manfred Kolster, *J. Amer. Math. Soc.*, **13**(1)(2000), 1-54.

[40] J. Rosenberg, Algebraic $K$-Theory and its Applications, GTM **147**, Springer-Verlag, 1994.

[41] J.-P. Serre, Local Fields, GTM **67**, Springer-Verlag, 1979.

[42] J. Tate, Relations between $K_2$ and Galois cohomology, *Invent. Math.*, **36**(1976), 257-274.

[43] J. Urbanowicz, On elements of given order in $K_2(F)$. *J. Pure App. Algebra*, **50**(1988), 298-307.

[44] A. Vazzana, 8-ranks of $K_2$ of rings in quadratic number fields, *J. Number Theory,***76**(1999), 248-264.

[45] V. Voevodsky, Motivic cohomology with $\mathbb{Z}/2$ coefficients, Publ. Math. IHES, **98** (2003), 1-57.

[46] V. Voevodsky, On Motivic cohomology with $\mathbb{Z}/l$ coefficients, preprint, 2003. Available at http:// www. math. uioc. edu/ K-theory/ 0639/.

[47] L. C. Washington, Introduction to Cyclotomic Fields, GTM **83**, Springer-Verlag, 1982.

[48] C. Weibel, 2007 Trieste Lectures on The Proof of the Bloch-Kato Conjecture, preprint, 2007.

[49] A. Wiles, The Iwasawa conjecture for totally real fields, *Ann. Math.*, **131**(1990), 493-540.

[50] K. J. Xu, On Browkin's conjecture about the elements of order five in $K_2(\mathbb{Q})$, *Science in China Ser A*, **50**(2007), 116-120.

[51] K. J. Xu, M. Liu, On the torsion in $K_2$ of a field, (preprint).

[52] K. J. Xu, H. R. Qin, Some elements of finite order in $K_2(\mathbb{Q})$, *Chinese Ann Math Ser A*, **22**(2001), 563-570.

[53] X. B. Yin, H. R. Qin, Q. S. Zhu, The structure of the tame kernels of quadratic number fields (II), *Acta Arith.*, **116**(3)(2005), 217-262.
[54] X. B. Yin, H. R. Qin, Q. S. Zhu, The structure of the tame kernels of quadratic number fields (III), to appear in *Comm. Algebra.*

*Cohomology of Groups and Algebraic K-theory*
ALM 12, pp. 413–478



# Higher Algebraic $K$-Theory: I

Daniel Quillen*

The purpose of this paper is to develop a higher $K$-theory for additive categories with exact sequences which extends the existing theory of the Grothendieck group in a natural way. To describe the approach taken here, let $\underline{\underline{M}}$ be an additive category embedded as a full subcategory of an abelian category $\underline{\underline{A}}$, and assume $\underline{\underline{M}}$ is closed under extensions in $\underline{\underline{A}}$. Then one can form a new category $Q(\underline{\underline{M}})$ having the same objects as $\underline{\underline{M}}$, but in which a morphism from $M'$ to $M$ is taken to be an isomorphism of $M'$ with a subquotient $M_1/M_0$ of $M$, where $M_0 \subset M_1$ are subobjects of $M$ such that $M_0$ and $M/M_1$ are objects of $\underline{\underline{M}}$. Assuming the isomorphism classes of objects of $\underline{\underline{M}}$ form a set, the category $Q(\underline{\underline{M}})$ has a classifying space $BQ(\underline{\underline{M}})$ determined up to homotopy equivalence. One can show that the fundamental group of this classifying space is canonically isomorphic to the Grothendieck group of $\underline{\underline{M}}$, which motivates defining a sequence of $K$-groups by the formula

$$K_i(\underline{\underline{M}}) = \pi_{i+1}(BQ(\underline{\underline{M}}), 0).$$

It is the goal of the present paper to show that this definition leads to an interesting theory.

The first part of the paper is concerned with the general theory of these $K$-groups. Section 1 contains various tools for working with the classifying space of a small category. It concludes with an important result which identifies the homotopy-theoretic fibre of the map of classifying spaces induced by a functor. In $K$-theory this is used to obtain long exact sequences of $K$-groups from the exact homotopy sequence of a map.

Section 2 is devoted to the definition of the $K$-groups and their elementary properties. One notes that the category $Q(\underline{\underline{M}})$ depends only on $\underline{\underline{M}}$ and the family of those short sequences $0 \to M' \to M \to M'' \to 0$ in $\underline{\underline{M}}$ which are exact in the ambient abelian category. In order to have an intrinsic object of study, it is convenient to introduce the notion of an *exact category*, which is an additive category equipped with a family of short sequences satisfying some standard conditions

*Supported in part by the National Science Foundation. Originally appeared in *Higher K-theories* (Proc. Conf., Battelle Memorial Inst., Seattle, Wash., 1972), pp. 85–147. Lecture Notes in Math., Vol. 341, Springer, Berlin 1973. Reprinted with permission from Springer.

(essentially those axiomatized in [Heller]). For an exact category $\underline{\underline{M}}$ with a set of isomorphism classes one has a sequence of $K$-groups $K_i(\underline{\underline{M}})$ varying functorially with respect to exact functors. Section 2 also contains the proof that $K_0(\underline{\underline{M}})$ is isomorphic to the Grothendieck group of $\underline{\underline{M}}$. It should be mentioned, however, that there are examples due to Gersten and Murthy showing that in general $K_1(\underline{\underline{M}})$ is not the same as the universal determinant group of Bass.

The next three sections contain four basic results which might be called the exactness, resolution, devissage, and localization theorems. Each of these generalizes a well-known result for the Grothendieck group ([Bass, Ch. VIII]), and, as will be apparent from the rest of the paper, they enable one to do a lot of $K$-theory.

The second part of the paper is concerned with applications of the general theory to rings and schemes. Given a ring (resp. a noetherian ring) $A$, one defines the groups $K_i(A)$ (resp. $K_i^{'}(A)$) to be the $K$-groups of the category of finitely generated projective $A$-modules (resp. the abelian category of finitely generated $A$-modules). There is a canonical map $K_i(A) \to K_i^{'}(A)$ which is an isomorphism for $A$ regular by the resolution theorem. Because the devissage and localization theorems apply only to abelian categories, the interesting results concern the groups $K_i^{'}(A)$. In section 6 we prove the formulas

$$K_i^{'}(A) = K_i^{'}(A[t]), \qquad K_i^{'}(A[t,t^{-1}]) = K_i^{'}(A) \oplus K_{i-1}^{'}(A)$$

for $A$ noetherian, which entail the corresponding results for $K$-groups when $A$ is regular. The first formula is proved more generally for a class of rings with increasing filtration, including some interesting non-commutative rings such as universal enveloping algebras. To illustrate the generality, the $K$-groups of certain skew fields are computed.

For a scheme (resp. noetherian) scheme $X$, the groups $K_i(X)$ (resp. $K_i^{'}(X)$) are defined using the category of vector bundles (resp. coherent sheaves) on $X$, and there is a canonical map $K_i(X) \to K_i^{'}(X)$ which is an isomorphism for $X$ regular. Section 7 is devoted to the $K'$-theory. Especially interesting is a spectral sequence

$$E_1^{pq} = \coprod_{cod(x)=p} K_{-p-q}(k(x)) \Rightarrow K_{-n}^{'}(X)$$

obtained by filtering the category of coherent sheaves according to the codimension of the support. In the case where $X$ is regular and of finite type over a field, we carry out a program proposed by Gersten at this conference ([Gersten 3]), which leads to a proof of *Bloch's formula*

$$A^P(X) = H^P(X, K_p(\underline{\underline{O}}_X))$$

proved by Bloch in particular cases ([Bloch]), where $A^P(X)$ is the group of codimension $p$ cycles modulo linear equivalence. One noteworthy feature of this formula is that the right side is clearly contravariant in $X$, which suggests rather strongly that higher $K$-theory might eventually provide a theory of the Chow ring for non-quasi-projective regular varieties.

Section 8 contains the computation of the $K$-groups of the projective bundle associated to a vector bundle over a scheme. This result generalizes the computation of the Grothendieck groups given in [SGA 6], and it may be viewed as a first step toward a higher $K$-theory for schemes, as opposed to the $K'$-theory of the preceding section. The proof, different from the one in [SGA 6], is based on the existence of canonical resolutions for regular sheaves on projective space, which may be of some independent interest. The method also permits one to determine the $K$-groups of a Severi-Brauer scheme in terms of the $K$-groups of the associated Azumaya algebra and its powers.

This paper contains proofs of all of the results announced in [Quillen 1], except for Theorem 1 of that paper, which asserts that the groups $K_i(A)$ here agree with those obtained by making $BGL(A)$ into an $H$-space (see [Gersten 5]). From a logical point of view, this theorem should have preceded the second part of the present paper, since it is used there a few times. However, I recently discovered that the ideas involved its proof could be applied to prove the expected generalization of the localization theorem and fundamental theorem for non-regular rings [Bass, p.494,663]. These results will appear in the next installment of this theory.

The proofs of Theorems A and B given in section 1 owe a great deal to conversations with Graeme Segal, to whom I am very grateful. One can derive these results in at least two other ways, using cohomology and the Whitehead theorem as in [Friedlander], and also by means of the theory of minimal fibrations of simplicial sets. The present approach, based on the Dold-Thom theory of quasi-fibrations, is quite a bit shorter than the others, although it is not as clear as I would have liked, since the main points are in the references. Someday these ideas will undoubtedly be incorporated into a general homotopy theory for topoi.

This paper was prepared with the editor's encouragement during the first two months of 1973. I mention this because the results in §7 on Gersten's conjecture and Bloch's formula, which were discovered at this time, directly affect the papers [Gersten 3, 4] and [Bloch] in this procedings, which were prepared earlier.

## CONTENTS

# §1. The classifying space of a small category

In the succeeding sections of this paper $K$-groups will be defined as the homotopy groups of the classifying space of a certain small category. In this rather long section we collect together the various facts about the classifying space functor we will need. All of these are fairly well-known, except for the important Theorem B which identifies the homotopy-fibre of the map of classifying spaces induced by a functor under suitable conditions. It will later be used to derive long exact sequences in $K$-theory from the homotopy exact sequence of a map.

Let $\underline{\underline{C}}$ be a small category. Its *nerve*, denoted $N\underline{\underline{C}}$, is the (semi-)simplicial set whose $p$-simplices are the diagrams in $\underline{\underline{C}}$ of the form

$$X_0 \to X_1 \to \cdots \to X_p.$$

The $i$-th face (resp. degeneracy) of this simplex is obtained by deleting the object $X_i$ (resp. replacing $X_i$ by id: $X_i \to X_i$) in the evident way. The *classifying space* of $\underline{\underline{C}}$, denoted $B\underline{\underline{C}}$, is the geometric realization of $N\underline{\underline{C}}$. It is a $CW$ complex whose $p$-cells are in one-one correspondence with the $p$-simplices of the nerve which are nondegenerate, i.e. such that none of the arrows is an identity map. (See [Segal 1],[Milnor 1].)

For example, let $J$ be a (partially) ordered set regarded as a category in the usual way. Then $BJ$ is the simplicial complex (with the weak topology) whose vertices are the elements of $J$ and whose simplices are the totally ordered non-empty finite subsets of $J$. Conversely, if $K$ is a simplicial complex and if $J$ is the ordered set of simplices of $K$, then the simplicial complex $BJ$ is the barycentric subdivision of $K$. Thus every simplicial complex (with the weak topology) is homeomorphic to the classifying space of some, and in fact many, ordered sets. Furthermore, since it is known that any $CW$ complex is homotopy equivalent to a simplicial complex, it follows that any interesting homotopy type is realized as the classifying space of an ordered set. (I am grateful to Graeme Segal for bringing these remarks to my attention.)

As another example, let a group $G$ be regarded as a category with one object in the usual way. Then $BG$ is a classifying space for the discrete group $G$ in the traditional sense. It is an Eilenberg-MacLane space of type $K(G,1)$, so few homotopy types occur in this way.

Let $X$ be an object of $\underline{\underline{C}}$. Using $X$ to denote also the corresponding 0-cell of $B\underline{\underline{C}}$, we have a family of homotopy groups $\pi_i(B\underline{\underline{C}}, X), i \geqslant 0$, which will be called the *homotopy groups* of $\underline{\underline{C}}$ with basepoint $X$ and denoted simply $\pi_i(\underline{\underline{C}}, X)$. Of course, $\pi_0(\underline{\underline{C}}, X)$ is not a group, but a pointed set, which can be described as the set $\pi_0\underline{\underline{C}}$ of components of the category $\underline{\underline{C}}$ pointed by the component containing $X$. In effect, connected components of $B\underline{\underline{C}}$ are in one-one correspondence with components of $\underline{\underline{C}}$.

We will see below that $\pi_1(\underline{\underline{C}}, X)$ and also the homology groups of $B\underline{\underline{C}}$ can be defined "algebraically" without the use of spaces or some closely related machine such as semi-simplicial homotopy theory, or simplicial complexes and subdivision. The existence of similar descriptions of the higher homotopy groups seems to be unlikely. because so far nobody has produced an "algebraic" definition of the homotopy groups of a simplicial complex.

**Coverings of $B\underline{\underline{C}}$ and the fundamental group**

Let $E$ be a covering space of $B\underline{\underline{C}}$. For any object $X$ of $\underline{\underline{C}}$, let $E(X)$ denote the fibre of $E$ over $X$ considered as a 0-cell of $B\underline{\underline{C}}$. If $u : X \to X'$ is a map in $\underline{\underline{C}}$, it determines a path from $X$ to $X'$ in $B\underline{\underline{C}}$, and hence gives rise to a bijection $E(u) : E(X) \xrightarrow{\sim} E(X')$. It is easy to see that $E(fg) = E(f)E(g)$, hence in this way we obtain a functor $X \mapsto E(X)$ from $\underline{\underline{C}}$ to Sets which is *morphism-inverting*, that is, it carries arrows into isomorphisms.

Conversely, given $F : \underline{\underline{C}} \to Sets$, let $F\backslash\underline{\underline{C}}$ denote the category of pairs $(X, x)$ with $X$ in $\underline{\underline{C}}$ and $x \in F(X)$, in which a morphism $(X, x) \to (X', x')$ is a map $u : X \to X'$ such that $F(u)x = x'$. The forgetful functor $F\backslash\underline{\underline{C}} \to C$ induces a map of classifying spaces $B(F\backslash\underline{\underline{C}}) \to B\underline{\underline{C}}$ having the fibre $F(X)$ over $X$ for each object $X$. Using [Gabriel-Zisman, App.I, 3.2] it is not difficult to see that when $F$ is morphism-inverting, the map $B(F\backslash\underline{\underline{C}}) \to B\underline{\underline{C}}$ is locally trivial, and hence $B(F\backslash\underline{\underline{C}})$ is a covering space of $B\underline{\underline{C}}$. It is clear that the two procedures just described are inverse to each other, whence we have an equivalence of categories

$$(\text{Coverings of } \quad B\underline{\underline{C}}) \simeq (\textit{Morph.-inv. } F : \underline{\underline{C}} \to \textit{Sets})$$

where the latter denotes the full subcategory of $Funct(\underline{\underline{C}}, Sets)$, the category of functors from $\underline{\underline{C}}$ to $Sets$, consisting of the morphism-inverting functors.

Let $\underline{\underline{G}} = \underline{\underline{C}}[(Ar\underline{\underline{C}})^{-1}]$ denote the groupoid obtained from $\underline{\underline{C}}$ by formally adjoining the inverses of all the arrows [Gabriel-Zisman, I, 1.1]. The canonical functor from $\underline{\underline{C}}$ to $\underline{\underline{G}}$ induces an equivalence of categories

$$Funct(\underline{\underline{G}}, Sets) = (\textit{Morph.-inv. } F : \underline{\underline{C}} \to \textit{Sets})$$

(*loc.cit.*, I, 1.2). Let $X$ be an object of $\underline{\underline{C}}$ and let $G_X$ be the group of its automorphisms as an object of $\underline{\underline{G}}$. When $\underline{\underline{C}}$ is connected, the inclusion functor $G_X \to \underline{\underline{G}}$ is an equivalence of categories, hence one has an equivalence

$$Funct(\underline{\underline{G}}, Sets) \xrightarrow{\sim} Funct(G_X, Sets) = (G_X - sets).$$

Therefore by combining the above equivalences, we obtain an equivalence of categories of the category of coverings of $B\underline{\underline{C}}$ with the category of $G_X - sets$ given by the functor $E \mapsto E(X)$. By the theory of covering spaces this implies that there is a canonical isomorphism: $\pi_1(\underline{\underline{C}}, X) \simeq G_X$. The same conclusion holds when $\underline{\underline{C}}$ is not connected, as both groups depend only on the component of $\underline{\underline{C}}$ containing $X$. Thus we have established the following.

**Proposition 1.** *The category of covering spaces of $B\underline{\underline{C}}$ is canonically equivalent to the category of morphism-inverting functors $F : \underline{\underline{C}} \to Sets$, or what amounts to the same thing, the category $Funct(\underline{\underline{G}}, Sets)$, where $\underline{\underline{G}} = C[(Ar\underline{\underline{C}})^{-1}]$ is the groupoid obtained by formally inverting the arrows of $\underline{\underline{C}}$. The fundamental group $\pi_1(\underline{\underline{C}}, X)$ is canonically isomorphic to the group of automorphisms of $X$ as an object of the groupoid $\underline{\underline{G}}$.*

It follows in particular that a local coefficient system $L$ of abelian groups on $B\underline{\underline{C}}$ may be identified with the morphism-inverting functor $X \mapsto L(X)$ from $\underline{\underline{C}}$ to abelian groups.

**The homology of** $B\underline{\underline{C}}$

It is well-known that the homology and cohomology of the classifying space of a discrete group coincide with the homology and cohomology of the group in the sense of homological algebra. We now describe the generalization of this fact for an arbitrary small category.

Let $A$ be a functor from $\underline{\underline{C}}$ to $Ab$, the category of abelian groups, and let $H_p(\underline{\underline{C}}, A)$ denote the homology of the simplicial abelian group

$$C_p(\underline{\underline{C}}, A) = \coprod_{X_0 \to \cdots \to X_p} A(X_0)$$

of chains on $N\underline{\underline{C}}$ with coefficients in $A$. (By the homology we mean the homology of the associated normalized chain complex.) Then there are canonical isomorphisms

$$H_p(\underline{\underline{C}}, A) = \varinjlim{}_p^{\underline{\underline{C}}}(A)$$

where $\varinjlim_*^{\underline{\underline{C}}}$ denotes the left derived functors of the right exact functor $\varinjlim$ from $Funct(\underline{\underline{C}}, Ab)$ to $Ab$. This is proved by showing that $A \mapsto H_*(\underline{\underline{C}}, A)$ is an exact $\partial$–functor which coincides with $\varinjlim$ in degree zero and is effaceable in positive degrees. (See [Gabriel-Zisman, App.II, 3.3] .)

Let $H_*(B\underline{\underline{C}}, L)$ denote the singular homology of $B\underline{\underline{C}}$ with coefficients in a local coefficient system $L$. Then there are canonical isomorphisms

$$H_p(B\underline{\underline{C}}, L) = H_p(\underline{\underline{C}}, L)$$

where we identify $L$ with a morphism-inverting functor as above. This may be proved by filtering the $CW$ complex $B\underline{\underline{C}}$ by means of its skeleta and considering the associated spectral sequence. One has $E^1_{pq} = 0$ for $q \neq 0$ and $E^1_{*0} =$ the normalized chain complex associated to $C_*(\underline{\underline{C}}, L)$. (Compare [Segal 1, 5.1].) The spectral sequence degenerates yielding the desired isomorphism.

Thus we have

$$H_p(B\underline{\underline{C}}, L) = \varinjlim_p^{\underline{\underline{C}}}(L) \tag{1}$$

and similarly we have a canonical isomorphism for cohomology

$$H^p(B\underline{\underline{C}}, L) = \varprojlim_{\underline{\underline{C}}}^p(L) \tag{2}$$

where $\varprojlim_{\underline{\underline{C}}}^*$ denotes the right derived functors of the left exact functor $\varprojlim$ from $Funct(\underline{\underline{C}}, Ab)$ to $Ab$.

**Properties of the classifying space functor**

From now on we use the letters $\underline{\underline{C}}$, $\underline{\underline{C}}'$, etc. to denote small categories. If $f : \underline{\underline{C}} \to \underline{\underline{C}}'$ is a functor, it induces a cellular map $Bf : B\underline{\underline{C}} \to B\underline{\underline{C}}'$. In this way we obtain a faithful functor from the category of small categories to the category of $CW$ complexes and cellular maps. This functor is of course not fully faithful. As a particularly interesting example, we note that there is an obvious canonical cellular homeomorphism

$$B\underline{\underline{C}} = B\underline{\underline{C}}^0 \tag{3}$$

where $\underline{\underline{C}}^0$ is the dual category, which is not realized by a functor from $\underline{\underline{C}}$ to $\underline{\underline{C}}^0$ except in very special cases, e.g. groups.

By the compatibility of geometric realization with products [Milnor 1], one knows that the canonical map

$$B(\underline{\underline{C}} \times \underline{\underline{C}}') \to B\underline{\underline{C}} \times B\underline{\underline{C}}' \tag{4}$$

is a homeomorphism if either $B\underline{\underline{C}}$ or $BC'$ is a finite complex, and also if the product is given the compactly generated topology. As pointed out in [Segal 1], this implies the following.

**Proposition 2.** *A natural transformation $\theta : f \to g$ of functors from $\underline{\underline{C}}$ to $\underline{\underline{C}}'$ induces a homotopy $B\underline{\underline{C}} \times I \to B\underline{\underline{C}}'$ between $Bf$ and $Bg$.*

In effect, the triple $(f, g, \theta)$ can be viewed as a functor $\underline{\underline{C}} \times 1 \to \underline{\underline{C}}'$, where 1 is the ordered set $\{0 < 1\}$, and $B1$ is the unit interval.

We will say that a functor is a *homotopy equivalence* if it induces a homotopy equivalence of classifying spaces, and that a category is *contractible* if its classifying space is.

**Corollary 1.** *If a functor $f$ has either a left or a right adjoint, then $f$ is a homotopy equivalence.*

For if $f'$ is say left adjoint to $f$, then there are natural transformations $f'f \to id, id \to ff'$, whence $Bf'$ is a homotopy inverse for $Bf$.

**Corollary 2.** *A category having either an initial or a final object is contractible.*

For then the functor from the category to the punctual category has an adjoint.

Let $I$ be a small category which is *filtering* (=non-empty+directed [Bass, p.41]) and let $i \mapsto \underline{\underline{C}}_i$ be a functor from $I$ to small categories. Let $\underline{\underline{C}}$ be the inductive limit of the $\underline{\underline{C}}_i$; because filtered inductive limits commute with finite projective limits, we have $Ob\underline{\underline{C}} = \varinjlim Ob\underline{\underline{C}}_i$, $Ar\underline{\underline{C}} = \varinjlim Ar\underline{\underline{C}}_i$, and more generally $N\underline{\underline{C}} = \varinjlim N\underline{\underline{C}}_i$. Let $X_i \in Ob\underline{\underline{C}}_i$ be a family of objects such that for every arrow $i \to i'$ in $I$, the induced functor $\underline{\underline{C}}_i \to \underline{\underline{C}}_{i'}$ carries $X_i$ to $X_{i'}$, whence we have an inductive system $\pi_n(\underline{\underline{C}}_i, X_i)$ indexed by $I$.

**Proposition 3.** *If $X$ is the common image of the $X_i$ in $\underline{\underline{C}}$, then*

$$\varinjlim \pi_n(\underline{\underline{C}}_i, X_i) = \pi_n(\underline{\underline{C}}, X).$$

*Proof*. Because $I$ is filtering and $N\underline{\underline{C}} = \varinjlim N\underline{\underline{C}}_i$, it follows that any simplicial subset of $N\underline{\underline{C}}$ with a finite number of nondegenerate simplices lifts to $N\underline{\underline{C}}_i$ for some $i$, and moreover the lifting is unique up to enlarging the index $i$ in the evident sense. As every compact subset of a $CW$ complex is contained in a finite subcomplex, we see that every compact subset of $B\underline{\underline{C}}$ lifts to $B\underline{\underline{C}}_i$ for some $i$, uniquely up to enlarging $i$. The proposition follows easily from this.

**Corollary 1.** *Suppose in addition that for every arrow $i \to i'$ in $I$ the induced functor $\underline{\underline{C}}_i \to \underline{\underline{C}}_{i'}$ is a homotopy equivalence. Then the functor $\underline{\underline{C}}_i \to \underline{\underline{C}}$ is a homotopy equivalence for each $i$.*

*Proof*. Replacing $I$ by the cofinal category $i\backslash I$ of objects under $i$, we can suppose $i$ is the initial object of $I$. It then follows from the proposition that the map of $CW$ complexes $B\underline{\underline{C}}_i \to B\underline{\underline{C}}$ induces isomorphisms on homotopy. Hence it is a homotopy equivalence by a well-known theorem of Whitehead.

**Corollary 2.** *Any filtering category is contractible.*

In effect, $I$ is the inductive limit of the functor $i \mapsto I/i$, and the category $I/i$ of objects over $i$ has a final object, hence is contractible.

**Sufficient conditions for a functor to be a homotopy equivalence**

Let $f : \underline{\underline{C}} \to \underline{\underline{C}}'$ be a functor and denote objects of $\underline{\underline{C}}$ by $X, X'$, etc. and objects of $\underline{\underline{C}}'$ by $Y, Y'$, etc. If $Y$ is a fixed object of $\underline{\underline{C}}'$, let $Y\backslash f$ denote the category consisting of pairs $(X, v)$ with $v : Y \to fX$, in which a morphism from $(X, v)$ to $(X', v')$ is a map $w : X \to X'$ such that $f(w)v = v'$. In particular, when $f$ is the identity functor of $\underline{\underline{C}}'$, we obtain the category $Y \backslash \underline{\underline{C}}'$ of objects under $Y$. Similarly one defines the category $f/Y$ consisting of pairs $(X, u)$ with $u : fX \to Y$.

**Theorem A.** *If the category $Y \backslash f$ is contractible for every object $Y$ of $\underline{\underline{C}}'$, then the functor $f$ is a homotopy equivalence.*

In view of (3), this result admits a dual formulation to the effect that $f$ is a homotopy equivalence when all of the categories $f/Y$ are contractible.

**Example.** Let $g : K \to K'$ be a simplicial map of simplicial complexes, and let $f : J \to J'$ be the induced map of ordered sets of simplices in $K$ and $K'$, so that $g$ is homeomorphic to $Bf$. If $\overline{\sigma}$ denotes the element of $J'$ corresponding to a simplex $\sigma$ of $K'$, then $f/\overline{\sigma}$ is the ordered set of simplices in $g^{-1}(\sigma)$. In this situation the theorem says that a simplicial map is a homotopy equivalence when the inverse image of each (closed) simplex is contractible.

Before proving the theorem we derive a corollary. First we recall the definition of fibred and cofibred categories [SGA 1, Exp. VI] in a suitable form. Let $f^{-1}(Y)$ denote the *fibre* of $f$ over $Y$, that is, the subcategory of $\underline{\underline{C}}$ whose arrows are those mapped to the identity of $Y$ by $f$. It is easily seen that $f$ makes $\underline{\underline{C}}$ a *prefibred category* over $\underline{\underline{C}}'$ in the sense of *loc. cit.* if and only if for every object $Y$ of $\underline{\underline{C}}'$ the functor

$$f^{-1}(Y) \to Y \backslash f, X \mapsto (X, id_Y)$$

has a right adjoint. Denoting the adjoint by $(X, v) \mapsto v^*X$, we obtain for any map $v : Y \to Y'$ a functor

$$v^* : f^{-1}(Y') \to f^{-1}(Y)$$

determined up to canonical isomorphism, called *base-change* by $v$. The prefibred category $\underline{\underline{C}}$ over $\underline{\underline{C}}'$ is a *fibred category* if for every pair $u, v$ of composable arrows in $\underline{\underline{C}}'$, the canonical morphism of functors $u^*v^* \to (vu)^*$ is an isomorphism. We will call such functors $f$ *prefibred* and *fibred* respectively.

Dually, $f$ makes $\underline{\underline{C}}$ into a *precofibred category* over $\underline{\underline{C}}'$ when the functors $f^{-1}(Y) \to f/Y$ have left adjoints $(X, v) \mapsto v_*X$. In this case the functor $v_* : f^{-1}(Y) \to f^{-1}(Y')$ induced by $v : Y \to Y'$ is called *cobase-change* by $v$, and $\underline{\underline{C}}$ is a *cofibred category* when $(vu)_* \xrightarrow{\sim} v_*u_*$ for all composable $u, v$. Such functors $f$ will be called *precofibred* and *cofibred* respectively.

**Corollary.** *Suppose that $f$ is either prefibred or precofibred, and that $f^{-1}(Y)$ is contractible for every $Y$. Then $f$ is a homotopy equivalence.*

This follows from Prop. 2, Cor. 1.

**Example.** Let $S(\underline{\underline{C}})$ be the category whose objects are the arrows of $\underline{\underline{C}}$, and in which a morphism from $u : X \to Y$ to $u' : X' \to Y'$ is a pair $v : X' \to X, w : Y \to Y'$ such that $u' = wuv$. (Thus $S(\underline{\underline{C}})$ is the cofibred category over $\underline{\underline{C}}^0 \times \underline{\underline{C}}$ with discrete fibres defined by the functor $(X, Y) \mapsto \mathrm{Hom}(X, Y)$.) One has functors

$$\underline{\underline{C}}^0 \xleftarrow{s} S(\underline{\underline{C}}) \xrightarrow{t} \underline{\underline{C}}$$

given by source and target, and it is easy to see that these functors are cofibred. The categories $s^{-1}(X) = X \backslash \underline{\underline{C}}$ and $t^{-1}(Y) = (\underline{\underline{C}}/Y)^0$ have initial objects, hence are contractible. Therefore $s$ and $t$ are homotopy equivalences by the corollary.

This construction provides the simplest way of realizing by means of functors the homotopy equivalence (3).

We now turn to the proof of Theorem A. We will need a standard fact about the realization of bisimplicial spaces which we now derive.

Let $\underline{\underline{Ord}}$ be the category of ordered sets $p = \{0 < 1 < \cdots < p\}, p \in N$, so that by definition simplicial objects are functors with domain $\underline{\underline{Ord}}^0$. The realization functor

$$(p \mapsto X_p) \mapsto |p \mapsto X_p|$$

from simplicial spaces to spaces ([segal 1]) may be defined as the functor left adjoint to the functor which associates to a space $Y$ the simplicial space $p \mapsto \underline{\text{Hom}}(\triangle^p, Y)$, where $\underline{\text{Hom}}$ denotes function space and $\triangle^p$ is the simplex having $p$ as its set of vertices. In particular the realization functor commutes with inductive limits.

Let $T : p, q \mapsto T_{pq}$ be a bisimplicial space, i.e. a functor from $\underline{\underline{Ord}}^0 \times \underline{\underline{Ord}}^0$ to spaces. Realizing with respect to $q$ keeping $p$ fixed, we obtain a simplicial space $p \mapsto |q \mapsto T_{pq}|$ which may then be realized with respect to $p$. Also, we may realize first in the $p$-direction and then in the $q$-direction, or we may realize the diagonal simplicial space $p \mapsto T_{pp}$. It is well-known (e,g. [ Tornehave] ) that these three procedures yield the same result:

**Lemma.** *There are homeomorphisms*

$$|p \mapsto T_{pp}| = |p \mapsto |q \mapsto T_{pq}|| = |q \mapsto |p \mapsto T_{pq}||$$

*which are functorial in the simplicial space* $T$.

*Proof*. Suppose first that $T$ is of the form

$$h^{rs} \times S : (p, q) \mapsto \text{Hom}(p, r) \times \text{Hom}(q, s) \times S$$

where $S$ is a given space. Then

$$|p \mapsto \text{Hom}(p, r) \times \text{Hom}(p, s) \times S| = \triangle^r \times \triangle^s \times S.$$

(This is the basic homeomorphism used to prove that geometric realization commutes with products [Milnor 1].) On the other hand, we have

$$|p \mapsto |q \mapsto \text{Hom}(p,r) \times \text{Hom}(q,s) \times S|| = |p \mapsto \text{Hom}(p,r) \times \triangle^s \times S| = \triangle^r \times \triangle^s \times S$$

and similarly for the double realization taken in the other order. Thus the required functorial homeomorphisms exist on the full subcategory of bisimplicial spaces of this form.

But any $T$ has a canonical presentation

$$\coprod_{(r,s)\to(r',s')} h^{r's'} \times T_{rs} \rightrightarrows \coprod_{(r,s)} h^{rs} \times T_{rs} \to T$$

which is exact in the sense that the right arrow is the cokernel of the pair of arrows. Since the three functors from bisimplicial spaces to spaces under consideration commute with inductive limits, the lemma follows.

*Proof of Theorem A.* Let $S(f)$ be the category whose objects are triples $(X, Y, v)$ with $X$ an object of $\underline{\underline{C}}$ and $v : Y \to fX$ a map in $\underline{\underline{C}}'$, and in which a morphism from $(X, Y, v)$ to $(X', Y', v')$ is a pair of arrows $u : X \to X'$, $w : Y' \to Y$ such that $v' = f(u)vw$. (Thus $S(f)$ is the cofibred category over $\underline{\underline{C}} \times \underline{\underline{C}}'^0$ defined by the functor $(X, Y) \mapsto \operatorname{Hom}(Y, fX)$.) We have functors

$$\underline{\underline{C}}'^0 \xleftarrow{p_2} S(f) \xrightarrow{p_1} \underline{\underline{C}}$$

given by $p_1(X, Y, v) = X, p_2(X, Y, v) = Y$.

Let $T(f)$ be the bisimplicial set such that an element of $T(f)_{pq}$ is a pair of diagrams

$$(Y_p \to \cdots \to Y_0 \to fX_0, X_0 \to \cdots \to X_q)$$

in $\underline{\underline{C}}'$ and $\underline{\underline{C}}$ respectively, and such that the $i$–th face in the $p$–(resp. $q$–) direction deletes the object $Y_i$ (resp $X_i$) in the obvious way. Forgetting the first component gives a map of bisimplicial sets

$$(*) \qquad T(f)_{pq} \to N\underline{\underline{C}}_q$$

where the latter is constant in the $p$–direction. Since the diagonal simplicial set of $T(f)$ is the nerve of the category $S(f)$, it is clear that the realization of $(*)$ is the map $Bp_1 : BS(f) \to B\underline{\underline{C}}$. (By the realization of a bisimplicial set we mean the space described in the above lemma, where the bisimplicial set is regarded as a bisimplicial space in the obvious way.) On the other hand, realizing $(*)$ with respect to $p$ gives a map of simplicial spaces

$$\coprod_{X_0 \to \cdots \to X_q} B(\underline{\underline{C}}'/fX_0)^0 \to \coprod_{X_0 \to \cdots \to X_q} pt = N\underline{\underline{C}}_q$$

which is a homotopy equivalence for each $q$ because the category $\underline{\underline{C}}'/fX_0$ has a final object. Applying a basic result of May and Tornehave ([Tornehave, A.3]), or the lemma below (Th. B), we see the realization of $(*)$ is a homotopy equivalence. Thus the functor $p_1$ is a homotopy equivalence.

Similarly there is a map of bisimplicial sets $T(f)_{pq} \to N(\underline{\underline{C}}'^0)_p$ whose realization is the map $Bp_2 : BS(f) \to B\underline{\underline{C}}'^0$. Realizing with respect to $q$, we obtain a map of simplicial spaces

$$(**) \qquad \coprod_{Y_0 \leftarrow \cdots \leftarrow Y_p} B(Y_0 \setminus f) \to \coprod_{Y_0 \leftarrow \cdots \leftarrow Y_p} pt = N(\underline{\underline{C}}'^0)_p$$

which is a homotopy equivalence for each $p$, because the categories $Y \setminus f$ are contractible by hypothesis. Thus we conclude that the functor $p_2$ is a homotopy equivalence.

But we have a commutative diagram of categories

$$\begin{array}{ccccc} \underline{\underline{C}}'^0 & \xleftarrow{p_2} & S(f) & \xrightarrow{p_1} & \underline{\underline{C}} \\ \| & & \downarrow{\scriptstyle f'} & & \downarrow{\scriptstyle f} \\ \underline{\underline{C}}'^0 & \xleftarrow{p_2} & S(id_{\underline{\underline{C}}'}) & \xrightarrow{p_1} & \underline{\underline{C}}' \end{array}$$

where $f'(X, Y, v) = (fX, Y, v)$. The horizontal arrows are homotopy equivalences by what has been proved, (note that $Y \setminus id_{\underline{\underline{C}}'} = Y \setminus \underline{\underline{C}}'$ is contractible as it has an initial object). Thus $f$ is a homotopy equivalence, whence the theorem.

**The exact homotopy sequence.**

Let $g : E \to B$ be a map of topological spaces and let $b$ be a point of $B$. The *homotopy-fibre* of $f$ over $b$ is the space

$$F(g, b) = E \times_B B^I \times_B \{b\}$$

consisting of pairs $(e, p)$ with $e$ a point of $E$ and $p$ a path joining $g(e)$ and $b$. For any $e$ in $g^{-1}(b)$ one has the *exact homotopy sequence* of $g$ with basepoint $e$

$$\to \pi_{i+1}(B, b) \to \pi_i(F(g, b), \bar{e}) \to \pi_i(E, e) \xrightarrow{g_*} \pi_i(B, b) \to \cdots$$

where $\bar{e} = (e, \bar{b}), \bar{b}$ denoting the constant path at $b$.

Let $f : \underline{\underline{C}} \to \underline{\underline{C}}'$ be a functor and $Y$ an object of $\underline{\underline{C}}'$. If $j : Y \setminus f \to \underline{\underline{C}}$ is the functor sending $(X, v : Y \to fX)$ to $X$, then $(X, v) \mapsto v : Y \to fX$ is a natural transformation from the constant functor with value $Y$ to $fj$. Hence by Prop. 2 the composite $B(Y \setminus f) \to B\underline{\underline{C}} \to B\underline{\underline{C}}'$ contracts canonically to the constant map with image $Y$, and so we obtain a canonical map

$$B(Y \setminus f) \to F(Bf, Y).$$

We want to know when this map is a homotopy equivalence, for then we have an exact sequence relating the homotopy groups of the categories $Y \setminus f, \underline{\underline{C}}$ and $\underline{\underline{C}}'$. Since the homotopy-fibres of a map over points connected by a path are homotopy equivalent, it is clearly necessary in order for the above map to be a homotopy equivalence for all $Y$, that the functor $Y' \setminus f \to Y \setminus f, (X, v) \mapsto (X, vu)$ induced by $u : Y \to Y'$ be a homotopy equivalence for every map $u$ in $\underline{\underline{C}}'$. We are going to show the converse is true.

Because homotopy-fibres are not classifying spaces of categories, and hence are somewhat removed from what we ultimately will work with, it is convenient to formulate things in terms of homotopy-cartesian squares. Recall that a commutative square of spaces

$$\begin{array}{ccc} E' & \xrightarrow{h'} & E \\ {\scriptstyle g'}\downarrow & & \downarrow{\scriptstyle g} \\ B' & \xrightarrow{h} & B \end{array}$$

is called *homotopy-cartesian* if the map

$$E' \to B' \times_B B^I \times_B E, e' \mapsto (g'(e'), \overline{hg'(b')}, h'(e'))$$

from $E'$ to the homotopy-fibre-product of $h$ and $g$ is a homotopy equivalence. When $B'$ is contractible, the map $F(g', b') \to E'$ is a homotopy equivalence for any $b'$ in $B'$, hence one has a map $E' \to F(g, h(b'))$ unique up to homotopy. In this case the square is easily seen to be homotopy-cartesian if and only if $E' \to F(g, h(b'))$ is a homotopy equivalence.

A commutative square of categories will be called homotopy-cartesian if the corresponding square of classifying spaces is. With this terminology we have the following generalization of Theorem A.

**Theorem B.** *Let $f : \underline{\underline{C}} \to \underline{\underline{C}}'$ be a functor such that for every arrow $Y \to Y'$ in $\underline{\underline{C}}'$, the induced functor $Y' \setminus f \to Y \setminus f$ is a homotopy equivalence. Then for any object $Y$ of $\underline{\underline{C}}'$ the cartesian square of categories*

$$\begin{array}{ccc} Y\backslash f & \xrightarrow{j} & \underline{\underline{C}} \\ {\scriptstyle f'}\downarrow & & \downarrow{\scriptstyle f} \\ Y\backslash \underline{\underline{C}}' & \xrightarrow{j'} & \underline{\underline{C}}' \end{array} \qquad \begin{array}{l} j(X,v) = X \\ f'(X,v) = (fX, v) \\ j'(Y',v) = Y' \end{array}$$

*is homotopy-cartesian. Consequently for any $X$ in $f^{-1}(Y)$ we have an exact sequence*

$$\to \pi_{i+1}(\underline{\underline{C}}', Y) \to \pi_i(Y \setminus f, \overline{X}) \xrightarrow{j_*} \pi_i(\underline{\underline{C}}, X) \xrightarrow{f_*} \pi_i(\underline{\underline{C}}', Y) \to \cdots$$

*where $\overline{X} = (X, id_Y)$.*

As with Theorem A, this result admits a dual formulation with the categories $f \setminus Y$.

**Corollary.** *Suppose $f : \underline{\underline{C}} \to \underline{\underline{C}}'$ is prefibred (resp. precofibred) and that for every arrow $u : Y \to Y'$ the base-change functor $u^* : f^{-1}(Y') \to f^{-1}(Y)$, (resp. the cobasechange functor $u_* : f^{-1}(Y) \to f^{-1}(Y')$) is a homotopy equivalence. Then for any $Y$ in $\underline{\underline{C}}'$, the category $f^{-1}(Y)$ is homotopy equivalent to the homotopy-fibre of $f$ over $Y$. (Precisely, the square*

$$\begin{array}{ccc} f^{-1}(Y) & \xrightarrow{i} & \underline{\underline{C}} \\ \downarrow & & \downarrow{\scriptstyle f} \\ pt & \xrightarrow{Y} & \underline{\underline{C}}' \end{array}$$

*where $i$ is the inclusion functor, is homotopy-cartesian.) Consequently for any $X$ in $f^{-1}(Y)$ we have an exact homotopy sequence*

$$\to \pi_{i+1}(\underline{\underline{C}}', Y) \to \pi_i(f^{-1}(Y), X) \xrightarrow{i_*} \pi_i(\underline{\underline{C}}, X) \xrightarrow{f_*} \pi_i(\underline{\underline{C}}', Y) \to .$$

This is clear, since $f^{-1}(Y) \to Y \setminus f$ is a homotopy equivalence for prefibred $f$.

For the proof of the theorem we will need a lemma based on the theory of quasi-fibrations [Dold-Lashof], which is a special case of a general result about the realization of a map of simplicial spaces [Segal 2]. A *quasi-fibration* is a map $g : E \to B$ of spaces such that the canonical map $g^{-1}(b) \to F(g, b)$ induces isomorphisms on homotopy for all $b$ in $B$. When $E, B$ are in the class $\underline{\underline{W}}$ of spaces having the homotopy type of a $CW$ complex, one knows from [Milnor 2] that $F(g, b)$ is in $\underline{\underline{W}}$. Thus if $g^{-1}(b)$ is also in $\underline{\underline{W}}$, and $g$ is a quasi-fibration, we, have that $g^{-1}(b) \to F(g,b)$ is a homotopy equivalence, i.e. the square

$$\begin{array}{ccc} g^{-1}(b) & \xrightarrow{i} & E \\ \downarrow & & \downarrow g \\ pt & \xrightarrow{b} & B \end{array}$$

is homotopy-cartesian.

**Lemma.** *Let $i \mapsto X_i$ be a functor from a small category $I$ to topological spaces, and let $g : X_I \to BI$ be the space over $BI$ obtained by realizing the simplicial space*

$$p \mapsto \coprod_{i_0 \to \cdots \to i_p} X_{i_0}.$$

*If $X_i \to X_{i'}$ is a homotopy equivalence for every arrow $i \to i'$ in $I$, then $g$ is a quasi-fibration.*

*Proof*. It suffices by Lemma 1.5 of [Dold-Lashof] to show that the restriction of $g$ to the $p$-skeleton $F_p$ of $BI$ is a quasi-fibration for all $p$. We have a map of cocartesian squares

$$\begin{array}{ccc} \coprod \partial\Delta^p & \subset & \coprod \Delta^p \\ \downarrow & & \downarrow \\ F_{p-1} & \subset & F_p \end{array} \quad \xleftarrow{g} \quad \begin{array}{ccc} \coprod X_{i_0} \times \partial\Delta^p & \subset & \coprod X_{i_0} \times \Delta^p \\ \downarrow & & \downarrow \\ g^{-1}(F_{p-1}) & \subset & g^{-1}(F_p) \end{array}$$

where the disjoint unions are taken over the nondegenerate $p$-simplices $i_0 \to \cdots \to i_p$ of $NI$. Let $U$ be the open set of $F_p$ obtained by removing the barycenters of the $p$-cells, and let $V = F_p - F_{p-1}$. It suffices by Lemma 1.4 of *loc. cit.* to show the restrictions of $g$ to $U, V$ and $U \cap V$ are quasi-fibrations. This is clear for $V$ and $U \cap V$, since over each $p$-cell $g$ is a product map.

We will apply Lemma 1.3 of *loc. cit.* to $g|U$, assuming as we may by induction that $g|F_{p-1}$ is a quasi-fibration, and using the evident fibre-preserving deformation $D$ of $g|U$ into $g|F_{p-1}$ provided by the radial deformation of $\Delta^p$ minus barycenter onto $\partial\Delta^p$. We have only to check that if $D$ carries $x \in U$ into $x' \in F_{p-1}$, then the map $g^{-1}(x) \to g^{-1}(x')$ induced by $D$ induces isomorphisms of homotopy groups. Supposing $x \notin F_{p-1}$ as we may, let $x$ come from an interior point $z$ of the copy of $\Delta^p$ corresponding to the simplex $s = (i_0 \to \cdots \to i_p)$, and let the radial deformation push $z$ into the open face of $\Delta^p$ with vertices $j_0 < \cdots < j_q$. Then it is easy to see that $g^{-1}(x) = X_{i_0}$ and $g^{-1}(x') = X_k, k = i_{j_0}$, and that the map in question is the one $X_{i_0} \to X_k$ induced by the face $i_0 \to k$ of $s$. As these induced maps are homotopy equivalences by hypothesis, the proof of the lemma is complete.

*Proof of Theorem B.* We return to the proof of Theorem A. The functor $p_1 : S(f) \to \underline{\underline{C}}$ is a homotopy equivalence as before, but not necessarily the functor $p_2$. The map $B_{p_2} : BS(f) \to B(\underline{\underline{C}}'^0)$ is the realization of the map (**). Thus applying the preceding lemma to the functor $Y \mapsto B(Y \setminus f)$ from $\underline{\underline{C}}'^0$ to spaces, we see that $B_{p_2}$ is a quasifibration, and hence the cartesian square

$$\begin{array}{ccc} Y\backslash f & \longrightarrow & S(f) \\ \downarrow & & \downarrow p_2 \\ pt & \xrightarrow{Y} & \underline{\underline{C}}'^{0} \end{array}$$

is homotopy-cartesian. Consider now the diagram

$$\begin{array}{ccccc} Y\backslash f & \longrightarrow & S(f) & \xrightarrow{\sim} & \underline{\underline{C}} \\ \downarrow & (1) & \downarrow f' & (2) & \downarrow f^{-} \\ Y\backslash \underline{\underline{C}}' & \longrightarrow & S(id_{\underline{\underline{C}}'}) & \underset{\sim}{\longrightarrow} & \underline{\underline{C}}' \\ \wr\downarrow & (3) & \wr\downarrow & & \\ pt & \underset{Y}{\longrightarrow} & \underline{\underline{C}}'^{0} & & \end{array}$$

in which the squares are cartesian, and in which the sign '$\sim$' denotes a homotopy equivalence. Since the square (1) + (3) is homotopy-cartesian, it follows that (1) is homotopy-cartesian, hence (1) + (2) is also, whence the theorem.

## §2. The $K$-groups of an exact category

**Exact categories.** Let $\underline{\underline{M}}$ be an additive category which is embedded as a full subcategory of an abelian category $\underline{\underline{A}}$, and suppose that $\underline{\underline{M}}$ is closed under extensions in $\underline{\underline{A}}$ in the sense that if an object $A$ of $\underline{\underline{A}}$ has a subobject $A'$ such that $A'$ and $A/A'$ are isomorphic to objects of $\underline{\underline{M}}$, then $A$ is isomorphic to an object of $\underline{\underline{M}}$. Let $\underline{\underline{E}}$ be the class of sequences

$$0 \to M' \xrightarrow{i} M \xrightarrow{j} M'' \to 0 \tag{1}$$

in $\underline{\underline{M}}$ which are exact in the abelian category $\underline{\underline{A}}$. We call a map in $\underline{\underline{M}}$ an *admissible monomorphism* (resp. *admissible epimorphism*) if it occurs as the map $i$ (resp. $j$) of some member (1) of $\underline{\underline{E}}$. Admissible monomorphisms and epimorphisms will sometimes be denoted $M' \rightarrowtail M$ and $M \twoheadrightarrow M''$, respectively.

The class $\underline{\underline{E}}$ clearly enjoys the following properties:

a) Any sequence in $\underline{\underline{M}}$ isomorphic to a sequence in $\underline{\underline{E}}$ is in $\underline{\underline{E}}$. For any $M', M''$ in $\underline{\underline{M}}$, the sequence

$$0 \to M' \xrightarrow{(id,0)} M' \oplus M'' \xrightarrow{pr_2} M'' \to 0 \tag{2}$$

is in $\underline{\underline{E}}$. For any sequence (1) in $\underline{\underline{E}}$, $i$ is a kernel for $j$ and $j$ is a cokernel for $i$ in the additive category $\underline{\underline{M}}$.

b) The class of admissible epimorphisms is closed under composition and under basechange by arbitrary maps in $\underline{\underline{M}}$. Dually, the class of admissible monomorphisms is closed under composition and under cobase-change by arbitrary maps in $\underline{\underline{M}}$.

c) Let $M \to M''$ be a map possessing a kernel in $\underline{\underline{M}}$. If there exists a map $N \to M$ in $\underline{\underline{M}}$ such that $N \to M \to M''$ is an admissible epimorphism, then $M \to M''$ is an admissible epimorphism. Dually for admissible monomorphisms.

For example, suppose given a sequence (1) in $\underline{\underline{E}}$ and a map $f : N \to M''$ in $\underline{\underline{M}}$. Form the diagram in $\underline{\underline{A}}$

$$\begin{array}{ccccccccc}
0 \to M' & \longrightarrow & M & \xrightarrow{j} & M'' & \longrightarrow 0 \\
\| & & \uparrow & & \uparrow f & \\
0 \longrightarrow M' & \longrightarrow & P & \longrightarrow & N & \longrightarrow 0
\end{array}$$

where $P$ is a fibre product of $f$ and $j$ in $\underline{\underline{A}}$. Because $\underline{\underline{M}}$ is closed under extensions in $\underline{\underline{A}}$, we can suppose $P$ is an object of $\underline{\underline{M}}$. Hence the basechange of $j$ by $f$ exists in $\underline{\underline{M}}$ and it is an admissible epimorphism.

**Definition.** An *exact category* is an additive category $\underline{\underline{M}}$ equipped with a family $\underline{\underline{E}}$ of sequences of the form (1), called the (short) *exact sequences* of $\underline{\underline{M}}$, such that the properties a), b), c) hold. An *exact functor* $F : \underline{\underline{M}} \to \underline{\underline{M}}'$ between exact categories is an additive functor carrying exact sequences in $\underline{\underline{M}}$ into exact sequences in $\underline{\underline{M}}'$.

**Examples.** Any abelian category is an exact category in an evident way. Any additive category can be made into an exact category in at least one way by taking $\underline{\underline{E}}$ to be the family of split exact sequences (2). A category which is 'abelian' in the sense of [Heller] is an exact category which is Karoubian (i.e. every projector has an image), and conversely.

Now suppose given an exact category $\underline{\underline{M}}$. Let $\underline{\underline{A}}$ be the additive category of additive contravariant functors from $\underline{\underline{M}}$ to abelian groups which are left exact, i.e. carry (1) to an exact sequence

$$0 \to F(M'') \to F(M) \to F(M').$$

(Precisely, choose a universe containing $\underline{\underline{M}}$, and let $\underline{\underline{A}}$ be the category of left exact functors whose values are abelian groups in the universe.) Following well-known ideas (e.g. [Gabriel]), one can prove $\underline{\underline{A}}$ is an abelian category, that the Yoneda functor $h$ embeds $\underline{\underline{M}}$ as a full subcategory of $\underline{\underline{A}}$ closed under extensions, and finally that a sequence (1) is in $\underline{\underline{E}}$ if and only if $h$ carries it into an exact sequence in $\underline{\underline{A}}$. The details will be omitted, as they are not really important for the sequel.

**The category $Q\underline{\underline{M}}$**

If $\underline{\underline{M}}$ is an exact category, we form a new category $Q\underline{\underline{M}}$ having the same objects as $\underline{\underline{M}}$ but with morphisms defined in the following way. Let $M$ and $M'$ be objects in $\underline{\underline{M}}$ and consider all diagrams

$$M \overset{j}{\twoheadleftarrow} N \overset{i}{\rightarrowtail} M' \tag{3}$$

where $j$ is an admissible epimorphism and $i$ is an admissible monomorphism. We consider isomorphisms of these diagrams which induce the identity on $M$ and $M'$, such isomorphisms being unique when they exist. A morphism from $M$ to $M'$ in the category $Q\underline{\underline{M}}$ is by definition an isomorphism class of these diagrams. Given a morphism from $M'$ to $M''$ represented by the diagram

$$M' \xleftarrow{j'} N' \xrightarrow{i'} M''$$

the composition of this morphism with the morphism from $M$ to $M'$ represented by (3) is the morphism represented by the pair $j \cdot pr_1, i' \cdot pr_2$ in the diagram

$$\begin{array}{ccccc}
N \times_{M'} N' & \xrightarrow{pr_2} & N' & \xrightarrow{i'} & M'' \\
{\scriptstyle pr_1}\downarrow & & \downarrow{\scriptstyle j'} & & \\
N & \xrightarrow[i]{} & M' & & \\
{\scriptstyle j}\downarrow & & & & \\
M & & & &
\end{array}$$

It is clear that composition is well-defined and associative. Thus when the isomorphism classes of diagrams (3) form a set (e.g. if every object of $\underline{\underline{M}}$ has a set of subobjects) then $Q\underline{\underline{M}}$ is a well-defined category. We assume this to be the case from now on.

It is useful to describe the preceding construction using admissible sub- and quotient objects. By an *admissible subobject* of $M$ we will mean an isomorphism class of admissible monomorphisms $M' \rightarrowtail M$, isomorphism being understood as isomorphism of objects over $M$. Admissible subobjects are in one-one correspondence with *admissible quotient* objects defined in the analogous way. The admissible subobjects of $M$ form an ordered set with the ordering: $M_1 \leqslant M_2$ if the unique map $M_1 \to M_2$ over $M$ is an admissible monomorphism. When $M_1 \leqslant M_2$, we call $(M_1, M_2)$ an *admissible layer* of $M$, and we call the cokernel $M_2/M_1$ an *admissible subquotient* of $M$.

With this terminology, it is clear that a morphism from $M$ to $M'$ in $Q\underline{\underline{M}}$ may be identified with a pair $((M_1, M_2), \theta)$ consisting of an admissible layer in $M'$ and an isomorphism $\theta : M \xrightarrow{\sim} M_2/M_1$. Composition is the obvious way of combining an isomorphism of $M$ with an admissible subquotient of $M'$ and an isomorphism of $M'$ with an admissible subquotient of $M''$ to get an isomorphism of $M$ with an admissible subquotient of $M''$.

For example, the morphisms from 0 to $M$ in $Q\underline{\underline{M}}$ are in one-one correspondence with the admissible subobjects of $M$. Isomorphisms from $M$ to $M'$ in $Q\underline{\underline{M}}$ are the same as isomorphisms from $M$ to $M'$ in $\underline{\underline{M}}$.

If $i : M' \rightarrowtail M$ is an admissible monomorphism, then it gives rise to a morphism from $M'$ to $M$ in $Q\underline{\underline{M}}$ which will be denoted

$$i_! : M' \to M.$$

Such morphisms will be called *injective*. Similarly, an admissible epimorphism $j : M \twoheadrightarrow M''$ gives rise to a morphism

$$j^! : M'' \to M$$

and these morphisms will be called *surjective*. By definition, any morphism $u$ in $Q\underline{\underline{M}}$ can be factored $u = i_! j^!$, and this factorization is unique up to unique isomorphism. If we form the bicartesian square

$$(4) \qquad \begin{array}{ccc} N & \overset{i}{\rightarrowtail} & M' \\ {\scriptstyle j}\big\downarrow & & \big\downarrow{\scriptstyle j'} \\ M & \underset{}{\overset{i'}{\rightarrowtail}} & N' \end{array}$$

then $u = j'^! i'_!$, and this injective-followed-by-surjective factorization is also unique up to unique isomorphism. A map which is both injective and surjective is an isomorphism and it is of the form $\theta_! = (\theta^{-1})^!$ for a unique isomorphism $\theta$ in $\underline{\underline{M}}$.

Injective and surjective maps in $Q\underline{\underline{M}}$ should not be confused with monomorphisms and epimorphisms in the categorical sense. Indeed, every morphism in $Q\underline{\underline{M}}$ is a monomorphism. In fact, the category $Q\underline{\underline{M}}/M$ is easily seen to be equivalent to the ordered set of admissible layers in $M$ with the ordering: $(M_0, M_1) \leqslant (M'_0, M'_1)$ if $M'_0 \leqslant M_0 \leqslant M_1 \leqslant M'_1$.

We can use the operations $i \mapsto i_!$ and $j \mapsto j^!$ to characterize the category $Q\underline{\underline{M}}$ by a *universal property*. First we note that these operations have the following properties:

a) If $i$ and $i'$ are composable admissible monomorphisms, then $(i'i)_! = i'_! i_!$. Dually, if $j$ and $j'$ are composable admissible epimorphisms then $(jj')^! = j'^! j^!$. Also $(id_M)_! = (id_M)^! = id_M$.

b) If (4) is a bicartesian square in which the horizontal (resp. vertical) maps are admissible monomorphisms (resp. epimorphisms), then $i_! j^! = j'^! i'_!$.

Now suppose given a category $\underline{\underline{C}}$ and for each object $M$ of $\underline{\underline{M}}$ an object $hM$ of $\underline{\underline{C}}$, and for each $i : M' \rightarrowtail M$ (resp. $j : M \twoheadrightarrow M''$)a map $i_! : hM' \to hM$ (resp. $j^! : hM'' \to hM$) such that the properties a), b) hold. Then it is clear that this data induces a unique functor $Q\underline{\underline{M}} \to \underline{\underline{C}}, M \mapsto hM$ compatible with the operations $i \mapsto i_!$ and $j \mapsto j^!$ in the two categories.

In particular, an exact functor $F : \underline{\underline{M}} \to \underline{\underline{M}}'$ between exact categories induces a functor $Q\underline{\underline{M}} \to Q\underline{\underline{M}}', M \mapsto FM, i_! \mapsto (Fi)_!, j^! \mapsto (Fj)^!$. We note also that if $\underline{\underline{M}}^0$ is the dual exact category, then we have an isomorphism of categories

$$(5) \qquad Q(\underline{\underline{M}}^0) = Q\underline{\underline{M}}$$

such that the injective arrows in the former correspond to surjective arrows in the latter and conversely.

**The fundamental group of $Q\underline{\underline{M}}$.** Suppose now that $\underline{\underline{M}}$ is a small exact category, so that the classifying space $B(Q\underline{\underline{M}})$ is defined. Let 0 be a given zero object of $\underline{\underline{M}}$.

**Theorem 1.** *The fundamental group $\pi_1(B(Q\underline{\underline{M}}), 0)$ is canonically isomorphic to the Grothendieck group $K_0\underline{\underline{M}}$.*

*Proof.* The Grothendieck group is by definition the abelian group with one generator $[M]$ for each object $M$ of $\underline{\underline{M}}$ and one relation $[M] = [M'][M'']$ for each exact sequence (1) in $\underline{\underline{M}}$. we note that it could also be defined as the not-necessarily-abelian group with the same generators and relations, because the relations $[M'][M''] = [M' \oplus M''] = [M''][M']$ force the group to be abelian.

According to Prop. 1, the category of covering spaces of $B(Q\underline{\underline{M}})$ is equivalent to the category $\underline{\underline{F}}$ of morphism-inverting functors $F : Q\underline{\underline{M}} \to Sets$. It suffices therefore to show the group $K_0\underline{\underline{M}}$ acts naturally on $F(0)$ for $F$ in $\underline{\underline{F}}$, and that the resulting functor from $\underline{\underline{F}}$ to $K_0\underline{\underline{M}} - sets$ is an equivalence of categories.

Let $i_M : 0 \rightarrowtail M$ and $j_M : M \twoheadrightarrow 0$ denote the obvious maps, and let $\underline{\underline{F}}'$ be the full subcategory of $\underline{\underline{F}}$ consisting of $F$ such that $F(M) = F(0)$ and $F(i_{M!}) = id_{F(0)}$ for all $M$. Clearly any $F$ is isomorphic to an object of $\underline{\underline{F}}'$, so it suffices to show $\underline{\underline{F}}'$ is equivalent to $K_0\underline{\underline{M}} - sets$.

Given a $K_0\underline{\underline{M}} - set$ $S$, let $F_S : Q\underline{\underline{M}} \to Sets$ be the functor defined by

$$F_S(M) = S, F_S(i_!) = id_S, F_S(j^!) = \text{ multiplication by } \quad [\operatorname{Ker} j] \text{ on } S,$$

using the universal property of $Q\underline{\underline{M}}$. Clearly $S \mapsto F_S$ is a functor from $K_0\underline{\underline{M}} - sets$ to $\underline{\underline{F}}'$. On the other hand if $F \in \underline{\underline{F}}'$, then given $i : M' \rightarrowtail M$ we have $i \cdot i_{M'} = i_M$, hence $F(i_!) = id_{F(0)}$. Given the exact sequence

$$0 \to M' \xrightarrow{i} M \xrightarrow{j} M'' \to 0$$

we have $j^! i_{M''!} = i_! j^!_{M'}$, hence $F(j^!) = F(j^!_{M'}) \in Aut(F(0))$. Also

$$F(j^!_M) = F(j^! j^!_{M''}) = F(j^!_{M'}) F(j^!_{M''})$$

so by the universal property of $K_0\underline{\underline{M}}$, there is a unique group homomorphism from $K_0\underline{\underline{M}}$ to $Aut(F(0))$ such that $[M] \mapsto F(j^!_M)$. Thus we have a natural action of $K_0\underline{\underline{M}}$ on $F(0)$ for any $F$ in $\underline{\underline{F}}'$. In fact, it is clear that the resulting functor $F \mapsto F(0)$ from $\underline{\underline{F}}'$ to $K_0\underline{\underline{M}} - sets$ is an isomorphism of categories with inverse $S \mapsto F_S$, so the proof of the theorem is complete.

**Higher $K$-groups.** The above theorem offers some motivation for the following definition of $K$-groups for a small exact category $\underline{\underline{M}}$.

**Definition.** $K_i\underline{\underline{M}} = \pi_{i+1}(B(Q\underline{\underline{M}}), 0)$.

Note first of all that the $K$-groups are independent of the choice of the zero object 0. Indeed, given another zero object $0'$, there is a unique map $0 \to 0'$ in $Q\underline{\underline{M}}$, hence there is a canonical path from 0 to $0'$ in the classifying space.

Secondly we note that *the preceding definition extends to exact categories having a set of isomorphism classes of objects.* We define $K_i\underline{\underline{M}}$ to be $K_i\underline{\underline{M}}'$, where $\underline{\underline{M}}'$ is a small subcategory equivalent to $\underline{\underline{M}}$, the choice of $\underline{\underline{M}}'$ being irrelevant by Prop. 2. From now on we will only consider exact categories whose isomorphism classes form a set, except when mentioned otherwise. In addition, when we apply the results of §1, it will be tacitly assumed that we have replaced any large exact category by an equivalent small one.

**Elementary properties of $K$-groups.** An exact functor $f : \underline{\underline{M}} \to \underline{\underline{M}}'$ induces a functor $Q\underline{\underline{M}} \to Q\underline{\underline{M}}'$, and hence a homomorphism of $K$-groups which will be denoted

(6) $$f_* : K_i\underline{\underline{M}} \to K_i\underline{\underline{M}}'.$$

In this way $K_i$ becomes a functor from exact categories and exact functors to abelian groups. Moreover, isomorphic functors induce the same map on $K$-groups by Prop. 2. From (5) we have

$$K_i(\underline{\underline{M}}^0) = K_i\underline{\underline{M}}. \tag{7}$$

The product $\underline{\underline{M}} \times \underline{\underline{M}}'$ of two exact categories is an exact category in which a sequence is exact when its projections in $\underline{\underline{M}}$ and $\underline{\underline{M}}'$ are. Clearly $Q(\underline{\underline{M}} \times \underline{\underline{M}}') = Q\underline{\underline{M}} \times Q\underline{\underline{M}}'$. Since the classifying space functor is compatible with products (§1, (4)), we have

$$K_i(\underline{\underline{M}} \times \underline{\underline{M}}') \simeq K_i\underline{\underline{M}} \oplus K_i\underline{\underline{M}}', x \mapsto pr_{1*}(x) + pr_{2*}(x). \tag{8}$$

The functor $\oplus : \underline{\underline{M}} \times \underline{\underline{M}} \to \underline{\underline{M}}, (M, M') \mapsto M \oplus M'$ is exact, so it induces a homomorphism

$$K_i\underline{\underline{M}} \oplus K_i\underline{\underline{M}} = K_i(\underline{\underline{M}} \times \underline{\underline{M}}) \xrightarrow{\oplus_*} K_i\underline{\underline{M}}.$$

This map coincides with the sum in the abelian group $K_i\underline{\underline{M}}$ because the functors $M \mapsto 0 \oplus M, M \mapsto M \oplus 0$ are isomorphic to the identity.

Let $j \mapsto \underline{\underline{M}}_j$ be a functor from a small filtering category to exact categories and functors, and let $\varinjlim \underline{\underline{M}}_j$ be the inductive limit of the $\underline{\underline{M}}_j$ in the sense of Prop. 3. Then $\varinjlim \underline{\underline{M}}_j$ is an exact category in a natural way, and $Q(\varinjlim \underline{\underline{M}}_j) = \varinjlim Q\underline{\underline{M}}_j$, hence from Prop. 3 we obtain an isomorphism

$$K_i(\varinjlim \underline{\underline{M}}_j) = \varinjlim K_i\underline{\underline{M}}_j. \tag{9}$$

**Example.** Let $A$ be a ring with 1 and let $\underline{\underline{P}}(A)$ denote the additive category of finitely generated projective (left) $A$-modules. We regard $\underline{\underline{P}}(A)$ as an exact category in which the exact sequences are those sequences which are exact in the category of all $A$-modules, and we define the $K$-groups of the ring $A$ by

$$K_iA = K_i(\underline{\underline{P}}(A)).$$

A ring homomorphism $A \to A'$ induces an exact functor $A'\otimes_A ? : \underline{\underline{P}}(A) \to \underline{\underline{P}}(A')$ which is defined up to canonical isomorphism, hence it induces a well-defined homomorphism

$$(A'\otimes_A ?)_* : K_iA \to K_iA'. \tag{10}$$

making $K_iA$ a covariant functor of $A$. From (8) we have

$$K_i(A \times A') = K_iA \oplus K_iA'. \tag{11}$$

If $j \mapsto A_j$ is a filtered inductive system of rings, we have from (9) an isomorphism

$$K_i(\varinjlim A_j) = \varinjlim K_iA_j. \tag{12}$$

(To apply (9), one replaces $\underline{\underline{P}}(A_j)$ by the equivalent category $\underline{\underline{P}}(A_j)'$ whose objects are the idempotent matrices over $A_j$, so that $\underline{\underline{P}}(\varinjlim A_j)' = \varinjlim \underline{\underline{P}}(A_j)'$.) Finally we note that $p \mapsto \mathrm{Hom}_A(P, A)$ is an equivalence of $\underline{\underline{P}}(A)$ with the dual category to $\underline{\underline{P}}(A^{op})$, where $A^{op}$ is the opposed ring to $A$, hence from (7) we get a canonical isomorphism

$$K_i(A) = K_i(A^{op}). \tag{13}$$

**Remarks.** It can be proved that the groups $K_iA$ defined here agree with those defined by making $BGL(A)$ into an $H$-space and taking homotopy groups (see for example [Gersten 5]). In particular, they coincide for $i = 1, 2$ with the groups defined by by Bass and Milnor, and with the $K$-groups computed for a finite field in [Quillen 2]. On the other hand, for a general exact category $\underline{\underline{M}}$, the group $K_1(\underline{\underline{M}})$ is not the same as the universal determinant group defined in [Bass, p.389]. There is a canonical homomorphism from the universal determinant group to $K_1(\underline{\underline{M}})$, but Gersten and Murthy have produced examples showing that it is neither surjective nor injective in general.

## §3. Characteristic exact sequences and filtrations

Let $\underline{\underline{M}}$ be an exact category and regard the family $\underline{\underline{E}}$ of short exact sequences in $\underline{\underline{M}}$ as an additive category in the obvious way. We denote objects of $\underline{\underline{E}}$ by $E, E'$, etc. and let $sE, tE, qE$ denote the sub-, total, and quotient objects of $E$, whence we have an exact sequence

$$0 \to sE \to tE \to qE \to 0$$

in $\underline{\underline{M}}$ associated to each object $E$ of $\underline{\underline{E}}$. A sequence in $\underline{\underline{E}}$ will be called exact if it gives rise to three exact sequences in $\underline{\underline{M}}$ on applying $s, t$, and $q$. With this notion of exactness, it is clear that $\underline{\underline{E}}$ is an exact category, and that $s, t$, and $q$ are exact functors from $\underline{\underline{E}}$ to $\underline{\underline{M}}$.

**Theorem 2.** *The functor $(s, q) : Q\underline{\underline{E}} \to Q\underline{\underline{M}} \times Q\underline{\underline{M}}$ is a homotopy equivalence.*

*Proof*. It suffices by Theorem A to show the category $(s, q)/(M, N)$ is contractible for any given pair $M, N$ of objects of $\underline{\underline{M}}$. Put $\underline{\underline{C}} = (s, q)/(M, N)$; it is the fibred category over $Q\underline{\underline{E}}$ consisting of triples $(E, u, v)$, where $u : sE \to M, v : qE \to N$ are maps in $Q\underline{\underline{M}}$. Let $\underline{\underline{C}}'$ be the full subcategory of $\underline{\underline{C}}$ consisting of the triples $(E, u, v)$ such that $u$ is surjective, and let $\underline{\underline{C}}''$ be the full subcategory of triples such that $u$ is surjective and $v$ is injective.

**Lemma.** *The inclusion functors $\underline{\underline{C}}' \to \underline{\underline{C}}$ and $\underline{\underline{C}}'' \to \underline{\underline{C}}'$ have left adjoints.*

Consider first the inclusion of $\underline{\underline{C}}'$ in $\underline{\underline{C}}$. Let $X = (E, u, v) \in \underline{\underline{C}}$; it suffices to show that there is a universal arrow $X \to \overline{X}$ in $\underline{\underline{C}}$ with $\overline{X}$ in $\underline{\underline{C}}'$.

Let $u = j^! i_!$ where $i : sE \rightarrowtail M', j : M \twoheadrightarrow M'$, and define the exact sequence $i_*E$ by 'pushout':

$$\begin{array}{rccccccccc}
E: & 0 & \longrightarrow & sE & \longrightarrow & tE & \longrightarrow & qE & \longrightarrow & 0 \\
 & & & \downarrow{\scriptstyle i} & & \downarrow & & \| & & \\
i_*E: & 0 & \longrightarrow & M' & \longrightarrow & T & \longrightarrow & qE & \longrightarrow & 0.
\end{array}$$

Let $\overline{X} = (i_* E, j^!, v)$; it belongs to $\underline{\underline{C}}'$ and there is a canonical arrow $X \to \overline{X}$ given by the evident injective map $E \to i_* E$.

Now suppose given $X \to X'$ with $X' = (E', j'^!, v')$ in $\underline{\underline{C}}'$. Represent the map $E \to E'$ by the pair $E \rightarrowtail E_0$, $E' \twoheadrightarrow E_0$. Since

$$sE \rightarrowtail sE_0 \twoheadleftarrow sE' \xleftarrow{j'} M$$

represents $u$, we can suppose $E_0$ chosen so that $sE \rightarrowtail sE_0$ is the map $i$, and $M \twoheadrightarrow sE_0$ is $j$. By the universal property of pushouts, the map $E \rightarrowtail E_0$ factors uniquely $E \rightarrowtail i_* E \rightarrowtail E_0$, so it is clear that we have a map $\overline{X} \to X'$ in $\underline{\underline{C}}'$ such that $X \to \overline{X} \to X'$ is the given map $X \to X'$.

It remains to show the uniqueness of the map $\overline{X} \to X'$. Consider factorizations $X \to X'' \to X'$ of $X \to X'$ such that $X''$ is in $\underline{\underline{C}}'$. Note that $\underline{\underline{C}}/X' = Q\underline{\underline{E}}/E'$ is equivalent to the ordered set of admissible layers in $E'$. Let $(E_0, E_1)$ be the layer corresponding to $X \to X'$ and $(E_0'', E_1'')$ the layer corresponding to $X'' \to X'$ so that $(E_0, E_1) \leqslant (E_0'', E_1'')$ and $sE_1'' = sE'$. There is a least such layer $(E_0'', E_1'')$ given by $tE_0'' = tE_0$, $tE_1'' = sE' + tE_1$, which is characterized by the fact that the map $E_1/E_0 \to E_1''/E_0''$ is injective and induces an isomorphism on quotient objects. Thus among the factorizations $X \to X'' \to X'$ there is a least one, unique up to canonical isomorphism, and characterized by the condition that $E \to E''$ should be injective and induce an isomorphism $qE \xrightarrow{\sim} qE''$. Since the factorization $X \to \overline{X} \to X'$ has this property, it is clear that the map $\overline{X} \to X'$ is uniquely determined. Thus $\underline{\underline{C}}' \to \underline{\underline{C}}$ has the left adjoint $X \mapsto \overline{X}$.

Next consider the inclusion of $\underline{\underline{C}}''$ in $\underline{\underline{C}}'$, and let $(E, u, v) \in \underline{\underline{C}}'$. Represent $v : qE \to N$ by the pair $j : N' \twoheadrightarrow qE$, $i : N' \rightarrowtail N$, and define $j^* E$ by pull-back:

$$\begin{array}{ccccccccc} 0 & \longrightarrow & sE & \longrightarrow & T & \longrightarrow & N' & \longrightarrow & 0 \\ & & \| & & \downarrow & & \downarrow & & \\ 0 & \longrightarrow & sE & \longrightarrow & tE & \longrightarrow & qE & \longrightarrow & 0. \end{array}$$

One verifies by an argument essentially dual to the preceding one that $(E, u, v) \mapsto (j^* E, u, i_!)$ is left adjoint to the inclusion of $\underline{\underline{C}}''$ in $\underline{\underline{C}}'$. This finishes the lemma.

By Prop. 2, Cor. 1, the categories $\underline{\underline{C}}$ and $\underline{\underline{C}}''$ are homotopy equivalent. Let $(E, j^!, i_!) \in \underline{\underline{C}}''$, and let $j_M : M \twoheadrightarrow 0$ and $i_N : 0 \rightarrowtail N$ be the obvious maps. A map from $(0, j_M^!, i_{N!})$ to $(E, j^!, i_!)$ may be identified with an admissible subobject $E'$ of $E$ such that $sE' = sE$ and $qE' = 0$. Clearly $E'$ is unique, so $(0, j_M^!, i_{N!})$ is an initial object of $\underline{\underline{C}}''$. Thus $\underline{\underline{C}}''$, and hence $\underline{\underline{C}}$ is contractible, which finishes the proof of the theorem.

**Corollary 1.** *Let $\underline{\underline{M}}'$ and $\underline{\underline{M}}$ be exact categories and let*

$$0 \longrightarrow F' \longrightarrow F \longrightarrow F'' \longrightarrow 0$$

*be an exact sequence of exact functors from $\underline{\underline{M}}'$ to $\underline{\underline{M}}$. Then*

$$F_* = F'_* + F''_* : K_i \underline{\underline{M}}' \to K_i \underline{\underline{M}}.$$

*Proof.* It clearly suffices to treat the case of the exact sequence

$$0 \longrightarrow s \longrightarrow t \longrightarrow q \longrightarrow 0$$

of functors from $\underline{\underline{E}}$ to $\underline{\underline{M}}$. Let $f : \underline{\underline{M}} \times \underline{\underline{M}} \to \underline{\underline{E}}$ be the exact functor sending $(M', M'')$ to the split exact sequence

$$0 \longrightarrow M' \longrightarrow M' \oplus M'' \longrightarrow M'' \longrightarrow 0.$$

The functors $tf$ and $\oplus(s, q)f$ are isomorphic, hence

$$t_* f_* = \oplus_*(s_*, q_*)f_* = (s_* + q_*)f_* : (K_i\underline{\underline{M}})^2 \to K_i\underline{\underline{M}}.$$

But $f_*$ is a section of $(s_*, q_*) : K_i\underline{\underline{E}} \to (K_i\underline{\underline{M}})^2$ which is an isomorphism by the theorem. Thus $t_* = s_* + q_*$, proving the corollary.

Note that the category of functors from a category $\underline{\underline{C}}$ to an exact category $\underline{\underline{M}}$ is an exact category in which a sequence of functors is exact if it is pointwise exact. We thus have the notion of an *admissible filtration* $0 = F_0 \subset F_1 \subset \cdots \subset F_n = F$ of a functor $F$. this means that $F_{p-1}(X) \to F_p(X)$ is an admissible monomorphism in $\underline{\underline{M}}$ for every $X$ in $\underline{\underline{C}}$, and it implies that there exist quotient functors $F_p/F_q$ for $q \leqslant p$, determined up to canonical isomorphism. It is easily seen that if $\underline{\underline{C}}$ is an exact category, and if the functors $F_p/F_{p-1}$ are exact for $1 \leqslant p \leqslant n$, then all the quotients $F_p/F_q$ are exact.

**Corollary 2.** (*Additivity for 'characteristic' filtrations*) *Let* $F : \underline{\underline{M}}' \to \underline{\underline{M}}$ *be an exact functor between exact categories equipped with an admissible filtration* $0 = F_0 \subset \cdots \subset F_n = F$ *such that the quotient functors* $F_p/F_{p-1}$ *are exact for* $1 \leqslant p \leqslant n$. *Then*

$$F_* = \sum_{p-1}^{n} (F_p/F_{p-1})_* : K_i\underline{\underline{M}}' \to K_i\underline{\underline{M}}.$$

**Corollary 3.** (*Additivity for 'characteristic' exact sequences*) *If*

$$0 \to F_0 \to \cdots \to F_n \to 0$$

*is an exact sequence of exact functors from* $\underline{\underline{M}}'$ *to* $\underline{\underline{M}}$, *then*

$$\sum_{p=0}^{n} (-1)^p (F_p)_* = 0 : K_i\underline{\underline{M}}' \to K_i\underline{\underline{M}}.$$

These result from Cor. 1 by induction.

**Applications.** we give two simple examples to illustrate the preceding results.

Let $X$ be a ringed space, and put $K_iX = K_i\underline{\underline{P}}(X)$, where $\underline{\underline{P}}(X)$ is the category of vector bundles on $X$,(i.e. sheaves of $\underline{\underline{O}}_X$-modules which are locally direct factors of $\underline{\underline{O}}_X^n$) equipped with the usual notion of exact sequence. Given $E$ in $\underline{\underline{P}}(X)$, we have an exact functor $E\otimes ? : \underline{\underline{P}}(X) \to \underline{\underline{P}}(X)$ which induces a homomorphism of $K$-groups $(E\otimes ?)_* : K_iX \to K_iX$. If $0 \to E' \to E \to E'' \to 0$

is an exact sequence of vector bundles, then Cor. 1 implies $(E\otimes ?)_* = (E'\otimes ?)_* + (E''\otimes ?)_*$. Thus we obtain products

(1) $$K_0X \otimes_{\mathbb{Z}} K_iX \to K_iX, [E] \otimes x \mapsto (E\otimes ?)_* x$$

which clearly make $K_iX$ into a module over $K_0X$. (Products $K_iX \otimes K_jX \to K_{i+j}X$ can also be defined, but this requires more machinery.)

**Graded rings.** Let $A = A_0 \oplus A_1 \oplus \cdots$ be a graded ring and denote by $\underline{\underline{P}}gr(A)$ the category of graded finitely generated projective $A$-modules $P = \oplus P_n, n \in \mathbb{Z}$. The group $K_i(\underline{\underline{P}}gr(A))$ is a $\mathbb{Z}[t, t^{-1}]$-module, where multiplication by $t$ is the automorphism induced by the translation functor $P \mapsto P(-1), P(-1)_n = P_{n-1}$.

**Proposition.** *There is a* $\mathbb{Z}[t, t^{-1}]$ *-module isomorphism*

$$\mathbb{Z}[t, t^{-1}] \otimes_{\mathbb{Z}} K_iA_0 \xrightarrow{\sim} K_i(\underline{\underline{P}}gr(A)), \quad 1 \otimes x \mapsto (A \otimes_{A_0} ?)_* x.$$

*Proof*. Given $P$ in $\underline{\underline{P}}gr(A)$, let $F_kP$ be the $A$-submodule of $P$ generated by $P_n$ for $n \leqslant k$, and let $\underline{\underline{P}}_q$ be the full subcategory of $\underline{\underline{P}}gr(A)$ consisting of those $P$ for which $F_{-q-1}P = 0$ and $F_qP = P$. We have an exact functor

$$T : \underline{\underline{P}}gr(A) \to \underline{\underline{P}}gr(A_0), \quad T(P) = A_0 \otimes_A P$$

where $A_0$ is considered as a graded ring concentrated in degree zero. It is known ([Bass], p.637) that $P$ is non-canonically isomorphic to

$$A \otimes_{A_0} T(P) = \coprod_n A(-n) \otimes_{A_0} T(P)_n.$$

It follows that $P \mapsto F_kP$ is an exact functor from $\underline{\underline{P}}gr(A)$ to itself, and that there is a canonical isomorphism of exact functors

$$F_nP/F_{n-1}P \simeq A(-n) \otimes_{A_0} T(P)_n.$$

Applying Cor. 2 to the identity functor of $\underline{\underline{P}}_q$ and the filtration $0 = F_{-q-1} \subset \cdots \subset F_q = id$, one sees that the homomorphism

$$\coprod_{-q \leqslant n \leqslant q} t^n \otimes K_iA_0 \to K_i\underline{\underline{P}}_q, \quad t^n \otimes x \mapsto (A(-n) \otimes_{A_0} ?)_* x$$

is an isomorphism with inverse given by the map with components $(T_n)_*, -q \leqslant n \leqslant q$. Since $\underline{\underline{P}}gr(A)$ is the union of the $\underline{\underline{P}}_q$, the proposition results from §2, (9).

## §4. Reduction by resolution

In this section $\underline{\underline{M}}$ denotes an exact category with a set of isomorphism classes, and $\underline{\underline{P}}$ a full subcategory *closed under extensions* in $\underline{\underline{M}}$ in the sense that $\underline{\underline{P}}$ contains a zero object and for any exact sequence in $\underline{\underline{M}}$

(1) $$0 \to M' \to M \to M'' \to 0$$

if $M'$ and $M''$ are isomorphic to objects of $\underline{\underline{P}}$, so is $M$. Such a $\underline{\underline{P}}$ is an exact category where a sequence is exact if and only if it is exact in $\underline{\underline{M}}$. The category $Q\underline{\underline{P}}$ is a subcategory of $Q\underline{\underline{M}}$ which is not usually a full subcategory, as $\underline{\underline{M}}$-admissible monomorphisms and epimorphisms need not be $\underline{\underline{P}}$-admissible.

In the following, letters $P, P'$, etc. will denote objects of $\underline{\underline{P}}$, and the symbols $\rightarrowtail, \twoheadrightarrow, \leqslant$ will always refer to $\underline{\underline{M}}$-admissible monomorphisms, epimorphisms and subobjects, respectively. The corresponding $\underline{\underline{P}}$-admissible notions will be specified explicitly. For example, $P \rightarrowtail P'$ denotes an $\underline{\underline{M}}$-admissible monomorphism between two objects of $\underline{\underline{P}}$; it is $\underline{\underline{P}}$-admissible iff the cokernel is isomorphic to an object of $\underline{\underline{P}}$.

We are interested in showing that the inclusion of $\underline{\underline{P}}$ in $\underline{\underline{M}}$ induces isomorphisms $K_i\underline{\underline{P}} \simeq K_i\underline{\underline{M}}$ when every object $M$ of $\underline{\underline{M}}$ has a finite $\underline{\underline{P}}$-resolution:

$$0 \to P_n \to \cdots \to P_0 \to M \to 0. \tag{2}$$

The standard proof for $K_0$ consists in defining an inverse map $K_0\underline{\underline{M}} \to K_0\underline{\underline{P}}$ by showing $\sum(-1)^n[P_n] \in K_0\underline{\underline{P}}$ depends only on $[M]$. By Cor. 3 of the preceding section, this method works when there exist resolutions (2) depending on $M$ in an exact functorial fashion. However, this situation occurs rarely, so we must proceed differently.

The following theorem handles the case where resolutions of length one exist. As an example, think of $\underline{\underline{M}}$ as modules of projective dimension $\leqslant n$, and $\underline{\underline{P}}$ as the subcategory of modules of projective dimension $< n$. The general case follows by induction (see Cor. 1).

**Theorem 3.** *Let $\underline{\underline{P}}$ be a full subcategory of an exact category $\underline{\underline{M}}$ which is closed under extensions and is such that*

i) *For any exact sequence* (1), *if $M$ is in $\underline{\underline{P}}$, then $M'$ is in $\underline{\underline{P}}$.*

ii) *For any $M''$ in $\underline{\underline{M}}$, there exists an exact sequence* (1) *with $M$ in $\underline{\underline{P}}$. Then the inclusion functor $Q\underline{\underline{P}} \to Q\underline{\underline{M}}$ is a homotopy equivalence, so $K_i\underline{\underline{P}} \xrightarrow{\sim} K_i\underline{\underline{M}}$.*

*Proof.* We factor $Q\underline{\underline{P}} \to Q\underline{\underline{M}}$. into two inclusion functors

$$Q\underline{\underline{P}} \xrightarrow{g} \underline{\underline{C}} \xrightarrow{f} Q\underline{\underline{M}}$$

where $\underline{\underline{C}}$ is the full subcategory of $Q\underline{\underline{M}}$ with the same objects as $Q\underline{\underline{P}}$. We will prove $g$ and $f$ are homotopy equivalences.

To show $g$ is a homotopy equivalence, it suffices by Theorem A to prove $g/P$ is contractible for any object $P$ in $\underline{\underline{C}}$. The category $g/P$ is easily seen to be equivalent to the ordered set $J$ of $\underline{\underline{M}}$-admissible layers $(M_0, M_1)$ in $P$ such that $M_1/M_0 \in \underline{\underline{P}}$, with the ordering $(M_0, M_1) \prec (M_0', M_1')$ iff $M_0' \leqslant M_0 \leqslant M_1 \leqslant M_1'$ and $M_0/M_0', M_1'/M_1 \in \underline{\underline{P}}$. By hypothesis i), one knows that $M_1$, and $M_0$ are in $\underline{\underline{P}}$ for every $(M_0, M_1)$ in $J$. Hence in $J$ we have arrows

$$(M_0, M_1) \prec (0, M_1) \succ (0, 0)$$

which can be viewed as natural transformations of functors from $J$ to $J$ joining the functor $(M_0, M_1) \mapsto (0, M_1)$ to the identity and to the constant functor with value $(0,0)$. Using Prop. 2, we see that $J$, hence $g/P$, is contractible, so $g$ is a homotopy equivalence.

To prove $f$ is a homotopy equivalence, we show $M \setminus f$ is contractible for any $M$ in $Q\underline{\underline{M}}$. Put $\underline{\underline{F}} = M \setminus f$; it is the cofibred category over $\underline{\underline{C}}$ consisting of pairs $(P, u)$ with $u : M \to P$ a map in $Q\underline{\underline{M}}$. Let $\underline{\underline{F}}'$ be the full subcategory consisting of $(P, u)$ with $u$ surjective. Given $X = (P, u)$ in $\underline{\underline{F}}$, write $u = i_! j^!$ with $j : \overline{P} \twoheadrightarrow M, i : \overline{P} \rightarrowtail P$. By hypothesis i), $\overline{P}$ is in $\underline{\underline{P}}$ as the notation suggests. Thus $\overline{X} = (\overline{P}, j^!)$ is an object of $\underline{\underline{F}}'$, and $i$ defines a map $\overline{X} \to X$. One verifies easily that $\overline{X} \to X$ is a universal arrow from an object of $\underline{\underline{F}}'$ to $X$, hence $X \mapsto \overline{X}$ is right adjoint to the inclusion of $\underline{\underline{F}}'$ in $\underline{\underline{F}}$. By Prop. 2, Cor. 1, we have only to prove that $\underline{\underline{F}}'$ is contractible.

The dual category $\underline{\underline{F}}'^{0}$ is the category whose objects are maps $P \twoheadrightarrow M$, and in which a morphism from $P \twoheadrightarrow M$ to $P' \twoheadrightarrow M$ is a map $P \twoheadrightarrow P'$ such that the obvious triangle commutes. By hypothesis ii), there is at least one such object $P_0 \twoheadrightarrow M$. Given another $P \twoheadrightarrow M$, the fibre product $P \times_M P_0$ is an object of $\underline{\underline{P}}$, as it is an extension of $P_0$ by Ker $(P \twoheadrightarrow M)$ which is in $\underline{\underline{P}}$ by hypothesis i). Hence in $\underline{\underline{F}}'^{0}$ we have arrows

$$(P \twoheadrightarrow M) \leftarrow (P \times_M P_0 \twoheadrightarrow M) \to (P_0 \twoheadrightarrow M)$$

which may be viewed as natural transformations from the functor $(P \twoheadrightarrow M) \mapsto (P \times_M P_0 \twoheadrightarrow M)$ to the constant functor with value $P_0 \twoheadrightarrow M$ and to the identity functor. Using Prop. 2, we conclude that $\underline{\underline{F}}'$ is contractible, finishing the proof of the theorem.

**Corollary 1.** *Assume $\underline{\underline{P}}$ is closed under extensions in $\underline{\underline{M}}$ and further that*

a) *For every exact sequence* (1), *if $M, M''$ are in $\underline{\underline{P}}$, then so is $M'$.*

b) *Given $j : M \twoheadrightarrow P$, there exists $j' : P' \twoheadrightarrow P$ and $f : P' \to M$ such that $jf = j'$.* (*This holds, for example, if for every $M$ there exists $P' \twoheadrightarrow M$.*) *Let $\underline{\underline{P}}_n$ be the full subcategory of $\underline{\underline{M}}$ consisting of $M$ having $\underline{\underline{P}}$-resolutions of length $\leqslant n$, i. e. such that there exists an exact sequence* (2), *and put $\underline{\underline{P}}_\infty = \bigcup \underline{\underline{P}}_n$. Then*

$$K_i\underline{\underline{P}} \xrightarrow{\sim} K_i\underline{\underline{P}}_1 \xrightarrow{\sim} \cdots \xrightarrow{\sim} K_i\underline{\underline{P}}_\infty .$$

That $\underline{\underline{P}}_n$ is closed under extensions in $\underline{\underline{M}}$, and hence the groups $K_i\underline{\underline{P}}_n$ are defined results from the following standard facts (compare [Bass, p.39]).

**Lemma.** *For any exact sequence* (1) *and integer $n \geqslant 0$, we have*

1) $M \in \underline{\underline{P}}_n, M'' \in \underline{\underline{P}}_{n+1} \Rightarrow M' \in \underline{\underline{P}}_n$

2) $M', M'' \in \underline{\underline{P}}_{n+1} \Rightarrow M \in \underline{\underline{P}}_{n+1}$

3) $M, M'' \in \underline{\underline{P}}_{n+1} \Rightarrow M' \in \underline{\underline{P}}_{n+1}$.

Assuming this, we apply Theorem 3 to the pair $\underline{\underline{P}}_n \subset \underline{\underline{P}}_{n+1}$. Hypothesis ii) is satisfied, for given $M \in \underline{\underline{P}}_{n+1}$, there exists an $\underline{\underline{M}}$-admissible epimorphism $P \twoheadrightarrow M$ with $P \in \underline{\underline{P}}$ ; and by 1) it is $\underline{\underline{P}}_{n+1}$-admissible. The other hypotheses are clear, so $K_i\underline{\underline{P}}_n \xrightarrow{\sim} K_i\underline{\underline{P}}_{n+1}$ for each $n$. The case of $\underline{\underline{P}}_\infty$ follows by passage to the limit (§2, (9)).

To prove the lemma, it suffices by a simple induction to treat the case $n = 0$.

1): Since $M'' \in \underline{\underline{P}}_1$, there exists a short exact sequence $P' \rightarrowtail P \twoheadrightarrow M''$, so we can form the diagram on the left with short exact rows and columns

$$\begin{array}{ccccc} 0 & \longrightarrow & P' & = & P' \\ \downarrow & & \downarrow & & \downarrow \\ M' & \longrightarrow & F & \longrightarrow & P \\ \| & & \downarrow & & \downarrow \\ M' & \longrightarrow & M & \longrightarrow & M'' \end{array} \qquad \begin{array}{ccccc} R' & \longrightarrow & R & \longrightarrow & R'' \\ \downarrow & & \downarrow & & \downarrow \\ P' & \longrightarrow & P' \oplus P'' & \longrightarrow & P'' \\ \downarrow & & \downarrow & & \downarrow \\ M' & \longrightarrow & M & \longrightarrow & M'' \end{array}$$

and with $F = M \times_{M''} P$. Since $P', M$ are in $\underline{\underline{P}}$ and $\underline{\underline{P}}$ is closed under extensions, we have $F \in \underline{\underline{P}}$. Since $F, P \in \underline{\underline{P}}$ we have from a) that $M' \in \underline{\underline{P}}$, proving 1).

2): Since $M'' \in \underline{\underline{P}}_1$, there exists $P \twoheadrightarrow M''$, so applying b) to $pr_1 : P \times_{M''} M \twoheadrightarrow P$, we can enlarge $P$ and find $P'' \twoheadrightarrow M$ factoring into $P'' \rightarrow M \rightarrow M''$. Thus we can form the above diagram on the right with short exact rows and columns, and with $P', R' \in \underline{\underline{P}}$ as $M' \in \underline{\underline{P}}_1$. Applying 1) we see that $R'' \in \underline{\underline{P}}$, so $R \in \underline{\underline{P}}$ and $M \in \underline{\underline{P}}_1$, proving 2).

3): Since $M \in \underline{\underline{P}}_1$, we can form the diagram with short exact rows and columns

$$\begin{array}{ccccc} P' & = & P' & \longrightarrow & 0 \\ \downarrow & & \downarrow & & \downarrow \\ K & \longrightarrow & P & \longrightarrow & M'' \\ \downarrow & & \downarrow & & \| \\ M' & \longrightarrow & M & \longrightarrow & M'' \end{array}$$

As $M'' \in \underline{\underline{P}}_1$, 1) implies $K \in \underline{\underline{P}}$, so $M' \in \underline{\underline{P}}_1$, proving 3). The lemma and Cor. 1 are done.

As an example of the corollary, take $\underline{\underline{P}} = \underline{\underline{P}}(A)$ and $\underline{\underline{M}} = Mod(A)$, the category of (left) $A$-modules. (Better, so that $\underline{\underline{M}}$ has a set of isomorphism classes, take $\underline{\underline{M}}$ to be the abelian category of all $A$-modules of cardinality $< \alpha$, where $\alpha$ is some infinite cardinal $> card(A)$.) Let $\underline{\underline{P}}_n(A)$ be the category of $A$-modules having $\underline{\underline{P}}$-resolutions of length $\leqslant n$, and $\underline{\underline{P}}_\infty(A) = \cup \underline{\underline{P}}_n(A)$. Then $\underline{\underline{P}}_n(A) = \underline{\underline{P}}_n$ as in the corollary, so we obtain

**Corollary 2.** *For $0 \leqslant n \leqslant \infty$, we have $K_i A \xrightarrow{\sim} K_i(\underline{\underline{P}}_n(A))$. In particular if $A$ is regular, then $K_i A \xrightarrow{\sim} K_i(Modf(A))$, where $Modf(A)$ is the category of finitely generated $A$-modules.*

We recall that a *regular* ring is a noetherian ring such that every (left) module has finite projective dimension. For such a ring $A$ we have $\underline{\underline{P}}_\infty(A) = Modf(A)$.

Similarly, Cor. 1 implies that for a regular noetherian separated scheme the $K$-groups of the category of coherent sheaves and the category of vector bundles are the same, since every coherent sheaf has a finite resolution by vector bundles [SGA 6, II, 2.2].

**Transfer maps.** Let $f : A \to B$ be a ring homomorphism such that as an $A$-module $B$ is in $\underline{\underline{P}}_\infty(A)$. Then restriction of scalars defines an exact functor from $\underline{\underline{P}}_\infty(B)$ to $\underline{\underline{P}}_\infty(A)$, hence by Cor. 2 it induces a homomorphism of $K$-groups which we will denote

$$f_* : K_i B \to K_i A \tag{3}$$

and call the *transfer* map with respect to $f$. Clearly given another homomorphism $g : B \to C$ with $C \in \underline{\underline{P}}_\infty(B)$, we have

$$(gf)_* = f_* g_* : K_i C \to K_i A. \tag{4}$$

We suppose now for simplicity that $A$ and $B$ are commutative, so that we have functors

$$\underline{\underline{P}}(A) \times \underline{\underline{P}}_n(A) \to \underline{\underline{P}}_n(A), (P, M) \mapsto P \otimes_A M.$$

for $0 \leqslant n \leqslant \infty$, which induce a product $K_0 A \otimes K_i A \to K_i A$, $[P] \otimes z \mapsto (P \otimes_A ?)_* z$, and similarly for $B$. Then if $f^* = (B \otimes_A ?)_* : K_i A \to K_i B$, we have the *projection formula*

$$f_*(f^* x \cdot y) = x \cdot f_* y \tag{5}$$

for $x \in K_0 A$ and $y \in K_i B$. This results immediately from the fact that for $X$ in $\underline{\underline{P}}(A)$ there is an isomorphism of exact functors

$$Y \mapsto (B \otimes_A X) \otimes_B Y = X \otimes_A Y$$

from $\underline{\underline{P}}_\infty(B)$ to $\underline{\underline{P}}_\infty(A)$.

**Corollary 3.** *Let $T = \{T_i, i \geqslant 1\}$ be an exact connected sequence of functors from an exact category $\underline{\underline{M}}$ to an abelian category $\underline{\underline{A}}$ (i.e. for every exact sequence (1), we have a long exact sequence*

$$\to T_2 M'' \to T_1 M' \to T_1 M \to T_1 M'').$$

*Let $\underline{\underline{P}}$ be the full subcategory of T-acyclic objects ($T_n M = 0$ for all $n \geqslant 1$), and assume for each $M$ in $\underline{\underline{M}}$ that there exists $P \twoheadrightarrow M$ with $P$ in $\underline{\underline{P}}$, and that $T_n M = 0$ for $n$ sufficiently large. Then $K_i \underline{\underline{P}} \xrightarrow{\sim} K_i \underline{\underline{M}}$.*

This results either from Cor. 1, or better by applying Theorem 3 directly to the inclusion $\underline{\underline{P}}_n \subset \underline{\underline{P}}_{n+1}$, where $\underline{\underline{P}}_n$ consists of $M$ such that $T_j M = 0$ for $j > n$.

Here is an application of this result. Put $K_i' A = K_i(Modf(A))$ for $A$ noetherian, and let $f : A \to B$ be a homomorphism of noetherian rings. If $B$ is flat as a right $A$-module, then we obtain a homomorphism of $K$-groups

$$(B \otimes_A ?)_* : K_i' A \to K_i' B \tag{6}$$

because $B \otimes_A ?$ is exact. But more generally if $B$ is of finite Tor-dimension as a right $A$-module, then applying Cor. 3 to $\underline{\underline{M}} = Modf(A)$ and $T_n M = Tor_n^A(B, M)$, we find that $K_i \underline{\underline{P}} \simeq K_i' A$, where $\underline{\underline{P}}$ is the full subcategory of $Modf(A)$ consisting of $M$ such that $T_n M = 0$ for $n > 0$. Since $B \otimes_A ?$ is exact on $\underline{\underline{P}}$, we obtain a homomorphism (6) in this more general situation.

# §5. Devissage and localization in abelian categories

In this section $\underline{\underline{A}}$ will denote an abelian category having a set of isomorphism classes of objects, and $\underline{\underline{B}}$ will be a non-empty full subcategory closed under taking subobjects, quotient objects, and finite products in $\underline{\underline{A}}$. Clearly $\underline{\underline{B}}$ is an abelian category and the inclusion functor $\underline{\underline{B}} \to \underline{\underline{A}}$ is exact. We regard $\underline{\underline{A}}$ and $\underline{\underline{B}}$ as exact categories in the obvious way, so that all monomorphisms and epimorphisms are admissible. Then $Q\underline{\underline{B}}$ is the full subcategory of $Q\underline{\underline{A}}$ consisting of those objects which are also objects of $\underline{\underline{B}}$.

**Theorem 4.** (Devissage) *Suppose that every object $M$ of $\underline{\underline{A}}$ has a finite filtration $0 = M_0 \subset M_1 \subset \cdots \subset M_n = M$ such that $M_j/M_{j-1}$ is in $\underline{\underline{B}}$ for each $j$. Then the inclusion functor $Q\underline{\underline{B}} \to Q\underline{\underline{A}}$ is a homotopy equivalence, so $K_i\underline{\underline{B}} \xrightarrow{\sim} K_i\underline{\underline{A}}$.*

*Proof.* Denoting the inclusion functor by $f$, it suffices by Theorem A to prove that $f/M$ is contractible for any object $M$ of $\underline{\underline{A}}$. The category $f/M$ is the fibred category over $Q\underline{\underline{B}}$ consisting of pairs $(N, u)$, where $N \in Q\underline{\underline{B}}$ and $u : N \to M$ is a map in $Q\underline{\underline{A}}$. By associating to $u$ what might be called its image, that is, the layer $(M_0, M_1)$ of $M$ such that $u$ is given by an isomorphism $N \simeq M_1/M_0$, it is clear that we obtain an equivalence of $f/M$ with the ordered set $J(M)$ consisting of layers $(M_0, M_1)$ in $M$ such that $M_1/M_0 \in \underline{\underline{B}}$, with the ordering $(M_0, M_1) \leqslant (M_0', M_1')$ iff $M_0' \subset M_0 \subset M_1 \subset M_1'$.

By virtue of the hypothesis that $M$ has a finite filtration with quotients in $\underline{\underline{B}}$, it will suffice to show the inclusion $i : J(M') \to J(M)$ is a homotopy equivalence whenever $M' \subset M$ is such that $M/M' \in \underline{\underline{B}}$. We define functors

$$r : J(M) \to J(M'), \quad (M_0, M_1) \mapsto (M_0 \cap M', M_1 \cap M')$$

$$s : J(M) \to J(M), \quad (M_0, M_1) \mapsto (M_0 \cap M', M_1).$$

These are well-defined because

$$M_1 \cap M'/M_0 \cap M' \subset M_1/M_0 \cap M' \subset M_1/M_0 \times M/M'$$

and because $\underline{\underline{B}}$ is closed under subobjects and products by assumption. Note that $ri = id_{J(M')}$ and that there are natural transformations $ir \to s \leftarrow id_{J(M)}$ represented by

$$(M_0 \cap M', M_1 \cap M') \leqslant (M_0 \cap M', M_1) \geqslant (M_0, M_1).$$

Hence by Prop. 2, $r$ is a homotopy inverse for $i$, so the proof is complete.

**Corollary 1.** *Let $\underline{\underline{A}}$ be an abelian category (with a set of isomorphism classes) such that every object has finite length. Then*

$$K_i\underline{\underline{A}} \simeq \coprod_{j \in J} K_i D_j$$

*where $\{X_j, j \in J\}$ is a set of representatives for the isomorphism classes of simple objects of $\underline{\underline{A}}$, and $D_j$ is the sfield $End(X_j)^{op}$.*

*Proof.* From the theorem we have $K_i\underline{\underline{B}} = K_i\underline{\underline{A}}$, where $\underline{\underline{B}}$ is the subcategory of semi-simple objects, so we reduce to the case where every object of $\underline{\underline{A}}$ is semi-simple. Using the fact that $K$-groups commute with products and filtered inductive limits (§2, (8),(9)) we reduce to the case where $\underline{\underline{A}}$ has a single simple object $X$ up to isomorphism. But then $M \mapsto \mathrm{Hom}(X, M)$ is an equivalence of $\underline{\underline{A}}$ with $\underline{\underline{P}}(D)$, $D = End(X)^{op}$, so the corollary follows.

**Corollary 2.** *If $I$ is a nilpotent two-sided ideal in a noetherian ring $A$, then $K_i'(A/I) \xrightarrow{\sim} K_i'A$, (notation as in §4, (6)).*

This results by applying the theorem to the inclusion $Modf(A/I) \subset Modf(A)$.

**Theorem 5.** (Localization) *Let $\underline{\underline{B}}$ be a Serre subcategory of $\underline{\underline{A}}$, let $\underline{\underline{A}}/\underline{\underline{B}}$ be the associated quotient abelian category (see for example* [Gabriel], [Swan]), *and let $e : \underline{\underline{B}} \to \underline{\underline{A}}$, $s : \underline{\underline{A}} \to \underline{\underline{A}}/\underline{\underline{B}}$ denote the canonical functors. Then there is a long exact sequence*

$$\cdots \xrightarrow{s_*} K_1(\underline{\underline{A}}/\underline{\underline{B}}) \to K_0\underline{\underline{B}} \xrightarrow{e_*} K_0\underline{\underline{A}} \xrightarrow{s_*} K_0(\underline{\underline{A}}/\underline{\underline{B}}) \to 0.$$

(It will be clear from the proof that this exact sequence is functorial for exact functors $(\underline{\underline{A}}, \underline{\underline{B}}) \to (\underline{\underline{A}}', \underline{\underline{B}}')$ . Unfortunately the proof does not shed much light on the nature of the boundary map $\partial : K_{i+1}(\underline{\underline{A}}/\underline{\underline{B}}) \to K_i(\underline{\underline{B}})$, and further work remains to be done in this direction.)

Before taking up the proof of the theorem, we give an example.

**Corollary.** *If $A$ is a Dedekind domain with quotient field $F$, there is a long exa sequence*

$$\to K_{i+1}F \to \coprod_m K_i(A/m) \to K_iA \to K_iF \to \cdots$$

*where $m$ runs over the maximal ideals of $A$.*

This follows by applying the theorem to $\underline{\underline{A}} = Modf(A)$, with $\underline{\underline{B}}$ the subcategory of torsion modules, whence $\underline{\underline{A}}/\underline{\underline{B}}$ is equivalent to $Modf(F) = \underline{\underline{P}}(F)$, (compare [Swan, p. 115]) We have $K_i\underline{\underline{A}} = K_iA$ by Cor. 2 of Theorem 3, and $K_i\underline{\underline{B}} = \coprod K_i(A/m)$ by Theorem 4, Cor. 1 Note that the map $K_iA \to K_iF$ in the exact sequence is the one induced by the homomorphism $A \to F$ as in §2, (10), and the map $K_i(A/m) \to K_iA$ is the transfer map associated to the homomorphism $A \to A/m$ in the sense of the preceding section.

*Proof of Theorem 5.* Fix a zero object 0 in $\underline{\underline{A}}$, and let 0 also denote its image in $\underline{\underline{A}}/\underline{\underline{B}}$. One knows that $\underline{\underline{B}}$ is the full subcategory of $\underline{\underline{A}}$ consisting of $M$ such that $sM \simeq 0$. Hence the composite of $Qe : Q\underline{\underline{B}} \to Q\underline{\underline{A}}$ with $Qs : Q\underline{\underline{A}} \to Q(\underline{\underline{A}}/\underline{\underline{B}})$ is isomorphic the constant functor with value 0, so $Qe$ factors

$$Q\underline{\underline{B}} \longrightarrow 0\backslash Qs \longrightarrow Q\underline{\underline{A}}$$

$$M \mapsto (M, 0 \simeq sM), (N, u) \mapsto N.$$

In view of Theorem B, §1, it suffices to establish the following assertions.

a) *For every $u : V' \to V$ in $Q(\underline{\underline{A}}/\underline{\underline{B}}), u^* : V \setminus Qs \to V' \setminus Qs$ is a homotopy equivalence.*

b) *The functor $Q\underline{\underline{B}} \longrightarrow 0 \setminus Qs$ is a homotopy equivalence.*

Factoring $u$ into injective and surjective maps, one sees that it suffices to prove a) when $u$ is either injective or surjective. On the other hand, replacing a category by its dual does not change the $Q$-category (§2, (7)). As surjective maps in $Q(\underline{\underline{A}}/\underline{\underline{B}})$ become injective in $Q((\underline{\underline{A}}/\underline{\underline{B}})^0) = Q(\underline{\underline{A}}^0/\underline{\underline{B}}^0)$, it is enough to prove a) when $u$ is injective, say $u = i_!, i : V' \rightarrowtail V$. Finally we have $i_! i_{V'!} = i_{V!}$, so it suffices to prove a) for the injective map $i_{V!}$ for any $V$ in $\underline{\underline{A}}/\underline{\underline{B}}$.

Let $\underline{\underline{F}}_V$ be the full subcategory of $V \setminus Qs$ consisting of pairs $(M, u)$ such that $u : V \xrightarrow{\sim} sM$ is an isomorphism. Clearly $\underline{\underline{F}}_0$ is isomorphic to $Q\underline{\underline{B}}$, so assertion b) results from the following.

**Lemma 1.** *The inclusion functor $\underline{\underline{F}}_V \longrightarrow V \setminus Qs$ is a homotopy equivalence.*

Denoting this functor by $f$, it suffices by Theorem A to show the category $f/(M, u)$ is contractible for any object $(M, u)$ of $V \setminus Qs$. Let the map $u : V \longrightarrow sM$ in $Q(\underline{\underline{A}}/\underline{\underline{B}})$ be represented by an isomorphism $V \simeq V_1/V_0$,where $(V_0, V_1)$ is a layer in $sM$. It is easily seen that the category $f/(M, u)$ is equivalent to the ordered set of layers $(M_0, M_1)$ in $M$ such that $(sM_0, sM_1) = (V_0, V_1)$, with the ordering $(M_0, M_1) \leqslant (M_0', M_1')$ iff $M_0' \subset M_0 \subset M_1 \subset M_1'$. This ordered set is directed because

$$(M_0, M_1) \leqslant (M_0 \cap M_0', M_1 + M_1') \geqslant (M_0', M_1').$$

It is non-empty because any subobject $V_1$ of $sM$ is of the form $sM_1$ for some $M_1 \subset M$. In effect, $V_1 = sN$ for some $N$ in $\underline{\underline{A}}$, and the map $V_1 \to sM$ can be represented as $s(g)s(i)^{-1}$ where $i : N' \rightarrowtail N$ has its cokernel in $\underline{\underline{B}}$ and $g : N' \to M$ is a map in $\underline{\underline{A}}$; then one can take $M_1$ to be the image of $g$. Thus $f/(M, u)$ is a filtering category, so it is contractible by Prop. 3, Cor. 2, proving the lemma.

The next four lemmas will be devoted to proving that the category $\underline{\underline{F}}_V$ is homotopy equivalent to $Q\underline{\underline{B}}$. To this end we introduce the following auxiliary categories. Let $N$ be a given object of $\underline{\underline{A}}$, and let $\underline{\underline{E}}_N$ be the category having as objects pairs $(M, h)$, where $h : M \to N$ is a mod-$\underline{\underline{B}}$ isomorphism, i.e. a map in $\underline{\underline{A}}$ whose kernel and cokernel are in $\underline{\underline{B}}$, or equivalently one which becomes an isomorphism in $\underline{\underline{A}}/\underline{\underline{B}}$. A morphism from $(M, h)$ to $(M', h')$ in $\underline{\underline{E}}_N$ is by definition a map $u : M \to M'$ in $Q\underline{\underline{A}}$ such that

$$(*)\qquad \begin{array}{ccc} M_1 & \overset{i}{\rightarrowtail} & M' \\ {\scriptstyle j}\downarrow & & \downarrow{\scriptstyle h'} \\ M & \xrightarrow{h} & N \end{array}$$

commutes if $u = i_! j^!$. To each $(M, h)$ in $\underline{\underline{E}}_N$ we associate $\mathrm{Ker}(h)$, which is an object of $\underline{\underline{B}}$ determined up to canonical isomorphism. To the map $(M, h) \to (M', h')$ represented by $(*)$ we associate the map in $Q\underline{\underline{B}}$ represented by the maps

$$\mathrm{Ker}(h) \twoheadleftarrow \mathrm{Ker}(hj) \rightarrowtail \mathrm{Ker}(h')$$

induced by $j$ and $i$ respectively. It is easily checked that in this way we obtain a functor

$$k_N : \underline{\underline{E}}_N \to Q\underline{\underline{B}}, \quad (M, h) \mapsto \mathrm{Ker}(h)$$

determined up to canonical isomorphism. We prove $k_N$ is a homotopy equivalence in two steps.

**Lemma 2.** *Let $\underline{\underline{E}}'_N$ be the full subcategory of $\underline{\underline{E}}_N$ consisting of pairs $(M,h)$ such that $h : M \to N$ is an epimorphism. Then the restriction $k'_N : \underline{\underline{E}}'_N \to Q\underline{\underline{B}}$ of $k_N$ is a homotopy equivalence.*

It suffices to prove $k'_N/T$ is contractible for any $T$ in $Q\underline{\underline{B}}$. Put $\underline{\underline{C}} = k'_N/T$; it is the fibred category over $\underline{\underline{E}}'_N$ consisting of pairs $((M,h),u)$, with $(M,h)$ in $\underline{\underline{E}}'_N$, and where $u : \mathrm{Ker}(h) \to T$ is a map in $Q\underline{\underline{B}}$. Let $\underline{\underline{C}}'$ be the full subcategory consisting of $((M,h),u)$ with $u$ surjective. Given $X = ((M,h),u)$ in $\underline{\underline{C}}$, write $u = j^! i_!$ with $i : \mathrm{Ker}(h) \rightarrowtail T_0, j : T \twoheadrightarrow T_0$ and define $(i_* M, \bar{h})$ by 'pushout':

$$\begin{array}{ccccc}
\mathrm{Ker}(h) & \rightarrowtail & M & \overset{h}{\twoheadrightarrow} & N \\
{\scriptstyle i}\downarrow & & \downarrow & & \| \\
T_0 & \rightarrowtail & i_* M & \overset{\bar{h}}{\twoheadrightarrow} & N.
\end{array}$$

Let $\overline{X} = ((i_* M, \bar{h}), j^!)$; it belongs to $\underline{\underline{C}}'$ and there is an evident map $X \to \overline{X}$. One verifies as in the proof of Theorem 3 that $X \to \overline{X}$ is a universal arrow from $X$ to an object of $\underline{\underline{C}}'$. Hence the inclusion $\underline{\underline{C}}' \to \underline{\underline{C}}$ has the left adjoint $X \mapsto \overline{X}$, so we have reduced to proving that $\underline{\underline{C}}'$ is contractible. But $\underline{\underline{C}}'$ has the initial object $((N, id_N), j^!_T)$, so this is clear, whence the lemma.

**Lemma 3.** *The functor $k_N : \underline{\underline{E}}_N \to Q\underline{\underline{B}}$ is a homotopy equivalence.*

Thanks to the preceding lemma, it suffices to show the inclusion $\underline{\underline{E}}'_N \to \underline{\underline{E}}_N$ is a homotopy equivalence. Let $\underline{\underline{I}}$ be the ordered set of subobjects $I$ of $N$ such that $N/I$ is in $\underline{\underline{B}}$, and consider the functor $f : \underline{\underline{E}}_N \to \underline{\underline{I}}$ sending $(M,h)$ to $\mathrm{Im}(h)$. One verifies easily that $f$ is fibred, the fibre over $I$ being $\underline{\underline{E}}'_I$, and the base change functor from $\underline{\underline{E}}'_I$ to $\underline{\underline{E}}'_J$ being $J \times_I ? : (M \twoheadrightarrow I) \mapsto (J \times_I M \twoheadrightarrow J)$. Since $J \times_I ?$ commutes with $k_I$ and $k_J$, it follows from Lemma 2 that $J \times_I ?$ is a homotopy equivalence for every arrow $J \subset I$ in $\underline{\underline{I}}$. From Theorem B, Cor., we conclude $\underline{\underline{E}}'_I$ is homotopy equivalent to the homotopy-fibre of $f$ over $I$. Since $\underline{\underline{I}}$ is contractible (it has $N$ for final object), one knows from homotopy theory that the inclusion $\underline{\underline{E}}'_I \to \underline{\underline{E}}_N$ is a homotopy equivalence for each $I$, proving the lemma.

We now want to show $\underline{\underline{F}}_V$ is homotopy equivalent to $\underline{\underline{E}}_N$ when $sN \simeq V$. First we note a simple consequence of the preceding.

**Lemma 4.** *Let $g : N \to N'$ be a map in $\underline{\underline{A}}$ which is a mod-$\underline{\underline{B}}$ isomorphism. Then the functor $g_* : \underline{\underline{E}}_N \to \underline{\underline{E}}_{N'}, (M,h) \mapsto (M, gh)$ is a homotopy equivalence.*

One verifies easily that by associating to $(M,h) \in \underline{\underline{E}}_N$ the obvious injective map $\mathrm{Ker}(h) \to \mathrm{Ker}(gh)$, one obtains a natural transformation from $k_N$ to $k_{N'} g_*$. (Observe: In 'lower' $K$-theory one calculates with matrices – in 'higher' $K$-theory with functors.) Thus $k_N$ and $k_{N'} g_*$ are homotopic, and since $k_N$ and $k_{N'}$ are homotopy equivalences, so is $g_*$, whence the lemma.

Now given $V$ in $\underline{\underline{A}}/\underline{\underline{B}}$, let $\underline{\underline{I}}_V$ be the category having as objects pairs $(N, \phi)$, where $N$ is in $\underline{\underline{A}}$ and $\phi : sN \xrightarrow{\sim} V$ is an isomorphism in $\underline{\underline{A}}/\underline{\underline{B}}$, in which a morphism $(N,\phi) \to (N', \phi')$ is a map $g : N \to N'$ such that $\phi' s(g) = \phi$. It is clear from the construction of $\underline{\underline{A}}/\underline{\underline{B}}$ that $\underline{\underline{I}}_V$ is a filtering category. For example, given two maps $g_1, g_2 : (N, \phi) \to (N', \phi')$ we have $s(g_1 - g_2) = 0$, so $\mathrm{Im}(g_1 - g_2) \in \underline{\underline{B}}$, hence we obtain a map $(N', \phi') \to (N'', \phi'')$ equalizing $g_1, g_2$ with $N'' = N'/\mathrm{Im}(g_1 - g_2)$.

We have a functor from $\underline{\underline{I}}_V$ to categories sending $(N, \phi)$ to $\underline{\underline{E}}_N$ and $g : (N, \phi) \to (N', \phi')$ to $g_* : \underline{\underline{E}}_N \longrightarrow \underline{\underline{E}}_{N'}$. Further, for each $(N, \phi)$ we have a functor

$$p_{(N,\phi)} : \underline{\underline{E}}_N \longrightarrow \underline{\underline{F}}_V, (M, h) \mapsto (M, s(h)^{-1}\phi^{-1} : V \underset{\rightarrow}{\sim} sN \underset{\rightarrow}{\sim} sM).$$

Since $p_{(N',\phi')}g_* = p_{(N,\phi)}$ for any map $g : (N, \phi) \to (N', \phi')$ in $\underline{\underline{I}}_V$, we obtain a functor

$$(**) \qquad \underset{\underline{\underline{I}}_V}{\underrightarrow{\lim}} \{(N, \phi) \mapsto \underline{\underline{E}}_N\} \underset{\rightarrow}{\sim} \underline{\underline{F}}_V$$

which we claim is an *isomorphism* of categories. In effect

$$(M, \theta : V \underset{\rightarrow}{\sim} sM) = p_{(M,\theta^{-1})}(M, id_M)$$

for any $(M, \theta)$ in $\underline{\underline{F}}_V$, showing that $(**)$ is surjective on objects. Also given $p_{(N,\phi)}(M, h) = p_{(N,\phi)}(M', h')$, then $M = M'$ and $s(h) = s(h')$. Letting $N' = N/\mathrm{Im}(h-h')$ we otain a map $g : (N, \phi) \to (N', \phi')$ such that $g_*(M, h) = g_*(M', h')$, showing that $(**)$ is injective on objects. The verification that $(**)$ is bijective on arrows is similar.

Applying Prop. 3, Cor. 1, we obtain from Lemma 4 and $(**)$ the following.

**Lemma 5.** *For any $\phi : sN \underset{\rightarrow}{\sim} V$, the functor $p_{(N,\phi)}$ is a homotopy equivalence.*

The end is now near. To finish the proof of the theorem, we have only to show $(i_{V!})^* : V \setminus Qs \to 0 \setminus Qs$ is a homotopy equivalence. Choose $(N, \phi)$ as in Lemma 5 and form the diagram

$$\begin{array}{ccc} \underline{\underline{E}}_N & \xrightarrow{p_{(N,\phi)}} & \underline{\underline{F}}_V \subset V \setminus Qs \\ {\scriptstyle k_N}\downarrow & & \downarrow{\scriptstyle (i_{V!})^*} \\ Q\underline{\underline{B}} & \xrightarrow{\sim} & \underline{\underline{F}}_0 \subset 0 \setminus Qs \end{array}$$

The diagram is not commutative, for the lower-left and upper-right paths are respectively the functors

$$(M, h) \mapsto (\mathrm{Ker}(h), 0 \simeq s(\mathrm{Ker}(h)))$$

$$(M, h) \mapsto (M, (i_{0M})_! : 0 \to sM).$$

However it is easily checked that by associating to $(M, h)$ the obvious injective map $\mathrm{Ker}(h) \to M$, one obtains a natural transformation between these two functors. Thus the diagram is homotopy commutative, and since all the arrows in the diagram are homotopy equivalences except possibly $(i_{V!})^*$ by Lemmas 1, 3, and 5, it follows that $(i_{V!})^*$ is one also. The proof of the localization theorem is now complete.

# §6. Filtered rings and the homotopy property for regular rings

This section contains some important applications of the preceding results to the groups $K'_iA = K_i(Modf(A))$ for $A$ noetherian. If $A$ is regular, we have $K_iA = K'_iA$ by the resolution theorem (Th. 3, Cor. 2), so we also obtain results about $K_iA$ for $A$ regular. In particular, we prove the homotopy theorem: $K_iA = K_i(A[t])$ for $A$ regular. According to [Gersten 1], this signifies that the groups $K_iA$ are the same as the $K$-groups of Karoubi and Villamayor for $A$ regular (assuming Theorem 1 of the announcement [Quillen 1] which asserts that the groups $K_iA$ are the same as the Quillen $K$-groups of [Gersten 1]).

**Graded rings.** Let $B = \coprod B_n, n \geqslant 0$ be a graded ring and put $k = B_0$. From now on we consider only graded $B$-modules $N = \coprod N_n$ with $n \geqslant 0$, unless specified otherwise. Put

$$T_i(N) = Tor_i^B(k, N)$$

where $k$ is regarded as a right $B$-module by means of the augmentation $B \to k$. Then $T_i(N)$ is a graded $k$-module in a natural way, e.g. $T_0(N)_n = N_n/(A_1N_{n-1} + \cdots + A_nN_0)$. Denote by $F_pN$ the submodule of $N$ generated by $N_n$ for $n \leqslant p$, so that we have $0 = F_{-1}N \subset F_0N \subset \cdots, \cup F_pN = N$. It is clear that

$$T_0(F_pN)_n = \begin{cases} 0, & n > p \\ T_0(N)_n, & n \leqslant p \end{cases} \tag{1}$$

and that there are canonical epimorphisms

$$B(-p) \otimes_k T_0(N)_p \twoheadrightarrow F_pN/F_{p-1}N \tag{2}$$

where $B(-p)_n = B_{n-p}$.

**Lemma 1.** *If $T_1(N) = 0$ and $Tor_i^k(B, T_0(N)) = 0$ for all $i > 0$, then (2) is an isomorphism for all $p$.*

*Proof.* For any $k$-module $X$ we have

$$Tor_i^k(B, X) = 0 \quad for \quad i > 0 \Longrightarrow T_i(B \otimes_k X) = 0 \quad for \quad i > 0. \tag{3}$$

In effect, if $P.$ is a $k$-projective resolution of $X$, then $B \otimes_k P.$ is a $B$-projective resolution of $B \otimes_k X$, and $T_i(B \otimes_k X) = H_i(k \otimes_B B \otimes_k P.) = H_i(P.) = 0$ for $i > 0$. In particular by the hypothesis on $T_0(N)$, we have

$$T_i(B \otimes_k T_0(N)) = 0 \quad for \quad i > 0. \tag{4}$$

Let $R^p$ be the kernel of (2). Since (2) clearly induces an isomorphism on $T_0$, we obtain from the Tor long exact sequence an exact sequence

$$T_1(B(-p) \otimes_k T_0(N)_p)_n \longrightarrow T_1(F_pN/F_{p-1}N)_n \xrightarrow{\partial} T_0(R^p)_n \longrightarrow 0.$$

The first group is zero by (4), so $\partial$ is an isomorphism.

Fix an integer $s$. we will show that (2) is an isomorphism in degrees $\leqslant s$ and also that $T_1(F_pN)_n = 0$ for $n \leqslant s$ by decreasing induction on $p$. For large $p$. this is true, because $T_1(F_pN)_n = T_1(N)_n$ for $p \geqslant n$, and because $T_1(N) = 0$ by hypothesis. Assuming $T_1(F_pN)_n = 0$ for $n \leqslant s$, we find from (1) and the exact sequence

$$T_1(F_pN)_n \to T_1(F_pN/F_{p-1}N)_n \to T_0(F_{p-1}N)_n \to T_0(F_pN)_n$$

that $T_1(F_pN/F_{p-1}N)_n = T_0(R^p)_n = 0$ for $n \leqslant s$. It follows that $R^p$ is zero in degrees $\leqslant s$, showing that (2) is an isomorphism in degrees $\leqslant s$ as claimed. In addition we find $0 = T_2(B(-p) \otimes_k T_0(N)_p)_n \simeq T_2(F_pN/F_{p-1}N)_n$ for $n \leqslant s$, whence from the exact sequence

$$T_2(F_pN/F_{p-1}N)_n \to T_1(F_{p-1}N)_n \to T_1(F_pN)_n$$

we have $T_1(F_{p-1}N)_n = 0$ for $n \leqslant s$, completing the induction. Since $s$ is arbitrary, the lemma is proved.

Suppose now that $B$ is (left) noetherian, and let $Modfgr(B)$ be the abelian category of finitely generated graded $B$-modules. Its $K$-groups are naturally modules over $\mathbb{Z}[t]$, where the action of $t$ is induced by the translation functor $N \mapsto N(-1)$. The ring $k$ is also noetherian, so if $B$ has finite Tor dimension as a right $k$-module, we have a homomorphism (§4, (6) )

$$(B\otimes_k ?)_* : K_i'k \to K_i(Modfgr(B)) \tag{5}$$

induced by the exact functor $B\otimes_k ?$ on the subcategory $\underline{\underline{F}}$ of $Modf(k)$ consisting of $k$-modules $F$ such that $Tor_i^k(B, F) = 0$ for $i > 0$.

**Theorem 6.** *Suppose $B$ is a graded noetherian ring such that $B$ has finite Tor dimension as a right $k$-module, and such that $k$ has finite Tor dimension as a right $B$-module. Then* (5) *extends to a $\mathbb{Z}[t]$-module isomorphism*

$$\mathbb{Z}[t] \otimes_{\mathbb{Z}} K_i'k \xrightarrow{\sim} K_i(Modfgr(B)).$$

(The hypothesis that $k$ be of finite Tor dimension over $B$ is very restrictive. For example, if $k$ is a field and $B$ is commutative,then $B$ has to be a polynomial ring over $k$. In all situations where this theorem is used, it happens that $B$ is flat over $k$. Does this follow from the assumpltion that $B$ and $k$ are of finite Tor dimension over each other?)

*Proof.* Let $\underline{\underline{N}}'$ be the full subcategory of $Modfgr(B)$ consisting of $N$ such that $T_i(N) = 0$ for $i > 0$, and let $\underline{\underline{N}}''$ be the full subcategory of $\underline{\underline{N}}'$ consisting of $N$ such that $T_0(N) \in \underline{\underline{F}}$. By the finite Tor dimension hypotheses and the resolution theorem (§4) one has isomorphisms $K_i\underline{\underline{F}} = K_i'k, K_i\underline{\underline{N}}'' = K_i\underline{\underline{N}}' = K_i(Modfgr(B))$. Let $\underline{\underline{N}}''_n$ be the full subcategory of $\underline{\underline{N}}''$ consisting of $N$ such that $F_nN = N$. We have homomorphisms

$$(K_i\underline{\underline{F}})^n = K_i(\underline{\underline{F}}^n) \xrightarrow{b} K_i(\underline{\underline{N}}''_n) \xrightarrow{c} (K_i\underline{\underline{F}})^n$$

induced by the exact functors $(F_j, 0 \leqslant j \leqslant n) \mapsto \coprod B(-j) \otimes_k F_j$ (this is in $\underline{\underline{N}}''$ by (3)) and $N \mapsto (T_0(N)_j)$ respectively. Clearly $cb = id$. On the other hand, by

Lemma 1 any $N$ in $\underline{\underline{N}}''_n$ has an exact characteristic filtration $0 \subset F_0 N \subset \cdots \subset F_n N = N$ with $F_p N / F_{p-1} N = B(-p) \otimes_k T_0(N)_p$, so applying Th. 2, Cor. 2, one finds that $bc = id$. Thus $b$ is an isomorphism, so by passing to the limit over $n$ we have $\mathbb{Z}[t] \otimes K_i \underline{\underline{F}} \xrightarrow{\sim} K_i \underline{\underline{N}}''$, which proves the theorem.

The following will be used in the proof of Theorem 7.

**Lemma 2.** *Suppose $B$ is noetherian, $k$ is regular, and that $k$ has finite Tor dimension as a right $B$-module. Then any $N$ in $Modfgr(B)$ has a finite resolution by finitely generated projective graded $B$-modules.*

*Proof.* Starting with $N_0 = N$, we recursively construct exact sequences in $Modfgr(B)$

$$0 \longrightarrow N_r \longrightarrow P_{r-1} \longrightarrow N_{r-1} \longrightarrow 0$$

where $P_{r-1}$ is projective. We have to show $N_r$ is projective for $r$ large. Since $T_i(N_r) = T_{i+1}(N_{r-1})$ for $i > 0$, it follows that $T_i(N_r) = 0$ for $i > 0$ and $r \geqslant d$, where $d$ is the Tor dimension of $k$ over $B$. Then for $r > d$ we have exact sequences

$$0 \longrightarrow T_0(N_r) \longrightarrow T_0(P_{r-1}) \longrightarrow T_0(N_{r-1}) \longrightarrow 0.$$

As $k$ is regular, $T_0(N_d)$ has finite projective dimension $s$, so $T_0(N_r)$ is projective for $r \geqslant d+s$. It follows from Lemma 1 that $N_{d+s}$ is projective, whence the lemma.

**Filtered rings.** Let $A$ be a ring equipped with an increasing filtration by subgroups $0 = F_{-1}A \subset F_0 A \subset F_1 A \subset \cdots$ such that $1 \in F_0 A, F_p A \cdot F_q A \subset F_{p+q} A$, and $\cup F_p A = A$. Let $B = gr(A) = \coprod F_p A / F_{p-1} A$ be the associated graded ring and put $k = F_0 A = B_0$. By a *filtered $A$-module $M$* we will mean an $A$-module equipped with an increasing filtration $0 = F_{-1} M \subset F_0 M \subset \cdots$ such that $F_p A \cdot F_q M \subset F_{p+q} M$ and $\cup F_p M = M$. Then $gr(M) = \coprod F_p M / F_{p-1} M$ is a graded $B$-module in a natural way.

**Lemma 3.** i) *If $gr(M)$ is a finitely generated $B$-module, then $M$ is a finitely generated $A$-module. In particular, if every graded left ideal in $B$ is finitely generated, then $A$ is noetherian.*

ii) *If $gr(M)$ is a projective $B$-module, then $M$ is a projective $A$-module.*

iii) *If $gr(M)$ has a resolution by finitely generated projective graded $B$-modules of length $\leqslant n$, then $M$ has a $\underline{\underline{P}}(A)$ -resolution of length $\leqslant n$.*

*Proof.* We use the following construction. Suppose given $k$-modules $L_j$ and maps of $k$-modules $L_j \longrightarrow F_j M$ for each $j \geqslant 0$ such that the composition

$$L_j \longrightarrow F_j M \longrightarrow gr_j(M) \longrightarrow T_0(gr(M))_j$$

is onto. Let $P$ be the filtered $A$-module with $F_n P = \coprod_j F_{n-j} A \otimes_k L_j$ and let $\phi : P \to M$ be such that $\phi$ restricted to $A \otimes_k L_j$ is the $A$-linear extension of the given map from $L_j$ to $F_j M$. Then $T_0(gr(P))_j = L_j$, and $\phi$ is a map of filtered $A$-modules such that $T_0(gr(\phi))$ is onto. It follows that $gr(\phi)$ is onto, hence $F_n(\phi)$ is onto for all $n$, and so $\phi$ is onto. Thus if $K = \mathrm{Ker}(\phi), F_n K = K \cap F_n M$, we have an exact sequence of $A$-modules

$$0 \longrightarrow K \longrightarrow P \longrightarrow M \longrightarrow 0$$

such that

$$0 \longrightarrow F_n K \longrightarrow F_n P \longrightarrow F_n M \longrightarrow 0$$

(6) $$0 \longrightarrow gr_n K \longrightarrow gr_n P \longrightarrow gr_n M \longrightarrow 0$$

are exact for all $n$.

i): If $gr(M)$ is a finitely generated $B$-module, then $T_0(gr(M))$ is a finitely generated $k$-module, hence we can take $L_j$ to be a free finitely generated $k$-module which is zero for large $j$. Then $P$ is a free finitely generated $A$-module, so $M$ is finitely generated, proving the first part of i). The second part follows by taking $M$ to be a left ideal of $A$ and endowing it with the induced filtration $F_n M = M \cap F_n A$.

ii): If $gr(M)$ is projective over $B$, then $T_0(gr(M))$ is projective over $k$, and we can take $L_j = T_0(gr(M))_j$. Then $T_0(gr(\phi))$ is an isomorphism, so from the exact sequence

$$T_1(gr(M)) \to T_0(gr(K)) \to T_0(gr(P)) \to T_0(gr(M))$$

we conclude that $T_0(gr(K)) = 0$. Then $gr(K) = 0$, so $K = 0, M = P$, and $M$ is projective over $A$, proving ii).

iii): We use induction on $n$, the case $n = 0$ being clear from i) and ii). Assuming $gr(M)$ has a resolution of length $\leqslant n$ by finitely generated graded projective $B$-modules, choose $P$ as in the proof of i), so that $gr(P)$ is a free finitely generated $B$-module. From the exact sequence (6), and the lemma after Th. 3, Cor. 1, (or Schanuel's lemma), we know that $gr(K)$ has a resolution of length $\leqslant n-1$ by finitely generated graded projective $B$-modules. Applying the induction hypothesis, it follows that $K$ has a $\underline{\underline{P}}(A)$-resolution of length $\leqslant n-1$, so $M$ has a $\underline{\underline{P}}(A)$-resolution of length $\leqslant n$, as was to be shown.

**Lemma 4.** *If $B$ is noetherian, $k$ is regular, and if $k$ has finite Tor dimension as a right $B$-module, then $A$ is regular.*

This is an immediate consequence of Lemma 2 and Lemma 3 iii).

We can now prove the main result of this section.

**Theorem 7.** *Let $A$ be a ring equipped with an increasing filtration $0 = F_{-1}A \subset F_0 A \subset F_1 A \subset \cdots$ such that $1 \in F_0 A$, $F_p A \cdot F_q A \subset F_{p+q} A$, and $\cup F_p A = A$. Suppose $B = gr(A)$ is noetherian and that $B$ is of finite Tor dimension as a right module over $B_0 = F_0 A$, (hence $F_0 A$ and $A$ are noetherian and $A$ is of finite Tor dimension as a right $F_0 A$-module). Suppose also that $F_0 A$ is of finite Tor dimension as a right $B$-module. Then the inclusion $F_0 A \subset A$ induces isomorphisms $K_i'(F_0 A) \simeq K_i' A$. If further $F_0 A$ is regular, then so is $A$, and we have isomorphisms $K_i(F_0 A) \xrightarrow{\sim} K_i A$.*

*Proof.* Put $k = F_0 A$. Since $B$ is noetherian, we know $A$ is also by Lemma 3 i). Also if $B$ has Tor dimension $d$ over $k$, then $F_n A / F_{n-1} A$ has Tor dimension $\leqslant d$ for each $n$, so the same is true for $F_n A$, and hence also for $A$. Thus the map $K_i' k \to K_i' A$ is defined, and we have only to prove that it is an isomorphism. Indeed,the last assertion of the theorem results from Lemma 4 and the fact that $K_i A = K_i' A$ for regular $A$ by the resolution theorem (Th. 3, Cor. 2).

Let $z$ be an indeterminate and let $A'$ be the subring $\coprod (F_n A) z^n$ of $A[Z]$. We show the graded ring $A'$ satisfies the hypotheses of Theorem 6. The fact that $A'$ has finite Tor dimension over $k$ is clear from the preceding paragraph. Since $z$ is a central non-zero-divisor in $A'$, we have that $B = A'/zA'$ is of Tor dimension one over $A'$. As $k$ has finite Tor dimension over $B$, it follows that $k$ has finite Tor dimension over $A'$. Finally to show $A'$ is noetherian, we filter $A'$ by letting $F_pA'$ consist of those polynomials whose coefficients are in $F_pA$ The ring

$$gr(A') = \coprod_{p \leqslant n} (gr_p A) z^n$$

is isomorphic to $gr(A)[z]$, which is noetherian, hence $A'$ is noetherian by Lemma 3 i).

Let $\underline{\underline{F}}$ be the full subcategory of $Modf(k)$ consisting of $F$ such that $Tor_i^k(B, F) = 0$ for $i > 0$, whence $K_i\underline{\underline{F}} = K_i'k$ by the resolution theorem (Th.3, Cor. 3). Applying Theorem 6 to $B$ and $A'$, we obtain $\mathbb{Z}[t]$-module isomorphisms

$$(7) \qquad \begin{array}{ll} \mathbb{Z}[t] \otimes K_i\underline{\underline{F}} \xrightarrow{\sim} K_i(Modf\,gr(B)), & 1 \otimes x \mapsto (B \otimes_k ?)_* x \\ \mathbb{Z}[t] \otimes K_i\underline{\underline{F}} \xrightarrow{\sim} K_i(Modf\,gr(A')), & 1 \otimes x \mapsto (A' \otimes_k ?)_* x. \end{array}$$

Let $\underline{\underline{B}}$ be the Serre subcategory of $\underline{\underline{A}} = Modf\,gr(A')$ consisting of modules on which $z$ is nilpotent. The functor

$$j : Modf\,gr(A') \to Modf(A), \quad M \mapsto M/(z-1)M$$

is exact and induces an equivalence of the quotient category $\underline{\underline{A}}/\underline{\underline{B}}$ with $Modf(A)$. (Compare [Swan, p.114, 130]; note that if $S = \{z^n\}$, then $S^{-1}A'$ is the Laurent polynomial ring $A[z, z^{-1}]$, and a graded module over $A[z, z^{-1}]$ is the same as a module over $A = A'/(z-1)A'$.) Since $A'/zA' = B$, we have an embedding

$$i : Modf\,gr(B) \to Modf\,gr(A')$$

identifying the former with the full subcategory of the latter consisting of modules killed by $z$. The devissage theorem implies that $K_i(Modf\,gr(B)) = K_i\underline{\underline{B}}$. Thus the exact sequence of the localization theorem for the pair $(\underline{\underline{A}}, \underline{\underline{B}})$ takes the form

$$(8) \qquad \to K_i(Modf\,gr(B)) \xrightarrow{i_*} K_i(Modf\,gr(A')) \xrightarrow{j_*} K_i'A \to .$$

We next compute $i_*$ with respect to the isomorphisms (7). Associating to $F$ in $\underline{\underline{F}}$ the exact sequence

$$0 \to A'(-1) \otimes_k F \xrightarrow{z} A' \otimes_k F \to B \otimes_k F \to 0$$

we obtain an exact sequence of exact functors from $\underline{\underline{F}}$ to $Modf\,gr(A')$. Applying Th. 2, Cor. 1, we conclude that the square of $\mathbb{Z}[t]$-module homomorphisms

$$\begin{array}{ccc} \mathbb{Z}[t] \otimes K_i\underline{\underline{F}} & \xrightarrow{\sim} & K_i(Modf\,gr(B)) \\ {\scriptstyle 1-t}\downarrow & & \downarrow{\scriptstyle i_*} \\ \mathbb{Z}[t] \otimes K_i\underline{\underline{F}} & \xrightarrow{\sim} & K_i(Modf\,gr(A')) \end{array}$$

is commutative. Since $1-t$ is injective with cokernel $K_i\underline{\underline{F}}$, we conclude from the exact sequence (8) that the composition

$$K_i\underline{\underline{F}} \longrightarrow K_i(Modf\,gr(A')) \xrightarrow{j_*} K_i'A$$

induced by $F \mapsto A' \otimes_k F \mapsto A \otimes_k F$ is an isomorphism. Since $K_i\underline{\underline{F}} = K'_i k$, this proves the theorem.

The preceding theorem enables one to compute the $K$-groups of some interesting non-commutative rings.

**Examples.** Let $\mathcal{J}$ be a finite dimensional Lie algebra over a field $k$, and let $U(\mathcal{J})$ be its universal enveloping algebra. The Poincare-Birkhoff-Witt theorem asserts that $U(\mathcal{J})$ is a filtered algrebra such that $gr(U(\mathcal{J}))$ is a polynomial ring over $k$. Thus Theorem 7 implies that $K_i k = K_i U(\mathcal{J})$. Similarly if $H_n$ is the Heisenberg-Weyl algebra over $k$ with generators $p_i, q_i, 1 \leqslant i \leqslant n$, subject to the relations $[p_i, p_j] = [q_i, q_j] = 0, [p_i, q_j] = \delta_{ij}$, then we have $K_i k = K_i H_n$.

**Theorem 8.** *If $A$ is noetherian, then there are canonical isomorphisms*

i) $K'_i(A[t]) \simeq K'_i A$

ii) $K'_i(A[t, t^{-1}]) \simeq K'_i A \oplus K'_{i-1} A$

*Proof.* i) follows immediately from the preceding theorem.

ii): Applying the localization theorem to the Serre subcategory $\underline{\underline{B}}$ of $Modf(A[t])$ consisting of modules on which $t$ is nilpotent, we get a long exact sequence

$$\begin{array}{ccccccc} \longrightarrow & K_i\underline{\underline{B}} & \longrightarrow & K'_i(A[t]) & \longrightarrow & K'_i(A[t,t^{-1}]) & \longrightarrow \\ & \uparrow{\scriptstyle s} & & \uparrow{\scriptstyle s} & \nearrow & & \\ & K'_i A & & K'_i A & & & \end{array}$$

where the first vertical isomorphism results from applying the devissage theorem to the embedding $Modf(A) = Modf(A[t]/tA[t]) \subset \underline{\underline{B}}$. The homomorphism $A[t, t^{-1}] \to A$ sending $t$ to 1 makes $A$ a right module of Tor dimension one over $A[t, t^{-1}]$, so it induces a map $K'_i(A[t, t^{-1}]) \to K'_i A$ left inverse to the oblique arrow. Thus the exact sequence breaks up into split short exact sequences proving ii).

**Corollary** (Fundamental theorem for regular rings). *If $A$ is regular, then there are canonical isomorphisms $K_i(A[t]) = K_i A$ and $K_i(A[t, t^{-1}]) = K_i A \oplus K_{i-1} A$.*

This is clear from Th. 3, Cor. 2, since $A[t]$ and $A[t, t^{-1}]$ are regular if $A$ is.

**Exercise.** Let $\phi$ be an automorphism of a noetherian ring $A$, and let $A_\phi[t]$, $A_\phi[t, t^{-1}]$ be the associated twisted polynomial and Laurent polynomial rings in which $t \cdot a = \phi(a) \cdot t$,([Farrell-Hsiang]). Show that $K'_i A = K'_i(A_\phi[t])$ and that there is a long exact sequence

$$(9) \qquad \to K'_i A \xrightarrow{1-\phi_*} K'_i A \to K'_i(A_\phi[t, t^{-1}]) \to K'_{i-1} A \to .$$

We finish this section by showing how the preceding results can be used to compute the $K$-groups of certain skew-fields. Keith Dennis points out that this has some interest already in the case of $K_2$, since a non-commutative generalization of Matsumoto's theorem is not known. (Here and in the computation to follow, we will be assuming Theorem 1 of the announcement [Quillen 1], which implies that the $K_2 A$ here is the same as Milnor's, and that the groups $K_i\mathbb{F}_q$ are the same as the ones computed in [Quillen 2].)

**Example 1.** Let $k$ be the algebraic closure of the finite field $\mathbb{F}_p$, and let $A$ be be the twisted polynomial ring $k_\phi[F]$ with $Fx = x^q F$ for $x$ in $k$, where $q = p^d$. Then $A$ is a non-commutative domain in which every left ideal is principal. Let $D$ be the quotient skew-field of $A$, whence $Modf(D) = Modf(A)/\underline{\underline{B}}$, where $\underline{\underline{B}}$ is the Serre subcategory consisting of $A$-modules which are torsion, or equivalently, which are finite dimensional over $k$. The localization theorem gives an exact sequence

$$(10) \qquad \to K_i\underline{\underline{B}} \xrightarrow{i_*} K_iA \to K_iD \to K_{i-1}\underline{\underline{B}} \to$$

($A$ and $D$ are regular), and we have $K_iA = K_ik$ by Theorem 7.

An object of $\underline{\underline{B}}$ is a finite dimensional vector space $V$ over $k$ equipped with an additive map $F : V \to V$ such that $F(xv) = x^q F(v)$ for $x$ in $k$ and $v$ in $V$. It is well-known that $V$ splits canonically: $V = V_0 \oplus V_1$, where $F$ is nilpotent on $V_0$ and bijective on $V_1$, and moreover that

$$k \otimes_{\mathbb{F}_q} V^F \xrightarrow{\sim} V_1$$

where $V^F = \{v \in V | Fv = v\}$ is a finite dimensional vector space over the subfield $\mathbb{F}_q$ of $k$ with $q$ elements. Thus we have an equivalence of categories

$$\underline{\underline{B}} \cong \bigcup_n Modf(A/AF^n) \times Modf(\mathbb{F}_q).$$

Applying the devissage theorem to the first factor, we obtain $K_i\underline{\underline{B}} = K_ik \oplus K_i\mathbb{F}_q$.

Let $\phi : k \longrightarrow k$ be the Frobenius automorphism: $\phi(x) = x^q$, and let $\phi(V)$ denote the base extension of the $k$-vector space $V$ with respect to $\phi$, i.e. $\phi(V) = k \otimes_k V$, where $k$ is regarded as a right $k$-module via $\phi$. If $V$ is regarded as an $A$-module killed by $F$, we have an exact sequence of $A$-modules

$$0 \to A \otimes_k \phi(V) \to A \otimes_k V \to V \to 0$$

$$a \otimes (x \otimes v) \mapsto axF \otimes v$$

On the other hand, if $W$ is a finite dimensional vector space over $\mathbb{F}_q$, we have an exact sequence of $A$-modules

$$0 \longrightarrow A \otimes_{\mathbb{F}_q} W \longrightarrow A \otimes_{\mathbb{F}_q} W \longrightarrow k \otimes_{\mathbb{F}_q} W \longrightarrow 0$$

$$a \otimes w \longmapsto a(F-1) \otimes w$$

where $F$ acts on the cokernel by $F(x \otimes w) = x^q \otimes w$. Applying Th. 2, Cor. 1, to these "characteristic" sequences, one easily deduces that the composite

$$K_ik \oplus K_i\mathbb{F}_q = K_i\underline{\underline{B}} \xrightarrow{i_*} K_iA = K_ik$$

is zero on the factor $K_i\mathbb{F}_q$ and the map $1 - \phi_*$ on $K_ik$. From [Quillen 2] one has exact sequences

$$0 \longrightarrow K_i\mathbb{F}_q \longrightarrow K_ik \xrightarrow{1-\phi_*} K_ik \longrightarrow 0$$

for $i > 0$. Combining this with (10) we obtain the formulas

$$\begin{aligned} &K_0D = \mathbb{Z}, K_1D = \mathbb{Z} \otimes \mathbb{Z} \\ &K_{2i}D = (K_{2i-1}\mathbb{F}_q)^2 = (\mathbb{Z}/(q^i - 1)\mathbb{Z})^2 \quad i > 0 \\ &K_{2i+1}D = (K_{2i}\mathbb{F}_q)^2 = 0 \qquad\qquad\quad\; i > 0. \end{aligned} \tag{11}$$

**Example 2.** Let $H$ be the Heisenberg-Weyl algebra with generators $p, q$ such that $pq - qp = 1$ over an algebraically closed field $k$, and let $D$ be the quotient skew-field of $H$. In this case, one can prove that the localization exact sequence associated to *Modf(H)* and the Serre subcategory of torsion modules breaks up into short exact sequences

$$0 \to K_ik \to K_iD \to \coprod K_{i-1}k \to 0$$

where the direct sum is taken over the set of isomorphism classes of simple $H$-modules. The proof is similar to the preceding, the essential points being a) torsion finitely generated $H$-modules are of finite length, because $H$ has no modules finite dimensional over $k$, and b) $k$ is the ring of endomorphisms of any simple $H$-module ([Quillen 3]).

## §7. $K'$-theory for schemes

**1.** If $X$ is a scheme, we put $K_qX = K_q\underline{\underline{P}}(X)$, where $\underline{\underline{P}}(X)$ is the category of vector bundles over $X$ (= locally free sheaves of $\underline{\underline{O}}_X$-modules of finite rank) equipped with the usual notion of exact sequence. If $X$ is a noetherian scheme, we put $K'_qX = K_q\underline{\underline{M}}(X)$, where $\underline{\underline{M}}(X)$ is the abelian category of coherent sheaves on $X$. The following theory concerns primarily the groups $K'_qX$, so for the rest of this section we will assume all schemes to be *noetherian* and *separated*, unless stated otherwise.

As the inclusion functor from $\underline{\underline{P}}(X)$ to $\underline{\underline{M}}(X)$ is exact, it induces a homomorphism

$$K_qX \to K'_qX. \tag{1.1}$$

*When $X$ is regular this is an isomorphism.* In effect, one knows that any coherent sheaf $F$ is a quotient of a vector bundle [SGA 6 II 2.2.3 - 2.2.7.1], hence it has a resolution by vector bundles, in fact a finite resolution as $X$ is regular and quasi-compact (see [SGA 2 VIII 2.4]). Thus 1.1 is an isomorphism by the resolution theorem (Th. 3, Cor. 1).

If $E$ is a vector bundle on $X$, then $F \mapsto E \otimes F$ is an exact functor from $\underline{\underline{M}}(X)$ to itself, hence as in §3, (1), we obtain pairings

$$K_0X \otimes K'_qX \longrightarrow K'_qX \tag{1.2}$$

making $K'_qX$ a module over the ring $K_0X$. (In a later paper I plan to extend this idea to define a graded anti-commutative ring structure on $K_*X$ such that $K'_*X$ is a graded module over $K_*X$.)

**2. Functorial behavior.** If $f : X \longrightarrow Y$ is a morphism of schemes (resp. a *flat* morphism), then the inverse image functor $f^* : \underline{\underline{P}}(Y) \longrightarrow \underline{\underline{P}}(X)$ (resp. $f^* : \underline{\underline{M}}(Y) \longrightarrow \underline{\underline{M}}(X)$ ) is exact, hence it induces a homomorphism of $K$-groups which will be denoted

$$f^* : K_qY \longrightarrow K_qX \quad (resp. f^* : K'_qY \longrightarrow K'_qX). \tag{2.1}$$

It is clear that in this way $K_q$ becomes a contravariant functor from schemes to abelian groups, and that $K'_q$ is a contravariant functor on the subcategory of schemes and flat morphisms.

**Proposition 2.2.** *Let $i \longmapsto X_i$ be a filtered projective system of schemes such that the transition morphisms $X_i \longrightarrow X_j$ are affine, and let $X = \varprojlim X_i$. Then*

$$K_qX = \varinjlim K_qX_i. \tag{2.3}$$

*If in addition the transition morphisms are flat, then*

$$K'_qX = \varinjlim K'_qX_i. \tag{2.4}$$

*Proof.* We wish to apply §2(9), using the fact that $\underline{\underline{P}}(X)$ is essentially the inductive limit of the $\underline{\underline{P}}(X_i)$ by [EGA IV 8.5]. In order to obtain an honest inductive system of categories, we replace $\underline{\underline{P}}(X_i)$ by an equivalent category using Giraud's method as follows. Let $I$ be the index category of the system $X_i$, and let $I'$ be the category obtained by adjoining an initial object $\phi$ to $I$. We extend the system $X_i$ to $I'$ by putting $X_\phi = X$, and let $\underline{\underline{P}}$ be the fibred category over $I'$ having the fibre $\underline{\underline{P}}(X_i)$ over $i$. Let $\underline{\underline{P}}_i$ be the category of cartesian sections of $\underline{\underline{P}}$ over $I'/i$. (An object of $\underline{\underline{P}}_i$ is a family of pairs $(E_j, \theta_j)$ with $E_j \in \underline{\underline{P}}(X_j)$ and $\theta_j$ an isomorphism $(j \to i) * E_i \simeq E_j$ for each object $j \longrightarrow i$ of $I'/i$.) Clearly $\underline{\underline{P}}_i$ is equivalent to $\underline{\underline{P}}(X_i)$ and $i \mapsto \underline{\underline{P}}_i$ is a functor from $I^0$ to categories. Using [EGA IV 8.5] it is not hard to see that we have an equivalence of categories

$$\varinjlim_I (i \mapsto \underline{\underline{P}}_i) \longrightarrow \underline{\underline{P}}(X)$$

such that a sequence is exact in $\underline{\underline{P}}(X)$ if and only if it comes from an exact sequence in some $\underline{\underline{P}}_i$. Thus from §2 (9) we have $K_q\underline{\underline{P}}(X) = \varinjlim K_q\underline{\underline{P}}_i$, proving 2.3. The proof of 2.4 is similar.

**2.5.** Suppose that $f : X \longrightarrow Y$ is a *morphism of finite Tor dimension* (i.e. $\underline{\underline{O}}_X$ is of finite Tor dimension as a module over $f^{-1}(\underline{\underline{O}}_Y)$), and let $\underline{\underline{P}}(Y, f)$ be the full subcategory of $\underline{\underline{M}}(Y)$ consisting of sheaves $F$ such that

$$Tor_i^{\underline{\underline{O}}_Y}(\underline{\underline{O}}_X, F) = 0 \text{ for } i > 0.$$

Assuming that every $F$ in $\underline{\underline{M}}(Y)$ is a quotient of a member of $\underline{\underline{P}}(Y, f)$, the resolution theorem (Th. 3, Cor. 3) implies that the inclusion $\underline{\underline{P}}(Y, f) \longrightarrow \underline{\underline{M}}(Y)$ induces isomorphisms on $K$-groups. Combining this isomorphism with the homomorphism induced by the exact functor $f^* : \underline{\underline{P}}(Y,f) \longrightarrow \underline{\underline{M}}(X)$, we obtain a homomorphism which will be denoted

$$f^* : K'_qY \longrightarrow K'_qX. \tag{2.6}$$

The assumption holds if either $f$ is flat (whence $\underline{\underline{P}}(Y,f) = \underline{\underline{M}}(Y)$), or if every coherent sheaf on $Y$ is the quotient of a vector bundle (e.g. if $Y$ has an ample line bundle). In both of these cases the formula $(fg)^* = g^*f^*$ is easily verified.

**2.7.** Let $f : X \longrightarrow Y$ be a *proper morphism*, so that the higher direct image functors $R^if_*$ carry coherent sheaves on $X$ to coherent sheaves on $Y$. Let $\underline{\underline{F}}(X,f)$ denote the full subcategory of $\underline{\underline{M}}(X)$ consisting of $F$ such that $R^if_*(F) = 0$ for $i > 0$. Since $R^if_* = 0$ for $i$ large [EGA III 1.4.12], we can apply Th. 3, Cor. 3 to the inclusion $\underline{\underline{F}}(X,f)^0 \longrightarrow \underline{\underline{M}}(X)^0$ to get an isomorphism $K_q\underline{\underline{F}}(X,f) \xrightarrow{\sim} K'_qX$, provided we assume that every coherent sheaf on $X$ can be embedded in a member of $\underline{\underline{F}}(X,f)$. Composing this isomorphism with the homomorphism of $K$-groups induced by the exact functor $f_* : \underline{\underline{F}}(X,f) \longrightarrow \underline{\underline{M}}(Y)$, we obtain a homomorphism which will be denoted

$$f_* : K'_qX \longrightarrow K'_qY. \tag{2.8}$$

The assumption is satisfied in the following cases:

i) When $f$ is finite, in particular, when $f$ is a closed immersion. In this case $R^if_* = 0$ for $i > 0$ [EGA III 1.3.2], so $\underline{\underline{F}}(X,f) = \underline{\underline{M}}(X)$.

ii) When $X$ has an ample line bundle [EGA II 4.5.3]. In effect if $L$ is ample on $X$, then it is ample when restricted to any open subset, and in particular, it is ample relative to $f$. Replacing $L$ by a high tensor power, we can suppose $L$ is very ample relative to $f$, and further that $L$ is generated by its global sections. Then for any $n$ we have an epimorphism $(\underline{\underline{O}}_X)^{rn} \longrightarrow L^{\otimes n}$, hence dualizing and tensoring with $L^{\otimes n}$, we obtain an exact sequence of vector bundles

$$0 \longrightarrow \underline{\underline{O}}_X \longrightarrow (L^{\otimes n})^{rn} \longrightarrow E \longrightarrow 0.$$

Hence for any coherent sheaf $F$ on $X$ we have an exact sequence

$$0 \longrightarrow F \longrightarrow F(n)^{rn} \longrightarrow F \otimes E \longrightarrow 0 \tag{2.9}$$

where $F(n) = F \otimes L^{\otimes n}$. But by Serre's theorem [EGA III 2.2.1], there is an $n_0$ such that $R^if_*(F(n)) = 0$ for $i > 0, n \geqslant n_0$, so $F(n) \in \underline{\underline{F}}(X,f)$ for $n \geqslant n_0$. Thus $F$ can be embedded in a member of $\underline{\underline{F}}(X,f)$ as asserted.

The verification of the formula $(fg)_* = f_*g_*$ in cases i) and ii) is straightforward and will be omitted.

**Proposition 2.10.** (Projection formula) *Suppose $f : X \longrightarrow Y$ proper and of finite Tor dimension, and assume $X$ and $Y$ have ample line bundles so that* 2.6 *and* 2.8 *are defined. Then for $x \in K_0X$ and $y \in K'_qY$ we have $f_*(x \cdot f^*y) = f_*(x) \cdot y$ in $K'_qY$, where $f_*(x)$ is the image of $x$ by the homomorphism $f_* : K_0X \longrightarrow K_0Y$ of* [SGA 6 2.12.3].

*Proof.* We recall that if $x = [E]$ is the class of a vector bundle $E$, then $f_*(x)$ is the class of the perfect complex $Rf_*(E)$. Arguing as in case ii) above, one sees that $K_0X$ is generated by the elements $[E]$ such that $R^if_*(E) = 0$ for $i > 0$. Then $Rf_*(E) \simeq f_*E$, and $f_*(x) = \sum(-1)^i[P_i] \in K_0X$, where $\{P.\}$ is a finite resolution of $f_*E$ by vector bundles on $Y$. Let $\underline{\underline{L}}$ denote the full subcategory of $\underline{\underline{M}}(Y)$ consisting of $F$ such that

$$Tor_i^{\underline{\underline{O}}_Y}(f_*E, F) = 0 = Tor_i^{\underline{\underline{O}}_Y}(\underline{\underline{O}}_X, F) \quad i > 0.$$

By the resolution theorem we have $K_q\underline{\underline{L}} = K'_qY$. Moreover, applying Th. 2, Cor. 3 to

$$0 \longrightarrow P_n \otimes F \longrightarrow \cdots \longrightarrow P_0 \otimes F \longrightarrow f_*E \otimes F \longrightarrow 0$$

for $F \in \underline{\underline{L}}$, one sees that $y \longmapsto f_*(x) \cdot y$ is the endomorphism of $K'_qY$ induced by the exact functor $F \longmapsto f_*E \otimes F$ from $\underline{\underline{L}}$ to $\underline{\underline{M}}(Y)$.

From the projection formula in the derived category: $Rf_*(E \overset{L}{\bigotimes}_Y F) = Rf_*(E) \overset{L}{\bigotimes}_Y F$ (see [SGA 6 III 2.7]), we find for $F$ in $\underline{\underline{L}}$ that

$$R^qf_*(E \otimes f^*E) = \begin{cases} 0 & q \neq 0 \\ f_*E \otimes F & q = 0. \end{cases}$$

Thus $E \otimes f^*F$ is in $\underline{\underline{F}}(X, f)$, so by the definition of 2.6 and 2.8, we have that $y \longmapsto f_*(x \cdot f^*y)$ is the endomorphism of $K'_qY$ induced by the exact functor $F \longmapsto f_*(E \otimes f^*F)$ from $\underline{\underline{L}}$ to $\underline{\underline{M}}(Y)$. Since we have an isomorphism $f_*(E \otimes f^*F) \simeq f_*E \otimes F$, the projection formula follows.

**Proposition 2.11.** *Let*

$$\begin{array}{ccc} X' & \xrightarrow{g'} & X \\ {\scriptstyle f'}\downarrow & & \downarrow{\scriptstyle f} \\ Y' & \xrightarrow{g} & Y \end{array}$$

*be a cartesian square of schemes having ample line bundles. Assume $f$ is proper, $g$ is of finite Tor dimension, and that $Y'$ and $X$ are Tor independent over $Y$, (i.e.*

$$Tor_i^{\underline{\underline{O}}_{Y,y}}(\underline{\underline{O}}_{Y',y'}, \underline{\underline{O}}_{X,x}) = 0 \ for \ i > 0$$

*for any $x \in X, y' \in Y', y \in Y$ such that $f(x) = y = g(y')$.) Then*

$$g^*f_* = f'_*g'^* : K'_qX \longrightarrow K'_qY'.$$

*Sketch of proof.* Set $\underline{\underline{L}} = \underline{\underline{P}}(X, g') \cap \underline{\underline{F}}(X, f)$. From the formula $Lg^*Rf_* = Rf'_*Lg'^*$ in the derived category [SGA 6 IV 3.1.0], one deduces that for $F \in \underline{\underline{L}}$ we have that $f_*F \in \underline{\underline{P}}(Y, g), g'^*F \in \underline{\underline{F}}(X', f')$, and that there is an isomorphism $g^*f_*(F) = f'_*g'^*(F)$. Thus everything comes to showing that $K_q\underline{\underline{L}} \xrightarrow{\sim} K'_qX$. Since $K_q\underline{\underline{P}}(X, g') \xrightarrow{\sim} K'_qX$, we have only to check that the inclusion $\underline{\underline{L}} \longrightarrow \underline{\underline{P}}(X, g')$ induces isomorphisms on $K$-groups. But this follows from the resolution theorem, because the exact sequence 2.9 shows that the functors $R^if_*$ on the category

$\underline{\underline{P}}(X, g')$ are effaceable for $i > 0$.

**3. Closed subschemes.** Let $Z$ be a closed subscheme of $X$, let $i : Z \longrightarrow X$ be the canonical immersion, and let $I$ be the coherent sheaf of ideals in $\underline{\underline{O}}_X$ defining $Z$. The functor $i_* : \underline{\underline{M}}(Z) \longrightarrow \underline{\underline{M}}(X)$ allows us to identify coherent sheaves on $Z$ with coherent sheaves on $X$ killed by $I$.

**Proposition 3.1.** *If $I$ is nilpotent, then $i_* : K'_q Z \longrightarrow K'_q X$ is an isomorphism. In particular, $K'_q(X_{red}) \xrightarrow{\sim} K'_q X$.*

This is an immediate consequence of Theorem 4.

**Proposition 3.2.** *Let $U$ be the complement of $Z$ in $X$, and $j : U \longrightarrow X$ the canonical open immersion. Then there is a long exact sequence*

$$(3.3) \qquad \longrightarrow K'_{q+1}U \longrightarrow K'_q Z \xrightarrow{i_*} K'_q X \xrightarrow{j^*} K'_q U \longrightarrow$$

*Proof.* One knows [Gabriel,Ch. V] that $j^* : \underline{\underline{M}}(X) \longrightarrow \underline{\underline{M}}(U)$ induces an equivalence of $\underline{\underline{M}}(U)$ with the quotient category $\underline{\underline{M}}(X)/\underline{\underline{B}}$, where $\underline{\underline{B}}$ is the Serre subcategory consisting of coherent sheaves with support in $Z$. Theorem 4 implies that $i_* : \underline{\underline{M}}(Z) \longrightarrow \underline{\underline{B}}$ induces isomorphisms on $K$-groups, so the desired exact sequence results from Theorem 5.

**Remark 3.4.** The exact sequence 3.3 has some evident naturality properties which follow from the fact that it is the homotopy exact sequence of the "fibration"

$$BQ(\underline{\underline{M}}(Z)) \to BQ(\underline{\underline{M}}(X)) \to BQ(\underline{\underline{M}}(U)).$$

For example, if $Z'$ is a closed subscheme of $X$ containing $Z$, then there is a map from the exact sequence of $(X, Z)$ to the one for $(X, Z')$. Also a flat map $f : X' \to X$ induces a map from the exact sequence for $(X, Z)$ to the one for $(X', f^{-1}Z)$.

**Remark 3.5.** From 3.3 one deduces in a well-known fashion a *Mayer-Vietoris sequence*

$$\longrightarrow K'_{q+1}(U \cap V) \longrightarrow K'_q(U \cup V) \longrightarrow K'_q U \oplus K'_q V \longrightarrow K'_q(U \cap V) \longrightarrow$$

for any two open sets $U$ and $V$ of $X$. Starting essentially from this point, Brown and Gersten (see their paper in this procedings) construct a spectral sequence

$$E_2^{pq} = H^p(X, \underline{\underline{K}}'_{-q}) \Longrightarrow K'_{-n} X$$

which reflects the fact that $K'$-theory is a sheaf of generalized cohomology theories in a certain sense. In connection with this, we mention that Gersten has proposed defining higher $K$-groups for regular schemes by piecing together the Karoubi-Villamayor theories belonging to the open affine subschemes (see [Gersten 2]). Using the above Mayer-Vietoris sequence and the fact that Karoubi-Villamayor

$K$-theory coincides with ours for regular rings, Gersten has shown that his method leads to the groups $K_qX = K'_qX$ studied here.

**4. Affine and projective space bundles.**

**Proposition 4.1.** (Homotopy property) *Let $f : P \longrightarrow X$ be a flat map whose fibres are affine spaces (for example, a vector bundle or a torsor under a vector bundle). Then $f^* : K'_qX \longrightarrow K'_qP$ is an isomorphism.*

*Proof.* If $Z$ is a closed subset of $X$ with complement $U$, then because $f$ is flat we have a map of exact sequences

$$\begin{array}{ccccccc} \longrightarrow K'_qZ & \longrightarrow & K'_qX & \longrightarrow & K'_qU & \longrightarrow \\ \downarrow & & \downarrow & & \downarrow & \\ \longrightarrow K'_qP_Z & \longrightarrow & K'_qP & \longrightarrow & K'_qP_U & \longrightarrow \end{array}$$

By the five lemma, the proposition is true for one of $X, Z$, and $U$ if it is true for the other two. Using noetherian induction we can assume the proposition holds for all closed subsets $Z \neq X$. We can suppose $X$ is irreducible, for if $X = Z_1 \cup Z_2$ with $Z_1, Z_2 \neq X$, then the proposition holds for $Z_1$ and $X - Z_1 = Z_2 - (Z_1 \cap Z_2)$, hence also for $X$. We can also suppose $X$ reduced by 3.1.

Now take the inductive limit in the above diagram as $Z$ runs over all closed subsets $\neq X$. Then by 2.4, $\varinjlim K'_qU = K'_q(k(x))$ and $\varinjlim K'_qP_U = K'_q(k(x) \times_X P)$, where $k(x)$ is the residue field at $x$, and where $x$ is the generic point of $X$. Thus we have reduced to the case where $X = Spec(k)$, $k$ a field, and we want to prove $K'_qk \xrightarrow{\sim} K'_q(k[t_1, \cdots, t_n])$. But this follows from §6 Th. 8, so the proof is complete.

**4.2.** *Jouanolou's device.* Jouanolou has shown that at least for a quasi-projective scheme $X$ over a field, there is a torsor $P$ over $X$ with group a vector bundle such that $P$ is an affine scheme. He defines higher $K$-groups for smooth $X$ by taking the Karoubi-Villamayor $K$-groups of the coordinate ring of $P$ and showing that these do not depend on the choice of $P$. From 4.1 it is clear that his method yields the groups $K_qX = K'_qX$ considered here.

**Proposition 4.3.** *Let $E$ be a vector bundle of rank $r$ over $X$, let $PE = Proj(SE)$ be the associated projective bundle, where $SE$ is the symmetric algebra of $E$, and let $f : PE \longrightarrow X$ be the structural map. Then we have a $K_0(PE)$-module isomorphism*

$$K_0(PE) \otimes_{K_0X} K'_qX \xrightarrow{\sim} K'_q(PE) \tag{4.4}$$

*given by $y \otimes x \longmapsto y \cdot f^*x$. Equivalently, if $z \in K_0(PE)$ is the class of the canonical line bundle $\underline{\underline{O}}(-1)$, then we have an isomorphism*

$$(K'_qX)^r \xrightarrow{\sim} K'_q(PE), (x_i)_{0 \leqslant i < r} \longmapsto \sum_{i=0}^{r-1} z^i \cdot f^*x_i. \tag{4.5}$$

*Sketch of proof.* The equivalence of 4.4 and 4.5 results from the fact that $K_0(PE)$ is a free $K_0X$-module with basis $1, \cdots, z^{r-1}$, [SGA 6 VI 1.1]. Using the exact sequence 3.3 as in the proof of 4.1, one reduces to the case where $X = Spec(k)$, $k$ a field. By the standard correspondence between coherent sheaves on $PE$ and finitely generated graded $SE$-modules, one knows that $\underline{\underline{M}}(PE)$ is equivalent to the quotient of $Modfgr(SE)$ by the subcategory of $M$ such that $M_n = 0$ for $n$ large. This subcategory has the same $K$-groups as the category Modfgr(k) by Theorem 4, where we view $k$-modules as $SE$-modules killed by the augmentation ideal. Thus from the localization theorem we have an exact sequence

$$(4.6) \qquad \longrightarrow K_q(Modfgr(k)) \xrightarrow{i_*} K_q(Modfgr(SE)) \xrightarrow{j_*} K'_q(PE) \longrightarrow$$

where $i$ is the inclusion and $j$ associates to a module $M$ the associated sheaf $\widetilde{M}$ on $PE$. From Theorem 6 we have the vertical isomorphisms in the square

$$\begin{array}{ccc} K_q(Modfgr(k)) & \xrightarrow{i_*} & K_q(Modfgr(SE)) \\ \uparrow \wr & & \uparrow \wr \\ \mathbb{Z}[t] \otimes K'_q k & \xrightarrow{h} & \mathbb{Z}[t] \otimes K'_q k \end{array}$$

Using the Koszul resolution

$$0 \longrightarrow SE(-r) \otimes \wedge^r E \otimes M \longrightarrow \cdots \longrightarrow SE \otimes M \longrightarrow M \longrightarrow 0$$

and Th. 2, Cor. 3, one shows that the map $h$ rendering the above square commutative is multiplication by $\lambda_{-t}(E) = \sum(-t)^i[\wedge^i E]$. Thus $i_*$ is injective, so from 4.6 we get an isomorphism

$$\coprod_{0 \leqslant i < r} t^i \otimes K'_q k \xrightarrow{\sim} K'_q(PE)$$

induced by the functors $M \mapsto \underline{\underline{O}}(-1)^{\otimes i} \otimes_k M, 0 \leqslant i < r$ from $Modf(k)$ to $\underline{\underline{M}}(PE)$. This gives the desired isomorphism 4.5.

The following generalizes 3.1.

**Proposition 4.7.** *Let $f : X' \longrightarrow X$ be a finite morphism which is radicial and surjective (i.e. for each $x$ in $X$ the fibre $f^{-1}(x)$ has exactly one point $x'$ and the residue field extension $k(x')/k(x)$ is purely inseparable). Let $S$ be the multiplicative system in $\mathbb{Z}$ generated by the degrees $[k(x') : k(x)]$ for all $x$ in $X$. Then $f_* : K'_q(X') \longrightarrow K'_q X$ induces an isomorphism $S^{-1}K'_q(X') \xrightarrow{\sim} S^{-1}K'_q X$.*

*Proof.* If $Z$ is a closed subscheme of $X$ with complement $U$, and, if $Z'$ and $U'$ are the respective inverse images of $Z$ and $U$ in $X'$, then we have a map of exact sequences

$$\begin{array}{ccccccc} \longrightarrow K'_q(Z') & \longrightarrow & K'_q(X') & \longrightarrow & K'_q(U') & \longrightarrow \\ \downarrow (f_Z)_* & & \downarrow f_* & & \downarrow (f_U)_* & \\ \longrightarrow K'_q Z & \longrightarrow & K'_q X & \longrightarrow & K'_q U & \longrightarrow \end{array}$$

Localizing with respect to $S$ and using the five lemma, we see that if the proposition holds for two of $f_Z, f, f_U$ it holds for the third. Thus arguing as in the proof of 4.1 we can reduce to the case where $X = Spec(k), k$ a field. By 3.1 we can suppose $X' = Spec(k')$, where $k'$ is a purely inseparable finite extension of $k$. Thus we have reduced to the following.

**Proposition 4.8.** *Let $f : k \longrightarrow k'$ be a purely inseparable finite extension of degree $p^d$. Then $f_*f^* =$ multiplication by $p^d$ on $K_q k$ and $f^*f_* =$ multiplication by $p^d$ on $K_q(k')$.*

*Proof.* The fact that $f_*f^* =$ multiplication by $[k' : k]$ is an immediate consequence of the projection formula §4(5) and does not use the purely inseparable hypothesis. The homomorphism $f^*f_*$ is induced by the exact functor

$$V \longmapsto k' \otimes_k V = (k' \otimes_k k') \otimes_{k'} V$$

from $\underline{\underline{P}}(k')$ to itself. Since $k'/k$ is purely inseparable, the augmentation ideal $I$ of $k' \otimes_k k'$ is nilpotent. Filtering by powers of $I$, one obtains a filtration of the above functor with

$$gr((k' \otimes_k k') \otimes_{k'} V) = \coprod_n (I^n/I^{n+1}) \otimes_{k'} V.$$

But because the two $k'$-module structures on $I^n/I^{n+1}$ coincide, this graded functor is isomorphic to the functor $V \longmapsto V^r$, where $r = dim_{k'}(gr(k' \otimes_k k')) = p^d$. Applying Th. 2, Cor. 2 to this filtration, we find $f^*f_* =$ multiplication by $p^d$, completing the proof.

**5. Filtration by support, Gersten's conjecture, and the Chow ring.** Let $\underline{\underline{M}}_p(X)$ denote the Serre subcategory of $\underline{\underline{M}}(X)$ consisting of those coherent sheaves whose support is of codimension $\geqslant p$. (The codimension of a closed subset $Z$ of $X$ is the infimum of the dimensions of the local rings $\underline{\underline{O}}_{X,z}$ where $z$ runs over the generic points of $Z$.) From §2 (9) and 3.1, it is clear that we have

$$K_q(\underline{\underline{M}}_p(X)) = \varinjlim K'_q Z \tag{5.1}$$

where $Z$ runs over the closed subsets of codimension $\geqslant p$. we also have

$$f^*(\underline{\underline{M}}_p(X)) \subset \underline{\underline{M}}_p(X') \quad if \quad f : X' \longrightarrow X \quad is \quad flat. \tag{5.2}$$

In effect, one has to show that if $Z$ has codimension $\geqslant p$ in $X$, then $f^{-1}Z$ has codimension $\geqslant p$ in $X'$. But if $z'$ is a generic point of $f^{-1}Z$, and $z = f(z')$, then the homomorphism $\underline{\underline{O}}_{X,z} \longrightarrow \underline{\underline{O}}_{X',z'}$ is a flat local homomorphism such that $rad(\underline{\underline{O}}_{X,z}) \cdot \underline{\underline{O}}_{X',z'}$ is primary for $rad(\underline{\underline{O}}_{X',z'})$; hence $dim(\underline{\underline{O}}_{X,z}) = dim(\underline{\underline{O}}_{X',z'})$ by [EGA IV 6.1.3], proving the assertion.

If $X = \varprojlim X_i$ where $i \longmapsto X_i$ is a filtered projective system with affine flat transition morphisms, then we have isomorphisms

$$K_q(\underline{\underline{M}}_p(X)) = \varinjlim K_q(\underline{\underline{M}}_p(X_i)). \tag{5.3}$$

In view of 5.1 this reduces to showing that any $Z$ of codimension $p$ in $X$ is of the form $f_i^{-1}(Z_i)$ for some $i$, where $Z_i$ is of codimension $p$ in $X_i$, and where $f_i : X \longrightarrow X_i$ denotes the canonical map. But for $i$ large enough, one has $Z = f_i^{-1}(Z_i)$ with $Z_i$ = the closure of $f_i(Z)$. Hence any generic point $z'$ of $Z_i$ is the image of a generic point $z$ of $Z$, so the local rings at $z'$ and $z$ have the same dimension by the result about dimension used above. Thus $Z_i$ also has codimension $p$, proving 5.3.

**Theorem 5.4.** *Let $X_p$ be the set of points of codimension $p$ in $X$. There is a spectral sequence*

(5.5)
$$E_1^{pq}(X) = \coprod_{x \in X_p} K_{-p-q} k(x) \Longrightarrow K'_{-n} X$$

*which is convergent when $X$ has finite (Krull) dimension. This spectral sequence is contravariant for flat morphisms. Furthermore, if $X = \varprojlim X_i$, where $i \longmapsto X_i$ is a filtered projective system with affine flat transition morphisms, then the spectral sequence for $X$ is the inductive limit of the spectral sequences for the $X_i$.*

In this spectral sequence we interpret $K_n$ as zero for $n < 0$. Thus the spectral sequence is concentrated in the range $p \geqslant 0, p + q \leqslant 0$.

*Proof*. We consider the filtration

$$\underline{\underline{M}}(X) = \underline{\underline{M}}_0(X) \supset \underline{\underline{M}}_1(X) \supset \cdots$$

of $\underline{\underline{M}}(X)$ by Serre subcategories. There is an equivalence

$$\underline{\underline{M}}_p(X)/\underline{\underline{M}}_{p+1}(X) \sim \coprod_{x \in X_p} \bigcup_n Modf(\underline{\underline{O}}_{X,x}/rad(\underline{\underline{O}}_{X,x})^n)$$

so from Th. 4, Cor. 1, one has an isomorphism

$$K_i(\underline{\underline{M}}_p(X)/\underline{\underline{M}}_{p+1}(X)) \simeq \coprod_{x \in X_p} K_i k(x)$$

where $k(x)$ is the residue field at $x$. From Th. 5 we get exact sequences

$$\longrightarrow K_i(\underline{\underline{M}}_{p+1}(X)) \longrightarrow K_i(\underline{\underline{M}}_p(X)) \longrightarrow \coprod_{x \in X_p} K_i k(x) \longrightarrow K_{i-1}(\underline{\underline{M}}_{p+1}(X)) \longrightarrow$$

which give rise to the desired spectral sequence in a standard way. The functorality assertions of the theorem follow immediately from 5.2 and 5.3.

We will now take up a line of investigation initiated by Gersten in his talk at this conference [Gersten 3].

**Proposition 5.6.** *The following conditions are equivalent:*

i) *For every* $p \geqslant 0$, *the inclusion* $\underline{\underline{M}}_{p+1}(X) \longrightarrow \underline{\underline{M}}_p(X)$ *induces zero on* $K$*-groups.*

ii) *For all* $q$, $E_2^{pq}(X) = 0$ *if* $p \neq 0$ *and the edge homomorphism* $K'_{-q}X \longrightarrow E_2^{0q}(X)$ *is an isomorphism.*

iii) *For every* $n$ *the sequence*

$$0 \longrightarrow K'_nX \xrightarrow{\ e\ } \coprod_{x \in X_0} K_nk(x) \xrightarrow{\ d_1\ } \coprod_{x \in X_1} K_{n-1}k(x) \xrightarrow{\ d_1\ } \cdots \tag{5.7}$$

*is exact. Here* $d_1$ *is the differential on* $E_1(X)$ *and* $e$ *is the map obtained by pulling-back with respect to the canonical morphisms Spec* $k(x) \longrightarrow X$.

This follows immediately from the spectral sequence 5.5 and its construction.

**Proposition 5.8.** (Gersten) *Let* $\underline{\underline{K}}'_n$ *denote the sheaf on* $X$ *associated to the presheaf* $U \longmapsto K'_nU$. *Assume that Spec*$(\underline{\underline{O}}_{X,x})$ *satisfies the equivalent conditions of* 5.6 *for all* $x$ *in* $X$. *Then there is a canonical isomorphism*

$$E_2^{pq}(X) = H^p(X, \underline{\underline{K}}'_{-q})$$

*with* $E_2^{pq}(X)$ *as in* 5.5.

*Proof*. We view the sequences 5.7 for the different open subsets of $X$ as a sequence of presheaves, and we sheafify to get a sequence of sheaves

$$0 \longrightarrow \underline{\underline{K}}'_n \longrightarrow \coprod_{x \in X_0} (i_x)_*(K_nk(x)) \longrightarrow \coprod_{x \in X_1} (i_x)_*(K_{n-1}k(x)) \longrightarrow \cdots \tag{5.9}$$

where $i_x :$ *Spec* $k(x) \longrightarrow X$ denotes the canonical map. The stalk of 5.9 over $x$ is the sequence 5.7 for *Spec*$(\underline{\underline{O}}_{X,x})$, because *Spec*$(\underline{\underline{O}}_{X,x}) = \varprojlim U$, where $U$ runs over the affine open neighborhoods of $x$, and because the spectral sequence 5.5 commutes with such projective limits. By hypothesis, 5.9 is exact, hence it is a flask resolution of $\underline{\underline{K}}'_n$, so

$$\begin{aligned} H^p(X, \underline{\underline{K}}'_n) &= H^p\{s \longmapsto \Gamma(X, \coprod_{x \in X_s} (i_x)_* K_{n-s}k(x))\} \\ &= H^p\{s \longmapsto E_1^{s,-n}(X)\} = E_2^{p,-n}(X) \end{aligned}$$

as asserted.

The following conjecture has been verified by Gersten in certain cases [Gersten 3].

**Conjecture 5.10** (Gersten). *The conditions of* 5.6 *are satisfied for the spectrum of a regular local ring.*

Actually, it seems reasonable to conjecture that the conditions of 5.6. hold more generally for semi-local regular rings, for in the cases where the conjecture has been proved, the arguments also apply to the corresponding semi-local situation. On the other hand there are examples suggesting that it is unreasonable to expect the conditions of 5.6 to hold for any general class of local rings besides the regular local rings.

We will now prove Gersten's conjecture in some important equi-characteristic cases.

**Theorem 5.11.** *Let $R$ be a finite type algebra over a field $k$, let $S$ be a finite set of primes in $R$ such that $R_p$ is regular for each $p$ in $S$, and let $A$ be the regular semi-local ring obtained by localizing $R$ with respect to $S$. Then Spec $A$ satisfies the conditions of* 5.6.

*Proof.* We first reduce to the case where $R$ is smooth over $k$. There exists a subfield $k'$ of $k$ finitely generated over the prime field, a finite type $k'$-algebra $R'$, and a finite subset $S'$ of Spec $R'$ such that $R = k \otimes_{k'} R'$ and such that the primes in $S$ are the base extensions of the primes in $S'$. If $A'$ is the localization of $R'$ with respect to $S'$, then $A = k \otimes_{k'} A'$ and $A'$ is regular. Letting $k_i$ run over the subfields of $k$ containing $k'$ and finitely generated over the prime field, we have $A = \varinjlim k_i \otimes_{k'} A'$ and $K_*(\underline{\underline{M}}_p(A)) = \varinjlim K_*(\underline{\underline{M}}_p(k_i \otimes_{k'} A'))$ by 5.3, where here and in the following we write $\underline{\underline{M}}_p(A)$ instead of $\underline{\underline{M}}_p(Spec\, A)$. Thus it suffices to prove the theorem when $k$ is finitely generated over the prime field. In this case $A$ is a localization of a finite type algebra over the prime field, so by changing $R$, we can suppose $k$ is the prime field. As prime fields are perfect, it follows that $R$ is smooth over $k$ at the points of $S$, hence also in an open neighborhood of $S$. Replacing $R$ by $R_f$ for some $f$ not vanishing at the points in $S$, we can suppose $R$ is smooth over $k$ as asserted.

We wish to prove that for any $p \geqslant 0$ the inclusion $\underline{\underline{M}}_{p+1}(A) \longrightarrow \underline{\underline{M}}_p(A)$ induces zero on $K$-groups. By 5.3 we have

$$K_*(\underline{\underline{M}}_{p+1}(A)) = \varinjlim K_*(\underline{\underline{M}}_{p+1}(R_f))$$

where $f$ runs over elements not vanishing at the points of $S$, hence replacing $R$ by $R_f$, we reduce to showing that the functor $\underline{\underline{M}}_{p+1}(R) \longrightarrow \underline{\underline{M}}_p(A)$ induces zero on $K$-groups. As

$$K_*(\underline{\underline{M}}_{p+1}(R)) = \varinjlim K_*(\underline{\underline{M}}_p(R/tR))$$

where $t$ runs over the regular elements of $R$, it suffices to show that given a regular element $t$, there exists an $f$, not vanishing at the points of $S$, such that the functor $M \longmapsto M_f$ from $\underline{\underline{M}}_p(R/tR)$ to $\underline{\underline{M}}_p(R)$ induces zero on $K$-groups.

We will need the following variant of the normalization lemma.

**Lemma 5.12.** *Let $R$ be a smooth finite type algebra of dimension $r$ over a field $k$, let $t$ be a regular element of $R$, and let $S$ be a finite subset of Spec $R$. Then there exist elements $x_1, \ldots, x_{r-1}$ of $R$ algebraically independent over $k$ such that if $B = k[x_1, \ldots, x_{r-1}] \subset R$, then* i) *$R/tR$ is finite over $B$, and* ii) *$R$ is smooth over $B$ at the points of $S$.*

Granting this for the moment, put $B' = R/tR$ and $R' = R \otimes_B B'$ so that we have arrows

$$\begin{array}{ccc} R' & \longrightarrow & R \\ \downarrow\!\uparrow u' & & \uparrow u \\ B' & \longrightarrow & B \end{array}$$

where the horizontal arrows are finite. Let $S'$ be the finite set of points of Spec $R'$ lying over the points in $S$. As $u$ is smooth of relative dimension one at the points of $S$, $u'$ is smooth of relative dimension one at the points of $S'$. One knows then [SGA 1 II 4.15] that the ideal $I = \text{Ker } (R' \longrightarrow B')$ is principal at the points of $S'$, hence principal in a neighborhood of $S'$. Since $R'/R$ is finite, this neighborhood contains the inverse image of a neighborhood of $S$ in Spec $R$. Thus we can find $f$ in $R$ not vanishing at the points of $S$ such that $I_f$ is isomorphic to $R'_f$ as an $R'_f$-module. We can also suppose $f$ chosen so that $R'_f$ is smooth, hence flat, over $B'$.

Then for any $B'$-module $M$ we have an exact sequence of $R_f$-modules

$$(*) \qquad 0 \longrightarrow I_f \otimes_{B'} M \longrightarrow R'_f \otimes_{B'} M \longrightarrow M_f \longrightarrow 0.$$

Since $R'_f$ is flat over $B'$, if $M$ is in $\underline{\underline{M}}_p(B')$, then $R'_f \otimes_{B'} M$ is in $\underline{\underline{M}}_p(R'_f)$, so viewed as an $R_f$-module, we have $R'_f \otimes_{B'} M$ is in $\underline{\underline{M}}_p(R_f)$. Thus $(*)$ is an exact sequence of exact functors from $\underline{\underline{M}}_p(B')$ to $\underline{\underline{M}}_p(R_f)$. Applying Th. 2, Cor. 1, and using the isomorphism $I_f \simeq R'_f$, we conclude that the functor from $\underline{\underline{M}}_p(B')$ to $\underline{\underline{M}}_p(R_f)$ induces the zero map on $K$-groups, as was to be shown.

*Proof of the lemma.* Choosing for each prime in $S$ a maximal ideal containing it, we can suppose $S$ is a finite set of maximal ideals of $R$. Let $\Omega^1$ be the module of Kahler differentials of $R$ over $k$. It is a projective $R$-module of rank $r$, and for $R$ to be smooth over $B = k[x_1, \ldots, x_{r-1}]$ at the points of $S$ means that the differentials $dx_i \in \Omega^1$ are independent at the points of $S$. Let $J$ be the intersection of the ideals in $S$. As $R/J^n = \prod R/m^n, m \in S$, is finite dimensional over $k$, we can find a finite dimensional $k$-subspace $V$ of $R$ such that for each $m$ in $S$, there exists $v_1, \ldots, v_r$ in $V$ whose differentials form a basis for $\Omega^1$ at $m$ vanishing at the other points of $S$. We can suppose also that $V$ generates $R$ as an algebra over $k$.

Define an increasing filtration of $R/tR$ by letting $F_n(R/tR)$ be the subspace spanned by the monomials of degree $\leqslant n$ in the elements of $V$. Then the associated graded ring $gr(R/tR)$ is of dimension $r - 1$. To see this, note that $Proj(\coprod F_n(R/tR))$ is the closure in projective space of the subscheme $Spec(R/tR)$ of the affine space $Spec\ S(V)$. Since $R/tR$ has dimension $r - 1$, the part of this $Proj$ at infinity, namely $Proj(gr(R/tR))$, is of dimension $r - 2$, so $gr(R/tR)$ has dimension $r - 1$ as asserted. Let $z_1, \ldots, z_{r-1}$ be a system of parameters for $gr(R/tR)$ such that each $z_i$ is homogeneous of degree $\geqslant 2$. Then $gr(R/tR)$ is finite over $k[z_1, \ldots, z_{r-1}]$, so if the $z_i$ are lifted to elements $x'_i$ of $R$, then $R/tR$ is finite over $k[x'_1, \ldots, x'_{r-1}]$.

By the choice of $V$, we can choose $v_1, \ldots, v_{r-1}$ in $V$ such that $x_i = x'_i + v_i, 1 \leqslant i < r$, have independent differentials at the points of $S$, whence condition ii) of the lemma is satisfied. On the other hand, the $x_i$ have the leading terms $z_i$ in $gr(R/tR)$, so $R/tR$ is finite over $k[x_1, \ldots, x_{r-1}]$. The proof of the lemma and

Theorem 5.11 is now complete.

**Theorem 5.13.** *The conditions of* 5.6 *hold for Spec $A$ when $A$ is the ring of formal power series $k[[X_1, \dots, X_n]]$ over a field $k$, and when $A$ is the ring of convergent power series in $X_1, \dots, X_n$ with coefficients in a field complete with respect to a non-trivial valuation.*

The proof is analogous to the preceding. Indeed, given $0 \neq t \in A = k[[X_1, \dots, X_n]]$, then after a change of coordinates, $A/tA$ becomes finite over $B = k[[X_1, \dots, X_{r-1}]]$ by the Weierstrass preparation theorem. Further, if we put $A' = A \otimes_B A/tA$, then $\mathrm{Ker}(A' \longrightarrow A/tA)$ is principal, so arguing as before, we can conclude that $\underline{\underline{K}}_p(A/tA) \longrightarrow \underline{\underline{K}}_p(A)$ induces zero on $K$-groups. The argument also works for convergent power series, since the preparation theorem is still available.

We now want to give an application of 5.11 to the Chow ring. We will assume known the fact that the $K_1A$ defined here is canonically isomorphic to the Bass $K_1$, and in particular that $K_1A$ is canonically isomorphic to the group of units $A^{\cdot}$, when $A$ is a local ring or a Euclidean domain.

**Proposition 5.14.** *Let $X$ be a regular scheme of finite type over a field. Then the image of*

$$d_1 : \coprod_{x \in X_{p-1}} k_1 k(x) \longrightarrow \coprod_{x \in X_p} K_0 k(x) = \coprod_{x \in X_p} \mathbb{Z}$$

*in the spectral sequence* 5.5 *is the subgroup of codimension $p$ cycles which are linearly equivalent to zero. Consequently $E_2^{p,-p}(X)$ is canonically isomorphic to the group $A^p(X)$ of cycles of codimension $p$ modulo linear equivalence.*

*Proof.* Let $P^1$ be the projective line over the ground field, and let $t$ denote tho canonical rational function on $P^1$. Let $C^p(X)$ denote the group of codimension $p$ cycles. The subgroup of cycles linearly equivalent to zero is generated by cycles of the form $W_0 - W_\infty$ where $W$ is an irreducible subvariety of $X \times P^1$ of codimension $p$ such that the intersections $W_0 = W \cap (X \times 0)$ and $W_\infty = W \cap (X \times \infty)$ are proper. We need a known formula for $W_0 - W_\infty$ which we now recall.

Let $Y$ be the image of $W$ under the projection $X \times P^1 \longrightarrow X$, so that $dim(Y) = dim(W)$ or $dim(W) - 1$. In the latter case we have $W = Y \times P^1$ and $W_0 - W_\infty = 0$, so we may assume $dim(W) = dim(Y)$, whence $Y$ has codimension $p-1$ in $X$. Let $y$ be the generic point of $Y$ and $w$ the generic point of $W$, so that $k(w)$ is a finite extension of $k(y)$. Let $t'$ be the non-zero element of $k(w)$ obtained by pulling $t$ back to $W$, and let $x$ be a point of codimension one in $Y$, whence $\underline{\underline{O}}_{Y,x}$ is a local domain of dimension one with quotient field $k(y)$. Then the formula we want is

$$\text{(5.15)} \qquad (\text{multiplicity of } x \text{ in } W_0 - W_\infty) = ord_{yx}(Norm_{k(w)/k(y)} t')$$

where $ord_{yx} : k(y)^{\cdot} \longrightarrow \mathbb{Z}$ is the unique homomorphism such that

$$ord_{yx}(f) = length(\underline{\underline{O}}_{Y,x} / f \underline{\underline{O}}_{Y,x})$$

for $f \in \underline{\underline{O}}_{Y,x}, f \neq 0$. For a proof of 5.15 see [Chevalley, p. 2-12].

From 5.15 it is clear that the subgroup of cycles linearly equivalent to zero is the image of the homomorphism

$$\phi : \coprod_{y \in X_{p-1}} k(y)^{\cdot} \longrightarrow \coprod_{x \in X_p} \mathbb{Z}x = C^p(X)$$

where if $f \in k(y)^{\cdot}$, then $\phi(f) = \sum ord_{yx}(f) \cdot x$ and we put $ord_{yx} = 0$ if $x \notin \overline{\{y\}}$. Since $K_1 k(y) = k(y)^{\cdot}$, we see $\phi$ is a map from $E_1^{p-1,-p}(X)$ to $E_1^{p,-p}(X)$, so all that remains to prove the proposition is to show that $\phi = d_1$.

Let $d_1$, have the components

$$(d_1)_{yx} : k(y)^{\cdot} = K_1 k(y) \longrightarrow K_0 k(x) = \mathbb{Z}$$

for $y$ in $X_{p-1}$ and $x$ in $X_p$. We want to show that $(d_1)_{yx} = ord_{yx}$. Fix $y$ in $X_{p-1}$ and let $Y$ be its closure. The closed immersion $Y \longrightarrow X$ carries $\underline{\underline{M}}_j(Y)$ to $\underline{\underline{M}}_{j+p-1}(X)$ for all $j$, hence it induces a map from the spectral sequence 5.5 for $Y$ to the one for $X$ augmenting the filtration by $p-1$. Thus we get a commutative diagram

$$\begin{array}{ccc} E_1^{p-1,-p}(X) & \xrightarrow{d_1} & E_1^{p,-p}(X) = C^p(X) \\ \uparrow & & \uparrow \\ K_1 k(y) = E_1^{0,-1}(Y) & \xrightarrow{d_1} & E_1^{1,-1}(Y) = C^1(Y) \end{array}$$

which shows that $(d_1)_{yx} = 0$ unless $x$ is in $Y$. On the other hand, if $x$ is of codimension one in $Y$, then the flat map $Spec(\underline{\underline{O}}_{Y,x}) \longrightarrow Y$ induces a map of spectral sequences, so we get a commutative diagram

$$\begin{array}{cccccc} K_1 k(y) = & E_1^{0,-1}(Y) & \xrightarrow{d_1} & E_1^{1,-1}(Y) = & C^1(Y) & \\ \| & \downarrow & & \downarrow & \downarrow & \text{multiplicity of } x \\ K_1 k(y) = & E_1^{0,-1}(\underline{\underline{O}}_{Y,x}) & \xrightarrow{d_1} & E_1^{1,-1}(\underline{\underline{O}}_{Y,x}) = & \mathbb{Z} & \end{array}$$

which shows that $(d_1)_{yx}$ is the map $d_1$ in the spectral sequence for $\underline{\underline{O}}_{Y,x}$. Therefore the equality $(d_1)_{yx} = ord_{yx}$ is a consequence of the following.

**Lemma 5.16.** *Let $A$ be an equi-characteristic local noetherian domain of dimension one with quotient field $F$ and residue field $k$, and let*

$$\longrightarrow K_1'A \longrightarrow K_1F \xrightarrow{\partial} K_0k \longrightarrow K_0'A \longrightarrow K_0F \longrightarrow 0$$

*be the exact sequence 3.3 associated to the closed set Spec $k$ of Spec $A$. Then $\partial : K_1F \longrightarrow K_0k$ is isomorphic to ord : $F^{\cdot} \longrightarrow \mathbb{Z}$, where ord is the homomorphism such that $ord(x) = length(A/xA)$ for $x$ in $A, x \neq 0$.*

*Proof.* We have isomorphisms $K_1F = F^{\cdot}$ and $K_1A = A^{\cdot}$ since $A$ and $F$ are local rings. We wish to show $\partial(x) = ord(x)$ for $x$ in $A, x \neq 0$. If $x$ is in $A^{\cdot}$, this is clear, as $\partial(X) = 0$ since $x$ is in the image of the map $K_1A \longrightarrow K_1'A \longrightarrow K_1F$. Thus we can suppose $x$ is not a unit. By hypothesis $A$ is an algebra over the prime subfield $k_0$ of $k$. If $x$ were algebraic over $k_0$, it would be a unit in $A$. Thus $x$ is not algebraic, so we have a *flat* homomorphism $k_0[t] \longrightarrow A$ sending the indeterminate $t$ to $x$. By naturality of the exact sequence 3.3 for flat maps, we get a commutative diagram

$$\begin{array}{ccccccc} \longrightarrow K_1k_0[t] \longrightarrow & K_1k_0[t,t^{-1}] & \xrightarrow{\partial} & K_0k_0 \longrightarrow \\ \downarrow & \downarrow u & & \downarrow v \\ \longrightarrow K_1'A \longrightarrow & K_1F & \xrightarrow{\partial} & K_0k \longrightarrow \end{array}$$

such that $u(t) = x$. The homomorphism $v$ is induced by sending a $k_0$-vector space $V$ to the $A$-module

$$A \otimes_{k_0[t]} V = A/xA \otimes_{k_0} V$$

and using devissage to identify the $K$-groups of the category of $A$-modules of finite length with those of $\underline{\underline{P}}(k)$. Thus with respect to the isomorphisms $K_0k_0 = K_0k = \mathbb{Z}$, $v$ is multiplication by $length(A/xA) = ord(x)$. Therefore it suffices to show that in the top row of the above diagram, one has $\partial(t) = \pm 1$. But this is easily verified by explicitly computing the top row, using the fact that $K_0R = \mathbb{Z}$ and $K_1R = R^{\cdot}$ for a Euclidean domain. q.e.d.

**Remark 5.17.** In another paper, along with the proof of Theorem 1 of [Quillen 1], I plan to justify the following description of the boundary map $\partial : K_nF \longrightarrow K_{n-1}k$ for a local noetherian domain $A$ of dimension one with quotient field $F$ and residue field $k$. By the universal property of the $K$-theory of a ring, such a map is defined by giving for every finite dimensional vector space $V$ over $F$ a homotopy class of maps

$$B(Aut(V)) \longrightarrow BQ(\underline{\underline{P}}(k)) \tag{5.18}$$

compatible with direct sums. To do this consider the set of $A$-lattices in $V$, i.e. finitely generated $A$-submodules $L$ such that $F \otimes_A L = V$. Let $X(V)$ be the ordered set of layers $(L_0, L_1)$ such that $L_1/L_0$ is killed by the maximal ideal of $A$, and put $G = Aut(V)$. Then $G$ acts on $X(V)$, so we can form a cofibred category $X(V)_G$ over $G$ with fibre $X(V)$. One can show that $X(V)$ is contractible (it is essentially a 'building'), hence the functor $X(V)_G \longrightarrow G$ is a homotopy. equivalence. On the other hand there is a functor $X(V)_G \longrightarrow Q(\underline{\underline{P}}(k))$ sending $(L_0, L_1)$ to $L_1/L_0$, hence we obtain the desired map 5.18.

It can be deduced from this description that the Lemma 5.16 is valid without the equi-characteristic hypothesis.

Combining 5.8, 5.11, and 5.14 we obtain the following.

**Theorem 5.19.** *For a regular scheme $X$ of finite type over a field, there is a canonical isomorphism*

$$H^p(X, \underline{\underline{K}}_p) = A^p(X).$$

For $p = 0$ and 1 this amounts to the trivial formulas $H^0(X, \mathbb{Z}) = C^0(X)$ and $H^1(X, \underline{\underline{O}}_X^{\cdot}) = Pic(X)$. For $p = 2$ this formula has been established by Spencer Bloch in certain cases (see his paper in this procedings).

One noteworthy feature about the formula 5.19 is that the left side is manifestly contravariant in $X$, which suggests that higher $K$-theory will eventually provide the tool for a theory of the Chow ring for non-projective nonsingular varieties.

## §8. Projective fibre bundles

The main result of this section is the computation of the $K$-groups of the projective bundle associated to a vector bundle over a scheme. It generalizes the theorem about Grothendieck groups in [SGA 6 VI] and may be considered as a first step toward a higher $K$-theory for schemes (as opposed to the $K'$-theory developed in the preceding section). The method of proof differs from that of [SGA 6] in that it uses the existence of canonical resolutions for sheaves on projective space which are regular in the sense of [Mumford, Lecture 14]. We also discuss two variants of this result proved by the same method. The first concerns the 'projective line' over a (not necessarily commutative) ring; it is one of the ingredients for a higher $K$ generalization of the 'Fundamental Theorem' of Bass to be presented in a later paper. The second is a formula relating the $K$-groups of a Severi-Brauer scheme with those of the associated Azumaya algebra and its powers, which was inspired by a calculation of Roberts.

**1. The canonical resolution of a regular sheaf on PE.** Let $S$ be a scheme (not necessarily noetherian or separated), let $E$ be a vector bundle of rank $r$ over $S$, and let $X = PE = Proj(SE)$ be the associated projective bundle, where $SE$ is the symmetric algebra of $E$ over $\underline{\underline{O}}_S$. Let $\underline{\underline{O}}_X(1)$ be the canonical line bundle on $X$ and $f : X \longrightarrow S$ the structural map. We will use the term "$X$-module" to mean a quasi-coherent sheaf of $\underline{\underline{O}}_X$-modules, unless specified otherwise.

The following lemma summarizes some standard facts about the higher direct image functors $R^q f_*$ we will need.

**Lemma 1.1.** a) *For any $X$-module $F$, $R^q f_*(F)$ is an $S$-module which is zero for $q \geqslant r$.*

b) *For any $X$-module $F$ and vector bundle $E'$ on $S$, one has*

$$R^q f_*(F) \otimes_S E' = R^q f_*(F \otimes_S E').$$

c) *For any $S$-module $N$, one has*

$$R^q f_*(\underline{\underline{O}}_X(n) \otimes_S N) = \begin{cases} 0 & q \neq 0, r-1 \\ S_n E \otimes_S N & q = 0 \\ (S_{r-n}E)R^{\vee} \otimes_S \wedge^r E \otimes_S N & q = r-1 \end{cases}$$

*where "$^\vee$" denotes the dual vector bundle.*

d) *If $F$ is an $X$-module of finite type (e.g. a vector bundle), and if $S$ is affine, then $F$ is a quotient of $(\underline{\underline{O}}_X(-1)^{\otimes n})^k$ for some $n, k$.*

Parts a),c) result from the standard Cech calculations of the cohomology of projective space [EGA III 2]. Part b) is obvious since locally $E'$ is a direct sum of finitely many copies of $\underline{\underline{O}}_S$. For d), see [EGA II 2.7.10].

Following Mumford, we call an $X$-module $F$ *regular* if $R^q f_*(F(-q)) = 0$ for $q > 0$, where as usual, $F(n) = \underline{\underline{O}}_X(1)^{\otimes n} \otimes_X F$. For example, we have $\underline{\underline{O}}_X(n) \otimes_S N$ is regular for $n \geqslant 0$ by c).

**Lemma 1.2.** *Let $0 \longrightarrow F' \longrightarrow F \longrightarrow F'' \longrightarrow 0$ be an exact sequence of $X$-modules.*

a) *If $F'(n)$ and $F''(n)$ are regular, so is $F(n)$.*

b) *If $F(n)$ and $F'(n+1)$ are regular, so is $F''(n)$.*

c) *If $F(n+1)$ and $F''(n)$ are regular, and if $f_*(F(n)) \longrightarrow f_*(F''(n))$ is onto, then $F'(n+1)$ is regular.*

*Proof.* This follows immediately from the long exact sequence

$$\begin{aligned} R^q f_*(F'(n-q)) \longrightarrow R^q f_*(F(n-q)) \longrightarrow R^q f_*(F''(n-q)) \\ \longrightarrow R^{q+1} f_*(F'(n-q)) \longrightarrow R^{q+1} f_*(F(n-q)). \end{aligned}$$

The following two lemmas appear in [Mumford, Lecture 14] and in [SGA 6 XIII 1.3], but the proof given here is slightly different.

**Lemma 1.3.** *If $F$ is regular, then $F(n)$ is regular for all $n \geqslant 0$.*

*Proof.* From the canonical epimorphism $\underline{\underline{O}}_X \otimes_S E \to \underline{\underline{O}}_X(1)$ one has an epimorphism

$$\underline{\underline{O}}_X(-1) \otimes_S E \longrightarrow \underline{\underline{O}}_X \tag{1.4}$$

so we get an exact sequence of vector bundles on $X$

$$0 \longrightarrow \underline{\underline{O}}_X(-r) \otimes_S \wedge^r E \longrightarrow \ldots \longrightarrow \underline{\underline{O}}_X(-1) \otimes_S E \longrightarrow \underline{\underline{O}}_X \longrightarrow 0 \tag{1.5}$$

by taking the exterior algebra of $\underline{\underline{O}}_X(-1) \otimes_S E$ with differential the interior product by 1.4. Tensoring with $F$ we obtain an exact sequence

$$0 \longrightarrow F(-r) \otimes_S \wedge^r E \longrightarrow \ldots \longrightarrow F(-1) \otimes_S E \longrightarrow F \longrightarrow 0. \tag{1.6}$$

Assuming $F$ to be regular, then $(F(-p) \otimes_S \wedge^p E)(p)$ is seen to be regular using 1.1 b). Thus if 1.6 is split into short exact sequences

$$0 \longrightarrow Z_p \longrightarrow F(-p) \otimes_S \wedge^p E \longrightarrow Z_{p-1} \longrightarrow 0$$

we can use 1.2 b) to show by decreasing induction on $p$ that $Z_p(p+1)$ is regular. Thus $Z_0(1) = F(1)$ is regular, so the lemma follows by induction on $n$.

**Lemma 1.7.** *If $F$ is regular, then the canonical map $O_X \otimes_S f_*(F) \longrightarrow F$ is surjective.*

*Proof.* From the preceding proof one has an exact sequence

$$0 \longrightarrow Z_1 \longrightarrow F(-1) \otimes_S E \longrightarrow F \longrightarrow 0$$

where $Z_1(2)$ is regular. Thus $R^1 f_*(Z_1(n)) = 0$ for $n \geqslant 1$, so we find that the canonical map $f_*(F(n-1)) \otimes_S E \longrightarrow f_*(F(n))$ is surjective for $n \geqslant 1$. Hence the canonical map of SE-modules

$$SE \otimes_S f_*(F) \longrightarrow \coprod_{n \geqslant 0} f_*(F(n))$$

is surjective. The lemma follows by taking associated sheaves.

Suppose now that $F$ is an $X$-module which admits a resolution

$$0 \longrightarrow \underline{\underline{O}}_X(-r+1) \otimes_S T_{r-1} \longrightarrow \dots \longrightarrow \underline{\underline{O}}_X \otimes_S T_0 \longrightarrow F \longrightarrow 0$$

where the $T_i$ are modules on $S$. Breaking this sequence up into short exact sequences and applying 1.2 b), one sees as in the proof of 1.3 that $F$ has to be regular. Moreover, the above exact sequence can be viewed as a resolution of the zero module by acyclic objects for the $\delta-$functor $R^q f_*(?(n))$, where $n$ is any fixed integer $\geqslant 0$. Thus on applying $f_*$ we get an exact sequence

$$0 \longrightarrow S_{n-r+1}E \otimes_S T_{r-1} \longrightarrow \dots \longrightarrow S_n E \otimes_S T_0 \longrightarrow f_*(F(n)) \longrightarrow 0$$

for each $n \geqslant 0$. In particular, we have exact sequences

$$(1.8) \qquad 0 \longrightarrow T_n \longrightarrow E \otimes_S T_{n-1} \longrightarrow \dots \longrightarrow f_*(F(n)) \longrightarrow 0$$

for $n = 0, \cdots, r-1$ which can be used to show recursively that the modules $T_n$ are determined by $F$ up to canonical isomorphism.

Conversely, given an $X$-module $F$, we inductively define a sequence of $X$-modules $Z_n = Z_n(F)$ and a sequence of $S$-modules $T_n = T_n(F)$ as follows. Starting with $Z_{-1} = F$, let $T_n = f_*(Z_{n-1}(n))$, and let $Z_n$ be the kernel of the canonical map $\underline{\underline{O}}_X(-n) \otimes_S T_n \to Z_{n-1}$. It is clear that $Z_n$ and $T_n$ are additive functors of $F$.

Supposing now that $F$ is regular, we show by induction that $Z_n(n+1)$ is regular, this being clear for $n = -1$. We have an exact sequence

$$(1.9) \qquad 0 \longrightarrow Z_n(n) \longrightarrow \underline{\underline{O}}_X \otimes_S T_n \stackrel{c}{\longrightarrow} Z_{n-1}(n) \longrightarrow 0$$

where the canonical map $c$ is surjective by 1.7 and the induction hypothesis. By 1.3, 1.2 c) we find that $Z_n(n+1)$ is regular, so the induction works. In addition we have

$$(1.10) \qquad f_*(Z_n(n)) = 0 \quad \textit{for} \quad n \geqslant 0$$

because $c$ induces an isomorphism after applying $f_*$.

From 1.9 and the fact that $f_*$ is exact on the category of regular $X$-modules, one concludes by induction that $F \longmapsto T_n(F)$ is an exact functor from regular $X$-modules to $S$-modules.

We next show that $Z_{r-1} = 0$. From 1.9 we get exact sequences

$$R^{q-1}f_*(Z_{n+q-1}(n)) \xrightarrow{\ \delta\ } R^q f_*(Z_{n+q}(n)) \longrightarrow R^q f_*(\underline{\underline{O}}_X(-q) \otimes_S T_{n+q})$$

which allow one to prove by induction on $q$, starting from 1.10, that $R^q f_*(Z_{n+q}(n)) = 0$ for $q, n \geqslant 0$. This shows that $Z_{r-1}(r-1)$ is regular, since $R^q f_*$ is zero for $q \geqslant r$. By 1.10 and 1.7 we have $Z_{r-1}(r-1) = 0$, so $Z_{r-1} = 0$ as was to be shown.

Combining the exact sequences 1.9 we obtain a canonical resolution of the regular sheaf $F$ of length $r-1$. Thus we have proved the following.

**Proposition 1.11.** *Any regular $X$-module $F$ has a resolution of the form*

$$0 \longrightarrow \underline{\underline{O}}_X(-r+1) \otimes_S T_{r-1}(F) \longrightarrow \dots \longrightarrow \underline{\underline{O}}_X \otimes_S T_0(F) \longrightarrow F \longrightarrow 0$$

*where the $T_i(F)$ are $S$-modules determined up to unique isomorphism by $F$. Moreover $F \longmapsto T_i(F)$ is an exact functor from the category of regular $X$-modules to the category of $S$-modules.*

The next three lemmas are concerned with the situation when $F$ is a vector bundle on $X$.

**Lemma 1.12.** *Assume $S$ is quasi-compact. Then for any vector bundle $F$ on $X$, there exists an integer $n_0$ such that for all $S$-modules $N$ and $n \geqslant n_0$, one has*

a) $R^q f_*(F(n) \otimes_S N) = 0$ *for* $q > 0$

b) $f_*(F(n)) \otimes_S N \xrightarrow{\sim} f_*(F(n) \otimes_S N)$

c) $f_*(F(n))$ *is a vector bundle on $S$.*

*Proof.* Because $S$ is the union of finitely many open affines, it suffices to prove the lemma when $S$ is affine. In this case $F$ is the quotient of $L = \underline{\underline{O}}_X(-n)^k$ for some $n$ and $k$ by 1.1 d). Thus for any vector bundle $F$ on $S$, there is an exact sequence of vector bundles

$$0 \longrightarrow F' \longrightarrow L \longrightarrow F \longrightarrow 0$$

such that the lemma is true for $L$ by 1.1. Since

$$0 \longrightarrow F'(n) \otimes_S N \longrightarrow L(n) \otimes_S N \longrightarrow F(n) \otimes_S N \longrightarrow 0$$

is exact, we have an exact sequence

$$R^q f_*(L(n) \otimes_S N) \longrightarrow R^q f_*(F(n) \otimes_S N) \longrightarrow R^{q+1} f_*(F'(n) \otimes_S N)$$

so part a) can be proved by decreasing induction on $q$, as in the proof of Serre's theorem [EGA III 2.2.1]. Using a) we have a diagram with exact rows

$$\begin{array}{ccccccc}
 & f_*(F'(n)) \otimes_S N & \longrightarrow & f_*(L(n)) \otimes_S N & \longrightarrow & f_*(F(n)) \otimes_S N & \longrightarrow 0 \\
 & \downarrow u' & & \downarrow \wr & & \downarrow u & \\
0 \longrightarrow & f_*(F'(n) \otimes_S N) & \longrightarrow & f_*(L(n) \otimes_S N) & \longrightarrow & f_*(F(n) \otimes_S N) & \longrightarrow 0
\end{array}$$

for $n \geqslant$ some $n_0$ and all $N$. Hence $u$ is surjective; applying this to the vector bundle $F'$, we see that $u'$ is surjective, hence $u$ is bijective for $n \geqslant$ some $n_0$ and all $N$, whence b). By a), $f_*(F(n) \otimes_S N)$ is exact as a functor of $N$ for sufficiently large $n$, whence using b) we see $f_*(F(n))$ is a flat $\underline{\underline{O}}_S$-module. On the other hand, $f_*(F(n))$ is a quotient of $f_*(L(n))$ for $n \geqslant$ some $n_0$, so $f_*(F(n))$ is of finite type.Applying this to $F'$ we see that $f_*(F(n))$ is of finite presentation for all sufficiently large $n$. But a flat module of finite presentation is a vector bundle, whence c).

**Lemma 1.13.** *If $F$ is a vector bundle on $X$ such that $R^q f_*(F(n)) = 0$ for $q \geqslant 0, n \geqslant 0$, then $f_*(F(n))$ is a vector bundle on $S$ for all $n \geqslant 0$.*

*Proof.* The assertion being local on $S$, one can suppose $S$ affine, whence $f_*(F(n))$ is a vector bundle on $S$ for large $n$ by 1.12 c). Consider the exact sequence

$$0 \longrightarrow F(n) \longrightarrow F(n+1) \otimes_S E^\vee \longrightarrow \dots \longrightarrow F(n+r) \otimes_S \wedge^r E^\vee \longrightarrow 0$$

obtained by tensoring $F(n)$ with the dual of the sequence 1.5. For $n \geqslant 0$, this is a resolution of the zero module by acyclic modules for the $\delta$-functor $R^q f_*$, hence one knows that on applying $f_*$ one gets an exact sequence

$$0 \longrightarrow f_*(F(n)) \longrightarrow \dots \longrightarrow f_*(F(n+r)) \otimes_S \wedge^r E \longrightarrow 0.$$

Therefore one can show $f_*(F(n))$ is a vector bundle for all $n \geqslant 0$ by decreasing induction on $n$.

**Lemma 1.14.** *If $F$ is a regular vector bundle on $X$, then $T_i(F)$ is a vector bundle on $S$ for each $i$.*

This follows by induction on $i$, using the exact sequences 1.8 and the lemma 1.13.

**2. The projective bundle theorem.** Recall that the $K$-groups of a scheme are naturally modules over $K_0$ by §3 (1). The following result generalizes [SGA 6 VI 1.1].

**Theorem 2.1.** *Let $E$ be a vector bundle of rank $r$ over a scheme $S$ and $X = Proj(SE)$ the associated projective scheme. If $S$ is quasi-compact, then one has isomorphisms*

$$(K_q S)^r \xrightarrow{\sim} K_q X, (a_i)_{0 \leqslant i < r} \longmapsto \sum_{i=0}^{r-1} z^i \cdot f^* a_i$$

*where $z \in K_0 X$ is the class of the canonical line bundle $\underline{\underline{O}}_X(-1)$ and $f : X \longrightarrow S$ is the structural map.*

*Proof.* Let $\underline{\underline{P}}_n$ denote the full subcategory of $\underline{\underline{P}}(X)$ consisting of vector bundles $F$ such that $R^q f_*(F(k)) = 0$ for $q \neq 0$ and $k \geqslant n$. Let $\underline{\underline{R}}_n$ denote the full subcategory of $\underline{\underline{P}}(X)$ consisting of $F$ such that $F(n)$ is regular. Each of these subcategories is closed under extensions, so its $K$-groups are defined.

**Lemma 2.2.** *For all $n$, one has isomorphisms: $K_q(\underline{\underline{R}}_n) \simeq K_q(\underline{\underline{P}}_n) \simeq K_q(\underline{\underline{P}}(X))$ induced by the inclusions $\underline{\underline{R}}_n \subset \underline{\underline{P}}_n \subset \underline{\underline{P}}(X)$.*

To prove the lemma, we consider the exact sequence

$$0 \longrightarrow F \longrightarrow F(1) \otimes_S E \longrightarrow \dots \longrightarrow F(r) \otimes_S \wedge^r E \longrightarrow 0. \tag{2.3}$$

For each $p > 0, F \longmapsto F(p) \otimes_S \wedge^p E$ is an exact functor from $\underline{\underline{P}}_n$ to $\underline{\underline{P}}_{n-1}$, hence it induces a homomorphism $u_p : K_q(\underline{\underline{P}}_n) \longrightarrow K_q(\underline{\underline{P}}_{n-1})$. From Th. 2, Cor. 3 it is clear that $\sum_{p>0}(-1)^{p-1}u_p$ is an inverse to the map induced by the inclusion of $\underline{\underline{P}}_{n-1}$ in $\underline{\underline{P}}_n$. Thus we have $K_q(\underline{\underline{P}}_{n-1}) \xrightarrow{\sim} K_q(\underline{\underline{P}}_n)$ for all $n$. By 1.12 a), $\underline{\underline{P}}(X)$ is the union of the $\underline{\underline{P}}_n$, so by §2(9) we have $K_q(\underline{\underline{P}}_n) \simeq K_q(\underline{\underline{P}}(X))$ for all $n$. The proof that $K_q(\underline{\underline{R}}_n) \simeq K_q(\underline{\underline{P}}(X))$ is similar, whence the lemma.

Put $U_n(N) = \underline{\underline{O}}_X(-n) \otimes_S N$ for $N$ in $\underline{\underline{P}}(S)$. For $0 \leqslant n < r, U_n$ is an exact functor from $\underline{\underline{P}}(S)$ to $\underline{\underline{P}}_0$ by 1.1 c), hence it induces a homomorphism $u_n : K_q(\underline{\underline{P}}(S)) \longrightarrow K_q(\underline{\underline{P}}_0)$. In view of 2.2, it suffices for the proof of the theorem to show that the homomorphism

$$u : K_q(\underline{\underline{P}}(S))^r \longrightarrow K_q(\underline{\underline{P}}_0), \quad (a_n)_{0 \leqslant n < r} \mapsto \sum_{n=0}^{r-1} u_n(a_n)$$

is an isomorphism.

From 1.13 we know that $V_n(F) = f_*(F(n))$ is an exact functor from $\underline{\underline{P}}_0$ to $\underline{\underline{P}}(S)$ for $n \geqslant 0$, hence we have a homomorphism

$$v : K_q(\underline{\underline{P}}_0) \longrightarrow K_q(\underline{\underline{P}}(S)), \quad x \mapsto (v_n(x))_{0 \leqslant n < r},$$

where $v_n$ is induced by $V_n$. Since

$$V_n U_m(N) = f_*(\underline{\underline{O}}_X(n-m) \otimes_S N) = S_{n-m}(E) \otimes_S N$$

by 1.1 c), it follows that the composition $vu$ is described by a triangular matrix with ones on the diagonal, Therefore $vu$ is an isomorphism, so $u$ is injective.

On the other hand, $T_n$ is an exact functor from $\underline{\underline{R}}_0$ to $\underline{\underline{P}}(S)$ by 1.11 and 1.14, hence we have a homomorphism

$$t : K_q(\underline{\underline{R}}_0) \longrightarrow K_q(\underline{\underline{P}}(S))^r, \quad x \mapsto ((-1)^n t_n(x))_{0 \leqslant n < r}$$

where $t_n$ is induced by $T_n$. Applying Th. 2, Cor. 3 to the exact sequence 1.11, we see that the composition ut is the map $K_q(\underline{\underline{R}}_0) \longrightarrow K_q(\underline{\underline{P}}_0)$ induced by the inclusion of $\underline{\underline{R}}_0$ in $\underline{\underline{P}}_0$. By 2.2, $ut$ is an isomorphism, so $u$ is surjective, concluding the proof.

**3. The projective line over a ring.** Let $A$ be a (not necessarily commutative) ring let $t$ be an indeterminate, and let

$$A[t] \xrightarrow{i_1} A[t, t^{-1}] \xleftarrow{i_2} A[t^{-1}]$$

denote the canonical homomorphisms. When $A$ is commutative, a quasi-coherent sheaf on $P^1_A = Proj(A[X_0, X_1])$ may be identified with a triple $F = (M^+, M^-, \theta)$, where $M^+ \in Mod(A[t]), M^- \in Mod(A[t^{-1}])$ and $\theta : i_1^*(M^+) \xrightarrow{\sim} i_2^*(M^-)$ is an isomorphism of $A[t,t^{-1}]$-modules. Following [Bass XII §9], we define $Mod(P_A^1)$ for $A$ non-commutative to be the abelian category of such triples, and we define the category of vector bundles on $P_A^1$, denoted $\underline{\underline{P}}(P_A^1)$, to be the full subcategory consisting of triples with $M^+ \in \underline{\underline{P}}(A[t]), M^- \in \underline{\underline{P}}(A[t^{-1}])$.

**Theorem 3.1.** *Let $h_n : \underline{\underline{P}}(A) \longrightarrow \underline{\underline{P}}(P_A^1)$ be the exact functor sending $P$ to the triple consisting of $P[t] = A[t] \otimes_A P, P[t^{-1}]$, and multiplication by $t^{-n}$ on $P[t,t^{-1}]$. Then one has isomorphisms*

$$(K_qA)^2 \xrightarrow{\sim} K_q(\underline{\underline{P}}(P_A^1)), \quad (x,y) \longmapsto (h_0)_*(x) + (h_1)_*(y)$$

*and the relations*

$$(h_n)_* - 2(h_{n-1})_* + (h_{n-2})_* = 0 \tag{3.2}$$

*for all $n$.*

When $A$ is commutative, this follows from 2.1, once one notices that $h_n(P)$ is the module $\underline{\underline{O}}_X(n) \otimes_S P$. For the non-commutative case, one modifies the proof of 2.1 in a straightforward way. For example, if $F = (M^+, M^-, \theta)$, we put $F(n) = (M^+, M^-, t^{-n}\theta)$, and let $X_0, X_1 : F(n-1) \longrightarrow F(n)$ be the homomorphisms given by $X_0 = 1$ on $M^+$ and $t^{-1}$ on $M^-$, $X_1 = t$ on $M^+$ and 1 on $M^-$. Then we have an exact sequence

$$0 \longrightarrow F(n-2) \xrightarrow{(X_1,-X_0)} F(n-1)^2 \xrightarrow{X_0pr_1+X_1pr_2} F(n) \longrightarrow 0$$

corresponding to 1.6, which leads to the relations 3.2. Also using the fact that $R^qf_*$ can be computed by means to the standard open affine covering of $P^1$, we can define $R^qf_*(F)$ in the non-commutative case to be the homology of the complex concentrated in degrees 0, 1 given by the map $d : M^+ \times M^- \longrightarrow i_2^*(M^-), d(x,y) = \theta(1 \otimes x) - 1 \otimes y$. One therefore has available all of the tools used in the proof of 2.1 in the non-commutative case; the rest is straightforward checking which will be omitted.

**4. Severi-Brauer schemes and Azumaya algebras.** Let $S$ be a scheme and let $X$ be a Severi-Brauer scheme over $S$ of relative dimension $r-1$. By definition $X$ is an $S$-scheme locally isomorphic to the projective space $P_S^{r-1}$ for the etale topology on $S$. (see [Grothendieck]), and it is essentially the same thing as an Azumaya algebra of rank $r^2$ over $S$. We propose now to generalize 2.1 to this situation.

When there exists a line bundle $L$ on $X$ which restricts to $0(-1)$ on each geometric fibre, one has $X = PE$, where $E$ is the vector bundle $f_*L^\vee$ on $S, f : X \longrightarrow S$ being the structural map of $X$. In general such a line bundle $L$ exists only locally for the etale topology on $X$. However, we shall now show that there is a canonical vector bundle of rank $r$ on $X$ which restricts to $0(-1)^r$ on each geometric fibre.

Let the group scheme $GL_{r,S}$ act on $\underline{\underline{O}}_S^r$ in the standard way, and put $Y = P_S^{r-1} = Proj(S(\underline{\underline{O}}_S^r))$. The induced action on $Y$ factors through the projective group $PGL_{r,S} = GL_{r,S}/G_{m,S}$. Since $G_{m,S}$ acts trivially on the vector bundle $\underline{\underline{O}}_Y(-1) \otimes_S \underline{\underline{O}}_S^r$, the group $PGL_{r,S}$ operates on this vector bundle compatibly with its action on $Y$. As $X$ is locally isomorphic to $Y$ for the etale topology on $S$ and $PGL_{r,S}$ is the group of automorphisms of $Y$ over $S$, one knows that $X$ is the bundle over $S$ with fibre $Y$ associated to torsor $T$ under $PGL_{r,S}$ locally trivial for the etale topology. Thus by faithfully flat descent, the bundle $\underline{\underline{O}}_Y(-1) \otimes_S \underline{\underline{O}}_S^r$ on $Y$ gives rise to a vector bundle $J$ on $X$ of rank $r$.

It is clear that the construction of $J$ is compatible with base change, and that $J = \underline{\underline{O}}_X(-1) \otimes_S E$ if $X = PE$. In the general case there is a cartesian square

$$\begin{array}{ccc} X' & \xrightarrow{g'} & X \\ {\scriptstyle f'}\downarrow & & \downarrow{\scriptstyle f} \\ S' & \xrightarrow{g} & S \end{array}$$

where $g$ is faithfully flat (e.g. an etale surjective map over which $T$ becomes trivial) such that $X' = PE$ for some vector bundle $E$ of rank $r$ on $S'$, and further

$$g'^*(J) = \underline{\underline{O}}_{X'}(-1) \otimes_{S'} E.$$

Let $A$ be the sheaf of (non-commutative) $\underline{\underline{O}}_S$-algebras given by

$$A = f_*(\underline{\underline{End}}_X(J))^{op}$$

where '*op*' denotes the opposed ring structure. As $g$ is flat, we have $g^*f_* = f'_*g'^*$. Hence we have

$$g^*(A)^{op} = f'_*(\underline{\underline{End}}_{X'}(\underline{\underline{O}}_{X'}(-1) \otimes_{S'} E)) = f'_*(\underline{\underline{O}}_{X'} \otimes_{S'} \underline{\underline{End}}_{S'}(E)) = \underline{\underline{End}}_{S'}(E),$$

hence $A$ is an Azumaya algebra of rank $r^2$ over $S$. Moreover one has

$$f^*A = \underline{\underline{End}}_X(J)^{op}$$

as one verifies by pulling back to $X'$.

Let $J_n$ (resp. $A_n$) be the $n$-fold tensor product of $J$ on $X$ (resp. $A$ on $S$), so that $A_n$ is an Azumaya algebra of rank $(r^n)^2$ such that

$$A_n = f_*(\underline{\underline{End}}_X(J_n))^{op}, f^*(A_n) = \underline{\underline{End}}_X(J_n)^{op}.$$

Let $\underline{P}(A_n)$ denote the category of vector bundles on $S$ which are left modules for $A_n$. Since $J_n$ is a right $f^*(A_n)$–module, which locally on $X$ is a direct summand of $f^*(A_n)$, we have an exact functor

$$J_n \otimes_{A_n} ? : \underline{P}(A_n) \longrightarrow \underline{P}(X), M \mapsto J_n \otimes_{f^*(A_n)} f^*(M).$$

and hence an induced map of $K$-groups.

**Theorem 4.1.** *If $S$ is quasi-compact, one has isomorphisms*

$$\coprod_{n=0}^{r-1} K_i(A_n) \xrightarrow{\sim} K_i(X), \quad (x_n) \mapsto \sum_{n=0}^{r-1} (J_n \otimes_{A_n} ?)_*(x_n).$$

This is actually a generalization of 2.1 because if two Azumaya algebras $A, B$ represent the same element of the Brauer group of $S$, then the categories $\underline{\underline{P}}(A)$, $\underline{\underline{P}}(B)$ are equivalent, and hence have isomorphic $K$-groups. Thus $K_i(\underline{\underline{P}}(A_n)) = K_i(S)$ for all $n$ if $X$ is the projective bundle associated to some vector bundle.

The proof of 4.1 is a modification of the proof of 2.1. One defines an $X$-module $F$ to be regular if its inverse image on $X' = PE$ is regular. For a regular $F$ one constructs a sequence

$$(4.2) \qquad 0 \longrightarrow J_{r-1} \otimes_{A_{r-1}} T_{r-1}(F) \longrightarrow \cdots \longrightarrow \underline{\underline{O}}_X \otimes_S T_0(F) \longrightarrow F \longrightarrow 0$$

recursively by

$$T_n(F) = f_*(\underline{\underline{\mathrm{Hom}}}_X(J_n, Z_{n-1}(F))); Z_n(F) = \mathrm{Ker}\{J_n \otimes_{A_n} T_n(F) \to Z_{n-1}(F)\}$$

starting with $Z_{-1}(F) = F$. It is easy to see this sequence when lifted to $X'$ coincides the the canonical resolution 1.11 for the inverse image of $F$ on $X'$. Since $X'$ is faithfully flat over $X$, 4.2 is a resolution of $F$.

We note also that there is a canonical epimorphism $J \twoheadrightarrow \underline{\underline{O}}_X$ obtained by descending 1.4, and hence a canonical vector bundle exact sequence

$$0 \longrightarrow \wedge^{r} J \longrightarrow \cdots \longrightarrow J \longrightarrow \underline{\underline{O}}_X \longrightarrow 0$$

on $X$ corresponding to 1.5. Therefore it should be clear that all of the tools used in the proof of 2.1 are available in the situation under consideration; the rest of the proof of 4.1 will be left to the reader.

**Example.** Let $X$ be a complete non-singular curve of genus zero over the field $k = H^0(X, \underline{\underline{O}}_X)$, and suppose $X$ has no rational point. Then $X$ is a Severi-Brauer scheme over $k$ of relative dimension one, and $J$ is the unique indecomposable vector bundle of rank 2 over $X$ with degree -2. The above theorem says

$$K_i(X) = K_i(k) \oplus K_i(A)$$

where $A$ is the skew-field of endomorphisms of $J$. This formula in low dimensions has been proved by Leslie Roberts ([Roberts]).

# References

H. Bass: Algebraic $K$-theory, Benjamin 1968.

S. Bloch: $K_2$ and algebraic cycles, these procedings.

K. Brown and S. Gersten: Algebraic $K$-theory as generalized sheaf cohomology, these procedings.

C. Chevalley: Les classes d'equivalence rationelle I, Exp. 2, Séminaire Chevalley 1958, Anneaux de Chow et applications, Secrétariat mathématique, Paris.

A. Dold and R. Lashof: Principal quasifibrations and fibre homotopy equivalence of bundles, III. J. Math. 3 (1959) 285-305.

F. T. Farrell and W. C. Hsiang: A formula for $K_1R_\alpha[T]$, Applications of categorical algebra, Procedings of symposia in pure mathematics XVIII (1970), Amer. Math. Soc.

E. Friedlander: Fibrations in etale homotopy theory, Publ. Math. I.H.E.S. 42 (1972).

P. Gabriel: Des categories abeliennes, Bull. Math. Soc. France 90 (1962) 323-448.

P. Gabriel and M. Zisman: Calculus of fractions and homotopy theory, Ergebnisse der Mathematik und ihrer Grenzgebiete, Band 35, Springer 1967.

S. Gersten 1: The relation between the $K$-theory of Karoubi and Villamayor and the $K$-theory of Quillen (preprint).

—— 2: $K$-theory of regular schemes, Bull. Amer. Math. Soc. (Jan. 1973).

—— 3: On some exact sequences in the higher $K$-theory of rings, these procedings.

—— 4: Problems about higher $K$-functors, these procedings.

—— 5: Higher $K$-theory of rings, these procedings.

A. Grothendieck: Le groupe de Brauer I, Dix exposés sur la cohomologie des schémas, North-Holland Publ. Co. 1968.

A. Heller: Homological algebra in abelian categories, Ann. of Math. 68 (1958) 484-525.

J. Milnor 1: The realization of a semi-simplicial complex, Ann. of Math. 65 (1957) 357-362.

—— 2: On spaces having the homotopy type of a CW-complex, Trans. Amer. Math. Soc. 90 (1959) 272-280.

D. Mumford: Lectures on curves on an algebraic surface, Annals of Math. Studies 59 (1966).

D. Quillen 1: Higher $K$-theory for categories with exact sequences, to appear in the procedings of the June 1972 Oxford symposium "New developments in topology".

—— 2: On the cohomology and $K$-theory of the general linear groups over a finite field, Ann. of Math. 96 (1972) 552-586.

—— 3: On the endomorphism ring of a simple module over an enveloping algebra, Proc. Amer. Math. Soc. 21 (1969) 171-172.

L. Roberts: Real quadrics and $K_1$ of a curve of genus zero, Mathematical Preprint No. 1971-60, Queen's University at Kingston.

G. Segal 1: Classifying spaces and spectral sequences, Publ. Math. I.H.E.S. 34 (1968) 105-112.

—— 2: Categories and cohomology theories, preprint, Oxford 1972.

R. G. Swan: Algebraic $K$-theory, Lecture notes in Math. 76 (1968).

J. Tornehave: On BSG and the symmetric groups (to appear).

EGA: Élements de Géométrie Algébrique, by A. Grothendieck and J. Dieudonné

EGA II: Publ. Math. I.H.E.S. 8 (1961)
EGA III (first part): —— 11 (1961)
EGA IV (third part): —— 28 (1966).

SGA: Séminaire de Géométrie Algébrique du Bois Marie, by A. Grothendieck and others

SGA 1: Lecture Notes in Math. 224 (1971)
SGA 6: —— 225 (1971)
SGA 2: Cohomologie locale des faisceaux cohérents et Théorèmes de Lefschetz locaux et globaux, North-Holland Publ. Co. 1968.

Massachusetts Institute of Technology

*Cohomology of Groups and Algebraic K-theory*
ALM 12, pp. 479–488



# Finite Generation of the Groups $K_i$ of Rings of Algebraic Integers

Daniel Quillen*

## 1. Statement of results

The title refers to:

**THEOREM 1:** *If $A$ is the ring of algebraic integers in a number field $F$ (finite over $\mathbb{Q}$) then $K_iA$ is a finitely generated group for all $i \geqslant 0$.*

**Remarks:** (1) The proof uses the definition of the groups $K_iA$, as $K_i\underline{\underline{P}}(A)$, given in [Quillen 1, §2]. Here $\underline{\underline{P}}(A)$ denotes the "exact category" of finitely generated projective $A$-modules.

(2) If $B$ is the ring of $S$-integers relative to some finite set $S$ of finite primes of $F$ then we have the localisation sequence [Quillen, 1, §5, Cor. to Thm. 5],

$$\cdots \to K_iA \to K_iB \to \coprod_{\mathscr{Y} \in S} K_{i-1}(A/\mathscr{Y}) \to K_{i-1}A \to \cdots$$

From [Quillen, 2, Thm. 8] one knows the $K$-groups of the finite fields $A/\mathscr{Y}$, whence one concludes that $K_iA \to K_iB$ has finite kernel (which is zero for $i$ even, $> 0$) and finite cokernel for $i > 1$. In particular $K_iB$ is finitely generated, and of the same rank as $K_iA$ for $i \neq 1$.

(3) The ranks of the groups $K_iA$ have been computed in [Borel, Cor. of Thm. 2]. Conjectures about the arithmetic, significance of the torsion subgroups of the $K_iA$ can be found in [Lichtenbaum].

(4) The proof of Theorem 1 yields the analogous result for a maximal order in a semi-simple $F$-algebra (cf. the end of §3).

To explain how Theorem 1 is deduced we must introduce the *building* of a vector space $V$, say of dimension $n$, over a field (or a division ring ) $F$. It is the

*Text prepared by H. Bass from Quillen's lecture at the Battelle Conference September, 1972. Originally appeared in *Higher K-theories* (Proc. Conf., Battelle Memorial Inst., Seattle, Wash., 1972), pp. 179–198. Lecture Notes in Math., Vol. 341, Springer, Berlin, 1973. Reprinted with permission from Springer.

simplicial complex, here denoted $\boxed{V}$, associated to the set of proper subspaces $W$ of $V (0 < W < V)$, (partially) ordered by inclusion. Thus a $p-$ simplex of $\boxed{V}$ is a chain $0 < W_0 < \cdots < W_p < V$ of proper subspaces $W_i$ of $V$. If $n \leqslant 1$ then $\boxed{V} = \varnothing$; if $n = 2$ then $\boxed{V}$ is the projective space $\mathbb{P}(V)$ of lines in $V$, as a discrete space.

**THEOREM 2** (Solomon-Tits, cf. [Solomon]): *Suppose* $n \geqslant 2$. *Then* $V$ *has the homotopy type of a bouquet of* $(n-2)-$ *spheres.*

We shall give a proof of this below, since some of the details are needed for other arguments.

It follows that $\boxed{V}$ has zero reduced (integral) homology except in dimension $n-2$, where we obtain a free $\mathbb{Z}$-module $H_{n-2}(\boxed{V})$ on which $GL(V)$ naturally acts. This is called the *Steinberg module* of $V$, here denoted $st(V)$. For $n = 1$ we agree to put $st(V) = \mathbb{Z}$, with trivial action of $GL(V)$.

Now let $A$ be a Dedekind ring with field of fractions $F$. For each $n \geqslant 0$ let $Q_n$ denote the full subcategory of $Q = Q\underline{P}(A)$ whose objects are the projective $A$-modules $P$ of rank $\leqslant n$. Thus $Q_0$ is equivalent to the trivial category, $Q_n \subset Q_{n+1}$ and $Q = U_n Q_n$. The main result to be proved below is the following.

**THEOREM 3:** Let $n \geqslant 1$. *The inclusion* $w : Q_{n-1} \to Q_n$ *induces a long exact sequence*

$$\cdots \to H_i Q_{n-1} \to H_i Q_n \to \coprod_{\alpha} H_{i-n}(GL(P_\alpha), st(V_\alpha)) \to H_{i-1} Q_{n-1} \to \cdots,$$

*where the* $P_\alpha$*'s represent the isomorphism classes of projective* $A$*-modules of rank* $=$ $n$, *and where* $V_\alpha = P_\alpha \otimes_A F$.

**Remarks:** (1) For any (essentially) small category $C$ we put $H_* C = H_*(BC, \mathbb{Z})$, where $BC$ is the classifying space ($=$ geometric realisation of the nerve) of $C$ in the sense of [Quillen, 1, §1].

(2) The structure theory of projective modules over Dedekind rings implies that $P_\alpha \mapsto \det P_\alpha = \bigwedge^n P_\alpha$ defines a bijection from the set of $P_\alpha$'s to $Pic(A)$.

(3) The proof we give of Theorem 3 applies, more generally, when $A$ is a maximal order, over a Dedekind domain, in a division algebra $F$. The corollaries of the theorem drawn below likewise apply in that generality.

**COROLLARY** ("Stability"). *The homomorphisms* $H_i Q_n \to H_i Q_{n+1}$ *are surjective for* $n \geqslant i$ *and injective for* $n \geqslant i+1$.

**Proof of Theorem 1 from Theorem 3.**

Suppose now that $F$ is a finite dimensional division algebra over $\mathbb{Q}$ and that $A$ is a maximal order in $F$.

Let $P \in \underline{P}(A)$ and put $V = P \otimes_A F = P \otimes_Z \mathbb{Q}$. Then $\Gamma = GL(P)$ is an arithmetic subgroup of $G(\mathbb{Q})$, where $G$ is the reductive algebraic group over $\mathbb{Q}$ whose rational points in a $\mathbb{Q}$-algebra $R$ form the group $G(R) = GL(V \otimes_{\mathbb{Q}} \mathbb{R})$ of $F \otimes_{\mathbb{Q}} \mathbb{R}$- automorphisms of $V \otimes_{\mathbb{Q}} \mathbb{R}$. Let $S$ denote the connected component of the kernel of the norm homomorphism $G \to \mathbb{G}_m$ (the norm being that of the $\mathbb{Q}$-algebra $End_F(V)$). Then $S$ is a connected reductive algebraic group, with no non-trivial characters, defined over $\mathbb{Q}$, and so it is subject to the result of [Borel-Serre]. Since $\Gamma \cap S(\mathbb{Q})$ has finite index in $\Gamma$ (the elements of $\Gamma$ having norm $\pm 1$) we can find a normal torsion free subgroup $\Gamma'$ of $\Gamma$ of finite index in $\Gamma \cap S(\mathbb{Q})$. According to Théorème 3 of [Borel-Serre] we have for any $\Gamma'$-module $M$ and any $i$, a duality isomorphism

$$H_i(\Gamma', I \otimes M) \cong H^{d-l-i}(\Gamma', M).$$

Here $l$ denotes the $\mathbb{Q}$-rank of $S$, $d$ the dimension of $S(\mathbb{R})$ modulo a maximal compact subgroup, and $I$ is the Steinberg module of the Tits building $T$ whose simplices correspond to the parabolic subgroups of $S$ defined over $\mathbb{Q}$. There is a natural isomorphism, in the present case, $\boxed{V} \to T$, such that the simplex $W_0 < \cdots < W_p$ of $\boxed{V}$ corresponds to its stabilizer in $S$, which is a parabolic subgroup defined over $\mathbb{Q}$. This isomorphism permits us to identify $I$ with $st(V)$, and so deduce isomorphisms

$$H_i(\Gamma', st(V)) \cong H^{d-l-i}(\Gamma', \mathbb{Z}). \tag{1}$$

Now according to [Raghunathan, Cor.3] the groups $H^j(\Gamma', M)$ are finitely generated for all $j$ whenever $M$ is finitely generated over $\mathbb{Z}$. Actually [Raghunathan] does not apply directly here because $S$ is not semi-simple. However there is an exact sequence $1 \to \Gamma_S \to \Gamma' \to \Gamma_t \to 1$ where $\Gamma_S$ is arithmetic in a semi-simple group and where $\Gamma_t$ is finitely generated abelian. Then the groups $H^q(\Gamma_S, M)$ are finitely generated by [Raghunathan] so the spectral sequence $H^p(\Gamma_t, H^q(\Gamma_S, M)) \Rightarrow H^{p+q}(\Gamma', M)$ gives finite generation of the latter. Taking $M = \mathbb{Z}$ we obtain from (1) the finite generation of the groups $H_i(\Gamma', st(V))$. Then the homology spectral sequence $H_p(\Gamma/\Gamma', H_q(\Gamma', st(V))) \Rightarrow H_{p+q}(\Gamma, st(V))$ yields, since $\Gamma/\Gamma'$ is finite, the finite generation of $H_i(\Gamma, st(V))$, which we now record:

(2) *If $P \subset \underline{\underline{P}}(A)$ and $V = P \otimes_A F$ then $H_i(GL(P), st(V))$ is a finitely generated group for all $i$.*

The Jordan-Zassenhaus theorem (see, for example, [Bass, Ch. X, Thm. (2.4)]) implies that the set $\{p_\alpha\}$, representing isomorphism classes of projective $A$-modules of rank $n$, is finite. Hence by (2), the groups

$$L_i = \coprod_\alpha H_i(GL(P_\alpha), st(V_\alpha)),$$

where $V_\alpha = P_\alpha \otimes_A F$, are finitely generated. For $n \geqslant 1$, Theorem 3 furnishes a long exact sequence

$$\cdots \to L_{i+1-n} \to H_iQ_{n-1} \to H_iQ_n \to L_{i-n} \to \cdots$$

Since $H_0Q_0 = \mathbb{Z}$ and $H_iQ_0 = 0$ for $i > 0$ we conclude by induction on $n$ that

(3) *$H_iQ_n$ is finitely generated for all* $i$ and $n$.

Fixing $i$ and letting $n \to \infty$ we obtain the finite generation of $H_iQ\underline{\underline{P}}(A)$.

Now $\otimes : \underline{\underline{P}}(A) \times \underline{\underline{P}}(A) \to \underline{\underline{P}}(A)$ gives to $BQ\underline{\underline{P}}(A)$ the structure of a homotopy associative and commutative $H$-space. Hence the finite generation of its homology implies that of its homotopy, whence $K_i A = \pi_{i+1}(BQ\underline{\underline{P}}(A), 0)$ is finitely generated, thus proving Theorem 1.

## 2. The Solomon-Tits theorem

We fix here a division ring $F$ and a (right) vector space $V$ of dimension $n < \infty$ over $F$. Let $\textcircled{V}$ denote the simplicial complex associated to the set of *all* subspaces of $V$, ordered by inclusion. It is contractible since, for example, $V$ has a least subspace 0 [Quillen, 1, §1, Cor. 2 of Prop. 2]. Its p-simplices are chains

$$W_0 < \cdots < W_p \tag{1}$$

of subspaces. We distinguish the following subcomplexes of $\textcircled{V}$ by the indicated restrictions on their simplices (1):

$$\boxed{V} : W_0 \neq 0 \text{ and } W_p \neq V$$

$$\overset{\triangle}{\boxed{V}} : W_0 \neq 0$$

$$\underset{\triangledown}{\boxed{V}} : W_p \neq V$$

$$\underset{\triangledown}{\overset{\triangle}{\boxed{V}}} : dim(W_p/W_0) < n$$

It is readily seen that, for $n \geqslant 2$,

$$\overset{\triangle}{\boxed{V}} \simeq Cone\ \boxed{V} \simeq \underset{\triangledown}{\boxed{V}}$$

and

$$\underset{\triangledown}{\overset{\triangle}{\boxed{V}}} \simeq Susp\ \boxed{V},$$

where "$\simeq$" denotes homotopy equivalence.

**Proof of Theorem 2:**

We argue by induction on $n \geqslant 2$. For $n = 2$ the discrete space $\boxed{V}$ is trivially a bouquet of 0-spheres, so assume $n \geqslant 3$. Fix a line $L$ in $V$ and let $\underline{\underline{H}}$ denote the set of hyperplanes $H$ of $V$ complementary to $L : V = H \oplus L$. Let $Y$ denote the full subcomplex of $\boxed{V}$ obtained by deleting the set of vertices $\underline{\underline{H}}$.

**Claim.** $Y$ is contractible.

In fact, let $q : V \to V/L$ be the canonical projection. It induces a simplicial map $q : Y \to \underset{\triangledown}{\boxed{V/L}}$, and the latter cone is contractible. Hence it suffices, by [Quillen, 1, §1, example following Theorem A], to show that $q^{-1}(\sigma)$ is contractible for each closed simplex $\sigma = (W_0/L < \cdots < W_p/L)$ of $\underset{\triangledown}{\boxed{V/L}}$. If $U$ is a vertex of $q^{-1}(\sigma)$ then, for some $i$, we have $qU = W_i/L$, i.e. $U + L = W_i$. Thus $U \mapsto U + L$ defines a simplicial map from $q^{-1}(\sigma)$ to the simplex with vertices $W_0 < \cdots < W_p$, and the relations $U \leqslant U + L$ show that this map is homotopic to the identity of $q^{-1}(\sigma)$ [Quillen, 1, §1, Prop. 2]. This proves the claim.

Now we have the following schematic picture of $\boxed{V}$:

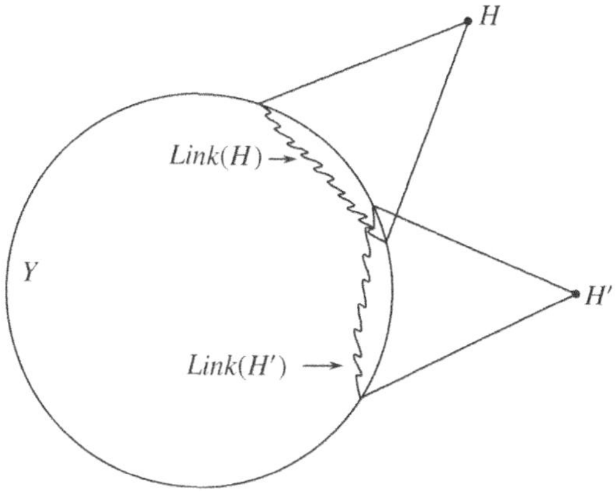


Let $H \in \underline{\underline{H}}$. Let $Link(H)$ denote the subcomplex of $\boxed{V}$ formed by simplices $\sigma$ such that $H \notin \sigma$ but $\sigma \cup \{H\}$ is a simplex. Evidently $Link(H) \subset Y$. Further $\boxed{V}$ is the union of $Y$ with the cones over these links, amalgamated along the links, as H varies over $\underline{\underline{H}}$. From the claim above we thus obtain

$$\begin{aligned} \boxed{V} &\simeq \boxed{V}/Y \\ &\simeq \bigvee_{H \in \underline{\underline{H}}} Susp(Link(H)). \end{aligned}$$

But clearly $Link(H) = \boxed{H}$ for $H \in \underline{\underline{H}}$. Thus the theorem follows by induction, since $dim\ H = n - 1$.

Let $J(V)$ denote the set of proper layers $(W_0, W_1)$ of $V$ $(0 \leqslant W_0 \leqslant W_1 \leqslant V$ and $dim(W_1/W_0) < n)$, ordered by $(W_0, W_1) \leqslant (W_0', W_1')$ if $W_0' \leqslant W_0$ and $W_1 \leqslant W_1'$. For $n = 1$ it is the "unrelated" set consisting of $(0, 0)$ and $(V, V)$, whence $BJ(V) = S^0$.

**PROPOSITION:** *Suppose $n \geqslant 2$. There is a $GL(V)$-equivariant homotopy equivalence*

$$\widehat{\boxed{V}} \to BJ(V)$$

Define a map $g$ from the set Simpl $\widehat{\boxed{V}}$ of simplices of $\widehat{\boxed{V}}$, ordered by inclusion to $J(V)$, by

$$g(W_0 < \cdots < W_p) = (W_0, W_p)$$

Clearly $g$ is order preserving, i.e. a functor (where we view ordered sets as categories) and it is also $GL(V)$-equivariant. Since $B\ Simpl(K)$ is canonically homeomorphic to the barycentric subdivision of $K$, for any simplicial complex $K$

(cf. [Quillen, 1, §1]) the proposition will follow if we show that $Bq$ is a homotopy equivalence. For this it suffices, by [Quillen, 1, §1, Theorem A], to show, for each $(U_0, U_1) \in J(V)$, that the category $g/(U_0, U_1)$ is contractible (i.e. that its classifying space is so). The objects of $g/(U_0, U_1)$ are simplices $W_0 < \cdots < W_p$ such that $U_0 \leqslant W_0$ and $W_p \leqslant U_1$; they are ordered by inclusion. Evidently $g/(U_0, U_1)$ is isomorphic to Simpl $U_1/U_0$ so indeed it is contractible.

**Remark:** The above proposition, and its proof, are purely combinatorial, in the following sense. Let $S$ be a partially ordered set with a least element, 0, and a greatest element, $V$. Set $S' = S - \{0, V\}$, and assume $S' = \varnothing$. Then $BS \cong Susp(BS')$. Let

$$J = \{(W_0, W_1) \in S \times S | W_0 \leqslant W_1, \text{ and } 0 < W_0 \text{ or } W_1 < V\}$$

Define $g : Simpl\ (BS) \to J$ by $g(W_0 < \cdots < W_p) = (W_0, W_p)$. Then $g/(U_0, U_1) = Simpl\ (B[U_0, U_1])$, where $[U_0, U_1] = \{W \in S | U_0 \leqslant W \leqslant U_1\}$. Moreover $B[U_0, U_1]$ is contractible since $[U_0, U_1]$ has a least (and greatest) element. Hence $Bg$ furnishes an $Aut\ (S)$-equivariant homotopy equivalence $BS \to BJ$.

**COROLLARY:** *Suppose $n \geqslant 1$. The reduced homology $\widetilde{H}_i(J(V))(= \widetilde{H}_i(BJ(V), \mathbb{Z}))$ vanishes for $i \neq n-1$. The $\mathbb{Z}$-module $\widetilde{H}_{n-1}(J(V))$ is free.*

**DEFINITION:** We call $\widetilde{H}_{n-1}(J(V))$, with the natural action of $GL(V)$ on it, the *Steinberg module* of $V$, and denote it $st(V)$.

In view of the proposition this definition accords with that given in §1 above.

## 3. Proof of the main theorem (theorem 3)

We begin by recalling some basic facts about the homology, $H_i C = H_i(BC, \mathbb{Z})$, of a small category $C$. A reference for this is [Gabriel-Zisman, Appendix II, §3] (cf. also [Quillen, 1, §1]).

Consider the abelian category $C - Ab$ of abelian group valued functors on $C$. The functor $\varinjlim : C - Ab \to Ab$ is right exact and has left derived functors $\varinjlim_p^C : C - Ab \to Ab$. For the constant functor $\mathbb{Z}$ we have

$$(1) \qquad H_p C = \varinjlim_p^C \mathbb{Z}$$

[Gabriel-Zisman, App. II, 3.3].

For example suppose $C$ is a group $G$, viewed as a category with one object. A functor $M : G \to Ab$ is just a $G$-module, and we have $\varinjlim^G M = H_0(G, M) = M/(\sum_{g \in G}(g-1)M)$. Similarly

$$(2) \qquad \varinjlim_p^G M = H_p(G, M),$$

the Eilenberg-MacLane homology of $G$.

Suppose $w : C' \to C$ is a functor between small categories. If $P \in C$ we have the functor $i_P : w/P \to C'$ sending $(P', u)$ to $P'$. Let $f : C' \to Ab$ be a functor. Then [Gabriel-Zisman, App. II, Thm. 3.6 and Remark 3.8] there is a spectral sequence

$$E^2_{p,q} = \varinjlim{}^C_p (P \mapsto \varinjlim{}^{(w/P)}_q f \circ i_P) \Rightarrow \varinjlim{}^{C'}_{p+q} f.$$

For the constant functor $f = \mathbb{Z}$ this takes the form (using (1)),

$$E^2_{p,q} = \varinjlim{}^C_p (P \to H_q(w/P)) \Rightarrow H_{p+q}(C'). \tag{3}$$

Now let $A$ be a Dedekind ring with field of fractions $F$, as in Theorem 3. Our arguments do not require these data to be commutative, so we may, more generally, allow $A$ to be a maximal order (over a Dedekind ring) in a division algebra $F$. The only feature we require is that if $P \in \underline{\underline{P}}(A)$ and $V = P \otimes_A F$ then $P' \mapsto P' \otimes_A F$ defines a bijection from the set of direct summands of $P$ to the set of all sub $F$-modules of $V$. We put $rk\ P = dim\ V$ and define $Q_n$ to be the full subcategory of $Q\underline{\underline{P}}(A)$ whose objects are the $P's$ of rank $\leqslant n$. Applying (3) to the inclusion functor $w : Q_{n-1} \to Q_n (n \geqslant 1)$ we obtain a spectral sequence

$$E^2_{p,q} = \varinjlim{}^{Q_n}_p (P \to H_q(w/P)) \Rightarrow H_{p+q}(Q_{n-1}). \tag{4}$$

Its analysis requires first the determination of the groups $H_q(w/P)$, for $P \in Q_n$.

Recall that an object of $w/P$ is a morphism $u : P' \mapsto P$ with $P' \in Q_{n-1}$. Up to isomorphism (over $P$) such an object is determined by an admissible layer $(P_0, P_1)$ of $P$ such that $u$ corresponds to an isomorphism $P' \to P_1/P_0$. Thus we see that $w/P$ is equivalent to the set $J$ of admissible layers $(P_0, P_1)$ of $P$ such that $rk(P_1, P_0) < n$, with the ordering $(P_0, P_1) \leqslant (P_0', P_1')$ if $P_0' \leqslant P_0$ and $P_1 \leqslant P_1'$.

If $rk\ P < n$, then $J$ has the maximal element $(0, P)$, so $w/P$ is contractible [Quillen, 1, §1, Cor. 2 to Prop. 2].

Suppose $rk\ P = n$. Then the map $P' \mapsto P' \otimes_A F \subset V = P \otimes_A F$ induces an isomorphism from $J$ to $J(V)$ (notation as in §2). Thus, in view of the proposition and corollary of §2, we can now describe the groups $H_q(w/P)$.

For $n = 1$ we have

$$\begin{aligned} &H_q(w/P) = 0 \quad \text{if } q > 0 \\ &H_0(w/P) = \begin{cases} \mathbb{Z}, & \text{if } P = 0 \\ \mathbb{Z} \oplus \mathbb{Z}, & \text{if } rk\ P = 1. \end{cases} \end{aligned} \tag{5}$$

For $n \geqslant 2$ we have

$$\begin{aligned} &H_0(w/P) = \mathbb{Z} \\ &H_q(w/P) = 0 \qquad \text{if } q \neq 0,\ n-1 \\ &H_{n-1}(w/P) = \begin{cases} 0 & \text{if } rk\ P < n \\ st(V) & \text{if } rk\ P = n \end{cases} \end{aligned} \tag{6}$$

Consider first the case $n \geqslant 2$. Then

$$\varinjlim{}^{Q_n}_p (P \to H_q(w/P)) = \begin{cases} H_p(Q_n) & \text{if } q = 0 \\ 0 & \text{if } q \neq 0, n-1 \end{cases}$$

For the case $q = n - 1$ we introduce the full subcategory $Q'$ of $Q_n$ whose objects are the $P$'s of rank $= n$. Then $h(P) = H_{n-1}(w/P)$ is, by (6) a functor in $Q_n$ which vanishes on $Q_{n-1}$, and hence on the source of every arrow whose target lies outside $Q'$. It follows that the complex in [Gabriel-Zisman, App. II, 3.2] use to compute $\varinjlim{}^{Q_n}_* h$ is isomorphic to that used to compute $\varinjlim{}^{Q'}_* (h|Q')$. Now $Q'$ is equivalent to the groupoid of projective modules of rank $n$ and their isomorphisms. It is equivalent to the full skeletal subcategory with one object $P_\alpha$ from each isomorphism class, and the latter category is just the groupoid $Q'' = \coprod_\alpha GL(P_\alpha)$. On $Q''$ the functor $P_\alpha \mapsto H_{n-1}(w/P_\alpha)$ corresponds to the family of $GL(P_\alpha)$ - modules $st(V_\alpha)$, where $V_\alpha = P_\alpha \otimes_A F$, by (6). In view of (2) therefore we have

$$\begin{aligned} &\varinjlim_p^{Q_n}(P \mapsto H_{n-1}(w/P)) \\ &= \coprod_\alpha H_p(GL(P_\alpha), st(V_\alpha)) \end{aligned}$$

Denoting this group by $L_p$ we can now display

$$E^2_{p,q} = \varinjlim_P^{Q_n}(P \mapsto H_q(w/P)).$$

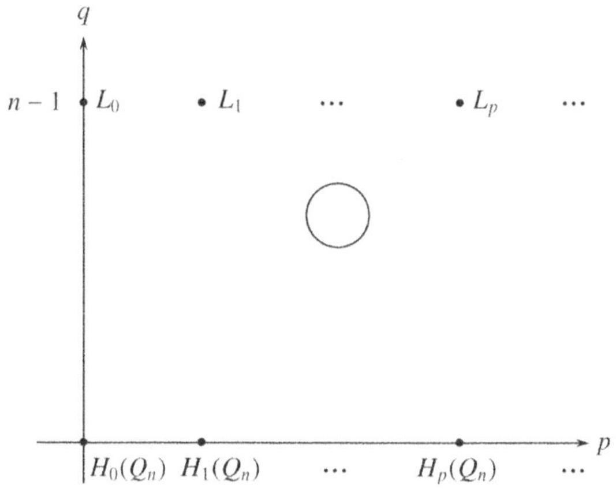


This picture remains in tact until the $E^n$-term, whose differentials furnish the horizontal sequences in the exact diagram

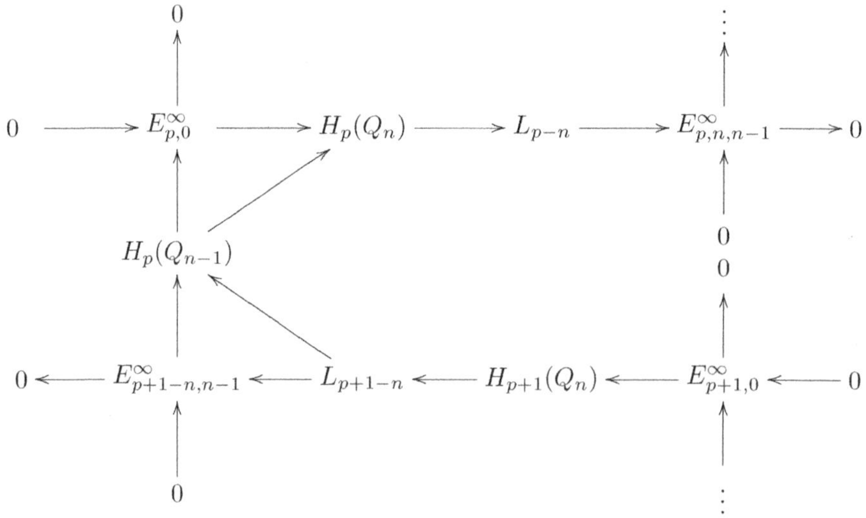

from which one extracts the exact sequence of Theorem 3.

Finally, consider the case $n = 1$. Then in view of (5) the spectral sequence (4) degenerates to an isomorphism

$$\varinjlim_p^{Q_1}(P \mapsto H_0(w/P) \cong H_p(Q_0) = \begin{cases} \mathbb{Z} & p = 0 \\ 0 & p > 0 \end{cases}$$

If $h$ denotes the functor $P \mapsto H_0(w/P)$ we have a functorial exact sequence

$$0 \to \widetilde{H_0}(w/P) \xrightarrow{d} H_0(w/P) \to \mathbb{Z} \to 0 \tag{7}$$

where $d$ is the antidiagonal map $\mathbb{Z} \to \mathbb{Z} \oplus \mathbb{Z}$ for $P$ of rank 1. Just as in the case $n \geqslant 2$ above we find that

$$\varinjlim_p^{Q_1}(P \mapsto \widetilde{H_0}(w/P)) = \coprod_\alpha H_p(GL(P_\alpha), st(V_\alpha))$$

a module we denote by $L_p$. (Recall that $st(V_\alpha) = \mathbb{Z}$ here since $n = 1$.) The homology exact sequence of $\varinjlim_*^{Q_1}$ for (7) thus takes the form

$$\cdots \to H_p(Q_0) \to H_p(Q_1) \to L_{p-1} \to H_{p-1}(Q_0) \to \cdots,$$

thus proving Theorem 3 for $n = 1$, and so completing its proof.

**Orders in semi-simple algebras.**

To apply this theorem to a maximal order $A$ in a semi-simple algebra we first note that $A$ is a finite product of maximal orders $A_j$ in simple algebras, and $K_iA$ is the direct sum of the $K_iA_j$'s. Further each $A_j$ is of the form $End_{B_j}(P_j)$ where $B_j$ is a maximal order in a division algebra and $P_j$ is a faithful projective left $B_j$-module. Then $P \mapsto P \otimes_{B_j} P_j$ is an equivalence of categories $\underline{\underline{P}}(B_j) \to \underline{\underline{P}}(A_j)$, whence isomorphisms $K_iB_j \to K_iA_j$. Since Theorem 3 above applies to the $B_j$'s we can use it to obtain information about $K_iA$, for example that $K_iA$ is finitely generated when $A$ is a maximal order in a semi-simple $\mathbb{Q}$-algebra.

In case $A$ is a not necessarily maximal order we can embed $A$ in a maximal order $B$. Then $A\left[\frac{1}{s}\right] = B\left[\frac{1}{s}\right]$ for some central non divisor of zero $s$ (in say the conductor, i.e. the annihilator of $B/A$). Then the localisation sequence takes the form $\cdots \to K_i(\mathfrak{m}) \to K_iA \to K_iB\left[\frac{1}{s}\right] \to \cdots$ where $\mathfrak{m}$ is the category of $s$-torsion finitely generated $A$-modules of finite homological dimension. Since $B\left[\frac{1}{s}\right]$ is a maximal order the groups $K_iB\left[\frac{1}{s}\right]$ can be treated by the methods above. Additional techniques seem to be required for the analysis of $K_i\mathfrak{m}$.

# References

[Bass] H. Bass, Algebraic $K$-theory, W.A. Benjamin, New York (1968).

[Borel] A. Borel, Cohomology réele stable des groupes $S$-arithmetiques classiques, C.R. Acad. Sci. Paris t. 274(24, 1972)A1700-A1702.

[Borel-Serre] A. Borel and J.-P. Serre, Adjonction de coins aux espaces symmétriques; applications à la cohomologie des groupes arithmétiques, C.R. Acad. Sci. Paris t. 271 (de. 1970) A1156-A1158.

[Gabriel-Zisman] P. Gabriel and M. Zisman, Calculus of fractions and homotopy theory, Springer-Verlag, Ergebnisse der Math. Band 35 (1967).

[Lichtenbaum] S. Lichtenbaum, Values of zeta-functions, étale cohomology, and algebraic $K$-theory, These Proceedings.

[Quillen 1] D. Quillen, Higher Algebraic $K$-theory I, These Proceedings.

[Quillen 2] ——, On the cohomology and $K$-theory of the general linear groups over a finite field, Ann. Math, Vol. 96, No. 3(1972) 552-586.

[Raghunathan] M. S. Raghunathan, A note on quotients of real algebraic groups by arithmetic subgroups, Invent. Math. 4 (1968), 318-335.

[Solomon] L. Solomon, The Steinberg character of a finite group with BN-pair, Theory of Finite Groups (Symposium edited by R. Brauer and C.-H. Sah), W.A. Benjamin, New York (1969), 213-221.

*Cohomology of Groups and*
*Algebraic K-theory*
ALM 12, pp. 489–501



# A User's Guide to Continuously Controlled Algebra

David Rosenthal*

**Abstract**

Continuously controlled algebra is an important tool for proving the Farrell-Jones conjecture and the Novikov conjecture. The purpose of this expository article is to present an accessible introduction to continuously controlled algebra and the continuously controlled version of assembly maps.

**2000 Mathematics Subject Classification:** 18F25

**Key words and Phrases:** $K$-theory, assembly map, controlled algebra

## 1. Introduction

Controlled algebra was first considered by Connell and Hollingsworth in [12], and its first major applications were developed by Quinn in [24]. Continuously controlled algebra was introduced by Anderson, Connolly, Ferry and Pedersen [1] and was used by Carlsson and Pedersen [10, 11] to analyze assembly maps in algebraic $K$- and $L$-theory. Carlsson and Pedersen used this theory to prove the Novikov conjecture for a large class of groups by showing that the assembly map was a split injection. Motivated by the groundbreaking work of Farrell and Jones [14], Bartels, Lück and Reich [5] have recently used the continuously controlled version of the assembly map in their proof of the $K$-theoretic Farrell-Jones conjecture for word hyperbolic groups, and Bartels and Lück have announced the analogous result in $L$-theory. These two isomorphism results combine to prove the Borel conjecture for these groups.

The aim of this article is to provide a friendly introduction to continuously controlled algebra, hopefully making the general framework of this proof technique a little more accessible. To achieve this modest goal, we focus on the continuously controlled algebra approach to equivariant homology theories and assembly maps.

*Department of Mathematics and Computer Science, St. John's University, 8000 Utopia Pkwy, Jamaica, NY 11439, USA. E-mail: rosenthd@stjohns.edu

The objects of study in this theory are additive categories of so-called "geometric modules." The notation used in the literature varies from paper to paper and is often quite involved. We hope the presentation here will serve to demystify these categories and to provide the reader with an understanding of the underlying concepts. Continuously controlled categories are introduced without group actions and are illustrated with several examples. Group actions are then incorporated into the definition, so that an equivariant homology theory and corresponding assembly map can be defined. The presentation focuses on algebraic $K$-theory. To study algebraic $L$-theory one has to be careful with involutions, but the treatment is virtually the same. We conclude with a general discussion of how this theory has been used to prove isomorphism and split injectivity results.

The author would like to thank Marco Varisco for useful comments and discussions. The author would also like to thank Erik K. Pedersen for his expertise and for his many invaluable insights.

## 2. Continuously controlled algebra

Let us begin by recalling some categorical terminology. Two functors $F$ and $G$ between categories $\mathcal{A}$ and $\mathcal{B}$ are *naturally equivalent* if there is a natural transformation from $F$ to $G$ that is an isomorphism for every object in $\mathcal{A}$. Two categories $\mathcal{A}$ and $\mathcal{B}$ are *equivalent* if there are functors $F : \mathcal{A} \to \mathcal{B}$ and $G : \mathcal{B} \to \mathcal{A}$ such that $FG$ is naturally equivalent to $\mathrm{id}_{\mathcal{B}}$ and $GF$ is naturally equivalent to $\mathrm{id}_{\mathcal{A}}$.

An *additive category* $\mathcal{A}$ is a small category in which every hom-set (i.e., the set of morphisms between two objects) is an abelian group, morphism composition is bilinear, there is a zero object, and for every finite collection of objects $A_1, \ldots, A_n$ in $\mathcal{A}$, the biproduct $A_1 \oplus \cdots \oplus A_n$ is an object of $\mathcal{A}$ (a biproduct is both a product and a coproduct). An *additive functor* between additive categories $\mathcal{A}$ and $\mathcal{B}$ is a functor that is a group homomorphism for every hom-set in $\mathcal{A}$. Two additive categories are equivalent when they are equivalent by additive functors.

Let $\mathcal{A}$ be a small additive category, and let $\mathbb{K}^{-\infty}(\mathcal{A})$ denote the non-connective $K$-theory spectrum associated with the symmetric monoidal category obtained from $\mathcal{A}$ by restricting to isomorphisms. That is, $\pi_n(\mathbb{K}^{-\infty}(\mathcal{A})) \cong K_n(\mathcal{A})$ for every integer $n$. In particular, if $\mathcal{F}_R$ is the category of finitely generated free $R$-modules, then $\pi_n(\mathbb{K}^{-\infty}(\mathcal{F}_R)) \cong K_n(R)$ for every integer $n$; $\mathbb{K}^{-\infty}$ is a functor from the category of small additive categories to the category of spectra. For a more detailed description of this functor, see [22, 9].

Let $X$ be a CW complex, $Y \subseteq X$ be a closed subcomplex, and $\mathcal{A}$ be a small additive category. A subset $K \subseteq X$ is called *relatively compact* if the closure of $K$ in $X$ is compact. It is called *locally finite* if for every point $x$ in $K$ there is a neighborhood $U$ of $x$ such that $U \cap K = \{x\}$.

**Definition 2.1.** The continuously controlled category $\mathcal{B}(X, Y; \mathcal{A})$ is an additive category whose objects $A = (A_x)$ are collections of objects of $\mathcal{A}$ indexed by the points in the space $X - Y$ such that the *support of* $A$, $\mathrm{supp}(A) = \{x \in X - Y \mid A_x \neq 0\}$, is

(1) locally finite in $X - Y$; and

(2) relatively compact in $X$.

A morphism $\phi : A \to B$ is a collection $\phi = (\phi_x^{x'} : A_x \to B_{x'})$ of morphisms in $\mathcal{A}$ such that

(1) for every $x \in X - Y$, the set $\{x' \in X - Y \mid \phi_x^{x'} \neq 0 \text{ or } \phi_{x'}^{x} \neq 0\}$ is finite; and
(2) $\phi : A \to B$ is *continuously controlled* at $Y$. That is, for every $y \in Y$ and every neighborhood $U \subseteq X$ of $y$, there is a neighborhood $V \subseteq X$ of $y$ such that $\phi_x^{x'} = 0$ and $\phi_{x'}^{x} = 0$ whenever $x \in V$ and $x' \notin U$.

The continuously controlled category depends functorially on the CW pair $(X, Y)$ (for a proof see, for example, [4, Section 3.3]). A map $f : (X, Y) \to (X', Y')$ induces a functor $f_* : \mathcal{B}(X, Y; \mathcal{A}) \to \mathcal{B}(X', Y'; \mathcal{A})$, defined by $f_*(A)_{x'} = \bigoplus_{x \in f^{-1}(x')} A_x$. Given a morphism $\phi : A \to B$ in $\mathcal{B}(X, Y; \mathcal{A})$, $f_*(\phi) : f_*(A) \to f_*(B)$ is defined in the obvious way.

**Example 2.2** (Forget-control). Consider the categories $\mathcal{B}(X \times [0,1], X \times \{1\}; \mathcal{A})$ and $\mathcal{B}(CX, *; \mathcal{A})$, where $CX = X \times [0,1]/X \times \{1\}$ is the cone on $X$ and $*$ is the cone point. The control spaces for these two categories are the same, namely $X \times [0,1)$, but the control conditions are different. In $\mathcal{B}(X \times [0,1], X \times \{1\}; \mathcal{A})$ the continuous control condition on morphisms is defined with respect to the boundary $X \times \{1\}$, whereas in $\mathcal{B}(CX, *; \mathcal{A})$ the continuous control condition is defined with respect to the one-point boundary. Therefore, the continuous control condition on morphisms in $\mathcal{B}(X \times [0,1], X \times \{1\}; \mathcal{A})$ is more restrictive than the one in $\mathcal{B}(CX, *; \mathcal{A})$. The quotient map $X \times [0,1] \to X \times [0,1]/X \times \{1\}$, induces a functor $\mathcal{B}(X \times [0,1], X \times \{1\}; \mathcal{A}) \to \mathcal{B}(CX, *; \mathcal{A})$ that is essentially the identity on objects and morphisms. For this reason, it is known as a *forget-control functor.* In addition, because of the relatively compact condition on objects, the map $CX \to pt \times [0,1]$, which collapses $X$ to a point, induces an equivalence of categories $\mathcal{B}(CX, *; \mathcal{A}) \cong \mathcal{B}([0,1], \{1\}; \mathcal{A})$. This helps to illustrate the difference between $\mathcal{B}(X \times [0,1], X \times \{1\}; \mathcal{A})$ and $\mathcal{B}(CX, *; \mathcal{A})$. (Compare with Example 2.3 below.)

An additive category $\mathcal{B}$ is called *flasque* if it admits an endofunctor $\Sigma : \mathcal{B} \to \mathcal{B}$ and a natural equivalence between $\mathrm{id}_{\mathcal{B}} \oplus \Sigma$ and $\Sigma$. By the Additivity Theorem, such a category has trivial $K$-theory. Using an argument of this type to prove that a category has trivial $K$-theory is referred to as an *Eilenberg swindle.*

**Example 2.3.** We will use an Eilenberg swindle to show that $\mathcal{B}([0,1], \{1\}; \mathcal{A})$ has trivial $K$-theory. The idea for constructing the necessary endofunctor is to shift the objects and morphisms towards 1. Let $A$ and $B$ be objects of $\mathcal{B}([0,1], \{1\}; \mathcal{A})$, and let $\phi : A \to B$ be a morphism. Let $S : \mathcal{B}([0,1], \{1\}; \mathcal{A}) \to \mathcal{B}([0,1], \{1\}; \mathcal{A})$ be the endofunctor defined by

$$S(A)_t := A_{2t-1},$$
$$S(\phi)_s^t := \phi_{2s-1}^{2t-1} : A_{2s-1} \to B_{2t-1}.$$

Now define the endofunctor $\Sigma : \mathcal{B}([0,1],\{1\};\mathcal{A}) \to \mathcal{B}([0,1],\{1\};\mathcal{A})$ by

$$\Sigma(A) := \bigoplus_{n\geq 1} S^n(A),$$
$$\Sigma(\phi) := \bigoplus_{n\geq 1} S^n(\phi).$$

For each object $A$ there is a continuously controlled isomorphism $U_A : A \to S(A)$, defined by

$$(U_A)_s^t := \begin{cases} \mathrm{id}_{A_s} & \text{if } s = 2t-1 \\ 0 & \text{otherwise} \end{cases}.$$

The desired natural equivalence, $\eta : \mathrm{id} \oplus \Sigma \to \Sigma$, is given by $\eta(A) := \bigoplus_{n\geq 0} U_{S^n(A)}$. Therefore, $\mathbb{K}^{-\infty}(\mathcal{B}([0,1],\{1\};\mathcal{A}))$ is weakly contractible (i.e., it has trivial homotopy groups).

At first glance it appears that $\mathcal{B}(X\times[0,1], X\times\{1\};\mathcal{A})$ admits an Eilenberg swindle for any $X$ by sliding the objects and morphisms towards 1, as was done in Example 2.3. But this does not work in general. Remember that the continuous control condition on morphisms says that the components of a morphism must become shorter and shorter as they approach 1. This must happen in the $X$-direction, as well as in the $[0,1)$-direction. For example, consider the two-point space $S^0 = \{-1,1\}$. By the continuous control condition, the components of a morphism in $\mathcal{B}(S^0\times[0,1], S^0\times\{1\};\mathcal{A})$ are zero between $(-1,s)$ and $(1,t)$, if $s$ or $t$ is sufficiently close to 1. Therefore, a morphism with a non-zero component between $(-1,s)$ and $(1,t)$ for some $s$ and $t$ away from 1, cannot be shifted arbitrarily close to $S^0\times\{1\}$.

The *support at infinity* of an object $A$ in $\mathcal{B}(X,Y;\mathcal{A})$ is the set of limit points of $\mathrm{supp}(A)$ in $Y$.

**Definition 2.4.** Let $C$ be a closed subset of $Y$. The category $\mathcal{B}(X,Y;\mathcal{A})_C$ is the full subcategory of $\mathcal{B}(X,Y;\mathcal{A})$ on objects whose support at infinity is contained in $C$.

**Example 2.5.** Consider $\mathcal{B}(X\times[0,1], X\times\{1\};\mathcal{A})_\emptyset$, the full subcategory of $\mathcal{B}(X\times[0,1], X\times\{1\};\mathcal{A})$ whose objects have empty support at infinity. Because of the relatively compact and locally finite conditions, the support of every object is finite. Furthermore, the continuous control condition is vacuous. Therefore, $\mathcal{B}(X\times[0,1], X\times\{1\};\mathcal{A})_\emptyset$ and $\mathcal{A}$ are equivalent additive categories.

**Definition 2.6.** Let $W$ be an open subset of $Y$. The *germ category* $\mathcal{B}(X,Y;\mathcal{A})^W$ has the same objects as $\mathcal{B}(X,Y;\mathcal{A})$, but morphisms are identified if they agree in a neighborhood of $W$. Specifically, $\phi,\psi : A\to B$ are identified if there is a neighborhood $U\subseteq X$ of $W$ such that $x\in U$ or $x'\in U$ implies that $\phi_x^{x'} = \psi_x^{x'}$.

Germ categories are very interesting. As noted above, $\mathcal{B}(S^0\times[0,1], S^0\times\{1\};\mathcal{A})$ is not equivalent to $\mathcal{B}([0,1],\{1\};\mathcal{A})\oplus\mathcal{B}([0,1],\{1\};\mathcal{A})$, since $\mathcal{B}([0,1],\{1\};\mathcal{A}) \oplus \mathcal{B}([0,1],\{1\};\mathcal{A})$ is flasque (because each summand is flasque), while $\mathcal{B}(S^0\times[0,1], S^0\times\{1\};\mathcal{A})$ is not. However, it is true that the germ categories $\mathcal{B}(S^0\times$

$[0,1], S^0 \times \{1\}; \mathcal{A})^{S^0\times\{1\}}$ and $\mathcal{B}([0,1],\{1\};\mathcal{A})^{\{1\}} \oplus \mathcal{B}([0,1],\{1\};\mathcal{A})^{\{1\}}$ are equivalent. The reason is that we are taking germs at 1, which implies that every morphism has a representative that does not "jump between levels." What is meant by this is the following. Let $\phi$ be a morphism in $\mathcal{B}(S^0 \times [0,1], S^0 \times \{1\}; \mathcal{A})$. The continuous control condition implies that there is a neighborhood $V \subseteq S^0 \times [0,1]$ of the point $(1,1)$ such that $\phi_{(1,t)}^{(-1,s)} = 0 = \phi_{(-1,s)}^{(1,t)}$ whenever $(1,t)$ is in $V$. Now define a morphism $\psi$ such that $\psi_{(1,t)}^{(-1,s)} = 0 = \psi_{(-1,s)}^{(1,t)}$ for all $s$ and $t$, and $\psi_x^{x'} = \phi_x^{x'}$ otherwise. Therefore, $\phi$ and $\psi$ represent the same morphism in the germ category $\mathcal{B}(S^0 \times [0,1], S^0 \times \{1\}; \mathcal{A})^{S^0\times\{1\}}$, and the components of $\psi$ are zero between points in $\{-1\} \times [0,1)$ and $\{1\} \times [0,1)$. More generally, we have:

**Lemma 2.7.** *If $T$ is a discrete space, then*

$$\mathcal{B}(T \times [0,1], T \times \{1\}; \mathcal{A})^{T\times\{1\}} \cong \bigoplus_T \mathcal{B}([0,1],\{1\};\mathcal{A})^{\{1\}}.$$

*Sketch of proof.* As with the two point space, since we are taking germs at 1, every morphism has a representative that does not jump between levels. The reason for the direct sum is the relative compactness condition, which implies that each object in $\mathcal{B}(T \times [0,1], T \times \{1\}; \mathcal{A})^{T\times\{1\}}$ can only be non-zero for finitely many points in $T$. □

The categories from Definitions 2.4 and 2.6 have a special relationship. In [20], Karoubi introduced the following notion.

**Definition 2.8.** Let $\mathcal{A}$ be a full subcategory of an additive category $\mathcal{U}$. Denote the objects of $\mathcal{A}$ by the letters $A$ through $F$ and the objects of $\mathcal{U}$ by the letters $U$ through $W$. Then $\mathcal{U}$ is said to be *$\mathcal{A}$-filtered* if every object $U$ has a family of decompositions $\{U = E_\alpha \oplus U_\alpha\}$ where

(i) the decomposition forms a filtered poset under the partial order in which $E_\alpha \oplus U_\alpha \le E_\beta \oplus U_\beta$ whenever $U_\beta \subseteq U_\alpha$ and $E_\alpha \subseteq E_\beta$;
(ii) every map $A \to U$ factors through $E_\alpha$ for some $\alpha$;
(iii) every map $U \to A$ factors through $E_\alpha$ for some $\alpha$;
(iv) for each $U$ and $V$, the filtration on $U \oplus V$ is equivalent to the sum of the filtrations $\{U = E_\alpha \oplus U_\alpha\}$ and $\{V = F_\beta \oplus V_\beta\}$; that is, $\{U \oplus V = (E_\alpha \oplus F_\beta) \oplus (U_\alpha \oplus V_\beta)\}$.

Karoubi also defined the *quotient category $\mathcal{U}/\mathcal{A}$* to have the same objects as $\mathcal{U}$, but morphisms $\phi$, $\psi : U \to V$ are identified if their difference, $\phi - \psi$, factors through $\mathcal{A}$. The key fact is that the induced sequence

$$\mathbb{K}^{-\infty}(\mathcal{A}) \to \mathbb{K}^{-\infty}(\mathcal{U}) \to \mathbb{K}^{-\infty}(\mathcal{U}/\mathcal{A})$$

is a homotopy fibration of spectra, which yields a long exact sequence of homotopy groups. For this reason Karoubi filtrations play a major role in the subject.

It is straightforward to check that

$$\mathcal{B}(X,Y;\mathcal{A})_C \to \mathcal{B}(X,Y;\mathcal{A})$$

is a Karoubi filtration, where $C$ is a closed subset of $Y$. Furthermore, the corresponding quotient category is precisely the germ category $\mathcal{B}(X, Y; \mathcal{A})^{Y-C}$. Therefore, the sequence

$$(1) \qquad \mathcal{B}(X, Y; \mathcal{A})_C \to \mathcal{B}(X, Y; \mathcal{A}) \to \mathcal{B}(X, Y; \mathcal{A})^{Y-C}$$

yields a homotopy fibration of spectra after applying $\mathbb{K}^{-\infty}$. Similarly, it is an exercise to show that

$$(2) \qquad \mathcal{B}(X, Y; \mathcal{A})_C^W \to \mathcal{B}(X, Y; \mathcal{A})^W \to \mathcal{B}(X, Y; \mathcal{A})^{W-C}$$

is a Karoubi filtration sequence, where $W$ is an open subset of $Y$.

**Example 2.9.** An instance of (1) that is useful for studying assembly maps is the sequence

$$\begin{aligned}\mathcal{B}(X \times [0,1], X \times \{1\}; \mathcal{A})_\emptyset &\to \mathcal{B}(X \times [0,1], X \times \{1\}; \mathcal{A}) \\ &\to \mathcal{B}(X \times [0,1], X \times \{1\}; \mathcal{A})^{X \times \{1\}}.\end{aligned}$$

When $X$ is a point, the associated long exact sequence implies that

$$K_{n+1}(\mathcal{B}([0,1], \{1\}; \mathcal{A})^{\{1\}}) \cong K_n(\mathcal{B}([0,1], \{1\}; \mathcal{A})_\emptyset) \cong K_n(\mathcal{A}),$$

since $\mathbb{K}^{-\infty}(\mathcal{B}([0,1], \{1\}; \mathcal{A}))$ is weakly contractible. Therefore, $\Omega\mathbb{K}^{-\infty}(\mathcal{B}([0,1], \{1\}; \mathcal{A})^{\{1\}})$ is weakly homotopy equivalent to $\mathbb{K}^{-\infty}(\mathcal{A})$.

The functor $\Omega\mathbb{K}^{-\infty}(\mathcal{B}(- \times [0,1], - \times \{1\}; \mathcal{A})^{- \times \{1\}})$ from the category of CW complexes to the category of spectra is homotopy invariant and excisive, thereby yielding the generalized homology theory $H_*(X; \mathbf{K}_\mathcal{A}) := \pi_*(\Omega\mathbb{K}^{-\infty}(\mathcal{B}(- \times [0,1], - \times \{1\}; \mathcal{A})^{- \times \{1\}}))$. By Example 2.9, its value at a point is the $K$-theory of $\mathcal{A}$. Karoubi filtrations are needed to establish the fact that $H_*(X; \mathbf{K}_\mathcal{A})$ is in fact an homology theory. Excision is proved by showing that the pushout diagram of CW complexes

$$\begin{array}{ccc} X \cap Y & \longrightarrow & X \\ \downarrow & & \downarrow \\ Y & \longrightarrow & X \cup Y \end{array}$$

induces a homotopy Cartesian square

$$(3) \qquad \begin{array}{ccc} \mathbb{K}^{-\infty}(\mathcal{B}(X \cap Y; \mathcal{A})^{(X \cap Y) \times \{1\}}) & \longrightarrow & \mathbb{K}^{-\infty}(\mathcal{B}(X; \mathcal{A})^{X \times \{1\}}) \\ \downarrow & & \downarrow \\ \mathbb{K}^{-\infty}(\mathcal{B}(Y; \mathcal{A})^{Y \times \{1\}}) & \longrightarrow & \mathbb{K}^{-\infty}(\mathcal{B}(X \cup Y; \mathcal{A})^{(X \cup Y) \times \{1\}}) \end{array}$$

where $\mathcal{B}(- \times [0,1], - \times \{1\}; \mathcal{A})$ is denoted by $\mathcal{B}(-; \mathcal{A})$. To prove this, consider the

following commutative diagram of Karoubi filtrations induced by inclusion.
(4)

$$\begin{array}{ccccc}
\mathcal{B}(X;\mathcal{A})^{X\times\{1\}}_{(X\cap Y)\times\{1\}} & \longrightarrow & \mathcal{B}(X;\mathcal{A})^{X\times\{1\}} & \longrightarrow & \mathcal{B}(X;\mathcal{A})^{(X-Y)\times\{1\}} \\
\downarrow & & \downarrow & & \downarrow \\
\mathcal{B}(X\cup Y;\mathcal{A})^{(X\cup Y)\times\{1\}}_{Y\times\{1\}} & \longrightarrow & \mathcal{B}(X\cup Y;\mathcal{A})^{(X\cup Y)\times\{1\}} & \longrightarrow & \mathcal{B}(X\cup Y;\mathcal{A})^{(X-Y)\times\{1\}}
\end{array}$$

It is an exercise to show that, when $C$ is a closed subset of $X$, the inclusion functor $\mathcal{B}(C;\mathcal{A})^{C\times\{1\}} \to \mathcal{B}(X;\mathcal{A})^{X\times\{1\}}_{C}$ produces an equivalence of categories. Therefore, $\mathcal{B}(X\cap Y;\mathcal{A})^{(X\cap Y)\times\{1\}} \cong \mathcal{B}(X;\mathcal{A})^{X\times\{1\}}_{(X\cap Y)\times\{1\}}$ and $\mathcal{B}(Y;\mathcal{A})^{Y\times\{1\}} \cong \mathcal{B}(X\cup Y;\mathcal{A})^{(X\cup Y)\times\{1\}}_{Y\times\{1\}}$. One also checks that $\mathcal{B}(X;\mathcal{A})^{(X-Y)\times\{1\}} \to \mathcal{B}(X\cup Y;\mathcal{A})^{(X-Y)\times\{1\}}$ has an inverse induced by projection. Thus, diagram (4) becomes
(5)

$$\begin{array}{ccccc}
\mathcal{B}(X\cap Y;\mathcal{A})^{(X\cap Y)\times\{1\}} & \longrightarrow & \mathcal{B}(X;\mathcal{A})^{X\times\{1\}} & \longrightarrow & \mathcal{B}(X;\mathcal{A})^{(X-Y)\times\{1\}} \\
\downarrow & & \downarrow & & \downarrow\cong \\
\mathcal{B}(Y;\mathcal{A})^{Y\times\{1\}} & \longrightarrow & \mathcal{B}(X\cup Y;\mathcal{A})^{(X\cup Y)\times\{1\}} & \longrightarrow & \mathcal{B}(X\cup Y;\mathcal{A})^{(X-Y)\times\{1\}}
\end{array}$$

Applying $\mathbb{K}^{-\infty}$ to (5) yields a commutative diagram of spectra in which each row is a fibration. Since fibration sequences are also cofibration sequences in the category of spectra, the two rows of the induced diagram are cofibration sequences with equivalent cofibers. Therefore, diagram (3) is a homotopy co-Cartesian square, which means it is a homotopy Cartesian square, since we are working in the category of spectra.

The above argument can be found in [17, Corollary 9.3], [15, p. 48] and [4, Proposition 4.3]. For a proof of homotopy invariance, the reader is referred to [4, Section 5]. The proof employs a trick, dating back to Pedersen and Weibel [22], that makes clever use of Eilenberg swindles.

## 3. Equivariant homology and the assembly map

Given a discrete group $G$ and a ring $R$, the classical assembly map in algebraic $K$-theory, introduced by Loday [21], is a map that relates the $K$-theory of the group ring $R[G]$ to the homology of B$G$ with coefficients in the $K$-theory spectrum of $R$. In the revolutionary work of Farrell and Jones [14], a more general assembly map relating the $K$-theory of $R[G]$ to a generalized $G$-equivariant homology theory was formulated using Quinn's "homology of simplicially stratified fibrations" [24]. Davis and Lück [13] used the "orbit category" to create an abstract approach to Farrell-Jones assembly, and Hambleton and Pedersen formalized the continuously controlled version of the Farrell-Jones assembly map in [15]. Recently, Bartels and Reich [7] used the Davis-Lück machinery to construct an assembly map with coefficients in an additive category on which $G$ acts. This more general formulation allows one to study, for example, twisted group rings. It also possesses interesting inheritance properties (see [7, 16]). In this section the assembly map *with coefficients* is defined using continuously controlled algebra. In order to do this the corresponding equivariant homology theory must be developed.

An *additive category with right $G$-action* $\mathcal{A}$ is an additive category together with a collection of covariant functors $\{g^*\colon \mathcal{A} \to \mathcal{A} \mid g \in G\}$ such that $(g \circ h)^* = h^* \circ g^*$ and $e^* = \mathrm{id}_{\mathcal{A}}$. A subset $S \subseteq X$ is called *$G$-compact* if $S = G \cdot K$ for some compact subset $K \subseteq X$. It is called *relatively $G$-compact* if its closure in $X$ is $G$-compact.

**Definition 3.1.** Let $G$ be a discrete group, $\mathcal{A}$ be an additive category with right $G$-action, $X$ be a $G$-CW complex and $Y \subseteq X$ be a closed $G$-invariant subcomplex of $X$. The category $\mathcal{D}(X, Y; \mathcal{A})$ has objects $A = (A_z)$, consisting of collections of objects of $\mathcal{A}$ indexed by $G \times (X - Y)$ such that the support of $A$, $\mathrm{supp}(A) = \{z \in G \times (X - Y) \mid A_z \neq 0\}$, is:

(1) locally finite in $G \times (X - Y)$; and
(2) relatively $G$-compact in $G \times X$.

A morphism $\phi : A \to B$ is a collection $\phi = (\phi_z^{z'} : A_z \to B_{z'})$ of morphisms in $\mathcal{A}$ such that:

(1) for every $z \in G \times (X - Y)$, the set $\{z' \in G \times (X - Y) \mid \phi_z^{z'} \neq 0 \text{ or } \phi_{z'}^{z} \neq 0\}$ is finite; and
(2) $\phi : A \to B$ is continuously controlled at $Y$, which means that for every $y \in Y$ and every $G_y$-invariant neighborhood $U \subseteq X$ of $y$, there is a $G_y$-invariant neighborhood $V \subseteq X$ of $y$ such that $\phi_z^{z'} = 0$ and $\phi_{z'}^{z} = 0$ whenever $z \in G \times V$ and $z' \notin G \times U$.

Adding a factor of $G$ to the control space, an idea due to Pedersen, creates an interesting action of $G$ on $\mathcal{D}(X, Y; \mathcal{A})$. The right $G$-action on $\mathcal{D}(X, Y; \mathcal{A})$ is induced by the diagonal action of $G$ on $G \times X$ and the right $G$-action on $\mathcal{A}$. It is given by

$$(g^* A)_z := g^*(A_{gz}),$$
$$(g^* \phi)_z^{z'} := g^*(\phi_{gz}^{gz'}).$$

In the corresponding fixed point category, $\mathcal{D}^G(X, Y; \mathcal{A})$, every object $A$ and every morphism $\phi$ satisfy $A_z = (g^* A)_z = g^*(A_{gz})$ and $\phi_z^{z'} = (g^*\phi)_z^{z'} = g^*(\phi_{gz}^{gz'})$ for every $g$ in $G$ and every $z, z'$ in $G \times (X - Y)$. If $\mathcal{A} = \mathcal{F}_R$ with the trivial $G$-action, then the objects of $\mathcal{D}^G(X \times [0,1); \mathcal{F}_R)$ are free $R[G]$-modules and the morphisms are $R[G]$-homomorphisms. Furthermore, the direct sum of the pieces of an object $A$ over an entire orbit, $\bigoplus_{g' \in G} A_{g'z} = \bigoplus_{g \in G} A_{(g, gx)}$ for some $x$ in $X - Y$, is a finitely generated free $R[G_x]$-module. Thus, we can think of objects in $\mathcal{D}^G(X, Y; \mathcal{F}_R)$ as being built out of finitely generated free $R[G_x]$-modules, where $x$ is a point in $X$. In particular, notice that $\mathcal{D}^G(\mathrm{pt}, \emptyset; \mathcal{F}_R)$ is equivalent to $\mathcal{F}_{R[G]}$, the category of finitely generated free $R[G]$-modules. For this reason the category $\mathcal{D}^G(\mathrm{pt}, \emptyset; \mathcal{A})$ is denoted by $\mathcal{A}[G]$.

As in the unequivariant case, $\mathcal{D}^G(X, Y; \mathcal{A})$ depends functorially on the $G$-CW pair $(X, Y)$ [4, Section 3.3]. In order to construct an equivariant homology theory and an assembly map, we will work with the category $\mathcal{D}^G(X \times [0,1], X \times \{1\}; \mathcal{A})$, where $X$ is a $G$-CW complex. For brevity, denote this category by $\mathcal{D}^G(X \times [0,1); \mathcal{A})$.

Let $\mathcal{D}^G(X \times [0,1); \mathcal{A})_\emptyset$ be the full subcategory of $\mathcal{D}^G(X \times [0,1); \mathcal{A})$ on objects $A$ such that the intersection of $X \times \{1\}$ with the closure of $\mathrm{supp}(A)$ in $X \times [0,1]$ is the empty set. Notice that, as in Example 2.9, the quotient category, which we denote by $\mathcal{D}^G(X \times [0,1); \mathcal{A})^{>0}$, is a germ category. The objects are the same as in $\mathcal{D}^G(X \times [0,1); \mathcal{A})$ but morphisms are identified if they agree close to $G \times X \times \{1\}$, i.e., on the complement of a neighborhood of $G \times X \times \{0\}$. (It is not difficult to check that taking germs and taking fixed categories commute.) Furthermore, there is a corresponding Karoubi filtration sequence

$$\mathcal{D}^G(X \times [0,1); \mathcal{A})_\emptyset \to \mathcal{D}^G(X \times [0,1); \mathcal{A}) \to \mathcal{D}^G(X \times [0,1); \mathcal{A})^{>0}.$$

**Example 3.2.** Let $A$ be an object in $\mathcal{D}^G(\mathrm{pt} \times [0,1); \mathcal{F}_R)_\emptyset$. Since objects have empty support at infinity, the local finiteness condition implies that there are only finitely many $t \in [0,1)$ with $A_{(e,t)} \neq 0$, where $e$ is the identity of $G$. By equivariance, $A_{(g,t)} = A_{(e,t)}$ for every $t \in [0,1)$. Therefore, $A$ is a finitely generated free $R[G]$-module. Since the continuous control condition is vacuous, equivariance tells us that the morphisms in $\mathcal{D}^G(\mathrm{pt} \times [0,1); \mathcal{F}_R)_\emptyset$ are just $R[G]$-homomorphisms. Thus, $\mathcal{D}^G(\mathrm{pt} \times [0,1); \mathcal{F}_R)_\emptyset$ is equivalent to $\mathcal{F}_{R[G]}$. (In fact, $\mathcal{D}^G(X \times [0,1); \mathcal{F}_R)_\emptyset$ is equivalent to $\mathcal{F}_{R[G]}$ for every $X$, because of the relatively $G$-compact condition on objects.) As in Example 2.9, since $\mathcal{D}^G(\mathrm{pt} \times [0,1); \mathcal{F}_R)$ is flasque, $\Omega\mathbb{K}^{-\infty}(\mathcal{D}^G(\mathrm{pt} \times [0,1); \mathcal{F}_R)^{>0})$ is weakly homotopy equivalent to $\mathbb{K}^{-\infty}(\mathcal{F}_{R[G]})$.

More generally, we have the following lemma about orbits $G/H$.

**Lemma 3.3.** $\Omega\mathbb{K}^{-\infty}(\mathcal{D}^G(G/H \times [0,1); \mathcal{A})^{>0}) \simeq \mathbb{K}^{-\infty}(\mathcal{A}[H])$.

*Proof.* Equivariance tells us that the objects of $\mathcal{D}^G(G/H \times [0,1); \mathcal{A})^{>0}$ are determined by their value over the points in $\{eH\} \times [0,1)$. Since the isotropy at every point in $\{eH\} \times [0,1)$ is $H$, the objects over $\{eH\} \times [0,1)$ are objects in $\mathcal{A}[H]$. Since $G/H$ is a discrete space and we are taking germs at 1, the morphisms in $\mathcal{D}^G(G/H \times [0,1); \mathcal{A})_\emptyset$ cannot jump between different points of $G/H$ (compare with Lemma 2.7). Combining this with the equivariance of morphisms, we see that $\mathcal{D}^G(G/H \times [0,1); \mathcal{A})^{>0}$ is equivalent to $\mathcal{D}^{\{e\}}(\mathrm{pt} \times [0,1); \mathcal{A}[H])^{>0}$. Since $\mathcal{D}^{\{e\}}(\mathrm{pt} \times [0,1); \mathcal{A}[H])$ is flasque, $\Omega\mathbb{K}^{-\infty}(\mathcal{D}^G(G/H \times [0,1); \mathcal{A})^{>0}) \simeq \mathbb{K}^{-\infty}(\mathcal{A}[H])$. □

The functor $\mathbf{K}(-; \mathcal{A})^G := \Omega\mathbb{K}^{-\infty}(\mathcal{D}^G(- \times [0,1); \mathcal{A})^{>0})$ from the category of $G$-CW complexes to the category of spectra is $G$-homotopy invariant and $G$-excisive [4, Section 5]. As in the unequivariant case, the proof of this fact makes use of Karoubi filtrations and Eilenberg swindles. In this sense $\mathbf{K}(-; \mathcal{A})^G$ defines a $G$-equivariant homology theory. The corresponding equivariant homology groups are denoted by $H_*^G(X; \mathbf{K}_\mathcal{A}) := \pi_*(\mathbf{K}(X; \mathcal{A})^G)$.

The collapse map, $X \to \mathrm{pt}$, induces the *continuously controlled assembly map*

$$A^G : \mathbf{K}(X; \mathcal{A})^G \to \mathbf{K}(\mathrm{pt}; \mathcal{A})^G,$$

which yields a map on homology, $A^G_* : H^G_*(X; \mathbf{K}_\mathcal{A}) \to H^G_*(\mathrm{pt}; \mathbf{K}_\mathcal{A}) = K_*(\mathcal{A}[G])$. In light of Example 2.2, this construction interprets assembly as a forget-control map. It is also worth noting that the continuously controlled assembly map is a map of fixed point spectra, since $\mathbb{K}^{-\infty}$ commutes with taking fixed sets. That is, if $G$ acts on an additive category $\mathcal{B}$, then there is an induced action of $G$ on $\mathbb{K}^{-\infty}(\mathcal{B})$, and the fixed point set of this spectrum, $\mathbb{K}^{-\infty}(\mathcal{B})^G$, is equivalent to the spectrum $\mathbb{K}^{-\infty}(\mathcal{B}^G)$. This implies that $A^G$ is the map induced by the equivariant map of spectra $\mathbf{K}(X; \mathcal{A}) \to \mathbf{K}(\mathrm{pt}; \mathcal{A})$.

The continuously controlled version of the assembly map is homotopy equivalent to the one defined by Bartels and Reich, who used the Davis-Lück machinery for constructing assembly maps [7, 16]. In the case $\mathcal{A} = \mathcal{F}_R$, Hambleton and Pedersen used Lemma 3.3 to identify the Davis-Lück assembly map with the continuously controlled assembly map [15]. They also showed that both versions were equivalent to the classical definition of the assembly map.

The *Farrell-Jones conjecture with coefficients* predicts that the assembly map is an isomorphism for all coefficients $\mathcal{A}$ when $X = \mathrm{E}_{\mathcal{VC}yc}G$, where $\mathcal{VC}yc$ is the family of virtually cyclic subgroups of $G$. [1] This is a very strong statement. For example, if the Farrell-Jones conjecture with coefficients is true for a group $G$, then the Fibered Farrell-Jones conjecture is true for $G$ [7]. Furthermore, it would imply that the Farrell-Jones conjecture with coefficients is true for every subgroup of $G$. It also behaves well with respect to extensions (see [16]).

By the universal property, there is a map $\mathrm{E}_{\mathcal{F}in}G \to \mathrm{E}_{\mathcal{VC}yc}G$, where $\mathcal{F}in$ is the family of finite subgroups of $G$. Bartels [2] proved that for every discrete group $G$, the induced map on homology, $H^G_*(\mathrm{E}_{\mathcal{F}in}G; \mathbf{K}_\mathcal{A}) \to H^G_*(\mathrm{E}_{\mathcal{VC}yc}G; \mathbf{K}_\mathcal{A})$, is a split injection. (He showed this for $\mathcal{A} = \mathcal{F}_R$, but the proof works in the general case as well.) As a result, the Farrell-Jones conjecture implies that the assembly map for the family of finite subgroups, $H^G_*(\mathrm{E}_{\mathcal{F}in}G; \mathbf{K}_\mathcal{A}) \to K_*(\mathcal{A}[G])$, should be a split injection for every discrete group $G$.

## 4. Proving isomorphism and injectivity results

In this section we present an alternate description of the continuously controlled assembly map that is useful for proofs.

Consider the following commutative diagram.

$$\begin{array}{ccccc}
\mathcal{D}^G(X \times [0,1); \mathcal{A})_\emptyset & \longrightarrow & \mathcal{D}^G(X \times [0,1); \mathcal{A}) & \longrightarrow & \mathcal{D}^G(X \times [0,1); \mathcal{A})^{>0} \\
\downarrow \cong & & \downarrow & & \downarrow A^G \\
\mathcal{D}^G(\mathrm{pt} \times [0,1); \mathcal{A})_\emptyset & \longrightarrow & \mathcal{D}^G(\mathrm{pt} \times [0,1); \mathcal{A}) & \longrightarrow & \mathcal{D}^G(\mathrm{pt} \times [0,1); \mathcal{A})^{>0}
\end{array}$$

[1] Recall that if $\mathcal{F}$ is a family of subgroups of $G$ that is closed under conjugation and taking subgroups, then the *universal space for $G$ with isotropy in $\mathcal{F}$*, $\mathrm{E}_\mathcal{F}G$, is a $G$-CW complex with the property that $(\mathrm{E}_\mathcal{F}G)^H$ is contractible if $H$ is in $\mathcal{F}$ and is empty otherwise. Such spaces are universal for $G$-actions with isotropy in $\mathcal{F}$, meaning that given any $G$-CW complex $Y$ whose isotropy groups belong to $\mathcal{F}$, there is a map $Y \to \mathrm{E}_\mathcal{F}G$ that is unique up to $G$-equivariant homotopy equivalence. Thus, $\mathrm{E}_\mathcal{F}G$ is unique up to $G$-equivariant homotopy equivalence.

Since $\mathcal{D}^G(\mathrm{pt} \times [0,1);\mathcal{A})$ is flasque, $\mathbb{K}^{-\infty}(\mathcal{D}^G(\mathrm{pt} \times [0,1);\mathcal{A}))$ is weakly contractible. Therefore, it is a simple diagram chase to verify that the assembly map, $A^G$, is homotopy equivalent to the connecting map $\Omega\mathbb{K}^{-\infty}(\mathcal{D}^G(X \times [0,1);\mathcal{A})^{>0}) \to \mathbb{K}^{-\infty}(\mathcal{D}^G(X \times [0,1);\mathcal{A})_\emptyset)$.

This variant of the continuously controlled assembly map shows us that in order to prove isomorphism results one needs to prove that the category $\mathcal{D}^G(X \times [0,1);\mathcal{A})$ has trivial $K$-theory. For this reason, Bartels-Lück-Reich call this category *the obstruction category.* In [5] they were able to prove the $K$-theoretic Farrell-Jones conjecture for all word hyperbolic groups and all coefficient categories by showing that the $K$-theory of this category vanishes when $X = \mathrm{E}_{\mathcal{VC}yc}G$. (Because of the nice inheritance properties of the assembly map with coefficients, their result proves the $K$-theoretic Farrell-Jones conjecture for all subgroups of finite products of word hyperbolic groups.) Recently Bartels and Lück have announced the corresponding result in algebraic $L$-theory. Combining these two theorems proves the Borel conjecture for all subgroups of finite products of word hyperbolic groups.

The ultimate result about the assembly map is that it is an isomorphism. Proving that the obstruction category vanishes achieves this goal, but it is a difficult task. However, with a bit less one can still prove that the assembly map is a split injection using a trick known as *the descent principle.* Using the fact that the continuously controlled assembly map is a map of fixed spectra, one can employ *homotopy fixed point sets* to prove the following theorem (a proof can be found in [8].)

**Theorem 4.1** (The Descent Principle). *Let $G$ be a discrete group, $\mathcal{A}$ be an additive $G$-category and $\mathrm{E}_{\mathcal{F}in}G$ be a finite dimensional $G$-CW complex. Assume that the $K$-theory of the category $\mathcal{D}^H(\mathrm{E}_{\mathcal{F}in}G \times [0,1);\mathcal{A})$ vanishes for every finite subgroup $H$ of $G$.*

*Then $A_*^G : H_*^G(\mathrm{E}_{\mathcal{F}in}G;\mathbb{K}_{\mathcal{A}}^{-\infty}) \to K_*(\mathcal{A}[G])$ is a split injection.*

Consider the case when $G$ is torsion-free. Then the descent principle says that one can prove the assembly map is a split injection by proving that it is an equivalence *unequivariantly.* This is the original version of the descent principle. It was used by Carlsson and Pedersen in [10] to prove the Novikov conjecture for torsion-free groups whose universal space $\mathrm{E}G$ satisfied certain geometric conditions. Since Carlsson and Pedersen's seminal work, the descent principle and continuously controlled algebra have been used to prove split injectivity results (see, for example, [11, 2, 3, 25, 8]). This method was recently used to prove that the assembly map is a split injection for all discrete subgroups of virtually connected Lie groups [8], and for all $S$-arithmetic subgroups of algebraic groups defined over global fields, regardless of rank, by Ji [19].

# References

[1] D. R. Anderson, F. Connolly, S. C. Ferry, and E. K. Pedersen, *Algebraic K-theory with continuous control at infinity*, J. Pure Appl. Algebra **94** (1994), 25–47.

[2] A. Bartels, *On the domain of the assembly map in algebraic K-theory*, Algebr. Geom. Topol. **3** (2003), 1037–1050.

[3] A. Bartels, *Squeezing and higher algebraic K-theory*, $K$-theory **28** (2003), 19–37.

[4] A. Bartels, F. T. Farrell, L. Jones, and H. Reich, *On the isomorphism conjecture in algebraic K-theory*, Topology **43**(1) (2004), 157–213.

[5] A. Bartels, W. Lück, and H. Reich, *The K-theoretic Farrell-Jones Conjecture for hyperbolic groups*. To appear in Invent. Math. arXiv: math.KT/0701434.

[6] A. Bartels, W. Lück, and H. Reich, *Equivariant covers for hyperbolic groups*, arXiv: math.GT/0609685.

[7] A. Bartels and H. Reich, *Coefficients for the Farrell-Jones Conjecture*, Adv. Math. **209**(1) (2007), 337–362.

[8] A. Bartels and D. Rosenthal, *On the algebraic K- and L-theory of groups with finite asymptotic dimension*. To appear in J. Reine Agnew. Math. arXiv: math.KT/0605088

[9] M. Cárdenas and E. K. Pedersen, *On the Karoubi filtration of a category*, $K$-theory **12** (1997), 165–191.

[10] G. Carlsson and E. K. Pedersen, *Controlled algebra and the Novikov conjectures for K- and L- theory*, Topology **34** (1995), 731–758.

[11] G. Carlsson and E. K. Pedersen, *Čech homology and the Novikov conjectures for K- and L- theory*, Math. Scand. **82** (1998), 5–47.

[12] E. H. Connell and J. Hollingsworth, *Geometric groups and Whitehead torsion*, Trans. Amer. Math. Soc. **140** (1969), 161–181.

[13] J. Davis and W. Lück, *Spaces over a category and assembly maps in isomorphism conjectures in K- and L-theory*, $K$-theory **15** (1998), 201-252.

[14] F. T. Farrell and L. E. Jones, *Isomorphism conjectures in algebraic K-theory*, J. Amer. Math. Soc. **6** (1993), no. 3, 249–297.

[15] I. Hambleton and E. K. Pedersen, *Identifying assembly maps in K- and L-theory*, Math. Ann. **328**, (2004), 27–57.

[16] I. Hambleton, E. K. Pedersen, and D. Rosenthal, *Assembly maps for group extensions in K- and L-theory*. arXiv:math.KT/07090437.

[17] N. Higson, E. K. Pedersen, and J. Roe, *$C^*$-algebras and controlled topology*, $K$-theory **11** (1997), 209–239.

[18] W. C. Hsiang, *Geometric applications of algebraic K-theory*, Proceedings of the ICM, Vols. 1, 2, Warsaw, 1983, P. W. N., Warsaw 1984, 99–118.

[19] L. Ji, *The integral Novikov conjectures for linear groups containing torsion elements*. Preprint.

[20] M. Karoubi, *Foncteur derivees et K-theorie*, Lecture Notes in Mathematics, vol. 136, Springer-Verlag, Berlin-New York, 1970.

[21] J.-L. Loday, *K-théorie algébrique et représentations de groupes*, Ann. Sci. École Norm. Sup. (4) **9** (1976), 309–377.

[22] E. K. Pedersen and C. Weibel, *A non-connective delooping of algebraic K-theory*, Algebraic and Geometric Topology (Rutgers, 1983), Lecture Notes in Mathematics, vol. 1126, Springer, Berlin (1985), 166–181.

[23] E. K. Pedersen and C. Weibel, *K-theory homology of spaces*, Algebraic Topology (Arcata, 1986), Lectures Notes in Mathematics, vol. 1370, Springer-

Verlag, Berlin-New York, 1989, 346–361.

[24] F. Quinn, *Ends of maps, II*, Invent. Math. **68** (1982), 353–424.

[25] D. Rosenthal, *Splitting with continuous control in algebraic $K$-theory*, $K$-theory **32** (2004), 139–166.

*Cohomology of Groups and*
*Algebraic K-theory*
ALM 12, pp. 503–517



# Higher $K$-Theory of Algebraic Integers and the Cohomology of Arithmetic Groups

Christophe Soulé*
(Notes by Marco Varisco)

**Abstract**

These lecture notes give an introduction to algebraic $K$-groups of the ring of integers of number fields.

**2000 Mathematics Subject Classification:** 19D50, 19F27

**Keywords and Phrases:** $K$-group, ring of integers, cohomology of arithmetic groups

## Lecture one: Two theorems of Armand Borel

Let $F$ be a number field, i.e., a finite field extension of $\mathbb{Q}$, and let $A = \mathcal{O}_F$ be its ring of integers, i.e., the integral closure of $\mathbb{Z}$ in $F$:

$$A = \mathcal{O}_F = \left\{ x \in F \mid x^n + a_1 x^{n-1} + \ldots + a_n = 0,\ a_i \in \mathbb{Z} \right\}.$$

Our goal in these lectures is to understand the algebraic $K$-theory of $A$.

First of all, observe that there is no negative $K$-theory because $A$ is regular.

**Proposition 1.** $K_0(A) \cong \mathbb{Z} \oplus \operatorname{Pic}(A)$.

Here $\operatorname{Pic}(A)$ is the ideal class group of $A$, i.e., the set of isomorphism classes of invertible $A$-modules with addition given by the tensor product. Proposition 1 is true more generally for any Dedekind domain $A$, since every projective module is a sum of ideals, each of which is projective and satisfies $I \oplus J \cong IJ \oplus A$, see [Mil71].

For $A = \mathcal{O}_F$ Dirichlet proved that $\operatorname{Pic}(A)$ is finite.

*Institut des Hautes Études Scientifiques, 35 route de Chartres, 91440 Bures-sur-Yvette, France. URL: http://www.ihes.fr/~soule/ E-mail address: soule@ihes.fr

**Proposition 2.** $K_1(A) = A^\times$.

In fact, Bass, Milnor, and Serre [BMS67] proved that $SK_1(A) = 0$, and for any commutative ring $A$ one has $K_1(A) = A^\times \times SK_1(A)$.

For $A = \mathcal{O}_F$ Dirichlet proved that

$$\dim_{\mathbb{Q}}(A^\times \otimes \mathbb{Q}) = r_1 + r_2 - 1 = d_1$$

where

$$r_1 = \#\{\text{ real places of } F\,\} = \#\{\,\sigma\colon F \hookrightarrow \mathbb{R}\,\},$$
$$r_2 = \#\{\text{ complex places of } F\,\} = \frac{1}{2}\#\{\,\sigma\colon F \hookrightarrow \mathbb{C},\ \sigma \neq \overline{\sigma}\,\},$$

(the resulting decomposition of $F \otimes_{\mathbb{Q}} \mathbb{R}$ then shows that $[F : \mathbb{Q}] = r_1 + 2r_2$), and for any $n \geq 1$ we put

$$d_n = \begin{cases} r_1 + r_2 - 1 & \text{if } n = 1, \\ r_1 + r_2 & \text{if } n \text{ is odd and } \geq 3, \\ r_2 & \text{if } n \text{ is even.} \end{cases}$$

More precisely, Dirichlet proved that $A^\times$ is the product of the finite cyclic group $\mu(F)$ of roots of unity in $F$ and a free abelian group of rank $r_1 + r_2 - 1 = d_1$.

**Theorem 3** (Quillen [Qui73]). *For all $m \geq 0$, $K_m(A)$ is finitely generated.*

**Theorem 4** (Borel [Bor74]). *For all $m > 0$,*

- *if $m$ is even then $K_m(A)$ is finite,*
- *if $m = 2n - 1$ then $\dim_{\mathbb{Q}}(K_m(A) \otimes \mathbb{Q}) = d_n$.*

These results generalize the aforementioned theorems by Dirichlet.

*Example* 5. If $F = \mathbb{Q}$, $A = \mathbb{Z}$ then $r_1 = 1$ and $r_2 = 0$, and hence for $m > 0$

$$K_m(\mathbb{Z}) = \begin{cases} \mathbb{Z} \oplus \text{finite} & m = 5, 9, 13, \ldots, \\ \text{finite} & \text{otherwise.} \end{cases}$$

We will not discuss the proof of Quillen's theorem 3 here.

As we will see below, Borel's theorem 4 follows from the following theorem.

**Theorem 6** (Borel). *Let $G = SL_N(\mathbb{R})^{r_1} \times SL_N(\mathbb{C})^{r_2} \supset \Gamma = SL_N(A)$.*

*Assume $q+1 \leq (N-1)/4$. Then the corestriction map $H^q_{\text{cont}}(G) \to H^q(\Gamma; \mathbb{R})$ is an isomorphism.*

Here $H^q_{\text{cont}}(G)$ is the continuous cohomology of $G$ with real coefficients. It can be defined as the cohomology of the complex

$$\cdots \longrightarrow C^q_{\text{cont}}(G)^G \xrightarrow{\ \delta\ } C^{q+1}_{\text{cont}}(G)^G \longrightarrow \cdots,$$

where $C^q_{\text{cont}}(G)$ is the real vector space of continuous maps from $G^{q+1}$ to $\mathbb{R}$ and $\delta$ is given by the formula

$$\delta\varphi(g_0,\dots,g_{q+1}) = \sum_{i=0}^{q+1}(-1)^i\,\varphi(g_0,\dots,\widehat{g_i},\dots,g_{q+1}).$$

Theorem 6 is a special case of the following result.

**Theorem 7** (Borel [Bor74]). *Let $\underline{G}$ be a semi-simple algebraic group over $\mathbb{Q}$ such that $G = \underline{G}(\mathbb{R})$ is connected and let $\Gamma < \underline{G}(\mathbb{Q})$ be an arithmetic group.*

*Assume $q + 1 \le \operatorname{rank}_{\mathbb{Q}}(G)/4$. Then the corestriction map $H^q_{\text{cont}}(G) \to H^q(\Gamma;\mathbb{R})$ is an isomorphism.*

*Proof that theorem 6 implies theorem 4.* Step 1: We first compute $H^*_{\text{cont}}(G)$ as follows. Consider the maximal compact subgroup $K$ of $G$, and the symmetric space $X = K\backslash G$.

*Example* 8. If $G = GL_N(\mathbb{R})$ then $K = O(N)$ and $X$ is the set of positive definite real quadratic forms. In fact, given $[g] \in X$ we can define $\varphi(x) = \|g(x)\|^2$ for $x \in \mathbb{R}^N$.

If $G = SL_N(\mathbb{R})$ then $K = SO(N)$ and $X$ is the set of positive definite real quadratic forms modulo the action of $\mathbb{R}^\times_{>0}$.

The manifold $X$ is contractible. Therefore the de Rham complex

$$0 \to \mathbb{R} \to \Omega^0(X) \to \Omega^1(X) \to \Omega^2(X) \to \cdots$$

is exact. This yields a "strong" resolution of $\mathbb{R}$ by "relatively" injective $G$-modules (this means that the resolution is "good" from the point of view of continuous cohomology, see [Gui80]). Hence

$$H^q_{\text{cont}}(G) = H^q(\Omega^*(X)^G).$$

Let $\mathfrak{g}$ and $\mathfrak{k}$ be the Lie algebras of $G$ and $K$ respectively. By restriction of differential forms at the origin we have

$$\Omega^q(X)^G = \hom_{\mathfrak{k}}(\Lambda^q(\mathfrak{g}/\mathfrak{k}),\mathbb{R}).$$

Consider the Cartan decomposition $\mathfrak{g} = \mathfrak{k}\oplus\mathfrak{p}$ and $\mathfrak{g}\otimes_{\mathbb{R}}\mathbb{C} \supset i\mathfrak{p}$. Then the so-called unitarian trick yields that $\mathfrak{k}\oplus i\mathfrak{p} = \operatorname{Lie}(G_u)$, where $G_u$ is a compact connected Lie group containing $K$. Then

$$\Omega^q(X)^G = \hom_{\mathfrak{k}}(\Lambda^q(\mathfrak{g}/\mathfrak{k}),\mathbb{R}) = \hom_{\mathfrak{k}}(\Lambda^q(\mathfrak{p}),\mathbb{R}) \cong \hom_{\mathfrak{k}}(\Lambda^q(i\mathfrak{p}),\mathbb{R}) = \Omega^q(K\backslash G_u)^{G_u}.$$

Since $G_u$ is compact and connected, integration on $G_u$ shows that the inclusion

$$\Omega^q(K\backslash G_u)^{G_u} \subseteq \Omega^q(K\backslash G_u)$$

is a homology equivalence [Gui80, rem. 7.1 and lemma E.2]. Therefore

$$H^q_{\text{cont}}(G) = H^q(\Omega^*(X)^G) = H^q(\Omega^*(K\backslash G_u)^{G_u}) = H^q(\Omega^*(K\backslash G_u)) = H^q(K\backslash G_u;\mathbb{R}).$$

*Example* 9. If $G = SL_N(\mathbb{R})$, $K = SO(N)$ then

$$\mathfrak{g} = \{ M \mid \operatorname{tr} M = 0 \}, \quad \mathfrak{k} = \{ M \in \mathfrak{g} \mid M^t = -M \}, \quad \mathfrak{p} = \{ M \in \mathfrak{g} \mid M^t = M \}$$

and therefore

$$\mathfrak{k} \oplus i\mathfrak{p} = \left\{ M \in \mathfrak{g} \otimes_{\mathbb{R}} \mathbb{C} \;\middle|\; \overline{M}^t = -M \right\} \cong \mathfrak{su}(N).$$

Hence $G_u = SU(N)$. Then we get

$$H^*_{\text{cont}}(SL_N(\mathbb{R})) \cong H^*(SO(N)\backslash SU(N); \mathbb{R}).$$

The right-hand side is known (by previous work of Borel) and gives

$$H^*_{\text{cont}}(SL_N(\mathbb{R})) \cong \Lambda^*(e_5, e_9, e_{13}, \dots, e_{4k+1})$$

with $e_q \in H^q(SO(N)\backslash SU(N); \mathbb{Z})$, $k = [\frac{N-1}{2}]$.

If $G = SL_N(\mathbb{C})$ then $K = SU(N)$ and $G_u = SU(N) \times SU(N)$. We get

$$H^*_{\text{cont}}(SL_N(\mathbb{C})) \cong H^*(SU(N); \mathbb{R}) \cong \Lambda^*(\varepsilon_3, \varepsilon_5, \varepsilon_7, \dots, \varepsilon_{2N-1})$$

with $\varepsilon_q \in H^q(SU(N); \mathbb{Z})$.

For $G = SL_N(\mathbb{R})^{r_1} \times SL_N(\mathbb{C})^{r_2}$ this yields

$$H^*_{\text{cont}}(G) \cong \Lambda^*(e_i)^{\otimes r_1} \otimes \Lambda^*(\varepsilon_j)^{\otimes r_2}.$$

Step 2: There is a homotopy equivalence

$$BSL(A)^+ \times B(A^\times) \xrightarrow{\simeq} BGL(A)^+$$

and hence for $m \geq 2$

$$K_m(A) \cong \pi_m BSL(A)^+.$$

For any CW-complex $X$ consider the Hurewicz map

$$h_m \colon \pi_m(X) \otimes \mathbb{R} \to (IH^m(X; \mathbb{R}))^\vee$$

where $E^\vee = \hom_{\mathbb{R}}(E, \mathbb{R})$ and $IH^m(X; \mathbb{R}) = H^m(X; \mathbb{R})/\{\text{cup products}\}$.

**Lemma 10.** *If $X$ is an H-space such that $\dim_{\mathbb{R}} H^m(X; \mathbb{R}) < \infty$ for all $m$, then $h_m$ is an isomorphism.*

*Proof.* To prove this lemma, we define

$$PH_m(X; \mathbb{R}) = \{ x \in H_m(X; \mathbb{R}) \mid \Delta_*(x) = x \otimes 1 + 1 \otimes x \}$$

where

$$\Delta_* \colon H_m(X; \mathbb{R}) \to H_m(X \times X; \mathbb{R}) \cong \bigoplus_{s+t=m} H_s(M; \mathbb{R}) \otimes H_t(M; \mathbb{R})$$

is induced by the diagonal map.

Then if $X$ is an H-space there is an isomorphism

$$\pi_m(X) \otimes \mathbb{R} \xrightarrow{\cong} PH_m(X; \mathbb{R})$$

[MM65, Appendix], and under the finite type assumption above $(IH^m(X; \mathbb{R}))^\vee \cong PH_m(X; \mathbb{R})$. □

Now $BSL(A)^+$ is an H-space satisfying the assumption of the previous lemma, because of (the proof of) Quillen's theorem 3, and therefore for $m \geq 2$ we get

$$\begin{aligned} K_m(A) \otimes \mathbb{R} \cong (IH^m(BSL(A)^+; \mathbb{R}))^\vee &= (IH^m(BSL(A); \mathbb{R}))^\vee \\ &= (IH^m(SL(A); \mathbb{R}))^\vee. \end{aligned}$$

Theorem 6 implies that for $N \gg m$

$$H^m(SL_N(A); \mathbb{R}) \cong H^m_{\text{cont}}(G) \cong (\Lambda^*(e_i)^{\otimes r_1} \otimes \Lambda^*(\varepsilon_j)^{\otimes r_2})^m$$

and therefore

$$H^m(SL(A); \mathbb{R}) \cong H^m(SL_N(A); \mathbb{R}).$$

This yields

$$\begin{aligned} (IH^m(SL(A); \mathbb{R}))^\vee \overset{N \gg m}{\cong} IH^m_{\text{cont}}(G)^\vee &\cong I(\Lambda^*(e_i)^{\otimes r_1} \otimes \Lambda^*(\varepsilon_j)^{\otimes r_2})^m \\ &= \begin{cases} \mathbb{R}^{d_n} & \text{if } m = 2n-1, \\ 0 & \text{otherwise,} \end{cases} \end{aligned}$$

completing the proof that Borel's theorem 4 follows from theorem 6. □

*Example* 11. If $F = \mathbb{Q}$ then $r_2 = 0$, $r_1 = 1$ and

$$I(\Lambda^*(e_5, e_9, e_{13}, \ldots))^m = \begin{cases} \mathbb{R} & \text{if } m = 5, 9, 13, \ldots, \\ 0 & \text{otherwise.} \end{cases}$$

*Sketch of proof of theorem 6.* For simplicity we only consider

$$G = SL_N(\mathbb{R}) \supset \Gamma = SL_N(\mathbb{Z}).$$

Recall that $H^q_{\text{cont}}(G) = H^q(\Omega^*(X)^G)$ where $X$ is the symmetric space $K\backslash G$.

**Lemma 12** (Cartan). *The differential* $d\colon \Omega^*(X)^G \to \Omega^{*+1}(X)^G$ *vanishes.*

*Proof of lemma 12.* Let $\theta\colon G \to G$ be the Cartan involution $\theta(g) = (g^{-1})^t$. It induces a map $\theta\colon X \to X$ and therefore a chain map $\theta^*\colon \Omega^*(X)^G \to \Omega^*(X)^G$.

Look at $\theta'\colon \mathfrak{g} \to \mathfrak{g}$, $\theta'(M) = -M^t$. Recall that $\Omega^q(X)^G = \hom_{\mathfrak{k}}(\Lambda^q\mathfrak{p}, \mathbb{R})$ and $\mathfrak{p} = \{\, M \in \mathfrak{g} \mid M^t = M \,\}$. If $x \in \Lambda^q\mathfrak{p}$ then $\theta'(x) = (-1)^q x$. Hence if $\alpha \in \Omega^q(X)^G$ we compute

$$(-1)^q(d\alpha) = d\theta^*(\alpha) = \theta^* d(\alpha) = (-1)^{q+1}(d\alpha)$$

and therefore $d\alpha = 0$. □

Now assume first that $\Gamma = \{\, \gamma \in SL_N(\mathbb{Z}) \mid \gamma \equiv 1 \pmod 3 \,\}$.

**Fact 13.** $\Gamma$ *is torsionfree.*

This fact (due to Minkowski) implies that $\Gamma$ is acting freely on $X = K\backslash G$, as we can see as follows. Let $\gamma \in \Gamma$ and $[g] \in K\backslash G$. If $[g]\gamma = [g]$, we get $g\gamma = kg$, i.e., $\gamma = g^{-1}kg \in g^{-1}Kg \cap \Gamma$. But $g^{-1}Kg \cap \Gamma$ is finite, being the intersection of a compact with a discrete group. Therefore $\gamma$ has finite order, but, since $\Gamma$ is torsionfree, this shows that $\gamma = 1$.

Since $X$ is contractible, $X/\Gamma$ is therefore a $K(\Gamma, 1)$-space. Then

$$H^q(\Gamma; \mathbb{R}) = H^q(X/\Gamma; \mathbb{R}) = H^q(\Omega^*(X/\Gamma)) = H^q(\Omega^*(X)^\Gamma)$$

and we have to study

$$\Omega^q(X)^G = H^q(\Omega^*(X)^G) \to H^q(\Omega^*(X)^\Gamma).$$

Fix a smooth $G$-invariant metric $h$ on $TX$, and define

- the volume form $\mu = \sqrt{det(h^{i,j})}\, dx_1 \cdots dx_n \in \Omega^n(X)$, where $n = \dim(X)$,
- the star operator $\star\colon \Omega^q(X) \to \Omega^{n-q}(X)$ by $\omega \wedge \star\omega = h(\omega, \omega)\mu$,
- the Laplace operator $\Delta = dd^* + d^*d$, where

$$d^* = (-1)^{n(q+1)-1} \star d\star\colon \Omega^q(X) \to \Omega^{q-1}(X).$$

Cartan's lemma 12 above shows that $\Omega^*(X)^G \subset \ker \Delta$.

Main idea: Do Hodge theory on $X/\Gamma$.

Main difficulty: $X/\Gamma$ is not compact, it has only finite volume.

First step:

$$H^q(\Omega^*(X)^\Gamma) = H^q(\Omega^*(X)^\Gamma_{\log})$$

where $\Omega^*(X)^\Gamma_{\log}$ is the complex of differential forms $\omega$ such that both $\omega$ and $d\omega$ have "logarithmic growth at infinity". For instance, when $G = SL_2(\mathbb{R})$, in which case $X$ is the Poincaré upper half-plane

$$X = G/K = \{\, z \in \mathbb{C} \mid \operatorname{Im}(z) > 0 \,\},$$

a form $\omega$ is said to have logarithmic growth at infinity when its restriction to a Siegel set

$$\mathfrak{S} = \big\{\, z \in X \;\big|\; |\operatorname{Re}(z)| \leq b,\ \operatorname{Im}(z) > t \,\big\}$$

can be written

$$\omega_{|\mathfrak{S}} = \sum_{I,J} a_{I,J}(z)(dx)^I \left(\frac{dy}{y}\right)^J$$

with

$$|a_{I,J}(z)| \leq C|\log(y)|^k$$

for some integer $k$.

The proof of this step relies upon a Poincaré lemma with logarithmic growth.

Next, assume $\omega \in \Omega^q(X)^\Gamma_{\log}$ and $q$ is small with respect to $N$. Then $\omega$ is $L^2$, i.e.,

$$\|\omega\|^2_{L^2} = \int_{X/\Gamma} h(\omega, \omega)\mu < \infty.$$

In other words, $\Omega^*(X)^\Gamma_{\log} \subset \Omega^*(X)^\Gamma_{L^2}$ for $q$ small.

Now we can do $L^2$-Hodge theory:

(a) If $\omega$ is $L^2$ and $d\omega = 0$ then $\omega = h + d\eta$ with $h$ harmonic and $L^2$.
(b) If $h$ is harmonic and $L^2$, and $h = d\eta$ where $\eta$ is $L^2$, then $h = 0$.

(E.g., in order to prove (b) compute

$$(h, h)_{L^2} = (h, d\eta)_{L^2} = (d^*h, \eta)_{L^2} = 0$$

and therefore $h = 0$.)

The next step is to show that $\Omega^*(X)^G \subset \Omega^*(X)^\Gamma_{L^2}$.

And then the crucial step, due essentially to Garland and Matsushima, is to prove that if $q$ is small and $h \in \Omega^q(X)^\Gamma_{L^2}$ with $\Delta(h) = 0$, then $h \in \Omega^q(X)^G$.

Putting all together we get

$$H^q(\Omega^*(X)^\Gamma) \cong \ker(\Delta) \cap \Omega^q(X)^\Gamma_{L^2} \cong \Omega^q(X)^G.$$

Finally, for $\Gamma_0 = SL_N(\mathbb{Z})$, since $\Gamma$ is a normal subgroup of $\Gamma_0$ with finite index, we have

$$H^q(\Gamma_0; \mathbb{R}) \cong H^q(\Gamma; \mathbb{R})^{\Gamma_0/\Gamma}$$

which is then equal to $\Omega^q(X)^G$ since the action of $\Gamma_0$ is trivial. □

## Lecture two: Regulators

Let $F$ be a number field and $A$ its ring of integers. Let $\mathfrak{a} \subset A$ be a non-zero ideal. The norm of $\mathfrak{a}$ is $N\mathfrak{a} = \#(A/\mathfrak{a}) < \infty$.

**Definition 14.** $\zeta_F(s) = \sum_{\mathfrak{a} \neq 0} \frac{1}{(N\mathfrak{a})^s}$.

*Example* 15. If $F = \mathbb{Q}$ then

$$\zeta_{\mathbb{Q}}(s) = \sum_{n \geq 1} \frac{1}{n^s}$$

is the classical zeta function $\zeta$.

**Fact 16.**

- *$\zeta_F(s)$ is absolutely convergent when $\Re(s) > 1$;*
- *$\zeta_F(s)$ has a meromorphic continuation to $\mathbb{C}$;*
- *$\zeta_F(s)$ has a pole of order 1 at $s = 1$;*
- *Let $\xi(s) = A^s \Gamma(s/2)^{r_1} \Gamma(s)^{r_2} \zeta_F(s)$, where $\Gamma(s)$ is the classical gamma function, $A = 2^{-r_2}\sqrt{|D|}\pi^{r_1+2r_2}$, and $D$ is the discriminant of $F$. Then $\xi$ satisfies the functional equation $\xi(1-s) = \xi(s)$.*

**Corollary 17.** *If $n \geq 1$ then $\zeta_F(s)$ has a zero of order $d_n$ at $s = 1 - n$.*

*Proof.* The gamma function $\Gamma(s)$ has poles of order 1 at $s = 0, -1, -2, -3, \ldots$, hence $\Gamma(s/2)^{r_1}\Gamma(s)^{r_2}$ has a pole of order $r_2$ (respectively $r_1 + r_2$) at $s = 1 - n$ when $n \geq 0$ is even (respectively, odd). Note that $\zeta_F(n) \neq 0$ hence $\xi(n) \neq 0$ when $n > 1$.

Now let $s \to n$ and consider $A^{1-s}\Gamma((1-s)/2)^{r_1}\Gamma(1-s)^{r_2}\zeta_F(1-s) = \xi(s)$, by the functional equation. Therefore $\zeta_F(s)$ has a zero of order $d_n$ at $s = 1 - n$. □

**Definition 18.** $\zeta_F^*(1-n) = \lim_{s \to 1-n} \dfrac{\zeta_F(s)}{(s+n-1)^{d_n}} \in \mathbb{R}^\times$.

**Theorem 19** (Dirichlet's class number formula).

$$\zeta_F^*(0) = -\frac{hR}{w}$$

*where*

$$\begin{aligned} h &= \#\operatorname{Pic}(A) = \#K_0(A)_{\text{tors}} \\ w &= \#\mu(F) \quad = \#K_1(A)_{\text{tors}} \end{aligned}$$

*and $R$ is the regulator defined below.*

**Definition 20** (of the regulator). If $u \in A^\times$ and $v$ is an archimedean place of $F$, put

$$\|u\|_v = \begin{cases} |\sigma(u)| & \text{if } v = \sigma\colon F \hookrightarrow \mathbb{R}, \\ |\sigma(u)|^2 & \text{if } v = \{\sigma, \overline{\sigma}\},\ \sigma\colon F \hookrightarrow \mathbb{C}. \end{cases}$$

There is the so-called product formula: $\prod_v \|u\|_v = 1$, where $v$ ranges over all archimedean places. This yields a map

$$\rho\colon A^\times \to \mathbb{R}^{d_1} = \mathbb{R}^{r_1+r_2-1} = \ker(\Sigma\colon \mathbb{R}^{r_1+r_2} \to \mathbb{R}) , \quad u \mapsto (\log \|u\|_v)_v .$$

**Fact 21** (Dirichlet). $\operatorname{im}(\rho)$ *is a lattice.*

Now endow $\mathbb{R}^{d_1}$ with the restriction of the Lebesgue measure on $\mathbb{R}^{r_1+r_2}$, and define

$$R = \operatorname{vol}\left(\frac{\mathbb{R}^{d_1}}{\rho(A^\times)}\right).$$

We now want to see how Dirichlet's class number formula 19 generalizes to higher $K$-theory.

Recall that $H^*_{\text{cont}}(SL_N(\mathbb{R})) \cong \Lambda^*(e_5, e_9, e_{13}, \ldots)$. Given $\sigma\colon F \hookrightarrow \mathbb{R}$, then, for $m = 4k+1, k \geq 1$, we get

$$\sigma^*(e_m) \in H^m(SL_N(A); \mathbb{R}) \overset{N \gg 0}{\cong} H^m(SL(A); \mathbb{R})$$

and hence a map

$$K_m(A) \to H^m(SL(A); \mathbb{R})^\vee \to \mathbb{R}$$

given by the composition of the Hurewicz homomorphism and the map sending $\varphi \in H^m(SL(A); \mathbb{R})^\vee$ to $2\pi\varphi(\sigma^*(e_m))$. Similarly for $\sigma\colon F \hookrightarrow \mathbb{C}$ and $m = 3, 5, 7, \ldots$.

In this way we get for $n > 1$ a map

$$\rho_n\colon K_{2n-1}(A) \to \mathbb{R}^{d_n}$$

called the *higher (Borel) regulator map.*

The proof of theorem 4 actually shows that $\operatorname{im}(\rho_n)$ is a lattice. Define

$$R_n = \operatorname{vol}\left(\frac{\mathbb{R}^{d_n}}{\rho_n(K_{2n-1}(A))}\right).$$

**Theorem 22** (Siegel; Borel [Bor77]). *For any $n > 1$ there is a $q_n \in \mathbb{Q}^\times$ such that*

$$\zeta_F^*(1-n) = q_n R_n.$$

*Example* 23. Assume $r_2 = 0$ and $n > 1$ is even. Then $d_n = 0$, and theorem 22 (in this case due to Siegel) gives

$$\zeta_F(1-n) \in \mathbb{Q}^\times.$$

For instance, if $F = \mathbb{Q}$ and $n > 1$ is even then

$$\zeta(1-n) = -\frac{b_n}{n}$$

where $b_n$ are the Bernoulli numbers defined by

$$\frac{te^t}{e^t - 1} = \sum_{n=0}^{\infty} b_n \frac{t^n}{n!}.$$

E.g.,

$$\zeta(1-n) = \begin{cases} -1/2 \\ 1/12 \\ 0 \\ 1/120 \\ 0 \\ \dots \\ 691/32760 \end{cases} \quad \text{if} \quad 1-n = \begin{cases} 0 \\ -1 \\ -2 \\ -3 \\ -4 \\ \dots \\ -11 \end{cases}.$$

# Lecture three: Étale cohomology

Let $E$ be a finite extension of the number field $F$ and let $B$ be the ring of integers of $E$. Let $p$ be a prime.

**Definition 24.** $E$ is said to be *unramified outside* $p$ when, for any prime ideal $\mathfrak{p} \subset A$ such that $\mathfrak{p} \nmid pA$, one has in $B$ that $\mathfrak{p}B = q_1 \cdots q_t$ with $q_1, \dots, q_t$ distinct prime ideals.

*Example* 25. For any integer $k \geq 1$ then $F(\mu_{p^k})|F$ is unramified outside $p$, where $\mu_{p^k}$ is the group of $p^k$-th roots of unity.

Define

$$\Phi = \bigcup_{\substack{E|F \\ \text{unramified} \\ \text{outside p}}} E$$

and notice that $\mu_{p^k} \subset \Phi^\times$ for all $k \geq 1$. Define a character

$$\epsilon\colon \operatorname{Gal}(\Phi|F) \to \mathbb{Z}_p^\times$$

by the equation $g(\xi) = \xi^{\epsilon(g)}$ for any $g \in \operatorname{Gal}(\Phi|F)$ and for any $\xi \in \mu_{p^k}$.

The abelian group $\mathbb{Z}_p$ carries then for any $n$ a new $\mathrm{Gal}(\Phi|F)$-module structure, denoted $\mathbb{Z}_p(n)$ and defined as

$$g \cdot \alpha = \epsilon(g)^n \alpha$$

for $g \in \mathrm{Gal}(\Phi|F)$ and $\alpha \in \mathbb{Z}_p(n)$.

Define *étale cohomology* as

$$H^q_{\text{ét}}\left( \operatorname{Spec} A\left[\frac{1}{p}\right]; \mathbb{Z}_p(n)\right) = H^q_{\text{cont}}(\mathrm{Gal}(\Phi|F); \mathbb{Z}_p(n))$$

and abbreviate these groups to $H^q(A; \mathbb{Z}_p(n))$.

The following theorem is in quotation marks because it depends on the proof of the so-called Bloch-Kato conjecture, announced by Voevodsky and Rost but not yet fully written-up (cf. [Wei05]). Notice however that the corresponding surjectivity statements were proved by Soulé [Sou79] and Dwyer-Friedlander [DF85].

**"Theorem" 26.** *If $p$ is odd and $n \geq 2$ there are canonical isomorphisms*

$$K_{2n-1}(A) \otimes \mathbb{Z}_p \xrightarrow{\cong} H^1(A; \mathbb{Z}_p(n)),$$
$$K_{2n-2}(A) \otimes \mathbb{Z}_p \xrightarrow{\cong} H^2(A; \mathbb{Z}_p(n)).$$

The natural maps in the theorem above were first constructed by Dwyer and Friedlander [DF85] using étale $K$-theory $K^{\text{ét}}_m(A; \mathbb{Z}_p)$, which can be thought of as $p$-adic topological $K$-theory of the étale homotopy type of $\operatorname{Spec} A[\frac{1}{p}]$. (There is also a more modern description using motivic cohomology instead of étale cohomology.) There is a natural map

$$K_m(A) \to K^{\text{ét}}_m(A; \mathbb{Z}_p)$$

and there is also a spectral sequence converging to $K^{\text{ét}}_{2n-q}(A; \mathbb{Z}_p)$ with

$$E_2^{q,-2n} = H^q(A; \mathbb{Z}_p(n)).$$

But, assuming that $p$ is odd, $H^q(A; \mathbb{Z}_p(n)) = 0$ if $n > 0$ and $q \neq 1$ or 2. So the spectral sequence degenerates, i.e., $E_2 = E_\infty$ and in any diagonal there is only one non-zero $E_2$-term, therefore

$$K^{\text{ét}}_{2n-1}(A; \mathbb{Z}_p) = H^1(A; \mathbb{Z}_p(n)),$$
$$K^{\text{ét}}_{2n-2}(A; \mathbb{Z}_p) = H^2(A; \mathbb{Z}_p(n)).$$

**Corollary 27.** *When $n \geq 2$ the group $H^2(A; \mathbb{Z}_p(n))$ is always finite, and it vanishes for almost all primes $p$; moreover $\dim_{\mathbb{Q}_p} H^1(A; \mathbb{Z}_p(n)) \otimes \mathbb{Q}_p = d_n$.*

This corollary is not in quotation marks because the surjectivity statements in theorem 26, combined with theorem 4, are enough for it.

**Theorem 28** (Wiles [Wil90]). *If $r_2 = 0$ (i.e., if $F$ is totally real) and $n \geq 2$ is even then*

$$|\zeta_F(1-n)| = 2^? \frac{\prod_{p>2} \# H^2(A; \mathbb{Z}_p(n))}{\prod_{p>2} \# H^1(A; \mathbb{Z}_p(n))}.$$

**"Corollary" 29.** *If $r_2 = 0$ and $n \geq 2$ is even then*

$$|\zeta_F(1-n)| = 2^? \frac{\# K_{2n-2}(A)}{\# K_{2n-1}(A)}.$$

*"Example"* 30. If $F = \mathbb{Q}$ and $n \geq 2$ is even then

$$|\zeta(1-n)| = 2 \frac{\# K_{2n-2}(\mathbb{Z})}{\# K_{2n-1}(\mathbb{Z})}.$$

Combining "theorem" 26 with work of Fleckinger, Kolster, and Nguyen Quang Do [KNQDF96] we get:

**"Theorem" 31.** *If $F$ is abelian and $n \geq 1$ then*

$$|\zeta_F^*(1-n)| = 2^? \frac{\# K_{2n-2}(A)}{\# K_{2n-1}(A)_{\text{tors}}} R_n.$$

**Conjecture 32** (Vandiver). *For an odd prime $p$ define*

$$C = \operatorname{Pic}(\mathbb{Q}(\mu_p)) \otimes \mathbb{Z}/p\mathbb{Z}\,, \quad C^+ = \{\, x \in C \mid \overline{x} = x \,\}\,.$$

*Then $C^+ = 0$.*

Recall that $p$ is called *regular* if $C = 0$ (and that Kummer proved Fermat for regular primes—Vandiver hoped that $C^+ = 0$ would also imply Fermat).

Using computers one can show that Vandiver's conjecture is true if $p < 10^7$.

**"Theorem" 33** ([Kur92]). *The Vandiver conjecture is true for all primes if and only if $K_{4k}(\mathbb{Z}) = 0$ for all $k \geq 1$.*

*Example* 34. $K_4(\mathbb{Z}) = 0$ by a theorem of Rognes [Rog00], but $K_8(\mathbb{Z})$ is still unknown.

*Proof.* For $i \in \mathbb{Z}$ denote

$$C^{(i)} = \left\{ x \in C \;\middle|\; g(x) = \epsilon(g)^i x \quad \forall g \in \operatorname{Gal}(\mathbb{Q}(\mu_p)|\mathbb{Q}) \overset{\epsilon}{\cong} (\mathbb{Z}/p\mathbb{Z})^\times \right\}.$$

Then $C^+ = \bigoplus_{i \text{ even}} C^{(i)}$. The last needed ingredient is the computation

$$H^2(\mathbb{Z}; \mathbb{Z}_p(n)) \otimes \mathbb{Z}/p\mathbb{Z} \cong C^{(p-n)},$$

which combined with "theorem" 26 finishes the proof. □

*Example* 35. Obviously $C^{(p-1)} = 0$; there is a surjection $K_4(\mathbb{Z}) \twoheadrightarrow C^{(p-3)}$, and therefore, as noticed by Kurihara [Kur92], $C^{(p-3)} = 0$.

Assuming the Bloch-Kato and Vandiver conjectures we get the following computation of $K_*(\mathbb{Z})$ (cf. [Wei05]; note that the 2-torsion is known [RW00]): setting

$$\begin{aligned} w_n &= \text{denominator of } \frac{1}{2}\zeta(1-n), \\ c_n &= \text{numerator} \quad \text{of } \frac{1}{2}\zeta(1-n), \\ k_m &= \left[1+\frac{m}{4}\right], \end{aligned}$$

then for all $m > 1$ we have

$$K_m(\mathbb{Z}) \cong \begin{cases} \mathbb{Z} \oplus \mathbb{Z}/2 & \text{if } m \equiv 1 \\ \mathbb{Z}/2c_{2k_m} & \text{if } m \equiv 2 \\ \mathbb{Z}/2w_{2k_m} & \text{if } m \equiv 3 \\ 0 & \text{if } m \equiv 4 \\ \mathbb{Z} & \text{if } m \equiv 5 \\ \mathbb{Z}/c_{2k_m} & \text{if } m \equiv 6 \\ \mathbb{Z}/w_{2k_m} & \text{if } m \equiv 7 \\ 0 & \text{if } m \equiv 8 \end{cases} \pmod 8.$$

## Lecture four: Perfect forms

Let $N \geq 2$ and let $\phi(x) = \sum_{1 \leq i,j \leq N} a_{ij} x_i x_j$ be a positive definite real quadratic form in $N$ variables, i.e., $a_{ij} = a_{ji} \in \mathbb{R}$, $\phi(x) \geq 0$, with equality if and only if $x = 0$.

Define $M(\phi) = \{\, x \in \mathbb{Z}^N - 0 \mid \phi(x) \text{ is minimal} \,\}$. This is a finite set.

**Definition 36.** We say that $\phi$ is *perfect* when $M(\phi)$ determines $\phi$ up to scalar multiplication.

*Example* 37. $M(x^2 + y^2) = \{(1,0), (-1,0), (0,1), (0,-1)\} = M(x^2 + \frac{1}{2}xy + y^2)$, hence these forms are not perfect. On the other hand $x^2 + xy + y^2$ is perfect, and

$$M(x^2 + xy + y^2) = \{(1,0), (-1,0), (0,1), (0,-1), (1,-1), (-1,1)\}.$$

The group $\Gamma = SL_N(\mathbb{Z})$ acts on forms by $(\phi\gamma)(x) = \phi(\gamma(x))$.

**Theorem 38** (Voronoï [Vor08]). *Modulo the action of* $\Gamma$ *and scalar multiplication there are only finitely many perfect forms of a given rank* $N$.

For small values of $N$ perfect forms have been classified and

$$\#\{\text{perfect forms in } N \text{ variables}\} = \begin{cases} 1 \\ 1 \\ 2 \\ 3 \\ 7 \\ 33 \\ 10916 \end{cases} \quad \text{if} \quad N = \begin{cases} 2 \\ 3 \\ 4 \\ 5 \\ 6 \\ 7 \\ 8 \end{cases}$$

(the last two numbers were obtained by computers).

Define

$$C_N^* = \left\{ \phi(x) = \sum_{1\le i,j\le N} a_{ij}x_ix_j \;\middle|\; \begin{array}{l} a_{ij} = a_{ji} \in \mathbb{R}\,, \quad \phi(x) \geq 0\,, \\ \exists\, V \subsetneq \mathbb{Q}^N : \ker(\phi) = V \otimes \mathbb{R} \end{array} \right\}$$

and $X_N^* = C_N^*/\mathbb{R}_{>0}^\times$, together with a projection $\pi\colon C_N^* \to X_N^*$.

For $v \in \mathbb{Z}^N - 0$ define $\widehat{v} \in C_N^*$ by $\widehat{v}(x) = (v|x)^2$ (where $(v|x)$ is the scalar product). If $\phi$ is perfect define

$$\sigma(\phi) = \pi\left(\left\{\sum_i \lambda_i \widehat{v}_i | \forall i\ \lambda_i \geq 0 \text{ and } v_i \in M(\phi)\right\}\right).$$

**Theorem 39** (Voronoï [Vor08]). *The family of cells $\sigma(\phi)$ for $\phi$ perfect and their intersections give a $\Gamma$-invariant cell decomposition of $X_N^*$.*

This can be used to compute $H^*(\Gamma; \mathbb{Z})$. Endow $X_N^*$ with the CW-topology coming from this cell decomposition (warning: this is different from the usual topology on matrices). Let $\partial X_N^* \subset X_N^*$ be the set of classes of forms with non-trivial kernel. Consider the integral equivariant homology of $(X_N^*, \partial X_N^*)$.

There is a first quadrant spectral sequence $E_{pq}^2 = H_p(\Gamma, H_q(X_N^*, \partial X_N^*; \mathbb{Z}))$ converging to $H_{p+q}^\Gamma(X_N^*, \partial X_N^*; \mathbb{Z})$.

**Proposition 40.** *The space $X_N^*$ is contractible and $\partial X_N^*$ is homotopy equivalent to the spherical Tits building of $SL_N(\mathbb{Q})$, i.e., has the homotopy type of a bouquet of infinitely many spheres of dimension $N-2$.*

Therefore

$$H_q(X_N^*, \partial X_N^*; \mathbb{Z}) = \widetilde{H}_{q-1}(\partial X_N^*; \mathbb{Z}) = \begin{cases} St_N & \text{if } q = N-1, \\ 0 & \text{otherwise,} \end{cases}$$

where $St_N$ is the Steinberg module, and so

$$H_m^\Gamma(X_N^*, \partial X_N^*; \mathbb{Z}) = H_{m-N+1}(\Gamma; St_N).$$

There is a second spectral sequence with $E_{pq}^1 = \bigoplus_{\dim(\sigma)=p} H_q(\Gamma_\sigma; \mathbb{Z}_\sigma)$ also converging to $H_{p+q}^\Gamma(X_N^*, \partial X_N^*; \mathbb{Z})$. Here $\Gamma_\sigma$ is the stabilizer of the cell $\sigma$ and $\mathbb{Z}_\sigma$ is its orientation module.

**Lemma 41.** *If a prime $p$ divides $\#\Gamma_\sigma$ then $p \leq N+1$.*

*Proof.* If $\gamma^p = 1$ then $\gamma^{p-1} + \gamma^{p-2} + \ldots + 1 = 0$, but since the minimal polynomial divides the characteristic polynomial we get that $p - 1 \leq N$. □

Denote by $\mathcal{S}_{N+1}$ the Serre subcategory of finite abelian groups $A$ such that if $p|\#A$ then $p \leq N+1$. We will now compute modulo $\mathcal{S}_{N+1}$.

If $q > 0$ then $\#\Gamma_\sigma$ annihilates $H_q(\Gamma_\sigma; \mathbb{Z}_\sigma)$, hence $E_{pq}^1 \equiv 0 \pmod{\mathcal{S}_{N+1}}$.

If $\Gamma_\sigma$ acts non-trivially on $\mathbb{Z}_\sigma$ then 2 annihilates $H_0(\Gamma_\sigma; \mathbb{Z}_\sigma)$.

Let $V_n$ be the free abelian group spanned by all $\Gamma$-orbits of cells $\sigma$ of dimension $n$ such that $\sigma$ meets $X_N^* - \partial X_N^*$ and $\Gamma_\sigma$ preserves the orientation of $\sigma$, and $V = (V_*, d^1)$. We get

$$H_n(V) \equiv H_{n-N+1}(\Gamma; St_N) \quad (\text{mod } \mathcal{S}_{N+1}).$$

According to Borel-Serre duality and an additional argument of Farrell

$$H_m(\Gamma; St_N) \equiv H^{d-m}(\Gamma; \mathbb{Z}) \quad (\text{mod } \mathcal{S}_{N+1})$$

where $d = N(N-1)/2$. One gets:

**Theorem 42.**

(a) $$H^m(SL_2(\mathbb{Z}); \mathbb{Z}) \equiv H^m(SL_3(\mathbb{Z}); \mathbb{Z}) \equiv \begin{cases} \mathbb{Z} & \textit{if } m = 0 \\ 0 & \textit{otherwise} \end{cases} \quad (\text{mod } \mathcal{S}_3);$$

(b) $$H^m(SL_4(\mathbb{Z}); \mathbb{Z}) \equiv \begin{cases} \mathbb{Z} & \textit{if } n = 0, 3 \\ 0 & \textit{otherwise} \end{cases} \quad (\text{mod } \mathcal{S}_5);$$

(c) $$H^m(SL_5(\mathbb{Z}); \mathbb{Z}) \equiv \begin{cases} \mathbb{Z} & \textit{if } n = 0, 5 \\ 0 & \textit{otherwise} \end{cases} \quad (\text{mod } \mathcal{S}_5);$$

(d) $$H^m(SL_6(\mathbb{Z}); \mathbb{Z}) \equiv \begin{cases} \mathbb{Z} & \textit{if } m = 0, 8, 9 \\ \mathbb{Z}^2 & \textit{if } n = 5 \\ 0 & \textit{otherwise} \end{cases} \quad (\text{mod } \mathcal{S}_7);$$

(e) $$H^m(SL_7(\mathbb{Z}); \mathbb{Q}) \cong \begin{cases} \mathbb{Q} & \textit{if } m = 0, 5, 11, 14, 15 \\ 0 & \textit{otherwise} \end{cases} .$$

Here part (b) is due to Lee and Szczarba, and (c)-(d)-(e) to Elbaz-Vincent, Gangl, and Soulé [EVGS02], involving computer calculations.

These results can be used, with some more work, to compute $K_m(\mathbb{Z}), m \leq 7$. The classification of perfect forms for $N = 8$ is also known, but the computation of the cohomology of $SL_8(\mathbb{Z})$ seems too complicated for today's computers.

# References


[BMS67] Hyman Bass, John Milnor, and Jean-Pierre Serre, *Solution of the congruence subgroup problem for $SL_n$ $(n \geq 3)$ and $Sp_{2n}$ $(n \geq 2)$*, Inst. Hautes Études Sci. Publ. Math., **33**, (1967), 59–137, MR0244257

[Bor74] Armand Borel, *Stable real cohomology of arithmetic groups*, Ann. Sci. École Norm. Sup. (4), **7**, (1974), 235–272, MR0387496

[Bor77] Armand Borel, *Cohomologie de $SL_n$ et valeurs de fonctions zeta aux points entiers*, Ann. Scuola Norm. Sup. Pisa Cl. Sci. (4), **4**, (1977), no. 4, 613–636, MR0506168

[DF85] William Dwyer, and Eric Friedlander, *Algebraic and etale $K$-theory*, Trans. Amer. Math. Soc., **292**, (1985), no. 1, 247–280, MR805962

[EVGS02] Philippe Elbaz-Vincent, Herbert Gangl, and Christophe Soulé, *Quelques calculs de la cohomologie de $GL_N(\mathbb{Z})$ et de la $K$-théorie de $\mathbb{Z}$*, C. R. Math. Acad. Sci. Paris, **335**, (2002), no. 4, 321–324, MR1931508

[Gui80] Alain Guichardet, *Cohomologie des groupes topologiques et des algèbres de Lie*, Textes Mathématiques, vol. 2, CEDIC, Paris, (1980), MR644979

[KNQDF96] Manfred Kolster, Thong Nguyen Quang Do, and Vincent Fleckinger, *Twisted $S$-units, $p$-adic class number formulas, and the Lichtenbaum conjectures*, Duke Math. J., **84**, (1996), no. 3, 679–717, MR1408541

[Kur92] Masato Kurihara, *Some remarks on conjectures about cyclotomic fields and $K$-groups of $\mathbb{Z}$*, Compositio Math., **81**, (1992), no. 2, 223–236, MR1145807

[Mil71] John Milnor, *Introduction to algebraic $K$-theory*, Annals of Mathematics Studies, vol. 72, Princeton University Press, Princeton, N.J., (1971), MR0349811

[MM65] John Milnor, and John Moore, *On the structure of Hopf algebras*, Ann. of Math. (2), **81**, (1965), 211–264, MR0174052

[Qui73] Daniel Quillen, *Finite generation of the groups $K_i$ of rings of algebraic integers*, Algebraic $K$-theory, I: Higher $K$-theories, (Proc. Conf., Battelle Memorial Inst., Seattle, Wash., 1972), Lecture Notes in Math., vol. 341, Springer, Berlin, 1973, pp. 179–198, MR0349812

[Rog00] John Rognes, *$K_4(\mathbb{Z})$ is the trivial group*, Topology **39** (2000), no. 2, 267–281, MR1722028

[RW00] John Rognes and Charles Weibel, *Two-primary algebraic $K$-theory of rings of integers in number fields*, J. Amer. Math. Soc. **13** (2000), no. 1, 1–54. Appendix A by Manfred Kolster. MR1697095

[Sou79] Christophe Soulé, *$K$-théorie des anneaux d'entiers de corps de nombres et cohomologie étale*, Invent. Math. **55**, (1979), no. 3, 251–295, MR553999

[Vor08] Georges Voronoï, *Nouvelles applications des paramètres continus à la théorie des formes quadratiques. Premier Mémoire: Sur quelques propriétés des formes quadratiques positives parfaites*, J. für Math. **133** (1908), 97–178.

[Wei05] Charles Weibel, *Algebraic $K$-theory of rings of integers in local and global fields*, Handbook of $K$-theory, vol. 1, Springer, Berlin, (2005), pp. 139–190, MR2181823

[Wil90] Andrew Wiles, *The Iwasawa conjecture for totally real fields*, Ann. of Math. (2) **131** (1990), no. 3, 493–540, MR1053488

Made in the USA
Middletown, DE
04 November 2026

36552539R00292